# The Environmental Regulatory Dictionary

# The Environmental Regulatory Dictionary

## Fourth Edition

*Compiled by*

**James J. King**

WILEY-INTERSCIENCE

A JOHN WILEY & SONS, INC., PUBLICATION

Published by John Wiley & Sons, Inc., Hoboken, New Jersey.
Published simultaneously in Canada.



***Library of Congress Cataloging-in-Publication Data:***

The environmental regulatory dictionary / edited by James L. King. — 4th ed.
p. cm.
Rev. ed. of : Environmental dictionary / compiled by James J. King. 3rd ed. 1995.
Designed to supplement the Code of federal regulations, Title 40, Protection of environment (40 CFR).
ISBN 0-471-70526-8 (cloth)
1. Environmental law—United States—Dictionaries. 2. Pollution—Law and legislation—United States—Dictionaries. I. King, James J. II. King, James J. Environmental dictionary. III. Code of federal regulations. Title 40, Protection of environment.
KF3775.A68K57 2005
344.7304'6'03—dc22 2004023474

Printed in the United States of America

10 9 8 7 6 5 4 3 2 1

# CONTENTS


Preface vii

Acknowledgements ix

Guide to Usage xi

Definitions from the *Code of Federal Regulations, Title 40, Protection of Environment (40 CFR)* 1

Appendix: Outline of 40 CFR 497

About the Author 508

# PREFACE

Thank you for choosing the 4th edition of *The Environmental Regulatory Dictionary.*

*The Environmental Regulatory Dictionary* is designed to be a supplement for researching environmental regulations in the *Code of Federal Regulations, Title 40, Protection of Environment* (40 CFR). 40 CFR is a maze of administrative and technical requirements, constructed with language that has very specific meanings. Locating correct terminology in the regulations can become complicated: it is not always obvious where to start looking for a term; and, the same tern may be defined differently for the various regulatory areas. This book will hopefully be useful not only as a dictionary, but also as a locator for the regulation(s) that applies to the term being researched.

*The Environmental Regulatory Dictionary* does not guarantee the correctness (regarding content, grammar, punctuation, spelling, etc.) or validity of the definitions contained in this dictionary; they are simply reproductions of those contained in the source publication.

All definitions in this dictionary have been reproduced as accurately as possible. However, definitions from this dictionary should not be quoted or cited. Please refer to 40 CFR for the final word. Please consult competent legal and technical professionals for guidance in interpreting the terms listed.

For more information, readers can visit Jim King's website at *www.JimKing.Biz.*

# ACKNOWLEDGEMENTS

I would like to thank several friends and associates for their help and encouragement to go forward with the 4th edition of *The Environmental Regulatory Dictionary.*

Thanks go to W. Corey Trench, author, business strategist, and good friend for his input for the need for an updated version of the Dictionary. To my friends and clients Bharat Arora, MISWACO, Halifax, Nova Scotia, David Blankenship, and John Babb, Jr. (my computer guru), TECO Coal Company, Corbin, Kentucky fellow auditor Dawn Henderlight, Knoxville, Tennessee and Richard Gentry, AER Manufacturing, Carrollton, Texas for reviewing my proposal and providing great insights into the need and shaping of this work. I wish to thank my wife JoAnne for continuing to be a great source of love and encouragement in all the endeavors of my life. Finally, I would like to thank Jonathan Rose of John Wiley and Sons, for his guidance and responsiveness in helping me get the finished product into production, and on-schedule.

JIM KING

*September 5, 2004*
*Morgan County, Tennessee*

# GUIDE TO USAGE

The definitions in the dictionary are taken from *The Code of Federal Regulations, Title 40, Protection of the Environment:* (40 CFR).

In using this book the listed term is followed by the section of the Code where the corresponding regulation is found. For example:

**A, as in "Cyanide A"** (Section 433.11) shall mean amenable to alkaline chlorination.

The section of the Code referenced in the definition takes the reader to that section of the Code where the definition in Section 433.11 is applicable. In this case the regulation is found in **Section 433: Metal finishing point source category** where the term **A, as in "Cyanide A"** is used.

The dictionary not only serves as a "dictionary" in the true sense, but also as a guide to finding the regulations.

My intent is to make your search for the applicable regulation you are researching easier than you may have experienced in the past.

# #

**0-bedroom dwelling** (Section 745.103) means any residential dwelling in which the living area is not separated from the sleeping area. The term includes efficiencies, studio apartments, dormitory housing, military barracks, and rentals of individual rooms in residential dwellings.

**10 year, 24 hour rainfall event** (Section 411.31) shall mean a rainfall event with a probable recurrence interval of once in ten years as defined by the National Weather Service in Technical Paper No. 40, "Rainfall Frequency Atlas of the United States," May 1961, and subsequent amendments, or equivalent regional or state rainfall probability information developed therefrom.

**10 year, 24 hour rainfall event and 25 year, 24 hour rainfall event** (Section 412.11) shall mean a rainfall event with a probable recurrence interval of once in ten years or twenty-five years, respectively, as defined by the National Weather Service in Technical Paper Number 40, "Rainfall Frequency Atlas of the United States", May 1961, and subsequent amendments, or equivalent regional or state rainfall probability information developed therefrom.

**10 year, 24/hour rainfall event** (Section 423.11) means a rainfall event with a probable recurrence interval of once in ten years as defined by the National Weather Service in Technical Paper No. 40. Rainfall Frequency Atlas of the United States, May 1961 or equivalent regional rainfall probability information developed therefrom.

**10-day LC50** (Section 435.11) as applied to the sediment toxicity BAT effluent limitations and NSPS means the concentration (milligrams/ kilogram dry sediment) of the base fluid in sediment that is lethal to 50 percent of the Leptocheirus plumulosus test organisms exposed to that concentration of the base fluids after ten days of constant exposure.

**10-year 24-hour precipitation event** (Section 436.21) shall mean the maximum 24-hour precipitation event with a probable reoccurrence interval of once in 10 years. This information is available in "Weather Bureau Technical Paper No. 40," May 1961 and "NOAA Atlas 2," 1973 for the 11 Western States, and may be obtained from the National Climatic Center of the Environmental Data Service, National Oceanic and Atmospheric Administration, U.S. Department of Commerce.

**10-year 24-hour precipitation event** (Section 438.181) shall mean the maximum 24 hour precipitation event with a probable reoccurrence interval of once in 10 years. This information is available in "Weather Bureau Technical Paper No. 40," May 1961 and "NOAA Atlas 2," 1973 for the 11 Western States, and may be obtained from the National Climatic Center of the Environmental Data Service, National Oceanic and Atmospheric Administration, U.S. Department of Commerce.

**10-year, 24-hour rainfall event and 25-year, 24-hour rainfall event** (Section 412.21) shall mean a rainfall event with a probable recurrence interval of once in ten years or twenty-five years, respectively, as defined by the National

**1-year, 24-hour precipitation event** (Section 434.11) means the maximum 24-hour precipitation event with a probable recurrence interval of once in one year as defined by the National Weather Service and Technical Paper No. 40, "Rainfall Frequency Atlas of the U.S.," May 1961, or equivalent regional or rainfall probability information developed therefrom.

**1-year, 2-year, and 10-year, 24-hour precipitation events** (Section 434.11) means the maximum 24-hour precipitation event with a probable recurrence interval of once in one, two, and ten years respectively as defined by the National Weather Service and Technical Paper No. 40, "Rainfall Frequency Altas of the U.S.," May 1961, or equivalent regional or rainfall probability information developed therefrom.

**25-year 24-hour rainfall event** (Section 418.11) shall mean the maximum 24-hour precipitation event with a probable recurrence interval of once in 25 years as defined by the National Weather Service in technical paper No. 40, "Rainfall Frequency Atlas of the United States", May 1961, and subsequent amendments in effect, as of the effective date of this regulation.

**25-year 24-hour rainfall event** (Section 422.41) shall mean the maximum precipitation event with a probable recurrence interval of once in 25 years as defined by the National Weather Service in technical paper no. 40, "Rainfall Frequency Atlas of the United States," May, 1961, and subsequent amendments or equivalent regional or State rainfall probability information developed therefrom.

**25-year 24-hour rainfall event** (Section 422.51) shall mean the maximum precipitation event with a probable recurrence interval of once in 25 years as defined by the National Weather Service in technical paper no. 40, "Rainfall Frequency Atlas of the United States," May 1961, and subsequent amendments or equivalent regional or State rainfall probability information developed therefrom.

**2-year, 24-hour precipitation event** (Section 434.11) means the maximum 24-hour precipitation event with a probable recurrence interval of once in two years as defined by the National Weather Service and Technical Paper No. 40, "Rainfall Frequency Atlas of the U.S., "May 1961, or equivalent regional or rainfall probability information developed therefrom.

**309 of the Act** (Section 108.2) any order issued by a State to secure compliance with a permit, or condition thereof, issued under a program approved pursuant to section 402 of the Act; or any order issued by a court in an action brought pursuant to section 309 or section 505 of the Act.

**30-day average** (Section 133.101) The arithmetic mean of pollutant parameter values of samples collected in a period of 30 consecutive days.

**37 States** (Section 86.1702-99) is the trading region comprised of the United States excluding California and the Northeast Trading Region.

**49 states** (Section 86.1702-99) is the region comprised of the United States excluding California.

**4-day LC50** (Section 435.11) as applied to the sediment toxicity BAT effluent limitations and NSPS means the concentration (milligrams/ kilogram dry sediment) of the drilling fluid in sediment that is lethal to 50 percent of the Leptocheirus plumulosus test organisms exposed to that concentration of the drilling fluids after four days of constant exposure.

**505 Cycle** (Section 86.000-2) means the test cycle that consists of the first 505 seconds (seconds 1 to 505) of the EPA Urban Dynamometer Driving Schedule, described in Sec. 86.115-00 and listed in appendix I, paragraph (a), of this part.

**505 Cycle** (Section 86.1803-01) means the test cycle that consists of the first 505 seconds (seconds 1 to 505) of the EPA Urban Dynamometer Driving Schedule, described in Sec. 86.115-00 and listed in Appendix I, paragraph (a), of this part.

**50-state engine family** (Section 86.1002-2001) means an engine family that meets both federal and California Air Resources Board motor vehicle emission control regulations and has received a federal certificate of conformity as well as an Executive Order.

**50-state engine family** (Section 86.1002-97) means an engine family that meets both federal and California Air Resources Board motor vehicle emission control regulations and has received a federal certificate of conformity as well as an Executive Order.

**50-state engine family** (Section 86.602-97) means an engine family that meets both federal and California Air Resources Board motor vehicle emission control regulations and has received a federal certificate of conformity as well as an Executive Order.

motor vehicle emission control regulations and has received a federal certificate of conformity as well as an Executive Order.

**7-day average** (Section 133.101) The arithmetic mean of pollutant parameter values for samples collected in a period of 7 consecutive days.

**866 Cycle** (Section 86.000-2) means the test cycle that consists of the last 866 seconds (seconds 505 to 1372) of the EPA Urban Dynamometer Driving Schedule, described in Sec. 86.115-00 and listed in appendix I, paragraph (a), of this part.

**866 Cycle** (Section 86.1803-01) means the test cycle that consists of the last 866 seconds (seconds 506 to 1372) of the EPA Urban Dynamometer Driving Schedule, described in Sec. 86.115-00 and listed in Appendix I, paragraph (a), of this part.

**96-hour LC50** (Section 435.11) means the concentration (parts per million) or percent of the suspended particulate phase (SPP) from a sample that is lethal to 50 percent of the test organisms exposed to that concentration of the SPP after 96 hours of constant exposure.

**96-hour LC50** (Section 435.41) means the concentration (parts per million) or percent of the suspended particulate phase (SPP) from a sample that is lethal to 50 percent of the test organisms exposed to that concentration of the SPP after 96 hours of constant exposure.

# A

**A, as in "Cyanide A"** (Section 433.11) shall mean amenable to alkaline chlorination.

**Abandoned mine** (Section 434.11) means a mine where mining operations have occurred in the past and (1) The applicable reclamation bond or financial assurance has been released or forfeited or (2) If no reclamation bond or other financial assurance has been posted, no mining operations have occurred for five years or more.

**Abandoned well** (Section 146.3) means a well whose use has been permanently discontinued or which is in a state of disrepair such that it cannot be used for its intended purpose or for observation purposes.

**Abatement** (Section 745.223) means any measure or set of measures designed to permanently eliminate lead-based paint hazards. Abatement includes, but is not limited to: (1) The removal of paint and dust, the permanent enclosure or encapsulation of lead-based paint, the replacement of painted surfaces or fixtures, or the removal or permanent covering of soil, when lead-based paint hazards are present in such paint, dust or soil; and (2) All preparation, cleanup, disposal, and post-abatement clearance testing activities associated with such measures. (3) Specifically, abatement includes, but is not limited to: (i) Projects for which there is a written contract or other documentation, which provides that an individual or firm will be conducting activities in or to a residential dwelling or child-occupied facility that: (A) Shall result in the permanent elimination of lead-based paint hazards; or (B) Are designed to permanently eliminate lead-based paint hazards and are described in paragraphs (1) and (2) of this definition. (ii) Projects resulting in the permanent elimination of lead-based paint hazards, conducted by firms or individuals certified in accordance with Sec. 745.226, unless such projects are covered by paragraph (4) of this definition; (iii) Projects resulting in the permanent elimination of lead-based paint hazards, conducted by firms or individuals who, through their company name or promotional literature, represent, advertise, or hold themselves out to be in the business of performing lead-based paint activities as identified and defined by this section, unless such projects are covered by paragraph (4) of this definition; or (iv) Projects resulting in the permanent elimination of lead-based paint hazards, that are conducted in response to State or local abatement orders. (4) Abatement does not include renovation, remodeling, landscaping or other activities, when such activities are not designed to permanently eliminate lead-based paint hazards, but, instead, are designed to repair, restore, or remodel a given structure or dwelling, even though these activities may incidentally result in a reduction or elimination of lead-based paint hazards. Furthermore, abatement does not include interim controls, operations and maintenance activities, or other measures and activities designed to temporarily, but not permanently, reduce lead-based paint hazards.

**Abnormally treated vehicle** (Section 86.1803-01) means any diesel light-duty vehicle or diesel light-duty truck that is operated for less than five miles in a 30 day period immediately prior to conducting a particulate emissions test.

**Above ground tank** (Section 260.10) means a device meeting the definition of "tank" in Sec. 260.10 and that is situated in such a way that the entire surface area of the tank is completely above the plane of the adjacent surrounding surface and the entire surface area of the tank (including the tank bottom) is able to be visually inspected.

**Aboveground release** (Section 280.12) means any release to the surface of the land or to surface water. This includes, but is not limited to, releases from the above-ground portion of an UST system and aboveground releases associated with overfills and transfer operations as the regulated substance moves to or from an UST system.

**Aboveground storage facility** (Section 113.3) means a tank or other container, the bottom of which is on a plane not more than 6 inches below the surrounding surface.

**Aboveground tank** (Section 279.1) means a tank used to store or process used oil that is not an underground storage tank as defined in Sec. 280.12 of this chapter.

**ac** (Section 401.11) means acre(s).

**AC1** (Section 86.000-2) means a test procedure as described in Sec. 86.162-00 which simulates testing with air conditioning operating in an environmental test cell by adding the air conditioning compressor load to the normal dynamometer forces.

**AC1** (Section 86.1803-01) means a test procedure as described in Sec. 86.162-00 which simulates testing with air conditioning operating in an environmental test cell by adding the air conditioning compressor load to the normal dynamometer forces.

**AC2** (Section 86.000-2) means a test procedure as described in Sec. 86.162-00 which simulates testing with air conditioning operating in an environmental test cell by adding a heat load to the passenger compartment.

**AC2** (Section 86.1803-01) means a test procedure as described in Sec. 86.162-00 which simulates testing with air conditioning operating in an environmental test cell by adding a heat load to the passenger compartment.

**Acceleration test procedure** (Section 205.151) means the measurement methodologies specified in Appendix I.

**Acceptable of a batch** (Section 204.51) means that the number of non-complying compressors in the batch sample is less than or equal to the acceptance number as determined by the appropriate sampling plan.

**Acceptable Quality Level (AQL)** (Section 204.51) means the maximum percentage of failing compressors that, for purposes of sampling inspection can be considered satisfactory as a process average.

**Acceptable quality level (AQL)** (Section 205.151) means the maximum allowable average percentage of vehicles or exhaust systems that can fail sampling inspection under a Selective Enforcement Audit.

**Acceptable quality level (AQL)** (Section 86.1002-2001) means the maximum percentage of failing engines or vehicles, that for purposes of sampling inspection, can be considered satisfactory as a process average.

**Acceptable Quality Level (AQL)** (Section 86.1002-84) means the maximum percentage of failing engines or vehicles, that for purposes of sampling inspection, can be considered satisfactory as a process average.

**Acceptable quality level (AQL)** (Section 86.1002-97) means the maximum percentage of failing engines or vehicles, that for purposes of sampling inspection, can be considered satisfactory as a process average.

**Acceptable Quality Level (AQL)** (Section 86.602-84) means the maximum percentage of failing vehicles that, for purposes of sampling inspection, can be considered satisfactory as a process average.

**Acceptable quality level (AQL)** (Section 89.502) means the maximum percentage of failing engines that can be considered a satisfactory process average for sampling inspections.

**Acceptable Quality Level** (Section 205.51) means the maximum percentage of failing vehicles that for purposes of sampling inspection, can be considered satisfactory as a process average.

**Acceptance date** (Section 195.2) means the date on which EPA enters the application into the data system.

**Acceptance of a batch** (Section 205.51) means that the number of noncomplying vehicles in the batch sample is less than or

**Acceptance of a batch sequence** (Section 204.51) means that the number of rejected batches in the sequence is less than or equal to the sequence acceptable number as determined by the appropriate sampling plan.

**Acceptance of a Batch sequence** (Section 205.51) means that the number of rejected batches in the sequence is less than or equal to the acceptance number as determined by the appropriate sampling plan.

**Acceptance of a compressor** (Section 204.51) means that the measured noise emissions of the compressor, when measured in accordance with the applicable procedure, conforms to the applicable standard.

**Acceptance of a vehicle** (Section 205.51) means that the measured emissions of the vehicle when measured in accordance with the applicable procedure, conforms to the applicable standard.

**Accepted application** (Section 195.2) refers to an application that has been entered into the data system.

**Accessible environment** (Section 191.12) means: (1) The atmosphere; (2) land surfaces; (3) surface waters; (4) oceans; and (5) all of the lithosphere that is beyond the controlled area.

**Accessible when referring to ACM** (Section 763.83) means that the material is subject to disturbance by school building occupants or custodial or maintenance personnel in the course of their normal activities.

**Accident** (Section 171.2) means an unexpected, undesirable event, caused by the use or presence of a pesticide that adversely affects man or the environment.

**Accidental release** (Section 1400.2) means an unanticipated emission of a regulated substance or other extremely hazardous substance into the ambient air from a stationary source.

**Accidental release** (Section 280.92) means any sudden or nonsudden release of petroleum from an underground storage tank that results in a need for corrective action and/or compensation for bodily injury or property damage neither expected nor intended by the tank owner or operator.

**Accidental release** (Section 68.3) means an unanticipated emission of a regulated substance or other extremely hazardous substance into the ambient air from a stationary source.

**Account certificate of representation** (Section 96.2) means the completed and signed submission required by subpart B of this part for certifying the designation of a NOX authorized account representative for a NOX Budget source or a group of identified NOX Budget sources who is authorized to represent the owners and operators of such source or sources and of the NOX Budget units at such source or sources with regard to matters under the NOX Budget Trading Program.

**Account number** (Section 72.2) means the identification number given by the Administrator to each Allowance Tracking System account pursuant to Sec. 73.31(d) of this chapter.

**Account number** (Section 96.2) means the identification number given by the Administrator to each NOX Allowance Tracking System account.

**Account number** (Section 97.2) means the identification number given by the Administrator to each NOX Allowance Tracking System account.

**Accredited or accreditation** (Section 763.83) when referring to a person or laboratory means that such person or laboratory is accredited in accordance with section 206 of Title II of the Act.

**Accredited training program** (Section 745.223) means a training program that has been accredited by EPA pursuant to Sec. 745.225 to provide training for individuals engaged in lead-based paint activities.

**Accuracy** (Section 86.082-2) means the difference between a measurement and true value.

**Accuracy** (Section 86.1803-01) means the difference between a measurement and true value.

**Accuracy** (Section 92.102) means the difference between the measured value and the true value, where the true value is determined from NIST traceable measurements where possible, or otherwise determined by good engineering practice.

**Acid or ferruginous mine drainage** (Section 434.11) means mine drainage which, before any treatment, either has a pH of less than 6.0 or a total iron concentration equal to or greater than 10 mg/l.

**Acid Rain compliance option** (Section 72.2) means one of the methods of compliance used by an affected unit under the Acid Rain Program as described in a compliance plan submitted and approved in accordance with subpart D of this part, part 74 of this chapter or part 76 of this chapter.

**Acid Rain emissions limitation** (Section 72.2) means: (1) For purposes of sulfur dioxide emissions: (i) The tonnage equivalent of the allowances authorized to be allocated to an affected unit for use in a calendar year under section 404(a)(1), (a)(3), and (h) of the Act, or the basic Phase II allowance allocations authorized to be allocated to an affected unit for use in a calendar year, or the allowances authorized to be allocated to an opt-in source under section 410 of the Act for use in a calendar year; (ii) As adjusted: (A) By allowances allocated by the Administrator pursuant to section 403, section 405 (a)(2), (a)(3), (b)(2), (c)(4), (d)(3), and (h)(2), and section 406 of the Act; (B) By allowances allocated by the Administrator pursuant to subpart D of this part; and thereafter © By allowance transfers to or from the compliance subaccount for that unit that were recorded or properly submitted for recordation by the allowance transfer deadline as provided in Sec. 73.35 of this chapter, after deductions and other adjustments are made pursuant to Sec. 73.34© of this chapter; and (2) For purposes of nitrogen oxides emissions, the applicable limitation under part 76 of this chapter.

**Acid Rain emissions limitation** (Section 96.2) means, as defined in Sec. 72.2 of this chapter, a limitation on emissions of sulfur dioxide or nitrogen oxides under the Acid Rain Program under title IV of the CAA.

**Acid Rain emissions limitation** (Section 97.2) means, as defined in Sec. 72.2 of this chapter, a limitation on emissions of sulfur dioxide or nitrogen oxides under the Acid Rain Program under title IV of the Clean Air Act.

**Acid Rain emissions reduction requirement** (Section 72.2) means a requirement under the Acid Rain Program to reduce the emissions of sulfur dioxide or nitrogen oxides from a unit to a specified level or by a specified percentage.

**Acid Rain permit or permit** (Section 72.2) means the legally binding written document or portion of such document, including any permit revisions, that is issued by a permitting authority under this part and specifies the Acid Rain Program requirements applicable to an affected source and to the owners and operators and the designated representative of the affected source or the affected unit.

**Acid Rain Program** (Section 72.2) means the national sulfur dioxide and nitrogen oxides air pollution control and emissions reduction program established in accordance with title IV of the Act, this part, and parts 73, 74, 75, 76, 77, and 78 of this chapter.

**Acid recovery** (Section 420.91) means those sulfuric acid pickling operations that include processes for recovering the unreacted acid from spent pickling acid solutions.

**Acid regeneration** (Section 420.91) means those hydrochloric acid pickling operations that include processes for regenerating acid from spent pickling acid solutions.

**Acoustic descriptor** (Section 211.102) means the numeric, symbolic, or narrative information describing a product's acoustic properties as they are determined according to the test methodology that the Agency prescribes.

**Acoustical Assurance Period (AAP)** (Section 205.151) means a specified period of time or miles driven after sale to the ultimate purchaser during which a newly manufactured vehicle or exhaust system, properly used and maintained, must continue in compliance with the Federal standard.

**Act** (Section 192) means the Uranium Mill Tailings Radiation Control Act of 1978, as amended.

**Act** (Section 201.1) means the Noise Control Act of 1972 (Pub. L. 92-574, 86 Stat. 1234).

**Act** (Section 230.3) means the Clean Water Act (also known as the Federal Water Pollution Control Act or FWPCA) Pub. L. 92-500, as amended by Pub. L. 95-217, 33 U.S.C. 1251, et seq.

**Act** (Section 50.1) means the Clean Air Act, as amended (42 U.S.C. 1857-18571, as amended by Pub. L. 91-604).

**Act** (Section 51.100) means the Clean Air Act (42 U.S.C. 7401 et seq., as amended by Pub. L. 91-604, 84 Stat. 1676 Pub. L. 95-95, 91 Stat., 685 and Pub. L. 95-190, 91 Stat., 1399.)

**Act** (Section 53.1) means the Clean Air Act (42 U.S.C. 1857-1857l), as amended. Administrator means the Administrator of the Environmental Protection Agency or the Administrator's authorized representative.

**Act** (Section 56.1) means the Clean Air Act as amended (42 U.S.C. 7401 et seq.).

**Act** (Section 57.103) means the Clean Air Act, as amended.

**Act** (Section 58.1) means the Clean Air Act as amended (42 U.S.C. 7401, et seq.).

**Act** (Section 60.2) means the Clean Air Act (42 U.S.C. 7401 et seq.)

**Act** (Section 600.022-85) means part I of title V of the Motor Vehicle Information and Cost Savings Act (15 U.S.C. 1901 et seq.).

**Act** (Section 61.02) means the Clean Air Act (42 U.S.C. 7401 et seq.).

**Act** (Section 63.2) means the Clean Air Act (42 U.S.C. 7401 et seq., as amended by Pub. L. 101-549, 104 Stat. 2399).

**Act** (Section 64.1) means the Clean Air Act, as amended by Pub.L. 101-549, 42 U.S.C. 7401, et seq.

**Act** (section 710.2) means the Toxic Substances Control Act, 15 U.S.C. 2601 et seq.

**Act** (Section 720.3) means the Toxic Substances Control Act, 15 U.S.C. 2601 et seq.

**Act** (Section 745.103) means the Residential Lead-Based Paint Hazard Reduction Act of 1992, 42 U.S.C. 4852d.

Act of 1986" and Title IV of SARA, also known as the "Radon Gas and Indoor Air Quality Research Act of 1986." Title V of SARA amending the Internal Revenue Code is also known as the "Superfund Revenue Act of 1986."

**Act or RCRA** (Section 247.3) means the Solid Waste Disposal Act, as amended by the Resource Conservation and Recovery Act, as amended, 42 U.S.C 6901 et seq;

**Action level** (Section 141.2)is the concentration of lead or copper in water specified in Sec. 141.80© which determines, in some cases, the treatment requirements contained in subpart I of this part that a water system is required to complete.

**Activation** (Section 300.5) means notification by telephone or other expeditious manner or, when required, the assembly of some or all appropriate members of the RRT or NRT.

**Active ingredient** (Section 152.3) means any substance (or group of structurally similar substances if specified by the Agency) that will prevent, destroy, repel or mitigate any pest, or that functions as a plant regulator, desiccant, or defoliant within the meaning of FIFRA sec. 2(a), except as provided in Sec. 174.3 of this chapter.

**Active ingredient** (Section 158.153) means any substance (or group of structurally similar substances, if specified by the Agency) that will prevent, destroy, repel or mitigate any pest, or that functions as a plant regulator, desiccant, or defoliant within the meaning of FIFRA sec. 2 (a).

**Active ingredient** (Section 174.3) means a pesticidal substance that is intended to be produced and used in a living plant, or in the produce thereof, and the genetic material necessary for the production of such a pesticidal substance.

**Active ingredient** (Section 455.10) means an ingredient of a pesticide which is intended to prevent, destroy, repel, or mitigate any pest.

**Active institutional control** (Section 191.12) means: (1) Controlling access to a disposal site by any means other than passive institutional controls; (2) performing maintenance operations or remedial actions at a site, (3) controlling or cleaning up releases from a site, or (4) monitoring parameters related to disposal system performance.

**Active life** (Section 258.2) means the period of operation beginning with the initial receipt of solid waste and ending at completion of closure activities in accordance with Sec. 258.60 of this part.

**Active life of a facility** (Section 260.10) means the period from the initial receipt of hazardous waste at the facility until the Regional Administrator receives certification of final closure.

**Active mining area** (Section 440.132) is a place where work or other activity related to the extraction, removal, or recovery of metal ore is being conducted, except, with respect to surface mines, any area of land on or in which grading has been completed to return the earth to desired contour and reclamation work has begun.

**Active mining area** (Section 434.11) means the area, on and beneath land, used or disturbed in activity related to the extraction, removal, or recovery of coal from its natural deposits. This term excludes coal preparation plants, coal preparation plant associated areas and post-mining areas.

**Active portion** (Section 258.2) means that part of a facility or unit that has received or is receiving wastes and that has not been closed in accordance with Sec. 258.60 of this part.

**Active portion** (Section 260.10) means that portion of a facility where treatment, storage, or disposal operations are being or have been conducted after the effective date of part 261 of this chapter and which is not a closed portion. (See also "closed portion" and "inactive portion".)

**Active range** (Section 266.201) means a military range that is currently in service and is being regularly used for range activities.

**Active sewage sludge unit** (Section 503.21) is a sewage sludge unit that has not closed.

**Active use** (Section 57.103) refers to an SO2 constant control system installed at a smelter before August 7, 1977 and not totally removed from regular service by that date.

**Actual emissions** (Section 63.2) is defined in subpart D of this part for the purpose of granting a compliance extension for an early reduction of hazardous air pollutants.

**Actual SO2 emissions rate** (Section 72.2) means the annual average sulfur dioxide emissions rate for the unit (expressed in lb/mmBtu), for the specified calendar year; provided that, if the unit is listed in the NADB, the "1985 actual SO2 emissions rate" for the unit shall be the rate specified by the Administrator in the NADB under the data field "SO2RTE."

**Acute dermal LD50** (Section 152.3) means a statistically derived estimate of the single dermal dose of a substance that would cause 50 percent mortality to the test population under specified conditions.

**Acute inhalation LC50** (Section 152.3) means a statistically derived estimate of the concentration of a substance that would cause 50 percent mortality to the test population under specified conditions.

**Acute oral LD50** (Section 152.3) means a statistically derived estimate of the single oral dose of a substance that would cause 50 percent mortality to the test population under specified conditions.

**Acute toxicity** (Section 132.2) is concurrent and delayed adverse effect(s) that results from an acute exposure and occurs within any short observation period which begins when the exposure begins, may extend beyond the exposure period, and usually does not constitute a substantial portion of the life span of the organism.

**Acute-chronic ratio (ACR)** (Section 132.2) is a standard measure of the acute toxicity of a material divided by an appropriate measure of the chronic toxicity of the same material under comparable conditions.

**Acutely toxic effects** (Section 721.3) A chemical substance produces acutely toxic effects if it kills within a short time period (usually 14 days): (1) At least 50 percent of the exposed mammalian test animals following oral administration of a single dose of the test substance at 25 milligrams or less per kilogram of body weight (LD50). (2) At least 50 percent of the exposed mammalian test animals following dermal administration of a single dose of the test substance at 50 milligrams or less per kilogram of body weight (LD50). (3) At least 50 percent of the exposed mammalian test animals following administration of the test substance for 8 hours or less by continuous inhalation at a steady concentration in air at 0.5 milligrams or less per liter of air (LC50).

**Added ingredients** (Section 407.81) shall mean the prepared sauces (prepared from items such as dairy products, starches, sugar, tomato sauce and concentrate, spices, and other related preprocessed ingredients) which are added during the canning and freezing of fruits and vegetables.

**Additional advance auction** (Section 72.2) means the auction of advance allowances that were offered the previous year for sale in an advance sale.

**Additive** (Section 79.2) means any substance, other than one composed solely of carbon and/or hydrogen, that is intentionally added to a fuel named in the designation (including any added to a motor vehicle's fuel system) and that is not intentionally removed prior to sale or use.

**Additive** (Section 790.3) means a chemical substance that is intentionally added to another chemical substance to improve its stability or impart some other desirable quality.

**Additive manufacturer** (Section 79.2) means any person who produces, manufactures, or imports an additive for use as an additive and/or sells or imports for sale such additive under the person's own name.

**Add-on control** (Section 72.2) means a pollution reduction control technology that operates independent of the combustion process.

**Adequate quality control** (Section 745.223) means a plan or design which ensures the authenticity, integrity, and accuracy of samples, including dust, soil, and paint chip or paint film samples. Adequate quality control also includes provisions for representative sampling.

**Adequate SO2 emission limitation** (Section 57.103) means a SIP emission limitation which was approved or promulgated by EPA as adequate to attain and maintain the NAAQS in the areas affected by the stack emissions without the use of any unauthorized dispersion technique.

**Adequate water quality funding** (Section 130.2) means that the State, Territory, or authorized Tribe has allocated existing water quality funds from any source to the implementation of the TMDL load allocations to the fullest extent practicable and in a manner consistent with the effective operation of its clean water program. In the event that existing funding is not adequate to fully implement the TMDL load allocations, you may satisfy the funding requirement of reasonable assurance by including an explanation of when adequate funds will become available and the schedule by which these funds will be used to implement the TMDL load allocations. When EPA establishes a TMDL, EPA must show there is adequate funding. It may do so by conditioning Clean Water Act grants to the fullest extent practicable and in a manner consistent with effective operation of other Clean Water Act programs.

**Adhesion promoter** (Section 59.101) means a coating designed to facilitate the bonding of a primer or topcoat on surfaces such as trim moldings, door locks, and door sills, where sanding is impracticable, and on plastic parts and the edges of sanded areas.

**Adjacent** (Section 230.3) means bordering, contiguous, or neighboring. Wetlands separated from other waters of the United States by man-made dikes or barriers, natural river berms, beach dunes, and the like are "adjacent wetlands."

**Adjustable parameter** (Section 89.2) means any device, system, or element of design which is physically capable of being adjusted (including those which are difficult to access) and which, if adjusted, may affect emissions or engine performance during emission testing.

**Adjustable parameter** (Section 90.3) means any device, system, or element of design which is physically capable of being adjusted (including those which are difficult to access) and which, if adjusted, may affect emissions or engine performance during emission testing or normal in-use operation.

**Adjustable parameter** (Section 91.3) means any device, system, or element of design which is physically capable of being adjusted (including those which are difficult to access) and which, if adjusted, may affect emissions or engine performance during emission testing or normal in-use operation.

**Adjustable Parameter** (Section 94.2) means any device, system, or element of design which is physically or electronically capable of being adjusted (including those which are difficult to access) and which, if adjusted, may affect emissions or engine performance during emission testing.

**Adjusted configuration** (Section 610.11) means the test configuration after adjustment of engine calibrations to the retrofit specifications, but excluding retrofit hardware installation.

**Adjusted Loaded Vehicle Weight** (Section 86.1803-01) means the numerical average of vehicle curb weight and gross vehicle weight rating (GVWR).

**Adjusted Loaded Vehicle Weight** (Section 88.102-94) is defined as the numerical average of the vehicle curb weight and the GVWR.

**Administrative controls** (Section 68.3) mean written procedural mechanisms used for hazard control.

**Administrative hearing** (Section 307.14) means an administrative adjudication required by section 112(b)(2) of CERCLA in the event a claimant contests a determination of his claim made by the U.S. Environmental Protection Agency (EPA).

**Administrative Law Judge** (Section 164.2) means an Administrative Law Judge appointed pursuant to 5 U.S.C. 3105 (see also 5 CFR part 930, as amended), and such term is synonymous with the term "Hearing Examiner" as used in the Act or in the United States Code.

**Administrative law judge** (Section 209.3) means an administrative law judge appointed under 5 U.S.C. 3105 (see also 5 CFR part 930, as amended by 37 FR 16787). "Administrative law judge" is synonymous with "hearing examiner" as used in Title 5 of the United States Code.

**Administrative Law Judge** (Section 57.103) means an administrative law judge appointed under 5 U.S.C. 3105 (see also 5 CFR part 930, as amended by 37 FR 16787), and is synonymous with the term "Hearing Examiner" as formerly used in Title 5 of the U.S. Code.

**Administrator** (Section 50.1) means the Administrator of the Environmental Protection Agency.

**Administrator** (Section 63.2) means the Administrator of the United States Environmental Protection Agency or his or her authorized representative (e.g., a State that has been delegated the authority to implement the provisions of this part).

**Administrator** (Section 68.3) means the administrator of the U.S. Environmental Protection Agency.

**Administrator and General Counsel** (Section 350.1) mean the EPA officers or employees occupying the positions so titled.

**Adsorbable organic halides (AOX)** (Section 430.0) A bulk parameter that measures the total mass of chlorinated organic matter in water and wastewater.

**Advance allowance** (Section 72.2) means an allowance that may be used for purposes of compliance with a unit's Acid Rain sulfur dioxide emissions limitation requirements beginning no earlier than seven years following the year in which the allowance is first offered for sale.

**Advance auction** (Section 72.2) means an auction of advance allowances.

**Advance sale** (Section 72.2) means a sale of advance allowances.

**Advanced air emission control devices** (Section 426.11) shall mean air pollution control equipment, such as electrostatic precipitators and high energy scrubbers, that are used to treat an air discharge which has been treated initially by equipment including knock-out chambers and low energy scrubbers.

**Adverse effect** (Section 132.2) is any deleterious effect to organisms due to exposure to a substance. This includes effects which are or may become debilitating, harmful or toxic to the normal functions of the organism, but does not include non-harmful effects such as tissue discoloration alone or the induction of enzymes involved in the metabolism of the substance.

**Adverse weather** (Section 112.2) means the weather conditions that make it difficult for response equipment and personnel to cleanup or remove spilled oil, and that will be considered when identifying response systems and equipment in a response plan for the applicable operating environment. Factors to consider include significant wave height as specified in Appendix E to this part, as appropriate, ice conditions, temperatures, weather-related visibility, and currents within the area in which the systems or equipment are intended to function.

**Advertised Engine Displacement** (Section 205.151) means the rounded off volumetric engine capacity used for marketing purposes by the motorcycle manufacturer.

**Aerobic digestion** (Section 503.31) is the biochemical decomposition of organic matter in sewage sludge into carbon dioxide and water by microorganisms in the presence of air.

**Aerometric Information Retrieval System (AIRS)-Air Quality Subsystem (AQS)** (Section 58.1) is EPA's computerized system for storing and reporting of information relating to ambient air quality data.

**Affected facility** (Section 60.2) means, with reference to a stationary source, any apparatus to which a standard is applicable.

**Affected Federal land manager** (Section 93.152) means the Federal agency or the Federal official charged with direct responsibility for management of an area designated as Class I under the Act (42 U.S.C. 7472) that is located within 100 km of the proposed Federal action.

**Affected source** (Section 70.2) shall have the meaning given to it in the regulations promulgated under title IV of the Act.

**Affected source** (Section 71.2) shall have the meaning given to it in 40 CFR 72.2.

**Affected source** (Section 72.2) means a source that includes one or more affected units.

**Affected source,** (Section 63.2) for the purposes of this part, means the collection of equipment, activities, or both within a single contiguous area and under common control that is included in a section 112© source category or subcategory for which a section 112(d) standard or other relevant standard is established pursuant to section 112 of the Act. Each relevant standard will define the "affected source," as defined in this paragraph unless a different definition is warranted based on a published justification as to why this definition would result in significant administrative, practical, or implementation problems and why the different definition would resolve those problems. The term "affected source," as used in this part, is separate and distinct from any other use of that term in EPA regulations such as those implementing title IV of the Act. Affected source may be defined differently for part 63 than affected facility and stationary source in parts 60 and 61, respectively. This definition of "affected source," and the procedures for adopting an alternative definition of "affected source," shall apply to each section 112(d) standard for which the initial proposed rule is signed by the Administrator after June 30, 2002.

**Affected States** (Section 70.2) are all States: (1) Whose air quality may be affected and that are contiguous to the State in which a part 70 permit, permit modification or permit renewal is being proposed; or (2) That are within 50 miles of the permitted source.

**Affected States** (Section 72.2) means any affected States as defined in part 71 of this chapter.

**Affected States are:** (Section 71.2) (1) All States and areas within Indian country subject to a part 70 or part 71 program whose air quality may be affected and that are contiguous to the State or the area within

Indian country in which the permit, permit modification, or permit renewal is being proposed; or that are within 50 miles of the permitted source. A Tribe shall be treated in the same manner as a State under this paragraph (1) only if EPA has determined that the Tribe is an eligible Tribe.(2) The State or area within Indian country subject to a part 70 or part 71 program in which a part 71 permit, permit modification, or permit renewal is being proposed. A Tribe shall be treated in the same manner as a State under this paragraph (2) only if EPA has determined that the Tribe is an eligible Tribe. (3) Those areas within the jurisdiction of the air pollution control agency for the area in which a part 71 permit, permit modification, or permit renewal is being proposed.

**Affected unit** (Section 70.2) shall have the meaning given to it in the regulations promulgated under title IV of the Act.

**Affected unit** (Section 71.2) shall have the meaning given to it in 40 CFR 72.2.

**Affected unit** (Section 72.2) means a unit that is subject to any Acid Rain emissions reduction requirement or Acid Rain emissions limitation under Sec. 72.6 or part 74 of this chapter.

**Affiliate** (Section 72.2) shall have the meaning set forth in section 2(a)(11) of the Public Utility Holding Company Act of 1935, 15 U.S.C. 79b(a)(11), as of November 15, 1990.

**Affiliated entity** (Section 66.3) means a person who directly, or indirectly through one or more intermediaries, controls, is controlled by, or is under common control with the owner or operator of a source.

**Aftermarket conversion certifier** (Section 85.502) means any company or individual which assembles the various aftermarket conversion hardware components into a particular combination or configuration and certifies that combination or configuration according to the provisions of this subpart.

**Aftermarket conversion installer** (Section 85.502) means any company or individual which installs an aftermarket conversion system on a light-duty vehicle, light-duty truck, heavy-duty vehicle, or heavy-duty engine with the effect of allowing the vehicle or engine to operate on a fuel other than the fuel which the vehicle or engine was originally certified to use.

**Aftermarket conversion system** (Section 85.502) means any combination of hardware, including but not limited to fuel storage and fuel metering hardware, which is installed on a light-duty vehicle, light-duty truck, heavy-duty vehicle, or heavy-duty engine with the effect of allowing the vehicle or engine to operate on a fuel other than the fuel which the vehicle or engine was originally certified to use. Components which do not affect the emissions performance of the converted vehicle or engine, as determined by the Administrator, are not included for the purposes of this subpart.

**Aftermarket Part** (Section 85.2113) means any part offered for sale for installation in or on a motor vehicle after such vehicle has left the vehicle manufacturer's production line.

**Aftermarket Part Manufacturer** (Section 85.2113) means: (1) A manufacturer of an aftermarket part or, (2) A party that markets aftermarket parts under its own brand name, or, (3) A rebuilder of original equipment or aftermarket parts, or (4) A party that licenses others to sell its parts.

**Aftertreatment** (Section 90.3) means the passage of exhaust gases through a device or system such as a catalyst whose purpose is to chemically alter the gases prior to their release to the atmosphere.

**Aftertreatment system or aftertreatment component or aftertreatment technology** (Section 92.2) means any system or component or technology mounted downstream of the exhaust valve or exhaust port whose design function is to reduce exhaust emissions.

**Aftertreatment system or aftertreatment component or aftertreatment technology** (Section 94.2) means any system or component or technology mounted downstream of the exhaust valve or exhaust port whose design function is to reduce exhaust emissions.

**Agency** (Section 50.1) means the Environmental Protection Agency.

**Agency** (Section 53.1) means the Environmental Protection Agency.

**Agent** (Section 745.103) means any party who enters into a contract with a seller or lessor, including any party who enters into a contract with a representative of the seller or lessor, for the purpose of selling or leasing target housing. This term does not apply to purchasers or any purchaser's representative who receives all compensation from the purchaser.

**Agreement State** (Section 191.02) means any State with which the Commission or the Atomic Energy Commission has entered into an effective agreement under subsection 274b of the Atomic Energy Act of 1954, as amended (68 Stat. 919).

**Agricultural commodity** (Section 171.2) means any plant, or part thereof, or animal, or animal product, produced by a person (including farmers, ranchers, vineyardists, plant propagators, Christmas tree growers, aquaculturists, floriculturists, orchardists, foresters, or other comparable persons) primarily for sale, consumption, propagation, or other use by man or animals.

**Agricultural employer** (Section 170.3) means any person who hires or contracts for the services of workers, for any type of compensation, to perform activities related to the production of agricultural plants, or any person who is an owner of or is responsible for the management or condition of an agricultural establishment that uses such workers.

**Agricultural establishment** (Section 170.3) means any farm, forest, nursery, or greenhouse.

**Agricultural land** (Section 503.11) is land on which a food crop, a feed crop, or a fiber crop is grown. This includes range land and land used as pasture.

**Agricultural plant** (Section 170.3) means any plant grown or maintained for commercial or research purposes and includes, but is not limited to, food, feed, and fiber plants; trees; turfgrass; flowers, shrubs; ornamentals; and seedlings.

**Agricultural solid** (Section 243.101) waste means the solid waste that is generated by the rearing of animals, and the producing and harvesting of crops or trees.

**Agricultural solid waste** (Section 246.101) means the solid waste that is generated by the rearing of animals, and the producing and harvesting of crops or trees.

**Agronomic rate** (Section 503.11) is the whole sludge application rate (dry weight basis) designed: (1) To provide the amount of nitrogen needed by the food crop, feed crop, fiber crop, cover crop, or vegetation grown on the land; and (2)To minimize the amount of nitrogen in the sewage sludge that passes below the root zone of the crop or vegetation grown on the land to the ground water.

**AIChE/CCPS** (Section 68.3) means the American Institute of Chemical Engineers/Center for Chemical Process Safety.

**Air (Section 240.101)** Over fire air means air, under control as to quantity and direction, introduced above or beyond a fuel bed by induced or forced draft. "Underfire air" means any forced or induced air, under control as to quantity and direction, that is supplied from beneath and which passes through the solid wastes fuel bed.

**Air compressor system** (Section 761.13) means air compressors, piping, receiver tanks, volume tanks and bottles, dryers, airlines, and related appurtenances.

**Air emissions** (Section 129.2) means the release or discharge of a toxic pollutant by an owner or operator into the ambient air either (1) by means of a stack or (2) as a fugitive dust, mist or vapor as a result inherent to the manufacturing or formulating process.

**Air erosion** (Section 763.83) means the passage of air over friable ACBM which may result in the release of asbestos fibers.

**Air pollution control system** (Section 62.1201) means the equipment used to reduce the release of particulate matter and other pollutants to the atmosphere.

**Air quality restricted operation of a spray tower** (Section 417.151) shall mean an operation utilizing formulations (e.g., those with high non-ionic content) which require a very high rate of wet scrubbing to maintain desirable quality of stack gases, and thus generate much greater quantities of waste water than can be recycled to process.

**Air stripping operation** (Section 264.1031) is a desorption operation employed to transfer one or more volatile components from a liquid mixture into a gas (air) either with or without the application of heat to the liquid. Packed towers, spray towers, and bubble-cap, sieve, or valve-type plate towers are among the process configurations used for contacting the air and a liquid.

**Aircraft** (Section 87.1) means any airplane for which a U.S. standard airworthiness certificate or equivalent foreign airworthiness certificate is issued.

**Aircraft engine** (Section 87.1) means a propulsion engine which is installed in or which is manufactured for installation in an aircraft.

**Aircraft gas turbine engine** (Section 87.1) means a turboprop, turbofan, or turbojet aircraft engine.

**Alcohol fuel** (Section 86.1702-99) means either methanol or ethanol as those terms are defined in this subpart.

**Alcohol fuel** (Section 92.2) means a fuel consisting primarily (more than 50 percent by weight) of one or more alcohols: e.g., methyl alcohol, ethyl alcohol.

**Alkaline cleaning** (Section 471.02) uses a solution (bath), usually detergent, to remove lard, oil, and other such compounds from a metal surface. Alkaline cleaning is usually followed by a water rinse. The rinse may consist of single or multiple stage rinsing. For the purposes of this part, an alkaline cleaning operation is defined as a bath followed by a rinse, regardless of the number of rinse stages. Each alkaline cleaning bath and rinse combination is entitled to a discharge allowance.

**Alkaline cleaning bath** (Section 468.02) shall mean a bath consisting of an alkaline cleaning solution through which a workpiece is processed.

**Alkaline cleaning rinse** (Section 468.02) shall mean a rinse following an alkaline cleaning bath through which a workpiece is processed. A rinse consisting of a series of rinse tanks is considered as a single rinse.

**Alkaline cleaning rinse for forged parts** (Section 468.02) shall mean a rinse following an alkaline cleaning bath through which a forged part is processed. A rinse consisting of a series of rinse tanks is considered as a single rinse.

**Alkaline, mine drainage** (Section 434.11) means mine drainage which, before any treatment, has a pH equal to or greater than 6.0 and total iron concentration of less than 10 mg/l.

**All States Trading Region (ASTR)** (Section 86.1702-99) means the region comprised of all states except the OTC States that have not opted into National LEV pursuant to the opt-in provisions at Sec. 86.1705 or that have opted out of National LEV and whose opt-outs have become effective, as provided at Sec. 86.1707; California; and any state outside the OTR with a Section 177 Program in effect that does not allow National LEV as a compliance alternative.

**Allegation** (Section 717.3) means a statement, made without formal proof or regard for evidence, that a chemical substance or mixture has caused a significant adverse reaction to health or the environment.

**All-electric range test** (Section 86.1702-99) means a test sequence used to determine the range of an electric vehicle or of a hybrid electric vehicle without the use of its auxiliary power unit. The All-Electric Range Test cycle is defined in Sec. 86.1770.

**Alley collection** (Section 243.101) means the collection of solid waste from containers placed adjacent to or in an alley.

**Allocate or allocation** (Section 72.2) means the initial crediting of an allowance by the Administrator to an Allowance Tracking System unit account or general account.

**Allocate or allocation** (Section 96.2) means the determination by the permitting authority or the Administrator of the number of NOX allowances to be initially credited to a NOX Budget unit or an allocation set-aside.

**Allocate or allocation** (Section 97.2) means, with regard to NOX allowances, the determination by the Administrator of the number of NOX allowances to be initially credited to a NOX Budget unit or an allocation set-aside.

**Allotment** (Section 35.102) EPA's calculation of the funds that may be available to an eligible recipient for an environmental program grant. An allotment is not an entitlement.

**Allowable SO2 emissions rate** (Section 72.2) means the most stringent federally enforceable emissions limitation for sulfur dioxide (in lb/ mmBtu) applicable to the unit or combustion source for the specified calendar year, or for such subsequent year as determined by the Administrator where such a limitation does not exist for the specified year; provided that, if a Phase I or Phase II unit is listed in the NADB, the "1985 allowable SO2 emissions rate" for the Phase I or Phase II unit shall be the rate specified by the Administrator in the NADB under the data field "1985 annualized boiler SO2 emission limit."

**Allowance** (Section 72.2) means an authorization by the Administrator under the Acid Rain Program to emit up to one ton of sulfur dioxide during or after a specified calendar year.

**Allowance deduction, or deduct when referring to allowances,** (Section 72.2) means the permanent withdrawal of allowances by the Administrator from an Allowance Tracking System compliance subaccount, or future year subaccount, to account for the number of tons of SO2 emissions from an affected unit for the calendar year, for tonnage emissions estimates calculated for periods of missing data as provided in part 75 of this chapter, or for any other allowance surrender obligations of the Acid Rain Program.

**Allowance reserve** (Section 72.2) means any bank of allowances established by the Administrator in the Allowance Tracking System pursuant to sections 404(a)(2) (Phase I extension reserve), 404(g) (energy conservation and renewable energy reserve), or 416(b) (special allowance reserve) of the Act, and implemented in accordance with part 73, subpart B of this chapter.

**Allowance Tracking System account** (Section 72.2) means an account in the Allowance Tracking System established by the Administrator for purposes of allocating, holding, transferring, and using allowances.

**Allowance Tracking System or ATS** (Section 72.2) means the Acid Rain Program system by which the Administrator allocates, records, deducts, and tracks allowances.

**Allowance transfer deadline** (Section 72.2) means midnight of March 1 (or February 29 in any leap year) or, if such day is not a business day, midnight of the first business day thereafter and is the deadline by which allowances may be submitted for recordation in an affected unit's compliance subaccount for the purposes of meeting the unit's Acid Rain emissions limitation requirements for sulfur dioxide for the previous calendar year.

**Allowances held or hold allowances** (Section 72.2) means the allowances recorded by the Administrator, or submitted to the Administrator for recordation in accordance with Sec. 73.50 of this chapter, in an Allowance Tracking System account.

**Alternative contemporaneous annual emission limitation** (Section 76.2) means the maximum allowable NOX emission rate (on a lb/mmBtu, annual average basis) assigned to an individual unit in a NOX emissions averaging plan pursuant to Sec. 76.10.

**Alternative effluent limitations or pretreatment standards** (Section 437.2) mean effluent limitations determined on a case-by-case basis under section 402(a)(1) of the CWA or pretreatment standards developed as local limits by the control authority under 40 CFR Sec. 403.6© that apply to the discharge of wastewater subject to this provision. The permit writer (or control authority) will calculate these limitations or standards using a "building block" approach or the "combined wastestream formula." Under this approach, the permit writer (or control authority) will develop flow-weighted effluent limitations or standards for the treated combined wastestream by applying the limitations or standards in 40 CFR subchapter N that would otherwise apply to a particular wastestream received from off-site if the wastestream were treated and discharged from the facility at which it was generated.

**Alternative emission limitation** (Section 63.2) means conditions established pursuant to sections 112(i) (5) or 112(i) (6) of the Act by the Administrator or by a State with an approved permit program.

**Alternative emission standard** (Section 63.2) means an alternative means of emission limitation that, after notice and opportunity for public comment, has been demonstrated by an owner or operator to the Administrator's satisfaction to achieve a reduction in emissions of any air pollutant at least equivalent to the reduction in emissions of such pollutant achieved under a relevant design, equipment, work practice, or operational emission standard, or combination thereof, established under this part pursuant to section 112(h) of the Act.

**Alternative Fueled Vehicle for the Federal Fleet, Guidance Document for Federal Agencies** (Section 88.302-94) shall apply.

**Alternative fuels** (Section 86.000-2) means any fuel other than gasoline and diesel fuels, such as methanol, ethanol, and gaseous fuels.

**Alternative fuels** (Section 86.1803-01) means any fuel other than gasoline and diesel fuels, such as methanol, ethanol, and gaseous fuels.

**Alternative method** (Section 60.2) means any method of sampling and analyzing for an air pollutant which is not a reference or equivalent method but which has been demonstrated to the Administrator's satisfaction to, in specific cases, produce results adequate for his determination of compliance.

**Alternative method** (Section 61.02) means any method of sampling and analyzing for an air pollutant which is not a reference method but which has been demonstrated to the Administrator's satisfaction to produce results adequate for the Administrator's determination of compliance.

**Alternative monitoring system** (Section 72.2) means a system or a component of a system designed to provide direct or indirect data of mass emissions per time period, pollutant concentrations, or volumetric flow, that is demonstrated to the Administrator as having the same precision, reliability, accessibility, and timeliness as the data provided by a certified CEMS or certified CEMS component in accordance with part 75 of this chapter.

**Alternative technology** (Section 76.2) means a control technology for reducing NOX emissions that is outside the scope of the definition of low NOX burner technology. Alternative technology does not include overfire air as applied to wall-fired boilers or separated overfire air as applied to tangentially fired boilers.

**Alternative test method** (Section 63.2) means any method of sampling and analyzing for an air pollutant that is not a test method in this chapter and that has been demonstrated to the Administrator's satisfaction, using Method 301 in Appendix A of this part, to produce results adequate for the Administrator's determination that it may be used in place of a test method specified in this part.

**Alternative water supplies** (Section 300.5) as defined by section 101(34) of CERCLA, includes, but is not limited to, drinking water and household water supplies.

**Alternator/generator efficiency** (Section 92.2) means the ratio of the electrical power output from the alternator/generator to the mechanical power input to the alternator/generator at the operating point.

**Alternator/generator input horsepower** (Section 92.2) means the mechanical horsepower input to the main alternator or generator of a locomotive. For the purpose of calculating brake horsepower, alternator/generator input horsepower does not include any power used to circulate engine coolant, circulate engine lubricant, or to supply fuel to the engine.

**Altitude performance adjustments** (Section 86.1602) are adjustments or modifications made to vehicle, engine, or emission control functions in order to improve emission control performance at altitudes other than those for which the vehicles were designed.

**Aluminum basis material** (Section 465.02) means aluminum, aluminum alloys and aluminum coated steels which are processed in coil coating. The term can means a container formed from sheet metal and consisting of a body and two ends or a body and a top.

**Aluminum casting** (Section 464.02) The remelting of aluminum or an aluminum alloy to form a cast intermediate or final product by pouring or forcing the molten metal into a mold, except for ingots, pigs, or other cast shapes related to nonferrous (primary and secondary) metals manufacturing (40 CFR part 421) and aluminum forming (40 CFR part 467). Processing operations following the cooling of castings not covered under aluminum forming, except for grinding scrubber operations which are covered here, are covered under the electroplating and metal finishing point source categories (40 CFR parts 413 and 433).

**Aluminum forming** (Section 467.02) is a set of manufacturing operations in which aluminum and aluminum alloys are made into semifinished products by hot or cold working.

**Aluminum forming subcategories** (Section 467.02) based on the drawing process. In the drawing with neat oils subcategory, the drawing process uses a pure or neat oil as a lubricant. In the drawing with emulsions or soaps subcategory, the drawing process uses an emulsion or soap solution as a lubricant.

**Ambient air** (Section 50.1) means that portion of the atmosphere, external to buildings, to which the general public has access.

**Ambient air** (Section 57.103) shall have the meaning given by 40 CFR 50.1(e), as that definition appears upon promulgation of this subpart, or as hereafter amended.

**Ambient air quality** (Section 57.103) refers only to concentrations of sulfur dioxide in the ambient air, unless otherwise specified.

**Ambient water criterion** (Section 129.2) means that concentration of a toxic pollutant in a navigable water that, based upon available data, will not result in adverse impact on important aquatic life, or on consumers of such aquatic life, after exposure of that aquatic life for periods of time exceeding 96 hours and continuing at least through one reproductive cycle; and will not result in a significant risk of adverse health effects in a large human population based on available information such as mammalian laboratory toxicity data, epidemiological studies of human occupational exposures, or human exposure data, or any other relevant data.

**Ammonia-N (or ammonia-nitrogen)** (Section 420.02) means the value obtained by manual distillation (at pH 9.5) followed by the Nesslerization method specified in 40 CFR 136.3.

**Amount of pesticidal product** (Section 167.3) means quantity, expressed in weight or volume of the product, and is to be reported in pounds for solid or semi-solid pesticides and active ingredients or gallons for liquid pesticides and active ingredients, or number of individual retail units for devices.

**Amount of pesticide or active ingredient** (Section 169.1) The term "amount of pesticide or active ingredient" means the weight or volume of the pesticide or active ingredient used in producing a pesticide expressed as weight for solid or semi-solid products and as weight or volume of liquid products.

**An approved measure** (Section 57.103) refers to one contained in an NSO which is in effect.

**Anaerobic digestion** (Section 503.31) is the biochemical decomposition of organic matter in sewage sludge into methane gas and carbon dioxide by microorganisms in the absence of air.

**Ancillary equipment** (Section 260.10) means any device including, but not limited to, such devices as piping, fittings, flanges, valves, and pumps, that is used to distribute, meter, or control the flow of hazardous waste from its point of generation to a storage or treatment tank(s), between hazardous waste storage and treatment tanks to a point of disposal onsite, or to a point of shipment for disposal off-site.

**Ancillary equipment** (Section 280.12) means any devices including, but not limited to, such devices as piping, fittings, flanges, valves, and pumps used to distribute, meter, or control the flow of regulated substances to and from an UST.

**Ancillary operation** (Section 467.02) is a manufacturing operation that has a large flow, discharges significant amounts of pollutants, and may not be present at every plant in a subcategory, but when present is an integral part of the aluminum forming process.

**Ancillary operation** (Section 467.11) shall mean any operation not previously included in the core, performed on-site, following or preceding the rolling operation. The ancillary operations shall include continuous rod casting, continuous sheet casting, solution heat treatment, cleaning or etching.

**Ancillary operation** (Section 467.31) shall mean any operation not previously included in the core, performed on-site, following or preceding the extrusion operation. The ancillary operations shall include direct chill casting, press or solution heat treatment, cleaning or etching, degassing, and extrusion press hydraulic fluid leakage.

**Ancillary operation** (Section 467.41) shall mean any operation not previously included in the core, performed on-site, following or preceding the forging operation. The ancillary operations shall include forging air pollution scrubbers, solution heat treatment, and cleaning or etching.

**Ancillary operation** (Section 467.51) shall mean any operation not previously included in the core, performed on-site, following or preceding the drawing operation. The ancillary operation shall include continuous rod casting, solution heat treatment, and cleaning or etching.

**Ancillary operation** (Section 468.02) shall mean any operation associated with a primary forming operation. These ancillary operations include surface and heat treatment, hydrotesting, sawing, and surface coating.

**Ancillary operations** (Section 461.2) means all of the operations specific to battery manufacturing and not included specifically within anode or cathode manufacture (ancillary operations are primarily associated with battery assembly and chemical production of anode or cathode active materials).

**Ancillary operations** (Section 461.2) means all of the operations specific to battery manufacturing and not included specifically within anode or cathode manufacture (ancillary operations are primarily associated with battery assembly and chemical production of anode or cathode active materials).

**Anhydrous product** (Section 417.11) shall mean the theoretical product that would result if all water were removed from the actual product.

**Anhydrous product** (Section 417.21) shall mean the theoretical product that would result if all water were removed from the actual product.

**Anhydrous product** (Section 417.31) shall mean the theoretical product that would result if all water were removed from the actual product.

**Anhydrous product** (Section 417.41) shall mean the theoretical product that would result if all water were removed from the actual product.

**Animal fat** (Section 112.2) means a non-petroleum oil, fat, or grease of animal, fish, or marine mammal origin.

**Animal feeding operation** (Section 122.2) is defined at Sec. 122.23.

**Animals** (Section 116.3) means appropriately sensitive animals which carry out respiration by means of a lung structure permitting gaseous exchange between air and the circulatory system;

**Annealing with oil** (Section 468.02) shall mean the use of oil to quench a workpiece as it passes from an annealing furnace.

**Annealing with water** (Section 468.02) shall mean the use of a water spray or bath, of which water is the major constituent, to quench a workpiece as it passes from an annealing furnace.

**Annual** (Section 704.3) means the corporate fiscal year.

**Annual average** (Section 407.61) shall mean the maximum allowable discharge of BOD5 or TSS as calculated by multiplying the total mass (kkg or 1000 lb) of each raw commodity processed for the entire processing season or calendar year by the applicable annual average limitation.

**Annual average** (Section 407.71) shall mean the maximum allowable discharge of BOD5 or TSS as calculated by multiplying the total mass (kkg or 1000 lb) of each raw commodity processed for the entire processing season or calendar year by the applicable annual average limitation.

**Annual average** (Section 407.81) shall mean the maximum allowable discharge of BOD5 or TSS, as calculated by multiplying the total mass (kkg or 1000 lb) of each final product produced for the entire processing season or calendar year by the applicable annual average limitation.

**Annual average** (Section 430.0) The mean concentration, mass loading or production-normalized mass loading of a pollutant over a period of 365 consecutive days (or such other period of time determined by the permitting authority to be sufficiently long to encompass expected variability of the concentration, mass loading, or production-normalized mass loading at the relevant point of measurement).

**Annual committed effective dose** (Section 191.12) means the committed effective dose resulting from one-year intake of radionuclides released plus the annual effective dose caused by direct radiation from facilities or activities subject to subparts B and C of this part.

**Annual committed effective dose equivalent** (Section 197.2) means the effective dose equivalent received by an individual in one year from radiation sources external to the individual plus the committed effective dose equivalent.

**Annual document log** (Section 761.13) means the detailed information maintained at the facility on the PCB waste handling at the facility.

**Annual pollutant loading rate** (Section 503.11) is the maximum amount of a pollutant that can be applied to a unit area of land during a 365 day period.

**Annual precipitation and annual evaporation** (Section 440.132) are the mean annual precipitation and mean annual lake evaporation, respectively, as established by the U.S. Department of Commerce, Environmental Science Services Administration, Environmental Data Services, or equivalent regional rainfall and evaporation data.

**Annual report** (Section 761.13) means the written document submitted each year by each disposer and commercial storer of PCB waste to the appropriate EPA Regional Administrator. The annual report is a brief summary of the information included in the annual document log.

**Annual State air monitoring report** (Section 58.1) is an annual report, prepared by control agencies and submitted to EPA for approval that consists of an annual data summary report for all pollutants and a detailed report describing any proposed changes to their air quality surveillance network.

**Annual whole sludge application rate** (Section 503.11) is the maximum amount of sewage sludge (dry weight basis) that can be applied to a unit area of land during a 365 day period.

**ANSI S3.19-1974** (Section 211.203) A revision of the ANSI Z24.22-1957 measurement procedure using one-third octave band stimuli presented under diffuse (reverberant) acoustic field conditions.

**ANSI Z24.22-1957** (Section 211.203) A measurement procedure published by the American National Standards Institute (ANSI) for obtaining hearing protector attenuation values at nine of the one-third octave band center frequencies by using pure tone stimuli presented to ten different test subjects under anechoic conditions.

**Antimony** (Section 415.661) shall mean the total antimony present in the process wastewater stream exiting the wastewater treatment system.

**Any collection** (Section 142.2) treatment, storage, and distribution facilities under control of the operator of such system and used primarily in connection with such system; and any collection or pretreatment storage facilities not under such control which are used primarily in connection with such system. Such term does not include any "special irrigation district." A public water system is either a "community water system" or a "noncommunity water system" as defined in Sec. 141.2.

**API** (Section 68.3) means the American Petroleum Institute.

**Appeals Officer** (Section 1601.3) means the person designated by the Chairperson to process appeals of denials of requests for CSB records under the FOIA.

**Applicable implementation plan** (Section 93.101) is defined in section 302(q) of the CAA and means the portion (or portions) of the implementation plan, or most recent revision thereof, which has been approved under section 110, or promulgated under section 110©, or promulgated or approved pursuant to regulations promulgated under section 301(d) and which implements the relevant requirements of the CAA.

**Applicable implementation plan or applicable SIP** (Section 93.152) means the portion (or portions) of the SIP or most recent revision thereof, which has been approved under section 110 of the Act, or promulgated under section 110© of the Act (Federal implementation plan), or promulgated or approved pursuant to regulations promulgated under section 301(d) of the Act and which implements the relevant requirements of the Act.

**Applicable legal requirements** (Section 66.3) means any of the following: (1) In the case of any major source, any emission limitation, emission standard, or compliance schedule under any EPA-approved State implementation plan (regardless of whether the source is subject to a Federal or State consent decree); (2) In the case of any source, an emission limitation, emission standard, standard of Performance, or other requirement (including, but not limited to, work practice standards) established under section 111 or 112 of the Act; (3) In the case of a source that is subject to a federal or federally approved state judicial consent decree or EPA approved extension, order, or suspension, any interim emission control requirement or schedule of compliance under that consent decree, extension, order or suspension; (4) In the case of a nonferrous smelter which has received a primary nonferrous smelter order issued or approved by EPA under Section 119 of the Act, any interim emission control requirement (including a requirement relating to the use of supplemental or intermittent controls) or schedule of compliance under that order.

**Applicable requirement** (Section 64.1) shall have the same meaning as provided under part 70 of this chapter.

**Applicable requirement** (Section 70.2) means all of the following as they apply to emissions units in a part 70 source (including requirements that have been promulgated or approved by EPA through rulemaking at the time of issuance but have future-effective compliance dates): (1) Any standard or other requirement provided for in the applicable implementation plan approved or promulgated by EPA through rulemaking under title I of the Act that implements the relevant requirements of the Act, including any revisions to that plan promulgated in part 52 of this chapter; (2) Any term or condition of any preconstruction permits issued pursuant to regulations approved or promulgated through rulemaking under title I, including parts C or D, of the Act; (3) Any standard or other requirement under section 111 of the Act, including section 111(d); (4) Any standard or other requirement under section 112 of the Act, including any requirement concerning accident prevention under section 112®(7) of the Act; (5) Any standard or other requirement of the acid rain program under title IV of the Act or the regulations promulgated thereunder; (6) Any requirements established pursuant to section 504(b) or section 114(a)(3) of the Act; (7) Any standard or other requirement governing solid waste incineration, under section 129 of the Act; (8) Any standard or other requirement for consumer and commercial products, under section 183(e) of the Act; (9 Any standard or other requirement for tank vessels under section 183(f) of the Act; (10) Any standard or other requirement of the program to control air pollution from outer continental shelf sources, under section 328 of the Act; (11) Any standard or other requirement of the regulations promulgated to protect stratospheric ozone under title VI of the Act, unless the Administrator has determined that such requirements need not be contained in a title V permit; and (12) Any national ambient air quality standard or increment or visibility requirement under part C of title I of the Act, but only as it would apply to temporary sources permitted pursuant to section 504(e) of the Act.

**Applicable requirement** (Section 71.2) means all of the following as they apply to emissions units in a part 71 source (including requirements that have been promulgated or approved by EPA through rulemaking at the time of issuance but have future compliance dates): (1) Any standard or other requirement provided for in the applicable implementation plan approved or promulgated by EPA through rulemaking under title I of the Act that implements the relevant requirements of the Act, including any revisions to that plan promulgated in part 52 of this chapter; (2) Any term or condition of any preconstruction permits issued pursuant to regulations approved or promulgated through rulemaking under title I, including parts C or D, of the Act; (3) Any standard or other requirement under section 111 of the Act, including section 111(d); (4) Any standard or other requirement under section 112 of the Act, including any requirement concerning accident prevention under section 112®(7) of the Act; (5) Any standard or other requirement of the acid rain program under title IV of the Act or 40 CFR parts 72 through 78; (6) Any requirements established pursuant to section 114(a)(3) or 504(b) of the Act; (7) Any standard or other requirement governing solid waste incineration, under section 129 of the Act; (8) Any standard or other requirement for consumer and commercial products, under section 183(e) of the Act; (9) Any standard or other requirement for tank vessels, under section 183(f) of the Act; (10) Any standard or other requirement of the program to control air pollution from outer continental shelf sources, under section 328 of the Act; (11) Any standard or other requirement of the regulations promulgated at 40 CFR part 82 to protect stratospheric ozone under title VI of the Act, unless the Administrator has determined that such requirements need not be contained in a title V permit; and (12) Any national ambient air quality standard or increment or visibility requirement under part C of title I of the Act, but only as it would apply to temporary sources permitted pursuant to section 504(e) of the Act.

**Applicable requirements** (Section 300.5) means those cleanup standards, standards of control, and other substantive requirements, criteria, or limitations promulgated under federal environmental or state environmental or facility siting laws that specifically address a hazardous substance, pollutant, contaminant, remedial action, location, or other circumstance found at a CERCLA site. Only those state standards that are identified by a state in a timely manner and that are more stringent than federal requirements may be applicable.

**Applicable standard** (Section 92.2) means a standard to which a locomotive or locomotive engine is subject; or, where a locomotive or locomotive engine is certified another standard or FEL, applicable standard means the other standard or FEL to which the locomotive or locomotive engine is certified, as allowed by Sec. 92.8. This definition does not apply to subpart D of this part.

**Applicable standard** (Section 92.302) means a standard that would have otherwise been applicable had the locomotive or locomotive engine not been certified under this subpart to an FEL different than that standard.

**Applicable standard** (Section 94.2) means a standard to which an engine is subject; or, where an engine is certified to another standard or FEL, applicable standard means the other standard or FEL to which the engine is certified, as allowed by Sec. 94.8. This definition does not apply to subpart D of this part.

**Applicable standard** (Section 94.302) means a standard that would have otherwise been applicable had the engine not been certified under this subpart to an FEL different than that standard.

**Applicable standards and limitations** (Section 122.2) means all State, interstate, and federal standards and limitations to which a "discharge," a "sewage sludge use or disposal practice," or a related activity is subject under the CWA, including "effluent limitations," water quality standards, standards of performance, toxic effluent standards or prohibitions, "best management practices," pretreatment standards, and "standards for sewage sludge use or disposal" under sections 301, 302, 303, 304, 306, 307, 308, 403 and 405 of CWA.

**Applicable water quality standards** (Section 110.1) means State water quality standards adopted by the State pursuant to section 303 of the Act or promulgated by EPA pursuant to that section;

**Applicant** (Section 105.2) means the person authorized to complete the application on behalf of an industrial organization or political subdivision of States.

**Applicant** (Section 152.3) means a person who applies for a registration, amended registration, or reregistration, under FIFRA sec. 3.

**Applicant** (Section 164.2) means any person who has made application to have a pesticide registered or classified pursuant to the provisions of the Act.

**Applicant** (Section 172.1) means any person who applies for an experimental use permit, pursuant to section 5 of the Act.

**Applicant** (Section 176.3) means any entity authorized under section 18 of FIFRA to request an emergency exemption that requests such an exemption under Sec. 166.20 of this chapter, or issues a crisis exemption under Sec. 166.40 of this chapter.

**Applicant** (Section 195.2) means an individual or organization that submits an application to the RMP program, including the individual proficiency component of the RMP program, or the RCP program. An applicant to the RMP program must submit a separate application for each location from which it provides radon measurement services. After the application is accepted by EPA, the applicant becomes a "participant" in the proficiency programs.

**Applicant** (Section 53.1) means a person or entity who submits an application for a reference or equivalent method determination under Sec. 53.4, or a person or entity who assumes the rights and obligations of an applicant under Sec. 53.7. Applicant may include a manufacturer, distributor, supplier, or vendor.

**Application** (Section 105.2) means a completed questionnaire, nomination form, or other documentation submitted to or by the States, EPA Regions or headquarters for consideration of a national CWA Recognition Award.

**Application** (Section 122.2) means the EPA standard national forms for applying for a permit, including any additions, revisions or modifications to the forms; or forms approved by EPA for use in "approved States," including any approved modifications or revisions.

**Application** (Section 124.2) means the EPA standard national forms for applying for a permit, including any additions, revisions or modifications to the forms; or forms approved by EPA for use in "approved States," including any approved modifications or revisions. For RCRA, application also includes the information required by the Director under Secs. 270.14 through 270.29 [contents of Part B of the RCRA application].

**Application** (Section 144.3) means the EPA standard national forms for applying for a permit, including any additions, revisions or modifications to the forms; or forms approved by EPA for use in approved States, including any approved modifications or revisions.

**Application** (Section 146.3) means the EPA standard national forms for applying for a permit, including any additions, revisions or modifications to the forms; or forms approved by EPA for use in approved States, including any approved modifications or revisions. For RCRA, application also includes the information required by the Director under Sec. 122.25 (contents of Part B of the RCRA application).

**Application** (Section 195.2) means the documents submitted to EPA by applicants to the RMP and RCP programs which request participation in a program.

**Application** (Section 232.2) means a form for applying for a permit to discharge dredged or fill material into waters of the United States.

**Application** (Section 270.2) means the EPA standard national forms for applying for a permit, including any additions, revisions or modifications to the forms; or forms approved by EPA for use in approved States, including any approved modifications or revisions. Application also includes the information required by the Director under Secs. 270.14 through 270.29 (contents of part B of the RCRA application).

**Application for research or marketing permit includes** (Section 160.3) (1) An application for registration, amended registration, or reregistration of a pesticide product under FIFRA sections 3, 4 or 24©. (2) An application for an experimental use permit under FIFRA section 5. (3) An application for an exemption under FIFRA section 18. (4) A petition or other request for establishment or modification of a tolerance, for an exemption for the need for a tolerance, or for other clearance under FFDCA section 408. (5) A petition or other request for establishment or modification of a food additive regulation or other clearance by EPA under FFDCA section 409. (6) A submission of data in response to a notice issued by EPA under FIFRA section 3© (2) (B). (7) Any other application, petition, or submission sent to EPA intended to persuade EPA to grant, modify, or leave unmodified a registration or other approval required as a condition of sale or distribution of a pesticide.

**Application means Form 9310-1** (Section 310.3) shown in Appendix III of this part, including all documentation and additional information you submit to support a request for reimbursement.

**Applications** (Section 85.2113) means all vehicle or engine configurations for which one part is being certified as set forth in the aftermarket part manufacturer's notification of intent to certify pursuant to Sec. 85.2115(a)(1).

**Applicator** (Section 82.3) means the person who applies methyl bromide.

**Approach angle** (Section 86.1803-01) means the smallest angle in a plan side view of an automobile, formed by the level surface on which the automobile is standing and a line tangent to the front tire static loaded radius arc and touching the underside of the automobile forward of the front tire.

**Appropriate** (Section 157.21) when used with respect to child-resistant packaging, means that the packaging is chemically compatible with the pesticide contained therein.

**Appropriate Act and regulations** (Section 124.2) means the Clean Water Act (CWA); the Solid Waste Disposal Act, as amended by the Resource Conservation Recovery Act (RCRA); or Safe Drinking Water Act (SDWA), whichever is applicable; and applicable regulations promulgated under those statutes. In the case of an "approved State program" appropriate Act and regulations includes program requirements.

**Appropriate Act and regulations** (Section 144.3) means the Solid Waste Disposal Act, as amended by the Resource Conservation and Recovery Act (RCRA); or Safe Drinking Water Act (SDWA), whichever is applicable; and applicable regulations promulgated under those statutes.

**Appropriate pollution control technology** (Section 455.10) means the wastewater treatment technology listed in Table 10 to this part 455 for a particular PAI(s) including an emulsion breaking step prior to the listed technology when emulsions are present in the wastewater to be treated.

**Appropriate treatment of the recycle water in subpart J, Sec. 440.104** (Section 440.132) includes, but is not limited to pH adjustment, settling and pH adjustment, settling, and mixed media filtration.

**Approval Authority** (Section 403.3) means the Director in an NPDES State with an approved State pretreatment program and the appropriate Regional Administrator in a non-NPDES State or NPDES State without an approved State pretreatment program.

**Approved clean coal technology demonstration project** (Section 76.2) means a project using funds appropriated under the Department of Energy's "Clean Coal Technology Demonstration Program," up to a total amount of $2,500,000,000 for commercial demonstration of clean coal technology, or similar projects funded through appropriations for the Environmental Protection Agency. The Federal contribution for a qualifying project shall be at least 20 percent of the total cost of the demonstration project.

**Approved permit program** (Section 60.2) means a State permit program approved by the Administrator as meeting the requirements of part 70 of this chapter or a Federal permit program established in this chapter pursuant to Title V of the Act (42 U.S.C. 7661).

**Approved permit program** (Section 61.02) means a State permit program approved by the Administrator as meeting the requirements of part 70 of this chapter or a Federal permit program established in this chapter pursuant to title V of the Act (42 U.S.C. 7661).

**Approved permit program** (Section 63.2) means a State permit program approved by the Administrator as meeting the requirements of part 70 of this chapter or a Federal permit program established in this chapter pursuant to title V of the Act (42 U.S.C. 7661).

**Approved permit program** (Section 65.2) means a State permit program approved by the Administrator as meeting the requirements of part 70 of this chapter or a Federal permit program established in this chapter pursuant to title V of the Act (42 U.S.C. 7661).

**Approved permit program or approved program** (Section 239.2) means a state Subtitle D permit program or other system of prior approval and conditions required under section 4005©(1)(B) of RCRA that has been determined to be adequate by EPA under this part.

**Approved POTW Pretreatment Program or Program or POTW Pretreatment Program** (Section 403.3) means a program administered by a POTW that meets the criteria established in this regulation (Secs. 403.8 and 403.9) and which has been approved by a Regional Administrator or State Director in accordance with Sec. 403.11 of this regulation.

**Approved program** (Section 232.2) means a State program which has been approved by the Regional Administrator under part 233 of this chapter or which is deemed approved under section 404(h)(3), 33 U.S.C. 1344(h)(3).

**Approved program or approved State** (Section 122.2) means a State or interstate program which has been approved or authorized by EPA under part 123.

**Approved program or approved State** (Section 270.2) means a State which has been approved or authorized by EPA under part 271.

**Approved Section 120 program** (Section 66.3) means a State program to assess and collect Section 120 penalties that has been approved by the Administrator.

**Approved state** (Section 239.2) means a state whose Subtitle D permit program or other system of prior approval and conditions required under section 4005©(1)(B) of RCRA has been determined to be adequate by EPA under this part.

**Approved State primacy program** (Section 142.2) consists of those program elements listed in Sec. 142.11(a) that were submitted with the initial State application for primary enforcement authority and approved by the EPA Administrator and all State program revisions thereafter that were approved by the EPA Administrator.

**Approved State Program** (Section 144.3) means a UIC program administered by the State or Indian Tribe that has been approved by EPA according to SDWA sections 1422 and/or 1425.

**Approved State program** (Section 501.2) means a State program which has received EPA approval under this part.

**Apricots** (Section 407.61) shall include the processing of apricots into the following product styles: Canned and frozen, pitted and unpitted, peeled and unpeeled, whole, halves, slices, nectar, and concentrate.

**Aquaculture project** (Section 122.2) is defined at Sec. 122.25.

**Aquatic animals** (Section 116.3) means appropriately sensitive wholly aquatic animals which carry out respiration by means of a gill structure permitting gaseous exchange between the water and the circulatory system;

**Aquatic environment and aquatic ecosystem** (Section 230.3) mean waters of the United States, including wetlands that serve as habitat for interrelated and interacting communities and populations of plants and animals.

**Aquatic flora** (Section 116.3) means plant life associated with the aquatic eco-system including, but not limited to, algae and higher plants;

**Aquifer** (Section 144.3) means a geological "formation," group of formations, or part of a formation that is capable of yielding a significant amount of water to a well or spring.

**Aquifer** (Section 146.3) means a geological formation, group of formations, or part of a formation that is capable of yielding a significant amount of water to a well or spring.

**Aquifer** (Section 147.2902) a geologic formation, group of formations, or part of a formation that is capable of yielding a significant amount of water to a well or spring.

**Aquifer** (Section 149.2) means a geological formation, group of formations, or part of a formation that is capable of yielding a significant amount of water to a well or spring.

**Aquifer** (Section 191.12) means an underground geological formation, group of formations, or part of a formation that is capable of yielding a significant amount of water to a well or spring.

**Aquifer** (Section 258.2) means a geological formation, group of formations, or portion of a formation capable of yielding significant quantities of ground water to wells or springs.

**Aquifer** (Section 260.10) means a geologic formation, group of formations, or part of a formation capable of yielding a significant amount of ground water to wells or springs.

**Aquifer** (Section 270.2) means a geological formation, group of formations, or part of a formation that is capable of yielding a significant amount of water to a well or spring.

**Aquifer** (Section 503.21) is a geologic formation, group of geologic formations, or a portion of a geologic formation capable of yielding ground water to wells or springs.

**Arbitrator** (Section 304.12) means the person appointed in accordance with Sec. 304.22 of this part and governed by the provisions of this part.

**Arch-fired boiler** (Section 76.2) means a dry bottom boiler with circular burners, or coal and air pipes, oriented downward and mounted on waterwalls that are at an angle significantly different from the horizontal axis and the vertical axis. This definition shall include only the following units: and (ii) Any fuel, except coal or solid or liquid coal-derived fuel, for the remaining heat input, if any. (2) For purposes of part 75 of this chapter, the combustion of: (i) Natural gas or other gaseous fuel (including coal-derived gaseous fuel) for at least 90.0 percent of the unit's average annual heat input during the previous three calendar years and for at least 85.0 percent of the annual heat input in each of those calendar years; and (ii) Fuel oil, for the remaining heat input, if any. (3) For purposes of part 75 of this chapter, a unit may initially qualify as gas-fired if the designated representative demonstrates to the satisfaction of the Administrator that the requirements of paragraph (2) of this definition are met, or will in the future be met, through one of the following submissions: (i) For a unit for which a monitoring plan has not been submitted under Sec. 75.62 of this chapter, the designated representative submits either: (A) Fuel usage data for the unit for the three calendar years immediately preceding the date of initial submission of the monitoring plan for the unit under Sec. 75.62; or (B) If a unit does not have fuel usage data for one or more of the three calendar years immediately preceding the date of initial submission of the monitoring plan for the unit under Sec. 75.62, the unit's designated fuel usage; all available fuel usage data (including the percentage of the unit's heat input derived from the combustion of gaseous fuels), beginning with the date on which the unit commenced commercial operation; and the unit's projected fuel usage. (ii) For a unit for which a monitoring plan has already been submitted under Sec. 75.62, that has not qualified as gas-fired under paragraph (3)(i) of this definition, and whose fuel usage changes, the designated representative submits either: (A) Three calendar years of data following the change in the unit's fuel usage, showing that no less than 90.0 percent of the unit's average annual heat input during the previous three calendar years, and no less than 85.0 percent of the unit's annual heat input during any one of the previous three calendar years, is from the combustion of gaseous fuels and the remaining heat input is from the combustion of fuel oil; or (B) A minimum of 720 hours of unit operating data following the change in the unit's fuel usage, showing that no less than 90.0 percent of the unit's heat input is from the combustion of gaseous fuels and the remaining heat input is from the combustion of fuel oil, and a statement that this changed pattern of fuel usage is considered permanent and is projected to continue for the foreseeable future. (iii) If a unit qualifies as gas-fired under paragraph (3)(i) or (ii) of this definition, the unit is classified as gas-fired as of the date of the submission under such paragraph. (4) For purposes of part 75 of this chapter, a unit that initially qualifies as gas-fired under paragraph (3)(i) or (ii) of this definition must meet the criteria in paragraph (2) of this definition each year in order to continue to qualify as gas-fired. If such a unit combusts only gaseous fuel and fuel oil but fails to meet such criteria for a given year, the unit no longer qualifies as gas-fired starting January 1 of the year after the first year for which the criteria are not met. If such a unit combusts fuel other than gaseous fuel or fuel oil and fails to meet such criteria in a given year, the unit no longer qualifies as gas-fired starting the day after the first day for which the criteria are not met. If a unit failing to meet the criteria in paragraph (2) of this definition initially qualified as a gas-fired unit under paragraph (3) of this definition, the unit may qualify as a gas-fired unit for a subsequent year only if the designated representative submits the data specified in paragraph (3)(ii)(A) of this definition.

**Area coated** (Section 466.02) means the area of basis material covered by each coating of enamel.

**Area Committee (AC) as provided for by CWA sections 311(a)(18) and (j)(4)** (Section 300.5) means the entity appointed by the President consisting of members from qualified personnel of federal, state, and local agencies with responsibilities that include preparing an area contingency plan for an area designated by the President.

**Area contingency plan (ACP) as provided for by CWA sections 311(a)(19) and (j)(4)** (Section 300.5) means the plan prepared by an Area Committee that is developed to be implemented in conjunction with the NCP and RCP, in part to address removal of a worst case discharge and to mitigate or prevent a substantial threat of such a discharge from a vessel, offshore facility, or onshore facility operating in or near an area designated by the President.

**Area of review** (Section 144.3) means the area surrounding an injection well described according to the criteria set forth in Sec. 146.06 or in the case of an area permit, the project area plus a circumscribing area the width of which is either \1/4\ of a mile or a number calculated according to the criteria set forth in Sec. 146.06.

**Area of review** (Section 146.3) means the area surrounding an injection well described according to the criteria set forth in Sec. 146.06 or in the case of an area permit, the project area plus a circumscribing area the width of which is either \1/4\ of a mile or a number calculated according to the criteria set forth in Sec. 146.06.

**Area of review** (Section 147.3001) For the purposes of this subpart, area of review means the area surrounding an injection well or project area described according to the criteria set forth in Sec. 147.3009 of this subpart.

**Area processed** (Section 465.02) means the area actually exposed to process solutions. Usually this includes both sides of the metal strip.

**Area processed** (Section 466.02) means the total basis material area exposed to processing solutions.

**Area source** (Section 51.100) means any small residential, governmental, institutional, commercial, or industrial fuel combustion operations; onsite solid waste disposal facility; motor vehicles, aircraft vessels, or other transportation facilities or other miscellaneous sources identified through inventory techniques similar to those described in the "AEROS Manual series, Vol. II AEROS User's Manual," EPA-450/2-76-029 December 1976.

**Area source** (Section 63.2) means any stationary source of hazardous air pollutants that is not a major source as defined in this part.

**Areawide agency** (Section 130.2) An agency designated under section 208 of the Act, which has responsibilities for WQM planning within a specified area of a State.

**Areawide air quality modeling analysis** (Section 93.152) means an assessment on a scale that includes the entire nonattainment or maintenance area which uses an air quality dispersion model to determine the effects of emissions on air quality.

**Arithmetic mean** (Section 745.63) means the algebraic sum of data values divided by the number of data values (e.g., the sum of the concentration of lead in several soil samples divided by the number of samples).

**Armed Forces vessel** (Section 1700.3) means a vessel owned or operated by the United States Department of Defense or the United States Coast Guard, other than vessels that are time or voyage chartered by the Armed Forces, vessels of the U.S. Army Corps of Engineers, or vessels that are memorials or museums.

**Arrest** (Section 303.11) Restraint of an arrestee's liberty or the equivalent through the service of judicial process compelling such a person to respond to a criminal accusation.

**Arsenic** (Section 415.671) shall mean the total arsenic present in the process wastewater stream exiting the wastewater treatment system.

**Article** (Section 372.3) means a manufactured item: (1) Which is formed to a specific shape or design during manufacture; (2) which has end use functions dependent in whole or in part upon its shape or design during end use; and (3) which does not release a toxic chemical under normal conditions of processing or use of that item at the facility or establishments.

**Article** (Section 68.3) means a manufactured item, as defined under 29 CFR 1910.1200(b), that is formed to a specific shape or design during manufacture, that has end use functions dependent in whole or in part upon the shape or design during end use, and that does not release or otherwise result in exposure to a regulated substance under normal conditions of processing and use.

**Article** (Section 704.3) means a manufactured item (1) which is formed to a specific shape or design during manufacture, (2) which has end use function(s) dependent in whole or in part upon its shape or design during end use, and (3) which has either no change of chemical composition during its end use or only those changes of composition which have no commercial purpose separate from that of the article, and that result from a chemical reaction that occurs upon end use of other chemical substances, mixtures, or articles; except that fluids and particles are not considered articles regardless of shape or design.

**Article** (Section 720.3) means a manufactured item (1) which is formed to a specific shape or design during manufacture, (2) which has end use function(s) dependent in whole or in part upon its shape or design during end use, and (3) which has either no change of chemical composition during its end use or only those changes of composition which have no commercial purpose separate from that of the article and that may occur as described in Sec. 720.36(g)(5), except that fluids and particles are not considered articles regardless of shape or design.

**Article 5 allowances** (Section 82.3) means the allowances apportioned under Sec. 82.9(a).

**Article is a manufactured item** (section 710.2) (1) Which is formed to a specific shape or design during manufacture, (2) which has end use function(s) dependent in whole or in part upon its shape or design during end use, and (3) which has either no change of chemical composition during its end use or only those changes of composition which have no commercial purpose separate from that of the article and that may occur as described in Sec. 710.4(d)(5); except that fluids and particles are not considered articles regardless of shape or design.

**As expeditiously as practicable considering technological feasibility** (Section 192.31) means as quickly as possible considering: the physical characteristics of the tailings and the site; the limits of available technology; the need for consistency with mandatory requirements of other regulatory programs; and factors beyond the control of the licensee. The phrase permits consideration of the cost of compliance only to the extent specifically provided for by use of the term "available technology."

**Asbestos** (Section 763.163) means the asbestiform varieties of: chrysotile (serpentine); crocidolite (riebeckite); amosite (cummingtonite-grunerite); tremolite; anthophyllite; and actinolite.

**Asbestos** (Section 763.83) means the asbestiform varieties of: Chrysotile (serpentine); crocidolite (riebeckite); amosite (cum- mingtonitegrunerite); anthophyllite; tremolite; and actinolite.

**Asbestos debris** (Section 763.83) means pieces of ACBM that can be identified by color, texture, or composition, or means dust, if the dust is determined by an accredited inspector to be ACM.

**Asbestos-containing building material (ACBM)** (Section 763.83) means surfacing ACM, thermal system insulation ACM, or miscellaneous ACM that is found in or on interior structural members or other parts of a school building.

**Asbestos-containing material (ACM)** (Section 763.83) when referring to school buildings means any material or product which contains more than 1 percent asbestos.

**Asbestos-containing product** (Section 763.163) means any product to which asbestos is deliberately added in any concentration or which contains more than 1.0 percent asbestos by weight or area.

**As-fired** (Section 72.2) means the taking of a fuel sample just prior to its introduction into the unit for combustion.

**ASME** (Section 68.3) means the American Society of Mechanical Engineers.

**As-received condition** (Section 86.1803-01) means the condition of an in-use vehicle procured for emission testing required by this subpart upon which no adjustments, maintenance, or component replacement has occurred subsequent to the vehicle's last routine operation by the vehicle's owner, lessee, or operator prior to procurement.

**Assistance agreement** (Section 307.14) means the legal instrument EPA uses to transfer money, property, services, or anything of value to a recipient to accomplish a public purpose. It is either a grant or cooperative agreement (see 40 CFR part 35) and will specify: budget and project periods; the Federal share of eligible project costs; a description of the work to be accomplished; and any special conditions.

**Assistant Administrator for Air and Radiation** (Section 57.103) means the Assistant Administrator for Air and Radiation of the U.S. Environmental Protection Agency.

**Association** (Section 304.12) means the organization offering arbitration services selected by EPA to conduct arbitrations pursuant to this part.

**ASTM** (Section 761.13) means American Society for Testing and Materials, 100 Barr Harbor Drive, West Conshohocken, PA 19428-2959.

**Atomization** (Section 471.02) is the process in which a stream of water or gas impinges upon a molten metal stream, breaking it into droplets which solidify as powder particles.

**At-the-source** (Section 421.31) means at or before the commingling of delacquering scrubber liquor blowdown with other process or nonprocess wastewaters.

**Attitudinal control** (Section 90.102) means the operator regulates either the horizontal or vertical position of the equipment, or both.

**Attorney General** (Section 1400.2) means the Attorney General of the United States or his or her designated representative.

**Auction subaccount** (Section 72.2) means a subaccount in the Special Allowance Reserve, as specified in section 416(b) of the Act, which contains allowances to be sold at auction in the amount of 150,000 per year from calendar year 1995 through 1999, inclusive, and 200,000 per year for each year begnning in calendar year 2000, subject to the adjustments noted in the regulations in part 73, subpart E of this chapter.

**Authorized account representative** (Section 72.2) means a responsible natural person who is authorized, in accordance with part 73 of this chapter, to transfer and otherwise dispose of allowances held in an Allowance Tracking System general account; or, in the case of a unit account, the designated representative of the owners and operators of the affected unit.

**Authorized representative** (Section 260.10) means the person responsible for the overall operation of a facility or an operational unit (i.e., part of a facility), e.g., the plant manager, superintendent or person of equivalent responsibility.

**Automated** (Section 53.1) method or analyzer means a method for measuring concentrations of an ambient air pollutant in which sample collection (if necessary), analysis, and measurement are performed automatically by an instrument.

**Automated continuous parameter monitoring system** (Section 65.2) means a continuous parameter monitoring system that automatically both records the measured data and calculates hourly averages.

**Automated data acquisition and handling system** (Section 72.2) means that component of the CEMS, COMS, or other emissions monitoring system approved by the Administrator for use in the Acid Rain Program, designed to interpret and convert individual output signals from pollutant concentration monitors, flow monitors, diluent gas monitors, opacity monitors, and other component parts of the monitoring system to produce a continuous record of the measured parameters in the measurement units required by part 75 of this chapter.

**Automated data acquisition and handling system or DAHS** (Section 96.2) means that component of the CEMS, or other emissions monitoring system approved for use under subpart H of this part, designed to interpret and convert individual output signals from pollutant concentration monitors, flow monitors, diluent gas monitors, and other component parts of the monitoring system to produce a continuous record of the measured parameters in the measurement units required by subpart H of this part.

**Automated data acquisition and handling system or DAHS** (Section 97.2) means that component of the CEMS, or other emissions monitoring system approved for use under subpart H of this part, designed to interpret and convert individual output signals from pollutant concentration monitors, flow monitors, diluent gas monitors, and other component parts of the monitoring system to produce a continuous record of the measured parameters in the measurement units required by subpart H of this part.

**Automated monitoring and recording system** (Section 65.2) means any means of measuring values of monitored parameters and creating a hard copy or computer record of the measured values that does not require manual reading of monitoring instruments and manual transcription of data values. Automated monitoring and recording systems include, but are not limited to, computerized systems, strip charts, and circular charts.

**Automatic waste feed cutoff (AWFCO) system** (Section 62.1201) means a system comprised of cutoff valves, actuator, sensor, data manager, and other necessary components and electrical circuitry designed, operated and maintained to stop the flow of hazardous waste to the combustion unit automatically and immediately (except as provided by Sec. 63.1206©(3)(viii)) when any operating requirement is exceeded.

**Automobile** (Section 59.101) means passenger cars, vans, motorcycles, trucks, and all other mobile equipment.

**Automobile** (Section 600.022-85) means: (i) Any four-wheel vehicle propelled by a combustion engine using onboard fuel or by an electric motor drawing current from rechargeable storage batteries or other portable energy storage devices (rechargeable using energy from a source off the vehicle such as residential electric service), (ii) Which is manufactured primarily for use on public streets, roads, or highways (except any vehicle operated on a rail or rails), (iii) Which is rated at not more than 8,500 pounds gross vehicle weight, which has a curb weight of not more than 6,000 pounds, and which has a basic vehicle frontal area of not more than 45 square feet, or (iv) Is a type of vehicle which the Secretary determines is substantially used for the same purposes.

**Automobile** (Section 610.11) means any four-wheeled vehicle propelled by fuel which is manufactured primarily for use on public streets, roads, and highways (except any vehicle operated exclusively on a rail or rails), and which is rated at 6,000 lbs. gross vehicle weight or less.

**Automobile refinish coating component** (Section 59.101) means any portion of a coating, such as a reducer or thinner, hardener, additive, etc., recommended (by its manufacturer or importer) to distributors or end-users for automobile refinishing. The raw materials used to produce the components that are mixed by the end-user to prepare a coating for application are not considered automobile refinish coating components. Any reference to automobile refinishing made by a manufacturer or importer on a container or in product literature constitutes a recommendation for automobile refinishing.

**Automobile refinish coating or coating component importer, or importer,** (Section 59.101) means any company, group, or individual that brings automobile refinish coatings or coating components from a location outside the United States into the United States for sale or distribution in the United States.

**Automobile refinish coating or coating component manufacturer, or manufacturer,** (Section 59.101) means any company, group, or individual that produces or packages automobile refinish coatings or coating components for sale or distribution in the United States, including an entity which produces or packages such coatings or coating components under a private label for another party.

**Automobile refinishing** (Section 59.101) means the process of coating automobiles or their parts, including partial body collision repairs that is subsequent to the original coating applied at an automobile original equipment manufacturing plant.

**Auxiliary Emission Control Device (AECD)** (Section 600.022-85) means an element of design as defined in part 86.

**Auxiliary Emission Control Device (AECD)** (Section 86.082-2) means any element of design which senses temperature, vehicle speed, engine RPM, transmission gear, manifold vacuum, or any other parameter for the purpose of activating, modulating, delaying, or deactivating the operation of any part of the emission control system.

**Auxiliary Emission Control Device (AECD)** (Section 86.1803-01) means any element of design which senses temperature, vehicle speed, engine RPM, transmission gear, manifold vacuum, or any other parameter for the purpose of activating, modulating, delaying, or deactivating the operation of any part of the emission control system.

**Auxiliary emission control device (AECD)** (Section 89.2) means any element of design that senses temperature, vehicle speed, engine RPM, transmission gear, or any other parameter for the purpose of activating, modulating, delaying, or deactivating the operation of any part of the emission control system.

**Auxiliary emission control device (AECD)** (Section 90.3) means any element of design that senses temperature, vehicle speed, engine RPM, transmission gear, or any other parameter for the purpose of activating, modulating, delaying, or deactivating the operation of any part of the emission control system.

**Auxiliary emission control device (AECD)** (Section 92.2) means any element of design which senses temperature, locomotive speed, engine RPM, atmospheric pressure, manifold pressure or vacuum, or any other parameter for the purpose of activating, modulating, delaying, or deactivating the operation of any part of the emission control system (including, but not limited to injection timing); or any other feature that causes in-use emissions to be higher than those measured under test conditions, except as allowed by this part.

**Auxiliary emission control device (AECD)** (Section 94.2) means any element of design which senses temperature, vessel speed, engine RPM, atmospheric pressure, manifold pressure or vacuum, or any other parameter for the purpose of activating, modulating, delaying, or deactivating the operation of any part of the emission control system (including, but not limited to injection timing); or any other feature that causes in-use emissions to be higher than those measured under test conditions.

**Auxiliary emission control device** (Section 91.3) means any element of design that senses temperature, engine speed, engine RPM, transmission gear, or any other parameter for the purpose of activating, modulating, delaying, or deactivating the operation of any part of the emission control system.

**Auxiliary engine** (Section 92.2) means a locomotive engine that provides hotel power, but does not provide power to propel the locomotive.

**Auxiliary engine** (Section 94.2) means a marine engine that is not a propulsion engine.

**Auxiliary marine diesel engine** (Section 89.2) means a marine diesel engine that is not a propulsion marine diesel engine.

**Auxiliary power** (Section 92.2) means the power provided by the main propulsion engine to operate accessories such as cooling fans.

**Available** (Section 745.103) means in the possession of or reasonably obtainable by the seller or lessor at the time of the disclosure.

**Available technology** (Section 192.31) means technologies and methods for emplacing a permanent radon barrier on uranium mill tailings piles or impoundments. This term shall not be construed to include extraordinary measures or techniques that would impose costs that are grossly excessive as measured by practice within the industry or one that is reasonably analogous, (such as, by way of illustration only, unreasonable overtime, staffing or transportation requirements, etc., considering normal practice in the industry; laser fusion, of soils, etc.), provided there is reasonable progress toward emplacement of a permanent radon barrier. To determine grossly excessive costs, the relevant baseline against which cost increases shall be compared is the cost estimate for tailings impoundment closure contained in the licensee's tailings closure plan, but costs beyond such estimates shall not automatically be considered grossly excessive.

**Average concentration as it relates to chlorine discharge** (Section 423.11) means the average of analyses made over a single period of chlorine release which does not exceed two hours.

**Average Fuel Economy** (Section 600.022-85) means the unique fuel economy value as computed under Sec. 600.510 for a specific class of automobiles produced by a manufacturer that is subject to average fuel economy standards.

**Average monthly discharge limitation** (Section 122.2) means the highest allowable average of "daily discharges" over a calendar month, calculated as the sum of all "daily discharges" measured during a calendar month divided by the number of "daily discharges" measured during that month.

**Average of daily values for 30 consecutive days** (Section 435.11) means the average of the daily values obtained during any 30 consecutive day period.

**Average of daily values for 30 consecutive days** (Section 435.41) means the average of the daily values obtained during any 30 consecutive day period.

**Average process water usage flow rate** (Section 463.31) of a finishing water process in liters per day is equal to the volume of process water (liters) used per year by a process divided by the number of days per year the process operates. The average process water usage flow rate for a plant with more than one plastics molding and forming process that uses finishing water is the sum of the average process water usage flow rates for the finishing processes.

**Average process water usage flow rate of a cleaning water process in liters per day** (Section 463.21) is equal to the volume of process water (liters) used per year by a process divided by the number of days per year the process operates. The average process water usage flow rate for a plant with more than one plastics molding and forming process that uses cleaning water is the sum of the average process water usage flow rates for the cleaning processes.

**Average volatile organic concentration or average VO concentration** (Section 265.1081) means the mass-weighted average volatile organic concentration of a hazardous waste as determined in accordance with the requirements of Sec. 265.1084 of this subpart.

**Average weekly discharge limitation** (Section 122.2) means the highest allowable average of "daily discharges" over a calendar week, calculated as the sum of all "daily discharges" measured during a calendar week divided by the number of "daily discharges" measured during that week.

**Averaging** (Section 91.1302) means the exchange of marine engine in-use emission credits among engine families within a given manufacturer's product line.

**Averaging** (Section 94.2) means the exchange of emission credits among engine families within a given manufacturer's product line.

**Averaging for chassis-bases heavy-duty vehicles** (Section 86.1803-01) means the exchange of NOX emission credits among test groups within a given manufacturer's product line.

**Averaging for clean-fuel vehicles** (Section 88.202-94) means the sale of clean-fuel vehicles that meet more stringent standards than required, which allows the manufacturer to sell fewer clean-fuel vehicles than would otherwise be required.

**Averaging for heavy-duty engines** (Section 86.090-2) means the exchange of NO<INF>X</INF> and particulate emission credits among engine families within a given manufacturer's product line.

**Averaging for locomotives and locomotive engines** (Section 92.2) means the exchange of emission credits among engine families within a given manufacturer's, or remanufacturer's, product line.

**Averaging set** (Section 86.090-2) means a subcategory of heavy-duty engines within which engine families can average and trade emission credits with one other.

**Averaging set** (Section 86.1803-01)means a subcategory of complete heavy-duty vehicles within which test groups can average and trade emission credits with one another.

**Averaging sets** (Section 86.1702-99) are the categories of LDVs and LDTs for which the manufacturer calculates a fleet average NMOG value. The four averaging sets for fleet average NMOG value calculation purposes are: (1) Class A delivered to a point of first sale in the Northeast Trading Region; (2) Class A delivered to a point of first sale in the All States Trading Region; (3) Class B delivered to a point of first sale in the Northeast Trading Region; (4) Class B delivered to a point of first sale in the All States Trading Region.

**Award** (Section 72.2) means the conditional set-aside by the Administrator, based on the submission of an early ranking application pursuant to subpart D of this part, of an allowance from the Phase I extension reserve, for possible future allocation to a Phase I extension applicant's Allowance Tracking System unit account.

**Awards and plaques** (Section 247.3) refers to free-standing statues and boardlike products generally used as wall-hangings.

**AWFCO (Section 62.1201)** means automatic waste feed cutoff.

**Axle clearance** (Section 86.1803-01) means the vertical distance from the level surface on which an automobile is standing to the lowest point on the axle differential of the automobile.

**Axle Ratio** (Section 600.022-85) means the number of times the input shaft to the differential (or equivalent) turns for each turn of the drive wheels.

**Axle ratio** (Section 86.1002-2001) means all ratios within [plusmn]3% of the axle ratio specified in the configuration in the test order.

**Axle ratio** (Section 86.1002-97) means all ratios within [plusmn]3% of the axle ratio specified in the configuration in the test order.

**Axle ratio** (Section 86.1702-99) means the number of times the input shaft to the differential (or equivalent) turns for each turn of the drive wheels.

**Axle Ratio** (Section 86.602-84) means all ratios within [plusmn]3% of the axle ratio specified in the configuration in the test order.

# B

**Baby foods** (Section 407.81) shall mean the processing of canned fresh fruits and vegetables, meats, eggs, fruit juices, cereal, formulated entrees, desserts and snacks using fresh, pre-processed, or any combination of these and other food ingredients necessary for the production of infant foods.

**Backup fuel** (Section 72.2) means a fuel for a unit where: (1) For purposes of the requirements of the monitoring exception of appendix E of part 75 of this chapter, the fuel provides less than 10.0 percent of the heat input to a unit during the three calendar years prior to certification testing for the primary fuel and the fuel provides less than 15.0 percent of the heat input to a unit in each of those three calendar years; or the Administrator approves the fuel as a backup fuel; and (2) For all other purposes under the Acid Rain Program, a fuel that is not the primary fuel (expressed in mmBtu) consumed by an affected unit for the applicable calendar year.

**Baler** (Section 246.101) means a machine used to compress solid wastes, primary materials, or recoverable materials, with or without binding, to a density or from which will support handling and transportation as a material unit rather than requiring a disposable or reusable container. This specifically excludes briquetters and stationary compaction equipment which is used to compact materials into disposable or reusable containers.

**Ballast** (Section 419.11) shall mean the flow of waters, from a ship, that is treated along with refinery wastewaters in the main treatment system.

**Banked credits** (Section 91.1302) refer to positive emission credits based on actual applicable production/sales volume as contained in the end of model year in-use testing reports submitted to EPA. Some or all of these banked credits may be revoked if EPA review of the end of model year in-use testing reports or any subsequent audit action(s) uncovers problems or errors.

**Banking** (Section 86.090-2) means the retention of heavy-duty engine NO<INF>X</INF> and particulate emission credits, by the manufacturer generating the emission credits, for use in future model year certification programs as permitted by regulation.

**Banking** (Section 86.1803-01) means the retention of NOX emission credits for complete heavy-duty vehicles by the manufacturer generating the emission credits, for use in future model year certification programs as permitted by regulation.

**Banking** (Section 88.202-94) means the retention of credits, by the manufacturer generating the emissions credits, for use in future model-year certification as permitted by regulation.

**Banking** (Section 91.1302) means the retention of marine engine in-use emission credits by the manufacturer generating the emission credits for use in future model year averaging or trading as permitted by these regulations.

**Banking** (Section 92.2) means the retention of emission credits by a credit holder for use in future calendar year averaging or trading as permitted by the regulations in this part.

**Banking** (Section 94.2) means the retention of emission credits by a credit holder for use in future calendar year averaging or trading as permitted by the regulations in this part.

**Bar, billet and bloom** (Section 420.91) means those acid pickling operations that pickle bar, billet or bloom products.

**Barrel** (Section 113.3) means 42 United States gallons at 60 degrees Fahrenheit.

**Barrier** (Section 191.12) means any material or structure that prevents or substantially delays movement of water or radionuclides toward the accessible environment. For example, a barrier may be a geologic structure, a canister, a waste form with physical and chemical characteristics that significantly decrease the mobility of radionuclides, or a material placed over and around waste, provided that the material or structure substantially delays movement of water or radionuclides.

**Base engine** (Section 94.2) means a land-based engine to be marinized, as configured prior to marinization.

**Base fluid** (Section 435.11) means the continuous phase or suspending medium of a drilling fluid formulation.

**Base fluid** (Section 435.41) means the continuous phase or suspending medium of a drilling fluid formulation.

**Base fluid retained on cuttings as applied to BAT effluent limitations and NSPS** (Section 435.11) refers to the American Petroleum Institute Recommended Practice 13B-2 supplemented with the specifications, sampling methods, and averaging method for retention values provided in Appendix 7 of Subpart A of this part.

**Base fluid retained on cuttings as applied to BAT effluent limitations and NSPS** (Section 435.41) refers to the American Petroleum Institute Recommended Practice 13B-2 supplemented with the specifications, sampling methods, and averaging method for retention values provided in Appendix 7 of subpart A of this part.

**Base Level** (Section 600.022-85) means a unique combination of basic engine inertia weight class and transmission class.

**Base Vehicle** (Section 600.022-85) means the lowest priced version of each body style that makes up a car line.

**Baseline** (Section 72.2) means the annual average quantity of fossil fuel consumed by a unit, measured in millions of British Thermal Units (expressed in mmBtu) for calendar years 1985 through 1987; provided that in the event that a unit is listed in the NADB, the baseline will be calculated for each unit-generator pair that includes the unit, and the unit's baseline will be the sum of such unit-generator baselines. The unit-generator baseline will be as provided in the NADB under the data field "BASE8587", as adjusted by the outage hours listed in the NADB under the data field "OUTAGEHR" in accordance with the following equation: Baseline = BASE8587 x {26280 / (26280 - OUTAGEHR)} x {36 / (36 - months not on line)} x 106 "Months not on line" is the number of months during January 1985 through December 1987 prior to the commencement of firing for units that commenced firing in that period, i.e., the number of months, in that period, prior to the on-line month listed under the data field "BLRMNONL" and the on-line year listed in the data field "BLRYRONL" in the NADB.

**Baseline BAT limitations or NSPS** (Section 430.21) means the BAT limitations specified in Sec. 430.24(a) (1) or (2), as applicable, and the NSPS specified in Sec. 430.25(b) (1) or (2), as applicable, that apply to any direct discharger that is not "enrolled" in the "Voluntary Advanced Technology Incentives Program."

**Baseline configuration** (Section 610.11) means the unretrofitted test configuration, tuned in accordance with the automobile manufacturer's specifications.

**Baseline consumption allowances** (Section 82.3) means the consumption allowances apportioned under Sec. 82.6.

**Baseline or trend assessment survey** (Section 228.2) means the planned sampling or measurement of parameters at set stations or in set areas in and near disposal sites for a period of time sufficient to provide synoptic data for determining water quality, benthic, or biological conditions as a result of ocean disposal operations. The minimum requirements for such surveys are given in Sec. 228.13.

**Baseline production allowances** (Section 82.3) means the production allowances apportioned under Sec. 82.5.

**Basic Engine** (Section 600.022-85) means a unique combination of manufacturer, engine displacement, number of cylinders, fuel system (as distinguished by number of carburetor barrels or use of fuel injection), catalyst usage, and other engine and emission control system characteristics specified by the Administrator. For electric vehicles, basic engine means a unique combination of manufacturer and electric traction motor, motor controller, battery configuration, electrical charging system, energy storage device, and other components as specified by the Administrator.

**Basic engine** (Section 86.082-2) means a unique combination of manufacturer, engine displacement, number of cylinders, fuel system (as distinguished by number of carburetor barrels or use of fuel injection), catalyst usage, and other engine and emission control system characteristics specified by the Administrator.

**Basic engine** (Section 86.1803-01) means a unique combination of manufacturer, engine displacement, number of cylinders, fuel system (as distinguished by number of carburetor barrels or use of fuel injection), catalyst usage, and other engine and emission control system characteristics specified by the Administrator.

**Basic oxygen furnace steelmaking** (Section 420.41) means the production of steel from molten iron, steel scrap, fluxes, and various combinations thereof, in refractory lined furnaces by adding oxygen.

**Basic Phase II allowance allocations** (Section 72.2) means: (1) For calendar years 2000 through 2009 inclusive, allocations of allowances made by the Administrator pursuant to section 403 and section 405 (b)(1), (3), and (4); (c)(1), (2), (3), and (5); (d)(1), (2), (4), and (5); (e); (f); (g)(1), (2), (3), (4), and (5); (h)(1); (i); and (j). (2) For each calendar year beginning in 2010, allocations of allowances made by the Administrator pursuant to section 403 and section 405 (b)(1), (3), and (4); (c)(1), (2), (3), and (5); (d)(1), (2), (4), and (5); (e); (f); (g)(1), (2), (3), (4), and (5); (h)(1) and (3); (i); and (j).

**Basic vehicle frontal area** (Section 86.082-2) means the area enclosed by the geometric projection of the basic vehicle along the longitudinal axis, which includes tires but excludes mirrors and air deflectors, onto a plane perpendicular to the longitudinal axis of the vehicle.

**Basic vehicle frontal area** (Section 86.1803-01) means the area enclosed by the geometric projection of the basic vehicle along the longitudinal axis, which includes tires but excludes mirrors and air deflectors, onto a plane perpendicular to the longitudinal axis of the vehicle.

**Basis material** (Section 465.02) means the coiled strip which is processed.

**Basis material** (Section 466.02) means the metal part or base onto which procelain enamel is applied.

**BAT** (Section 467.02) means the best available technology economically achievable under section 304(b)(2)(B) of the Act.

**Batch** (Section 160.3) means a specific quantity or lot of a test, control, or reference substance that has been characterized according to Sec. 160.105(a).

**Batch** (Section 169.1) The term "batch" means a quantity of a pesticide product or active ingredient used in producing a pesticide made in one operation or lot or if made in a continuous or semi-continuous process or cycle, the quantity produced during an interval of time to be specified by the producer.

**Batch** (Section 204.51) means the collection of compressors of the same category or configuration, as designated by the Administrator in a test request, from which a batch sample is to be randomly drawn and inspected to determine conformance with the acceptability criteria.

**Batch** (Section 205.51) means the collection of vehicles of the same category, configuration or subgroup thereof as designated by the Administrator in a test request, from which a batch sample is to be drawn, and inspected to determine conformance with the acceptability criteria.

**Batch** (Section 420.111) means those alkaline cleaning operations which process steel products such as coiled wire, rods, and tubes in discrete batches or bundles.

**Batch** (Section 420.81) means those descaling operations in which the products are processed in discrete batches.

**Batch** (Section 420.91) means those pickling operations which process steel products such as coiled wire, rods, and tubes in discrete batches or bundles.

**Batch** (Section 792.3) means a specific quantity or lot of a test, control, or reference substance that has been characterized according to Sec. 792.105(a).

**Batch of gasoline** (Section 80.2) means a quantity of gasoline that is homogeneous with regard to those properties that are specified for conventional or reformulated gasoline.

**Batch of motor vehicle diesel fuel** (Section 80.2) means a quantity of diesel fuel which is homogeneous with regard to those properties that are specified for motor vehicle diesel fuel under subpart I of this part.

**Batch process** (Section 65.2) means a process in which the equipment is fed intermittently or discontinuously. Processing then occurs in this equipment after which the equipment is generally emptied. Examples of industries that use batch processes include pharmaceutical production and pesticide production.

**Batch product-process equipment train** (Section 65.2) means the collection of equipment (for example, connectors, reactors, valves, pumps) configured to produce a specific product or intermediate by a batch process.

**Batch sample** (Section 204.51) means the collection of compressors that are drawn from a batch.

**Batch sample** (Section 205.51) means the collection of vehicles of the same category, configuration or subgroup thereof which are drawn from a batch and from which test samples are drawn.

**Batch sample size** (Section 204.51) means the number of compressors of the same category or configuration which is randomly drawn from the batch sample and which will receive emissions tests.

**Batch sample size** (Section 205.51) means the number of vehicles of the same category or configuration in a batch sample.

**Batch size** (Section 204.51) means the number, as designated by the Administrator in the test request, of compressors of the same category or configuration in a batch.

**Batch size** (Section 205.51) means the number as designated by the Administrator in the test request of vehicles of the same category or configuration in a batch.

**Batch, pipe and tube** (Section 420.81) means those descaling operations that remove surface scale from pipe and tube products in batch processes.

**Batch, rod and wire** (Section 420.81) means those descaling operations that remove surface scale from rod and wire products in batch processes.

**Batch, sheet and plate** (Section 420.81) means those descaling operations that remove surface scale from sheet and plate products in batch processes.

**Battery** (Section 260.10) means a device consisting of one or more electrically connected electrochemical cells which is designed to receive, store, and deliver electric energy. An electrochemical cell is a system consisting of an anode, cathode, and an electrolyte, plus such connections (electrical and mechanical) as may be needed to allow the cell to deliver or receive electrical energy. The term battery also includes an intact, unbroken battery from which the electrolyte has been removed.

**Battery** (Section 273.9) means a device consisting of one or more electrically connected electrochemical cells which is designed to receive, store, and deliver electric energy. An electrochemical cell is a system consisting of an anode, cathode, and an electrolyte, plus such connections (electrical and mechanical) as may be needed to allow the cell to deliver or receive electrical energy. The term battery also includes an intact, unbroken battery from which the electrolyte has been removed.

**Battery** (Section 461.2) means a modular electric power source where part or all of the fuel is contained within the unit and electric power is generated directly from a chemical reaction rather than indirectly through a heat cycle engine. In this regulation there is no differentiation between a single cell and a battery.

**Battery** (Section 461.2) means a modular electric power source where part or all of the fuel is contained within the unit and electric power is generated directly from a chemical reaction rather than indirectly through a heat cycle engine. In this regulation there is no differentiation between a single cell and a battery.

**Battery Configuration** (Section 600.022-85) means the electrochemical type, voltage, capacity (in Watt-hours at the c/3 rate), and physical characteristics of the battery used as the tractive energy storage device.

**Battery manufacturing operations** (Section 461.2) means all of the specific processes used to produce a battery including the manufacture of anodes and cathodes and associated ancillary operations. These manufacturing operations are excluded from regulation under any other point source category.

**Battery manufacturing operations** (Section 461.2) means all of the specific processes used to produce a battery including the manufacture of anodes and cathodes and associated ancillary operations. These manufacturing operations are excluded from regulation under any other point source category.

**Battery pack** (Section 86.1702-99) means any electrical energy storage device consisting of any number of individual battery modules which is used to propel electric or hybrid electric vehicles.

**Bauxite** (Section 421.11) shall mean ore containing alumina monohydrate or alumina trihydrate which serves as the principal raw material for the production of alumina by the Bayer process or by the combination process.

**BCT** (Section 467.02) means the best conventional pollutant control technology, under section 304(b)(4) of the Act.

**Beehive cokemaking** (Section 420.11) means those operations in which coal is heated with the admission of air in controlled amounts for the purpose of producing coke. There are no by-product recovery operations associated with beehive cokemaking operations.

**Beets** (Section 407.71) shall include the processing of beets into the following product styles: Canned and peeled, whole, sliced, diced, French style, sections, irregular, and other cuts but not dehydrated beets.

**Belowground release** (Section 280.12) means any release to the subsurface of the land and to ground water. This includes, but is not limited to, releases from the belowground portions of an underground storage tank system and belowground releases associated with overfills and transfer operations as the regulated substance moves to or from an underground storage tank.

**Belowground storage facility** (Section 113.3) means a tank or other container located other than as defined as "Aboveground".

**Bench-scale operation** (Section 439.1) means the laboratory testing of materials, methods, or processes on a small scale, such as on a laboratory worktable.

**Beneath the surface of the ground** (Section 280.12) means beneath the ground surface or otherwise covered with earthen materials.

**Beneficial organism** (Section 166.3) means any pollinating insect, or any pest predator, parasite, pathogen or other biological control agent which functions naturally or as part of an integrated pest management program to control another pest.

**Beneficiation** (Section 372.3) means the preparation of ores to regulate the size (including crushing and grinding) of the product, to remove unwanted constituents, or to improve the quality, purity, or grade of a desired product.

**Benzene (or priority pollutant No. 4)** (Section 420.02) means the value obtained by the standard method Number 602 specified in 44 FR 69464, 69570 (December 3, 1979).

**Benzo(a)pyrene (or priority pollutant No. 73)** (Section 420.02) means the value obtained by the standard method Number 610 specified in 44 FR 69464, 69570 (December 3, 1979).

**Beryllium copper alloy** (Section 468.02) shall mean any copper alloy that is alloyed to contain 0.10 percent or greater beryllium.

**Best available technology or BAT** (Section 141.2) means the best technology, treatment techniques, or other means which the Administrator finds, after examination for efficacy under field conditions and not solely under laboratory conditions, are available (taking cost into consideration). For the purposes of setting MCLs for synthetic organic chemicals, any BAT must be at least as effective as granular activated carbon.

**Best Management Practice (BMP)** (Section 130.2) Methods, measures or practices selected by an agency to meet its nonpoint source control needs. BMPs include but are not limited to structural and nonstructural controls and operation and maintenance procedures. BMPs can be applied before, during and after pollution-producing activities to reduce or eliminate the introduction of pollutants into receiving waters.

**Best management practices ("BMPs")** (Section 122.2) means schedules of activities, prohibitions of practices, maintenance procedures, and other management practices to prevent or reduce the pollution of "waters of the United States." BMPs also include treatment requirements, operating procedures, and practices to control plant site runoff, spillage or leaks, sludge or waste disposal, or drainage from raw material storage.

**Best management practices (BMPs)** (Section 232.2) means schedules of activities, prohibitions of practices, maintenance procedures, and other management practices to prevent or reduce the pollution of waters of the United States from discharges of dredged or fill material. BMPs include methods, measures, practices, or design and performance standards which facilitate compliance with the section 404(b)(1) Guidelines (40 CFR part 230), effluent limitations or prohibitions under section 307(a), and applicable water quality standards.

**BIA** (Section 147.2902) the "Bureau of Indian Affairs," United States Department of Interior.

**Bias** (Section 72.2) means systematic error, resulting in measurements that will be either consistently low or high relative to the reference value.

**Bi-directional control** (Section 86.094-2) means the capability of a diagnostic tool to send messages on the data bus that temporarily overrides the module's control over a sensor or actuator and gives control to the diagnostic tool operator. Bi-directional controls do not create permanent changes to engine or component calibrations.

**Bi-directional control** (Section 86.1803-01) means the capability of a diagnostic tool to send messages on the data bus that temporarily overrides the module's control over a sensor or actuator and gives control to the diagnostic tool operator. Bi-directional controls do not create permanent changes to engine or component calibrations.

**Bin or emission bin** (Section 86.1803-01) means a set of emission standards applicable to exhaust pollutants measured on the Federal Test Procedure (FTP). A bin is equivalent to a horizontal row of FTP standards in Tables S04-1 and S04-2 shown in this subpart. Manufacturers are generally free to choose the bin of standards that will apply to a certain test group of vehicles, provided that on a sales weighted average of those bins, all of their vehicles meet a specified fleet average standard for a particular pollutant.

**Bioaccumulation** (Section 132.2) is the net accumulation of a substance by an organism as a result of uptake from all environmental sources.

**Bioaccumulation factor (BAF)** (Section 132.2) is the ratio (in L/kg) of a substance's concentration in tissue of an aquatic organism to its concentration in the ambient water, in situations where both the organism and its food are exposed and the ratio does not change substantially over time.

**Bioaccumulative chemical of concern (BCC)** (Section 132.2) is any chemical that has the potential to cause adverse effects which, upon entering the surface waters, by itself or as its toxic transformation product, accumulates in aquatic organisms by a human health bioaccumulation factor greater than 1000, after considering metabolism and other physicochemical properties that might enhance or inhibit bioaccumulation, in accordance with the methodology in appendix B of this part. Chemicals with half-lives of less than eight weeks in the water column, sediment, and biota are not BCCs. The minimum BAF information needed to define an organic chemical as a BCC is either a field-measured BAF or a BAF derived using the BSAF methodology. The minimum BAF information needed to define an inorganic chemical, including an organometal, as a BCC is either a field-measured BAF or a laboratory-measured BCF. BCCs include, but are not limited to, the pollutants identified as BCCs in section A of Table 6 of this part.

**Bioconcentration** (Section 132.2) is the net accumulation of a substance by an aquatic organism as a result of uptake directly from the ambient water through gill membranes or other external body surfaces.

**Bioconcentration factor (BCF)** (Section 132.2) is the ratio (in L/kg) of a substance's concentration in tissue of an aquatic organism to its concentration in the ambient water, in situations where the organism is exposed through the water only and the ratio does not change substantially over time.

**Biodegradation rate** (Section 435.41) as applied to BAT effluent limitations and NSPS for drilling fluids and drill cuttings refers to the ISO 11734:1995 method: "Water quality—Evaluation of the 'ultimate' anaerobic biodegradability of organic compounds in digested sludge—Method by measurement of the biogas production (1995 edition)" (Available from the American National Standards Institute, 11 West 42nd Street, 13th Floor, New York, NY 10036) supplemented with modifications in Appendix 4 of subpart A of this part.

**Biodegradation rate as applied to BAT effluent limitations and NSPS for drilling fluids and drill cuttings** (Section 435.11) refers to the ISO 11734:1995 method: "Water quality—Evaluation of the 'ultimate' anaerobic biodegradability of organic compounds in digested sludge—Method by measurement of the biogas production (1995 edition)" supplemented with modifications in Appendix 4 of 40 CFR part 435, subpart A. This incorporation by reference was approved by the Director of the Federal Register in accordance with 5 U.S.C. 552(a) and 1 CFR part 51. Copies may be obtained from the American National Standards Institute, 11 West 42nd Street, 13th Floor, New York, NY 10036. Copies may be inspected at the Office of the Federal Register, 800 North Capitol Street, NW., Suite 700, Washington, DC. A copy may also be inspected at EPA's Water Docket, 401 M Street SW., Washington, DC 20460.

**Biological control agent** (Section 152.3) means any living organism applied to or introduced into the environment that is intended to function as a pesticide against another organism declared to be a pest by the Administrator.

**Bioremediation agents** (Section 300.5) means microbiological cultures, enzyme additives, or nutrient additives that are deliberately introduced into an oil discharge and that will significantly increase the rate of biodegradation to mitigate the effects of the discharge.

**Biota-sediment accumulation factor (BSAF)** (Section 132.2) is the ratio (in kg of organic carbon/kg of lipid) of a substance's lipid-normalized concentration in tissue of an aquatic organism to its organic carbon-normalized concentration in surface sediment, in situations where the ratio does not change substantially over time, both the organism and its food are exposed, and the surface sediment is representative of average surface sediment in the vicinity of the organism.

**Bitterns** (Section 415.161) shall mean the saturated brine solution remaining after precipitation of sodium chloride in the solar evaporation process.

**Blanket insulation** (Section 247.3) means relatively flat and flexible insulation in coherent sheet form, furnished in units of substantial area. Batt insulation is included in this term;

**Bleach plant** (Section 430.0) All process equipment used for bleaching beginning with the first application of bleaching agents (e.g., chlorine, chlorine dioxide, ozone, sodium or calcium hypochlorite, or peroxide), each subsequent extraction stage, and each subsequent stage where bleaching agents are applied to the pulp. For mills in subpart E of this part producing specialty grades of pulp, the bleach plant includes process equipment used for the hydrolysis or extraction stages prior to the first application of bleaching agents. Process equipment used for oxygen delignification prior to the application of bleaching agents is not part of the bleach plant.

**Bleach plant effluent** (Section 430.0) The total discharge of process wastewaters from the bleach plant from each physical bleach line operated at the mill, comprising separate acid and alkaline filtrates or the combination thereof.

**Blend fertilizer** (Section 418.71) shall mean a mixture of dry, straight and mixed fertilizer materials.

**Blowdown** (Section 401.11) means the minimum discharge of recirculating water for the purpose of discharging materials contained in the water, the further buildup of which would cause concentration in amounts exceeding limits established by best engineering practice.

**Blowdown** (Section 423.11) means the minimum discharge of recirculating water for the purpose of discharging materials contained in the water, the further buildup of which would cause concentration in amounts exceeding limits established by best engineering practices.

**Blue Sky Series engine** (Section 89.2) means a nonroad engine meeting the requirements of Sec. 89.112(f).

**Blue Sky Series engine** (Section 94.2) means an engine meeting the requirements of Sec. 94.7(e).

**BMPs** (Section 122.2) means "best management practices."

**Board insulation** (Section 247.3) means semi-rigid insulation preformed into rectangular units having a degree of suppleness, particularly related to their geometrical dimensions;

**BOD** (Section 133.101) The five day measure of the pollutant parameter biochemical oxygen demand (BOD).

**BOD5** (Section 401.11) means five-day biochemical oxygen demand.

**BOD5 input** (Section 405) shall mean the biochemical oxygen demand of the materials entered into process. It can be calculated by multiplying the fats, proteins and carbohydrates by factors of 0.890, 1.031 and 0.691 respectively. Organic acids (e.g., lactic acids) should be included as carbohydrates. Composition of input materials may be based on either direct analyses or generally accepted published values.

**BOD7** (Section 417.151) shall mean the biochemical oxygen demand as determined by incubation at 20 degrees C for a period of 7 days using an acclimated seed. Agitation employing a magnetic stirrer set at 200 to 500 rpm may be used.

**Bodily injury** (Section 280.92) shall have the meaning given to this term by applicable state law; however, this term shall not include those liabilities which, consistent with standard insurance industry practices, are excluded from coverage in liability insurance policies for bodily injury.

**Body Style** (Section 600.022-85) means a level of commonality in vehicle construction as defined by number of doors and roof treatment (e.g., sedan, convertible, fastback, hatchback) and number of seats (i.e., front, second, or third seat) requiring seat belts pursuant to National Highway Traffic Safety Administration safety regulations. Station wagons and light trucks are identified as car lines.

**Body style** (Section 86.082-2) means a level of commonality in vehicle construction as defined by number of doors and roof treatment (e.g., sedan, convertible, fastback, hatchback).

**Body style** (Section 86.1803-01) means a level of commonality in vehicle construction as defined by number of doors and roof treatment (e.g., sedan, convertible, fastback, hatchback).

**Body type** (Section 86.082-2) means a name denoting a group of vehicles that are either in the same car line or in different car lines provided the only reason the vehicles qualify to be considered in different car lines is that they are produced by a separate division of a single manufacturer.

**Body type** (Section 86.1803-01) means a name denoting a group of vehicles that are either in the same car line or in different car lines provided the only reason the vehicles qualify to be considered in different car lines is that they are produced by a separate division of a single manufacturer.

**Boiler** (Section 260.10) means an enclosed device using controlled flame combustion and having the following characteristics: (1) (i) The unit must have physical provisions for recovering and exporting thermal energy in the form of steam, heated fluids, or heated gases; and (ii) The unit's combustion chamber and primary energy recovery sections(s) must be of integral design. To be of integral design, the combustion chamber and the primary energy recovery section(s) (such as waterwalls and superheaters) must be physically formed into one manufactured or assembled unit. A unit in which the combustion chamber and the primary energy recovery section(s) are joined only by ducts or connections carrying flue gas is not integrally designed; however, secondary energy recovery equipment (such as economizers or air preheaters) need not be physically formed into the same unit as the combustion chamber and the primary energy recovery section. The following units are not precluded from being boilers solely because they are not of integral design: process heaters (units that transfer energy directly to a process stream), and fluidized bed combustion units; and (iii) While in operation, the unit must maintain a thermal energy recovery efficiency of at least 60 percent, calculated in terms of the recovered energy compared with the thermal value of the fuel; and (iv) The unit must export and utilize at least 75 percent of the recovered energy, calculated on an annual basis. In this calculation, no credit shall be given for recovered heat used internally in the same unit. (Examples of internal use are the preheating of fuel or combustion air, and the driving of induced or forced draft fans or feedwater pumps); or (2) The unit is one which the Regional Administrator has determined, on a case-by-case basis, to be a boiler, after considering the standards in Sec. 260.32.

**Boiler** (Section 372.3) means an enclosed device using controlled flame combustion and having the following characteristics: (1)(i) The unit must have physical provisions for recovering and exporting thermal energy in the form of steam, heated fluids, or heated gases; and (ii) The unit's combustion chamber and primary energy recovery sections(s) must be of integral design. To be of integral design, the combustion chamber and the primary energy recovery section(s) (such as waterwalls and superheaters) must be physically formed into one manufactured or assembled unit. A unit in which the combustion chamber and the primary energy recovery section(s) are joined only by ducts or connections carrying flue gas is not integrally designed; however, secondary energy recovery equipment (such as economizers or air preheaters) need not be physically formed into the same unit as the combustion chamber and the primary energy recovery section. The following units are not precluded from being boilers solely because they are not of integral design: process heaters (units that transfer energy directly to a process stream), and fluidized bed combustion units; and (iii) While in operation, the unit must maintain a thermal energy recovery efficiency of at least 60 percent, calculated in terms of the recovered energy compared with the thermal value of the fuel; and (iv) The unit must export and utilize at least 75 percent of the recovered energy, calculated on an annual basis. In this calculation, no credit shall be given for recovered heat used internally in the same unit. (Examples of internal use are the preheating of fuel or combustion air, and the driving of induced or forced draft fans or feedwater pumps); or (2) The unit is one which the Regional Administrator has determined, on a case-by-case basis, to be a boiler, after considering the standards in Sec. 260.32 of this chapter.

**Boiler** (Section 65.2) means any enclosed combustion device that extracts useful energy in the form of steam and is not an incinerator or a process heater. Boiler also means any industrial furnace as defined in 40 CFR 260.10.

**Boiler** (Section 72.2) means an enclosed fossil or other fuel-fired combustion device used to produce heat and to transfer heat to recirculating water, steam, or any other medium.

**Boiler** (Section 96.2) means an enclosed fossil or other fuel-fired combustion device used to produce heat and to transfer heat to recirculating water, steam, or other medium.

**Boiler** (Section 97.2) means an enclosed fossil or other fuel-fired combustion device used to produce heat and to transfer heat to recirculating water, steam, or other medium.

**Bond release** (Section 434.11) means the time at which the appropriate regulatory authority returns a reclamation or performance bond based upon its determination that reclamation work (including, in the case of underground mines, mine sealing and abandonment procedures) has been satisfactorily completed.

**Borrower, debtor, or obligor** (Section 280.200) is a person whose UST or UST system or facility or property on which the UST or UST system is located is encumbered by a security interest. These terms may be used interchangeably.

**Bottom ash** (Section 240.101) means the solid material that remains on a hearth or falls off the grate after thermal processing is complete.

**Bottom ash** (Section 423.11) means the ash that drops out of the furnace gas stream in the furnace and in the economizer sections. Economizer ash is included when it is collected with bottom ash.

**Bottoms receiver** (Section 264.1031) means a container or tank used to receive and collect the heavier bottoms fractions of the distillation feed stream that remain in the liquid phase.

**Bottoms receiver** (Section 65.2) means a tank that collects distillation bottoms before the stream is sent for storage or for further downstream processing.

**BPT** (Section 467.02) means the best practicable control technology currently available under section 304(b)(1) of the Act.

**Brake horsepower** (Section 92.2) means the sum of the alternator/generator input horsepower and the mechanical accessory horsepower, excluding any power used to circulate engine coolant, circulate engine lubricant, or to supply fuel to the engine.

**Breakover angle** (Section 86.1803-01) means the supplement of the largest angle, in the plan side view of an automobile, that can be formed by two lines tangent to the front and rear static loaded radii arcs and intersecting at a point on the underside of the automobile.

**Bridging crosses between plants** (Section 174.3) means the utilization of an intermediate plant in a cross to produce a viable zygote between the intermediate plant and a first plant, in order to cross the plant resulting from that zygote with a third plant that would not otherwise be able to produce viable zygotes from the fusion of its gametes with those of the first plant. The result of the bridging cross is the mixing of genetic material of the first and third plant through the formation of an intermediate zygote.

**Broccoli** (Section 407.71) shall include the processing of broccoli into the following product styles: Frozen, chopped, spears, and miscellaneous cuts.

**Broker** (Section 92.302) means any entity that facilitates a trade between a buyer and seller.

**Broker** (Section 94.302) means any entity that facilitates a trade between a buyer and seller.

**Brushing and grubbing area** (Section 434.80) means the area where woody plant materials that would interfere with soil salvage operations have been removed or incorporated into the soil that is being salvaged.

**Building insulation** (Section 247.3) means a material, primarily designed to resist heat flow, which is installed between the conditioned volume of a building and adjacent unconditioned volumes or the outside. This term includes but is not limited to insulation products such as blanket, board, spray-in-place, and loose-fill that are used as ceiling, floor, foundation, and wall insulation:

**Bulk container** (Section 246.101) means a large container that can either be pulled or lifted mechanically onto a service vehicle or emptied mechanically into a service vehicle.

**Bulk sewage sludge** (Section 503.11) is sewage sludge that is not sold or given away in a bag or other container for application to the land.

**Bulky waste** (Section 243.101) means large items of solid waste such as household appliances, furniture, large auto parts, trees, branches, stumps, and other oversize wastes whose large size precludes or complicates their handling by normal solid wastes collection, processing, or disposal methods.

**Burning agents** (Section 300.5) means those additives that, through physical or chemical means, improve the combustibility of the materials to which they are applied.

**Burnishing** (Section 471.02) nis a surface finishing process in which minute surface irregularities are displaced rather than removed.

**Bus** (Section 86.093-2) means a heavy heavy-duty diesel-powered passenger-carrying vehicle with a load capacity of fifteen or more passengers that is centrally fueled, and all urban buses. This definition only applies in the context of Secs. 86.093-11 and 86.093-35.

**Business confidentiality or confidential business information** (Section 350.1) includes the concept of trade secrecy and other related legal concepts which give (or may give) a business the right to preserve the confidentiality of business information and to limit its use or disclosure by others in order that the business may obtain or retain business advantages it derives from its right in the information. The definition is meant to encompass any concept which authorizes a Federal agency to withhold business information under 5 U.S.C. 552(b) (4), as well as any concept which requires EPA to withhold information from the public for the benefit of a business under 18 U.S.C. 1905.

**Business submitter** (Section 1601.3) means any person or entity which provides confidential business information, directly or indirectly, to the CSB and who has a proprietary interest in the information.

**Buyer** (Section 92.302) means the entity that receives credits as a result of trade or transfer.

**Buyer** (Section 94.302) means the entity that receives credits as a result of trade.

**By compound** (Section 65.2) means by individual stream components, not carbon equivalents.

**Bypass** (Section 122.2) is defined at Sec. 122.41(m).

**By-pass duct** (Section 62.1201) means a device which diverts a minimum of 10 percent of a cement kiln's off gas, or a device which the Administrator determines on a case-by-case basis diverts a sample of kiln gas that contains levels of carbon monoxide or hydrocarbons representative of the levels in the kiln.

**Bypass operating quarter** (Section 72.2) means a calendar quarter during which emissions pass through a stack, duct or flue that bypasses add-on emission controls.

**By-pass stack** (Section 72.2) means any duct, stack, or conduit through which emissions from an affected unit may or do pass to the atmosphere, which either augments or substitutes for the principal stack exhaust system or ductwork during any portion of the unit's operation.

**Byproduct** (Section 704.3) means a chemical substance produced without a separate commercial intent during the manufacture, processing, use, or disposal of another chemical substance(s) or mixture(s).

**Byproduct** (section 710.2) means a chemical substance produced without separate commercial intent during the manufacture or processing of another chemical substance(s) or mixture(s).

**Byproduct** (Section 712.3) means any chemical substance or mixture produced without a separate commercial intent during the manufacture, processing, use, or disposal of another chemical substance or mixture.

**Byproduct** (Section 716.3) means a chemical substance produced without a separate commercial intent during the manufacture, processing, use, or disposal of another chemical substance(s) or mixture(s).

**Byproduct** (Section 720.3) means a chemical substance produced without a separate commercial intent during the manufacture, processing, use, or disposal of another chemical substance or mixture.

**Byproduct** (Section 761.13) means a chemical substance produced without separate commercial intent during the manufacturing or processing of another chemical substance(s) or mixture(s).

**Byproduct** (Section 791.3) refers to a chemical substance produced without a separate commercial intent during the manufacture, processing, use or disposal of another chemical substance or mixture.

**By-product cokemaking** (Section 420.11) means those cokemaking operations in which coal is heated in the absence of air to produce coke. In this process, by-products may be recovered from the gases and liquids driven from the coal during cokemaking.

**Byproduct material, source material, and special nuclear material** (Section 720.3) have the meanings contained in the Atomic Energy Act of 1954, 42 U.S.C 2014 et seq. and the regulations issued under it.

# C

**C12-C14 ester and C8 ester** (Section 435.11) means the fatty acid/2-ethylhexyl esters with carbon chain lengths ranging from 8 to 16 and represented by the Chemical Abstracts Service (CAS) No. 135800-37-2. (Properties available from the Chemical Abstracts Service, 2540 Olentangy River Road, PO Box 3012, Columbus, OH, 43210)

**C16-C18 internal olefin** (Section 435.11) means a 65/35 blend, proportioned by mass, of hexadecene and octadecene, respectively. Hexadecene is an unsaturated hydrocarbon with a carbon chain length of 16, an internal double carbon bond, and is represented by the Chemical Abstracts Service (CAS) No. 26952-14-7. Octadecene is an unsaturated hydrocarbon with a carbon chain length of 18, an internal double carbon bond, and is represented by the Chemical Abstracts Service (CAS) No. 27070-58-2. (Properties available from the Chemical Abstracts Service, 2540 Olentangy River Road, PO Box 3012, Columbus, OH, 43210).

**C16-C18 internal olefin drilling fluid** (Section 435.11) means a C16-C18 internal olefin drilling fluid formulated as specified in Appendix 8 of subpart A of this part.

**Cab over axle or cab over engine** (Section 205.51) means the cab which contains the operator/passenger compartment is directly above the engine and front axle and the entire cab can be tilted forward to permit access to the engine compartment.

**Cadmium** (Section 415.641) shall mean the total cadmium present in the process wastewater stream exiting the wastewater treatment system.

**Calcium sulfate storage pile runoff** (Section 418.11) shall mean the calcium sulfate transport water runoff from or through the calcium sulfate pile, and the precipitation which falls directly on the storage pile and which may be collected in a seepage ditch at the base of the outer slopes of the storage pile, provided such seepage ditch is protected from the incursion of surface runoff from areas outside of the outer perimeter of the seepage ditch.

**Calculated level** (Section 82.3) means the weighted amount of a controlled substance determined by multiplying the amount (in kilograms) of the controlled substance by that substance's ozone depletion potential (ODP) weight listed in appendix A or appendix B to this subpart.

**Calibrating gas** (Section 86.082-2) means a gas of known concentration which is used to establish the response curve of an analyzer.

**Calibration** (Section 600.022-85) means the set of specifications, including tolerances, unique to a particular design, version of application of a component, or component assembly capable of functionally describing its operation over its working range.

**Calibration** (Section 86.082-2) means the set of specifications, including tolerances, unique to a particular design, version, or application of a component or components assembly capable of functionally describing its operation over its working range.

**Calibration** (Section 86.1803-01) means the set of specifications, including tolerances, unique to a particular design, version, or application of a component or components assembly capable of functionally describing its operation over its working range.

**Calibration** (Section 92.102) means the act of calibrating an analytical instrument using known standards.

**Calibration** (Section 92.2) means the set of specifications, including tolerances, unique to a particular design, version, or application of a component, or components, or assembly capable of functionally describing its operation over its working range. This definition does apply to subpart B of this part.

**Calibration** (Section 94.2) means the set of specifications, including tolerances, specific to a particular design, version, or application of a component, or components, or assembly capable of functionally describing its operation over its working range.

**Calibration error** (Section 72.2) means the difference between: (1) The response of gaseous monitor to a calibration gas and the known concentration of the calibration gas; (2) The response of a flow monitor to a reference signal and the known value of the reference signal; or (3) The response of a continuous opacity monitoring system to an attenuation filter and the known value of the filter after a stated period of operation during which no unscheduled maintenance, repair, or adjustment took place.

**Calibration gas** (Section 72.2) means: (1) A standard reference material; (2) A standard reference material-equivalent compressed gas primary reference material; (3) A NIST traceable reference material; (4) NIST/EPA-approved certified reference materials; (5) A gas manufacturer's intermediate standard; (6) An EPA protocol gas; (7) Zero air material; or (8) A research gas mixture.

**Calibration gas** (Section 86.1803-01) means a gas of known concentration which is used to establish the response curve of an analyzer.

**Calibration gas** (Section 92.102) means a gas of known concentration which is used to establish the response curve of an analyzer.

**Calibration of equipment** (Section 171.2) means measurement of dispersal or output of application equipment and adjustment of such equipment to control the rate of dispersal, and droplet or particle size of a pesticide dispersed by the equipment.

**CalLEV II or California LEV II** (Section 86.1803-01) refers to California's second phase of its low emission vehicle (LEV) program. This program was adopted at the hearing of the California Air Resources Board held on November 5, 1998 and became effective on November 27, 1999.

**Can be centrally fueled** (Section 88.302-94) means the sum of those vehicles that are centrally fueled and those vehicles that are capable of being centrally fueled.

**Candidate** (Section 53.1) method means a method for measuring the concentration of an air pollutant in the ambient air for which an application for a reference method determination or an equivalent method determination is submitted in accordance with Sec. 53.4, or a method tested at the initiative of the Administrator in accordance with Sec. 53.7.

**Candidate in-use vehicle** (Section 86.1803-01) means an in-use vehicle which would be eligible to participate in the in-use verification program in accordance with Sec. 86.1845-01.

**Canmaking** (Section 465.02) means the manufacturing process or processes used to manufacture a can from a basic metal.

**Canned meat processor** (Section 432.91) shall mean an operation which prepares and cans meats (such as stew, sandwich spreads, or similar products) alone or in combination with other finished products at rates greater than 2730 kg (6000 lb.) per day.

**Canned onions** (Section 407.71) shall mean the processing of onions into the following product styles: Canned, frozen, and fried (canned), peeled, whole, sliced, and any other piece size but not including frozen, battered onion rings or dehydrated onions.

**Capable of being centrally fueled** (Section 88.302-94) means a fleet, or that part of a fleet, consisting of vehicles that could be refueled 100 percent of the time at a location that is owned, operated, or controlled by the covered fleet operator, or is under contract with the covered fleet operator. The fact that one or more vehicles in a fleet is/are not capable of being centrally fueled does not exempt an entire fleet from the program.

**Capable of Transportation of Property on a street or highway** (Section 205.51) means that the vehicle: (i) Is self propelled and is capable of transporting any material or fixed apparatus, or is capable of drawing a trailer or semi-trailer; (ii) Is capable of maintaining a cruising speed of at least 25 mph over level, paved surface; (iii) Is equipped or can readily be equipped with features customarily associated with practical street or highway use, such features including but not being limited to: A reverse gear and a differential, fifth wheel, cargo platform or cargo enclosure, and (iv) Does not exhibit features which render its use on a street or highway impractical, or highly unlikely, such features including, but not being limited to, tracked road means, an inordinate size or features ordinarily associated with combat or tactical vehicles.

**Capacitor** (Section 761.13) means a device for accumulating and holding a charge of electricity and consisting of conducting surfaces separated by a dielectric. Types of capacitors are as follows: (1) Small capacitor means a capacitor which contains less than 1.36 kg (3 lbs.) of dielectric fluid. The following assumptions may be used if the actual weight of the dielectric fluid is unknown. A capacitor whose total volume is less than 1,639 cubic centimeters (100 cubic inches) may be considered to contain less than 1.36 kgs (3 lbs.) of dielectric fluid and a capacitor whose total volume is more than 3,278 cubic centimeters (200 cubic inches) must be considered to contain more than 1.36 kg (3 lbs.) of dielectric fluid. A capacitor whose volume is between 1,639 and 3,278 cubic centimeters may be considered to contain less then 1.36 kg (3 lbs.) of dielectric fluid if the total weight of the capacitor is less than 4.08 kg (9 lbs.). (2) Large high voltage capacitor means a capacitor which contains 1.36 kg (3 lbs.) or more of dielectric fluid and which operates at 2,000 volts (a.c. or d.c.) or above. (3) Large low voltage capacitor means a capacitor which contains 1.36 kg (3 lbs.) or more of dielectric fluid and which operates below 2,000 volts (a.c. or d.c.).

**Capacity factor** (Section 51.100) means the ratio of the average load on a machine or equipment for the period of time considered to the capacity rating of the machine or equipment.

**Capacity factor** (Section 72.2) means either: (1) the ratio of a unit's actual annual electric output (expressed in MWe-hr) to the unit's nameplate capacity times 8760 hours, or (2) the ratio of a unit's annual heat input (in million British thermal units or equivalent units of measure) to the unit's maximum design heat input (in million British thermal units per hour or equivalent units of measure) times 8,760 hours.

**Capital expenditure** (Section 60.2) means an expenditure for a physical or operational change to an existing facility which exceeds the product of the applicable "annual asset guideline repair allowance percentage" specified in the latest edition of Internal Revenue Service (IRS) Publication 534 and the existing facility's basis, as defined by section 1012 of the Internal Revenue Code. However, the total expenditure for a physical or operational change to an existing facility must not be reduced by any "excluded additions" as defined in IRS Publication 534, as would be done for tax purposes.

**Capital expenditure** (Section 61.02) means an expenditure for a physical or operational change to a stationary source which exceeds the product of the applicable "annual asset guideline repair allowance percentage" specified in the latest edition of Internal Revenue Service (IRS) Publication 534 and the stationary source's basis, as defined by section 1012 of the Internal Revenue Code. However, the total expenditure for a physical or operational change to a stationary source must not be reduced by any "excluded additions" as defined for stationary sources constructed after December 31, 1981, in IRS Publication 534, as would be done for tax purposes. In addition, "annual asset guideline repair allowance" may be used even though it is excluded for tax purposes in IRS Publication 534.

**Capture system** (Section 64.1) means the equipment (including but not limited to hoods, ducts, fans, and booths) used to contain, capture and transport a pollutant to a control device.

**Car Coupling Sound** (Section 201.1) means a sound which is heard and identified by the observer as that of car coupling impact, and that causes a sound level meter indicator (FAST) to register an increase of at least ten decibels above the level observed immediately before hearing the sound.

**Car Line** (Section 600.022-85) means a name denoting a group of vehicles within a make or car division which has a degree of commonality in construction (e.g., body, chassis). Car line does not consider any level of decor or opulence and is not generally distinguished by characteristics as roof line, number of doors, seats, or windows, except for station wagons or light-duty trucks.

Station wagons and light-duty trucks are considered to be different car lines than passenger cars.

**Car line** (Section 86.082-2) means a name denoting a group of vehicles within a make or car division which has a degree of commonality in construction (e.g., body, chassis). Car line does not consider any level of decor or opulence and is not generally distinguished by characteristics as roofline, number of doors, seats, or windows except for station wagons or light-duty trucks. Station wagons and light-duty trucks are considered to be different car lines than passenger cars.

**Car line** (Section 86.1803-01) means a name denoting a group of vehicles within a make or car division which has a degree of commonality in construction (e.g., body, chassis). Car line does not consider any level of decor or opulence and is not generally distinguished by characteristics as roofline, number of doors, seats, or windows except for station wagons or light-duty trucks. Station wagons, light-duty trucks, and complete heavy-duty vehicles are considered to be different car lines than passenger cars.

**Carbon hot forming operation (or "carbon") (Section 420.71)** means those hot forming operations which produce a majority, on a tonnage basis, of carbon steel products.

**Carbon regeneration unit** (Section 260.10) means any enclosed thermal treatment device used to regenerate spent activated carbon.

**Carbon steel** (Section 420.71) means those steel products other than specialty steel products.

**Carcinogen** (Section 132.2) is a substance which causes an increased incidence of benign or malignant neoplasms, or substantially decreases the time to develop neoplasms, in animals or humans. The classification of carcinogens is discussed in section II.A of appendix C to part 132.

**Carpet cushion, also known as carpet underlay** (Section 247.3) is padding placed beneath carpet to reduce carpet wear caused by foot traffic or furniture indentation, enhance comfort, and prolong appearance.

**Carrier** (Section 160.3) means any material, including but not limited to feed, water, soil, nutrient media, with which the test substance is combined for administration to a test system.

**Carrier** (Section 201.1) means a common carrier by railroad, or partly by railroad and partly by water, within the continental United States, subject to the Interstate Commerce Act, as amended, excluding street, suburban, and interurban electric railways unless operated as a part of a general railroad system of transportation.

**Carrier** (Section 792.3) means any material, including but not limited to, feed, water, soil, and nutrient media, with which the test substance is combined for administration to a test system.

**Carrier** (Section 80.2) means any distributor who transports or stores or causes the transportation or storage of gasoline or diesel fuel without taking title to or otherwise having any ownership of the gasoline or diesel fuel, and without altering either the quality or quantity of the gasoline or diesel fuel.

**Carrier of contaminant** (Section 230.3) means dredged or fill material that contains contaminants.

**Carrots** (Section 407.71) shall include the processing of carrots into the following product styles: Canned and frozen, peeled, whole, sliced, diced, nuggets, crinkle cut, julienne, shoestrings, chunks, chips and other irregular cuts, and juices but not dehydrated carrots.

**Carry** (Section 90.102) means the operator completely bears the weight of the equipment, including the engine.

**Carrying Case** (Section 211.203) The container used to store reusable hearing protectors.

**Carryout collection** (Section 243.101) means collection of solid waste from a storage area proximate to the dwelling unit(s) or establishment.

**Carry-over engine family** (Section 91.1302) means an engine family which undergoes certification using carryover test data from previous model years. See Sec. 91.118©.

**Car-seal** (Section 65.2) means a seal that is placed on a device that is used to change the position of a valve (for example, from opened to closed) in such a way that the position of the valve cannot be changed without breaking the seal.

**CAS** (Section 62.1201) means chemical abstract services registry.

**CAS** (Section 68.3) means the Chemical Abstracts Service.

**CAS** (Section 704.3) Number means Chemical Abstracts Service Registry Number.

**CAS** (Section 721.3) Number means Chemical Abstracts Service Registry Number assigned to a chemical substance on the Inventory.

**Casing** (Section 146.3) means a pipe or tubing of appropriate material, of varying diameter and weight, lowered into a borehole during or after drilling in order to support the sides of the hole and thus prevent the walls from caving, to prevent loss of drilling mud into porous ground, or to prevent water, gas, or other fluid from entering or leaving the hole.

**Casing** (Section 147.2902) a pipe or tubing of varying diameter and weight, lowered into a borehole during or after drilling in order to support the sides of the hole and, thus, prevent the walls from caving, to prevent loss of drilling mud into porous ground, or to prevent water, gas, or other fluid from entering the hole.

**Casting** (Section 471.02) is pouring molten metal into a mold to produce an object of desired shape.

**Catastrophic collapse** (Section 146.3) means the sudden and utter failure of overlying "strata" caused by removal of underlying materials.

**Catastrophic release** (Section 68.3) means a major uncontrolled emission, fire, or explosion, involving one or more regulated substances that presents imminent and substantial endangerment to public health and the environment.

**Category** (Section 204.51) means a group of compressor configurations which are identical in all aspects with respect to the parameters listed in paragraph © (1) (i) of Sec. 204.55-2.

**Category** (Section 205.151) means a group of vehicle configurations which are identical in all material aspects with respect to the parameters listed in Sec. 205.157-2 of this subpart.

**Category** (Section 205.165) means a group of exhaust systems which are identical in all material aspects with respect to the parameters listed in Sec. 205.168 of this subpart.

**Category** (Section 205.51) means a group of vehicle configurations which are identical in all material aspects with respect to the parameters listed in Sec. 205.55-2.

**Category** (Section 211.203) A group of hearing protectors which are identical in all aspects to the parameters listed in Sec. 211.210-2©.

**Category 1** (Section 94.2) means relating to a marine engine with a rated power greater than or equal to 37 kilowatts and a specific engine displacement less than 5.0 liters per cylinder.

**Category 2** (Section 94.2) means relating to a marine engine with a specific engine displacement greater than or equal to 5.0 liters per cylinder but less than 30 liters per cylinder.

**Category 3** (Section 94.2) means relating to a marine engine with a specific engine displacement greater than or equal to 30 liters per cylinder.

**Cathode ray tubes** (Section 469.31) means electronic devices in which electrons focus through a vacuum to generate a controlled image on a luminescent surface. This definition does not include receiving and transmitting tubes.

**Cathodic protection** (Section 280.12) is a technique to prevent corrosion of a metal surface by making that surface the cathode of an electrochemical cell. For example, a tank system can be cathodically protected through the application of either galvanic anodes or impressed current.

**Cathodic protection tester** (Section 280.12) means a person who can demonstrate an understanding of the principles and measurements of all common types of cathodic protection systems as applied to buried or submerged metal piping and tank systems. At a minimum, such persons must have education and experience in soil resistivity, stray current, structure-to-soil potential, and component electrical isolation measurements of buried metal piping and tank systems.

**Cause or contribute to a new violation** (Section 93.152) means a Federal action that: (1) Causes a new violation of a national ambient air quality standard (NAAQS) at a location in a nonattainment or maintenance area which would otherwise not be in violation of the standard during the future period in question if the Federal action were not taken; or (2) Contributes, in conjunction with other reasonably foreseeable actions, to a new violation of a NAAQS at a location in a nonattainment or maintenance area in a manner that would increase the frequency or severity of the new violation.

**Cause or contribute to a new violation for a project** (Section 93.101) means: (1) To cause or contribute to a new violation of a standard in the area substantially affected by the project or over a region which would otherwise not be in violation of the standard during the future period in question, if the project were not implemented; or (2) To contribute to a new violation in a manner that would increase the frequency or severity of a new violation of a standard in such area.

**Caused by** (Section 93.152) as used in the terms "direct emissions" and "indirect emissions," means emissions that would not otherwise occur in the absence of the Federal action.

**CBOD5** (Section 133.101) The five day measure of the pollutant parameter carbonaceous biochemical oxygen demand (CBOD5).

**Cell burner boiler** (Section 76.2) means a wall-fired boiler that utilizes two or three circular burners combined into a single vertically oriented assembly that results in a compact, intense flame. Any low NOX retrofit of a cell burner boiler that reuses the existing cell burner, close-coupled wall opening configuration would not change the designation of the unit as a cell burner boiler.

**Cell fusion** (Section 174.3) means the fusion in vitro of two or more cells or protoplasts.

**Cellulose fiber loose-fill** (Section 247.3) means a basic material of recycled wood-based cellulosic fiber made from selected paper, paperboard stock, or ground wood stock, excluding contaminated materials which may reasonably be expected to be retained in the finished product, with suitable chemicals introduced to provide properties such as flame resistance, processing and handling characteristics. The basic cellulosic material may be processed into a form suitable for installation by pneumatic or pouring methods;

**Cementing** (Section 146.3) means the operation whereby a cement slurry is pumped into a drilled hole and/or forced behind the casing.

**Cementing** (Section 147.2902) the operation whereby a cement slurry is pumped into a drilled hole and/or forced behind the casing.

**CEMS** (Section 62.1201) means continuous emissions monitoring system.

**CEMS precision or precision as applied to the monitoring requirements of part 75 of this chapter,** (Section 72.2) means the closeness of a measurement to the actual measured value expressed as the uncertainty associated with repeated measurements of the same sample or of different samples from the same process (e.g., the random error associated with simultaneous measurements of a process made by more than one instrument). A measurement technique is determined to have increasing "precision" as the variation among the repeated measurements decreases.

**Centralized waste treatment (CWT) facility** (Section 437.2) means any facility that treats (for disposal, recycling or recovery of material) any hazardous or non-hazardous industrial wastes, hazardous or non-hazardous industrial wastewater, and/or used material received from off-site. "CWT facility" includes both a facility that treats waste received exclusively from off-site and a facility that treats wastes generated on-site as well as waste received from off-site. For example, an organic chemical manufacturing plant may, in certain circumstances, be a CWT facility if it treats industrial wastes received from offsite as well as industrial waste generated at the organic chemical manufacturing plant. CWT facilities may also include re-refiners and may be owned by the federal government.

**Centralized waste treatment wastewater** (Section 437.2) means any wastewater generated as a result of CWT activities. CWT wastewater sources may include, but are not limited to: liquid waste receipts, solubilization water, used oil emulsion-breaking wastewater, tanker truck/drum/roll-off box washes, equipment washes, air pollution control scrubber blow-down, laboratory-derived wastewater, on-site landfill wastewaters, and contaminated storm water.

**Centrally fueled** (Section 88.302-94) means a fleet, or that part of a fleet, consisting of vehicles that are fueled 100 percent of the time at a location that is owned, operated, or controlled by the covered fleet operator, or is under contract with the covered fleet operator. Any vehicle that is under normal operations garaged at home at night but that is, in fact, centrally fueled 100 percent of the time shall be considered to be centrally fueled for the purpose of this definition. The fact that one or more vehicles in a fleet is/are not centrally fueled does not exempt an entire fleet from the program. The fact that a vehicle is not centrally fueled does not mean it could not be centrally fueled in accordance with the definition of "capable of being centrally fueled."

**Centrally fueled bus** (Section 86.093-2) means a bus that is refueled at least 75 percent of the time at one refueling facility that is owned, operated, or controlled by the bus operator.

**Centroidal area** (Section 72.2) means a representational concentric area that is geometrically similar to the stack or duct cross section, and is not greater than 1 percent of the stack or duct cross-sectional area.

**CERCLA** (Section 280.12) means the Comprehensive Environmental Response, Compensation, and Liability Act of 1980, as amended.

**CERCLA** (Section 300.5) is the Comprehensive Environmental Response, Compensation, and Liability Act of 1980, as amended by the Superfund Amendments and Reauthorization Act of 1986.

**CERCLA** (Section 761.13) means the Comprehensive Environmental Response, Compensation, and Liability Act (42 U.S.C. 9601-9657).

**CERCLA Hazardous Substance** (Section 355.20) means a substance on the list defined in section 101(14) of CERCLA. Note: Listed CERCLA hazardous substances appear in table 302.4 of 40 CFR part 302.

**CERCLIS** (Section 300.5) is the abbreviation of the CERCLA Information System, EPA's comprehensive data base and data management system that inventories and tracks releases addressed or needing to be addressed by the Superfund program. CERCLIS contains the official inventory of CERCLA sites and supports EPA's site planning and tracking functions. Sites that EPA decides do not warrant moving further in the site evaluation process are given a "No Further Response Action Planned" (NFRAP) designation. This means that no additional federal steps under CERCLA will be taken at the site unless future information so warrants. Sites given a NFRAP designation are placed in a separate archival data base. Inclusion of a specific site or area in the CERCLIS data base does not represent a determination of any party's liability, nor does it represent a finding that any response action is necessary.

**Certificate holder** (Section 85.1502) The entity in whose name the certificate of conformity for a class of motor vehicles or motor vehicle engines has been issued.

**Certificate of conformity** (Section 85.1502) The document issued by the Administrator under section 206(a) of the Act.

**Certificate of conformity** (Section 89.602) The document issued by the Administrator under section 213 and section 206(a) of the Act.

**Certificate of conformity** (Section 91.702) The document issued by the Administrator under section 213 and section 206(a) of the Act.

**Certificate of representation** (Section 72.2) means the completed and signed submission required by Sec. 72.20, for certifying the appointment of a designated representative for an affected source or a group of identified affected sources authorized to represent the owners and operators of such source(s) and of the affected units at such source(s) with regard to matters under the Acid Rain Program.

**Certification** (Section 163.2) means a certification by the Director that a pesticide chemical is useful for the purpose for which a tolerance or exemption is sought under the act.

**Certification** (Section 171.2) means the recognition by a certifying agency that a person is competent and thus authorized to use or supervise the use of restricted use pesticides.

**Certification** (Section 194.2) means any action taken by the Administrator pursuant to section 8(d) (1) of the WIPP LWA.

**Certification** (Section 260.10) means a statement of professional opinion based upon knowledge and belief.

**Certification** (Section 761.13) means a written statement regarding a specific fact or representation that contains the following language: Under civil and criminal penalties of law for the making or submission of false or fraudulent statements or representations (18 U.S.C. 1001 and 15 U.S.C. 2615), I certify that the information contained in or accompanying this document is true, accurate, and complete. As to the identified section(s) of this document for which I cannot personally verify truth and accuracy, I certify as the company official having supervisory responsibility for the persons who, acting under my direct instructions, made the verification that this information is true, accurate, and complete.

**Certification** (Section 89.2) means, with respect to new nonroad engines, obtaining a certificate of conformity for an engine family complying with the nonroad engine emission standards and requirements specified in this part.

**Certification** (Section 90.3) means, with respect to new nonroad engines, obtaining a certificate of conformity for an engine family complying with the nonroad engine emission standards and requirements specified in this part.

**Certification** (Section 91.3) means, with respect to new SI marine engines, obtaining a certificate of conformity for an engine family complying with the marine SI engine emission standards and requirements specified in this part.

**Certification level** (Section 86.1702-99) means the official exhaust emission result from an emission-data vehicle which has been adjusted by the applicable mass deterioration factor and is submitted to the Administrator for use in determining compliance with an emission standard for the purpose of certifying a particular engine family. For those engine families which are certified using reactivity adjustment factors developed by the manufacturer pursuant to appendix XVII of this part, the exhaust NMOG certification level shall include adjustment by the ozone deterioration factor.

**Certification Short Test (CST)** (Section 86.1803-01) means the test, for gasoline-fueled Otto-cycle light-duty vehicles and light-duty trucks, performed in accordance with the procedures contained in 40 CFR part 86, subpart O.

**Certification Short Test** (Section 86.096-2) means the test, for gasoline-fueled Otto-cycle light-duty vehicles and light-duty trucks, performed in accordance with the procedures contained in 40 CFR part 86 subpart O.

**Certification Vehicle** (Section 600.022-85) means a vehicle which is selected under 40 CFR 86.084-24(b)(1) or 40 CFR 86.1824-01 as applicable, and is used to determine compliance under 40 CFR 86.084-30 or 40 CFR 86.1844-01 as applicable for issuance of an original certificate of conformity.

**Certification Vehicle Emission Margin for a certified engine family** (Section 85.2113) means the difference between the EPA emission standards and the average FTP emission test results of that engine family's emission-data vehicles at the projected applicable useful life mileage point (i.e., useful life mileage for light-duty vehicles is 50,000 miles and for light-duty trucks is 120,000 miles for 1985 and later model years or 50,000 miles for 1984 and earlier model years).

**Certified abatement worker** (Section 745.223) means an individual who has been trained by an accredited training program, as defined by this section, and certified by EPA pursuant to Sec. 745.226 to perform abatements.

**Certified Aftermarket Part** (Section 85.2113) means any aftermarket part which has been certified pursuant to this subpart.

**Certified applicator** (Section 171.2) means any individual who is certified to use or supervise the use of any restricted use pesticides covered by his certification.

**Certified firm** (Section 745.223) means a company, partnership, corporation, sole proprietorship, association, or other business entity that performs lead-based paint activities to which EPA has issued a certificate of approval pursuant to Sec. 745.226(f).

**Certified inspector** (Section 745.223) means an individual who has been trained by an accredited training program, as defined by this section, and certified by EPA pursuant to Sec. 745.226 to conduct inspections. A certified inspector also samples for the presence of lead in dust and soil for the purposes of abatement clearance testing.

**Certified Part** (Section 85.2102) means a part certified in accordance with the aftermarket part certification regulations contained in this subpart.

**Certified project designer** (Section 745.223) means an individual who has been trained by an accredited training program, as defined by this section, and certified by EPA pursuant to Sec. 745.226 to prepare abatement project designs, occupant protection plans, and abatement reports.

**Certified risk assessor** (Section 745.223) means an individual who has been trained by an accredited training program, as defined by this section, and certified by EPA pursuant to Sec. 745.226 to conduct risk assessments. A risk assessor also samples for the presence of lead in dust and soil for the purposes of abatement clearance testing.

**Certified supervisor** (Section 745.223) means an individual who has been trained by an accredited training program, as defined by this section, and certified by EPA pursuant to Sec. 745.226 to supervise and conduct abatements, and to prepare occupant protection plans and abatement reports.

**Certifying agency** (Section 121.1) means the person or agency designated by the Governor of a State, by statute, or by other governmental act, to certify compliance with applicable water quality standards. If an interstate agency has sole authority to so certify for the area within its jurisdiction, such interstate agency shall be the certifying agency. Where a State agency and an interstate agency have concurrent authority to certify, the State agency shall be the certifying agency. Where water quality standards have been promulgated by the Administrator pursuant to section 10©(2) of the Act, or where no State or interstate agency has authority to certify, the Administrator shall be the certifying agency.

**Certifying official, for purposes of part 73 of this chapter,** (Section 72.2) means: (1) For a corporation, a president, secretary, treasurer, or vice-president of the corporation in charge of a principal business function, or any other person who performs similar policy or decision-making functions for the corporation; (2) For partnership or sole proprietorship, a general partner or the proprietor, respectively; and (3) For a local government entity or State, Federal, or other public agency, either a principal executive officer or ranking elected official.

**Cesspool** (Section 144.3) means a "drywell" that receives untreated sanitary waste containing human excreta, and which sometimes has an open bottom and/or perforated sides.

**Cesspool** (Section 146.3) means a "drywell" that receives untreated sanitary waste containing human excreta, and which sometimes has an open bottom and/or perforated sides.

**Cetane index or "Calculated cetane index"** (Section 80.2) is a number representing the ignition properties of diesel fuel oils from API gravity and mid-boiling point as determined by ASTM standard method D 976-80, entitled "Standard Methods for Calculated Cetane Index of Distillate Fuels". ASTM test method D 976-80 is incorporated by reference. This incorporation by reference was approved by the Director of the Federal Register in accordance with 5 U.S.C. 552(a) and 1 CFR part 51. A copy may be obtained from the American Society for Testing and Materials, 1916 Race Street, Philadelphia, PA 19103. A copy may be inspected at the Air Docket Section (A-130), Room M-1500, U.S. Environmental Protection Agency, Docket No. A-86-03, 401 M Street SW., Washington, DC 20460 or at the Office of the Federal Register, 800 North Capitol Street, NW., suite 700, Washington, DC.

**Chairperson** (Section 1601.3) means the Chairperson of the CSB (including, in the absence of a Chairperson, the Board Member supervising personnel matters) or his or her designee.

**Channelizers** (Section 247.3) means highly visible barrels or drums that can be positioned to direct traffic through detours;

**Chemical** (Section 790.3) means a chemical substance or mixture.

**Chemical agents** (Section 300.5) means those elements, compounds, or mixtures that coagulate, disperse, dissolve, emulsify, foam, neutralize, precipitate, reduce, solubilize, oxidize, concentrate, congeal, entrap, fix, make the pollutant mass more rigid or viscous, or otherwise facilitate the mitigation of deleterious effects or the removal of the pollutant from the water. Chemical agents include biological additives, dispersants, sinking agents, miscellaneous oil spill control agents, and burning agents, but do not include sorbents.

**Chemical agents and munitions** (Section 266.201) are defined as in 50 U.S.C. section 1521(j)(1).

**Chemical cargos** (Section 442.2) mean, but are not limited to, the following: latex, rubber, plastics, plasticizers, resins, soaps, detergents, surfactants, agricultural chemicals and pesticides, hazardous waste, organic chemicals including: alcohols, aldehydes, formaldehydes, phenols, peroxides, organic salts, amines, amides, other nitrogen compounds, other aromatic compounds, aliphatic organic chemicals, glycols, glycerines, and organic polymers; refractory organic compounds including: ketones, nitriles, organo-metallic compounds containing chromium, cadmium, mercury, copper, zinc; and inorganic chemicals including: aluminum sulfate, ammonia, ammonium nitrate, ammonium sulfate, and bleach. Cargos which are not considered to be food grade or petroleum cargos are considered to be chemical cargos.

**Chemical composition** (Section 79.2) means the name and percentage by weight of each compound in an additive and the name and percentage by weight of each element in an additive.

**Chemical metal cleaning waste** (Section 423.11) means any wastewater resulting from the cleaning of any metal process equipment with chemical compounds, including, but not limited to, boiler tube cleaning.

**Chemical name** (Section 721.3) means the scientific designation of a chemical substance in accordance with the nomenclature system developed by the International Union of Pure and Applied Chemistry or the Chemical Abstracts Service's rules of nomenclature, or a name which will clearly identify a chemical substance for the purpose of conducting a hazard evaluation.

**Chemical oxygen demand (COD)** (Section 430.0) A bulk parameter that measures the oxygen-consuming capacity of organic and inorganic matter present in water or wastewater. It is expressed as the amount of oxygen consumed from a chemical oxidant in a specific test.

**Chemical protective clothing** (Section 721.3) means items of clothing that provide a protective barrier to prevent dermal contact with chemical substances of concern. Examples can include, but are not limited to: full body protective clothing, boots, coveralls, gloves, jackets, and pants.

**Chemical structure** (Section 79.2) means the molecular structure of a compound in an additive.

**Chemical substance** (section 710.2) means any organic or inorganic substance of a particular molecular identity, including any combination of such substances occurring in whole or in part as a result of a chemical reaction or occurring in nature, and any chemical element or uncombined radical; except that "chemical substance" does not include: (1) Any mixture, (2) Any pesticide when manufactured, processed, or distributed in commerce for use as a pesticide, (3) Tobacco or any tobacco product, but not including any derivative products, (4) Any source material, special nuclear material, or byproduct material, (5) Any pistol, firearm, revolver, shells, and cartridges, and (6) Any food, food additive, drug, cosmetic, or device, when manufactured, processed, or distributed in commerce for use as a food, food additive, drug, cosmetic, or device.

**Chemical substance** (Section 720.3) means any organic or inorganic substance of a particular molecular identity, including any combination of such substances occurring in whole or in part as a result of a chemical reaction or occurring in nature, and any chemical element or uncombined radical, except that "chemical substance" does not include: (1) Any mixture. (2) Any pesticide when manufactured, processed, or distributed in commerce for use as a pesticide. (3) Tobacco or any tobacco product. (4) Any source material, special nuclear material, or byproduct material. (5) Any pistol, firearm, revolver, shells, or cartridges. (6) Any food, food additive, drug, cosmetic, or device, when manufactured, processed, or distributed in commerce for use as a food, food additive, drug, cosmetic, or device.

**Chemical substance** (Section 761.13) (1) except as provided in paragraph (2) of this definition, means any organic or inorganic substance of a particular molecular identity, including: Any combination of such substances occurring in whole or part as a result of a chemical reaction or occurring in nature, and any element or uncombined radical. (2) Such term does not include: Any mixture; any pesticide (as defined in the Federal Insecticide, Fungicide, and Rodenticide Act) when manufactured, processed, or distributed in commerce for use as a pesticide; tobacco or any tobacco product; any source material, special nuclear material, or byproduct material (as such terms are defined in the Atomic Energy Act of 1954 and regulations issued under such Act); any article the sale of which is subject to the tax imposed by section 4181 of the Internal Revenue Code of 1954 (determined without regard to any exemptions from such tax provided by section 4182 or section 4221 or any provisions of such Code); and any food, food additive, drug, cosmetic, or device (as such terms are defined in section 201 of the Federal Food, Drug, and Cosmetic Act) when manufactured, processed, or distributed in commerce for use as a food, food additive, drug, cosmetic, or device.

**Chemical substance** (Section 763.163) has the same meaning as in section 3 of the Act (TSCA).

**Chemical synthesis** (Section 439.31) means using one or a series of chemical reactions in the manufacturing process of a specified product.

**Chemical waste landfill** (Section 761.13) means a landfill at which protection against risk of injury to health or the environment from migration of PCBs to land, water, or the atmosphere is provided from PCBs and PCB Items deposited therein by locating, engineering, and operating the landfill as specified in Sec. 761.75.

**Chemigation** (Section 170.3) means the application of pesticides through irrigation systems.

**Cherries, brined** (Section 407.61) shall include the processing of all varieties of cherries into the following brined product styles: Canned, bottled and bulk, sweet and sour, pitted and unpitted, bleached, sweetened, colored and flavored, whole, halved and chopped.

**Cherries, sour** (Section 407.61) shall include the processing of all sour varieties of cherries into the following products styles: Frozen and canned, pitted and unpitted, whole, halves, juice and concentrate.

**Cherries, sweet** (Section 407.61) shall include the processing of all sweet varieties of cherries into the following products styles: Frozen and canned, pitted and unpitted, whole, halves, juice and concentrate.

**Chewable surface** (Section 745.63) means an interior or exterior surface painted with lead-based paint that a young child can mouth or chew. A chewable surface is the same as an "accessible surface" as defined in 42 U.S.C. 4851b(2)). Hard metal substrates and other materials that cannot be dented by the bite of a young child are not considered chewable.

**Chief Executive Officer of the tribe** (Section 350.1) means the person who is recognized by the Bureau of Indian Affairs as the chief elected administrative officer of the tribe.

**Chief Executive Officer of the tribe** (Section 355.20) means the person who is recognized by the Bureau of Indian Affairs as the chief elected administrative officer of the tribe.

**Chief Executive Officer of the tribe** (Section 370.2) means the person who is recognized by the Bureau of Indian Affairs as the chief elected administrative officer of the tribe.

**Chief Executive Officer of the tribe** (Section 372.3) means the person who is recognized by the Bureau of Indian Affairs as the chief elected administrative officer of the tribe.

**Chief Financial Officer** (Section 280.92) in the case of local government owners and operators, means the individual with the overall authority and responsibility for the collection, disbursement, and use of funds by the local government.

**Child-occupied facility** (Section 745.223) means a building, or portion of a building, constructed prior to 1978, visited regularly by the same child, 6 years of age or under, on at least two different days within any week (Sunday through Saturday period), provided that each day's visit lasts at least 3 hours and the combined weekly visit lasts at least 6 hours, and the combined annual visits last at least 60 hours. Child-occupied facilities may include, but are not limited to, day-care centers, preschools and kindergarten classrooms.

**Child-resistant packaging** (Section 157.21) means packaging that is designed and constructed to be significantly difficult for children under 5 years of age to open or obtain a toxic or harmful amount of the substance contained therein within a reasonable time, and that is not difficult for normal adults to use properly.

**Chips, corn** (Section 407.81) shall mean the processing of fried corn, made by soaking, rinsing, milling and extruding into a fryer without toasting. In terms of finished corn chips, 1 kg (lb) of finished product is equivalent to 0.9 kg (lb) of raw material.

**Chips, potato** (Section 407.81) shall mean the processing of fried chips, made from fresh or stored white potatoes, all varieties. In terms of finished potato chips, 1 kg (lb) of finished product is equivalent to 4 kg (lb) of raw material.

**Chips, tortilla** (Section 407.81) shall mean the processing of fried corn, made by soaking, rinsing, milling, rolling into sheets, toasting and frying. In terms of finished tortilla chips, 1 kg (lb) of finished product is equivalent to 0.9 kg (lb) of raw material.

**Chlorine** (Section 415.661) shall mean the total residual chlorine present in the process wastewater stream exiting the wastewater treatment system.

**Chrome pigments** (Section 415.341) means chrome yellow, chrome orange, molybdate chrome orange, anhydrous and hydrous chromium oxide, chrome green, and zinc yellow.

**Chrome tan** (Section 425.02) means the process of converting hide into leather using a form of chromium.

**Chromium** (Section 415.661) shall mean the total chromium present in the process wastewater stream exiting the wastewater treatment system.

**Chromium** (Section 420.02) means total chromium and is determined by the method specified in 40 CFR 136.3.

**Chromium** (Section 428.101) shall mean total chromium.

**Chronic toxicity** (Section 132.2) is concurrent and delayed adverse effect(s) that occurs only as a result of a chronic exposure.

**City Fuel Economy** (Section 600.022-85) means the fuel economy determined by operating a vehicle (or vehicles) over the driving schedule in the Federal emission test procedure.

**Cladding or metal cladding** (Section 471.02) is the art of producing a composite metal containing two or more layers that have been metallurgically bonded together by roll bonding (co-rolling), solder application (or brazing), or explosion bonding.

**Claim** (Section 211.203) An assertion made by a manufacturer regarding the effectiveness of his product.

**Claim (Section 304.12)** means the amount sought by EPA as recovery of response costs incurred and to be incurred by the United States at a facility, which does not exceed $500,000, excluding interest.

**Claim** (Section 305.3) means a demand in writing for a sum certain, which is presented to the Fund in accordance with CERCLA sections 111 and 112.

**Claim** (Section 307.14) means a demand in writing for a sum certain presented to the Fund in accordance with sections 111 and 112 of CERCLA.

**Claim for purposes of a release under CERCLA** (Section 300.5) means a demand in writing for a sum certain; for purposes of a discharge under CWA, it means a request, made in writing for a sum certain, for compensation for damages or removal costs resulting from an incident.

**Claimant** (Section 305.3) means any person who presents a claim to the Fund for reimbursement under CERCLA section 112(b)(1).

**Claimant** (Section 307.14) means any person who presents a claim to the Fund for reimbursement under section 112(b)(1) of CERCLA.

**Claimant** (Section 350.1) means a person submitting a claim of trade secrecy to EPA in connection with a chemical otherwise required to be disclosed in a report or other filing made under Title III.

**Claimant as defined by section 1001 of the OPA** (Section 300.5) means any person or government who presents a claim for compensation under Title I of the OPA.

**Claims Official** (Section 305.3) means the Assistant Administrator or the Regional Administrator or his delegatee who makes the initial decision awarding or denying a claim in whole or in part.

**Class** (Section 205.151) means a group of vehicles which are identical in all material aspects with respect to the parameters listed in Sec. 205.155 of this subpart.

**Class A** (Section 86.1702-99) comprises LDVs and LDTs 0-3750 lbs LVW that are subject to the provisions of this subpart.

**Class B** (Section 86.1702-99) comprises LDTs 3751-5750 lbs LVW that are subject to the provisions of this subpart.

**Class I** (Section 82.3) refers to the controlled substances listed in appendix A to this subpart.

**Class I equivalent method** (Section 53.1) means an equivalent method for PM2.5 which is based on a sampler that is very similar to the sampler specified for reference methods in appendix L of this part, with only minor deviations or modifications, as determined by EPA.

**Class I freight railroad** (Section 92.2) means a Class I railroad that primarily transports freight rather than passengers.

**Class I railroad** (Section 92.2) means a railroad that has been classified as a Class I railroad by the Surface Transportation Board.

**Class I sludge management facility** (Section 122.2) means any POTW identified under 40 CFR 403.8(a) as being required to have an approved pretreatment program (including such POTWs located in a State that has elected to assume local program responsibilities pursuant to 40 CFR 403.10(e)) and any other treatment works treating domestic sewage classified as a Class I sludge management facility by the Regional Administrator, or, in the case of approved State programs, the Regional Administrator in conjunction with the State Director, because of the potential for its sludge use or disposal practices to adversely affect public health and the environment.

**Class I sludge management facility** (Section 501.2) means any POTW identified under 40 CFR 403.8(a) as being required to have an approved pretreatment program (including such POTWs located in a State that has elected to assume local program responsibilities pursuant to 40 CFR 403.10(e)) and any other treatment works treating domestic sewage classified as a Class I sludge management facility by the Regional Administrator in conjunction with the State Program Director because of the potential for its sludge use or disposal practices to adversely affect public health or the environment.

**Class II** (Section 82.3) refers to the controlled substances listed in appendix B to this subpart.

**Class II equivalent method** (Section 53.1) means an equivalent method for PM2.5 that utilizes a PM2.5 sampler in which an integrated PM2.5 sample is obtained from the atmosphere by filtration and is subjected to a subsequent filter conditioning process followed by a gravimetric mass determination, but which is not a Class I equivalent method because of substantial deviations from the design specifications of the sampler specified for reference methods in appendix L of part 50 of this chapter, as determined by EPA.

**Class II railroad** (Section 92.2) means a railroad that has been classified as a Class II railroad by the Surface Transportation Board.

**Class II Wells** (Section 147.2902) wells which inject fluids: (a) Which are brought to the surface in connection with conventional oil or natural gas production and may be commingled with waste waters from gas plants which are an integral part of production operations, unless those waters would be classified as a hazardous waste at the time of injection; (b) For enhanced recovery of oil or natural gas; and (c) For storage of hydrocarbons which are liquid at standard temperature and pressure.

**Class III equivalent method** (Section 53.1) means an equivalent method for PM2.5 that has been determined by EPA not to be a Class I or Class II equivalent method. This fourth type of PM2.5 method includes alternative equivalent method samplers and continuous analyzers, based on designs and measurement principles different from those specified for reference methods (e.g., a means for estimating aerosol mass concentration other than by conventional integrated filtration followed by equilibration and gravimetric analysis. These samplers (or monitors) are those deemed to be substantially different from reference method samplers and are likely to use components and methods other than those specified for reference method samplers.

**Class III railroad** (Section 92.2) means a railroad that has been classified as a Class III railroad by the Surface Transportation Board.

**Class T3** (Section 87.1) means all aircraft gas turbine engines of the JT3D model family.

**Class T8** (Section 87.1) means all aircraft gas turbine engines of the JT8D model family.

**Class TF** (Section 87.1) means all turbofan or turbojet aircraft engines or aircraft engines designed for applications that otherwise would have been fulfilled by turbojet and turbofan engines except engines of class T3, T8, and TSS.

**Class TP** (Section 87.1) means all aircraft turboprop engines.

**Class TSS** (Section 87.1) means all aircraft gas turbine engines employed for propulsion of aircraft designed to operate at supersonic flight speeds.

**Classification of Railroads** (Section 201.1) means the division of railroad industry operating companies by the Interstate Commerce Commission into three categories. As of 1978, Class I railroads must have annual revenues of $50 million or greater, Class II railroads must have annual revenues of between $10 and $50 million, and Class III railroads must have less than $10 million in annual revenues.

**Classified information** (Section 68.3) means "classified information" as defined in the Classified Information Procedures Act, 18 U.S.C. App. 3, section 1(a) as "any information or material that has been determined by the United States Government pursuant to an executive order, statute, or regulation, to require protection against unauthorized disclosure for reasons of national security."

**Classified Waste** (Section 246.101) means waste material that has been given security classification in accordance with 50 U.S.C. 401 and Executive Order 11652.

**Clean Air Act or Act** (Section 49.2) means those statutory provisions in the United States Code at 42 U.S.C. 7401, et seq.

**Clean coal technology demonstration project** (Section 60.2) means a project using funds appropriated under the heading 'Department of Energy-Clean Coal Technology', up to a total amount of $2,500,000,000 for commercial demonstrations of clean coal technology, or similar projects funded through appropriations for the Environmental Protection Agency.

**Clean data** (Section 93.101) means air quality monitoring data determined by EPA to meet the requirements of 40 CFR part 58 that indicate attainment of the national ambient air quality standard.

**Clean-fuel vehicle aftermarket conversion certifier** (Section 88.302-94) means the business or entity that obtains a certificate of conformity with the clean-fuel vehicle standards and requirements for a vehicle/engine conversion configuration pursuant to the requirements of 40 CFR part 86 and this part 88.

**Cleaning or etching** (Section 467.02) is a chemical solution bath and a rinse or series of rinses designed to produce a desired surface finish on the workpiece. This term includes air pollution control scrubbers which are sometimes used to control fumes from chemical solution baths. Conversion coating and anodizing when performed as an integral part of the aluminum forming operations are considered cleaning or etching operations. When conversion coating or anodizing are covered here they are not subject to regulation under the provisions of 40 CFR Part 433, Metal Finishing.

**Cleaning water** (Section 463.2) is process water used to clean the surface of an intermediate or final plastic product or to clean the surfaces of equipment used in plastics molding and forming that contact an intermediate or final plastic product. It includes water used in both the detergent wash and rinse cycles of a cleaning process.

**Cleanup site** (Section 761.13) means the areal extent of contamination and all suitable areas in very close proximity to the contamination necessary for implementation of a cleanup of PCB remediation waste, regardless of whether the site was intended for management of waste.

**Clearance levels** (Section 745.223) are values that indicate the maximum amount of lead permitted in dust on a surface following completion of an abatement activity.

**Closed course competition event** (Section 205.151) means any organized competition event covering an enclosed, repeated or confined route intended for easy viewing of the entire route by all spectators. Such events include short track, dirt track, drag race, speedway, hill climb, ice race, and the Bonneville Speed Trials.

**Closed portion** (Section 260.10) means that portion of a facility which an owner or operator has closed in accordance with the approved facility closure plan and all applicable closure requirements. (See also "active portion" and "inactive portion".)

**Closed vent system** (Section 65.2) means a system that is not open to the atmosphere and is composed of piping, ductwork, connections, and, if necessary, flow inducing devices that transport gas or vapor from an emission point to a control device. A closed vent system does not include the vapor collection system that is part of any tank truck or railcar or the loading arm or hose that is used for vapor return. For transfer racks, the closed vent system begins at, and includes, the first block valve on the downstream side of the loading arm or hose used to convey displaced vapors.

**Closed vent system shutdown** (Section 65.2) means a work practice or operational procedure that stops production from a process unit or part of a process unit during which it is technically feasible to clear process material from a closed vent system or part of a closed vent system consistent with safety constraints and during which repairs can be effected. An unscheduled work practice or operational procedure that stops production from a process unit or part of a process unit for less than 24 hours is not a closed vent system shutdown. An unscheduled work practice or operational procedure that would stop production from a process unit or part of a process unit for a shorter period of time than would be required to clear the closed vent system or part of the closed vent system of materials and start up the unit, and would result in greater emissions than delay of repair of leaking components until the next scheduled closed vent system shutdown, is not a closed vent system shutdown. The use of spare equipment and technically feasible bypassing of equipment without stopping production are not closed vent system shutdowns.

**Closed-loop system** (Section 65.2) means an enclosed system that returns process fluid to a process.

**Closed-purge system** (Section 65.2) means a system or combination of systems and portable containers to capture purged liquids. Containers must be covered or closed when not being filled or emptied.

**Closed-top hopper** (Section 442.2) means a completely enclosed storage vessel used to transport dry bulk cargos, either by truck, rail, or barge. Closed-top hoppers are not designed or constructed to carry liquid cargos and are typically used to transport grain, soybeans, soy meal, soda ash, lime, fertilizer, plastic pellets, flour, sugar, and similar commodities or cargos. The cargos transported come in direct contact with the hopper interior. Closed-top hoppers are also commonly referred to as dry bulk hoppers.

**Closed-vent system** (Section 264.1031) means a system that is not open to the atmosphere and that is composed of piping, connections, and, if necessary, flow-inducing devices that transport gas or vapor from a piece or pieces of equipment to a control device.

**Closing rpm** (Section 205.151) means the engine speed in Figure 2 of Appendix I.

**Closure** (Section 270.2) means the act of securing a Hazardous Waste Management facility pursuant to the requirements of 40 CFR part 264.

**Closure device** (Section 265.1081) means a cap, hatch, lid, plug, seal, valve, or other type of fitting that blocks an opening in a cover such that when the device is secured in the closed position it prevents or reduces air pollutant emissions to the atmosphere. Closure devices include devices that are detachable from the cover (e.g., a sampling port cap), manually operated (e.g., a hinged access lid or hatch), or automatically operated (e.g., a spring-loaded pressure relief valve).

**Closure period** (Section 192.31) means the period of time beginning with the cessation, with respect to a waste impoundment, of uranium ore processing operations and ending with completion of requirements specified under a closure plan.

**Closure plan** (Section 192.31) means the plan required under Sec. 264.112 of this chapter.

**cm** (Section 401.11) means centimeter.

**CMS** (Section 62.1201) means continuous monitoring system.

**CN,A** (Section 413.02) shall mean cyanide amenable to chlorination as defined by 40 CFR 136.

**CN,T** (Section 413.02) shall mean cyanide, total.

**CO** (Section 58.1) means carbon monoxide.

**Coagulation** (Section 141.2) means a process using coagulant chemicals and mixing by which colloidal and suspended materials are destabilized and agglomerated into flocs.

**Coal** (Section 72.2) means all solid fuels classified as anthracite, bituminous, subbituminous, or lignite by the American Society for Testing and Materials Designation ASTM D388-92 "Standard Classification of Coals by Rank" (as incorporated by reference in Sec. 72.13).

**Coal extraction** (Section 372.3) means the physical removal or exposure of ore, coal, minerals, waste rock, or overburden prior to beneficiation, and encompasses all extraction-related activities prior to beneficiation. Extraction does not include beneficiation (including coal preparation), mineral processing, in situ leaching or any further activities.

**Coal pile runoff** (Section 423.11) means the rainfall runoff from or through any coal storage pile.

**Coal preparation plant** (Section 434.11) means a facility where coal is subjected to cleaning, concentrating, or other processing or preparation in order to separate coal from its impurities and then is loaded for transit to a consuming facility.

**Coal preparation plant associated areas** (Section 434.11) means the coal preparation plant yards, immediate access roads, coal refuse piles and coal storage piles and facilities.

**Coal preparation plant water circuit** (Section 434.11) means all pipes, channels, basins, tanks, and all other structures and equipment that convey, contain, treat, or process any water that is used is coal preparation processes within a coal preparation plant.

**Coal refuse disposal pile** (Section 434.11) means any coal refuse deposited on the earth and intended as permanent dispoal or long-term storage (greater than 180 days) of such material, but does not include coal refuse deposited within the active mining area or coal refuse never removed from the active mining area.

**Coal remining operation** (Section 434.70) means a coal mining operation at a site on which coal mining was previously conducted and where the site has been abandoned or the performance bond has been forfeited.

**Coal-derived fuel** (Section 72.2) means any fuel, whether in a solid, liquid, or gaseous state, produced by the mechanical, thermal, or chemical processing of coal (e.g., pulverized coal, coal refuse, liquified or gasified coal, washed coal, chemically cleaned coal, coal-oil mixtures, and coke).

**Coal-fired** (Section 72.2) means the combustion of fuel consisting of coal or any coal-derived fuel (except a coal-derived gaseous fuel that meets the definition of "very low sulfur fuel" in this section), alone or in combination with any other fuel, where: (1) For purposes of the requirements of part 75 of this chapter, a unit is "coal-fired" independent of the percentage of coal or coal-derived fuel consumed in any calendar year (expressed in mmBtu); and (2) For all other purposes under the Acid Rain Program, except for purposes of applying part 76 of this chapter, a unit is "coal-fired" if it uses coal or coal-derived fuel as its primary fuel (expressed in mmBtu); provided that, if the unit is listed in the NADB, the primary fuel is the fuel listed in the NADB under the data field "PRIMEFUEL".

**Coal-fired utility unit** (Section 76.2) means a utility unit in which the combustion of coal (or any coal-derived fuel) on a Btu basis exceeds 50.0 percent of its annual heat input during the following calendar year: for Phase I units, in calendar year 1990; and, for Phase II units, in calendar year 1995 or, for a Phase II unit that did not combust any fuel that resulted in the generation of electricity in calendar year 1995, in any calendar year during the period 1990-1995. For the purposes of this part, this definition shall apply notwithstanding the definition in Sec. 72.2 of this chapter.

**Coast Guard District Response Group (DRG) as provided for by CWA sections 311(a)(20) and (j)(3)** (Section 300.5) means the entity established by the Secretary of the department in which the USCG is operating, within each USCG district, and shall consist of: the combined USCG personnel and equipment, including marine firefighting equipment, of each port in the district; additional prepositioned response equipment; and a district response advisory team.

**Coastal waters for the purposes of classifying the size of discharges** (Section 300.5) means the waters of the coastal zone except for the Great Lakes and specified ports and harbors on inland rivers.

**Coastal zone as defined for the purpose of the NCP** (Section 300.5) means all United States waters subject to the tide, United States waters of the Great Lakes, specified ports and harbors on inland rivers, waters of the contiguous zone, other waters of the high seas subject to the NCP, and the land surface or land substrata, ground waters, and ambient air proximal to those waters. The term coastal zone delineates an area of federal responsibility for response action. Precise boundaries are determined by EPA/USCG agreements and identified in federal regional contingency plans.

**Coating operations** (Section 466.02) means all of the operations associated with preparation and application of the vitreous coating. Usually this includes ballmilling, slip transport, application of slip to the workpieces, cleaning and recovery of faulty parts, and firing (fusing) of the enamel coat.

**Cobalt** (Section 415.651) shall mean the total cobalt present in the process wastewater stream exiting the wastewater treatment system.

**COD** (Section 401.11) means chemical oxygen demand; (3) TOC means total organic carbon.

**COD** (Section 427.61) shall mean COD added to the process waste water.

**COD** (Section 427.71) shall mean COD added to the process waste water.

**Cogeneration unit** (Section 72.2) means a unit that has equipment used to produce electric energy and forms of useful thermal energy (such as heat or steam) for industrial, commercial, heating or cooling purposes, through the sequential use of energy.

**Cogeneration unit** (Section 72.2) means a unit that produces electric energy and useful thermal energy for industrial, commercial, or heating or cooling purposes, through the sequential use of the original fuel energy.

**Coil** (Section 465.02) means a strip of basis material rolled into a roll for handling.

**Coil coating** (Section 465.02) means the process of converting basis material strip into coated stock. Usually cleaning, conversion coating, and painting are performed on the basis material. This regulation covers processes which perform any two or more of the three operations.

**Cold rolling** (Section 468.02) shall mean the process of rolling a workpiece below the recrystallization temperature of the copper or copper alloy.

**Cold worked pipe and tube** (Section 420.101) means those cold forming operations that process unheated pipe and tube products using either water or oil solutions for cooling and lubrication.

**Collection** (Section 243.101) means the act of removing solid waste (or materials which have been separated for the purpose of recycling) from a central storage point.

**Collection** (Section 246.101) means the act of removing solid waste (or materials which have been separated for the purpose of recycling) from a central storage point.

**Collection frequency** (Section 243.101) means the number of times collection is provided in a given period of time.

**Collocated** (Section 53.1) describes two or more air samplers, analyzers, or other instruments which sampler the ambient air that are operated simultaneously while located side by side, separated by a distance that is large enough to preclude the air sampled by any of the devices from being affected by any of the other devices, but small enough so that all devices obtain identical or uniform ambient air samples that are equally representative of the general area in which the group of devices is located.

**Combination** (Section 420.101) means those cold rolling operations which include recirculation of rolling solutions at one or more mill stands, and once-through use of rolling solutions at the remaining stand or stands.

**Combination acid pickling** (Section 420.91) means those operations in which steel products are immersed in solutions of more than one acid to chemically remove scale and oxides, and those rinsing steps associated with such immersions.

**Combination heavy-duty vehicle** (Section 88.302-93) means a vehicle with a GVWR greater than 8,500 pounds (3,900 kilograms) which is comprised of a truck-tractor and one or more pieces of trailered equipment. The truck-tractor is a self-propelled motor vehicle built on one chassis which encompasses the engine, passenger compartment, and a means of coupling to a cargo carrying trailer(s). The truck-tractor itself is not designed to carry cargo.

**Combined cycle system** (Section 96.2) means a system comprised of one or more combustion turbines, heat recovery steam generators, and steam turbines configured to improve overall efficiency of electricity generation or steam production.

**Combined cycle system** (Section 97.2) means a system comprised of one or more combustion turbines, heat recovery steam generators, and steam turbines configured to improve overall efficiency of electricity generation or steam production.

**Combined Fuel Economy** (Section 600.022-85) means the fuel economy value determined for a vehicle (or vehicles) by harmonically averaging the city and highway fuel economy values, weighted 0.55 and 0.45 respectively, for gasoline-fueled and diesel vehicles. For electric vehicles, the term means the equivalent petroleum-based fuel economy value as determined by the calculation procedure promulgated by the Secretary of Energy.

**Combined Metals** (Section 421.261) shall mean the total of gold, platinum and palladium.

**Combustibles** (Section 240.101) means materials that can be ignited at a specific temperature in the presence of air to release heat energy.

**Combustion chamber** (Section 62.1201) means the area in which controlled flame combustion of hazardous waste occurs.

**Combustion controls** (Section 76.2) means technology that minimizes NOX formation by staging fuel and combustion air flows in a boiler. This definition shall include low NOX burners, overfire air, or low NOX burners with overfire air.

**Combustion device** (Section 65.2) means an individual unit of equipment, such as a flare, incinerator, process heater, or boiler, used for the combustion of organic emissions.

**Combustion source** (Section 72.2) means a stationary fossil fuel fired boiler, turbine, or internal combustion engine that has submitted or intends to submit an opt-in permit application under Sec. 74.14 of this chapter to enter the Opt-in Program.

**Combustion turbine** (Section 96.2) means an enclosed fossil or other fuel-fired device that is comprised of a compressor, a combustor, and a turbine, and in which the flue gas resulting from the combustion of fuel in the combustor passes through the turbine, rotating the turbine.

**Combustion turbine** (Section 97.2) means an enclosed fossil or other fuel-fired device that is comprised of a compressor, a combustor, and a turbine, and in which the flue gas resulting from the combustion of fuel in the combustor passes through the turbine, rotating the turbine.

**Commence commercial operation** (Section 72.2) means to have begun to generate electricity for sale, including the sale of test generation.

**Commence commercial operation** (Section 96.2) means, with regard to a unit that serves a generator, to have begun to produce steam, gas, or other heated medium used to generate electricity for sale or use, including test generation. Except as provided in Sec. 96.5, for a unit that is a NOX Budget unit under Sec. 96.4 on the date the unit commences commercial operation, such date shall remain the unit's date of commencement of commercial operation even if the unit is subsequently modified, reconstructed, or repowered. Except as provided in Sec. 96.5 or subpart I of this part, for a unit that is not a NOX Budget unit under Sec. 96.4 on the date the unit commences commercial operation, the date the unit becomes a NOX Budget unit under Sec. 96.4 shall be the unit's date of commencement of commercial operation.

**Commence commercial operation** (Section 97.2) means, with regard to a unit that serves a generator, to have begun to produce steam, gas, or other heated medium used to generate electricity for sale or use, including test generation. Except as provided in Sec. 97.4 (b), Sec. 97.5, or subpart I of this part, for a unit that is a NOX Budget unit under Sec. 97.4(a) on the date the unit commences commercial operation, such date shall remain the unit's date of commencement of commercial operation even if the unit is subsequently modified, reconstructed, or repowered. Except as provided in Sec. 97.4 (b), Sec. 97.5, or subpart I of this part, for a unit that is not a NOX Budget unit under Sec. 97.4 (a) on the date the unit commences commercial operation, the date the unit becomes a NOX Budget unit under Sec. 97.4 (a) shall be the unit's date of commencement of commercial operation.

**Commence construction** (Section 72.2) means that an owner or operator has either undertaken a continuous program of construction or has entered into a contractual obligation to undertake and complete, within a reasonable time, a continuous program of construction.

**Commence operation** (Section 72.2) means to have begun any mechanical, chemical, or electronic process, including start-up of an emissions control technology or emissions monitor or of a unit's combustion chamber.

**Commence operation** (Section 96.2) means to have begun any mechanical, chemical, or electronic process, including, with regard to a unit, start-up of a unit's combustion chamber. Except as provided in Sec. 96.5, for a unit that is a NOX Budget unit under Sec. 96.4 on the date of commencement of operation, such date shall remain the unit's date of commencement of operation even if the unit is subsequently modified, reconstructed, or repowered. Except as provided in Sec. 96.5 or subpart I of this part, for a unit that is not a NOX Budget unit under Sec. 96.4 on the date of commencement of operation, the date the unit becomes a NOX Budget unit under Sec. 96.4 shall be the unit's date of commencement of operation.

**Commence operation** (Section 97.2) means to have begun any mechanical, chemical, or electronic process, including, with regard to a unit, start-up of a unit's combustion chamber. Except as provided in Sec. 97.4 (b), Sec. 97.5, or subpart I of this part for a unit that is a NOX Budget unit under Sec. 97.4 (a) on the date of commencement of operation, such date shall remain the unit's date of commencement of operation even if the unit is subsequently modified, reconstructed, or repowered. Except as provided in Sec. 97.4 (b), Sec. 97.5, or subpart I of this part, for a unit that is not a NOX Budget unit under Sec. 97.4 (a) on the date of commencement of operation, the date the unit becomes a NOX Budget unit under Sec. 97.4(a) shall be the unit's date of commencement of operation.

**Commenced** (Section 60.2) means, with respect to the definition of new source in section 111(a) (2) of the Act, that an owner or operator has undertaken a continuous program of construction or modification or that an owner or operator has entered into a contractual obligation to undertake and complete, within a reasonable time, a continuous program of construction or modification.

**Commenced** (Section 61.02) means, with respect to the definition of "new source" in section 111(a)(2) of the Act, that an owner or operator has undertaken a continuous program of construction or modification or that an owner or operator has entered into a contractual obligation to undertake and complete, within a reasonable time, a continuous program of construction or modification.

**Commenced** (Section 63.2) means, with respect to construction or reconstruction of an affected source, that an owner or operator has undertaken a continuous program of construction or reconstruction or that an owner or operator has entered into a contractual obligation to undertake and complete, within a reasonable time, a continuous program of construction or reconstruction.

**Commerce** (Section 205.2) means trade, traffic, commerce, or transportation: (i) Between a place in a State and any place outside thereof, or (ii) Which affects trade, traffic, commerce, or transportation described in paragraph (a)(17)(i) of this section.

**Commerce** (section 710.2) means trade, traffic, transportation, or other commerce: (1) Between a place in a State and any place outside of such State, or (2) which affects trade, traffic, transportation, or commerce described in paragraph (i)(1) of this section.

**Commerce** (Section 720.3) means trade, traffic, transportation, or other commerce (1) between a place in a State and any place outside of such State, or (2) which affects trade, traffic, transportation, or commerce between a place in a State and any place outside of such State.

**Commerce** (Section 761.13) means trade, traffic, transportation, or other commerce: (1) Between a place in a State and any place outside of such State, or (2) Which affects trade, traffic, transportation, or commerce described in paragraph (1) of this definition.

**Commerce** (Section 763.163) has the same meaning as in section 3 of the Act (TSCA).

**Commercial** (Section 85.1602) means an activity engaged in as a vocation.

**Commercial aircraft engine** (Section 87.1) means any aircraft engine used or intended for use by an "air carrier," (including those engaged in "intrastate air transportation") or a "commercial operator" (including those engaged in "intrastate air transportation") as these terms are defined in the Federal Aviation Act and the Federal Aviation Regulations.

**Commercial aircraft gas turbine engine** (Section 87.1) means a turboprop, turbofan, or turbojet commercial aircraft engine.

**Commercial applicator** (Section 171.2) means a certified applicator (whether or not he is a private applicator with respect to some uses) who uses or supervises the use of any pesticide which is classified for restricted use for any purpose or on any property other than as provided by the definition of "private applicator."

**Commercial establishment** (Section 246.101) means stores, offices, restaurants, warehouses and other non-manufacturing activities.

**Commercial hazardous waste combustor** (Section 444.11) means any thermal unit, except a cement kiln, that is subject either to 40 CFR part 264, subpart O; 40 CFR part 265, subpart O; or 40 CFR part 266, subpart H if the thermal unit burns RCRA hazardous wastes received from off-site for a fee or other remuneration in the following circumstances. The thermal unit is a commercial hazardous waste combustor if the off-site wastes are generated at a facility not under the same corporate structure or subject to the same ownership as the thermal unit and (1) The thermal unit is burning wastes that are not of a similar nature to wastes being burned from industrial processes on site or (2) There are no wastes being burned from industrial processes on site. Examples of wastes of a "similar nature" may include the following: Wastes generated in industrial operations whose wastewaters are subject to the same provisions in 40 CFR Subchapter N or wastes burned as part of a product stewardship activity. The term commercial hazardous waste combustor includes the following facilities: a facility that burns exclusively waste received from off-site; and, a facility that burns both wastes generated on-site and wastes received from off-site. Facilities that may be commercial hazardous waste combustors include hazardous waste incinerators, rotary kiln incinerators, lime kilns, lightweight aggregate kilns, and boilers. A facility not otherwise a commercial hazardous waste combustor is not a commercial hazardous waste combustor if it burns RCRA hazardous waste for charitable organizations, as a community service or as an accommodation to local, state or government agencies so long as the waste is burned for no fee or other remuneration.

**Commercial hazardous waste combustor wastewater** (Section 444.11) means wastewater attributable to commercial waste combustion operations, but includes only wastewater from air pollution control systems and water used to quench flue gas or slag generated as a result of commercial hazardous waste combustor operations.

**Commercial marine engine** (Section 94.2) means a marine engine that is not a recreational marine engine.

**Commercial paper** (Section 763.163) means an asbestos-containing product which is made of paper intended for use as general insulation paper or muffler paper. Major applications of commercial papers are insulation against fire, heat transfer, and corrosion in circumstances that require a thin, but durable, barrier.

**Commercial pesticide handling establishment** (Section 170.3) means any establishment, other than an agricultural establishment, that: (1) Employs any person, including a self-employed person, to apply on an agricultural establishment, pesticides used in the production of agricultural plants. (2) Employs any person, including a self-employed person, to perform on an agricultural establishment, tasks as a crop advisor.

**Commercial Property** (Section 201.1) means any property that is normally accessible to the public and that is used for any of the purposes described in the following standard land use codes (reference Standard Land Use Coding Manual. U.S. DOT/FHWA, reprinted March 1977): 53-59, Retail Trade; 61-64, Finance, Insurance, Real Estate, Personal, Business and Repair Services; 652-659, Legal and other professional services; 671, 672, and 673 Governmental Services; 692 and 699, Welfare, Charitable and Other Miscellaneous Services; 712 and 719, Nature exhibitions and other Cultural Activities; 721, 723, and 729, Entertainment, Public and other Public Assembly; and 74-79, Recreational, Resort, Park and other Cultural Activities.

**Commercial solid waste** (Section 243.101) means all types of solid wastes generated by stores, offices, restaurants, warehouses, and other non-manufacturing activities, excluding residential and industrial wastes.

**Commercial solid waste** (Section 246.101) means all types of solid wastes generated by stores, offices, restaurants, warehouses and other non-manufacturing activities, and non-processing wastes such as office and packing wastes generated at industrial facilities.

**Commercial solid waste** (Section 258.2) means all types of solid waste generated by stores, offices, restaurants, warehouses, and other nonmanufacturing activities, excluding residential and industrial wastes.

**Commercial storer of PCB waste** (Section 761.13) means the owner or operator of each facility that is subject to the PCB storage unit standards of Sec. 761.65(b)(1) or (c)(7) or meets the alternate storage criteria of Sec. 761.65(b)(2), and who engages in storage activities involving either PCB waste generated by others or that was removed while servicing the equipment owned by others and brokered for disposal. The receipt of a fee or any other form of compensation for storage services is not necessary to qualify as a commercial storer of PCB waste. A generator who only stores its own waste is subject to the storage requirements of Sec. 761.65, but is not required to obtain approval as a commercial storer. If a facility's storage of PCB waste generated by others at no time exceeds a total of 500 gallons of liquid and/or non-liquid material containing PCBs at regulated levels, the owner or operator is a commercial storer but is not required to seek EPA approval as a commercial storer of PCB waste. Storage of one company's PCB waste by a related company is not considered commercial storage. A "related company" includes, but is not limited to: a parent company and its subsidiaries; sibling companies owned by the same parent company; companies owned by a common holding company; members of electric cooperatives; entities within the same Executive agency as defined at 5 U.S.C. 105; and a company having a joint ownership interest in a facility from which PCB waste is generated (such as a jointly owned electric power generating station) where the PCB waste is stored by one of the co-owners of the facility. A "related company" does not include another voluntary member of the same trade association. Change in ownership or title of a generator's facility, where the generator is storing PCB waste, does not make the new owner of the facility a commercial storer of PCB waste.

**Commercial use** (Section 721.3) means the use of a chemical substance or any mixture containing the chemical substance in a commercial enterprise providing saleable goods or a service to consumers (e.g., a commercial dry cleaning establishment or painting contractor).

**Commercial-use requester** (Section 1601.3) means requesters seeking information for a use or purpose that furthers the commercial, trade, or profit interests of the requester or the person on whose behalf the request is made. In determining whether a requester properly belongs in this category, the CSB shall determine, whenever reasonably possible, the use to which a requester will put the documents requested. Where the CSB has reasonable cause to doubt the use to which a requester will put the records sought, or where that use is not clear from the request itself, the CSB shall seek additional clarification before assigning the request to a specific category.

**Commission** (Section 191, 192) means the Nuclear Regulatory Commission.

**Commission** (Section 350, 355, 370) means the emergency response commission for the State in which the facility is located except where the facility is located in Indian Country, in which case, commission means the emergency response commission for the tribe under whose jurisdiction the facility is located. In the absence of an emergency response commission, the Governor and the chief executive officer, respectively, shall be the commission. Where there is a cooperative agreement between a State and a Tribe, the commission shall be the entity identified in the agreement.

**Commission scouring** (Section 410.11) shall mean the scouring of wool, 50 percent or more of which is owned by others, in mills that are 51 percent or more independent (i.e., only a minority ownership by company(ies) with greige or integrated operations); the mills must process 20 percent or more of their commissioned production through batch, noncontinuous processing operations.

**Committed effective dose equivalent** (Section 197.2) means the effective dose equivalent received over a period of time (e.g., 30 years,), as determined by NRC, by an individual from radionuclides internal to the individual following a one-year intake of those radionuclides.

**Committee** (Section 164.2) means a group of qualified scientists designated by the National Academy of Sciences according to agreement under the Act to submit an independent report to the Administrative Law Judge on questions of scientific fact referred from a hearing under subpart B of this part.

**Committee or Local emergency planning committee** (Section 355.20) means the local emergency planning committee appointed by the emergency response commission.

**Committee or local emergency planning committee** (Section 370.2) means the local emergency planning committee appointed by the emergency response commission.

**Commodity owner, shipper or their agent** (Section 82.3) means the person requesting that an applicator use methyl bromide for quarantine or preshipment applications.

**Common area** (Section 745.103) means a portion of a building generally accessible to all residents/users including, but not limited to, hallways, stairways, laundry and recreational rooms, playgrounds, community centers, and boundary fences.

**Common area** (Section 745.223) means a portion of a building that is generally accessible to all occupants. Such an area may include, but is not limited to, hallways, stairways, laundry and recreational rooms, playgrounds, community centers, garages, and boundary fences.

**Common area group** (Section 745.63) means a group of common areas that are similar in design, construction, and function. Common area groups include, but are not limited to hallways, stairwells, and laundry rooms.

**Common carrier by motor vehicle** (Section 202.10) means any person who holds himself out to the general public to engage in the transportation by motor vehicle in interstate or foreign commerce of passengers or property or any class or classes thereof for compensation, whether over regular or irregular routes.

**Common exposure route** (Section 171.2) means a likely way (oral, dermal, respiratory) by which a pesticide may reach and/or enter an organism.

**Common name** (Section 721.3) means any designation or identification such as code name, code number, trade name, brand name, or generic chemical name used to identify a chemical substance other than by its chemical name.

**Common pipe** (Section 72.2) means an oil or gas supply line through which the same type of fuel is distributed to two or more affected units.

**Common pipe operating time** (Section 72.2) means the portion of a clock hour during which fuel flows through a common pipe. The common pipe operating time, in hours, is expressed as a decimal fraction, with valid values ranging from 0.00 to 1.00.

**Common stack** (Section 72.2) means the exhaust of emissions from two or more units through a single flue.

**Common stack** (Section 96.2) means a single flue through which emissions from two or more units are exhausted.

**Common stack** (Section 97.2) means a single flue through which emissions from two or more units are exhausted.

**Community Monitoring Zone (CMZ)** (Section 58.1) means an optional averaging area with established, well defined boundaries, such as county or census block, within a MPA that has relatively uniform concentrations of annual PM2.5 as defined by appendix D of this part. Two or more core SLAMS and other monitors within a CMZ that meet certain requirements as set forth in Appendix D of this part may be averaged for making comparisons to the annual PM2.5 NAAQS.

**Community relations** (Section 300.5) means EPA's program to inform and encourage public participation in the Superfund process and to respond to community concerns. The term "public" includes citizens directly affected by the site, other interested citizens or parties, organized groups, elected officials, and potentially responsible parties (PRPs).

**Community relations coordinator** (Section 300.5) means lead agency staff who work with the OSC/RPM to involve and inform the public about the Superfund process and response actions in accordance with the interactive community relations requirements set forth in the NCP.

**Community water system** (Section 141.2) means a public water system which serves at least 15 service connections used by year-round residents or regularly serves at least 25 year-round residents.

**Compactor collection vehicle** (Section 243.101) means a vehicle with an enclosed body containing mechanical devices that convey solid waste into the main compartment of the body and compress it into a smaller volume of greater density.

**Compartmentalized vehicle** (Section 246.101) means a collection vehicle which has two or more compartments for placement of solid wastes or recyclable materials. The compartments may be within the main truck body or on the outside of that body as in the form of metal racks.

**Compatibility** (Section 171.2) means that property of a pesticide which permits its use with other chemicals without undesirable results being caused by the combination.

**Compatible** (Section 280.12) means the ability of two or more substances to maintain their respective physical and chemical properties upon contact with one another for the design life of the tank system under conditions likely to be encountered in the UST.

**Compensating unit** (Section 72.2) means an affected unit that is not otherwise subject to Acid Rain emissions limitation or Acid Rain emissions reduction requirements during Phase I and that is designated as a Phase I unit in a reduced utilization plan under Sec. 72.43; provided that an opt-in source shall not be a compensating unit.

**Competent** (Section 171.2) means properly qualified to perform functions associated with pesticide application, the degree of capability required being directly related to the nature of the activity and the associated responsibility.

**Competent authorities** (Section 262.81) means the regulatory authorities of concerned countries having jurisdiction over transfrontier movements of wastes destined for recovery operations.

**Complainant** (Section 209.3) means the Agency acting through any person authorized by the Administrator to issue a complaint to alleged violators of the Act. The complainant shall not be the judicial officer or the Administrator.

**Complete heavy-duty vehicle** (Section 86.1803-01) means any Otto-cycle heavy-duty vehicle of 14,000 pounds Gross Vehicle Weight Rating or less that has the primary load carrying device or container attached at the time the vehicle leaves the control of the manufacturer of the engine.

**Completely destroy** (Section 82.3) means to cause the expiration of a controlled substance at a destruction efficiency of 98 percent or greater, using one of the destruction technologies approved by the Parties.

**Complex** (Section 112.2) means a facility possessing a combination of transportation-related and non-transportation-related components that is subject to the jurisdiction of more than one Federal agency under section 311(j) of the Clean Water Act.

**Complex manufacturing operation** (Section 410.41) shall mean "simple" unit processes (desizing, fiber preparation and dyeing) plus any additional manufacturing operations such as printing, water proofing, or applying stain resistance or other functional fabric finishes.

**Complex manufacturing operation** (Section 410.61) shall mean "simple" unit processes (fiber preparation, dyeing and carpet backing) plus any additional manufacturing operations such as printing or dyeing and printing.

**Complex slaughterhouse** (Section 432.21) shall mean a slaughterhouse that accomplishes extensive by-product processing, usually at least three of such operations as rendering, paunch and viscera handling, blood processing, hide processing, or hair processing.

**Compliance account** (Section 96.2) means a NOX Allowance Tracking System account, established by the Administrator for a NOX Budget unit under subpart F of this part, in which the NOX allowance allocations for the unit are initially recorded and in which are held NOX allowances available for use by the unit for a control period for the purpose of meeting the unit's NOX Budget emissions limitation.

**Compliance account** (Section 97.2) means a NOX Allowance Tracking System account, established by the Administrator for a NOX Budget unit under subpart F of this part, in which the NOX allowance allocations for the unit are initially recorded and in which are held NOX allowances available for use by the unit for a control period for the purpose of meeting the unit's NOX Budget emissions limitation.

**Compliance application(s)** (Section 194.2) means the compliance certification application submitted to the Administrator pursuant to section 8(d)(1) of the WIPP LWA or any compliance re-certification applications submitted to the Administrator pursuant to section 8(f) of the WIPP LWA.

**Compliance assessment(s)** (Section 194.2) means the analysis conducted to determine compliance with Sec. 191.15, and part 191, subpart C of this chapter.

**Compliance certification** (Section 72.2) means a submission to the Administrator or permitting authority, as appropriate, that is required by this part, by part 73, 74, 75, 76, 77, or 78 of this chapter, to report an affected source or an affected unit's compliance or non-compliance with a provision of the Acid Rain Program and that is signed and verified by the designated representative in accordance with subparts B and I of this part and the Acid Rain Program regulations generally.

**Compliance certification** (Section 96.2) means a submission to the permitting authority or the Administrator, as appropriate, that is required under subpart D of this part to report a NOX Budget source's or a NOX Budget unit's compliance or noncompliance with this part and that is signed by the NOX authorized account representative in accordance with subpart B of this part.

**Compliance cycle** (Section 141.2) means the nine-year calendar year cycle during which public water systems must monitor. Each compliance cycle consists of three three-year compliance periods. The first calendar year cycle begins January 1, 1993 and ends December 31, 2001; the second begins January 1, 2002 and ends December 31, 2010; the third begins January 1, 2011 and ends December 31, 2019.

**Compliance date** (Section 63.2) means the date by which an affected source is required to be in compliance with a relevant standard, limitation, prohibition, or any federally enforceable requirement established by the Administrator (or a State with an approved permit program) pursuant to section 112 of the Act.

**Compliance date** (Section 65.2) means the date by which a regulated source is required to be in compliance with a relevant standard, limitation, prohibition, or any federally enforceable requirement established by the Administrator (or a State with an approved permit program) pursuant to the Act.

**Compliance date** (Section 94.2) means the date on which compliance with a standard becomes mandatory. For example, the compliance date for standards which first apply to the 2004 model year, is January 1, 2004.

**Compliance level** (Section 86.1002-2001) means an emission level determined during a Production Compliance Audit pursuant to subpart L of this part.

**Compliance level** (Section 86.1002-84) means an emission level determined during a Production Compliance Audit pursuant to subpart L of this part.

**Compliance level** (Section 86.1002-97) means an emission level determined during a Production Compliance Audit pursuant to subpart L of this part.

**Compliance level for an engine family** (Section 91.1302) is determined by averaging the in-use test results from each engine.

**Compliance period** (Section 141.2) means a three-year calendar year period within a compliance cycle. Each compliance cycle has three three-year compliance periods. Within the first compliance cycle, the first compliance period runs from January 1, 1993 to December 31, 1995; the second from January 1, 1996 to December 31, 1998; the third from January 1, 1999 to December 31, 2001.

**Compliance plan, for the purposes of the Acid Rain Program,** (Section 72.2) means the document submitted for an affected source in accordance with subpart C of this part or subpart E of part 74 of this chapter, or part 76 of this chapter, specifying the method(s) (including one or more Acid Rain compliance options as provided under subpart D of this part or subpart E of part 74 of this chapter, or part 76 of this chapter by which each affected unit at the source will meet the applicable Acid Rain emissions limitation and Acid Rain emissions reduction requirements.

**Compliance schedule** (Section 51.100) means the date or dates by which a source or category of sources is required to comply with specific emission limitations contained in an implementation plan and with any increments of progress toward such compliance.

**Compliance schedule** (Section 61.02) means the date or dates by which a source or category of sources is required to comply with the standards of this part and with any steps toward such compliance which are set forth in a waiver of compliance under Sec. 61.11.

**Compliance schedule** (Section 63.2) means: (1) In the case of an affected source that is in compliance with all applicable requirements established under this part, a statement that the source will continue to comply with such requirements; or (2) In the case of an affected source that is required to comply with applicable requirements by a future date, a statement that the source will meet such requirements on a timely basis and, if required by an applicable requirement, a detailed schedule of the dates by which each step toward compliance will be reached; or (3) In the case of an affected source not in compliance with all applicable requirements established under this part, a schedule of remedial measures, including an enforceable sequence of actions or operations with milestones and a schedule for the submission of certified progress reports, where applicable, leading to compliance with a relevant standard, limitation, prohibition, or any federally enforceable requirement established pursuant to section 112 of the Act for which the affected source is not in compliance. This compliance schedule shall resemble and be at least as stringent as that contained in any judicial consent decree or administrative order to which the source is subject. Any such schedule of compliance shall be supplemental to, and shall not sanction noncompliance with, the applicable requirements on which it is based.

**Compliance subaccount** (Section 72.2) means the subaccount in an affected unit's Allowance Tracking System account, established pursuant to Sec. 73.31 (a) or (b) of this chapter, in which are held, from the date that allowances for the current calendar year are recorded under Sec. 73.34(a) until December 31, allowances available for use in the current calendar year and, after December 31 until the date that deductions are made under Sec. 73.35(b), allowances available for use by the unit in the preceding calendar year, for the purpose of meeting the Acid Rain emissions limitation for sulfur dioxide.

**Compliance use date** (Section 72.2) means the first calendar year for which an allowance may be used for purposes of meeting a unit's Acid Rain emissions limitation for sulfur dioxide.

**Compliance with any "submission," "service," or "mailing" deadline** (Section 96.2) shall be determined by the date of dispatch, transmission, or mailing and not the date of receipt.

**Compliance with any "submission," "service," or "mailing" deadline** (Section 97.2) shall be determined by the date of dispatch, transmission, or mailing and not the date of receipt.

**Complying with the Protocol,** (Section 82.3) when referring to a foreign state not Party to the 1987 Montreal Protocol, the London Amendments, or the Copenhagen Amendments, means that the non-Party has been determined as complying with the Protocol, as indicated in appendix C to this subpart, by a meeting of the Parties as noted in the records of the directorate of the United Nations Secretariat.

**Component** (Section 260.10) means either the tank or ancillary equipment of a tank system.

**Component** (Section 270.2) means any constituent part of a unit or any group of constituent parts of a unit which are assembled to perform a specific function (e.g., a pump seal, pump, kiln liner, kiln thermocouple).

**Component or building component** (Section 745.223) means specific design or structural elements or fixtures of a building, residential dwelling, or child-occupied facility that are distinguished from each other by form, function, and location. These include, but are not limited to, interior components such as: ceilings, crown molding, walls, chair rails, doors, door trim, floors, fireplaces, radiators and other heating units, shelves, shelf supports, stair treads, stair risers, stair stringers, newel posts, railing caps, balustrades, windows and trim (including sashes, window heads, jambs, sills or stools and troughs), built in cabinets, columns, beams, bathroom vanities, counter tops, and air conditioners; and exterior components such as: painted roofing, chimneys, flashing, gutters and downspouts, ceilings, soffits, fascias, rake boards, cornerboards, bulkheads, doors and door trim, fences, floors, joists, lattice work, railings and railing caps, siding, handrails, stair risers and treads, stair stringers, columns, balustrades, window sills or stools and troughs, casings, sashes and wells, and air conditioners.

**Composite NO<INF>X</INF> standard** (Section 86.088-2) for a manufacturer which elects to average light-duty trucks subject to the NO<INF>X</INF> standard of Sec. 86.088-9(a)(iii)(A) together with those subject to the NO<INF>X</INF> standard of Sec. 86.088-9(a)(iii)(B) in the light-duty truck NO<INF>X</INF> averaging program, means that standard calculated according to the following equation and rounded to the nearest one-tenth gram per mile: [GRAPHIC] [TIFF OMITTED] TR06OC93.006 Where: PROD<INF>A</INF> = The manufacturer's total light-duty truck production for those engine families subject to the standard of Sec. 86.088-9(a)(iii)(A) and included in the average for a given model year,STD<INF>A</INF> = The NO<INF>X</INF> standard of Sec. 86.088-9(a)(iii)(A), PROD<INF>B</INF> = The manufacturer's total light-duty truck production for those engine families subject to the standard of Sec. 86.088-9(a)(iii)(B) and included in the average for a given model year, and STD<INF>B</INF> = The NO<INF>X</INF> standard of Sec. 86.088-9(a)(iii)(B).

**Composite particulate standard** (Section 86.085-2) for a manufacturer which elects to average diesel light-duty vehicles and diesel light-duty trucks together in the particulate averaging program, means that standard calculated according to the following equation and rounded to the nearest hundredth gram-per-mile: [GRAPHIC] [TIFF OMITTED] TR06OC93.004 Where: PROD<INF>LDV</INF> represents the manufacturer's total diesel light-duty vehicle production for those engine families being included in the average for a given model year. STD<INF>LDV</INF> represents the light-duty vehicle particulate standard. PROD<INF>LDT</INF> represents the manufacturer's total diesel light-duty truck production for those engine families being included in the average for a given model year. STD<INF>LDT</INF> represents the light-duty truck particulate standard.

**Composite particulate standard** (Section 86.087-2) for a manufacturer which elects to average diesel light-duty vehicles and diesel light-duty trucks with a loaded vehicle weight equal to or less than 3,750 lbs (LDDT1s) together in the particulate averaging program, means that standard calculated according to the following equation and rounded to the nearest hundredth gram per mile: [GRAPHIC] [TIFF OMITTED] TR06OC93.005 Where: PROD<INF>LDV</INF> represents the manufacturer's total light-duty vehicle production for those engine families being included in the average for a given model year. STD<INF>LDV</INF> represents the light-duty vehicle particulate standard. PROD<INF>LDDT1</INF> represents the manufacturer's total diesel light-duty truck production for those engine families with a loaded vehicle weight equal to or less than 3,750 lbs which are being included in the average for a given model year. STD<INF>LDDT1</INF> represents the light-duty truck particulate standard for diesel light-duty trucks with a loaded vehicle weight equal to or less than 3,750 lbs.

**Composite particulate standard** (Section 86.090-2) for a manufacturer which elects to average light-duty vehicles and light-duty trucks together in either the petroleum-fueled or methanol-fueled light-duty particulate averaging program, means that standards calculated using the following equation and rounded to the nearest one-hundredth (0.01) of a gram per mile: [GRAPHIC] [TIFF OMITTED] TR06OC93.007 Where: PROD<INF>LDV</INF> represents the manufacturer's total petroleum-fueled diesel or methanol-fueled diesel light-duty vehicle production for those engine families being included in the appropriate average for a given model year. STD<INF>LDV</INF> represents the light-duty vehicle particulate standard. PROD<INF>LDT</INF> represents the manufacturer's total petroleum-fueled diesel or methanol-fueled diesel light-duty truck production for those engine families being included in the appropriate average for a given model year. STD<INF>LDT</INF> represents the light-duty truck particulate standard.

**Composite sample** (Section 471.02) is a sample composed of no less than eight grab samples taken over the compositing period.

**Compost** (Section 247.3) made from yard trimmings, leaves, grass clippings, and/or food wastes is a thermophilic converted product with high humus content. Compost can be used as a soil amendment and can also be used to prevent or remediate pollutants in soil, air, and storm water run-off.

**Comprehensive performance evaluation (CPE)** (Section 141.2) is a thorough review and analysis of a treatment plant's performance-based capabilities and associated administrative, operation and maintenance practices. It is conducted to identify factors that may be adversely impacting a plant's capability to achieve compliance and emphasizes approaches that can be implemented without significant capital improvements. For purpose of compliance with subparts P and T of this part, the comprehensive performance evaluation must consist of at least the following components: Assessment of plant performance; evaluation of major unit processes; identification and prioritization of performance limiting factors; assessment of the applicability of comprehensive technical assistance; and preparation of a CPE report.

**Compression-ignition** (Section 89.2) means relating to a type of engine with operating characteristics significantly similar to the theoretical Diesel combustion cycle. The non-use of a throttle to regulate intake air flow for controlling power during normal operation is indicative of a compression-ignition engine. This definition is applicable beginning January 1, 2000.

**Compression-ignition** (Section 94.2) means relating to a type of engine with operating characteristics significantly similar to the theoretical Diesel combustion cycle. The non-use of a throttle to regulate intake air flow for controlling power during normal operation is indicative of a compression-ignition engine.

**Compressor configuration** (Section 204.51) means the basic classification unit of a manufacturer's product line and is comprised of compressor lines, models or series which are identical in all material respects with regard to the parameters listed in Sec. 204.55-3.

**Computer program** (Section 66.3) means the computer program used to calculate noncompliance penalties under section 120 of the Clean Air Act. This computer program appears as appendix C to these regulations.

**Concentrated animal feeding operation** (Section 122.2) is defined at Sec. 122.23.

**Concentrated aquatic animal feeding operation** (Section 122.2) is defined at Sec. 122.24.

**Concentration** (Section 745.63) means the relative content of a specific substance contained within a larger mass, such as the amount of lead (in micrograms per gram or parts per million by weight) in a sample of dust or soil.

**Concerned countries** (Section 262.81) means the exporting and importing OECD member countries and any OECD member countries of transit.

**Condensate** (Section 68.3) means hydrocarbon liquid separated from natural gas that condenses due to changes in temperature, pressure, or both, and remains liquid at standard conditions.

**Condenser** (Section 264.1031) means a heat-transfer device that reduces a thermodynamic fluid from its vapor phase to its liquid phase.

**Conditionally valid data** (Section 72.2) means data from a continuous monitoring system that are not quality assured, but which may become quality assured if certain conditions are met. Examples of data that may qualify as conditionally valid are: data recorded by an uncertified monitoring system prior to its initial certification; or data recorded by a certified monitoring system following a significant change to the system that may affect its ability to accurately measure and record emissions. A monitoring system must pass a probationary calibration error test, in accordance with section 2.1.1 of appendix B to part 75 of this chapter, to initiate the conditionally valid data status. In order for conditionally valid emission data to become quality assured, one or more quality assurance tests or diagnostic tests must be passed within a specified time period in accordance with Sec. 75.20(b)(3).

**Confer** (Section 82.3) means to shift the essential-use allowances obtained under Sec. 82.4(u) from the holder of the unexpended essential-use allowance to a person for the production of a specified controlled substance, or to shift the destruction and transformation credits obtained under Sec. 82.9(f) from the holder of the unexpended destruction and transformation credits to a person for the production of a specified controlled substance.

**Confidential business information** (Section 154.3) means trade secrets or confidential commercial or financial information under FIFRA section 10(b) or 5 U.S.C. 552(b) (3) or (4).

**Confidential business information** (Section 155.23) means trade secrets or confidential commercial or financial information under FIFRA sec. 10(b) or 5 U.S.C. 552(b) (3) or (4).

**Confidential business information** (Section 1601.3) means records provided to the government by a submitter that arguably contain material exempt from disclosure under Exemption 4 of the FOIA, because disclosure could reasonably be expected to cause substantial competitive harm.

**Confidential business information or CBI** (Section 305.3) means business information for which a person has made a "business confidentiality claim" as defined in 40 CFR 2.201(h) and in accordance with all applicable provisions in 40 CFR part 2, subpart B, except insofar as the Administrator has denied the claim pursuant to the procedures in 40 CFR part 2, subpart B.

**Configuration** (Section 86.602-98) when used for LDV refueling emissions testing, means a subclassification of an evaporative/refueling emission family on the basis of evaporative and refueling control system and other parameters which may be designated by the Administrator.

**Configuration** (Section 205.151) means the basic classification unit of a manufacturer's product line and is comprised of all vehicle designs, models or series which are identical in all material aspects with respect to the parameters listed in Sec. 205.157-3 of this subpart.

**Configuration** (Section 205.51) means the basic classification unit of a manufacturer's product line and is comprised of all vehicle designs, models or series which are identical in material aspects with respect to the parameters listed in Sec. 205.55-3. equal to the acceptance number as determined by the appropriate sampling plan.

**Configuration** (Section 610.11) means the mechanical arrangement, calibration and condition of a test automobile, with particular respect to carburetion, ignition timing, and emission control systems.

**Configuration** (Section 86.082-2) means a subclassification of an engine-system combination on the basis of engine code, inertia weight class, transmission type and gear ratios, final drive ratio, and other parameters which may be designated by the Administrator.

**Configuration** (Section 86.1002-2001) means a subclassification, if any, of a heavy-duty engine family for which a separate projected sales figure is listed in the manufacturer's Application for Certification and which can be described on the basis of emission control system, governed speed, injector size, engine calibration and other parameters which may be designated by the Administrator, or for light-duty trucks a subclassification of a light-duty truck engine family/emission control system combination on the basis of engine code, inertia weight class, transmission type and gear ratios, axle ratio, and other parameters which may be designated by the Administrator and/or a subclassification of a light-duty truck evaporative/refueling emission family/emission control system.

**Configuration** (Section 86.1002-84) means a subclassification, if any, of a heavy-duty engine family for which a separate projected sales figure is listed in the manufacturer's Application for Certification and which can be described on the basis of emission control system, governed speed, injector size, engine calibration, and other parameters which may be designated by the Administrator, or a subclassification of a light-duty truck engine family/emission control system combination on the basis of engine code, inertia weight class, transmission type and gear ratios, axle ratio, and other parameters which may be designated by the Administrator.

**Configuration** (Section 86.1002-97) means a subclassification, if any, of a heavy-duty engine family for which a separate projected sales figure is listed in the manufacturer's Application for Certification and which can be described on the basis of emission control system, governed speed, injector size, engine calibration, and other parameters which may be designated by the Administrator, or a subclassification of a light-duty truck engine family/emission control system combination on the basis of engine code, inertia weight class, transmission type and gear rations, axle ratio, and other parameters which may be designated by the Administrator.

**Configuration** (Section 86.1803-01) means a subclassification within a test group which is based on engine code, inertia weight class, transmission type and gear ratios, final drive ratio, and other parameters which may be designated by the Administrator.

**Configuration** (Section 86.602-84) means a subclassification of an engine-system combination on the basis of engine code, inertia weight class, transmission type and gear ratios, axle ratio, and other parameters which may be designated by the Administrator.

**Configuration** (Section 86.602-98) when used for LDV exhaust emissions testing, means a subclassification of an engine-system combination on the basis of engine code, inertia weight class, transmission type and gear ratios, axle ratio, and other parameters which may be designated by the Administrator.

**Configuration** (Section 89.502) means any subclassification of an engine family which can be described on the basis of gross power, emission control system, governed speed, injector size, engine calibration, and other parameters as designated by the Administrator.

**Configuration** (Section 90.702) means any subclassification of an engine family which can be described on the basis of gross power, emission control system, governed speed, injector size, engine calibration, and other parameters as designated by the Administrator.

**Configuration** (Section 92.2) means any subclassification of an engine family which can be described on the basis of gross power, emission control system, governed speed, injector size, engine calibration, and other parameters as designated by the Administrator.

**Configuration** (Section 94.2) means any subclassification of an engine family which can be described on the basis of gross power, emission control system, governed speed, injector size, engine calibration, and other parameters as designated by the Administrator.

**Confined aquifer** (Section 260.10) means an aquifer bounded above and below by impermeable beds or by beds of distinctly lower permeability than that of the aquifer itself; an aquifer containing confined ground water.

**Confining bed** (Section 146, 147) means a body of impermeable or distinctly less permeable material stratigraphically adjacent to one or more aquifers.

**Confining zone** (Section 146, 147) means a geological formation, group of formations, or part of a formation that is capable of limiting fluid movement above an injection zone.

**Confluent growth** (Section 141.2) means a continuous bacterial growth covering the entire filtration area of a membrane filter, or a portion thereof, in which bacterial colonies are not discrete.

**Congener** (Section 766.3) means any one particular member of a class of chemical substances. A specific congener is denoted by unique chemical structure, for example 2,3,7,8-tetrachlorodibenzofuran.

**Connected piping** (Section 280.12) means all underground piping including valves, elbows, joints, flanges, and flexible connectors attached to a tank system through which regulated substances flow. For the purpose of determining how much piping is connected to any individual UST system, the piping that joins two UST systems should be allocated equally between them.

**Connecting channels of the Great Lakes** (Section 132.2) are the Saint Mary's River, Saint Clair River, Detroit River, Niagara River, and Saint Lawrence River to the Canadian Border.

**Connector** (Section 264.1031) means flanged, screwed, welded, or other joined fittings used to connect two pipelines or a pipeline and a piece of equipment. For the purposes of reporting and recordkeeping, connector means flanged fittings that are not covered by insulation or other materials that prevent location of the fittings.

**Connector** (Section 65.2) means flanged, screwed, or other joined fittings used to connect two pipelines or a pipeline and a piece of equipment. A common connector is a flange. Joined fittings welded completely around the circumference of the interface are not considered connectors for the purpose of this regulation. For the purpose of reporting and recordkeeping, connector means joined fittings that are not inaccessible, ceramic, or ceramic-lined (for example, porcelain, glass, or glass-lined) as described in Sec. 65.108(e)(2).

**Conservation Verification Protocol** (Section 72.2) means a methodology developed by the Administrator for calculating the kilowatt hour savings from energy conservation measures and improved unit efficiency measures for the purposes of title IV of the Act.

**Consignee** (Section 262.51) means the ultimate treatment, storage or disposal facility in a receiving country to which the hazardous waste will be sent.

**Consignee** (Section 262.81) means the person to whom possession or other form of legal control of the waste is assigned at the time the waste is received in the importing country.

**Consolidated grant** (Section 35.102) A single grant made to a recipient consolidating funds from more than one environmental grant program. After the award is made, recipients must account for grant funds in accordance with the funds' original environmental program sources. Consolidated grants are not Performance Partnership Grants.

**Consolidated Metropolitan Statistical Area (CMSA)** (Section 58.1) means the most recent area as designated by the U.S. Office of Management and Budget and population figures from the Bureau of the Census. The Department of Commerce provides that within metropolitan complexes of 1 million or more population, separate component areas are defined if specific criteria are met. Such areas are designated primary metropolitan statistical areas (PMSAs; and any area containing PMSAs is designated CMSA.

**Consolidated microbial commercial activity notice or consolidated MCAN** (Section 700.43) means any MCAN submitted to EPA that covers more than one microorganism (each being assigned a separate MCAN number by EPA) as a result of a prenotice agreement with EPA.

**Consolidated microbial commercial activity notice or consolidated MCAN** (Section 725.3) means any MCAN submitted to EPA that covers more than one microorganism (each being assigned a separate MCAN number by EPA) as a result of a prenotice agreement with EPA.

**Consolidated premanufacture notice or consolidated PMN** (Section 700.43) means any PMN submitted to EPA that covers more than one chemical substance (each being assigned a separate PMN number by EPA) as a result of a prenotice agreement with EPA (See 48 FR 21734).

**Consortium** (Section 790.3) means an association of manufacturers and/or processors who have made an agreement to jointly sponsor testing.

**Constant controls, control technology, and continuous emission reduction technology** (Section 57.103) means systems which limit the quantity, rate, or concentration, excluding the use of dilution, and emissions of air pollutants on a continuous basis.

**Constant-speed engine** (Section 89.2) means an engine that is governed to operate only at rated speed.

**Constant-speed engine** (Section 94.2) means an engine that is governed to operate only at a single rated speed.

**Construction** (Section 129.2) means any placement, assembly, or installation of facilities or equipment (including contractual obligations to purchase such facilities or equipment) at the premises where such equipment will be used, including preparation work at such premises.

**Construction** (Section 60.2) means fabrication, erection, or installation of an affected facility.

**Construction** (Section 61.02) means fabrication, erection, or installation of an affected facility.

**Construction** (Section 63.2) means the on-site fabrication, erection, or installation of an affected source. Construction does not include the removal of all equipment comprising an affected source from an existing location and reinstallation of such equipment at a new location. The owner or operator of an existing affected source that is relocated may elect not to reinstall minor ancillary equipment including, but not limited to, piping, ductwork, and valves. However, removal and reinstallation of an affected source will be construed as reconstruction if it satisfies the criteria for reconstruction as defined in this section. The costs of replacing minor ancillary equipment must be considered in determining whether the existing affected source is reconstructed.

**Construction** (Section 72.2) means fabrication, erection, or installation of a unit or any portion of a unit.

**Construction and demolition waste** (Section 243.101) means the waste building materials, packaging, and rubble resulting from construction, remodeling, repair, and demolition operations on pavements, houses, commercial buildings, and other structures.

**Construction and demolition waste** (Section 246.101) means the waste building materials, packaging, and rubble resulting from construction, remodeling, repair, and demolition operations on pavements, houses, commercial buildings and other structures.

**Construction equipment or vehicle** (Section 85.1602) means any internal combustion engine-powered machine primarily used in construction and located on commercial construction sites.

**Consumer** (Section 721.3) means a private individual who uses a chemical substance or any product containing the chemical substance in or around a permanent or temporary household or residence, during recreation, or for any personal use or enjoyment.

**Consumer Price Index or CPI** (Section 72.2) means, for purposes of the Acid Rain Program, the U.S. Department of Labor, Bureau of Labor Statistics unadjusted Consumer Price Index for All Urban Consumers for the U.S. city average, for All Items on the latest reference base, or if such index is no longer published, such other index as the Administrator in his or her discretion determines meets the requirements of the Clean Air Act Amendments of 1990.

**Consumer product** (Section 721.3) means a chemical substance that is directly, or as part of a mixture, sold or made available to consumers for their use in or around a permanent or temporary household or residence, in or around a school, or in recreation.

**Consumption** (Section 82.3) means the production plus imports minus exports of a controlled substance (other than transhipments, or used controlled substances).

**Consumption allowances** (Section 82.3) means the privileges granted by this subpart to produce and import class I controlled substances; however, consumption allowances may be used to produce class I controlled substances only in conjunction with production allowances. A person's consumption allowances are the total of the allowances obtained under Secs. 82.6 and 82.7 and 82.10, as may be modified under Sec. 82.12 (transfer of allowances).

**Consumptive use with respect to heating oil** (Section 280.12) means consumed on the premises.

**Contact cooling and heating water** (Section 463.2) is process water that contacts the raw materials or plastic product for the purpose of heat transfer during the plastics molding and forming process.

**Contact cooling water** (Section 467.02) is any wastewater which contacts the aluminum workpiece or the raw materials used in forming aluminum.

**Contact cooling water** (Section 471.02) is any wastewater which contacts the metal workpiece or the raw materials used in forming metals for the purpose of removing heat from the metal.

**Container** (Section 260.10) means any portable device in which a material is stored, transported, treated, disposed of, or otherwise handled.

**Container** (Section 279.1) means any portable device in which a material is stored, transported, treated, disposed of, or otherwise handled.

**Container** (Section 59.101) means the individual receptacle that holds a coating or coating component for storage and distribution.

**Containment** (Section 745.223) means a process to protect workers and the environment by controlling exposures to the lead-contaminated dust and debris created during an abatement.

**Containment and/or inactivation controls** (Section 725.3) means any combination of engineering, mechanical, procedural, or biological controls designed and operated to restrict environmental release of viable microorganisms from a structure.

**Containment building** (Section 260.10) means a hazardous waste management unit that is used to store or treat hazardous waste under the provisions of subpart DD of parts 264 or 265 of this chapter.

**Contaminant** (Section 141.2) means any physical, chemical, biological, or radiological substance or matter in water.

**Contaminant** (Section 142.2) means any physical, chemical, biological, or radiological substance or matter in water.

**Contaminant** (Section 143.2) means any physical, chemical, biological, or radiological substance or matter in water.

**Contaminant** (Section 144.3) means any physical, chemical, biological, or radiological substance or matter in water.

**Contaminant** (Section 146.3) means any physical, chemical, biological, or radiological substance or matter in water.

**Contaminant** (Section 147.2902) any physical, chemical, biological, or radiological substance or matter in water.

**Contaminant** (Section 230.3) means a chemical or biological substance in a form that can be incorporated into, onto or be ingested by and that harms aquatic organisms, consumers of aquatic organisms, or users of the aquatic environment, and includes but is not limited to the substances on the 307(a) (1) list of toxic pollutants promulgated on January 31, 1978 (43 FR 4109).

**Contaminate an aquifer** (Section 503.21) means to introduce a substance that causes the maximum contaminant level for nitrate in 40 CFR 141.62(b) to be exceeded in the ground water or that causes the existing concentration of nitrate in ground water to increase when the existing concentration of nitrate in the ground water exceeds the maximum contaminant level for nitrate in 40 CFR 141.62(b).

**Contaminated ground water** (Section 445.2) means water below the land surface in the zone of saturation which has been contaminated by activities associated with waste disposal.

**Contaminated non-process waste water** (Section 422) shall mean any water including precipitation runoff, which during manufacturing or processing, comes into incidental contact with any raw material, intermediate product, finished product, by-product or waste product by means of: (1) Precipitation runoff, (2) accidental spills, (3) accidental leaks caused by the failure of process equipment and which are repaired or the discharge of pollutants therefrom contained or terminated within the shortest reasonable time which shall not exceed 24 hours after discovery or when discovery should reasonably have been made, whichever is earliest, and (4) discharges from safety showers and related personal safety equipment, and from equipment washings for the purpose of safe entry, inspection and maintenance; Provided, That all reasonable measures have been taken to prevent, reduce, eliminate and control to the maximum extent feasible such contact and provided further that all reasonable measures have been taken that will mitigate the effects of such contact once it has occurred.

**Contaminated nonprocess wastewater** (Section 415) shall mean any water which, during manufacturing or processing, comes into incidental contact with any raw material, intermediate product, finished product, by-product or waste product by means of (1) rainfall runoff; (2) accidental spills; (3) accidental leaks caused by the failure of process equipment, which are repaired within the shortest reasonable time not to exceed 24 hours after discovery; and (4) discharges from safety showers and related personal safety equipment: Provided, that all reasonable measures have been taken (i) to prevent, reduce and control such contact to the maximum extent feasible; and (ii) to mitigate the effects of such contact once it has occurred.

**Contaminated non-process wastewater** (Section 418) shall mean any water including precipitation runoff which, during manufacturing or processing, comes into incidental contact with any raw material, intermediate product, finished product, by-product or waste product by means of: (1) Precipitation runoff; (2) accidental spills; (3) accidental leaks caused by the failure of process equipment and which are repaired or the discharge of pollutants therefrom contained or terminated within the shortest reasonable time which shall not exceed 24 hours after discovery or when discovery should reasonably have been made, whichever is earliest; and (4) discharges from safety showers and related personal safety equipment, and from equipment washings for the purpose of safe entry, inspection and maintenance; provided that all reasonable measures have been taken to prevent, reduce, eliminate and control to the maximum extent feasible such contact and provided further that all reasonable measures have been taken that will mitigate the effects of such contact once it has occurred.

**Contaminated runoff** (Section 419.11) shall mean runoff which comes into contact with any raw material, intermediate product, finished product, by-product or waste product located on petroleum refinery property.

**Contaminated storm water** (Section 437.2) means storm water which comes in direct contact with CWT wastes, the waste handling and treatment areas, or other centralized waste treatment wastewater as defined in paragraph (d) of this section.

**Contaminated storm water** (Section 445.2) means storm water which comes in direct contact with landfill wastes, the waste handling and treatment areas, or landfill wastewater as defined in paragraph (f) of this section. Some specific areas of a landfill that may produce contaminated storm water include (but are not limited to): the open face of an active landfill with exposed waste (no cover added); the areas around wastewater treatment operations; trucks, equipment or machinery that has been in direct contact with the waste; and waste dumping areas.

**Contiguous zone** (Section 116.3) means the entire zone established or to be established by the United States under article 24 of the Convention of the Territorial Sea and the Contiguous Zone;

**Contiguous zone** (Section 117, 122) means the entire zone established by the United States under Article 24 of the Convention on the Territorial Sea and Contiguous Zone.

**Contiguous zone** (Section 300.5) means the zone of the high seas, established by the United States under Article 24 of the Convention on the Territorial Sea and Contiguous Zone, which is contiguous to the territorial sea and which extends nine miles seaward from the outer limit of the territorial sea.

**Contingency plan** (Section 260.10) means a document setting out an organized, planned, and coordinated course of action to be followed in case of a fire, explosion, or release of hazardous waste or hazardous waste constituents which could threaten human health or the environment.

**Continually regenerating trap oxidizer system** (Section 86.1702-99) means a trap oxidizer system that does not utilize an automated regeneration mode during normal driving conditions for cleaning the trap.

**Continuous** (Section 420.111) means those alkaline cleaning operations which process steel products other than in discrete batches or bundles.

**Continuous** (Section 420.81) means those descaling operations that remove surface scale from the sheet or wire products in continuous processes.

**Continuous** (Section 420.91) means those pickling operations which process steel products other than in discrete batches or bundles.

**Continuous casting** (Section 467.02) is the production of sheet, rod, or other long shapes by solidifying the metal while it is being poured through an open-ended mold using little or no contact cooling water. Continuous casting of rod and sheet generates spent lubricants and rod casting also generates contact cooling water.

**Continuous casting** (Section 471.02) is the production of sheet, rod, or other long shapes by solidifying the metal while it is being poured through an open-ended mold.

**Continuous compliance determination method** (Section 64.1) means a method, specified by the applicable standard or an applicable permit condition, which: (1) Is used to determine compliance with an emission limitation or standard on a continuous basis, consistent with the averaging period established for the emission limitation or standard; and (2) Provides data either in units of the standard or correlated directly with the compliance limit.

**Continuous discharge** (Section 122.2) means a "discharge" which occurs without interruption throughout the operating hours of the facility, except for infrequent shutdowns for maintenance, process changes, or other similar activities.

**Continuous emission monitoring system (CEMS)** (Section 63.2) means the total equipment that may be required to meet the data acquisition and availability requirements of this part, used to sample, condition (if applicable), analyze, and provide a record of emissions.

**Continuous emission monitoring system or CEMS** (Section 72, 96, 97) means the equipment required by part 75 of this chapter used to sample, analyze, measure, and provide, by readings taken at least once every 15 minutes, a permanent record of emissions, expressed in pounds per hour (lb/hr) for sulfur dioxide and in pounds per million British thermal units (lb/ mmBtu) for nitrogen oxides. The following systems are component parts included in a continuous emission monitoring system: (1) Sulfur dioxide pollutant concentration monitor; (2) Flow monitor; (3) Nitrogen oxides pollutant concentration monitors; (4) Diluent gas monitor (oxygen or carbon dioxide); (5) A continuous moisture monitor when such monitoring is required by part 75 of this chapter; and (6) A data acquisition and handling system.

**Continuous emission monitoring system or CEMS** (Section 72.2) means the equipment required by part 75 of this chapter used to sample, analyze, measure, and provide, by means of readings recorded at least once every 15 minutes (using an automated data acquisition and handling system (DAHS)), a permanent record of SO2, NOX, or CO2 emissions or stack gas volumetric flow rate. The following are the principal types of continuous emission monitoring systems required under part 75 of this chapter. Sections 75.10 through 75.18 and Sec. 75.71(a) of this chapter indicate which type(s) of CEMS is required for specific applications: (1) A sulfur dioxide monitoring system, consisting of an SO2 pollutant concentration monitor and an automated DAHS. An SO2 monitoring system provides a permanent, continuous record of SO2 emissions in units of parts per million (ppm); (2) A flow monitoring system, consisting of a stack flow rate monitor and an automated DAHS. A flow monitoring system provides a permanent, continuous record of stack gas volumetric flow rate, in units of standard cubic feet per hour (scfh); (3) A nitrogen oxides (NOX) emission rate (or NOX-diluent) monitoring system, consisting of a NOX pollutant concentration monitor, a diluent gas (CO2 or

$O_2$) monitor, and an automated DAHS. A NOX-diluent monitoring system provides a permanent, continuous record of: NOX concentration in units of parts per million (ppm), diluent gas concentration in units of percent $O_2$ or $CO_2$ (% $O_2$ or $CO_2$), and NOX emission rate in units of pounds per million British thermal units (lb/mmBtu); (4) A nitrogen oxides concentration monitoring system, consisting of a NOX pollutant concentration monitor and an automated DAHS. A NOX concentration monitoring system provides a permanent, continuous record of NOX emissions in units of parts per million (ppm). This type of CEMS is used only in conjunction with a flow monitoring system to determine NOX mass emissions (in lb/hr) under subpart H of part 75 of this chapter; (5) A carbon dioxide monitoring system, consisting of a $CO_2$ pollutant concentration monitor (or an oxygen monitor plus suitable mathematical equations from which the $CO_2$ concentration is derived) and the automated DAHS. A carbon dioxide monitoring system provides a permanent, continuous record of $CO_2$ emissions in units of percent $CO_2$ (% $CO_2$); and (6) A moisture monitoring system, as defined in Sec. 75.11(b)(2) of this chapter. A moisture monitoring system provides a permanent, continuous record of the stack gas moisture content, in units of percent $H_2O$ (% $H_2O$)

**Continuous monitor** (Section 62.1201) means a device which continuously samples the regulated parameter specified in Sec. 63.1209 without interruption, evaluates the detector response at least once every 15 seconds, and computes and records the average value at least every 60 seconds, except during allowable periods of calibration and except as defined otherwise by the CEMS Performance Specifications in appendix B, part 60 of this chapter.

**Continuous monitoring system (CMS)** (Section 63.2) is a comprehensive term that may include, but is not limited to, continuous emission monitoring systems, continuous opacity monitoring systems, continuous parameter monitoring systems, or other manual or automatic monitoring that is used for demonstrating compliance with an applicable regulation on a continuous basis as defined by the regulation.

**Continuous monitoring system** (Section 60.2) means the total equipment, required under the emission monitoring sections in applicable subparts, used to sample and condition (if applicable), to analyze, and to provide a permanent record of emissions or process parameters.

**Continuous opacity monitoring system (COMS)** (Section 63.2) means a continuous monitoring system that measures the opacity of emissions.

**Continuous opacity monitoring system or COMS** (Section 72.2) means the equipment required by part 75 of this chapter to sample, measure, analyze, and provide, with readings taken at least once every 6 minutes, a permanent record of opacity or transmittance. The following systems are component parts included in a continuous opacity monitoring system: (1) Opacity monitor; and (2) A data acquisition and handling system.

**Continuous operations** (Section 471.02) means that the industrial user introduces regulated wastewaters to the POTW throughout the operating hours of the facility, except for infrequent shutdowns for maintenance, process changes, or other similar activities.

**Continuous parameter monitoring system** (Section 63.2) means the total equipment that may be required to meet the data acquisition and availability requirements of this part, used to sample, condition (if applicable), analyze, and provide a record of process or control system parameters.

**Continuous parameter monitoring system or CPMS** (Section 65.2) means the total equipment that may be required to meet the data acquisition and availability requirements of this part used to sample, condition (if applicable), analyze, and provide a record of process or control system parameters.

**Continuous record** (Section 65.2) means documentation, either in hard copy or computer-readable form, of data values measured at least once every 15 minutes and recorded at the frequency specified in Sec. 65.161(a).

**Continuous recorder** (Section 264.1031) means a data-recording device recording an instantaneous data value at least once every 15 minutes.

**Continuous seal** (Section 265.1081) means a seal that forms a continuous closure that completely covers the space between the edge of the floating roof and the wall of a tank. A continuous seal may be a vapor-mounted seal, liquid-mounted seal, or metallic shoe seal. A continuous seal may be constructed of fastened segments so as to form a continuous seal.

**Continuous seal** (Section 65.2) means a seal that is designed to form a continuous closure that completely covers the space between the wall of the storage vessel and the edge of the floating roof. A continuous seal may be a vapor-mounted, liquid-mounted, or metallic shoe seal. A continuous seal may be constructed of fastened segments so as to form a continuous seal.

**Contract carrier by motor vehicle** (Section 202.10) means any person who engages in transportation by motor vehicle of passengers or property in interstate or foreign commerce for compensation (other than transportation referred to in paragraph (b) of this section) under continuing contracts with one person or a limited number of persons either (1) for the furnishing of transportation services through the assignment of motor vehicles for a continuing period of time to the exclusive use of each person served or (2) for the furnishing of transportation services designed to meet the distinct need of each individual customer.

**Contract for the purchase and sale of residential real property** (Section 745.103) means any contract or agreement in which one party agrees to purchase an interest in real property on which there is situated one or more residential dwellings used or occupied, or intended to be used or occupied, in whole or in part, as the home or residence of one or more persons.

**Contract or other approved** (Section 112.2) means: (1) A written contractual agreement with an oil spill removal organization(s) that identifies and ensures the availability of the necessary personnel and equipment within appropriate response times; and/or (2) A written certification by the owner or operator that the necessary personnel and equipment resources, owned or operated by the facility owner or operator, are available to respond to a discharge within appropriate response times; and/or (3) Active membership in a local or regional oil spill removal organization(s) that has identified and ensures adequate access through such membership to necessary personnel and equipment to respond to a discharge within appropriate response times in the specified geographic areas; and/or (4) Other specific arrangements approved by the Regional Administrator upon request of the owner or operator.

**Contractor claim** (Section 307.14) means the disputed portion of a written demand or written assertion by any contractor who has contracted with a person (i.e., the owner) for the conduct of a preauthorized response action, seeking as a matter of right, the payment of money, adjustment, or interpretation of contract terms, or other relief, arising under or related to a contract, which has been finally rejected or not acted upon by the owner and which is subsequently settled by the owner or is awarded by a third party in accordance with the disputes clause of the contract document.

**Control (including the terms controlling, controlled by, and under common control with)** (Section 66.3) means the power to direct or cause the direction of the management and policies of a person or organization, whether by the ownership of stock, voting rights, by contract, or otherwise.

**Control** (Section 192) means any remedial action intended to stabilize, inhibit future misuse of,

or reduce emissions or effluents from residual radioactive materials.

**Control** (Section 192.31) means any action to stabilize, inhibit future misuse of, or reduce emissions or effluents from uranium byproduct materials.

**Control** (Section 88.302-94) means: (1) When it is used to join all entities under common management, means any one or a combination of the following: (i) A third person or firm has equity ownership of 51 percent or more in each of two or more firms; (ii) Two or more firms have common corporate officers, in whole or in substantial part, who are responsible for the day-to-day operation of the companies. (iii) One firm leases, operates, supervises, or in 51 percent or greater part owns equipment and/or facilities used by another person or firm, or has equity ownership of 51 percent or more of another firm.

**Control area** (Section 80.2) means a geographic area in which only oxygenated gasoline under the oxygenated gasoline program may be sold or dispensed, with boundaries determined by section 211(m) of the Act.

**Control authority** (Section 413, 466, 471) is defined as the POTW if it has an approved pretreatment program; in the absence of such a program, the NPDES State if it has an approved pretreatment program or EPA if the State does not have an approved program.

**Control device** (Section 264.1031) means an enclosed combustion device, vapor recovery system, or flare. Any device the primary function of which is the recovery or capture of solvents or other organics for use, reuse, or sale (e.g., a primary condenser on a solvent recovery unit) is not a control device.

**Control device** (Section 64.1) means equipment, other than inherent process equipment, that is used to destroy or remove air pollutant(s) prior to discharge to the atmosphere. The types of equipment that may commonly be used as control devices include, but are not limited to, fabric filters, mechanical collectors, electrostatic precipitators, inertial separators, afterburners, thermal or catalytic incinerators, adsorption devices (such as carbon beds), condensers, scrubbers (such as wet collection and gas absorption devices), selective catalytic or non-catalytic reduction systems, flue gas recirculation systems, spray dryers, spray towers, mist eliminators, acid plants, sulfur recovery plants, injection systems (such as water, steam, ammonia, sorbent or limestone injection), and combustion devices independent of the particular process being conducted at an emissions unit (e.g., the destruction of emissions achieved by venting process emission streams to flares, boilers or process heaters). For purposes of this part, a control device does not include passive control measures that act to prevent pollutants from forming, such as the use of seals, lids, or roofs to prevent the release of pollutants, use of low-polluting fuel or feedstocks, or the use of combustion or other process design features or characteristics. If an applicable requirement establishes that particular equipment which otherwise meets this definition of a control device does not constitute a control device as applied to a particular pollutant-specific emissions unit, then that definition shall be binding for purposes of this part.

**Control device** (Section 65.2) means any combustion device, recovery device, or any combination of these devices used to comply with this part. Such equipment or devices include, but are not limited to, absorbers, carbon adsorbers, condensers, incinerators, flares, boilers, and process heaters. For process vents (as defined in this section), recovery devices are not considered control devices except for the recovery devices specified in Sec. 65.63(a)(2)(ii). A fuel gas system is not a control device. For a steam stripper, a primary condenser is not considered a control device.

**Control device shutdown** (Section 264.1031) means the cessation of operation of a control device for any purpose.

**Control period** (Section 80.2) means the period during which oxygenated gasoline must be sold or dispensed in any control area, pursuant to section 211(m)(2) of the Act.

**Control period** (Section 82.3) means the period from January 1, 1992 through December 31, 1992, and each twelve-month period from January 1 through December 31, thereafter.

**Control period** (Section 96, 97) means the period beginning May 1 of a year and ending on September 30 of the same year, inclusive.

**Control strategy** (Section 51.100) means a combination of measures designated to achieve the aggregate reduction of emissions necessary for attainment and maintenance of national standards including, but not limited to, measures such as: (1) Emission limitations. (2) Federal or State emission charges or taxes or other economic incentives or disincentives. (3) Closing or relocation of residential, commercial, or industrial facilities. (4) Changes in schedules or methods of operation of commercial or industrial facilities or transportation systems, including, but not limited to, short-term changes made in accordance with standby plans. (5) Periodic inspection and testing of motor vehicle emission control systems, at such time as the Administrator determines that such programs are feasible and practicable. (6) Emission control measures applicable to in-use motor vehicles, including, but not limited to, measures such as mandatory maintenance, installation of emission control devices, and conversion to gaseous fuels. (7) Any transportation control measure including those transportation measures listed in section 108(f) of the Clean Air Act as amended. (8) Any variation of, or alternative to any measure delineated herein. (9) Control or prohibition of a fuel or fuel additive used in motor vehicles, if such control or prohibition is necessary to achieve a national primary or secondary air quality standard and is approved by the Administrator under section 211©(4)© of the Act.

**Control strategy implementation plan revision** (Section 93.101) is the implementation plan which contains specific strategies for controlling the emissions of and reducing ambient levels of pollutants in order to satisfy CAA requirements for demonstrations of reasonable further progress and attainment (CAA sections 182(b)(1), 182(c)(2)(A), 182(c)(2)(B), 187(a)(7), 189(a)(1)(B), and 189(b)(1)(A); and sections 192(a) and 192(b), for nitrogen dioxide).

**Control substance** (Section 160, 792) means any chemical substance or mixture, or any other material other than a test substance, feed, or water, that is administered to the test system in the course of a study for the purpose of establishing a basis for comparison with the test substance for known chemical or biological measurements.

**Control system** (Section 65.2) means the combination of the closed vent system and the control devices used to collect and control vapors or gases from a regulated source.

**Control unit** (Section 72.2) means a unit employing a qualifying Phase I technology in accordance with a Phase I extension plan under Sec. 72.42.

**Controlled area** (Section 191.12) means: (1) A surface location, to be identified by passive institutional controls, that encompasses no more than 100 square kilometers and extends horizontally no more than five kilometers in any direction from the outer boundary of the original location of the radioactive wastes in a disposal system; and (2) the subsurface underlying such a surface location.

**Controlled product** (Section 82.3) means a product that contains a controlled substance listed as a Class I, Group I or II substance in appendix A to this subpart. Controlled products include, but are not limited to, those products listed in appendix D to this subpart. Controlled products belong to one or more of the following six categories of products: (1) Automobile and truck air conditioning units (whether incorporated in vehicles or not); (2) Domestic and commercial refrigeration and air-conditioning/heat pump equipment (whether containing controlled substances as a refrigerant and/or in insulating material of the product), e.g. Refrigerators, Freezers, Dehumidifiers, Water coolers, Ice machines, Air-conditioning and heat pump units; (3) Aerosol products, except medical aerosols; (4) Portable fire extinguishers; (5) Insulation boards, panels and pipe covers; (6) Pre-polymers.

**Controlled substance** (Section 82.3) means any substance listed in appendix A or appendix B to this subpart, whether existing alone or in a mixture, but excluding any such substance or mixture that is in a manufactured product other than a container used for the transportation or storage of the substance or mixture. Thus, any amount of a listed substance in appendix A or appendix B to this subpart that is not part of a use system containing the substance is a controlled substance. If a listed substance or mixture must first be transferred from a bulk container to another container, vessel, or piece of equipment in order to realize its intended use, the listed substance or mixture is a "controlled substance." The inadvertent or coincidental creation of insignificant quantities of a listed substance in appendix A or appendix B to this subpart; during a chemical manufacturing process, resulting from unreacted feedstock, from the listed substance's use as a process agent present as a trace quantity in the chemical substance being manufactured, or as an unintended byproduct of research and development applications, is not deemed a controlled substance. Controlled substances are divided into two classes, Class I in appendix A to this subpart, and Class II listed in appendix B to this subpart. Class I substances are further divided into seven groups, Group I, Group II, Group III, Group IV, Group V, Group VI, and Group VII, as set forth in appendix A to this subpart.

**Controlled surface mine drainage** (Section 434.11) means any surface mine drainage that is pumped or siphoned from the active mining area.

**Controlling interest** (Section 280.92) means direct ownership of at least 50 percent of the voting stock of another entity.

**Conveniently available service facility and spare parts for small-volume manufacturers** (Section 86.092-2, 86.1803-01) means that the vehicle manufacturer has a qualified service facility at or near the authorized point of sale or delivery of its vehicles and maintains an inventory of all emission-related spare parts or has made arrangements for the part manufacturers to supply the parts by expedited shipment (e.g., utilizing overnight express delivery service, UPS, etc.).

**Conventional breeding of plants** (Section 174.3) means the creation of progeny through either: The union of gametes, i.e., syngamy, brought together through processes such as pollination, including bridging crosses between plants and wide crosses, or vegetative reproduction. It does not include use of any of the following technologies: Recombinant DNA; other techniques wherein the genetic material is extracted from an organism and introduced into the genome of the recipient plant through, for example, micro-injection, macro-injection, micro-encapsulation; or cell fusion.

**Conventional filtration treatment** (Section 141.2) means a series of processes including coagulation, flocculation, sedimentation, and filtration resulting in substantial particulate removal.

**Conventional gasoline** (Section 80.2) means any gasoline which has not been certified under Sec. 80.40.

**Conventional gasoline** (Section 86.1702-99) means any certification gasoline which meets the specifications of Sec. 86.113(a). The ozone-forming potential of conventional gasoline vehicle emissions shall be determined by using the methods and gasoline specifications contained in appendix XVII of this part.

**Conventional mine** (Section 146.3) means an open pit or underground excavation for the production of minerals.

**Conversion configuration** (Section 88.302-94) means any combination of vehicle/engine conversion hardware and a base vehicle of a specific engine family.

**Conviction** (Section 303.11) A judgment of guilt entered in U.S. District Court, upon a verdict rendered by the court or petit jury or by a plea of guilty, including a plea of nolo contendere.

**Cook Inlet** (Section 435.41) refers to coastal locations north of the line between Cape Douglas on the West and Port Chatham on the east.

**Cooperative agreement** (Section 300.5) is a legal instrument EPA uses to transfer money, property, services, or anything of value to a recipient to accomplish a public purpose in which substantial EPA involvement is anticipated during the performance of the project.

**Cooperator** (Section 172.1) means any person who grants permission to a permittee or a permittee's designated participant for the use of an experimental use pesticide at an application site owned or controlled by the cooperator.

**Copenhagen Amendments** (Section 82.3) means the Montreal Protocol on Substances That Deplete the Ozone Layer, as amended at the Fourth Meeting of the Parties to the Montreal Protocol in Copenhagen in 1992.

**Copper** (Section 415) shall mean the total copper present in the process wastewater stream exiting the wastewater treatment system.

**Copper** (Section 420.02) means total copper and is determined by the method specified in 40 CFR 136.3.

**Copper casting** (Section 464.02) The remelting of copper or a copper alloy to form a cast intermediate or final product by pouring or forcing the molten metal into a mold, except for ingots, pigs, or other cast shapes related to nonferrous (primary and secondary) metals manufacturing (40 CFR part 421). Also excluded are casting of beryllium alloys in which beryllium is present at 0.1 or greater percent by weight and precious metals alloys in which the precious metal is present at 30 or greater percent by weight. Except for grinding scrubber operations which are covered here, processing operations following the cooling of castings are covered under the electroplating and metal finishing point source categories (40 CFR parts 413 and 433).

**Coproduct** (Section 704.3) means a chemical substance produced for a commercial purpose during the manufacture, processing, use, or disposal of another chemical substance or mixture.

**Co-product** (Section 716.3) means a chemical substance produced for a commercial purpose during the manufacture, processing, use, or disposal of another chemical substance(s) or mixture(s).

**Copy of study** (Section 716.3) means the written presentation of the purpose and methodology of a study and its results.

**Core of the drawing with neat oils subcategory** (Section 467.51) shall include drawing using neat oils, stationary casting, artificial aging, annealing, degreasing, sawing, and swaging.

**Core of the extrusion subcategory** (Section 467.31) shall include extrusion die cleaning, dummy block cooling, stationary casting, artificial aging, annealing, degreasing, and sawing.

**Core of the rolling with neat oils subcategory** (Section 467.11) shall include rolling using neat oils, roll grinding, sawing, annealing, stationary casting, homogenizing artificial aging, degreasing, and stamping.

**Core PM2.5 SLAMS** (Section 58.1) means community-oriented monitoring sites representative of community-wide exposures that are the basic component sites of the PM2.5 SLAMS regulatory network. Core PM2.5 SLAMS include community-oriented SLAMS monitors, and sites collocated at PAMS.

**Core Stable Standards** (Section 86.1702-99) means the standards and other requirements listed in Sec. 86.1707(d)(9)(i)(A) through (F).

**Corn** (Section 406.11) shall mean the shelled corn delivered to a plant before processing.

**Corn, canned** (Section 407.71) shall mean the processing of corn into the following product styles: Canned, yellow and white, whole kernel, cream style, and on-the-cob.

**Corn, frozen** (Section 407.71) shall mean the processing of corn into the following product styles: Frozen, yellow and white, whole kernel and whole cob.

**Corps** (Section 233.2) means the U.S. Army Corps of Engineers.

**Corrected concentration** (Section 58.1) pertains to the result of accuracy or precision assessment test of an open path analyzer in which a high-concentration test or audit standard gas contained in a short test cell is inserted into the optical measurement beam of the instrument. When the pollutant concentration measured by the analyzer in such a test includes both the pollutant concentration in the test cell and the concentration in the atmosphere, the atmospheric pollutant concentration must be subtracted from the test measurement to obtain the corrected concentration test result. The corrected concentration is equal to the measured concentration minus the average of the atmospheric pollutant concentrations measured (without the test cell) immediately before and immediately after the test.

**Corrective Action Management Unit or CAMU** (Section 270.2) means an area within a facility that is designated by the Regional Administrator under part 264 subpart S, for the purpose of implementing corrective action requirements under Sec. 264.101 and RCRA section 3008(h). A CAMU shall only be used for the management of remediation wastes pursuant to implementing such corrective action requirements at the facility.

**Correlated acceptable continuous (CAC) PM analyzer** (Section 58.1) means an optional fine particulate matter analyzer that can be used to supplement a PM2.5 reference or equivalent sampler, in accordance with the provisions of Sec. 58.13(f).

**Corresponding Onshore Area (COA)** (Section 55.2) means, with respect to any existing or proposed OCS source located within 25 miles of a State's seaward boundary, the onshore area that is geographically closest to the source or another onshore area that the Administrator designates as the COA, pursuant to Sec. 55.5 of this part.

**Corrosion expert** (Section 260.10) means a person who, by reason of his knowledge of the physical sciences and the principles of engineering and mathematics, acquired by a professional education and related practical experience, is qualified to engage in the practice of corrosion control on buried or submerged metal piping systems and metal tanks. Such a person must be certified as being qualified by the National Association of Corrosion Engineers (NACE) or be a registered professional engineer who has certification or licensing that includes education and experience in corrosion control on buried or submerged metal piping systems and metal tanks.

**Corrosion expert** (Section 280.12) means a person who, by reason of thorough knowledge of the physical sciences and the principles of engineering and mathematics acquired by a professional education and related practical experience, is qualified to engage in the practice of corrosion control on buried or submerged metal piping systems and metal tanks. Such a person must be accredited or certified as being qualified by the National Association of Corrosion Engineers or be a registered professional engineer who has certification or licensing that includes education and experience in corrosion control of buried or submerged metal piping systems and metal tanks.

**Corrosion inhibitor** (Section 141.2) means a substance capable of reducing the corrosivity of water toward metal plumbing materials, especially lead and copper, by forming a protective film on the interior surface of those materials.

**Corrugated box** (Section 246.101) means a container for goods which is composed of an inner fluting of material (corrugating medium) and one or two outer liners of material (linerboard).

**Corrugated container waste** (Section 246.101) means discarded corrugated boxes.

**Corrugated paper** (Section 763.163) means an asbestos-containing product made of corrugated paper, which is often cemented to a flat backing, may be laminated with foils or other materials, and has a corrugated surface. Major applications of asbestos corrugated paper include: thermal insulation for pipe coverings; block insulation; panel insulation in elevators; insulation in appliances; and insulation in low-pressure steam, hot water, and process lines.

**Corrugating medium furnish subdivision mills** (Section 430.101) are mills where only recycled corrugating medium is used in the production of paperboard.

**Cotton fiber furnish subdivision mills** (Section 430.111) are those mills where significant quantities of cotton fibers (equal to or greater than 4 percent of the total product) are used in the production of fine papers.

**Council** (Section 1517.2) shall mean the Council on Environmental Quality established under title II of the National Environmental Policy Act of 1969 (42 U.S.C. 4321 through 4347).

**Country of transit** (Section 262.81) means any designated OECD country in Sec. 262.58(a) (1) and (a) (2) other than the exporting or importing country across which a transfrontier movement of wastes is planned or takes place.

**Course agenda** (Section 745.223) means an outline of the key topics to be covered during a training course, including the time allotted to teach each topic.

**Course test** (Section 745.223) means an evaluation of the overall effectiveness of the training which shall test the trainees' knowledge and retention of the topics covered during the course.

**Course test blue print** (Section 745.223) means written documentation identifying the proportion of course test questions devoted to each major topic in the course curriculum.

**Cover** (Section 265.1081) means a device that provides a continuous barrier over the hazardous waste managed in a unit to prevent or reduce air pollutant emissions to the atmosphere. A cover may have openings (such as access hatches, sampling ports, gauge wells) that are necessary for operation, inspection, maintenance, and repair of the unit on which the cover is used. A cover may be a separate piece of equipment which can be detached and removed from the unit or a cover may be formed by structural features permanently integrated into the design of the unit.

**Cover** (Section 503.21) is soil or other material used to cover sewage sludge placed on an active sewage sludge unit.

**Covered area** (Section 80.2) means each of the geographic areas specified in Sec. 80.70 in which only reformulated gasoline may be sold or dispensed to ultimate consumers.

**Covered fleet operator** (Section 88.302-94) means a person who operates a fleet of at least ten covered fleet vehicles (as defined in section 241(6) of the Act) and that fleet is operated in a single covered area (even if the covered fleet vehicles are garaged outside of it). For purposes of this definition, the vehicle types described in the definition of covered fleet (section 241(5) of the Act) as exempt from the program will not be counted toward the ten-vehicle criterion.

**Covered process** (Section 68.3) means a process that has a regulated substance present in more than a threshold quantity as determined under Sec. 68.115.

**Covered state** (Section 86.1702-99) means a state that meets the conditions specified under Sec. 86.1705(d).

**Covered vehicle or engine** (Section 86.1702-99) means a vehicle specified in Sec. 86.1701(a), or an engine in such a vehicle, that is manufactured by a covered manufacturer.

**CPI (1990)** (Section 72.2) means the CPI for all urban consumers for the month of August 1989. The "CPI (1990)" is 124.6 (with 1982-1984=100). Beginning in the month for which a new reference base is established, "CPI (1990)" will be the CPI value for August 1989 on the new reference base.

**CPI (year)** (Section 72.2) means the CPI for all urban consumers for the month of August of the previous year.

**Cr(+6)** (Section 415.171) shall mean hexavalent chromium.

**Cr(T)** (Section 415.171) shall mean total chromium.

**Cr,VI** (Section 413.02) shall mean hexavalent chromium.

**Cranberries** (Section 407.61) shall include the processing of the following berries: Canned and frozen blackberries, blueberries, boysenberries, currants, gooseberries, loganberries, ollalieberries, raspberries, and any other similar cane or bushberry but not strawberries or cranberries.

**Cranberries** (Section 407.61) shall mean the processing of cranberries into the following product styles: Canned, bottled, and frozen, whole, sauce, jelly, juice and concentrate.

**Crankcase emissions** (Section 86.082-2, 86.402-78, 89.2) means airborne substances emitted to the atmosphere from any portion of the engine crankcase ventilation or lubrication systems.

**Crankcase emissions** (Section 86.1803-01) means airborne substances emitted to the atmosphere from any portion of the engine crankcase ventilation or lubrication systems.

**Crankcase emissions** (Section 92.2) means emissions to the atmosphere from any portion of the crankcase ventilation or engine lubrication systems.

**Crankcase emissions** (Section 94.2) means airborne substances emitted to the atmosphere from any portion of the engine crankcase ventilation or engine lubrication system.

**Credits** (Section 86.1702-99) means fleet average NMOG credits as calculated from the amount that the manufacturer's applicable fleet average NMOG value is below the applicable fleet average NMOG standard, times the applicable production for a given model year. NMOG credits have units of g/mi.

**Crisis exemption** (Section 176.3) means an exemption authorized under FIFRA section 18, in accordance with Sec. Sec. 166.40 through 166.53 of this chapter.

**Criteria** (Section 131.3) are elements of State water quality standards, expressed as constituent concentrations, levels, or narrative statements, representing a quality of water that supports a particular use. When criteria are met, water quality will generally protect the designated use.

**Criteria** (Section 220.2) means the criteria set forth in part 227 of this subchapter H.

**Criteria** (Section 256.06) means the "Criteria for Classification of Solid Waste Disposal Facilities", 40 CFR Part 257, promulgated under section 4004(a) of the Act.

**Criteria pollutant or standard** (Section 93.152) means any pollutant for which there is established a NAAQS at 40 CFR part 50.

**Criterion continuous concentration (CCC)** (Section 132.2) is an estimate of the highest concentration of a material in the water column to which an aquatic community can be exposed indefinitely without resulting in an unacceptable effect.

**Criterion maximum concentration (CMC)** (Section 132.2) is an estimate of the highest concentration of a material in the water column to which an aquatic community can be exposed briefly without resulting in an unacceptable effect.

**Critical emission-related components** (Section 86.088-2) are those components which are designed primarily for emission control, or whose failure may result in a significant increase in emissions accompanied by no significant impairment (or perhaps even an improvement) in performance, driveability, and/or fuel economy as determined by the Administrator. Critical emission-related maintenance means that maintenance to be performed on critical emission-related components. Emission-related maintenance means that maintenance which does substantially affect emissions or which is likely to affect the emissions deterioration of the vehicle or engine during normal in-use operation, even if the maintenance is performed at some time other than that which is recommended.

**Critical emission-related components** (Section 86.1803-01) are those components which are designed primarily for emission control, or whose failure may result in a significant increase in emissions accompanied by no significant impairment (or perhaps even an improvement) in performance, driveability, and/or fuel economy as determined by the Administrator.

**Critical emission-related maintenance** (Section 86.1803-01) means that maintenance to be performed on critical emission-related components.

**Critical organ** (Section 191.02) means the most exposed human organ or tissue exclusive of the integumentary system (skin) and the cornea.

**Crop advisor** (Section 170.3) means any person who is assessing pest numbers or damage, pesticide distribution, or the status or requirements of agricultural plants. The term does not include any person who is performing hand labor tasks.

**Crude intermediate plastic material** (Section 463.2) is plastic material formulated in an on-site polymerization process.

**Crude oil** (Section 68.3) means any naturally occurring, unrefined petroleum liquid.

**CSB incident report** (Section 1611.2) means the report containing the CSB's determinations, including the probable cause of an incident, issued either as a narrative report or in a computer format. Pursuant to 42 U.S.C. 7412®(6)(G), no part of the conclusions, findings or recommendations of the CSB relating to an accidental release or the investigation thereof, may be admitted as evidence or used in any suit or action for damages growing out of any matter mentioned in such report.

**CT or CTcalc** (Section 141.2) is the product of "residual disinfectant concentration" © in mg/l determined before or at the first customer, and the corresponding "disinfectant contact time" (T) in minutes, i.e., "C" x "T". If a public water system applies disinfectants at more than one point prior to the first customer, it must determine the CT of each disinfectant sequence before or at the first customer to determine the total percent inactivation or "total inactivation ratio." In determining the total inactivation ratio, the public water system must determine the residual disinfectant concentration of each disinfection sequence and corresponding contact time before any subsequent disinfection application point(s). "CT99.9" is the CT value required for 99.9 percent (3-log) inactivation of Giardia lamblia cysts. CT99.9 for a variety of disinfectants and conditions appear in tables 1.1-1.6, 2.1, and 3.1 of Sec. 141.74(b)(3).

**cu ft** (Section 401.11) means cubic foot (feet).

**cu m** (Section 401.11) means cubic meter(s).

**Cubic feet or cubic meters of production in subpart A** (Section 429.11) means the cubic feet or cubic meters of logs from which bark is removed.

**Cullet** (Section 426.21) shall mean any broken glass generated in the manufacturing process.

**Cullet water** (Section 426.11) shall mean that water which is exclusively and directly applied to molten glass in order to solidify the glass.

**Cumulative pollutant loading rate** (Section 503.11) is the maximum amount of an inorganic pollutant that can be applied to an area of land.

**Curb collection** (Section 243.101) means collection of solid waste placed adjacent to a street.

**Curb mass** (Section 86.402-78) means the actual or manufacturer's estimated mass of the vehicle with fluids at nominal capacity and with all equipment specified by the Administrator.

**Curb weight** (Section 86.1803-01) means the actual or the manufacturer's estimated weight of the vehicle in operational status with all standard equipment, and weight of fuel at nominal tank capacity, and the weight of optional equipment computed in accordance with Sec. 86.1832-01; incomplete light-duty trucks shall have the curb weight specified by the manufacturer.

**Curb-idle** (Section 86.1803-01) means, for manual transmission code motor vehicles, the engine speed with the transmission in neutral or with the clutch disengaged and with the air conditioning system, if present, turned off. For automatic transmission code motor vehicles, curb-idle means the engine speed with the automatic transmission in the park position (or neutral position if there is no park position), and with the air conditioning system, if present, turned off.

**Curb-idle for manual transmission code heavy-duty engines** (Section 86.082-2) means the manufacturer's recommended engine speed with the transmission in neutral or with the clutch disengaged. For automatic transmission code heavy-duty engines, curb-idle means the manufacturer's recommended engine speed with the automatic transmission in gear and the output shaft stalled.

**Curie (Ci)** (Section 190.02) means that quantity of radioactive material producing 37 billion nuclear transformations per second. (One millicurie (mCi) =0.001 Ci.)

**Curie (Ci)** (Section 192) means the amount

of radioactive material that produces 37 billion nuclear transformation per second. One picocurie (pCi) = $10^{-12}$ Ci.

**Current production |sales or distribution|** (Section 167.3) means amount of planned production in the calendar year in which the pesticides report is submitted, including new pesticidal products not previously sold or distributed.

**Current year subaccount** (Section 72.2) means the subaccount in an Allowance Tracking System general account, established pursuant to Sec. 73.31© of this chapter, in which are held allowances that may be transferred to a unit's compliance subaccount for use for the purpose of meeting the Acid Rain sulfur dioxide emissions limitation.

**Currently valid certificate of conformity (Section 89.602)** A certificate of conformity for which the current date is within the effective period as specified on the certificate of conformity, and which has not been withdrawn, superseded, voided, suspended, revoked, or otherwise rendered invalid.

**Custom blender** (Section 167.3) means any establishment which provides the service of mixing pesticides to a customer's specifications, usually a pesticide(s)-fertilizer(s), pesticide-pesticide, or a pesticide-animal feed mixture, when: (1) The blend is prepared to the order of the customer and is not held in inventory by the blender; (2) the blend is to be used on the customer's property (including leased or rented property); (3) the pesticide(s) used in the blend bears end-use labeling directions which do not prohibit use of the product in such a blend; (4) the blend is prepared from registered pesticides; (b) the blend is delivered to the end-user along with a copy of the end-use labeling of each pesticide used in the blend and a statement specifying the composition of mixture; and (6) no other pesticide production activity is performed at the establishment.

**Customer** (Section 704.3) means any person to whom a manufacturer, importer, or processor directly distributes any quantity of a chemical substance, mixture, mixture containing the substance or mixture, or article containing the substance or mixture, whether or not a sale is involved.

**Customer** (Section 72.2) means a purchaser of electricity not for the purposes of retransmission or resale. For generating rural electrical cooperatives, the customers of the distribution cooperatives served by the generating cooperative will be considered customers of the generating cooperative.

**Customer** (Section 721.3) means any person to whom a manufacturer, importer, or processor distributes any quantity of a chemical substance, or of a mixture containing the chemical substance, whether or not a sale is involved.

**Custom-molded device** (Section 211.203) A hearing protective device that is made to conform to a specific ear canal. This is usually accomplished by using a moldable compound to obtain an impression of the ear and ear canal. The compound is subsequently permanently hardened to retain this shape.

**Customs territory of the United States** (Section 372.3) means the 50 States, the District of Columbia, and Puerto Rico.

**Customs territory of the United States** (Section 720.3) means the 50 States, Puerto Rico, and the District of Columbia.

**Customs territory of the United States** (Section 763.163) means the 50 States, Puerto Rico, and the District of Columbia.

**Cut-in, or jambing, clearcoat** (Section 59.101) means a fast-drying, ready-to-spray clearcoat applied to surfaces such as door jambs and trunk and hood edges to allow for quick closure.

**Cutout or by-pass or similar devices** (Section 202.10) means devices which vary the exhaust system gas flow so as to discharge the exhaust gas and acoustic energy to the atmosphere without passing through the entire length of the exhaust system, including all exhaust system sound attenuation components.

**CWA** (Section 122.2) means the Clean Water Act (formerly referred to as the Federal Water Pollution Control Act or Federal Water Pollution Control Act Amendments of 1972) Public Law 92-500, as amended by Public Law 95-217, Public Law 95-576, Public Law 96-483 and Public Law 97-117, 33 U.S.C. 1251 et seq.

**CWA** (Section 270.2) means the Clean Water Act (formerly referred to as the Federal Water Pollution Control Act or Federal Water Pollution Control Act amendments of 1972) Pub. L. 92-500, as amended by Pub. L. 92-217 and Pub. L. 95-576; 33 U.S.C. 1251 et seq.

**CWA** (Section 501.2) means the Clean Water Act (formerly referred to as the Federal Water Pollution Control Act or Federal Water Pollution Control Act Amendments of 1972), Pub. L. 92-500, as amended by Pub. L. 95-217, Pub. L. 95-576, Pub. L.96-483, Pub. L. 97-117, and Pub. L. 100-4, 33 U.S.C. 1251 et seq.

**CWA and regulations** (Section 122.2) means the Clean Water Act (CWA) and applicable regulations promulgated thereunder. In the case of an approved State program, it includes State program requirements.

**Cwt** (Section 406.61) shall mean hundred weight.

**Cyanide** (Section 420.02) means total cyanide and is determined by the method specified in 40 CFR 136.3.

**Cyanide (T)** (Section 439.1) means the parameter total cyanide.

**Cyanide A** (Section 415.421) means those cyanides amenable to chlorination and is determined by the methods specified in 40 CFR 136.3.

**Cyanide A** (Section 415.91) shall mean those cyanides amenable to chlorination and is determined by the methods specified in 40 CFR 136.3.

**Cyclone boiler** (Section 76.2) means a boiler with one or more water-cooled horizontal cylindrical chambers in which coal combustion takes place. The horizontal cylindrical chamber(s) is (are) attached to the bottom of the furnace. One or more cylindrical chambers are arranged either on one furnace wall or on two opposed furnace walls. Gaseous combustion products exiting from the chamber(s) turn 90 degrees to go up through the boiler while coal ash exits the bottom of the boiler as a molten slag.

# D

**Daily discharge** (Section 122.2) means the "discharge of a pollutant" measured during a calendar day or any 24-hour period that reasonably represents the calendar day for purposes of sampling. For pollutants with limitations expressed in units of mass, the "daily discharge" is calculated as the total mass of the pollutant discharged over the day. For pollutants with limitations expressed in other units of measurement, the "daily discharge" is calculated as the average measurement of the pollutant over the day.

**Daily maximum limitation** (Section 429.11) is a value that should not be exceeded by any one effluent measurement. The 30-day limitation is a value that should not be exceeded by the average of daily measurements taken during any 30-day period.

**Daily values as applied to produced water effluent limitations and NSPS** (Section 435.11) means the daily measurements used to assess compliance with the maximum for any one day.

**Daily values as applied to produced water effluent limitations and NSPS** (Section 435.41) means the daily measurements used to assess compliance with the maximum for any one day.

**Damaged friable miscellaneous ACM** (Section 763.83) means friable miscellaneous ACM which has deteriorated or sustained physical injury such that the internal structure (cohesion) of the material is inadequate or, if applicable, which has delaminated such that its bond to the substrate (adhesion) is inadequate or which for any other reason lacks fiber cohesion or adhesion qualities. Such damage or deterioration may be illustrated by the separation of ACM into layers; separation of ACM from the substrate; flaking, blistering, or crumbling of the ACM surface; water damage; significant or repeated water stains, scrapes, gouges, mars or other signs of physical injury on the ACM. Asbestos debris originating from the ACBM in question may also indicate damage.

**Damaged friable surfacing ACM** (Section 763.83) means friable surfacing ACM which has deteriorated or sustained physical injury such that the internal structure (cohesion) of the material is inadequate or which has delaminated such that its bond to the substrate (adhesion) is inadequate, or which, for any other reason, lacks fiber cohesion or adhesion qualities. Such damage or deterioration may be illustrated by the separation of ACM into layers; separation of ACM from the substrate; flaking, blistering, or crumbling of the ACM surface; water damage; significant or repeated water stains, scrapes, gouges, mars or other signs of physical injury on the ACM. Asbestos debris originating from the ACBM in question may also indicate damage.

**Damaged or significantly damaged thermal system insulation ACM** (Section 763.83) means thermal system insulation ACM on pipes, boilers, tanks, ducts, and other thermal system insulation equipment where the insulation has lost its structural integrity, or its covering, in whole or in part, is crushed, water-stained, gouged, punctured, missing, or not intact such that it is not able to contain fibers. Damage may be further illustrated by occasional punctures, gouges or other signs of physical injury to ACM; occasional water damage on the protective coverings/jackets; or exposed ACM ends or joints. Asbestos debris originating from the ACBM in question may also indicate damage.

**Damages as defined by section 1001 of the OPA (Section 300.5)** means damages specified in section 1002(b) of the Act, and includes the cost of assessing these damages.

**Data** (Section 64.1) means the results of any type of monitoring or method, including the results of instrumental or non-instrumental monitoring, emission calculations, manual sampling procedures, recordkeeping procedures, or any other form of information collection procedure used in connection with any type of monitoring or method.

**Data fleet** (Section 610.11) means a fleet of automobiles tested at zero device-miles in baseline configuration, the retrofitted configuration and in some cases the adjusted configuration, in order to determine the changes in fuel economy and exhaust emissions due to the retrofitted configuration, and where applicable the changes due to the adjusted configuration, as compared to the fuel economy and exhaust emissions of the baseline configuration.

**Data gap** (Section 152.83) means the absence of any valid study or studies in the Agency's files which would satisfy a specific data requirement for a particular pesticide product.

**Data stream information** (Section 86.094-2) means information (i.e., messages and parameters) originated within the vehicle by a module or intelligent sensors (i.e., a sensor that contains and is controlled by itsown module) and transmitted between a network of modules and/or intelligent sensors connected in parallel with either one or two communication wires. The information is broadcast over the communication wires for use by other modules (e.g., chassis, transmission, etc.) to conduct normal vehicle operation or for use by diagnostic tools. Data stream information does not include engine calibration related information.

**Data stream information** (Section 86.1803-01) means information (i.e., messages and parameters) originated within the vehicle by a module or intelligent sensors (i.e., a sensor that contains and is controlled by its own module) and transmitted between a network of modules and/or intelligent sensors connected in parallel with either one or two communication wires. The information is broadcast over the communication wires for use by other modules (e.g., chassis, transmission, etc.) to conduct normal vehicle operation or for use by diagnostic tools. Data stream information does not include engine calibration related information.

**Data Submitters List** (Section 152.83) means the current Agency list, entitled "Pesticide Data Submitters by Chemical," of persons who have submitted data to the Agency.

**Date of completion** (Section 310.3) means the date when you have completed all field work and you have received all deliverables (such as lab results, technical expert reports, or invoices) due under a contract or other agreement.

**Day** (Section 65.2) means a calendar day.

**Day-night Sound Level** (Section 201.1) means the 24-hour time of day weighted equivalent sound level, in decibels, for any continuous 24-hour period, obtained after addition of ten decibels to sound levels produced in the hours from 10 p.m. to 7 a.m. (2200-0700). It is abbreviated as Ldn.

**Days** (Section 1603.3) means calendar days, except where noted otherwise.

**dB (A)** (Section 202.10) means the standard abbreviation for A-weighted sound level in decibels.

**dB (A)** is an abbreviation meaning A-weighted sound level in decibels, reference: 20 micropascals.

**dB(A)** (Section 205.2) means the standard abbreviation for A-weighted sound levels in decibels.

**dBA** (Section 204.2) is the standard abbreviation for A-weighted sound level in decibels.

**Dealer** (Section 600.022-85) means a person who resides or is located in the United States, any territory of the United States, or the District of Columbia and who is engaged in the sale or distribution of new automobiles to the ultimate purchaser.

**Dealer demonstration vehicle** (Section 88.302-94) means any vehicle that is operated by a motor vehicle dealer (as defined in section 216(4) of the Act) solely for the purpose of promoting motor vehicle sales, either on the sales lot or through other marketing or sales promotions, or for permitting potential purchasers to drive the vehicle for pre-purchase or pre-lease evaluation.

**Dealership** (Section 171.2) means any site owned or operated by a restricted use pesticide retail dealer where any restricted use pesticide is made available for use, or where the dealer offers to make available for use any such pesticide. Limited. The following definitions apply only to dealers, dealerships and transactions in States or on Indian Reservations where EPA conducts a Federal Pesticide Applicator Certification Program.

**Debits** (Section 86.1702-99) means fleet average NMOG debits as calculated from the amount that the manufacturer's applicable fleet average NMOG value is above the applicable fleet average NMOG standard, times the applicable production for a given model year. NMOG debits have units of g/mi.

**Debris** (Section 268.2) means solid material exceeding a 60 mm particle size that is intended for disposal and that is: A manufactured object; or plant or animal matter; or natural geologic material. However, the following materials are not debris: Any material for which a specific treatment standard is provided in Subpart D, Part 268, namely lead acid batteries, cadmium batteries, and radioactive lead solids; Process residuals such as smelter slag and residues from the treatment of waste, wastewater, sludges, or air emission residues; and Intact containers of hazardous waste that are not ruptured and that retain at least 75% of their original volume. A mixture of debris that has not been treated to the standards provided by Sec. 268.45 and other material is subject to regulation as debris if the mixture is comprised primarily of debris, by volume, based on visual inspection.

**Debris** (Section 429.11) means woody material such as bark, twigs, branches, heartwood or sapwood that will not pass through a 2.54 cm (1.0 in) diameter round opening and is present in the discharge from a wet storage facility.

**Decibel** (Section 201.1) means the unit measure of sound level, abbreviated as dB.

**Decisional body** (Section 72.2) means any EPA employee who is or may reasonably be expected to act in a decision-making role in a proceeding under part 78 of this chapter, including the Administrator, a member of the Environmental Appeals Board, and a Presiding Officer, and any staff of any such person who are participating in the decisional process.

**Deck drainage** (Section 435.11) means any waste resulting from deck washings, spillage, rainwater, and runoff from gutters and drains including drip pans and work areas within facilities subject to this Subpart.

**Deck drainage** (Section 435.41) means any waste resulting from deck washings, spillage, rainwater, and runoff from gutters and drains including drip pans and work areas within facilities subject to this Subpart.

**Dedicated ethanol vehicle** (Section 86.1702-99) means any ethanol-fueled motor vehicle that is engineered and designed to be operated solely on ethanol.

**Dedicated methanol vehicle** (Section 86.1702-99) means any methanol-fueled motor vehicle that is engineered and designed to be operated solely on methanol.

**Dedicated vehicle (or engine)** (Section 86.090-2) means any motor vehicle (or motor vehicle engine) engineered and designed to be operated using a single fuel. Flexible fuel vehicles and multi-fuel vehicles are not dedicated vehicles.

**Dedicated vehicle** (Section 86.1803-01) means any motor vehicle engineered and designed to be operated using a single fuel. Flexible fuel vehicles and multi-fuel vehicles are not dedicated vehicles.

**Deep drilling** (Section 194.2) means those drilling events in the Delaware Basin that reach or exceed a depth of 2,150 feet below the surface relative to where such drilling occurred.

**Defeat device** (Section 86.004-2) means an auxiliary emission control device (AECD) that reduces the effectiveness of the emission control system under conditions which may reasonably be expected to be encountered in normal vehicle operation and use, unless: (1) Such conditions are substantially included in the applicable Federal emission test procedure for heavy-duty vehicles and heavy-duty engines described in subpart N of this part; (2) The need for the AECD is justified in terms of protecting the vehicle against damage or accident; or (3) The AECD does not go beyond the requirements of engine starting.

**Defeat Device** (Section 86.082-2) means an AECD that reduces the effectiveness of the emission control system under conditions which may reasonably be expected to be encountered in normal urban vehicle operation and use, unless (1) such conditions are substantially included in the Federal emission test procedure, (2) the need for the AECD is justified in terms of protecting the vehicle against damage or accident, or (3) the AECD does not go beyond the requirements of engine starting.

**Defeat device** (Section 86.094-2) means an auxilary emission control device (AECD) that reduces the effectiveness of the emission control system under conditions which may reasonably be expected to be encountered in normal vehicle operation and use, unless: (1) Such conditions are substantially included in the Federal emission test procedure; (2) The need for the AECD is justified in terms of protecting the vehicle against damage or accident; or (3) The AECD does not go beyond the requirements of engine starting.

**Defeat device** (Section 86.1803-01) means an auxiliary emission control device (AECD) that reduces the effectiveness of the emission control system under conditions which may reasonably be expected to be encountered in normal vehicle operation and use, unless: (1) Such conditions are substantially included in the Federal emission test procedure; (2) The need for the AECD is justified in terms of protecting the vehicle against damage or accident; or (3) The AECD does not go beyond the requirements of engine starting.

**Defeat device** (Section 92.2) means an AECD or other control feature that reduces the effectiveness of the emission control system under conditions which may reasonably be expected to be encountered in normal locomotive operation and use, unless the AECD or other control feature has been identified by the certifying manufacturer or remanufacturer in the application for certification, and: (1) Such conditions are substantially represented by the portion of the federal test procedure during which the applicable emission rates are measured; (2) The need for the AECD is justified in terms of protecting the locomotive or locomotive engine against damage or accident; or (3) The AECD does not go beyond the requirements of engine starting.

**Defeat device** (Section 94.2) means an AECD or other control feature that reduces the effectiveness of the emission control system under conditions which may reasonably be expected to be encountered in normal engine operation and use, unless the AECD or other control feature has been identified by the manufacturer in the application for certification, and: (1) Such conditions are substantially represented by the portion of the applicable duty cycle of Sec. 94.105 during which the applicable emission rates are measured; (2) The need for the AECD or other control feature is justified in terms of protecting the engine or vessel against damage or accident; or (3) The AECD or other control feature does not go beyond the requirements of engine starting.

**Degassing** (Section 467.02) is the removal of dissolved hydrogen from the molten aluminum prior to casting. Chemicals are added and gases are bubbled through the molten aluminum. Sometimes a wet scrubber is used to remove excess chlorine gas.

**Degreasing** (Section 471.02) is the removal of oils and greases from the surface of the metal workpiece. This process can be accomplished with detergents as in alkaline cleaning or by the use of solvents.

**Dehydrated onions and garlic** (Section 407.71) shall mean the processing of dehydrated onions and garlic into the following product styles: Air, vacuum, and freeze dried, all varieties, diced, strips, and other piece sizes ranging from large sliced to powder but not including green onions, chives, or leeks.

**Dehydrated vegetables** (Section 407.71) shall mean the processing of dehydrated vegetables in the following product styles: Air, vacuum and freeze dried, blanched and unblanched, peeled and unpeeled, beets, bell peppers, cabbage, carrots, celery, chili pepper, horseradish, turnips, parsnips, parsley, asparagus, tomatoes, green beans, corn, spinach, green onion tops, chives, leeks, whole, diced, and any other piece size ranging from sliced to powder.

**Delaware Basin** (Section 194.2) means those surface and subsurface features which lie inside the boundary formed to the north, east and west of the disposal system by the innermost edge of the Capitan Reef, and formed, to the south, by a straight line drawn from the southeastern point of the Davis Mountains to the most southwestern point of the Glass Mountains.

**Delegate agency** (Section 71.2) means the State air pollution control agency, local agency, other State agency, Tribal agency, or other agency authorized by the Administrator pursuant to Sec. 71.10 to carry out all or part of a permit program under part 71.

**Delegated agency** (Section 55.2) means any agency that has been delegated authority to implement and enforce requirements of this part by the Administrator, pursuant to Sec. 55.11 of this part. It can refer to a State agency, a local agency, or an Indian tribe, depending on the delegation status of the program.

**Delineator** (Section 247.3) means a highly visible pavement marker that can be positioned to direct traffic or define boundaries;

**Demand-side measure** (Section 72.2) means a measure: (1) To improve the efficiency of consumption of electricity from a utility by customers of the utility; or (2) To reduce the amount of consumption of electricity from a utility by customers of the utility without increasing the use by the customer of fuel other than: Biomass (i.e., combustible energy-producing materials from biological sources, which include wood, plant residues, biological wastes, landfill gas, energy crops, and eligible components of municipal solid waste), solar, geothermal, or wind resources; or industrial waste gases where the party making the submission involved certifies that there is no net increase in sulfur dioxide emissions from the use of such gases. "Demand-side measure" includes the measures listed in part 73, appendix A, section 1 of this chapter.

**Demonstration period** (Section 76.2) means a period of time not less than 15 months, approved under Sec. 76.10, for demonstrating that the affected unit cannot meet the applicable emission limitation under Sec. 76.5, 76.6, or 76.7 and establishing the minimum NOX emission rate that the unit can achieve during long-term load dispatch operation.

**Density of microorganisms** (Section 503.31) is the number of microorganisms per unit mass of total solids (dry weight) in the sewage sludge.

**Deny or restrict the use of any defined area for specification** (Section 231.2) is to deny or restrict the use of any area for the present or future discharge of any dredged or fill material.

**Department** (Section 191.02) means the Department of Energy.

**Department** (Section 194.2) means the United States Department of Energy.

**Departure angle** (Section 86.1803-01) means the smallest angle, in a plan side view of a motor vehicle, formed by the level surface on which the motor vehicle is standing and a line tangent to the rear tire static loaded radius arc and touching the underside of the motor vehicle rearward of the rear tire.

**Depository site** (Section 192) means a site (other than a processing site) selected under Section 104(b) or 105(b) of the Act.

**Design capacity** (Section 240.101) means the weight of solid waste of a specified gross calorific value that a thermal processing facility is designed to process in 24 hours of continuous operation; usually expressed in tons per day.

**Design concept** (Section 93.101) means the type of facility identified by the project, e.g., freeway, expressway, arterial highway, grade-separated highway, reserved right-of-way rail transit, mixed-traffic rail transit, exclusive busway, etc.

**Design scope** (Section 93.101) means the design aspects which will affect the proposed facility's impact on regional emissions, usually as they relate to vehicle or person carrying capacity and control, e.g., number of lanes or tracks to be constructed or added, length of project, signalization, access control including approximate number and location of interchanges, preferential treatment for high-occupancy vehicles, etc.

**Designated agency** (Section 68.3) means the state, local, or Federal agency designated by the state under the provisions of Sec. 68.215(d).

**Designated facility** (Section 260.10) means a hazardous waste treatment, storage, or disposal facility which (1) has received a permit (or interim status) in accordance with the requirements of parts 270 and 124 of this chapter, (2) has received a permit (or interim status) from a State authorized in accordance with part 271 of this chapter, or (3) is regulated under Sec. 261.6©(2) or subpart F of part 266 of this chapter, and (4) that has been designated on the manifest by the generator pursuant to Sec. 260.20. If a waste is destined to a facility in an authorized State which has not yet obtained authorization to regulate that particular waste as hazardous, then the designated facility must be a facility allowed by the receiving State to accept such waste.

**Designated facility** (Section 761.13) means the off-site disposer or commercial storer of PCB waste designated on the manifest as the facility that will receive a manifested shipment of PCB waste.

**Designated management agency (DMA)** (Section 130.2) An agency identified by a WQM plan and designated by the Governor to implement specific control recommendations. [50 FR 1779, Jan. 11, 1985, as amended at 54 FR 14359, Apr. 11, 1989] Effective Date Note: At 65 FR 43662, July 13, 2000, Sec. 130.2 was amended by revising paragraphs © through (j) and (m), and by adding paragraphs (o) through ®, effective 30 days after the date that Congress allows EPA to implement this regulation. At 66 FR 53048, Oct. 18, 2001 this amendment was made effective April 30, 2003. For the convenience of the user, the revised and added text is set forth as follows:

**Designated Officer** (Section 94.2) means the person designated by the Director of the Office of Mobile Sources to act as the Designated Officer under the provisions of this part. For marine engines, the address for the Designated Officer is: Group Manager, Engine Compliance Group, U.S. EPA (mail code 6403J), 1200 Pennsylvania Ave., NW., Washington, DC 20460.

**Designated representative** (Section 70.2) shall have the meaning given to it in section 402(26) of the Act and the regulations promulgated thereunder.

**Designated representative** (Section 71.2) shall have the meaning given to it in section 402(26) of the Act and 40 CFR 72.2.

**Designated representative** (Section 72.2) means a responsible natural person authorized by the owners and operators of an affected source and of all affected units at the source or by the owners and operators of a combustion source or process source, as evidenced by a certificate of representation submitted in accordance with subpart B of this part, to represent and legally bind each owner and operator, as a matter of Federal law, in matters pertaining to the Acid Rain Program. Whenever the term "responsible official" is used in part 70 of this chapter, in any other regulations implementing title V of the Act, or in a State operating permit program, it shall be deemed to refer to the "designated representative" with regard to all matters under the Acid Rain Program.

**Designated State Agency** (Section 172.21) means the State agency designated by State law or other authority to be responsible for registering pesticides to meet special local needs.

**Designated uses** (Section 131.3) are those uses specified in water quality standards for each water body or segment whether or not they are being attained.

**Designated Volatility Attainment Area** (Section 80.2) means an area not designated as being in nonattainment with the National Ambient Air Quality Standard for ozone pursuant to rulemaking under section 107(d)(4)(A)(ii) of the Clean Air Act.

**Designated Volatility Nonattainment Area** (Section 80.2) means any area designated as being in nonattainment with the National Ambient Air Quality Standard for ozone pursuant to rulemaking under section 107(d)(4)(A)(ii) of the Clean Air Act.

**Desizing facilities** (Section 410.41) shall mean those facilities that desize more than 50 percent of their total production. These facilities may also perform other processing such as fiber preparation, scouring, mercerizing, functional finishing, bleaching, dyeing and printing. NSPS (Sec. 410.45)

**Destination facility** (Section 260.10) means a facility that treats, disposes of, or recycles a particular category of universal waste, except those management activities described in paragraphs (a) and (c) of Secs. 273.13 and 273.33 of this chapter. A facility, at which a particular category of universal waste is only accumulated, is not a destination facility for purposes of managing that category of universal waste.

**Destination facility** (Section 273.9) means a facility that treats, disposes of, or recycles a particular category of universal waste, except those management activities described in Sec. 273.13 (a) and (c) and Sec. 273.33 (a) and (c). A facility at which a particular category of universal waste is only accumulated, is not a destination facility for purposes of managing that category of universal waste.

**Destruction** (Section 82.3) means the expiration of a controlled substance to the destruction efficiency actually achieved, unless considered completely destroyed as defined in this section. Such destruction does not result in a commercially useful end product and uses one of the following controlled processes approved by the Parties to the Protocol: (1) Liquid injection incineration; (2) Reactor cracking; (3) Gaseous/fume oxidation; (4) Rotary kiln incineration; (5) Cement kiln; (6) Radio frequency plasma; or (7) Municipal waste incinerators only for the destruction of foams.

**Destruction Credits** (Section 82.3) means those privileges that may be obtained under Sec. 82.9 to produce controlled substances.

**Desulfurization** (Section 72.2) refers to various procedures whereby sulfur is removed from petroleum during or apart from the refining process. "Desulfurization" does not include such processes as dilution or blending of low sulfur content diesel fuel with high sulfur content diesel fuel from a diesel refinery not eligible under 40 CFR part 73, subpart G.

**Detection limit** (Section 136.2) means the minimum concentration of an analyte (substance) that can be measured and reported with a 99% confidence that the analyte concentration is greater than zero as determined by the procedure set forth at appendix B of this part.

**Deteriorated paint** (Section 745.223) means paint that is cracking, flaking, chipping, peeling, or otherwise separating from the substrate of a building component.

**Deteriorated paint** (Section 745.63) means any interior or exterior paint or other coating that is peeling, chipping, chalking or cracking, or any paint or coating located on an interior or exterior surface or fixture that is otherwise damaged or separated from the substrate.

**Deterioration factor** (Section 92.2) means the difference between exhaust emissions at the end of useful life and exhaust emissions at the low mileage test point expressed as either: the ratio of exhaust emissions at the end of useful life to exhaust emissions at the low mileage test point (for multiplicative deterioration factors); or the difference between exhaust emissions at the end of useful life exhaust emissions at the low mileage test point (for additive deterioration factors).

**Deterioration factor** (Section 94.2) means the difference between exhaust emissions at the end of useful life and exhaust emissions at the low hour test point expressed as either: the ratio of exhaust emissions at the end of useful life to exhaust emissions at the low hour test point (for multiplicative deterioration factors); or the difference between exhaust emissions at the end of useful life and exhaust emissions at the low hour test point (for additive deterioration factors).

**Development facility** (Section 435.11) is any fixed or mobile structure subject to this subpart that is engaged in the drilling of productive wells.

**Development facility** (Section 435.41) means any fixed or mobile structure subject to this Subpart that is engaged in the drilling of productive wells.

**Device** (Section 167.3) means any device or class of devices as defined by the Act and determined by the Administrator pursuant to section 25© to be subject to the provisions of section 7 of the Act.

**Device** (Section 169.1) The term "device" means any device or class of device as defined by the Act and determined by the Administrator to be subject to the provisions of the Act.

**Device integrity** (Section 610.11) means the durability of a device and effect of its malfunction on vehicle safety or other parts of the vehicle system.

**Device/measurement device** (Section 195.2) means a unit, component, or system designed to measure radon gas or radon decay products.

**Dewatering effluent** (Section 435.41) means wastewater from drilling fluids and drill cuttings dewatering activities (including but not limited to reserve pits or other tanks or vessels, and chemical or mechanical treatment occurring during the drilling solids separation/recycle/ disposal process).

**DF or df** (Section 90.3) means deterioration factor.

**Diatomaceous earth filtration** (Section 141.2) means a process resulting in substantial particulate removal in which (1) a precoat cake of diatomaceous earth filter media is deposited on a support membrance (septum), and (2) while the water is filtered by passing through the cake on the septum, additional filter media known as body feed is continuously added to the feed water to maintain the permeability of the filter cake.

**Dibenzofuran** (Section 766.3) means any of a family of compounds which has as a nucleus a triple-ring structure consisting of two benzene rings connected through a pair of bridges between the benzene rings. The bridges are a carbon-carbon bridge and a carbon-oxygen-carbon bridge at both substitution positions.

**Dibenzo-p-dioxin or dioxin** (Section 766.3) means any of a family of compounds which has as a nucleus a triple-ring structure consisting of two benzene rings connected through a pair of oxygen atoms.

**Dielectric material** (Section 280.12) means a material that does not conduct direct electrical current. Dielectric coatings are used to electrically isolate UST systems from the surrounding soils. Dielectric bushings are used to electrically isolate portions of the UST system (e.g., tank from piping).

**Diesel** (Section 86.090-2) means type of engine with operating characteristics significantly similar to the theoretical Diesel combustion cycle. The non-use of a throttle during normal operation is indicative of a diesel engine.

**Diesel** (Section 86.1803-01) means a type of engine with operating characteristics significantly similar to the theoretical Diesel combustion cycle. The non-use of a throttle during normal operation is indicative of a diesel engine.

**Diesel engine** (Section 86.1702-99) means any engine powered with diesel fuel, gaseous fuel, or alcohol fuel for which diesel engine speed/torque characteristics and vehicle applications are retained.

**Diesel fuel** (Section 72.2) means a low sulfur fuel oil of grades 1-D or 2-D, as defined by the American Society for Testing and Materials standard ASTM D975-91, "Standard Specification for Diesel Fuel Oils," grades 1-GT or 2-GT, as defined by ASTM D2880-90a, "Standard Specification for Gas Turbine Fuel Oils," or grades 1 or 2, as defined by ASTM D396-90a, "Standard Specification for Fuel Oils" (incorporated by reference in Sec. 72.13).

**Diesel fuel** (Section 80.2) means any fuel sold in any state or Territory of the United States and suitable for use in diesel motor vehicles, diesel motor vehicle engines or diesel nonroad engines, and which is commonly or commercially known or sold as diesel fuel.

**Diesel fuel** (Section 92.2) means any fuel suitable for use in diesel engines, and which is commonly or commercially known or sold as diesel fuel.

**Diesel fuel** (Section 94.2) means any fuel suitable for use in diesel engines which is commonly or commercially known or sold as diesel fuel.

**Diesel oil** (Section 435.11) refers to the grade of distillate fuel oil, as specified in the American Society for Testing and Materials Standard Specification for Diesel Fuel Oils D975-91, that is typically used as the continuous phase in conventional oil-based drilling fluids. This incorporation by reference was approved by the Director of the Federal Register in accordance with 5 U.S.C. 552(a) and 1 CFR part 51. Copies may be obtained from the American Society for Testing and Materials, 100 Barr Harbor Drive, West Conshohocken, PA, 19428. Copies may be inspected at the Office of the Federal Register, 800 North Capitol Street, NW., Suite 700, Washington, DC. A copy may also be inspected at EPA's Water Docket, 1200 Pennsylvania Ave., NW., Washington, DC 20460.

**Diesel oil** (Section 435.41) refers to the grade of distillate fuel oil, as specified in the American Society for Testing and Materials Standard Specification for Diesel Fuel Oils D975-91, that is typically used as the continuous phase in conventional oil-based drilling fluids. This incorporation by reference was approved by the Director of the Federal Register in accordance with 5 U.S.C. 552(a) and 1 CFR part 51. Copies may be obtained from the American Society for Testing and Materials, 1916 Race Street, Philadelphia, PA 19103. Copies may be inspected at the Office of the Federal Register, 800 North Capitol Street, NW., Suite 700, Washington, DC. A copy may also be inspected at EPA's Water Docket, 401 M Street SW., Washington, DC 20460.

**Diesel reciprocating engine unit** (Section 72.2) means an internal combustion engine that combusts only diesel fuel and that thereby generates electricity through the operation of pistons, rather than by heating steam or water.

**Diesel-fired unit** (Section 72.2) means, for the purposes of part 75 of this chapter, an oil-fired unit that combusts diesel fuel as its fuel oil, where the supplementary fuel, if any, shall be limited to natural gas or gaseous fuels containing no more sulfur than natural gas.

**Dietary LC50** (Section 152.161) means a statistically derived estimate of the concentration of a test substance in the diet that would cause 50 percent mortality to the test population under specified conditions.

**Dike** (Section 260.10) means an embankment or ridge of either natural or man-made materials used to prevent the movement of liquids, sludges, solids, or other materials.

**Diluent cap value** (Section 72.2) means a default value of percent CO2 or O2 which may be used to calculate the hourly NOX emission rate, CO2 mass emission rate, or heat input rate, when the measured hourly average percent CO2 is below the default value or when the measured hourly average percent O2 is above the default value. The diluent cap values for boilers are 5.0 percent CO2 and 14.0 percent O2. For combustion turbines, the diluent cap values are 1.0 percent CO2 and 19.0 percent O2.

**Diluent gas** (Section 72.2) means a major gaseous constituent in a gaseous pollutant mixture, which in the case of emissions from fossil fuel-fired units are carbon dioxide and oxygen.

**Diluent gas monitor** (Section 72.2) means that component of the continuous emission monitoring system that measures the diluent gas concentration in a unit's flue gas.

**Dioxin/furan and dioxins and furans** (Section 62.1201) mean tetra-, penta-, hexa-, hepta-, and octa-chlorinated dibenzo dioxins and furans.

**Dioxins and furans (D/F)** (Section 260.10) means tetra, penta, hexa, hepta, and octa-chlorinated dibenzo dioxins and furans.

**Direct application** (Section 420.101) means those cold rolling operations which include once-through use of rolling solutions at all mill stands.

**Direct chill casting** (Section 467.02) is the pouring of molten aluminum into a water-cooled mold. Contact cooling water is sprayed onto the aluminum as it is dropped into the mold, and the aluminum ingot falls into a water bath at the end of the casting process.

**Direct chill casting** (Section 471.02) is the pouring of molten nonferrous metal into a water-cooled mold. Contact cooling water is sprayed onto the metal as it is dropped into the mold, and the metal ingot falls into a water bath at the end of the casting process.

**Direct costs** (Section 1601.3) means those expenditures by the CSB actually incurred in searching for and duplicating records to respond to a FOIA request. Direct costs include the salary of the employee or employees performing the work (the basic rate of pay for the employee plus a percentage of that rate to cover benefits) and the cost of operating duplicating machinery. Direct costs do not include overhead expenses, such as the cost of space and heating or lighting of the facility in which the records are stored.

**Direct discharge** (Section 122.2) means the "discharge of a pollutant."

**Direct Dischargers (Section 455.41)**: Pollution prevention (P2) allowable discharge for this subpart means the quantity of concentrations of pollutants in PFPR process wastewaters that remain after a facility has demonstrated that it is using the specified practices of the Pollution Prevention Alternative as listed in Table 8 to this part 455 and that have been treated using appropriate pollution control technologies, as defined in Sec. 455.10(g), or a pesticide manufacturer's treatment system, or an equivalent system, used individually, or in any combination to achieve a sufficient level of pollutant reduction. The facility must demonstrate that the appropriate pollution control technology is properly maintained and operated.

**Direct emissions** (Section 93.152) means those emissions of a criteria pollutant or its precursors that are caused or initiated by the Federal action and occur at the same time and place as the action.

**Direct filtration** (Section 141.2) means a series of processes including coagulation and filtration but excluding sedimentation resulting in substantial particulate removal.

**Direct public utility ownership** (Section 72.2) means direct ownership of equipment and facilities by one or more corporations, the principal business of which is sale of electricity to the public at retail. Percentage ownership of such equipment and facilities shall be measured on the basis of book value.

**Direct Sale Subaccount** (Section 72.2) means a subaccount in the Special Allowance Reserve, as specified in section 416(b) of the Act, which contains Phase II allowances to be sold in the amount of 25,000 per year, from calendar year 1993 to 1999, inclusive, and of 50,000 per year for each year beginning in calendar year 2000, subject to the adjustments noted in the regulations at part 73, subpart E of this chapter.

**Director** (Section 122.2) means the Regional Administrator or the State Director, as the context requires, or an authorized representative. When there is no "approved State program," and there is an EPA administered program, "Director" means the Regional Administrator. When there is an approved State program, "Director" normally means the State Director. In some circumstances, however, EPA retains the authority to take certain actions even when there is an approved State program. (For example, when EPA has issued an NPDES permit prior to the approval of a State program, EPA may retain jurisdiction over that permit after program approval, see Sec. 123.1.) In such cases, the term "Director" means the Regional Administrator and not the State Director.

**Director** (Section 124.2) means the Regional Administrator, the State director or the Tribal director as the context requires, or an authorized representative. When there is no approved State or Tribal program, and there is an EPA administered program, Director means the Regional Administrator. When there is an approved State or Tribal program, "Director" normally means the State or Tribal director. In some circumstances, however, EPA retains the authority to take certain actions even when there is an approved State or Tribal program. (For example, when EPA has issued an NPDES permit prior to the approval of a State program, EPA may retain jurisdiction over that permit after program approval; see Sec. 123.1) In such cases, the term "Director" means the Regional Administrator and not the State or Tribal director.

**Director** (Section 124.2) means the State Director or Regional Administrator and is used when the accompanying provision is required of EPA-administered programs and of State programs under Secs. 123.25 (NPDES), 145.11 (UIC), 233.26 (404), and 271.14 (RCRA). The term Regional Administrator is used when the accompanying provision applies exclusively to EPA-issued permits and is not applicable to State programs under these sections. While States are not required to implement these latter provisions, they are not precluded from doing so, notwithstanding use of the term "Regional Administrator."

**Director** (Section 136.2) means the Director of the State Agency authorized to carry out an approved National Pollutant Discharge Elimination System Program under section 402 of the Act.

**Director** (Section 144.3) means the Regional Administrator, the State director or the Tribal director as the context requires, or an authorized representative. When there is no approved State or Tribal program, and there is an EPA administered program, "Director" means the Regional Administrator. When there is an approved State or Tribal program, "Director" normally means the State or Tribal director. In some circumstances, however, EPA retains the authority to take certain actions even when there is an approved State or Tribal program. In such cases, the term "Director" means the Regional Administrator and not the State or Tribal director.

**Director** (Section 146.3) means the Regional Administrator, the State director or the Tribal director as the context requires, or an authorized representative. When there is no approved State or Tribal program, and there is an EPA administered program, "Director" means the Regional Administrator. When there is an approved State or Tribal program, "Director" normally means the State or Tribal director. In some circumstances, however, EPA retains the authority to take certain actions even when there is an approved State or Tribal program. (For example, when EPA has issued an NPDES permit prior to the approval of a State program, EPA may retain jurisdiction over that permit after program approval; see Sec. 123.69). In such cases, the term Director means the Regional Administrator and not the State or Tribal director.

**Director** (Section 1517.2) means the Chairman of the Council on Environmental Quality acting as the head of the Office of Environmental Quality pursuant to the Environmental Quality Improvement Act of 1970, Pub. L. 91-224, 42 U.S.C. 4371 through 4374.

**Director** (Section 163.2) means the Director of the Pesticides Regulation Division, Environmental Protection Agency, Washington, D.C.

**Director** (Section 266.201) is as defined in 40 CFR 270.2.

**Director** (Section 270.2) means the Regional Administrator or the State Director, as the context requires, or an authorized representative. When there is no approved State program, and there is an EPA administered program, Director means the Regional Administrator. When there is an approved State program, Director normally means the State Director. In some circumstances, however, EPA retains the authority to take certain actions even when there is an approved State program. In such cases, the term Director means the Regional Administrator and not the State Director.

**Director** (Section 403.3) means the chief administrative officer of a State or Interstate water pollution control agency with an NPDES permit program approved pursuant to section 402(b) of the Act and an approved State pretreatment program.

**Director** (Section 720.3) means the Director of the EPA Office of Pollution Prevention and Toxics.

**Director** (Section 725.3) means the Director of the EPA Office of Pollution Prevention and Toxics.

**Director of an Approved State** (Section 258.2) means the chief administrative officer of a state agency responsible for implementing the state permit program that is deemed to be adequate by EPA under regulations published pursuant to sections 2002 and 4005 of RCRA.

**Director of the Implementing Agency** (Section 280.92) means the EPA Regional Administrator, or, in the case of a state with a program approved under section 9004, the Director of the designated state or local agency responsible for carrying out an approved UST program.

**Director of the Office of Pollution Prevention and Toxics** (Section 721.3) means the Director of the EPA Office of Pollution Prevention and Toxics or any EPA employee delegated by the Office Director to carry out the Office Director's functions under this part.

**Discarded material** (Section 261.2) is any material which is: (i) Abandoned, as explained in paragraph (b) of this section; or (ii) Recycled, as explained in paragraph © of this section; or (iii) Considered inherently waste-like, as explained in paragraph (d) of this section; or (iv) A military munition identified as a solid waste in 40 CFR 266.202. (b) Materials are solid waste if they are abandoned by being: (1) Disposed of; or (2) Burned or incinerated; or (3) Accumulated, stored, or treated (but not recycled) before or in lieu of being abandoned by being disposed of, burned, or incinerated. © Materials are solid wastes if they are recycled—or accumulated, stored, or treated before recycling—as specified in paragraphs ©(1) through (4) of this section. (1) Used in a manner constituting disposal. (i) Materials noted with a "*" in Column 1 of Table I are solid wastes when they are: (A) Applied to or placed on the land in a manner that constitutes disposal; or (B) Used to produce products that are applied to or placed on the land or are otherwise contained in products that are applied to or placed on the land (in which cases the product itself remains a solid waste). (ii) However, commercial chemical products listed in Sec. 261.33 are not solid wastes if they are applied to the land and that is their ordinary manner of use. (2) Burning for energy recovery. (i) Materials noted with a "*" in column 2 of Table 1 are solid wastes when they are: (A) Burned to recover energy; (B) Used to produce a fuel or are otherwise contained in fuels (in which cases the fuel itself remains a solid waste). (ii) However, commercial chemical products listed in Sec. 261.33 are not solid wastes if they are themselves fuels. (3) Reclaimed. Materials noted with a "*" in column 3 of Table 1 are solid wastes when reclaimed (except as provided under Sec. 261.4(a)(17)). Materials noted with a "—"in column 3 of Table 1 are not solid wastes when reclaimed. (4) Accumulated speculatively. Materials noted with a "*" in column 4 of Table 1 are solid wastes when accumulated speculatively.

**Discharge** (Section 109.2) includes, but is not limited to, any spilling, leaking, pumping, pouring, emitting, emptying, or dumping.

**Discharge** (Section 113.3) includes, but is not limited to any spilling, leaking, pumping, pouring, emitting, emptying or dumping.

**Discharge** (Section 122.2) when used without qualification means the "discharge of a pollutant."

**Discharge** (Section 140.1) includes, but is not limited to, any spilling, leaking, pumping, pouring, emitting, emptying, or dumping;

**Discharge** (Section 240.101) means water-borne pollutants released to a receiving stream directly or indirectly or to a sewerage system.

**Discharge allowance** (Section 461.2) means the amount of pollutant (mg per kg of production unit) that a plant will be permitted to discharge. For this category the allowances are specific to battery manufacturing operations.

**Discharge allowance** (Section 461.2) means the amount of pollutant (mg per kg of production unit) that a plant will be permitted to discharge. For this category the allowances are specific to battery manufacturing operations.

**Discharge as defined by section 311(a)(2) of the CWA** (Section 300.5) includes, but is not limited to, any spilling, leaking, pumping, pouring, emitting, emptying, or dumping of oil, but excludes discharges in compliance with a permit under section 402 of the CWA, discharges resulting from circumstances identified and reviewed and made a part of the public record with respect to a permit issued or modified under section 402 of the CWA, and subject to a condition in such permit, or continuous or anticipated intermittent discharges from a point source, identified in a permit or permit application under section 402 of the CWA, that are caused by events occurring within the scope of relevant operating or treatment systems. For purposes of the NCP, discharge also means substantial threat of discharge.

**Discharge incidental to the normal operation of a vessel** (Section 1700.3) means a discharge, including, but not limited to: graywater, bilgewater, cooling water, weather deck runoff, ballast water, oil water separator effluent, and any other pollutant discharge from the operation of a marine propulsion system, shipboard maneuvering system, crew habitability system, or installed major equipment, such as an aircraft carrier elevator or a catapult, or from a protective, preservative, or absorptive application to the hull of a vessel; and a discharge in connection with the testing, maintenance, and repair of any of the aforementioned systems whenever the vessel is waterborne, including pierside. A discharge incidental to normal operation does not include: (1) Sewage; (2) A discharge of rubbish, trash, or garbage; (3) A discharge of air emissions resulting from the operation of a vessel propulsion system, motor driven equipment, or incinerator; (4) A discharge that requires a National Pollutant Discharge Elimination System (NPDES) permit under the Clean Water Act; or (5) A discharge containing source, special nuclear, or byproduct materials regulated by the Atomic Energy Act.

**Discharge includes** (Section 112.2) but is not limited to, any spilling, leaking, pumping, pouring, emitting, emptying or dumping. For purposes of this part, the term discharge shall not include any discharge of oil which is authorized by a permit issued pursuant to section 13 of the River and Harbor Act of 1899 (30 Stat. 1121, 33 U.S.C. 407), or sections 402 or 405 of the FWPCA Amendments of 1972 (86 Stat. 816 et seq., 33 U.S.C. 1251 et seq.).

**Discharge includes** (Section 116.3) but is not limited to, any spilling, leaking, pumping, pouring, emitting, emptying or dumping, but excludes (A) discharges in compliance with a permit under section 402 of this Act, (B) discharges resulting from circumstances identified and reviewed and made a part of the public record with respect to a permit issued or modified under section 402 of this Act, and subject to a condition in such permit, and © continuous or anticipated intermittent discharges from a point source, identified in a permit or permit application under section 402 of this Act, which are caused by events occurring within the scope of relevant operating or treatment systems;

**Discharge Monitoring Report ("DMR")** (Section 122.2) means the EPA uniform national form, including any subsequent additions, revisions, or modifications for the reporting of self-monitoring results by permittees. DMRs must be used by "approved States" as well as by EPA. EPA will supply DMRs to any approved State upon request. The EPA national forms may be modified to substitute the State Agency name, address, logo, and other similar information, as appropriate, in place of EPA's.

**Discharge of a pollutant** (Section 122.2) means: (a) Any addition of any "pollutant" or combination of pollutants to "waters of the United States" from any "point source," or (b) Any addition of any pollutant or combination of pollutants to the waters of the "contiguous zone" or the ocean from any point source other than a vessel or other floating craft which is being used as a means of transportation. This definition includes additions of pollutants into waters of the United States from: surface runoff which is collected or channelled by man; discharges through pipes, sewers, or other conveyances owned by a State, municipality, or other person which do not lead to a treatment works; and discharges through pipes, sewers, or other conveyances, leading into privately owned treatment works. This term does not include an addition of pollutants by any "indirect discharger."

**Discharge of dredged material** (Section 232.2) (1) Except as provided below in paragraph (3), the term discharge of dredged material means any addition of dredged material into, including redeposit of dredged material other than incidental fallback within, the waters of the United States. The term includes, but is not limited to, the following: (i) The addition of dredged material to a specified discharge site located in waters of the United States; (ii) The runoff or overflow, associated with a dredging operation, from a contained land or water disposal area; and (iii) Any addition, including redeposit other than incidental fallback, of dredged material, including excavated material, into waters of the United States which is incidental to any activity, including mechanized land clearing, ditching, channelization, or other excavation. (2)(i) The Corps and EPA regard the use of mechanized earth-moving equipment to conduct land clearing, ditching, channelization, in-stream mining or other earth-moving activity in waters of the United States as resulting in a discharge of dredged material unless project-specific evidence shows that the activity results in only incidental fallback. This paragraph (i) does not and is not intended to shift any burden in any administrative or judicial proceeding under the CWA.

**Discharge of dredged material** (Section 232.2) does not include the following: (i) Discharges of pollutants into waters of the United States resulting from the onshore subsequent processing of dredged material that is extracted for any commercial use (other than fill). These discharges are subject to section 402 of the Clean Water Act even though the extraction and deposit of such material may require a permit from the Corps or applicable state. (ii) Activities that involve only the cutting or removing of vegetation above the ground (e.g., mowing, rotary cutting, and chain sawing) where the activity neither substantially disturbs the root system nor involves mechanized pushing, dragging, or other similar activities that redeposit excavated soil material. (iii) Incidental fallback. Section 404 authorization is not required for the following: (i) Any incidental addition, including redeposit, of dredged material associated with any activity that does not have or would not have the effect of destroying or degrading an area of waters of the U.S. as defined in paragraphs (5) and (6) of this definition; however, this exception does not apply to any person preparing to undertake mechanized land clearing, ditching, channelization and other excavation activity in a water of the United States, which would result in a redeposit of dredged material, unless the person demonstrates to the satisfaction of the Corps, or EPA as appropriate, prior to commencing the activity involving the discharge, that the activity would not have the effect of destroying or degrading any area of waters of the United States, as defined in paragraphs (5) and (6) of this definition. The person proposing to undertake mechanized land clearing, ditching, channelization or other excavation activity bears the burden of demonstrating that such activity would not destroy or degrade any area of waters of the United States. (ii) Incidental movement of dredged material occurring during normal dredging operations, defined as dredging for navigation in navigable waters of the United States, as that term is defined in 33 CFR part 329, with proper authorization from the Congress or the Corps pursuant to 33 CFR part 322; however, this exception is not applicable to dredging activities in wetlands, as that term is defined at Sec. 232.2® of this chapter. (iii) Certain discharges, such as those associated with normal farming, silviculture, and ranching activities, are not prohibited by or otherwise subject to regulation under Section 404. See 40 CFR 232.3 for discharges that do not require permits. (5) For purposes of this section, an activity associated with a discharge of dredged material destroys an area of waters of the United States if it alters the area in such a way that it would no longer be a water of the United States. Note: Unauthorized discharges into waters of the United States do not eliminate Clean Water Act jurisdiction, even where such unauthorized discharges have the effect of destroying waters of the United States. (6) For purposes of this section, an activity associated with a discharge of dredged material degrades an area of waters of the United States if it has more than a de minimis (i.e., inconsequential) effect on the area by causing an identifiable individual or cumulative adverse effect on any aquatic function. Discharge of fill material. (1) The term discharge of fill material means the addition of fill material into waters of the United States. The term generally includes, without limitation, the following activities: Placement of fill that is necessary for the construction of any structure or infrastructure in a water of the United States; the building of any structure, infrastructure, or impoundment requiring rock, sand, dirt, or other material for its construction; site-development fills for recreational, industrial, commercial, residential, or other uses; causeways or road fills; dams and dikes; artificial islands; property protection and/or reclamation devices such as riprap, groins, seawalls, breakwaters, and revetments; beach nourishment; levees; fill for structures such as sewage treatment facilities, intake and outfall pipes associated with power plants and subaqueous utility lines; placement of fill material for construction or maintenance of any liner, berm, or other infrastructure associated with solid waste landfills; placement of overburden, slurry, or tailings or similar mining-related materials;" after the words "utility lines; and artificial reefs. (2) In addition, placement of pilings in waters of the United States constitutes a discharge of fill material and requires a Section 404 permit when such placement has or would have the effect of a discharge of fill material. Examples of such activities that have the effect of a discharge of fill material include, but are not limited to, the following: Projects where the pilings are so closely spaced that sedimentation rates would be increased; projects in which the pilings themselves effectively would replace the bottom of a water body; projects involving the placement of pilings that would reduce the reach or impair the flow or circulation of waters of the United States; and projects involving the placement of pilings which would result in the adverse alteration or elimination of aquatic functions. (i) Placement of pilings in waters of the United States that does not have or would not have the effect of a discharge of fill material shall not require a Section 404 permit. Placement of pilings for linear projects, such as bridges, elevated walkways, and power line structures, generally does not have the effect of a discharge of fill material. Furthermore, placement of pilings in waters of the United States for piers, wharves, and an individual house on stilts generally does not have the effect of a discharge of fill material. All pilings, however, placed in the navigable waters of the United States, as that term is defined in 33 CFR part 329, require authorization under section 10 of the Rivers and Harbors Act of 1899 (see 33 CFR part 322).(ii) [Reserved]

**Discharge of pollutant(s)** (Section 401.11) means: (1) The addition of any pollutant to navigable waters from any point source and (2) any addition of any pollutant to the waters of the contiguous zone or the ocean from any point source, other than from a vessel or other floating craft. The term "discharge" includes either the discharge of a single pollutant or the discharge of multiple pollutants.

**Discharge or hazardous waste discharge** (Section 260.10) means the accidental or intentional spilling, leaking, pumping, pouring, emitting, emptying, or dumping of hazardous waste into or on any land or water.

**Discharge point** (Section 230.3) means the point within the disposal site at which the dredged or fill material is released.

**Discharger** (Section 437.2) means a facility that discharges wastewater directly to waters of the United States or introduces wastewater to a publicly-owned treatment works.

**Discipline** (Section 745.223) means one of the specific types or categories of lead-based paint activities identified in this subpart for which individuals may receive training from accredited programs and become certified by EPA. For example, "abatement worker" is a discipline.

**Discrete Wet Scrubbing Device** (Section 464.31) A discrete wet scrubbing device is a distinct, stand-alone device that removes particulates and fumes from a contaminated gas stream by bringing the gas stream into contact with a scrubber liquor, usually water, and from which there is a wastewater discharge. Examples of discrete wet scrubbing devices are: Spray towers and chambers, venturi scrubbers (fixed and variable), wet caps, packed bed scrubbers, quenchers, and orifice scrubbers. Semi-wet scrubbing devices where water is added and totally evaporates prior to dry air pollution control are not considered to be discrete wet scrubbing devices. Ancillary scrubber operations such as fan washes and backwashes are not considered to be discrete wet scrubber devices. These ancillary operations are covered by the mass limitations of the associated scrubber. Aftercoolers are not considered to be discrete wet scrubbing devices, and water discharges from aftercooling are not regulated as a process wastewater in this category.

**Disinfectant** (Section 141.2) means any oxidant, including but not limited to chlorine, chlorine dioxide, chloramines, and ozone added to water in any part of the treatment or distribution process, that is intended to kill or inactivate pathogenic microorganisms.

**Disinfectant contact time ("T" in CT calculations)** (Section 141.2) means the time in minutes that it takes for water to move from the point of disinfectant application or the previous point of disinfectant residual measurement to a point before or at the point where residual disinfectant concentration ("C") is measured. Where only one "C" is measured, "T" is the time in minutes that it takes for water to move from the point of disinfectant application to a point before or at where residual disinfectant concentration ("C") is measured. Where more than one "C" is measured, "T" is (a) for the first measurement of "C", the time in minutes that it takes for water to move from the first or only point of disinfectant application to a point before or at the point where the first "C" is measured and (b) for subsequent measurements of "C", the time in minutes that it takes for water to move from the previous "C" measurement point to the "C" measurement point for which the particular "T" is being calculated. Disinfectant contact time in pipelines must be calculated based on "plug flow" by dividing the internal volume of the pipe by the maximum hourly flow rate through that pipe. Disinfectant contact time within mixing basins and storage reservoirs must be determined by tracer studies or an equivalent demonstration.

**Disinfection** (Section 141.2) means a process which inactivates pathogenic organisms in water by chemical oxidants or equivalent agents.

**Disinfection profile** (Section 141.2) is a summary of Giardia lamblia inactivation through the treatment plant. The procedure for developing a disinfection profile is contained in Sec. 141.172 (Disinfection profiling and benchmarking) in subpart P and Secs. 141.530-141.536 (Disinfection profile) in subpart T of this part.

**Dispatch** (Section 72.2) means the assignment within a dispatch system of generating levels to specific units and generators to effect the reliable and economical supply of electricity, as customer demand rises or falls, and includes: (1) The operation of high-voltage lines, substations, and related equipment; and (2) The scheduling of generation for the purpose of supplying electricity to other utilities over interconnecting transmission lines.

**Dispensed fuel temperature** (Section 86.098-2) means the temperature (deg.F or deg.C may be used) of the fuel being dispensed into the tank of the test vehicle during a refueling test.

**Dispensed fuel temperature** (Section 86.1803-01) means the temperature (deg. F or deg. C may be used) of the fuel being dispensed into the tank of the test vehicle during a refueling test.

**Dispenser** (Section 211.203) The permanent (intended to be refilled) or disposable (discarded when empty) container designed to hold more than one complete set of hearing protector(s) for the express purpose of display to promote sale or display to promote use or both.

**Dispersants** (Section 300.5) means those chemical agents that emulsify, disperse, or solubilize oil into the water column or promote the surface spreading of oil slicks to facilitate dispersal of the oil into the water column.

**Dispersion technique** (Section 51.100) means any technique which attempts to affect the concentration of a pollutant in the ambient air by: (i) Using that portion of a stack which exceeds good engineering practice stack height: (ii) Varying the rate of emission of a pollutant according to atmospheric conditions or ambient concentrations of that pollutant; or (iii) Increasing final exhaust gas plume rise by manipulating source process parameters, exhaust gas parameters, stack parameters, or combining exhaust gases from several existing stacks into one stack; or other selective handling of exhaust gas streams so as to increase the exhaust gas plume rise. (2) The preceding sentence does not include: (i) The reheating of a gas stream, following use of a pollution control system, for the purpose of returning the gas to the temperature at which it was originally discharged from the facility generating the gas stream; (ii) The merging of exhaust gas streams where: (A) The source owner or operator demonstrates that the facility was originally designed and constructed with such merged gas streams; (B) After July 8, 1985 such merging is part of a change in operation at the facility that includes the installation of pollution controls and is accompanied by a net reduction in the allowable emissions of a pollutant. This exclusion from the definition of dispersion techniques shall apply only to the emission limitation for the pollutant affected by such change in operation; or © Before July 8, 1985, such merging was part of a change in operation at the facility that included the installation of emissions control equipment or was carried out for sound economic or engineering reasons. Where there was an increase in the emission limitation or, in the event that no emission limitation was in existence prior to the merging, an increase in the quantity of pollutants actually emitted prior to the merging, the reviewing agency shall presume that merging was significantly motivated by an intent to gain emissions credit for greater dispersion. Absent a demonstration by the source owner or operator that merging was not significantly motivated by such intent, the reviewing agency shall deny credit for the effects of such merging in calculating the allowable emissions for the source; (iii) Smoke management in agricultural or silvicultural prescribed burning programs; (iv) Episodic restrictions on residential woodburning and open burning; or (v) Techniques under Sec. 51.100(hh)(1)(iii) which increase final exhaust gas plume rise where the resulting allowable emissions of sulfur dioxide from the facility do not exceed 5,000 tons per year.

**Displacement** (Section 503.21) is the relative movement of any two sides of a fault measured in any direction.

**Displacement, and Displacement Class, see Sec. 86.419.**\

**Disposable Device** (Section 211.203) A hearing protective device that is intended to be discarded after one period of use.

**Disposal** (Section 191.02) means permanent isolation of spent nuclear fuel or radioactive waste from the accessible environment with no intent of recovery, whether or not such isolation permits the recovery of such fuel or waste. For example, disposal of waste in a mined geologic repository occurs when all of the shafts to the repository are backfilled and sealed.

**Disposal** (Section 260.10) means the discharge, deposit, injection, dumping, spilling, leaking, or placing of any solid waste or hazardous waste into or on any land or water so that such solid waste or hazardous waste or any constituent thereof may enter the environment or be emitted into the air or discharged into any waters, including ground waters.

**Disposal** (Section 270.2) means the discharge, deposit, injection, dumping, spilling, leaking, or placing of any hazardous waste into or on any land or water so that such hazardous waste or any constituent thereof may enter the environment or be emitted into the air or discharged into any waters, including ground water.

**Disposal** (Section 272.2) means the discharge, deposit, injection, dumping, spilling, leaking, or placing of any solid waste or hazardous waste into or on any land or water so that such solid waste or hazardous waste or any constituent thereof may enter the environment or be emitted into the air or discharged into any waters, including ground waters.

**Disposal** (Section 372.3) means any underground injection, placement in landfills/ surface impoundments, land treatment, or other intentional land disposal.

**Disposal** (Section 373.4) means the discharge, deposit, injection, dumping, spilling, leaking or placing of any hazardous substance into or on any land or water so that such hazardous substance or any constituent thereof may enter the environment or be emitted into the air or discharged into any waters, including groundwater.

**Disposal** (Section 761.13) means intentionally or accidentally to discard, throw away, or otherwise complete or terminate the useful life of PCBs and PCB Items. Disposal includes spills, leaks, and other uncontrolled discharges of PCBs as well as actions related to containing, transporting, destroying, degrading, decontaminating, or confining PCBs and PCB Items.

**Disposal area** (Section 192.31) means the region within the perimeter of an impoundment or pile containing uranium by product materials to which the post-closure requirements of Sec. 192.32(b) (1) of this subpart apply.

**Disposal facility** (Section 260.10) means a facility or part of a facility at which hazardous waste is intentionally placed into or on any land or water, and at which waste will remain after closure. The term disposal facility does not include a corrective action management unit into which remediation wastes are placed.

**Disposal facility** (Section 270.2) means a facility or part of a facility at which hazardous waste is intentionally placed into or on the land or water, and at which hazardous waste will remain after closure. The term disposal facility does not include a corrective action management unit into which remediation wastes are placed.

**Disposal period** (Section 192) means the period of time beginning March 7, 1983 and ending with the completion of all subpart A requirements specified under a plan for remedial action except those specified in Sec. 192.03 and Sec. 192.04. (o) Plan for remedial action means a written plan (or plans) for disposal and cleanup of residual radioactive materials associated with a processing site that incorporates the results of site characterization studies, environmental assessments or impact statements, and engineering assessments so as to satisfy the requirements of subparts A and B of this part. The plan(s) shall be developed in accordance with the provisions of Section 108(a) of the Act with the concurrence of the Commission and in consultation, as appropriate, with the Indian Tribe and the Secretary of Interior.

**Disposal regulations** (Section 194.2) means part 191, subparts B and C of this chapter.

**Disposal site** (Section 192) means the region within the smallest perimeter of residual radioactive material (excluding cover materials) following completion of control activities.

**Disposal site** (Section 192.31) means a site selected pursuant to section 83 of the Act.

**Disposal site** (Section 228.2) means an interim or finally approved and precise geographical area within which ocean dumping of wastes is permitted under conditions specified in permits issued under sections 102 and 103 of the Act. Such sites are identified by boundaries established by (1) coordinates of latitude and longitude for each corner, or by (2) coordinates of latitude and longitude for the center point and a radius in nautical miles from that point. Boundary coordinates shall be identified as precisely as is warranted by the accuracy with which the site can be located with existing navigational aids or by the implantation of transponders, buoys or other means of marking the site.

**Disposal site** (Section 230.3) means that portion of the "waters of the United States" where specific disposal activities are permitted and consist of a bottom surface area and any overlying volume of water. In the case of wetlands on which surface water is not present, the disposal site consists of the wetland surface area.

**Disposal site designation study** (Section 228.2) means the collection, analysis and interpretation of all available pertinent data and information on a proposed disposal site prior to use, including but not limited to, that from baseline surveys, special purpose surveys of other Federal agencies, public data archives, and social and economic studies and records of areas which would be affected by use of the proposed site.

**Disposal site evaluation study** (Section 228.2) means the collection, analysis, and interpretation of all pertinent information available concerning an existing disposal site, including but not limited to, data and information from trend assessment surveys, monitoring surveys, special purpose surveys of other Federal agencies, public data archives, and social and economic studies and records of affected areas.

**Disposal system** (Section 191.12) means any combination of engineered and natural barriers that isolate spent nuclear fuel or radioactive waste after disposal.

**Disposal well** (Section 146.3) means a well used for the disposal of waste into a subsurface stratum.

**Disposal well** (Section 147.2902) a well used for the disposal of waste into a subsurface stratum.

**Disposer of PCB waste** (Section 761.13) as the term is used in subparts J and K of this part, means any person who owns or operates a facility approved by EPA for the disposal of PCB waste which is regulated for disposal under the requirements of subpart D of this part.

**Dispute** (Section 791.3) refers to a present controversy between parties subject to a test rule over the amount or method of reimbursement for the cost of developing health and environmental data on the test chemical.

**Distance piece** (Section 65.2) means an open or enclosed casing through which the piston rod travels, separating the compressor cylinder from the crankcase.

**Distillate receiver** (Section 264.1031) means a container or tank used to receive and collect liquid material (condensed) from the overhead condenser of a distillation unit and from which the condensed liquid is pumped to larger storage tanks or other process units.

**Distillation operation** (Section 264.1031) means an operation, either batch or continuous, separating one or more feed stream(s) into two or more exit streams, each exit stream having component concentrations different from those in the feed stream(s). The separation is achieved by the redistribution of the components between the liquid and vapor phase as they approach equilibrium within the distillation unit.

**Distinct painting history** (Section 745.223) means the application history, as indicated by its visual appearance or a record of application, over time, of paint or other surface coatings to a component or room.

**Distribute in commerce** (Section 205.2) means sell in, offer for sale in, or introduce or deliver for introduction into, commerce.

**Distribute in commerce** (Section 720.3) means to sell in commerce, to introduce or deliver for introduction into commerce, or to hold after introduction into commerce.

**Distribute in commerce** (Section 763.163) has the same meaning as in section 3 of the Act (TSCA), but the term does not include actions taken with respect to an asbestos-containing product (to sell, resale, deliver, or hold) in connection with the end use of the product by persons who are users (persons who use the product for its intended purpose after it is manufactured or processed). The term also does not include distribution by manufacturers, importers, and processors, and other persons solely for purposes of disposal of an asbestos-containing product.

**Distribute in commerce and distribution in commerce** (section 710.2) when used to describe an action taken with respect to a chemical substance or mixture or article containing a substance or mixture, mean to sell or the sale of, the substance, mixture, or article in commerce; to introduce or deliver for introduction into commerce, or the introduction or delivery for introduction into commerce of, the substance, mixture, or article; or to hold, or the holding of, the substance, mixture, or article after its introduction into commerce.

**Distribute in commerce and Distribution in Commerce** (Section 761.13) when used to describe an action taken with respect to a chemical substance, mixture, or article containing a substance or mixture means to sell, or the sale of, the substance, mixture, or article in commerce; to introduce or deliver for introduction into commerce, or the introduction or delivery for introduction into commerce of the substance, mixture, or article; or to hold or the holding of, the substance, mixture, or article after its introduction into commerce.

**Distribute or sell and other grammatical variations of the term such as "distributed or sold" and "distribution or sale,"** (Section 152.3) means the acts of distributing, selling, offering for sale, holding for sale, shipping, holding for shipment, delivering for shipment, or receiving and (having so received) delivering or offering to deliver, or releasing for shipment to any person in any State.

**Distributor** (Section 80.2) means any person who transports or stores or causes the transportation or storage of gasoline or diesel fuel at any point between any gasoline or diesel fuel refinery or importer's facility and any retail outlet or wholesale purchaser-consumer's facility.

**Distributor of methyl bromide** (Section 82.3) means the person directly selling a class I, Group VI controlled substance to an applicator.

**Diurnal breathing losses** (Section 86.082-2) means evaporative emissions as a result of the daily range in temperature.

**Diurnal breathing losses** (Section 86.096-2) means diurnal emissions.

**Diurnal breathing losses** (Section 86.1803-01) means diurnal emissions.

**Diurnal emissions** (Section 86.096-2) means evaporative emissions resulting from the daily cycling of ambient temperatures.

**Diurnal emissions** (Section 86.1803-01) means evaporative emissions resulting from the daily cycling of ambient temperatures.

**DMR** (Section 122.2) means "Discharge Monitoring Report."

**Documented methodologies** (Section 745.223) are methods or protocols used to sample for the presence of lead in paint, dust, and soil.

**DOE** (Section 197.2) means the Department of Energy.

**Do-it-yourselfer used oil collection center** (Section 279.1) means any site or facility that accepts/aggregates and stores used oil collected only from household do-it-yourselfers.

**Domestic** (Section 704.3) means within the geographical boundaries of the 50 United States, including the District of Columbia, the Commonwealth of Puerto Rico, the Virgin Islands, Guam, American Samoa, the Northern Mariana Islands, and any other territory or possession of the United States.

**Domestic or other non-distribution system plumbing problem** (Section 141.2) means a coliform contamination problem in a public water system with more than one service connection that is limited to the specific service connection from which the coliform-positive sample was taken.

**Domestic septage** (Section 272.2) is either liquid or solid material removed from a septic tank, cesspool, portable toilet, Type III marine sanitation device, or similar treatment works that receives only domestic sewage. Domestic septage does not include liquid or solid material removed from a septic tank, cesspool, or similar treatment works that receives either commercial wastewater or industrial wastewater and does not include grease removed from a grease trap at a restaurant.

**Domestic waste** (Section 435.11) means materials discharged from sinks, showers, laundries, safety showers, eye-wash stations, hand-wash stations, fish cleaning stations, and galleys located within facilities subject to this Subpart.

**Domestic waste** (Section 435.41) means the materials discharged from sinks, showers, laundries, safety showers, eye-wash stations, hand-wash stations, fish cleaning stations, and galleys located within facilities subject to this Subpart.

**Dose equivalent** (Section 141.2) means the product of the absorbed dose from ionizing radiation and such factors as account for differences in biological effectiveness due to the type of radiation and its distribution in the body as specified by the International Commission on Radiological Units and Measurements (ICRU).

**Dose equivalent** (Section 190.02) means the product of absorbed dose and appropriate factors to account for differences in biological effectiveness due to the quality of radiation and its spatial distribution in the body. The unit of dose equivalent is the "rem." (One millirem (mrem) = 0.001 rem.)

**Dose equivalent** (Section 191.12) means the product of absorbed dose and appropriate factors to account for differences in biological effectiveness due to the quality of radiation and its spatial distribution in the body; the unit of dose equivalent is the "rem" ("sievert" in SI units).

**DOT** (Section 68.3) means the United States Department of Transportation.

**DOT** (Section 761.13) means the United States Department of Transportation.

**DOT** (Section 93.101) means the United States Department of Transportation.

**Double block and bleed system** (Section 264.1031) means two block valves connected in series with a bleed valve or line that can vent the line between the two block valves.

**Double block and bleed system** (Section 65.2) means two block valves connected in series with a bleed valve or line that can vent the line between the two block valves.

**Double wash/rinse** (Section 761.123) means a minimum requirement to cleanse solid surfaces (both impervious and nonimpervious) two times with an appropriate solvent or other material in which PCBs are at least 5 percent soluble (by weight). A volume of PCB-free fluid sufficient to cover the contaminated surface completely must be used in each wash/rinse. The wash/rinse requirement does not mean the mere spreading of solvent or other fluid over the surface, nor does the requirement mean a once-over wipe with a soaked cloth. Precautions must be taken to contain any runoff resulting from the cleansing and to dispose properly of wastes generated during the cleansing.

**Draft Acid Rain permit or draft permit** (Section 72.2) means the version of the Acid Rain permit, or the Acid Rain portion of an operating permit, that a permitting authority offers for public comment.

**Draft permit** (Section 122.2) means a document prepared under Sec. 124.6 indicating the Director's tentative decision to issue or deny, modify, revoke and reissue, terminate, or reissue a "permit." A notice of intent to terminate a permit, and a notice of intent to deny a permit, as discussed in Sec. 124.5, are types of "draft permits." A denial of a request for modification, revocation and reissuance, or termination, as discussed in Sec. 124.5, is not a "draft permit." A "proposed permit" is not a "draft permit."

**Draft permit** (Section 124.2) means a document prepared under Sec. 124.6 indicating the Director's tentative decision to issue or deny, modify, revoke and reissue, terminate, or reissue a "permit." A notice of intent to terminate a permit and a notice of intent to deny a permit as discussed in Sec. 124.5, are types of "draft permits." A denial of a request for modification, revocation and reissuance or termination, as discussed in Sec. 124.5, is not a "draft permit." A "proposal permit" is not a "draft permit."

**Draft permit** (Section 144.3) means a document prepared under Sec. 124.6 indicating the Director's tentative decision to issue or deny, modify, revoke and reissue, terminate, or reissue a "permit." A notice of intent to terminate a permit, and a notice of intent to deny a permit, as discussed in Sec. 124.5 are types of "draft permits." A denial of a request for modification, revocation and reissuance, or termination, as discussed in Sec. 124.5 is not a "draft permit."

**Draft permit** (Section 270.2) means a document prepared under Sec. 124.6 indicating the Director's tentative decision to issue or deny, modify, revoke and reissue, terminate, or reissue a permit. A notice of intent to terminate a permit, and a notice of intent to deny a permit, as discussed in Sec. 124.5, are types of draft permits. A denial of a request for modification, revocation and reissuance, or termination, as discussed in Sec. 124.5 is not a "draft permit." A proposed permit is not a draft permit.

**Draft permit** (Section 70.2) means the version of a permit for which the permitting authority offers public participation under Sec. 70.7(h) or affected State review under Sec. 70.8 of this part.

**Draft permit** (Section 71.2) means the version of a permit for which the permitting authority offers public participation under Sec. 71.7 or Sec. 71.11 and affected State review under Sec. 71.8.

**Drawing** (Section 467.02) is the process of pulling metal through a die or succession of dies to reduce the metal's diameter or alter its shape.

**Drawing** (Section 468.02) shall mean pulling the workpiece through a die or succession of dies to reduce the diameter or alter its shape.

**Drawing** (Section 471.02) is the process of pulling a metal through a die or succession of dies to reduce the metal's diameter or alter its cross-sectional shape.

**DRE** (Section 62.1201) means destruction and removal efficiency.

**Dredged material** (Section 232.2) means material that is excavated or dredged from waters of the United States.

**Dredged Material Permit** (Section 220.2) means a permit issued by the Corps of Engineers under section 103 of the Act (see 33 CFR 209.120) and any Federal projects reviewed under section 103(e) of the Act (see 33 CFR 209.145).

**Dresser** (Section 94.2) means any entity that modifies a land-based engine for use in a marine vessel, in compliance with the provisions of Sec. 94.907. This means that dressers may not modify the engine in a way that would affect emissions.

**Dried fruit** (Section 407.61) shall mean the processing of various fruits into the following products styles: Air, vacuum, and freeze dried, pitted and unpitted, blanched and unblanched, whole, halves, slices and other similar styles of apples, apricots, figs, peaches, pears, prunes, canned extracted prune juice and pulp from rehydrated and cooked dehydrated prunes; but not including dates or raisins.

**Drill cuttings** (Section 435.11) means the particles generated by drilling into subsurface geologic formations and carried out from the wellbore with the drilling fluid. Examples of drill cuttings include small pieces of rock varying in size and texture from fine silt to gravel. Drill cuttings are generally generated from solids control equipment and settle out and accumulate in quiescent areas in the solids control equipment or other equipment processing drilling fluid (i.e., accumulated solids).

**Drill cuttings** (Section 435.41) means the particles generated by drilling into subsurface geologic formations and carried out from the wellbore with the drilling fluid. Examples of drill cuttings include small pieces of rock varying in size and texture from fine silt to gravel. Drill cuttings are generally generated from solids control equipment and settle out and accumulate in quiescent areas in the solids control equipment or other equipment processing drilling fluid (i.e., accumulated solids).

**Drilling fluid** (Section 435.11) means the circulating fluid (mud) used in the rotary drilling of wells to clean and condition the hole and to counterbalance formation pressure. Classes of drilling fluids are:

**Drilling fluid** (Section 435.41) means the circulating fluid (mud) used in the rotary drilling of wells to clean and condition the hole and to counterbalance formation pressure. Classes of drilling fluids are: (1) Water-based drilling fluid means the continuous phase and suspending medium for solids is a water-miscible fluid, regardless of the presence of oil. (2) Non-aqueous drilling fluid means the continuous phase and suspending medium for solids is a water-immiscible fluid, such as oleaginous materials (e.g., mineral oil, enhanced mineral oil, paraffinic oil, C16-C18 internal olefins, and C8-C16 fatty acid/2-ethylhexyl esters).

**Drilling mud** (Section 144.3) means a heavy suspension used in drilling an "injection well," introduced down the drill pipe and through the drill bit.

**Drinking water supply as defined by section 101(7) of CERCLA** (Section 300.5) means any raw or finished water source that is or may be used by a public water system (as defined in the Safe Drinking Water Act (42 U.S.C. 300 et seq.) or as drinking water by one or more individuals.

**Drip pad** (Section 260.10) is an engineered structure consisting of a curbed, free-draining base, constructed of non-earthen materials and designed to convey preservative kick-back or drippage from treated wood, precipitation, and surface water run-on to an associated collection system at wood preserving plants.

**Dripline** (Section 745.63) means the area within 3 feet surrounding the perimeter of a building.

**Drive System** (Section 600.022-85) is determined by the number and location of drive axles (e.g., front wheel drive, rear wheel drive, four wheel drive) and any other feature of the drive system if the Administrator determines that such other features may result in a fuel economy difference.

**Drive train configuration** (Section 86.082-2) means a unique combination of engine code, transmission configuration, and axle ratio.

**Drive train configuration** (Section 86.1803-01) means a unique combination of engine code, transmission configuration, and axle ratio.

**Drums** (Section 442.2) mean metal or plastic cylindrical containers with either an open-head or a tight-head (also known as bung-type top) used to hold liquid, solid, or gaseous commodities or cargos which are in direct contact with the container interior. Drums typically range in capacity from 30 to 55 gallons.

**Dry** (Section 437.2) means not producing a wastewater.

**Dry beans** (Section 407.71) shall mean the production of canned pinto, kidney, navy, great northern, red, pink or related type, with and without formulated sauces, meats and gravies.

**Dry bottom** (Section 76.2) means the boiler has a furnace bottom temperature below the ash melting point and the bottom ash is removed as a solid.

**Dry drill cuttings** (Section 435.11) means the residue remaining in the retort vessel after completing the retort procedure specified in appendix 7 of subpart A of this part.

**Dry drill cuttings** (Section 435.41) means the residue remaining in the retort vessel after completing the retort procedure specified in Appendix 7 of subpart A of this part.

**Dry lot** (Section 412.21) shall mean a confinement facility for growing ducks in confinement with a dry litter floor cover and no access to swimming areas.

**Dry weight** (Section 761.13) means the weight of the sample, excluding the weight of the water in the sample. Prior to chemical analysis the water may be removed by any reproducible method that is applicable to measuring PCBs in the sample matrix at the concentration of concern, such as air drying at ambient temperature, filtration, decantation, heating at low temperature followed by cooling in the presence of a desiccant, or other processes or combinations of processes which would remove water but not remove PCBs from the sample. Analytical procedures which calculate the dry weight concentration by adjusting for moisture content may also be used.

**Drywell** (Section 144.3) means a well, other than an improved sinkhole or subsurface fluid distribution system, completed above the water table so that its bottom and sides are typically dry except when receiving fluids.

**Drywell** (Section 146.3) means a well, other than an improved sinkhole or subsurface fluid distribution system, completed above the water table so that its bottom and sides are typically dry except when receiving fluids.

**Dual fuel vehicle (or engine)** (Section 86.090-2) means any motor vehicle (or motor vehicle engine) engineered and designed to be operated on two different fuels, but not on a mixture of fuels.

**Dual Fuel Vehicle (or Engine)** (Section 88.102-94) means any motor vehicle (or motor vehicle engine) engineered and designed to be operated on two different fuels, but not on a mixture of the fuels.

**Dual fuel vehicle** (Section 86.1803-01) means any motor vehicle engineered and designed to be operated on two different fuels, but not on a mixture of the fuels.

**Dual-fuel reciprocating engine unit** (Section 72.2) means an internal combustion engine that combusts any combination of natural gas and diesel fuel and that thereby generates electricity through the operation of pistons, rather than by heating steam or water.

**Ductile Iron** (Section 464.31) A cast iron that has been treated while molten with a master alloy containing an element such as magnesium or cerium to induce the formation of free graphite as nodules or spherules, which imparts a measurable degree of ductility to the cast metal.

**Ductwork** (Section 65.2) means a conveyance system such as those commonly used for heating and ventilation systems. It is often made of sheet metal and often has sections connected by screws or crimping. Hard-piping is not ductwork.

**Dumping** (Section 220.2) means a disposition of material: Provided, That it does not mean a disposition of any effluent from any outfall structure to the extent that such disposition is regulated under the provisions of the FWPCA, under the provisions of section 13 of the River and Harbor Act of 1899, as amended (33 U.S.C. 407), or under the provisions of the Atomic Energy Act of 1954, as amended (42 U.S.C. 2011), nor does it mean a routine discharge of effluent incidental to the propulsion of, or operation of motor-driven equipment on, vessels: Provided further, That it does not mean the construction of any fixed structure or artificial island nor the intentional placement of any device in ocean waters or on or in the submerged land beneath such waters, for a purpose other than disposal, when such construction or such placement is otherwise regulated by Federal or State law or occurs pursuant to an authorized Federal or State program; And provided further, That it does not include the deposit of oyster shells, or other materials when such deposit is made for the purpose of developing, maintaining, or harvesting fisheries resources and is otherwise regulated by Federal or State law or occurs pursuant to an authorized Federal or State program.

**Duplication** (Section 1601.3) refers to the process of making a copy of a document necessary to fulfill a FOIA request. Such copies can take the form of, among other things, paper copy, microform, audio-visual materials, or machine-readable documentation. The copies provided shall be in a form that is reasonably usable by requesters.

**Durability data vehicle** (Section 86.1803-01) means a vehicle used to generate durability data as required in this subpart.

**Durability fleet** (Section 610.11) means a fleet of automobiles operated for mileage accumulation used to assess deterioration effects associated with the retrofit device.

**Durability group** (Section 86.1803-01) means the basic classification unit of a manufacturer's product line used for the purpose of selecting a vehicle configuration to demonstrate durability and predict deterioration in accordance with Sec. 86.1822-01.

**Durability useful life** (Section 86.094-2) means the highest useful life mileage out of the set of all useful life mileages that apply to a given vehicle. The durability useful life determines the duration of service accumulation on a durability data vehicle. The determination of durability useful life shall reflect any alternative useful life mileages approved by the Administrator under Sec. 86.094-21(f). The determination of durability useful life shall exclude any standard and related useful life mileage for which the manufacturer has obtained a waiver of emission data submission requirements under Sec. 86.094-23© Element of design means any control system (i.e., computer software, electronic control system, emission control system, computer logic), and/or control system calibrations, and/or the results of systems interaction, and/or hardware items on a motor vehicle or motor vehicle engine.

**Durability useful life** (Section 86.1803-01) means the highest useful life mileage out of the set of all useful life mileages that apply to a given vehicle. The durability useful life determines the duration of service accumulation on a durability data vehicle. The determination of durability useful life shall reflect any light-duty truck or complete heavy-duty vehicle alternative useful life periods approved by the Administrator under Sec. 86.1805-01©. The determination of durability useful life shall exclude any standard and related useful life mileage for which the manufacturer has obtained a waiver of emission data submission requirements under Sec. 86.1829-01.

**Dye penetrant testing** (Section 471.02) is a nondestructive method for finding discontinuities that are open to the surface of the metal. A dye is applied to the surface of metal and the excess is rinsed off. Dye that penetrates surface discontinuities will not be rinsed away thus marking these discontinuities.

**Dynamometer-idle for automatic transmission code heavy-duty engines** (Section 86.082-2) means the manufacturer's recommended engine speed without a transmission that simulates the recommended engine speed with a transmission and with the transmission in neutral.

# E

**Ear Insert Device** (Section 211.203) A hearing protective device that is designed to be inserted into the ear canal, and to be held in place principally by virtue of its fit inside the ear canal.

**Ear Muff Device** (Section 211.203) A hearing protective device that consists of two acoustic enclosures which fit over the ears and which are held in place by a spring-like headband to which the enclosures are attached.

**Early entry** (Section 170.3) means entry by a worker into a treated area on the agricultural establishment after a pesticide application is complete, but before any restricted-entry interval for the pesticide has expired.

**EC50** (Section 132.2) is a statistically or graphically estimated concentration that is expected to cause one or more specified effects in 50 percent of a group of organisms under specified conditions.

**Economic poison** (Section 163.2) shall have the same meaning as it has under the Federal Insecticide, Fungicide, and Rodenticide Act (7 U.S.C. 135-135k) and the regulations issued thereunder.

**Economizer** (Section 76.2) means the lowest temperature heat exchange section of a utility boiler where boiler feed water is heated by the flue gas.

**Educational institution** (Section 1601.3) refers to a preschool, a public or private elementary or high school, an institution of undergraduate higher education, an institution of graduate higher education, an institution of professional education, and an institution of vocational education, which operates a program of scholarly research.

**Effective concentration** (Section 58.1) pertains to testing an open path analyzer with a high-concentration calibration or audit standard gas contained in a short test cell inserted into the optical measurement beam of the instrument. Effective concentration is the equivalent ambient-level concentration that would produce the same spectral absorbance over the actual atmospheric monitoring path length as produced by the high-concentration gas in the short test cell. Quantitatively, effective concentration is equal to the actual concentration of the gas standard in the test cell multiplied by the ratio of the path length of the test cell to the actual atmospheric monitoring path length.

**Effective corrosion inhibitor residual** (Section 141.2) for the purpose of subpart I of this part only, means a concentration sufficient to form a passivating film on the interior walls of a pipe.

**Effective date** (Section 61.02) is the date of promulgation in the Federal Register of an applicable standard or other regulation under this part.

**Effective date** (Section 63.2) means: (1) With regard to an emission standard established under this part, the date of promulgation in the Federal Register of such standard; or (2) With regard to an alternative emission limitation or equivalent emission limitation determined by the Administrator (or a State with an approved permit program), the date that the alternative emission limitation or equivalent emission limitation becomes effective according to the provisions of this part.

**Effective date of a UIC program** (Section 146.3) means the date that a State UIC program is approved or established by the Administrator.

**Effective date of an NSO** (Section 57.103) means the effective date listed in the Federal Register publication of EPA's issuance or approval of an NSO.

**Effective dose** (Section 191.12) means the sum over specified tissues of the products of the dose equivalent received following an exposure of, or an intake of radionuclides into, specified tissues of the body, multiplied by appropriate weighting factors. This allows the various tissue-specific health risks to be summed into an overall health risk. The method used to calculate effective dose is described in appendix B of this part.

**Effective dose equivalent** (Section 197.2) means the sum of the products of the dose equivalent received by specified tissues following an exposure of, or an intake of radionuclides into, specified tissues of the body, multiplied by appropriate weighting factors.

**Effluent** (Section 232.2) means dredged material or fill material, including return flow from confined sites.

**Effluent concentrations consistently achievable through proper operation and maintenance** (Section 133.101) (1) For a given pollutant parameter, the 95$^{th}$ percentile value for the 30-day average effluent quality achieved by a treatment works in a period of at least two years, excluding values attributable to upsets, bypasses, operational errors, or other unusual conditions, and (2) a 7-day average value equal to 1.5 times the value derived under paragraph (f)(1) of this section.

**Effluent limitation** (Section 108.2) means any effluent limitation which is established as a condition of a permit issued or proposed to be issued by a State or by the Environmental Protection Agency pursuant to section 402 of the Act; any toxic or pretreatment effluent standard established under section 307 of the Act; any standard of performance established under section 306 of the Act; and any effluent limitation established under section 302, section 316, or section 318 of the Act.

**Effluent limitation** (Section 122.2) means any restriction imposed by the Director on quantities, discharge rates, and concentrations of "pollutants" which are "discharged" from "point sources" into "waters of the United States," the waters of the "contiguous zone," or the ocean.

**Effluent limitation** (Section 401.11) means any restriction established by the Administrator on quantities, rates, and concentrations of chemical, physical, biological and other constituents which are discharged from point sources, other than new sources, into navigable waters, the waters of the contiguous zone or the ocean.

**Effluent limitations guidelines** (Section 122.2) means a regulation published by the Administrator under section 304(b) of CWA to adopt or revise "effluent limitations."

**Effluent limitations guidelines** (Section 401.11) means any effluent limitations guidelines issued by the Administrator pursuant to section 304(b) of the Act.

**Effluent standard** (Section 104.2) means any effluent standard or limitation, which may include a prohibition of any discharge, established or proposed to be established for any toxic pollutant under section 307(a) of the Act.

**Effluent standard** (Section 129.2) means, for purposes of section 307, the equivalent of effluent limitation as that term is defined in section 502(11) of the Act with the exception that it does not include a schedule of compliance.

**Elastomeric coating** (Section 59.101) means a coating designed for application over flexible parts, such as elastomeric bumpers.

**Electric arc furnace steelmaking** (Section 420.41) means the production of steel principally from steel scrap and fluxes in refractory lined furnaces by passing an electric current through the scrap or steel bath.

**Electric Traction Motor** (Section 600.022-85) means an electrically powered motor which provides tractive energy to the wheels of a vehicle.

**Electric utility steam generating unit** (Section 60.2) means any steam electric generating unit that is constructed for the purpose of supplying more than one-third of its potential electric output capacity and more than 25 MW electrical output to any utility power distribution system for sale. Any steam supplied to a steam distribution system for the purpose of providing steam to a steam-electric generator that would produce electrical energy for sale is also considered in determining the electrical energy output capacity of the affected facility.

**Electric vehicle** (Section 86.1702-99) means any vehicle which operates solely by use of a battery or battery pack. This definition also includes vehicles which are powered mainly through the use of an electric battery or battery pack, but which use a flywheel that stores energy produced by the electric motor or through regenerative braking to assist in vehicle operation.

**Electrical Charging System** (Section 600.022-85) means a device to convert 60Hz alternating electric current, as commonly available in residential electric service in the United States, to a proper form for recharging the energy storage device.

**Electrical equipment** (Section 280.12) means underground equipment that contains dielectric fluid that is necessary for the operation of equipment such as transformers and buried electrical cable.

**Electricity for sale under firm contract to the grid** (Section 97.2) means electricity for sale where the capacity involved is intended to be available at all times during the period covered by a guaranteed commitment to deliver, even under adverse conditions.

**Electrocoating** (Section 471.02) is the electrodeposition of a metallic or nonmetallic coating onto the surface of a workpiece.

**Electrolyses plating** (Section 413.71) shall mean the deposition of conductive material from an autocatalytic plating solution without application of electrical current.

**Electronic crystals** (Section 469.22) means crystals or crystalline material which because of their unique structural and electronic properties are used in electronic devices. Examples of these crystals are crystals comprised of quartz, ceramic, silicon, gallium arsenide, and idium arsenide.

**Electroplating process wastewater** (Section 413.02) shall mean process wastewater generated in operations which are subject to regulation under any of subparts A through H of this part.

**Element of design** (Section 86.1702-99) means any control system (i.e., computer software, electronic control system, emission control system, computer logic), and/or control system calibrations and/or the results of systems interaction, and/or hardware items on a motor vehicle or motor vehicle engine.

**Element of design** (Section 86.1803-01) means any control system (i.e., computer software, electronic control system, emission control system, computer logic), and/or control system calibrations, and/or the results of systems interaction, and/ or hardware items on a motor vehicle or motor vehicle engine.

**Elemental chlorine-free (ECF)** (Section 430.0)Any process for bleaching pulps in the absence of elemental chlorine and hypochlorite that uses exclusively chlorine dioxide as the only chlorine-containing bleaching agent.

**Elementary neutralization unit** (Section 260.10) means a device which: (1) Is used for neutralizing wastes that are hazardous only because they exhibit the corrosivity characteristic defined in Sec. 261.22 of this chapter, or they are listed in subpart D of part 261 of the chapter only for this reason; and (2) Meets the definition of tank, tank system, container, transport vehicle, or vessel in Sec. 260.10 of this chapter.

**Elementary neutralization unit** (Section 270.2) means a device which: (a) Is used for neutralizing wastes only because they exhibit the corrosivity characteristic defined in Sec. 261.22 of this chapter, or are listed in subpart D of part 261 of this chapter only for this reason; and (b) Meets the definition of tank, tank system, container, transport vehicle, or vessel in Sec. 260.10 of this chapter.

**Elevated blood lead level (EBL)** (Section 745.223) means an excessive absorption of lead that is a confirmed concentration of lead in whole blood of 20 [mu] g/dl (micrograms of lead per deciliter of whole blood) for a single venous test or of 15-19 [mu] g/dl in two consecutive tests taken 3 to 4 months apart.

**Eligible claim** (Section 307.14) means any claim that has satisfied the requirements set forth in Sec. 307.21(b).

**Eligible Indian Tribe** (Section 144.3) is a Tribe that meets the statutory requirements established at 42 U.S.C. 300j-11(b)(1).

**Eligible Indian tribe** (Section 72.2) means any eligible Indian tribe as defined in part 71 of this chapter.

**Eligible Indian Tribe or eligible Tribe** (Section 71.2) means a Tribe that has been determined by EPA to meet the criteria for being treated in the same manner as a State, pursuant to the regulations implementing section 301(d)(2) of the Act.

**Eligible production or U.S. production** (Section 90.3) means Phase 2 engines produced for purposes of being used in the United States, and includes any engine exported and subsequently imported in a new piece of equipment, but excludes any engine introduced into commerce, by itself or in a piece of equipment, for use in a state that has established its own emission requirements applicable to such engines pursuant to a waiver granted by EPA under section 209(e) of the Clean Air Act.

**ELWK (equipment live weight killed)** (Section 432.41) shall mean the total weight of the total number of animals slaughtered at locations other than the slaughterhouse or packinghouse, which animals provide hides, blood, viscera or renderable materials for processing at that slaughterhouse, in addition to those derived from animals slaughtered on-site.

**ELWK (equivalent live weight killed)** (Section 432.11) shall mean the total weight of the total number of animals slaughtered at locations other than the slaughterhouse or packinghouse, which animals provide hides, blood, viscera or renderable materials for processing at that slaughterhouse, in addition to those derived from animals slaughtered on site.

**ELWK (equivalent live weight killed)** (Section 432.21) shall mean the total weight of the total number of animals slaughtered at locations other than the slaughterhouse or packinghouse, which animals provide hides, blood, viscera or renderable materials for processing at that slaughterhouse, in addition to those derived from animals slaughtered on site.

**Emergency** (Section 93.152) means a situation where extremely quick action on the part of the Federal agencies involved is needed and where the timing of such Federal activities makes it impractical to meet the requirements of this subpart, such as natural disasters like hurricanes or earthquakes, civil disturbances such as terrorist acts and military mobilizations.

**Emergency condition** (Section 166.3) means an urgent, non-routine situation that requires the use of a pesticide(s) and shall be deemed to exist when:

**Emergency exemption** (Section 176.3) means a specific, quarantine, or public health exemption authorized under FIFRA section 18 and the regulations at Sec. Sec. 166.20 through 166.35 of this chapter.

**Emergency fuel** (Section 72.2) means either: (1) (Section 72.2) (Section 72.2) (Section 72.2) (2) For purposes of the requirement for stack testing for an excepted monitoring system under appendix E of part 75 of this chapter, the fuel identified in a federally-enforceable permit for a plant and identified by the designated representative in the unit's monitoring plan as the fuel which is combusted only during emergencies where the primary fuel is not available.

**Emergency fuel** (Section 72.2) means either: (1) For purposes of the requirements for a fuel flowmeter used in an excepted monitoring system under appendix D or E of part 75 of this chapter, the fuel identified by the designated representative in the unit's monitoring plan as the fuel which is combusted only during emergencies where the primary fuel is not available; or (2) For purposes of the requirement for stack testing for an excepted monitoring system under appendix E of part 75 of this chapter, the fuel identified in the State, local, or Federal permit for a plant and is identified by the designated representative in the unit's monitoring plan as the fuel which is combusted only during emergencies where the primary fuel is not available, as established in a petition under Sec. 75.66 of this chapter.

**Emergency permit** (Section 144.3) means a UIC "permit" issued in accordance with Sec. 144.34.

**Emergency permit** (Section 270.2) means a RCRA permit issued in accordance with Sec. 270.61.

**Emergency Planning and Community Right-to-Know Act of 1986** (Section 310.3) means Title III—Emergency Planning and Community Right-to-Know Act of the Superfund Amendments and Reauthorization Act of 1986 (EPCRA) (Pub. L. 99-499, 42 U.S.C. 11000-11050).

**Emergency renovation operations** (Section 745.83) means renovation activities, such as operations necessitated by non-routine failures of equipment, that were not planned but result from a sudden, unexpected event that, if not immediately attended to, presents a safety or public health hazard, or threatens equipment and/or property with significant damage.

**Emergency vehicle** (Section 88.302-94) means any vehicle that is legally authorized by a governmental authority to exceed the speed limit to transport people and equipment to and from situations in which speed is required to save lives or property, such as a rescue vehicle, fire truck, or ambulance.

**Emission** (Section 240.101) means gas-borne pollutants released to the atmosphere.

**Emission control system** (Section 86.1803-01) is a unique group of emission control devices, auxiliary emission control devices, engine modifications and strategies, and other elements of design designated by the Administrator used to control exhaust emissions of a vehicle.

**Emission control system** (Section 89.2) means any device, system, or element of design which controls or reduces the emission of substances from an engine.

**Emission control system** (Section 90.3) means any device, system, or element of design which controls or reduces the emission of substances from an engine.

**Emission control system** (Section 91.3) means any device, system, or element of design which controls or reduces the emission of substances from an engine.

**Emission control system** (Section 92.2) means those devices, systems or elements of design which control or reduce the emission of substances from an engine. This includes, but is not limited to, mechanical and electronic components and controls, and computer software.

**Emission control system** (Section 94.2) means those devices, systems or elements of design which control or reduce the emission of substances from an engine. This includes, but is not limited to, mechanical and electronic components and controls, and computer software.

**Emission credits** (Section 86.090-2) mean the amount of emission reductions or exceedances, by a heavy-duty engine family, below or above the emission standard, respectively. Emission credits below the standard are considered as "positive credits," while emission credits above the standard are considered as "negative credits." In addition, "projected credits" refer to emission credits based on the projected U.S. production volume of the engine family. "Reserved credits" are emission credits generated within a model year waiting to be reported to EPA at the end of the model year. "Actual credits" refer to emission credits based on actual U.S. production volumes as contained in the end-of-year reports submitted to EPA. Some or all of these credits may be revoked if EPA review of the end of year reports or any subsequent audit actions uncover problems or errors.

**Emission credits** (Section 86.1803-01) mean the amount of emission reductions or exceedances, by a complete heavy-duty vehicle test group, below or above the emission standard, respectively. Emission credits below the standard are considered as "positive credits," while emission credits above the standard are considered as "negative credits." In addition, "projected credits" refer to emission credits based on the projected U.S. production volume of the test group. "Reserved credits" are emission credits generated within a model year waiting to be reported to EPA at the end of the model year. "Actual credits" refer to emission credits based on actual U.S. production volumes as contained in the end-of-year reports submitted to EPA. Some or all of these credits may be revoked if EPA review of the end of year reports or any subsequent audit actions uncover problems or errors.

**Emission credits** (Section 92.2) represent the amount of emission reduction or exceedance, by a locomotive engine family, below or above the emission standard, respectively. Emission reductions below the standard are considered as "positive credits," while emission exceedances above the standard are considered as "negative credits." In addition, "projected credits" refer to emission credits based on the projected applicable production/sales volume of the engine family. "Reserved credits" are emission credits generated within a calendar year waiting to be reported to EPA at the end of the calendar year. "Actual credits" refer to emission credits based on actual applicable production/sales volume as contained in the end-of-year reports submitted to EPA.

**Emission credits** (Section 94.2) means the amount of emission reduction or exceedance, by an engine family, below or above the emission standard, respectively, as calculated under subpart D of this part. Emission reductions below the standard are considered as "positive credits," while emission exceedances above the standard are considered as "negative credits." In addition, "projected credits" refer to emission credits based on the projected applicable production sales volume of the engine family. "Reserved credits" are emission credits generated within a calendar year waiting to be reported to EPA at the end of the calendar year. "Actual credits" refer to emission credits based on actual applicable production/sales volume as contained in the end-of-year reports submitted to EPA.

**Emission credits or in-use credits** (Section 91.1302) represent the amount of emission reduction or exceedance, for each regulated pollutant, by a marine engine family below or above, respectively, the applicable certification family emission limit (FEL) to which the engine family is certified. Emission reductions below the FEL are considered "positive credits," while emission exceedances above the FEL are considered "negative or required credits."

**Emission limitation and emission standard** (Section 51.100) mean a requirement established by a State, local government, or the Administrator which limits the quantity, rate, or concentration of emissions of air pollutants on a continuous basis, including any requirements which limit the level of opacity, prescribe equipment, set fuel specifications, or prescribe operation or maintenance procedures for a source to assure continuous emission reduction.

**Emission limitation or standard** (Section 64.1) means any applicable requirement that constitutes an emission limitation, emission standard, standard of performance or means of emission limitation as defined under the Act. An emission limitation or standard may be expressed in terms of the pollutant, expressed either as a specific quantity, rate or concentration of emissions (e.g., pounds of SO2 per hour, pounds of SO2 per million British thermal units of fuel input, kilograms of VOC per liter of applied coating solids, or parts per million by volume of SO2) or as the relationship of uncontrolled to controlled emissions (e.g., percentage capture and destruction efficiency of VOC or percentage reduction of SO2). An emission limitation or standard may also be expressed either as a work practice, process or control device parameter, or other form of specific design, equipment, operational, or operation and maintenance requirement. For purposes of this part, an emission limitation or standard shall not include general operation requirements that an owner or operator may be required to meet, such as requirements to obtain a permit, to operate and maintain sources in accordance with good air pollution control practices, to develop and maintain a malfunction abatement plan, to keep records, submit reports, or conduct monitoring.

**Emission measurement system** (Section 87.1) means all of the equipment necessary to transport and measure the level of emissions. This includes the sample system and the instrumentation system.

**Emission Performance Warranty** (Section 85.2102) means that warranty given pursuant to this subpart and section 207(b) of the Act.

**Emission point** (Section 65.2) means an individual process vent, storage vessel, transfer rack, wastewater stream, or equipment leak.

**Emission Related Parts** (Section 85.1402) means those parts installed for the specific purpose of controlling emissions or those components, systems, or elements of design which must function properly to assure continued emission compliance.

**Emission Related Parts** (Section 85.2102) means those parts installed for the specific purpose of controlling emissions or those components, systems, or elements of design which must function properly to assure continued vehicle emission compliance.

**Emission standard** (Section 63.2) means a national standard, limitation, prohibition, or other regulation promulgated in a subpart of this part pursuant to sections 112(d), 112(h), or 112(f) of the Act.

**Emission Vehicle Program (October, 1996)** (Section 86.1702-99) which is incorporated by reference (see Sec. 86.1); and has nationwide sales of light-duty vehicles and light light-duty trucks less than or equal to 40,000 units per model year based on the average number of vehicles sold by the manufacturer for each of the three most recent model years. For manufacturers certifying for the first time, model-year sales shall be based on projected sales.

**Emission Warranty** (Section 85.2113) means those warranties given by vehicle manufacturers pursuant to section 207 of the Act.

**Emission-Critical Parameters** (Section 85.2113) means those critical parameters and tolerances which, if equivalent from one part to another, will not cause the vehicle to exceed applicable emission standards with such parts installed.

**Emission-data engine** (Section 92.2) means an engine which is tested for purposes of emission certification or production line testing.

**Emission-data engine** (Section 94.2) means an engine which is tested for purposes of emission certification or production line testing.

**Emission-data locomotive** (Section 92.2) means a locomotive which is tested for purposes of emission certification or production line testing.

**Emission-related component** (Section 86.1803-01) means any component which can affect emissions.

**Emission-related defect** (Section 85.1902) shall mean a defect in design, materials, or workmanship in a device, system, or assembly described in the approved Application for Certification (required by 40 CFR 86.1843-01 and 86.1844-01, 40 CFR 86.098-22 and like provisions of subpart A of this part and 40 CFR part 86) which affects any parameter or specification enumerated in Appendix VIII of this part.

**Emission-related defect** (Section 90.802) means a defect in design, materials, or workmanship in a device, system, or assembly described in the approved application for certification which affects any applicable parameter or specification enumerated in 40 CFR part 85, Appendix VIII.

**Emission-related defect** (Section 92.2) means a defect in design, materials, or workmanship in a device, system, or assembly described in the approved Application for certification which affects any parameter or specification enumerated in Appendix I of this part.

**Emission-related defect** (Section 94.2) means a defect in design, materials, or workmanship in a device, system, or assembly which affects any parameter or specification enumerated in Appendix I of this part.

**Emission-related maintenance** (Section 86.1803-01) means that maintenance which does substantially affect emissions or which is likely to affect the emissions deterioration of the vehicle during normal in-use operation, even if the maintenance is performed at some time other than that which is recommended.

**Emission-related maintenance** (Section 92.2) means that maintenance which substantially affects emissions or which is likely to affect the deterioration of the locomotive or engine with respect to emissions, as described in an approved Application for certification.

**Emission-related maintenance** (Section 94.2) means that maintenance which substantially affects emissions or which is likely to affect the deterioration of the engine or vessel with respect to emissions.

**Emissions** (Section 72.2) means air pollutants exhausted from a unit or source into the atmosphere, as measured, recorded, and reported to the Administrator by the designated representative and as determined by the Administrator, in accordance with the emissions monitoring requirements of part 75 of this chapter.

**Emissions** (Section 96.2) means air pollutants exhausted from a unit or source into the atmosphere, as measured, recorded, and reported to the Administrator by the NOX authorized account representative and as determined by the Administrator in accordance with subpart H of this part.

**Emissions** (Section 97.2) means air pollutants exhausted from a unit or source into the atmosphere, as measured, recorded, and reported to the Administrator by the NOX authorized account representative and as determined by the Administrator in accordance with subpart H of this part.

**Emissions allowable under the permit** (Section 70.2) means a federally enforceable permit term or condition determined at issuance to be required by an applicable requirement that establishes an emissions limit (including a work practice standard) or a federally enforceable emissions cap that the source has assumed to avoid an applicable requirement to which the source would otherwise be subject.

**Emissions allowable under the permit** (Section 71.2) means a federally enforceable permit term or condition determined at issuance to be required by an applicable requirement that establishes an emissions limit (including a work practice standard) or a federally enforceable emissions cap that the source has assumed to avoid an applicable requirement to which the source would otherwise be subject.

**Emissions averaging** (Section 63.2) is a way to comply with the emission limitations specified in a relevant standard, whereby an affected source, if allowed under a subpart of this part, may create emission credits by reducing emissions from specific points to a level below that required by the relevant standard, and those credits are used to offset emissions from points that are not controlled to the level required by the relevant standard.

**Emissions budgets** (Section 93.152) are those portions of the applicable SIP's projected emission inventories that describe the levels of emissions (mobile, stationary, area, etc.) that provide for meeting reasonable further progress milestones, attainment, and/or maintenance for any criteria pollutant or its precursors.

**Emissions offsets** (Section 93.152) for purposes of Sec. 93.158, are emissions reductions which are quantifiable, consistent with the applicable SIP attainment and reasonable further progress demonstrations, surplus to reductions required by, and credited to, other applicable SIP provisions, enforceable at both the State and Federal levels, and permanent within the timeframe specified by the program.

**Emissions that a Federal agency has a continuing program responsibility for** (Section 93.152) means emissions that are specifically caused by an agency carrying out its authorities, and does not include emissions that occur due to subsequent activities, unless such activities are required by the Federal agency. When an agency, in performing its normal program responsibilities, takes actions itself or imposes conditions that result in air pollutant emissions by a non-Federal entity taking subsequent actions, such emissions are covered by the meaning of a continuing program responsibility.

**Emissions unit** (Section 64.1) shall have the same meaning as provided under part 70 of this chapter.

**Emissions unit** (Section 70.2) means any part or activity of a stationary source that emits or has the potential to emit any regulated air pollutant or any pollutant listed under section 112(b) of the Act. This term is not meant to alter or affect the definition of the term "unit" for purposes of title IV of the Act.

**Emissions unit** (Section 71.2) means any part or activity of a stationary source that emits or has the potential to emit any regulated air pollutant or any pollutant listed under section 112(b) of the Act. This term is not meant to alter or affect the definition of the term "unit" for purposes of title IV of the Act.

**Employee** (Section 1611.2) for the purpose of this part and part 1612 of this chapter, refers to current or former CSB Board Members or employees, including student interns, and contractors, contract employees, or consultants (and their employees). This definition does not include persons who are no longer employed by or under contract to the CSB, and who are retained or hired as expert witnesses or agree to testify about matters that do not involve their work for the CSB.

**Employee in Sec. 311.1** (Section 311.2) is defined as a compensated or non-compensated worker who is controlled directly by a State or local government, as contrasted to an independent contractor.

**Employer** (Section 721.3) means any manufacturer, importer, processor, or user of chemical substances or mixtures.

**Empty or emptying** (Section 65.2) means the removal of the stored liquid from a storage vessel. Storage vessels where stored liquid is left on the walls, as bottom clingage, or in pools due to bottom irregularities are considered empty. Lowering of the stored liquid level, so that the floating roof is resting on its legs, as necessitated by normal vessel operation (for example, when changing stored material or when transferring material out of the vessel for shipment) is not considered emptying.

**Emulsions** (Section 467.02) are stable dispersions of two immiscible liquids. In the aluminum forming category this is usually an oil and water mixture.

**Emulsions** (Section 471.02) are stable dispersions of two immiscible liquids. In the Nonferrous Metals Forming and Metal Powders Point Source category, this is usually an oil and water mixture.

**Encapsulant** (Section 745.223) means a substance that forms a barrier between lead-based paint and the environment using a liquid-applied coating (with or without reinforcement materials) or an adhesively bonded covering material.

**Encapsulation** (Section 745.223) means the application of an encapsulant.

**Encapsulation** (Section 763.83) means the treatment of ACBM with a material that surrounds or embeds asbestos fibers in an adhesive matrix to prevent the release of fibers, as the encapsulant creates a membrane over the surface (bridging encapsulant) or penetrates the material and binds its components together (penetrating encapsulant).

**Enclosed process** (Section 704.3) means a manufacturing or processing operation that is designed and operated so that there is no intentional release into the environment of any substance present in the operation. An operation with fugitive, inadvertent, or emergency pressure relief releases remains an enclosed process so long as measures are taken to prevent worker exposure to and environmental contamination from the releases.

**Enclosure** (Section 265.1081) means a structure that surrounds a tank or container, captures organic vapors emitted from the tank or container, and vents the captured vapors through a closed-vent system to a control device.

**Enclosure** (Section 745.223) means the use of rigid, durable construction materials that are mechanically fastened to the substrate in order to act as a barrier between lead-based paint and the environment.

**Enclosure** (Section 763.83) means an airtight, impermeable, permanent barrier around ACBM to prevent the release of asbestos fibers into the air.

**End of the pipe** (Section 430.0) The point at which final mill effluent is discharged to waters of the United States or introduced to a POTW.

**End use product** (Section 152.3) means a pesticide product whose labeling (1) Includes directions for use of the product (as distributed or sold, or after combination by the user with other substances) for controlling pests or defoliating, desiccating, or regulating the growth of plants, and (2) Does not state that the product may be used to manufacture or formulate other pesticide products.

**End use product** (Section 158.153) means a pesticide product whose labeling (1) Includes directions for use of the product (as distributed or sold, or after combination by the user with other substances) for controlling pests or defoliating, desiccating or regulating growth of plants, and (2) Does not state that the product may be used to manufacture or formulate other pesticide products.

**Endangered or threatened species** (Section 132.2) are those species that are listed as endangered or threatened under section 4 of the Endangered Species Act.

**Energy Average Level** (Section 201.1) means a quantity calculated by taking ten times the common logarithm of the arithmetic average of the antilogs of one-tenth of each of the levels being averaged. The levels may be of any consistent type, e.g. maximum sound levels, sound exposure levels, and day-night sound levels.

**Energy Information Administration** (Section 96.2) means the Energy Information Administration of the United States Department of Energy.

**Energy Information Administration** (Section 97.2) means the Energy Information Administration of the United States Department of Energy.

**Energy Storage Device** (Section 600.022-85) means a rechargeable means of storing tractive energy on board a vehicle such as storage batteries or a flywheel.

**Energy Summation of Levels** (Section 201.1) means a quantity calculated by taking ten times the common logarithm of the sum of the antilogs of one-tenth of each of the levels being summed. The levels may be of any consistent type, e.g., day-night sound level or equivalent sound level.

**Engine** (Section 89.2) as used in this part, refers to nonroad engine.

**Engine** (Section 90.3) as used in this part, refers to nonroad engine.

**Engine** (Section 91.3) as used in this part, refers to marine SI engine.

**Engine Code** (Section 600.022-85) means, for gasoline-fueled and diesel vehicles, a unique combination, within an engine-system combination (as defined in part 86 of this chapter), of displacement, carburetor (or fuel injection) calibration, distributor calibration, choke calibration, auxiliary emission control devices, and other engine and emission control system components specified by the Administrator. For electric vehicles, engine code means a unique combination of manufacturer, electric traction motor, motor configuration, motor controller, and energy storage device.

**Engine code** (Section 86.082-2) means a unique combination, within an engine-system combination, of displacement, carburetor (or fuel injection) calibration, choke calibration, distributor calibration, auxiliary emission control devices, and other engine and emission control system components specified by the Administrator.

**Engine code** (Section 86.1803-01) means a unique combination within a test group of displacement, fuel injection (or carburetor) calibration, choke calibration, distributor calibration, auxiliary emission control devices, and other engine and emission control system components specified by the Administrator.

**Engine Configuration** (Section 85.1402) means the set of components, tolerances, specifications, design parameters, and calibrations related to the emissions performance of the engine and specific to a subset of an engine family having a unique combination of displacement, fuel injection calibration, auxiliary emission control devices and emission control system components.

**Engine displacement** (Section 205.151) means volumetric engine capacity as defined in Sec. 205.153.

**Engine Family** (Section 85.2113) means the basic classification unit of a vehicle's product line for a single model year used for the purpose of emission-data vehicle or engine selection and as determined in accordance with 40 CFR 86.078-24.

**Engine family** (Section 86.082-2) means the basic classification unit of a manufacturer's product line used for the purpose of test fleet selection and determined in accordance with Sec. 86.082-24.

**Engine family** (Section 86.402-78) means the basic classification unit of a manufacturer's product line used for the purpose of test fleet selection and determined in accordance with Sec. 86.420. Engine-displacement-system combination means an engine family-displacement-emission control system combination. EPA Enforcement Officer means any officer or employee of the Environmental Protection Agency so designated in writing by the Administrator (or by his designee).

**Engine family** (Section 90.3) means a group of engines, as specified in Sec. 90.116.

**Engine family** (Section 91.3) means a group of engines, as specified in Sec. 91.115.

**Engine family** (Section 92.2) means a group of locomotive or locomotive engine configurations which are expected to have similar emission characteristics throughout the useful lives of the locomotives and engines (see Sec. 92.204), and which are (or were) covered (or requested to be covered) by a specific certificate of conformity.

**Engine family** (Section 94.2) means a group of engine configurations that are expected to have similar emission characteristics throughout the useful lives of the engines (see Sec. 94.204), and that are (or were) covered (or requested to be covered) by a specific certificate of conformity.

**Engine family group** (Section 86.082-2) means a combination of engine families for the purpose of determining a minimum deterioration factor under the Alternative Durability Program.

**Engine lubricating oils** (Section 247.3) means petroleum-based oils used for reducing friction in engine parts;

**Engine manufacturer** (Section 89.2) means any person engaged in the manufacturing or assembling of new nonroad engines or importing such engines for resale, or who acts for and is under the control of any such person in connection with the distribution of such engines. Engine manufacturer does not include any dealer with respect to new nonroad engines received by such person in commerce.

**Engine manufacturer** (Section 90.3) means any person engaged in the manufacturing or assembling of new nonroad engines or the importing of such engines for resale, or who acts for and is under the control of any such person in connection with the distribution of such engines. Engine manufacturer does not include any dealer with respect to new nonroad engines received by such person in commerce.

**Engine Model** (Section 87.1) means all commercial aircraft turbine engines which are of the same general series, displacement, and design characteristics and are usually approved under the same type certificate.

**Engine Rebuild** (Section 85.1402) means an activity, occurring over one or more maintenance events, involving: (1) Disassembly of the engine including the removal of the cylinder head(s); and (2) The replacement or reconditioning of more than one major cylinder component in more than half of the cylinders.

**Engine Replacement** (Section 85.1402) means the removal of an engine from the coach followed by the installation of another engine.

**Engine used in a locomotive** (Section 85.1602) means either an engine placed in the locomotive to move other equipment, freight, or passenger traffic, or an engine mounted on the locomotive to provide auxiliary power.

**Engine used in a locomotive** (Section 89.2) means either an engine placed in the locomotive to move other equipment, freight, or passenger traffic, or an engine mounted on the locomotive to provide auxiliary power.

**Engine used in a locomotive** (Section 92.2) means an engine incorporated into a locomotive or intended for incorporation into a locomotive.

**Engine warm-up cycle** (Section 86.094-2) means sufficient vehicle operation such that the coolant temperature has risen by at least 40 deg.F from engine starting and reaches a minimum temperature of 160 deg.F. Enhanced service and repair information means information which is specific for an original equipment manufacturer's brand of tools and equipment.

**Engine warm-up cycle** (Section 86.1803-01) means sufficient vehicle operation such that the coolant temperature has risen by at least 40 deg. F from engine starting and reaches a minimum temperature of 160 deg. F.

**Engineering analysis** (Section 92.2) means a summary of scientific and/or engineering principles and facts that support a conclusion made by a manufacturer or remanufacturer, with respect to compliance with the provisions of this part.

**Engineering analysis** (Section 94.2) means a summary of scientific and/or engineering principles and facts that support a conclusion made by a manufacturer, with respect to compliance with the provisions of this part.

**Engine-system combination** (Section 86.082-2) means an engine family-exhaust emission control system combination.

**Enhanced coagulation** (Section 141.2) means the addition of sufficient coagulant for improved removal of disinfection byproduct precursors by conventional filtration treatment.

**Enhanced mineral oil** (Section 435.41) as applied to enhanced mineral oil-based drilling fluid means a petroleum distillate which has been highly purified and is distinguished from diesel oil and conventional mineral oil in having a lower polycyclic aromatic hydrocarbon (PAH) content. Typically, conventional mineral oils have a PAH content on the order of 0.35 weight percent expressed as phenanthrene, whereas enhanced mineral oils typically have a PAH content of 0.001 or lower weight percent PAH expressed as phenanthrene.

**Enhanced mineral oil as applied to enhanced mineral oil-based drilling fluid** (Section 435.11) means a petroleum distillate which has been highly purified and is distinguished from diesel oil and conventional mineral oil in having a lower polycyclic aromatic hydrocarbon (PAH) content. Typically, conventional mineral oils have a PAH content on the order of 0.35 weight percent expressed as phenanthrene, whereas enhanced mineral oils typically have a PAH content of 0.001 or lower weight percent PAH expressed as phenanthrene.

**Enhanced mineral oil-based** (Section 435.11) means the continuous phase of the drilling fluid is enhanced mineral oil.

**Enhanced mineral oil-based** (Section 435.41) means the continuous phase of the drilling fluid is enhanced mineral oil.

**Enhanced softening** (Section 141.2) means the improved removal of disinfection byproduct precursors by precipitative softening.

**Enroll** (Section 430.21) means to notify the permitting authority that a mill intends to participate in the "Voluntary Advanced Technology Incentives Program." A mill can enroll by indicating its intention to participate in the program either as part of its application for a National Pollutant Discharge Elimination System (NPDES) permit, or through separate correspondence to the permitting authority as long as the mill signs the correspondence in accordance with 40 CFR 122.22.

**Environment** (Section 171.2) means water, air, land, and all plants and man and other animals living therein, and the interrelationships which exist among them.

**Environment** (Section 302.3) means (1) the navigable waters, the waters of the contiguous zone, and the ocean waters of which the natural resources are under the exclusive management authority of the United States under the Fishery Conservation and Management Act of 1976, and (2) any other surface water, ground water, drinking water supply, land surface or subsurface strata, or ambient air within the United States or under the jurisdiction of the United States;

**Environment** (Section 355.20) includes water, air, and land and the interrelationship which exists among and between water, air, and land and all living things.

**Environment** (Section 370.2) includes water, air, and land and the interrelationship that exists among and between water, air, and land and all living things.

**Environment as defined by section 101(8) of CERCLA** (Section 300.5) means the navigable waters, the waters of the contiguous zone, and the ocean waters of which the natural resources are under the exclusive management authority of the United States under the Magnuson Fishery Conservation and Management Act (16 U.S.C. 1801 et seq.); and any other surface water, ground water, drinking water supply, land surface or subsurface strata, or ambient air within the United States or under the jurisdiction of the United States.

**Environmental Appeals Board** (Section 124.2) shall mean the Board within the Agency described in Sec. 1.25(e) of this title. The Administrator delegates authority to the Environmental Appeals Board to issue final decisions in RCRA, PSD, UIC, or NPDES permit appeals filed under this subpart, including informal appeals of denials of requests for modification, revocation and reissuance, or termination of permits under Section 124.5(b). An appeal directed to the Administrator, rather than to the Environmental Appeals Board, will not be considered. This delegation does not preclude the Environmental Appeals Board from referring an appeal or a motion under this subpart to the Administrator when the Environmental Appeals Board, in its discretion, deems it appropriate to do so. When an appeal or motion is referred to the Administrator by the Environmental Appeals Board, all parties shall be so notified and the rules in this subpart referring to the Environmental Appeals Board shall be interpreted as referring to the Administrator.

**Environmental Appeals Board** (Section 164.2) shall mean the Board within the Agency described in Sec. 1.25 of this title. The Administrator delegates authority to the Environmental Appeals Board to issue final decisions in appeals filed under subparts B and C of this part. An appeal directed to the Administrator, rather than to the Environmental Appeals Board, will not be considered. This delegation does not preclude the Environmental Appeals Board from referring an appeal or a motion under subparts B and C to the Administrator when the Environmental Appeals Board, in its discretion, deems it appropriate to do so. When an appeal or motion is referred to the Administrator, all of the parties shall be so notified and the rules in subparts B and C referring to the Environmental Appeals Board shall be interpreted as referring to the Administrator.

**Environmental Appeals Board** (Section 209.3) means the Board within the Agency described in Sec. 1.25 of this title. The Administrator delegates authority to the Environmental Appeals Board to issue final decisions in appeals filed under this part. An appeal directed to the Administrator, rather than to the Environmental Appeals Board, will not be considered. This delegation of authority to the Environmental Appeals Board does not preclude the Environmental Appeals Board from referring an appeal or a motion filed under this part to the Administrator for decision when the Environmental Appeals Board, in its discretion, deems it appropriate to do so. When an appeal or motion is referred to the Administrator, all parties shall be so notified and the rules in this part referring to the Environmental Appeals Board shall be interpreted as referring to the Administrator.

**Environmental Appeals Board** (Section 66.3) shall mean the Board within the Agency described in Sec. 1.25 of this title. The Administrator delegates authority to the Environmental Appeals Board to issue final decisions in appeals filed under this part. Appeals directed to the Administrator, rather than to the Environmental Appeals Board, will not be considered. This delegation of authority to the Environmental Appeals Board does not preclude the Environmental Appeals Board from referring an appeal or a motion filed under this part to the Administrator for decision when the Environmental Appeals Board, in its discretion, deems it appropriate to do so. When an appeal or motion is referred to the Administrator, all parties shall be so notified and the rules in this part referring to the

Environmental Appeals Board shall be interpreted as referring to the Administrator.

**Environmental Appeals Board** (Section 72.2) means the three-member board established pursuant to Sec. 1.25(e) of this chapter and authorized to hear appeals pursuant to part 78 of this chapter.

**Environmental education and environmental education and training** (Section 47.105) mean educational activities and training activities involving elementary, secondary, and postsecondary students, as such terms are defined in the State in which they reside, and environmental education personnel, but does not include technical training activities directed toward environmental management professionals or activities primarily directed toward the support of noneducational research and development.

**Environmental noise** (Section 205.2) means the intensity, duration, and the character of sounds from all sources.

**Environmental program** (Section 35.102) A program for which EPA awards grants under the authorities listed in Sec. 35.101. The grants are subject to the requirements of this subpart.

**Environmental receptor** (Section 68.3) means natural areas such as national or state parks, forests, or monuments; officially designated wildlife sanctuaries, preserves, refuges, or areas; and Federal wilderness areas, that could be exposed at any time to toxic concentrations, radiant heat, or overpressure greater than or equal to the endpoints provided in Sec. 68.22(a), as a result of an accidental release and that can be identified on local U. S. Geological Survey maps.

**Environmental test cell** (Section 86.000-2) means a test cell capable of wind-speed, solar thermal load, ambient temperature, and humidity control or simulation which meets the requirements of Sec. 86.161-00 for running emission tests with the air conditioning operating.

**Environmental test cell** (Section 86.1803-01) means a test cell capable of wind-speed, solar thermal load, ambient temperature, and humidity control or simulation which meets the requirements of Sec. 86.161-00 for running emission tests with the air conditioning operating.

**Environmentally transformed** (Section 721.3) A chemical substance is "environmentally transformed" when its chemical structure changes as a result of the action of environmental processes on it.

**EPA** (Section 197.2) means the Environmental Protection Agency.

**EPA Acknowledgement of Consent** (Section 262.51) means the cable sent to EPA from the U.S. Embassy in a receiving country that acknowledges the written consent of the receiving country to accept the hazardous waste and describes the terms and conditions of the receiving country's consent to the shipment.

**EPA Enforcement Officer** (Section 86.082-2) means any officer or employee of the Environmental Protection Agency so designated in writing by the Administrator (or by his designee).

**EPA Enforcement Officer** (Section 86.1803-01) means any officer or employee of the Environmental Protection Agency so designated in writing by the Administrator (or by his/her designee).

**EPA enforcement officer** (Section 89.2) means any officer or employee of the Environmental Protection Agency so designated in writing by the Administrator (or by his or her designee).

**EPA enforcement officer** (Section 90.3) means any officer, employee, or authorized representative of the U.S. Environmental Protection Agency so designated in writing by the Administrator (or by his or her designee).

**EPA enforcement officer** (Section 91.3) means any officer, employee, or authorized representative of the U.S. Environmental Protection Agency so designated in writing by the Administrator (or by his or her designee).

**EPA Enforcement Officer** (Section 92.2) means any officer or employee of the Environmental Protection Agency so designated in writing by the Administrator or his/her designee.

**EPA Enforcement Officer** (Section 94.2) means any officer or employee of the Environmental Protection Agency so designated in writing by the Administrator or his/her designee.

**EPA hazardous waste number** (Section 260.10) means the number assigned by EPA to each hazardous waste listed in part 261, subpart D, of this chapter and to each characteristic identified in part 261, subpart C, of this chapter.

**EPA identification number** (Section 260.10) means the number assigned by EPA to each generator, transporter, and treatment, storage, or disposal facility.

**EPA identification number** (Section 761.13) means the 12-digit number assigned to a facility by EPA upon notification of PCB waste activity under Sec. 761.205.

**EPA protocol gas** (Section 72.2) means a calibration gas mixture prepared and analyzed according to section 2 of the "EPA Traceability Protocol for Assay and Certification of Gaseous Calibration Standards," September 1997, EPA-600/R-97/121 or such revised procedure as approved by the Administrator.

**EPA region** (Section 260.10) means the states and territories found in any one of the following ten regions: **Region I**—Maine, Vermont, New Hampshire, Massachusetts, Connecticut, and Rhode Island. **Region II**—New York, New Jersey, Commonwealth of Puerto Rico, and the U.S. Virgin Islands. **Region III**—Pennsylvania, Delaware, Maryland, West Virginia, Virginia, and the District of Columbia. **Region IV**—Kentucky, Tennessee, North Carolina, Mississippi, Alabama, Georgia, South Carolina, and Florida. **Region V**—Minnesota, Wisconsin, Illinois, Michigan, Indiana and Ohio. **Region VI**—New Mexico, Oklahoma, Arkansas, Louisiana, and Texas. **Region VII**—Nebraska, Kansas, Missouri, and Iowa. **Region VIII**—Montana, Wyoming, North Dakota, South Dakota, Utah, and Colorado. **Region IX**—California, Nevada, Arizona, Hawaii, Guam, American Samoa, Commonwealth of the Northern Mariana Islands. **Region X**—Washington, Oregon, Idaho, and Alaska. Equivalent method means any testing or analytical method approved by the Administrator under Secs. 260.20 and 260.21.

**EPA trial staff** (Section 72.2) means an employee of EPA, whether temporary or permanent, who has been designated by the Administrator to investigate, litigate, and present evidence, arguments, and positions of EPA in any evidentiary hearing under part 78 of this chapter. Any EPA or permitting authority employee, consultant, or contractor who is called as a witness in the evidentiary hearing by EPA trial staff shall be deemed to be "EPA trial staff".

**Equipment** (Section 264.1031) means each valve, pump, compressor, pressure relief device, sampling connection system, open-ended valve or line, or flange or other connector, and any control devices or systems required by this subpart.

**Equipment** (Section 65.2) means each of the following that is subject to control under the referencing subpart: pump, compressor, agitator, pressure relief device, sampling connection system, open-ended valve or line, valve, connector, and instrumentation system; and any control devices or systems used to comply with subpart F of this part.

**Equipment manufacturer** (Section 90.3) means a manufacturer of equipment using engines covered by the provisions of this Part who does not also manufacture engines covered by the provisions of this Part.

**Equivalence data** (Section 790.3) means chemical data or biological test data intended to show that two substances or mixtures are equivalent.

**Equivalent** (Section 790.3) means that a chemical substance or mixture is able to represent or substitute for another in a test or series of tests, and that the data from one substance can be used to make scientific and regulatory decisions concerning the other substance.

**Equivalent diameter** (Section 72.2) means a value, calculated using the equation in paragraph 2.1 of Method 1 in part 60, appendix A of this chapter, and used to determine the upstream and downstream distances for locating CEMS or CEMS components in flues or stacks with rectangular cross sections.

**Equivalent emission limitation** (Section 63.2) means any maximum achievable control technology emission limitation or requirements which are applicable to a major source of hazardous air pollutants and are adopted by the Administrator (or a State with an approved permit program) on a case-by-case basis, pursuant to section 112(g) or (j) of the Act.

**Equivalent method** (Section 50.1) means a method of sampling and analyzing the ambient air for an air pollutant that has been designated as an equivalent method in accordance with part 53 of this chapter; it does not include a method for which an equivalent method designation has been cancelled in accordance with Sec. 53.11 or Sec. 53.16 of this chapter.

**Equivalent method** (Section 53.1) means a method for measuring the concentration of an air pollutant in the ambient air that has been designated as an equivalent method in accordance with this part; it does not include a method for which an equivalent method designation has been canceled in accordance with Sec. 53.11 or Sec. 53.16.

**Equivalent method** (Section 58.1) means a method of sampling and analyzing the ambient air for an air pollutant that has been designated as an equivalent method in accordance with part 53 of this chapter; it does not include a method for which an equivalent method designation has been canceled in accordance with Sec. 53.11 or Sec. 53.16 of this chapter.

**Equivalent method** (Section 60.2) means any method of sampling and analyzing for an air pollutant which has been demonstrated to the Administrator's satisfaction to have a consistent and quantitatively known relationship to the reference method, under specified conditions.

**Equivalent method** (Section 65.2) means any method of sampling and analyzing for an air pollutant that has been demonstrated to the Administrator's satisfaction to have a consistent and quantitatively known relationship to the reference method under specified conditions.

**Equivalent P<INF>2</INF>O<INF>5</INF> feed** (Section 63.601) means the quantity of phosphorus, expressed as phosphorous pentoxide, fed to the process.

**Equivalent Sound Level** (Section 201.1) means the level, in decibels, of the mean-square A-weighted sound pressure during a stated time period, with reference to the square of the standard reference sound pressure of 20 micropascals. It is the level of the sound exposure divided by the time period and is abbreviated as Leq.

**Equivalent system** (Section 455.10) means a wastewater treatment system that is demonstrated in literature, treatability tests or self-monitoring data to remove a similar level of pesticide active ingredient (PAI) or priority pollutants as the applicable appropriate pollution control technology listed in Table 10 to this part 455.

**Equivalent test weight** (Section 86.094-2) means the weight, within an inertia weight class, which is used in the dynamometer testing of a vehicle and which is based on its loaded vehicle weight or adjusted loaded vehicle weight in accordance with the provisions of subparts A and B of this part. Gaseous fuel means natural gas or liquefied petroleum gas. Generic service and repair information means information which is not specific for an original equipment manufacturer's brand of tools and equipment.

**Equivalent test weight** (Section 86.1803-01) means the weight, within an inertia weight class, which is used in the dynamometer testing of a vehicle and which is based on its loaded vehicle weight or adjusted loaded vehicle weight in accordance with the provisions of this Part.

**Equivalent treatment** (Section 437.2) means a wastewater treatment system that achieves comparable pollutant removals to the applicable treatment technology selected as the basis for the limitations and pretreatment standards. Comparable removals may be demonstrated through literature, treatability tests, or self-monitoring data.

**Essential Metered Dose Inhaler (Essential MDI)** (Section 82.3) means metered dose inhalers for the treatment of asthma and chronic obstructive pulmonary disease, approved by the Food and Drug Administration or by another Party's analogous health authority before December 31, 2000, and considered to be essential by the Party where the MDI product will eventually be sold. In addition, if the MDI product is to be sold in the U.S., the active moiety contained in the MDI must be listed as essential at 21 CFR 2.125(e).

**Essential-Use Allowances** (Section 82.3) means the privileges granted by Sec. 82.4(t) to produce class I substances, as determined by allocation decisions made by the Parties to the Montreal Protocol and in accordance with the restrictions delineated in the Clean Air Act Amendments of 1990.

**Essential-Use Chlorofluorocarbons (Essential-use CFCs)** (Section 82.3) are the CFCs (CFC-11, CFC-12, or CFC-114) produced under the authority of essential-use allowances and not the allowances themselves. Essential-use CFCs include CFCs imported or produced by U.S. entities under the authority of essential-use allowances for use in essential metered dose inhalers, as well as CFCs imported or produced by non-U.S. entities under the authority of privileges granted by the Parties and the national authority of another country for use in essential metered dose inhalers.

**Essential-Uses** (Section 82.3) means those uses of controlled substances designated by the Parties to the Protocol to be necessary for the health and safety of, or critical for the functioning of, society; and for which there are no available technically and economically feasible alternatives or substitutes that are acceptable from the standpoint of environment and health. Beginning January 1, 2000 (January 1, 2002 for methyl chloroform) the essential use designations for class I substances must be made in accordance with the provisions of the Clean Air Act Amendments of 1990.

**Established level** (Section 159.153) means a tolerance, temporary tolerance, food additive regulation, action level, or other limitation on pesticide residues imposed by law, regulation, or other authority.

**Establishment** (Section 167.3) means any site where a pesticidal product, active ingredient, or device is produced, regardless of whether such site is independently owned or operated, and regardless of whether such site is domestic and producing a pesticidal product for export only, or whether the site is foreign and producing any pesticidal product for import into the United States.

**Establishment** (Section 372.3) means an economic unit, generally at a single physical location, where business is conducted or where services or industrial operations are performed.

**Ethanol** (Section 86.1702-99) means any fuel for motor vehicles and motor vehicle engines that is composed of either commercially available or chemically pure ethanol (CH3CH2OH) and gasoline as specified in Sec. 86.1771 (Fuel Specifications). The required fuel blend is based on the type of ethanol-fueled vehicle being certified and the particular aspect of the certification procedure being conducted.

**Ethanol** (Section 92.2) means a fuel that contains at least 50 percent ethanol (ethyl alcohol, (C2H5OH)) by volume.

**Ethanol blender** (Section 80.2) means any person who owns, leases, operates, controls, or supervises an ethanol blending plant.

**Ethanol blending plant** (Section 80.2) means any refinery at which gasoline is produced solely through the addition of ethanol to gasoline, and at which the quality or quantity of gasoline is not altered in any other manner.

**Ethanol vehicle** (Section 86.1702-99) means any motor vehicle that is engineered and designed to be operated using ethanol as a fuel.

**Ethnic foods** (Section 407.81) shall mean the production of canned and frozen Chinese and Mexican specialties utilizing fresh and pre-processed bean sprouts, bamboo shoots, water chestnuts, celery, cactus, tomatoes, and other similar vegetables necessary for the production of the various characteristic product styles.

**Evaluation** (Section 745.103) means a risk assessment and/or inspection.

**Evaluation program or program** (Section 610.11) means the sequence of analyses and tests prescribed by the Administrator as described in Sec. 610.13 in order to evaluate the performance of a retrofit device.

**Evaporative cooling tower** (Section 63.601) means an open water recirculating device that uses fans or natural draft to draw or force ambient air through the device to remove heat from process water by direct contact.

**Evaporative emission code** (Section 86.082-2) means a unique combination, in an evaporative emission family-evaporative emission control system combination, of purge system calibrations, fuel tank and carburetor bowl vent calibrations and other fuel system and evaporative emission control system components and calibrations specified by the Administrator.

**Evaporative emissions** (Section 86.082-2) means hydrocarbons emitted into the atmosphere from a motor vehicle, other than exhaust and crankcase emissions.

**Evaporative emissions** (Section 86.1803-01) means hydrocarbons emitted into the atmosphere from a motor vehicle, other than exhaust and crankcase emissions.

**Evaporative vehicle configuration** (Section 86.082-2) means a unique combination of basic engine, engine code, body type, and evaporative emission code.

**Evaporative/refueling control system** (Section 86.1803-01) means a unique combination within an evaporative/refueling family of canister adsorptive material, purge system configuration, purge strategy, and other parameters determined by the Administrator to affect evaporative and refueling emission control system durability or deterioration factors.

**Evaporative/refueling emission code** (Section 86.1803-01) means a unique combination, in an evaporative/refueling family-evaporative emission control system combination, of purge system calibrations, fuel tank and carburetor bowl vent calibrations and other fuel system and evaporative emission control system components and calibrations specified by the Administrator.

**Evaporative/refueling emission control system** (Section 86.098-2) means a unique combination within an evaporative/refueling family of canister adsorptive material, purge system configuration, purge strategy, and other parameters determined by the Administrator to affect evaporative and refueling emission control system durability or deterioration factors.

**Evaporative/refueling emission family** (Section 86.098-2) means the basic classification unit of a manufacturers' product line used for the purpose of evaporative and refueling emissions test fleet selection and determined in accordance with Sec. 86.098-24.

**Evaporative/refueling family** (Section 86.1803-01) means the basic classification unit of a manufacturers' product line used for the purpose of evaporative and refueling emissions test fleet selection and determined in accordance with Sec. 86.1821-01.

**Evaporative/refueling vehicle configuration** (Section 86.1803-01) means a unique combination of basic engine, engine code, body type, and evaporative emission code.

**Ex parte communication** (Section 304.12) means any communication, written or oral, relating to the merits of the arbitral proceeding, between the Arbitrator and any interested person, which was not originally filed or stated in the administrative record of the proceeding. Such communication is not ex parte communication if all parties to the proceeding have received prior written notice of the proposed communication and have been given the opportunity to be present and to participate therein.

**Ex parte communication** (Section 72.2) means any communication, written or oral, relating to the merits of an adjudicatory proceeding under part 78 of this chapter, that was not originally included or stated in the administrative record, in a pleading, or in an evidentiary hearing or oral argument under part 78 of this chapter, between the decisional body and any interested person outside EPA or any EPA trial staff. Ex parte communication shall not include: (1) Communication between EPA employees other than between EPA trial staff and a member of the decisional body; or (2) Communication between the decisional body and interested persons outside the Agency, or EPA trial staff, where all parties to the proceeding have received prior written notice of the proposed communication and are given an opportunity to be present and to participate therein.

**Excavation zone** (Section 280.12) means the volume containing the tank system and backfill material bounded by the ground surface, walls, and floor of the pit and trenches into which the UST system is placed at the time of installation.

**Exceedance** (Section 63.601) means a departure from an indicator range established under this subpart, consistent with any averaging period specified for averaging the results of the monitoring.

**Exceedance** (Section 64.1) shall mean a condition that is detected by monitoring that provides data in terms of an emission limitation or standard and that indicates that emissions (or opacity) are greater than the applicable emission limitation or standard (or less than the applicable standard in the case of a percent reduction requirement) consistent with any averaging period specified for averaging the results of the monitoring.

**Excepted monitoring system** (Section 72.2) means a monitoring system that follows the procedures and requirements of Sec. 75.19 of this chapter or of appendix D or E to part 75 for approved exceptions to the use of continuous emission monitoring systems.

**Excess emissions** (Section 51.100) means emissions of an air pollutant in excess of an emission standard.

**Excess emissions** (Section 72.2) means: (1) Any tonnage of sulfur dioxide emitted by an affected unit during a calendar year that exceeds the Acid Rain emissions limitation for sulfur dioxide for the unit; and (2) Any tonnage of nitrogen oxide emitted by an affected unit during a calendar year that exceeds the annual tonnage equivalent of the Acid Rain emissions limitation for nitrogen oxides applicable to the affected unit taking into account the unit's heat input for the year.

**Excess emissions** (Section 96.2) means any tonnage of nitrogen oxides emitted by a NOX Budget unit during a control period that exceeds the NOX Budget emissions limitation for the unit.

**Excess emissions** (Section 97.2) means any tonnage of nitrogen oxides emitted by a NOX Budget unit during a control period that exceeds the NOX Budget emissions limitation for the unit.

**Excess emissions and continuous monitoring system performance report** (Section 63.2) is a report that must be submitted periodically by an affected source in order to provide data on its compliance with relevant emission limits, operating parameters, and the performance of its continuous parameter monitoring systems.

**Excess Emissions and Monitoring Systems Performance Report** (Section 60.2) is a report that must be submitted periodically by a source in order to provide data on its compliance with stated emission limits and operating parameters, and on the performance of its monitoring systems.

**Excessive concentration** (Section 51.100) is defined for the purpose of determining good engineering practice stack height under Sec. 51.100(ii)(3) and means: (1) For sources seeking credit for stack height exceeding that established under Sec. 51.100(ii)(2) a maximum ground-level concentration due to emissions from a stack due in whole or part to downwash, wakes, and eddy effects produced by nearby structures or nearby terrain features which individually is at least 40 percent in excess of the maximum concentration experienced in the absence of such downwash, wakes, or eddy effects and which contributes to a total concentration due to emissions from all sources that is greater than an ambient air quality standard. For sources subject to the prevention of significant deterioration program (40 CFR 51.166 and 52.21), an excessive concentration alternatively means a maximum ground-level concentration due to emissions from a stack due in whole or part to downwash, wakes, or eddy effects produced by nearby structures or nearby terrain features which individually is at least 40 percent in excess of the maximum concentration experienced in the absence of such downwash, wakes, or eddy effects and greater than a prevention of significant deterioration increment. The allowable emission rate to be used in making demonstrations under this part shall be prescribed by the new source performance standard that is applicable to the source category unless the owner or operator demonstrates that this emission rate is infeasible. Where such demonstrations are approved by the authority administering the State implementation plan, an alternative emission rate shall be established in consultation with the source owner or operator. (2) For sources seeking credit after October 11, 1983, for increases in existing stack heights up to the heights established under Sec. 51.100(ii)(2), either (i) a maximum ground-level concentration due in whole or part to downwash, wakes or eddy effects as provided in paragraph (kk)(1) of this section, except that the emission rate specified by any applicable State implementation plan (or, in the absence of such a limit, the actual emission rate) shall be used, or (ii) the actual presence of a local nuisance caused by the existing stack, as determined by the authority administering the State implementation plan; and (3) For sources seeking credit after January 12, 1979 for a stack height determined under Sec. 51.100(ii)(2) where the authority administering the State implementation plan requires the use of a field study or fluid model to verify GEP stack height, for sources seeking stack height credit after November 9, 1984 based on the aerodynamic influence of cooling towers, and for sources seeking stack height credit after December 31, 1970 based on the aerodynamic influence of structures not adequately represented by the equations in Sec. 51.100(ii)(2), a maximum ground-level concentration due in whole or part to downwash, wakes or eddy effects that is at least 40 percent in excess of the maximum concentration experienced in the absence of such downwash, wakes, or eddy effects. (ll)-(mm) [Reserved]

**Excluded manufacturing process** (Section 761.13) means a manufacturing process in which quantities of PCBs, as determined in accordance with the definition of inadvertently generated PCBs, calculated as defined, and from which releases to products, air, and water meet the requirements of paragraphs (1) through (5) of this definition, or the importation of products containing PCBs as unintentional impurities, which products meet the requirements of paragraphs (1) and (2) of this definition. (1) The concentration of inadvertently generated PCBs in products leaving any manufacturing site or imported into the United States must have an annual average of less than 25 ppm, with a 50 ppm maximum. (2) The concentration of inadvertently generated PCBs in the components of detergent bars leaving the manufacturing site or imported into the United States must be less than 5 ppm. (3) The release of inadvertently generated PCBs at the point at which emissions are vented to ambient air must be less than 10 ppm. (4) The amount of inadvertently generated PCBs added to water discharged from a manufacturing site must be less than 100 micrograms per resolvable gas chromatographic peak per liter of water discharged. (5) Disposal of any other process wastes above concentrations of 50 ppm PCB must be in accordance with subpart D of this part.

**Excluded PCB products** (Section 761.13) means PCB materials which appear at concentrations less than 50 ppm, including but not limited to: (1) Non-Aroclor inadvertently generated PCBs as a byproduct or impurity resulting from a chemical manufacturing process. (2) Products contaminated with Aroclor or other PCB materials from historic PCB uses (investment casting waxes are one example). (3) Recycled fluids and/or equipment contaminated during use involving the products described in paragraphs (1) and (2) of this definition (heat transfer and hydraulic fluids and equipment and other electrical equipment components and fluids are examples). (4) Used oils, provided that in the cases of paragraphs (1) through (4) of this definition: (i) The products or source of the products containing $<$ 50 ppm concentration PCBs were legally manufactured, processed, distributed in commerce, or used before October 1, 1984. (ii) The products or source of the products containing $<$ 50 ppm concentrations PCBs were legally manufactured, processed, distributed in commerce, or used, i.e., pursuant to authority granted by EPA regulation, by exemption petition, by settlement agreement, or pursuant to other Agency-approved programs; (iii) The resulting PCB concentration (i.e. below 50 ppm) is not a result of dilution, or leaks and spills of PCBs in concentrations over 50 ppm.

**Exclusive economic zone** (Section 300.5) as defined by OPA section 1001, means the zone established by Presidential Proclamation Numbered 5030, dated March 10, 1983, including the ocean waters of the areas referred to as "eastern special areas" in Article 3(1) of the Agreement between the United States of America and the Union of Soviet Socialist Republics on the Maritime Boundary, signed June 1, 1990.

**Exclusive use study** (Section 152.83) means a study that meets each of the following requirements: (1) The study pertains to a new active ingredient (new chemical) or new combination of active ingredients (new combination) first registered after September 30, 1978; (2) The study was submitted in support of, or as a condition of approval of, the application resulting in the first registration of a product containing such new chemical or new combination (first registration), or an application to amend such registration to add a new use; and (3)The study was not submitted to satisfy a data requirement imposed under FIFRA section 3©(2)(B); Provided that, a study is an exclusive use study only during the 10-year period following the date of the first registration.

**Excursion** (Section 64.1) shall mean a departure from an indicator range established for monitoring under this part, consistent with any averaging period specified for averaging the results of the monitoring.

**Executive Officer** (Section 86.1002-2001) means the Executive Officer of the California Air Resources Board or his or her authorized representative.

**Executive Officer** (Section 86.1002-97) means the Executive Officer of the California Air Resources Board or his or her authorized representative.

**Executive Officer** (Section 86.602-97) means the Executive Officer of the California Air Resources Board or his or her authorized representative.

**Executive Officer** (Section 86.602-98) means the Executive Officer of the California Air Resources Board or his or her authorized representative.

**Executive Officer of the California Air Resources Board (ARB)** (Section 86.1702-99) as used in the referenced materials listed in Sec. 86.1 and appendix XIII of this part, means the Administrator of the Environmental Protection Agency (EPA).

**Executive Order** (Section 86.1002-2001) means the document the Executive Officer grants a manufacturer for an engine family that certifies the manufacturer has verified the engine family complies with all applicable standards and requirements pursuant to Title 13 of the California Code of Regulations.

**Executive Order** (Section 86.1002-97) means the document the Executive Officer grants a manufacturer for an engine family that certifies the manufacturer has verified the engine family complies with all applicable standards and requirements pursuant to Title 13 of the California Code of Regulations.

**Executive Order** (Section 86.602-97) means the document the Executive Officer grants a manufacturer for an engine family that certifies the manufacturer has verified that the engine family complies with all applicable standards and requirements pursuant to Title 13 of the California Code of Regulations.

**Executive Order** (Section 86.602-98) means the document the Executive Officer grants a manufacturer for an engine family that certifies the manufacturer has verified that the engine family complies with all applicable standards and requirements pursuant to Title 13 of the California Code of Regulations.

**Exempt compounds** (Section 59.101) means specific organic compounds that are not considered volatile organic compounds due to negligible photochemical reactivity. The exempt compounds are specified in Sec. 51.100 (s) of this chapter.

**Exempted aquifer** (Section 144.3) means an "aquifer" or its portion that meets the criteria in the definition of "underground source of drinking water" but which has been exempted according to the procedures in Sec. 144.7.

**Exempted aquifer** (Section 146.3) means an aquifer or its portion that meets the criteria in the definition of "underground source of drinking water" but which has been exempted according to the procedures of Sec. 144.8(b).

**Exemption** (Section 790.3) means an exemption from a testing requirement of a test rule promulgated under section 4 of the Act and part 799 of this chapter.

**Exemption** (Section 90.902) means exemption from the prohibitions of Sec. 90.1003.

**Exemption** (Section 91.1002) means exemption from the prohibitions of Sec. 91.1103.

**Exemption application** (Section 700.43) means any application submitted to EPA under section 5(h)(2) of the Act.

**Exemption holder** (Section 791.3) refers to a manufacturer or processor, subject to a test rule, that has received an exemption under sections 4©(1) or 4©(2) of TSCA from the requirement to conduct a test and submit data.

**Exemption notice** (Section 700.43) means any notice submitted to EPA under Sec. 723.175 of this chapter.

**Exemption request** (Section 725.3) means any application submitted to EPA under subparts E, F, or G of this part.

**Exhaust emissions** (Section 86.082-2) means substances emitted to the atmosphere from any opening downstream from the exhaust port of a motor vehicle engine.

**Exhaust emissions** (Section 86.1803-01) means substances emitted to the atmosphere from any opening downstream from the exhaust port of a motor vehicle engine.

**Exhaust emissions** (Section 86.402-78) means substances emitted to the atmosphere from any opening downstream from the exhaust port of a motor vehicle engine. Fuel system means the combination of fuel tank, fuel pump, fuel lines, oil injection metering system, and carburetor or fuel injection components, and includes all fuel system vents. Loaded vehicle mass means curb mass plus 80 kg (176 lb.), average driver mass.

**Exhaust emissions** (Section 87.1) means substances emitted to the atmosphere from the exhaust discharge nozzle of an aircraft or aircraft engine.

**Exhaust emissions** (Section 90.3) means matter emitted into the atmosphere from any opening downstream from the exhaust port of a nonroad engine.

**Exhaust emissions** (Section 91.3) means matter emitted into the atmosphere from any opening downstream from the exhaust port of a marine engine.

**Exhaust emissions** (Section 92.2) means substances (i.e., gases and particles) emitted to the atmosphere from any opening downstream from the exhaust port or exhaust valve of a locomotive engine.

**Exhaust emissions** (Section 94.2) means substances (i.e., gases and particles) emitted to the atmosphere from any opening downstream from the exhaust port or exhaust valve of an engine.

**Exhaust gas recirculation** (Section 89.2) means an emission control technology that reduces emissions by routing exhaust gases that had been exhausted from the combustion chamber(s) back into the engine to be mixed with incoming air prior to or during combustion. The use of valve timing to increase the amount of residual exhaust gas in the combustion chamber(s) that is mixed with incoming air prior to or during combustion is not considered to be exhaust gas recirculation for the purposes of this part.

**Exhaust gas recirculation** (Section 94.2) means an emission control technology that reduces emissions by routing gases that had been exhausted from the combustion chamber(s) back into the engine to be mixed with incoming air prior to or during combustion. The use of valve timing to increase the amount of residual exhaust gas in the combustion chamber(s) that is mixed with incoming air prior to or during combustion is not considered to be exhaust gas recirculation for the purposes of this part.

**Exhaust gas recirculation valve** (Section 86.1803-01) means a device which directs a portion of the exhaust gas into the intake air stream for the purpose of controlling emissions.

**Exhaust header pipe** (Section 205.165) means any tube of constant diameter which conducts exhaust gas from an engine exhaust port to other exhaust system components which provide noise attenuation. Tubes with cross connections or internal baffling are not considered to be "exhaust header pipes."

**Exhaust system** (Section 202.10) means the system comprised of a combination of components which provides for enclosed flow of exhaust gas from engine parts to the atmosphere.

**Exhaust system** (Section 205.151) means the combination of components which provides for the enclosed flow of exhaust gas from the engine exhaust port to the atmosphere. "Exhaust system" further means any constituent components of the combination which conduct exhaust gases and which are sold as separate products. "Exhaust System" does not mean any of the constituent components of the combination, alone, which do not conduct exhaust gases, such as brackets and other mounting hardware.

**Exhaust System (Section 205.51)** means the system comprised of a combination of components which provides for enclosed flow of exhaust gas from engine exhaust port to the atmosphere.

**Existing Class II Wells** (Section 147.2902) wells that were authorized by BIA and constructed and completed before the effective date of this program.

**Existing effluent quality** (Section 430.21) means the level at which the pollutants identified in Sec. 430.24(a)(1) are present in the effluent of a mill "enrolled" in the "Voluntary Advanced Technology Incentives Program."

**Existing facility** (Section 60.2) means, with reference to a stationary source, any apparatus of the type for which a standard is promulgated in this part, and the construction or modification of which was commenced before the date of proposal of that standard; or any apparatus which could be altered in such a way as to be of that type.

**Existing Great Lakes discharger** (Section 132.2) is any building, structure, facility, or installation from which there is or may be a "discharge of pollutants" (as defined in 40 CFR 122.2) to the Great Lakes System, that is not a new Great Lakes discharger.

**Existing hazardous waste management (HWM) facility or existing facility** (Section 260.10) means a facility which was in operation or for which construction commenced on or before November 19, 1980. A facility has commenced construction if: (1) The owner or operator has obtained the Federal, State and local approvals or permits necessary to begin physical construction; and either (2) (i) A continuous on-site, physical construction program has begun; or (ii) The owner or operator has entered into contractual obligations-which cannot be cancelled or modified without substantial loss—for physical construction of the facility to be completed within a reasonable time.

**Existing hazardous waste management (HWM) facility or existing facility** (Section 270.2) means a facility which was in operation or for which construction commenced on or before November 19, 1980. A facility has commenced construction if: (a) The owner or operator has obtained the Federal, State and local approvals or permits necessary to begin physical construction; and either (b)(1) A continuous on-site, physical construction program has begun; or

**Existing indirect dischargers** (Section 420.31) means only those two iron blast furnace operations with discharges to publicly owned treatment works prior to May 27, 1982.

**Existing injection well** (Section 144.3) means an "injection well" other than a "new injection well."

**Existing injection well** (Section 146.3) means an "injection well" other than a "new injection well."

**Existing MSWLF unit** (Section 258.2) means any municipal solid waste landfill unit that is receiving solid waste as of the appropriate dates specified in Sec. 258.1(e).

**Existing portion** (Section 192.31) means that land surface area of an existing surface impoundment on which significant quantities of uranium byproduct materials have been placed prior to promulgation of this standard.

**Existing portion** (Section 260.10) means that land surface area of an existing waste management unit, included in the original Part A permit application, on which wastes have been placed prior to the issuance of a permit.

**Existing source** (Section 129.2) means any source which is not a new source as defined above.

**Existing source** (Section 61.02) means any stationary source which is not a new source.

**Existing source** (Section 62.1201) means any affected source that is not a new source.

**Existing source** (Section 63.2) means any affected source that is not a new source.

**Existing source or existing OCS** (Section 55.2) source shall have the meaning given in the applicable requirements incorporated into Secs. 55.13 and 55.14 of this part, except that for two years following the date of promulgation of this part the definition given in Sec. 55.3 of this part shall apply for the purpose of determining the required date of compliance with this part.

**Existing tank** (Section 279.1) means a tank that is used for the storage or processing of used oil and that is in operation, or for which installation has commenced on or prior to the effective date of the authorized used oil program for the State in which the tank is located.

**Existing tank system** (Section 280.12) means a tank system used to contain an accumulation of regulated substances or for which installation has commenced on or before December 22, 1988. Installation is considered to have commenced if: (a) The owner or operator has obtained all federal, state, and local approvals or permits necessary to begin physical construction of the site or installation of the tank system; and if, (b) (1) Either a continuous on-site physical construction or installation program has begun; or, (2) The owner or operator has entered into contractual obligations— which cannot be cancelled or modified without substantial loss—for physical construction at the site or installation of the tank system to be completed within a reasonable time.

**Existing tank system or existing component** (Section 260.10) means a tank system or component that is used for the storage or treatment of hazardous waste and that is in operation, or for which installation has commenced on or prior to July 14, 1986. Installation will be considered to have commenced if the owner or operator has obtained all Federal, State, and local approvals or permits necessary to begin physical construction of the site or installation of the tank system and if either (1) a continuous on-site physical construction or installation program has begun, or (2) the owner or operator has entered into contractual obligations—which cannot be canceled or modified without substantial loss—for physical construction of the site or installation of the tank system to be completed within a reasonable time.

**Existing technology OB/PWC** (Section 91.3) means an outboard engine or a personal watercraft engine which was in production for the 1997 or any previous model years and that did not utilize newer technologies such as four-stroke technology, direct-injection two-stroke technology, catalyst technology, or other technology used to comply with emission standards which the Administrator determines is a new type of OB/PWC technology.

**Existing unit** (Section 72.2) means a unit (including a unit subject to section 111 of the Act) that commenced commercial operation before November 15, 1990 and that on or after November 15, 1990 served a generator with nameplate capacity of greater than 25 MWe. "Existing unit" does not include simple combustion turbines or any unit that on or after November 15, 1990 served only generators with a nameplate capacity of 25 MWe or less. Any "existing unit" that is modified, reconstructed, or repowered after November 15, 1990 shall continue to be an "existing unit."

**Existing uses** (Section 131.3) are those uses actually attained in the water body on or after November 28, 1975, whether or not they are included in the water quality standards.

**Existing vessel** (Section 140.1) refers to any vessel on which construction was initiated before January 30, 1975;

**Existing ZEV Mandate** (Section 86.1702-99) means any state regulation or other law that imposes (or purports to impose) obligations on auto manufacturers to produce, deliver for sale, or sell a certain number or percentage of ZEVs and that was adopted prior to December 16, 1997.

**Expedited Hearing** (Section 164.2) means a hearing commenced as the result of the issuance of a notice of intention to suspend or the suspension of a registration of a pesticide by an emergency order, and is limited to a consideration as to whether a pesticide presents an imminent hazard which justifies such suspension.

**Experimental animals** (Section 172.1) means individual animals or groups of animals, regardless of species, intended for use and used solely for research purposes and does not include animals intended to be used for any food purposes.

**Experimental start date** (Section 160.3) means the first date the test substance is applied to the test system.

**Experimental start date** (Section 792.3) means the first date the test substance is applied to the test system.

**Experimental technology** (Section 146.3) means a technology which has not been proven feasible under the conditions in which it is being tested.

**Experimental termination date** (Section 160.3) means the last date on which data are collected directly from the study.

**Experimental termination date** (Section 792.3) means the last date on which data are collected directly from the study.

**Exploratory facility** (Section 435.11) means any fixed or mobile structure subject to this Subpart that is engaged in the drilling of wells to determine the nature of potential hydrocarbon reservoirs.

**Exploratory facility** (Section 435.41) means any fixed or mobile structure subject to this Subpart that is engaged in the drilling of wells to determine the nature of potential hydrocarbon reservoirs.

**Exploratory source or exploratory OCS source** (Section 55.2) means any OCS source that is a temporary operation conducted for the sole purpose of gathering information. This includes an operation conducted during the exploratory phase to determine the characteristics of the reservoir and formation and may involve the extraction of oil and gas.

**Explosives or munitions emergency** (Section 260.10) means a situation involving the suspected or detected presence of unexploded ordnance (UXO), damaged or deteriorated explosives or munitions, an improvised explosive device (IED), other potentially explosive material or device, or other potentially harmful military chemical munitions or device, that creates an actual or potential imminent threat to human health, including safety, or the environment, including property, as determined by an explosives or munitions emergency response specialist. Such situations may require immediate and expeditious action by an explosives or munitions emergency response specialist to control, mitigate, or eliminate the threat.

**Explosives or munitions emergency** (Section 266.201) is as defined in 40 CFR 260.10.

**Explosives or munitions emergency response** (Section 260.10) means all immediate response activities by an explosives and munitions emergency response specialist to control, mitigate, or eliminate the actual or potential threat encountered during an explosives or munitions emergency. An explosives or munitions emergency response may include in-place render-safe procedures, treatment or destruction of the explosives or munitions and/or transporting those items to another location to be rendered safe, treated, or destroyed. Any reasonable delay in the completion of an explosives or munitions emergency response caused by a necessary, unforeseen, or uncontrollable circumstance will not terminate the explosives or munitions emergency. Explosives and munitions emergency responses can occur on either public or private lands and are not limited to responses at RCRA facilities.

**Explosives or munitions emergency response** (Section 266.201) is as defined in 40 CFR 260.10.

**Explosives or munitions emergency response specialist** (Section 260.10) means an individual trained in chemical or conventional munitions or explosives handling, transportation, render-safe procedures, or destruction techniques. Explosives or munitions emergency response specialists include Department of Defense (DOD) emergency explosive ordnance disposal (EOD), technical escort unit (TEU), and DOD-certified civilian or contractor personnel; and other Federal, State, or local government, or civilian personnel similarly trained in explosives or munitions emergency responses.

**Explosives or munitions emergency response specialist** (Section 266.201) is as defined in 40 CFR 260.10.

**Export** (Section 82.3) means the transport of virgin or used controlled substances from inside the United States or its territories to persons outside the United States or its territories, excluding United States military bases and ships for on-board use.

**Export exemption** (Section 204.2) means an exemption from the prohibitions of section 10 (a) (1), (2), (3), and (4) of the Act, granted by statute under section 10(b) (2) of the Act for the purpose of exporting regulated products.

**Export exemption** (Section 205.2) means an exemption from the prohibitions of section 10(a) (1), (2), (3), and (4) of the Act, granted by statute under section 10(b)(2) of the Act for the purpose of exporting regulated products.

**Export exemption** (Section 211.102) means an exemption from the prohibitions of section 10(a) (3) and (4) of the Act; this type of exemption is granted by statute under section 10(b) (2) of the Act for the purpose of exporting regulated products.

**Export exemption** (Section 85.1702) means an exemption granted by statute under section 203(b)(3) of the Act for the purpose of exporting new motor vehicles or new motor vehicle engines.

**Export exemption** (Section 90.902) means an exemption granted under Sec. 90.1004 (b) for the purpose of exporting new nonroad engines.

**Export exemption** (Section 91.1002) means an exemption granted under Sec. 91.1104 (b) for the purpose of exporting new marine SI engines.

**Exporter** (Section 707.63) means the person who, as the principal party in interest in the export transaction, has the power and responsibility for determining and controlling the sending of the chemical substance or mixture to a destination out of the customs territory of the United States.

**Exporter** (Section 82.3) means the person who contracts to sell controlled substances for export or transfers controlled substances to his affiliate in another country.

**Exporting country** (Section 262.81) means any designated OECD member country in Sec. 262.58(a) (1) from which a transfrontier movement of wastes is planned or has commenced.

**External floating roof** (Section 265.1081) means a pontoon-type or double-deck type cover that rests on the surface of the material managed in a tank with no fixed roof.

**External floating roof or EFR** (Section 65.2) means a pontoon-type (noncontact) or double-deck-type (contact) roof that is designed to rest on the stored liquid surface in a storage vessel with no fixed roof.

**Extraction** (Section 439.21) means process operations that derive pharmaceutically active ingredients from natural sources such as plant roots and leaves, animal glands, and parasitic fungi by chemical and physical extraction.

**Extraction site** (Section 230.3) means the place from which the dredged or fill material proposed for discharge is to be removed.

**Extremely hazardous substance** (Section 355.20) means a substance listed in appendices A and B of this part.

**Extremely hazardous substance** (Section 370.2) means a substance listed in the appendices to 40 CFR part 355, Emergency Planning and Notification.

**Extrusion** (Section 467.02) is the application of pressure to a billet of aluminum, forcing the aluminum to flow through a die orifice. The extrusion subcategory is based on the extrusion process.

**Extrusion** (Section 468.02) shall mean the application of pressure to a copper workpiece, forcing the copper to flow through a die orifice.

**Extrusion** (Section 471.02) is the application of pressure to a billet of metal, forcing the metal to flow through a die orifice.

**Extrusion die cleaning** (Section 467.31) shall mean the process by which the steel dies used in extrusion of aluminum are cleaned. The term includes a dip into a concentrated caustic bath to dissolve the aluminum followed by a water rinse. It also includes the use of a wet scrubber with the die cleaning operation.

**Extrusion heat treatment** (Section 468.02) shall mean the spray application of water to a workpiece immediately following extrusions for the purpose of heat treatment.

**Exudate** (Section 174.3) means a substance gradually discharged or secreted across intact cellular membranes or cell walls and present in the intercellular spaces or on the exterior surfaces of the plant.

# F

**Facilities eligible for treatment equivalent to secondary treatment** (Section 133.101) Treatment works shall be eligible for consideration for effluent limitations described for treatment equivalent to secondary treatment (Sec. 133.105), if:

**Facility** (Section 240.101) means all thermal processing equipment, buildings, and grounds at a specific site.

**Facility** (Section 256.06) refers to any resource recovery system or component thereof, any system, program or facility for resource conservation, and any facility for collection, source separation, storage, transportation, transfer, processing, treatment or disposal of solid waste, including hazardous waste, whether such facility is associated with facilities generating such wastes or not.

**Facility** (Section 258.2) means all contiguous land and structures, other appurtenances, and improvements on the land used for the disposal of solid waste.

**Facility** (Section 272.2) means all contiguous land and structures, other appurtenances, and improvements on the land used for the disposal of solid waste.

**Facility** (Section 302.3) means (1) any building, structure, installation, equipment, pipe or pipeline (including any pipe into a sewer or publicly owned treatment works), well, pit, pond, lagoon, impoundment, ditch, landfill, storage container, motor vehicle, rolling stock, or aircraft, or (2) any site or area where a hazardous substance has been deposited, stored, disposed of, or placed, or otherwise come to be located; but does not include any consumer product in consumer use or any vessel;

**Facility** (Section 350.1) means all buildings, equipment, structure, and other stationary items that are located on a single site or on contiguous or adjacent sites and which are owned or operated by the same person (or by any person which controls, is controlled by, or under common control with, such person). Facility shall include man-made structures as well as all natural structures in which chemicals are purposefully placed or removed through human means such that it functions as a containment structure for human use. For purposes of emergency release notification, the term includes motor vehicles, rolling stock, and aircraft.

**Facility** (Section 355.20) means all buildings, equipment, structure, and other stationary items that are located on a single site or on contiguous or adjacent sites and which are owned or operated by the same person (or by any person which controls, is controlled by, or under common control with, such person). Facility shall include manmade structures in which chemicals are purposefully placed or removed through human means such that it functions as a containment structure for human use. For purposes of emergency release notification, the term includes motor vehicles, rolling stock, and aircraft.

**Facility** (Section 370.2) means all buildings, equipment, structure, and other stationary items that are located on a single site or on contiguous or adjacent sites and which are owned or operated by the same person (or by any person which controls, is controlled by, or under common control with, such person). Facility shall include manmade structures as well as all natural structures in which chemicals are purposefully placed or removed through human means such that it functions as a containment structure for human use. For purposes of emergency release notification, the term includes motor vehicles, rolling stock, and aircraft.

**Facility** (Section 372.3) means all buildings, equipment, structures, and other stationary items which are located on a single site or on contiguous or adjacent sites and which are owned or operated by the same person (or by any person which controls, is controlled by, or under common control with such person). A facility may contain more than one establishment.

**Facility** (Section 445.2) means all contiguous property owned, operated, leased or under the control of the same person or entity.

**Facility** (Section 72.2) means any institutional, commercial, or industrial structure, installation, plant, source, or building.

**Facility** (Section 721.3) means all buildings, equipment, structures, and other stationary items which are located on a single site or on contiguous or adjacent sites and which are owned or operated by the same person (or by any person which controls, is controlled by, or under common control with such person).

**Facility** (Section 761.13) means all contiguous land, and structures, other appurtenances, and improvements on the land, used for the treatment, storage, or disposal of PCB waste. A facility may consist of one or more treatment, storage, or disposal units.

**Facility** (Section 82.3) means any process equipment (e.g., reactor, distillation column) used to convert raw materials or feedstock chemicals into controlled substances or consume controlled substances in the production of other chemicals.

**Facility as defined by section 101(9) of CERCLA** (Section 300.5) means any building, structure, installation, equipment, pipe or pipeline (including any pipe into a sewer or publicly owned treatment works), well, pit, pond, lagoon, impoundment, ditch, landfill, storage container, motor vehicle, rolling stock, or aircraft, or any site or area, where a hazardous substance has been deposited, stored, disposed of, or placed, or otherwise come to be located; but does not include any consumer product in consumer use or any vessel. As defined by section 1001 of the OPA, it means any structure, group of structures, equipment, or device (other than a vessel) which is used for one or more of the following purposes: Exploring for, drilling for, producing, storing, handling, transferring, processing, or transporting oil. This term includes any motor vehicle, rolling stock, or pipeline used for one or more of these purposes.

**Facility as defined by section 101(9) of CERCLA** (Section 300) means any: (1) Building, structure, installation, equipment, pipe or pipeline (including any pipe into sewer or publicly owned treatment works), well, pit, pond, lagoon, impoundment, ditch, landfill, storage container, motor vehicle, rolling stock, or aircraft; or (2) Any site or area where a hazardous substance has been deposited, stored, disposed of, or placed, or otherwise come to be located; but does not include any consumer product in consumer use or any vessel.

**Facility mailing list** (Section 270.2) means the mailing list for a facility maintained by EPA in accordance with 40 CFR 124.10© (1) (ix).

**Facility** (Section 260.10) means: (1) All contiguous land, and structures, other appurtenances, and improvements on the land, used for treating, storing, or disposing of hazardous waste. A facility may consist of several treatment, storage, or disposal operational units (e.g., one or more landfills, surface impoundments, or combinations of them). (2) For the purpose of implementing corrective action under Sec. 264.101, all contiguous property under the control of the owner or operator seeking a permit under subtitle C of RCRA. This definition also applies to facilities implementing corrective action under RCRA Section 3008(h). (3) Notwithstanding paragraph (2) of this definition, a remediation waste management site is not a facility that is subject to 40 CFR 264.101, but is subject to corrective action requirements if the site is located within such a facility.

**Facility or activity** (Section 122.2) means any NPDES "point source" or any other facility or activity (including land or appurtenances thereto) that is subject to regulation under the NPDES program.

**Facility or activity** (Section 124.2) means any "HWM facility," UIC "injection well," NPDES "point source" or "treatment works treating domestic sewage" or State 404 dredge or fill activity, or any other facility or activity (including land or appurtenances thereto) that is subject to regulation under the RCRA, UIC, NPDES, or 404 programs.

**Facility or activity** (Section 144.3) means any UIC "injection well," or an other facility or activity that is subject to regulation under the UIC program.

**Facility or activity** (Section 146.3) means any "HWM facility," UIC "injection well," NPDES "point source," or State 404 dredge and fill activity, or any other facility or activity (including land or appurtenances thereto) that is subject to regulation under the RCRA, UIC, NPDES, or 404 programs.

**Facility or activity** (Section 270.2) means any HWM facility or any other facility or activity (including land or appurtenances thereto) that is subject to regulation under the RCRA program.

**Factors beyond the control of the licensee** (Section 192.31) means factors proximately causing delay in meeting the schedule in the applicable license for timely emplacement of the permanent radon barrier notwithstanding the good faith efforts of the licensee to achieve compliance. These factors may include, but are not limited to, physical conditions at the site; inclement weather or climatic conditions; an act of God; an act of war; a judicial or administrative order or decision, or change to the statutory, regulatory, or other legal requirements applicable to the licensee's facility that would preclude or delay the performance of activities required for compliance; labor disturbances; any modifications, cessation or delay ordered by state, Federal or local agencies; delays beyond the time reasonably required in obtaining necessary governmental permits, licenses, approvals or consent for activities described in the tailings closure plan (radon) proposed by the licensee that result from agency failure to take final action after the licensee has made a good faith, timely effort to submit legally sufficient applications, responses to requests (including relevant data requested by the agencies), or other information, including approval of the tailings closure plan by NRC or the affected Agreement State; and an act or omission of any third party over whom the licensee has no control.

**Failing compressor** (Section 204.51) means that the measured noise emissions of the compressor, when measured in accordance with the applicable procedure, exceeds the applicable standard.

**Failing exhaust system** (Section 205.165) means that, when installed on any Federally regulated motorcycle for which it is designed and marketed, that motorcycle and exhaust system exceed the applicable standards.

**Failing vehicle** (Section 205.151) means a vehicle whose noise level is in excess of the applicable standard.

**Failing vehicle** (Section 205.51) means that the measured emissions of the vehicle, when measured in accordance with the applicable procedure, exceeds the applicable standard.

**Failure, EFR (referred to as EFR failure)** (Section 65.2) is defined as any time the external floating roof's primary seal has holes, tears, or other openings in the shoe, seal fabric, or seal envelope; or the secondary seal has holes, tears, or other openings in the seal or the seal fabric; or the gaskets no longer close off the stored liquid surface from the atmosphere; or a slotted membrane has more than 10 percent open area.

**Failure, internal floating roof type A (referred to as IFR type A failure)** (Section 65.2) means any time, as determined during visual inspection through roof hatches, in which the internal floating roof is not resting on the surface of the stored liquid inside the storage vessel and is not resting on the leg supports; or there is stored liquid on the floating roof; or there are holes, tears, or other openings in the seal or seal fabric; or there are visible gaps between the seal and the wall of the storage vessel.

**Failure, internal floating roof type B (referred to as IFR type B failure)** (Section 65.2) means any time, as determined during internal inspections, the internal floating roof's primary seal has holes, tears, or other openings in the seal or the seal fabric; or the secondary seal (if one has been installed) has holes, tears, or other openings in the seal or the seal fabric; or the gaskets no longer close off the stored liquid surface from the atmosphere; or a slotted membrane has more than 10 percent open area.

**Family emission limit (FEL)** (Section 86.090-2) means an emission level declared by the manufacturer which serves in lieu of an emission standard for certification purposes in any of the averaging, trading, or banking programs. FELs must be expressed to the same number of decimal places as the applicable emission standard. The FEL for an engine family using NO<INF>X</INF> or particulate NCPs must equal the value of the current NO<INF>X</INF> or particulate emission standard.

**Family emission limit (FEL)** (Section 86.1803-01) means an emission level declared by the manufacturer which serves in lieu of an emission standard for certification purposes in the averaging, trading and banking program. FELs must be expressed to the same number of decimal places as the applicable emission standard.

**Family emission limit (FEL)** (Section 89.2) means an emission level that is declared by the manufacturer to serve in lieu of an emission standard for certification purposes and for the averaging, banking, and trading program. A FEL must be expressed to the same number of decimal places as the applicable emission standard.

**Family Emission Limit (FEL)** (Section 91.3) means an emission level that is declared by the manufacturer to serve in lieu of an emission standard for certification and for the averaging, banking, and trading program. A FEL must be expressed to the same number of decimal places as the applicable emission standard.

**Family Emission Limit (FEL)** (Section 94.2) means an emission level declared by the certifying manufacturer to serve in lieu of an otherwise applicable emission standard for certification and compliance purposes in the averaging, banking and trading program. FELs are expressed to the same number of decimal places as the applicable emission standard.

**Family Emission Limit** (Section 92.2) means an emission level declared by the certifying manufacturer or remanufacturer to serve in lieu of an otherwise applicable emission standard for certification and compliance purposes in the averaging, banking and trading program. FELs are expressed to the same number of decimal places as the applicable emission standard.

**Family Emission Limit or FEL** (Section 90.3) means an emission level that is declared by the manufacturer to serve in lieu of an emission standard for the purposes of certification, production line testing, and Selective Enforcement Auditing for engines participating in the averaging, banking and trading program. A declared FEL will also serve in lieu of an emission standard where the manufacturer elects to perform voluntary in-use testing under this part. An FEL must be expressed to the same number of decimal places as the applicable emission standard.

**Family NO<INF>X</INF> emission limit** (Section 86.088-2) means the O<INF>X</INF>emission level to which an engine family is certified in the light-duty truck NO<INF>X</INF> averaging program, expressed to one-tenth of a gram per mile accuracy.

**Family particulate emission limit** (Section 86.085-2) means the diesel particulate emission level to which an engine family is certified in the particulate averaging program, expressed to an accuracy of one hundredth gram-per-mile.

**Farm** (Section 170.3) means any operation, other than a nursery or forest, engaged in the outdoor production of agricultural plants.

**Farm equipment or vehicle** (Section 85.1602) means any internal combustion engine-powered machine primarily used in the commercial production and/or commercial harvesting of food, fiber, wood, or commercial organic products or for the processing of such products for further use on the farm.

**Farm tank** (Section 280.12) is a tank located on a tract of land devoted to the production of crops or raising animals, including fish, and associated residences and improvements. A farm tank must be located on the farm property. "Farm" includes fish hatcheries, rangeland and nurseries with growing operations.

**Fast Meter Response** (Section 201.1) means that the "fast" response of the sound level meter shall be used. The fast dynamic response shall comply with the meter dynamic characteristics in paragraph 5.3 of the American National Standard Specification for Sound Level Meters. ANSI S1.4-1971. This publication is available from the American National Standards Institute, Inc., 1430 Broadway, New York, New York 10018.

**Fast meter response** (Section 202.10) means that the fast dynamic response of the sound level meter shall be used. The fast dynamic response shall comply with the meter dynamic characteristics in paragraph 5.3 of the American National Standard Specification for Sound Level Meters, ANSI S1. 4-1971. This publication is available from the American National Standards Institute, Inc., 1420 Broadway, New York, New York 10018.

**Fast Meter Response** (Section 205.2) means that the fast dynamic response of the sound level meter shall be used. The fast dynamic response shall comply with the meter dynamic characteristics in paragraph 5.3 of the American National Standard Specification for Sound Level Meters, ANSI SI.4-1971. This publication is available from the American National Standards Institute, Inc., 1430 Broadway, New York, New York 10018.

**Fast turnaround operation of a spray drying tower** (Section 417.151) shall mean operation involving more than 6 changes of formulation in a 30 consecutive day period that are of such degree and type (e.g., high phosphate to no phosphate) as to require cleaning of the tower to maintain minimal product quality.

**Fault** (Section 146.3) means a surface or zone of rock fracture along which there has been displacement.

**Fault** (Section 147.2902) a surface or zone of rock fracture along which there has been displacement.

**Fault** (Section 503.21) is a fracture or zone of fractures in any materials along which strata on one side are displaced with respect to strata on the other side.

**FDA** (Section 160.3) means the U.S. Food and Drug Administration.

**FDA** (Section 792.3) means the U.S. Food and Drug Administration.

**Feasibility study (FS)** (Section 300.5) means a study undertaken by the lead agency to develop and evaluate options for remedial action. The FS emphasizes data analysis and is generally performed concurrently and in an interactive fashion with the remedial investigation (RI), using data gathered during the RI. The RI data are used to define the objectives of the response action, to develop remedial action alternatives, and to undertake an initial screening and detailed analysis of the alternatives. The term also refers to a report that describes the results of the study.

**Fecal coliform bacteria** (Section 140.1) are those organisms associated with the intestines of warm-blooded animals that are commonly used to indicate the presence of fecal material and the potential presence of organisms capable of causing human disease.

**Federal Act** (Section 109.2) means the Federal Water Pollution Control Act, as amended, 33 U.S.C. 1151 et seq.

**Federal action** (Section 93.152) means any activity engaged in by a department, agency, or instrumentality of the Federal government, or any activity that a department, agency or instrumentality of the Federal government supports in any way, provides financial assistance for, licenses, permits, or approves, other than activities related to transportation plans, programs, and projects developed, funded, or approved under title 23 U.S.C. or the Federal Transit Act (49 U.S.C. 1601 et seq.). Where the Federal action is a permit, license, or other approval for some aspect of a non-Federal undertaking, the relevant activity is the part, portion, or phase of the non-Federal undertaking that requires the Federal permit, license, or approval.

**Federal agency** (Section 142.2) means any department, agency, or instrumentality of the United States.

**Federal Agency** (Section 205.2) means an executive agency (as defined in section 105 of title 5, United States Code) and includes the United States Postal Service.

**Federal agency** (Section 247.3) means any department, agency, or other instrumentality of the Federal government; any independent agency or establishment of the Federal government including any government corporation; and the Government Printing Office;

**Federal agency** (Section 260.10) means any department, agency, or other instrumentality of the Federal Government, any independent agency or establishment of the Federal Government including any Government corporation, and the Government Printing Office.

**Federal agency** (Section 93.152) means, for purposes of this subpart, a Federal department, agency, or instrumentality of the Federal government.

**Federal agency or agency of the United States** (Section 47.105) means any department, agency or other instrumentality of the Federal Government, any independent agency or establishment of the Federal Government including any Government corporation.

**Federal Compliance Testing sequence (FCT)** (Section 85.1502) The testing sequence that incorporates all of the testing requirements of part 86 applicable at the time of an emissions test conducted pursuant to this subpart.

**Federal Emission Test Procedure** (Section 600.022-85) refers to the dynamometer driving schedule, dynamometer procedure, and sampling and analytical procedures described in part 86 for the respective model year, which are used to derive city fuel economy data for gasoline-fueled or diesel vehicles.

**Federal facility** (Section 243.101) means any building, installation, structure, land, or public work owned by or leased to the Federal Government. Ships at sea, aircraft in the air, land forces on maneuvers, and other mobile facilities are not considered "Federal facilities" for the purpose of these guidelines. United States Government installations located on foreign soil or on land outside the jurisdiction of the United States Government are not considered "Federal facilities" for the purpose of these guidelines.

**Federal facility** (Section 246.101) means any building, installation, structure, land, or public work owned by or leased to the Federal Government. Ships at sea, aircraft in the air, land forces on maneuvers, and other mobile facilities are not considered Federal facilities for the purpose of these guidelines. United States Government installations located on foreign soil or on land outside the jurisdiction of the United States Government are not considered Federal facilities for the purpose of these guidelines.

**Federal Government** (Section 203.1) includes the legislative, executive, and judicial branches of the Government of the United States, and the government of the District of Columbia.

**Federal government official** (Section 1400.2) means—(1) An officer or employee of the United States; and (2) An officer or employee of an agent or contractor of the Federal government.

**Federal Highway Fuel Economy Test Procedure** (Section 600.022-85) refers to the dynamometer driving schedule, dynamometer procedure, and sampling and analytical procedures described in subpart B of this part and which are used to derive highway fuel economy data for gasoline-fueled or diesel vehicles.

**Federal Indian reservation (in the case of NPDES)** (Section 124.2) means all land within the limits of any Indian reservation under the jurisdiction of the United States Government, notwithstanding the issuance of any patent, and including rights-of-way running through the reservation.

**Federal Indian reservation** (Section 122.2) means all land within the limits of any Indian reservation under the jurisdiction of the United States Government, notwithstanding the issuance of any patent, and including rights-of-way running through the reservation.

**Federal Indian reservation** (Section 232.2) means all land within the limits of any Indian reservation under the jurisdiction of the United States Government, notwithstanding the issuance of any patent, and including rights-of-way running through the reservation.

**Federal Indian reservation** (Section 233.2) means all land within the limits of any Indian reservation under the jurisdiction of the United States Government, notwithstanding the issuance of any patent, and including rights-of-way running through the reservation.

**Federal Indian reservation** (Section 501.2) means all land within the limits of any Indian reservation under the jurisdiction of the United States Government, notwithstanding the issuance of any patent, and including rights-of-way running through the reservation.

**Federal Indian Reservation, Indian Reservation or Reservation** (Section 49.2) means all land within the limits of any Indian reservation under the jurisdiction of the United States government, notwithstanding the issuance of any patent, and including rights-of-way running through the reservation.

**Federal Indian reservation, Indian reservation or reservation** (Section 71.2) means all land within the limits of any Indian reservation under the jurisdiction of the United States Government, notwithstanding the issuance of any patent, and including rights-of-way running through the reservation.

**Federal Indian Reservation, Indian Reservation, or Reservation** (Section 131.3) means all land within the limits of any Indian reservation under the jurisdiction of the United States Government, notwithstanding the issuance of any patent, and including rights-of-way running through the reservation."

**Federal Indian reservation, Indian reservation, or reservation** (Section 132.2) means all land within the limits of any Indian reservation under the jurisdiction of the United States Government, notwithstanding the issuance of any patent, and including rights-of-way running through the reservation.

**Federal Radiological Emergency Response Plan (FRERP)** (Section 300.5) means the inter-agency agreement for coordinating the response of various agencies, under a variety of statutes, to a large radiological accident. The Lead Federal Agency (LFA), defined by the FRERP, activates the FRERP for any peacetime radiological emergency which, based upon its professional judgment, is expected to have a significant radiological effect within the United States, its territories, possessions, or territorial waters and that could require a response by several federal agencies.

**Federal Response Plan (FRP)** (Section 300.5) means the agreement signed by 27 federal departments and agencies in April 1987 and developed under the authorities of the Earthquake Hazards Reduction Act of 1977 (42 U.S.C. 7701 et seq.) and the Disaster Relief Act of 1974 (42 U.S.C. 3231 et seq.), as amended by the Stafford Disaster Relief Act of 1988.

**Federal standards** (Section 205.165) means, for the purpose of this subpart, the standards specified in Sec. 205.152(a) (1), (2) and (3).

**Federal Test Procedure or City Fuel Economy Test** (Section 610.11) means the test procedures specified in 40 CFR part 86, except as those procedures are modified in these protocols.

**Federal Test Procedure, or FTP** (Section 86.000-2) means the test procedure as described in Sec. 86.130-00 (a) through (d) and (f) which is designed to measure urban driving tail pipe exhaust emissions and evaporative emissions over the Urban Dynamometer Driving Schedule as described in appendix I to this part.

**Federal Test Procedure, or FTP** (Section 86.1803-01) means the test procedure as described in Sec. 86.130-00(a) through (d) and (f) which is designed to measure urban driving tail pipe exhaust emissions and evaporative emissions over the Urban Dynamometer Driving Schedule as described in Appendix I to this part.

**Federal, State and local approvals or permits necessary to begin physical construction** (Section 260.10) means permits and approvals required under

Federal, State or local hazardous waste control statutes, regulations or ordinances.

**Federal, State and local approvals or permits necessary to begin physical construction** (Section 270.2) means permits and approvals required under Federal, State or local hazardous waste control statutes, regulations or ordinances.

**Federally enforceable** (Section 63.2) means all limitations and conditions that are enforceable by the Administrator and citizens under the Act or that are enforceable under other statutes administered by the Administrator. Examples of federally enforceable limitations and conditions include, but are not limited to: (1) Emission standards, alternative emission standards, alternative emission limitations, and equivalent emission limitations established pursuant to section 112 of the Act as amended in 1990; (2) New source performance standards established pursuant to section 111 of the Act, and emission standards established pursuant to section 112 of the Act before it was amended in 1990; (3) All terms and conditions in a title V permit, including any provisions that limit a source's potential to emit, unless expressly designated as not federally enforceable; (4) Limitations and conditions that are part of an approved State Implementation Plan (SIP) or a Federal Implementation Plan (FIP); (5) Limitations and conditions that are part of a Federal construction permit issued under 40 CFR 52.21 or any construction permit issued under regulations approved by the EPA in accordance with 40 CFR part 51; (6) Limitations and conditions that are part of an operating permit where the permit and the permitting program pursuant to which it was issued meet all of the following criteria: (i) The operating permit program has been submitted to and approved by EPA into a State implementation plan (SIP) under section 110 of the CAA; (ii) The SIP imposes a legal obligation that operating permit holders adhere to the terms and limitations of such permits and provides that permits which do not conform to the operating permit program requirements and the requirements of EPA's underlying regulations may be deemed not "federally enforceable" by EPA; (iii) The operating permit program requires that all emission limitations, controls, and other requirements imposed by such permits will be at least as stringent as any other applicable limitations and requirements contained in the SIP or enforceable under the SIP, and that the program may not issue permits that waive, or make less stringent, any limitations or requirements contained in or issued pursuant to the SIP, or that are otherwise "federally enforceable"; (iv) The limitations, controls, and requirements in the permit in question are permanent, quantifiable, and otherwise enforceable as a practical matter; and (v) The permit in question was issued only after adequate and timely notice and opportunity for comment for EPA and the public. (7) Limitations and conditions in a State rule or program that has been approved by the EPA under subpart E of this part for the purposes of implementing and enforcing section 112; and (8) Individual consent agreements that the EPA has legal authority to create.

**Federally registered** (Section 162.151) means currently registered under sec. 3 of the Act, after having been initially registered under the Federal Insecticide, Fungicide, and Rodenticide Act of 1947 (Pub. L. 86-139; 73 Stat. 286; June 25, 1947) by the Secretary of Agriculture or under FIFRA by the Administrator.

**Federally regulated motorcycle** (Section 205.165) means, for the purpose of this subpart, any motorcycle subject to the noise standards of subpart D of this part.

**Federally-recognized Indian Tribe** (Section 310.3) as defined by section 101(36) of CERCLA, means any Indian Tribe, band, nation, or other organized group or community, including any Alaska Native village but not including any Alaska Native regional or village corporation, which is recognized as eligible for the special programs and services provided by the United States to Indians because of their status as Indians.

**Feedlot** (Section 412.11) shall mean a concentrated, confined animal or poultry growing operation for meat, milk or egg production, or stabling, in pens or houses wherein the animals or poultry are fed at the place of confinement and crop or forage growth or production is not sustained in the area of confinement.

**Feedlot** (Section 412.21) shall mean a concentrated, confined animal or poultry growing operation for meat, milk or egg production, or stabling, in pens or houses wherein the animals or poultry are fed at the place of confinement and crop or forage production or growth is not sustained in the area of confinement.

**Feedrate operating limits** (Section 62.1201) means limits on the feedrate of materials (e.g., metals, chlorine) to the combustor that are established based on comprehensive performance testing. The limits are established and monitored by knowing the concentration of the limited material (e.g., chlorine) in each feedstream and the flowrate of each feedstream.

**Feedstock** (Section 419.11) shall mean the crude oil and natural gas liquids fed to the topping units.

**Feedstream** (Section 62.1201) means any material fed into a hazardous waste combustor, including, but not limited to, any pumpable or nonpumpable solid, liquid, or gas.

**Fermentation** (Section 439.11) means process operations that utilize a chemical change induced by a living organism or enzyme, specifically, bacteria, or the microorganisms occurring in unicellular plants such as yeast, molds, or fungi to produce a specified product.

**Ferromanganese blast furnace** (Section 420.31) means those blast furnaces which produce molten iron containing more than fifty percent manganese.

**Ferrous casting** (Section 464.02) The remelting of ferrous metals to form a cast intermediate or finished product by pouring the molten metal into a mold. Except for grinding scrubber operations which are covered here, processing operations following the cooling of castings are covered under the electroplating and metal finishing point source categories (40 CFR parts 413 and 433).

**FFDCA** (Section 160.3) means the Federal Food, Drug and Cosmetic Act, as amended (21 U.S.C. 321 et seq).

**FHWA** (Section 93.101) means the Federal Highway Administration of DOT.

**FHWA/FTA project** (Section 93.101) for the purpose of this subpart, is any highway or transit project which is proposed to receive funding assistance and approval through the Federal-Aid Highway program or the Federal mass transit program, or requires Federal Highway Administration (FHWA) or Federal Transit Administration (FTA) approval for some aspect of the project, such as connection to an interstate highway or deviation from applicable design standards on the interstate system.

**Fiber** (Section 410.21) shall mean the dry wool and other fibers as received at the wool finishing mill for processing into wool and blended products.

**Fiber line** (Section 430.0)A series of operations employed to convert wood or other fibrous raw material into pulp. If the final product is bleached pulp, the fiber line encompasses pulping, de-knotting, brownstock washing, pulp screening, centrifugal cleaning, and multiple bleaching and washing stages.

**Fiber release episode** (Section 763.83) means any uncontrolled or unintentional disturbance of ACBM resulting in visible emission.

**Fiberglass insulation** (Section 247.3) means insulation which is composed principally of glass fibers, with or without binders;

**Field gas** (Section 68.3) means gas extracted from a production well before the gas enters a natural gas processing plant.

**FIFRA** (Section 152.3) means the Federal Insecticide, Fungicide, and Rodenticide Act, as amended (7 U.S.C. 136-136y).

**FIFRA** (Section 160.3) means the Federal Insecticide, Fungicide and Rodenticide Act as amended (7 U.S.C. 136 et seq).

**FIFRA** (Section 176.3) means the Federal Insecticide, Fungicide, and Rodenticide Act (7 U.S.C 136 et seq.).

**FIFRA** (Section 273.9) means the Federal Insecticide, Fungicide, and Rodenticide Act (7 U.S.C. 136-136y).

**Fifteen working day hold period** (Section 89.602) The period of time between a request for final admission and the automatic granting of final admission (unless EPA intervenes) for a nonconforming nonroad engine conditionally imported pursuant to Sec. 89.605 or Sec. 89.609. Day one of the hold period is the first working day (see definition for "working day" in this section) after the Engine Programs and Compliance Division of EPA receives a complete and valid application for final admission.

**File** (Section 72.2) means to send or transmit a document, information, or correspondence to the official custody of the person specified to take possession in accordance with the applicable regulation. Compliance with any "filing" deadline shall be determined by the date that person receives the document, information, or correspondence.

**Fill material** (Section 232.2) (1) Except as specified in paragraph (3) of this definition, the term fill material means material placed in waters of the United States where the material has the effect of: (i) Replacing any portion of a water of the United States with dry land; or (ii) Changing the bottom elevation of any portion of a water of the United States. (2) Examples of such fill material include, but are not limited to: rock, sand, soil, clay, plastics, construction debris, wood chips, overburden from mining or other excavation activities, and materials used to create any structure or infrastructure in the waters of the United States. (3) The term fill material does not include trash or garbage.

**Fill or filling** (Section 65.2) means the introduction of liquids into a storage vessel, but not necessarily to complete capacity.

Filter profile is a graphical representation of individual filter performance, based on continuous turbidity measurements or total particle counts versus time for an entire filter run, from startup to backwash inclusively, that includes an assessment of filter performance while another filter is being backwashed.

**Filtration** (Section 141.2) means a process for removing particulate matter from water by passage through porous media.

**Final acute value (FAV)** (Section 132.2) is (a) a calculated estimate of the concentration of a test material such that 95 percent of the genera (with which acceptable acute toxicity tests have been conducted on the material) have higher GMAVs, or (b) the SMAV of an important and/or critical species, if the SMAV is lower than the calculated estimate.

**Final approval** (Section 281.12) means the approval received by a state program that meets the requirements in Sec. 281.11(b).

**Final authorization** (Section 270.2) means approval by EPA of a State program which has met the requirements of section 3006(b) of RCRA and the applicable requirements of part 271, subpart A.

**Final chronic value (FCV)** (Section 132.2) is (a) a calculated estimate of the concentration of a test material such that 95 percent of the genera (with which acceptable chronic toxicity tests have been conducted on the material) have higher GMCVs, (b) the quotient of an FAV divided by an appropriate acute-chronic ratio, or © the SMCV of an important and/or critical species, if the SMCV is lower than the calculated estimate or the quotient, whichever is applicable.

**Final closure** (Section 260.10) means the closure of all hazardous waste management units at the facility in accordance with all applicable closure requirements so that hazardous waste management activities under parts 264 and 265 of this chapter are no longer conducted at the facility unless subject to the provisions in Sec. 262.34.

**Final cover** (Section 503.21) is the last layer of soil or other material placed on a sewage sludge unit at closure.

**Final order** (Section 305.3) means the decision of the Review Officer which has become final in accordance with Sec. 305.4(a), or of the Presiding Officer, or in the case of a voluntary agreement (see Sec. 305.25) of the parties, disposing of all legal and factual matters presented in the Request for a Hearing. A final order made by the Review Officer or the Presiding Officer shall contain findings of fact, conclusions of law, as well as the reasons therefore, and an order for an award of a sum certain, or an explanation of why no award is granted. The final order may consist of one or more of the following documents: the findings of fact, conclusions of law, and order of the Review Officer or the Presiding Officer; a voluntary agreement; an accelerated order; or a default order, if the default order provides for dismissal of the Request for a Hearing with prejudice. A final order is the final administrative decision of the Agency and (with the exception of a voluntary agreement) is appealable to the Federal district court for the district where the release or threat of release took place.

**Final permit** (Section 70.2) means the version of a part 70 permit issued by the permitting authority that has completed all review procedures required by Secs. 70.7 and 70.8 of this part.

**Final permit** (Section 71.2) means the version of a part 71 permit issued by the permitting authority that has completed all review procedures required by Secs. 71.7, 71.8, and 71.11.

**Final plant value (FPV)** (Section 132.2) is the lowest plant value that was obtained with an important aquatic plant species in an acceptable toxicity test for which the concentrations of the test material were measured and the adverse effect was biologically important.

**Final printed labeling** (Section 152.3) means the label or labeling of the product when distributed or sold. Final printed labeling does not include the package of the product, unless the labeling is an integral part of the package.

**Final product** (Section 700.43) means a new chemical substance (as "new chemical substance" is defined in Sec. 720.3 of this chapter) that is manufactured by a person for distribution in commerce, or for use by the person other than as an intermediate.

**Financial reporting year** (Section 280.92) may thus comprise a fiscal or a calendar year period.

**Financial reporting year** (Section 280.92) means the latest consecutive twelve-month period for which any of the following reports used to support a financial test is prepared: (1) a 10-K report submitted to the SEC; (2) an annual report of tangible net worth submitted to Dun and Bradstreet; or (3) annual reports submitted to the Energy Information Administration or the Rural Electrification Administration.

**Finished product** (Section 432.51) shall mean the final manufactured product as fresh meat cuts, hams, bacon or other smoked meats, sausage, luncheon meats, stew, canned meats or related products.

**Finished product** (Section 432.61) shall mean the final manufactured product as fresh meat cuts including, but not limited to, steaks, roasts, chops, or boneless meats.

**Finished product** (Section 432.71) shall mean the final manufactured product as fresh meat cuts including steaks, roasts, chops or boneless meat, bacon or other smoked meats (except hams) such as sausage, bologna or other luncheon meats, or related products (except canned meats).

**Finished products** (Section 432.81) shall mean the final manufactured product as fresh meat cuts including steaks, roasts, chops or boneless meat, smoked or cured hams, bacon or other smoked meats, sausage, bologna or other luncheon meats (except canned meats).

**Finishing water** (Section 463.2) is processed water used to remove waste plastic material generated during a finishing process or to lubricate a plastic product during a finishing process. It includes water used to machine or to assemble intermediate or final plastic products.

**Firm or company** (Section 717.3) means any person, that is subject to this part, as defined in Sec. 717.5.

**First attempt at repair** (Section 264.1031) means to take rapid action for the purpose of stopping or reducing leakage of organic material to the atmosphere using best practices.

**First attempt at repair, for the purposes of subparts F and G of this part,** (Section 65.2) means to take action for the purpose of stopping or reducing leakage of organic material to the atmosphere, followed by monitoring as specified in Sec. 65.104(b) and Sec. 65.143(c), as appropriate, to verify whether the leak is repaired, unless the owner or operator determines by other means that the leak is not repaired.

**First draw sample** (Section 141.2) means a one-liter sample of tap water, collected in accordance with Sec. 141.86(b)(2), that has been standing in plumbing pipes at least 6 hours and is collected without flushing the tap.

**First federal official** (Section 300.5) means the first federal representative of a participating agency of the National Response Team to arrive at the scene of a discharge or a release. This official coordinates activities under the NCP and may initiate, in consultation with the OSC, any necessary actions until the arrival of the predesignated OSC. A state with primary jurisdiction over a site covered by a cooperative agreement will act in the stead of the first federal official for any incident at the site.

**First food use** (Section 166.3) refers to the use of a pesticide on a food or in a manner which otherwise would be expected to result in residues in a food, if no permanent tolerance, exemption from the requirement of a tolerance, or food additive regulation for residues of the pesticide on any food has been established for the pesticide under section 408 (d) or (e) or 409 of the Federal Food, Drug, and Cosmetic Act.

**Fish and wildlife and sensitive environments** (Section 112.2) means areas that may be identified by either their legal designation or by evaluations of Area Committees (for planning) or members of the Federal On-Scene Coordinator's spill response structure (during responses). These areas may include wetlands, National and State parks, critical habitats for endangered/threatened species, wilderness and natural resource areas, marine sanctuaries and estuarine reserves, conservation areas, preserves, wildlife areas, wildlife refuges, wild and scenic rivers, recreational areas, national forests, Federal and State lands that are research national areas, heritage program areas, land trust areas, and historical and archeological sites and parks. These areas may also include unique habitats such as: aquaculture sites and agricultural surface water intakes, bird nesting areas, critical biological resource areas, designated migratory routes, and designated seasonal habitats.

**Fixed capital cost** (Section 63.2) means the capital needed to provide all the depreciable components of an existing source.

**Fixed liquid level gauge** (Section 86.098-2) means a type of liquid level gauge used on liquefied petroleum gas-fueled vehicles which uses a relatively small positive shutoff valve and is designed to indicate when the liquid level in the fuel tank being filled reaches the proper fill level. The venting of fuel vapor and/or liquid fuel to the atmosphere during the refueling event is generally associated with the use of the fixed liquid level gauge.

**Fixed liquid level gauge** (Section 86.1803-01) means a type of liquid level gauge used on liquefied petroleum gas-fueled vehicles which uses a relatively small positive shutoff valve and is designed to indicate when the liquid level in the fuel tank being filled reaches the proper fill level. The venting of fuel vapor and/or liquid fuel to the atmosphere during the refueling event is generally associated with the use of the fixed liquid level gauge.

**Fixed roof** (Section 265.1081) means a cover that is mounted on a unit in a stationary position and does not move with fluctuations in the level of the material managed in the unit.

**Fixed roof** (Section 65.2) means a roof that is mounted (for example, permanently affixed) on a storage vessel in a stationary manner and that does not move with fluctuations in stored liquid level.

**Flame zone** (Section 264.1031) means the portion of the combustion chamber in a boiler occupied by the flame envelope.

**Flame zone** (Section 65.2) means the portion of the combustion chamber in a boiler or process heater occupied by the flame envelope.

**Flat mill** (Section 420.71) means those steel hot forming operations that reduce heated slabs to plates, strip and sheet, or skelp.

**Fleet average NMOG value** (Section 86.1702-99) is the fleet average NMOG value calculated for a particular averaging set, based upon the applicable production for that averaging set.

**Fleet average NOX standard** (Section 86.1803-01) means, for light-duty vehicles, light-duty trucks and medium-duty passenger vehicles, a NOX standard imposed over an individual manufacturer's total U.S. sales (or a fraction of total U.S. sales during phase-in years), as 'U.S. sales" is defined in this subpart, of a given model year. Manufacturers determine their compliance with such a standard by averaging, on a sales weighted basis, the individual NOX standards they choose for the fleet of light-duty vehicles, light-duty trucks and medium-duty passenger vehicles they sell of that model year.

**Flexible delineator** (Section 247.3) means a highly visible marker that can be positioned to direct traffic or define boundaries and that will flex if struck by a vehicle to prevent damage to the vehicle or the delineator;

**Flexible fuel vehicle (or engine)** (Section 86.090-2) means any motor vehicle (or motor vehicle engine) engineered and designed to be operated on any mixture of two or more different fuels.

**Flexible Fuel Vehicle (or Engine)** (Section 88.102-94) means any motor vehicle (or motor vehicle engine) engineered and designed to be operated on any mixture of two or more different fuels.

**Flexible fuel vehicle** (Section 86.1803-01) means any motor vehicle engineered and designed to be operated on a petroleum fuel, a methanol fuel, or any mixture of the two. Methanol-fueled vehicles that are only marginally functional when using gasoline (e.g., the engine has a drop in rated horsepower of more than 80 percent) are not flexible fuel vehicles.

**Floating membrane cover** (Section 265.1081) means a cover consisting of a synthetic flexible membrane material that rests upon and is supported by the hazardous waste being managed in a surface impoundment.

**Floating roof** (Section 265.1081) means a cover consisting of a double deck, pontoon single deck, or internal floating cover which rests upon and is supported by the material being contained, and is equipped with a continuous seal.

**Floating roof** (Section 65.2) means a roof consisting of an external floating roof or an internal floating roof that is designed to rest upon and is supported by the stored liquid and is equipped with a continuous seal.

**Flocculation** (Section 141.2) means a process to enhance agglomeration or collection of smaller floc particles into larger, more easily settleable particles through gentle stirring by hydraulic or mechanical means.

**Flooring felt** (Section 763.163) means an asbestos-containing product which is made of paper felt intended for use as an underlayer for floor coverings, or to be bonded to the underside of vinyl sheet flooring.

**Flow indicator** (Section 264.1031) means a device that indicates whether gas flow is present in a vent stream.

**Flow indicator** (Section 65.2) means a device that indicates whether gas flow is present in a line, or whether the valve position would allow gas flow to be present in a line.

**Flow meter accuracy** (Section 72.2) means the closeness of the measurement made by a flow meter to the reference value of the fuel flow being measured, expressed as the difference between the measurement and the reference value.

**Flow monitor** (Section 72.2) means a component of the continuous emission monitoring system that measures the volumetric flow of exhaust gas.

**Flow proportional composite sample** (Section 471.02) is composed of grab samples collected continuously or discretely in proportion to the total flow at time of collection or to the total flow since collection of the previous grab sample. The grab volume or frequency of grab collection may be varied in proportion to flow.

**Flow rate** (Section 146.3) means the volume per time unit given to the flow of gases or other fluid substance which emerges from an orifice, pump, turbine or passes along a conduit or channel.

**Flowable fill** (Section 247.3) is a low strength material that is mixed to a wet, flowable slurry and used as an economical fill or backfill material in place of concrete, compacted soils, or sand.

**Flowrate** (Section 62.1201) means the rate at which a feedstream is fed into a hazardous waste combustor.

**Flow-through process tank** (Section 280.12) is a tank that forms an integral part of a production process through which there is a steady, variable, recurring, or intermittent flow of materials during the operation of the process. Flow-through process tanks do not include tanks used for the storage of materials prior to their introduction into the production process or for the storage of finished products or by-products from the production process.

**Flue** (Section 72.2) means a conduit or duct through which gases or other matter are exhausted to the atmosphere.

**Flue gas** (Section 76.2) means the combustion products arising from the combustion of fossil fuel in a utility boiler.

**Flue gas desulfurization system** (Section 72.2) means a type of add-on emission control used to remove sulfur dioxide from flue gas, commonly referred to as a "scrubber."

**Fluid** (Section 144.3) means any material or substance which flows or moves whether in a semisolid, liquid, sludge, gas, or any other form or state.

**Fluid** (Section 146.3) means material or substance which flows or moves whether in a semisolid, liquid, sludge, gas, or any other form or state.

**Fluid** (Section 147.2902) material or substance which moves or flows whether in a semisolid, liquid, sludge, gas or any other form or state.

**Fluorescent light ballast** (Section 761.13) means a device that electrically controls fluorescent light fixtures and that includes a capacitor containing 0.1 kg or less of dielectric.

**Fly ash** (Section 240.101) means suspended particles, charred paper, dust, soot, and other partially oxidized matter carried in the products of combustion.

**Fly ash** (Section 423.11) means the ash that is carried out of the furnace by the gas stream and collected by mechanical precipitators, electrostatic precipitators, and/or fabric filters. Economizer ash is included when it is collected with fly ash.

**Foam-in-place insulation** (Section 247.3) is rigid cellular foam produced by catalyzed chemical reactions that hardens at the site of the work. The term includes spray-applied and injected applications such as spray-in-place foam and pour-in-place;

**FOIA Officer** (Section 1601.3) means the person designated to process requests for CSB documents under the FOIA.

**Food** (Section 166.3) means any article used for food or drink for man or animals.

**Food** (Section 174.3) includes articles used for food or drink by humans or other animals.

**Food additive** (Section 177.3) means any substance the intended use of which results or may reasonably be expected to result, directly or indirectly, in its becoming a component of or otherwise affecting the characteristics of any food (including any such substance intended for use in producing, manufacturing, packing, processing, preparing, treating, packaging, transporting, or holding food), except that such term does not include: (1) A pesticide chemical in or on a raw agricultural commodity. (2) A pesticide chemical to the extent that it is intended for use or is used in the production, storage, or transportation of any raw agricultural commodity. (3) A color additive. (4) Any substance used in accordance with a sanction or approval granted prior to September 6, 1958, pursuant to the FFDCA, the Poultry Products Inspection Act, or the Federal Meat Inspection Act. (5) A new animal drug. (6) A substance that is generally recognized, among experts qualified by scientific training and experience to evaluate its safety, as having been adequately shown through scientific procedures (or, in the case of a substance used in food prior to January 1, 1958, through either scientific procedures or experience based on common use in food) to be safe under the conditions of its intended use.

**Food additive regulation** (Section 177.3) means a regulation issued pursuant to FFDCA section 409 that states the conditions under which a food additive may be safely used. A food additive regulation under this part ordinarily establishes a tolerance for pesticide residues in or on a particular processed food or a group of such foods. It may also specify: (1) The particular food or classes of food in or on which a food additive may be used. (2) The maximum quantity of the food additive which may be used in or on such food. (3) The manner in which the food additive may be added to or used in or on such food. (4) Directions or other labeling or packaging requirements for the food additive.

**Food commodity** (Section 180.1) means: (1) Any raw agricultural commodity (food or feed) as defined in section 201® of the Federal Food, Drug, and Cosmetic Act (FFDCA); and (2) Any processed food or feed as defined in section 201(gg) of the FFDCA.

**Food grade cargos** (Section 442.2) mean edible and non-edible food products. Specific examples of food grade cargos include, but are not limited to, the following: alcoholic beverages, animal by-products, animal fats, animal oils, caramel, caramel coloring, chocolate, corn syrup and other corn products, dairy products, dietary supplements, eggs, flavorings, food preservatives, food products that are not suitable for human consumption, fruit juices, honey, lard, molasses, non-alcoholic beverages, sweeteners, tallow, vegetable oils, and vinegar.

**Food plant** (Section 174.3) means a plant which either in part or in toto, is used as food.

**Food uses** (Section 180.2003) are the uses of a pesticide chemical that are likely to yield residues in food or feed crops, meat, milk, poultry or eggs.

**Food waste** (Section 243.101) means the organic residues generated by the handling, storage, sale, preparation, cooking, and serving of foods, commonly called garbage.

**Food waste** (Section 246.101) means the organic residues generated by the handling, storage, sale, preparation, cooking, and serving of foods; commonly called garbage.

**Food-chain crops** (Section 260.10) means tobacco, crops grown for human consumption, and crops grown for feed for animals whose products are consumed by humans.

**Forced outage** (Section 72.2) means the removal of a unit from service due to an unplanned component failure or other unplanned condition that requires such removal immediately or within 7 days from the onset of the unplanned component failure or condition. For purposes of Secs. 72.43, 72.91, and 72.92, "forced outage" also includes a partial reduction in the heat input or electrical output due to an unplanned component failure or other unplanned condition that requires such reduction immediately or within 7 days from the onset of the unplanned component failure or condition.

**Forecast period** (Section 93.101) with respect to a transportation plan is the period covered by the transportation plan pursuant to 23 CFR part 450.

**Foreclosure** (Section 745.103) means any of the various methods, statutory or otherwise, known in different jurisdictions, of enforcing payment of a debt, by the taking and selling of real property.

**Foreign** (Section 82.3) state means an entity which is recognized as a sovereign nation or country other than the United States of America. \1\ \1\ Taiwan is not considered a foreign state.

**Foreign state not Party to or Non-Party** (Section 82.3) means a foreign state that has not deposited instruments of ratification, acceptance, or other form of approval with the Directorate of the United Nations Secretariat, evidencing the foreign state's ratification of the provisions of the 1987 Montreal Protocol, the London Amendments, or of the Copenhagen Amendments, as specified.

**Foreign vessel** (Section 94.2) means a vessel of foreign registry or a vessel operated under the authority of a country other than the United States.

**Forest** (Section 170.3) means any operation engaged in the outdoor production of any agricultural plant to produce wood fiber or timber products.

**Forest** (Section 171.2) means a concentration of trees and related vegetation in non-urban areas sparsely inhabited by and infrequently used by humans; characterized by natural terrain and drainage patterns.

**Forest** (Section 503.11) is a tract of land thick with trees and underbrush.

**Forging** (Section 467.02) is the exertion of pressure on dies or rolls surrounding heated aluminum stock, forcing the stock to change shape and in the case where dies are used to take the shape of the die. The forging subcategory is based on the forging process.

**Forging** (Section 471.02) is deforming metal, usually hot, with compressive force into desired shapes, with or without dies. Where dies are used, the metal is forced to take the shape of the die.

**Formal Review** (Section 159.153) means Special Review, Rebuttable Presumption Against Registration (RPAR), FIFRA section 6© suspension proceeding, or FIFRA section 6(b) cancellation proceeding, whether completed or not.

**Formation** (Section 144.3) means a body of consolidated or unconsolidated rock characterized by a degree of lithologic homogeneity which is prevailingly, but not necessarily, tabular and is mappable on the earth's surface or traceable in the subsurface.

**Formation** (Section 146.3) means a body of rock characterized by a degree of lithologic homogeneity which is prevailingly, but not necessarily, tabular and is mappable on the earth's surface or traceable in the subsurface.

**Formation** (Section 147.2902) a body of rock characterized by a degree of lithologic homogeneity which is prevailingly, but not necessarily, tabular and is mappable on the earth's surface or traceable in the subsurface.

**Formation fluid** (Section 144.3) means "fluid" present in a "formation" under natural conditions as opposed to introduced fluids, such as "drilling mud."

**Formation fluid** (Section 146.3) means "fluid" present in a "formation" under natural conditions as opposed to introduced fluids, such as drilling mud.

**Formation oil** (Section 435.11) means the oil from a producing formation which is detected in the drilling fluid, as determined by the GC/MS compliance assurance method specified in appendix 5 of subpart A of this part when the drilling fluid is analyzed before being shipped offshore, and as determined by the RPE method specified in appendix 6 of subpart A of this part when the drilling fluid is analyzed at the offshore point of discharge. Detection of formation oil by the RPE method may be confirmed by the GC/MS compliance assurance method, and the results of the GC/MS compliance assurance method shall supercede those of the RPE method.

**Formation oil** (Section 435.41) means the oil from a producing formation which is detected in the drilling fluid, as determined by the GC/MS compliance assurance method specified in Appendix 5 of subpart A of this part when the drilling fluid is analyzed before being shipped offshore, and as determined by the RPE method specified in Appendix 6 of subpart A of this part when the drilling fluid is analyzed at the offshore point of discharge. Detection of formation oil by the RPE method may be confirmed by the GC/MS compliance assurance method, and the results of the GC/MS compliance assurance method shall supercede those of the RPE method.

**Forming** (Section 471.02) is a set of manufacturing operations in which metals and alloys are made into semifinished products by hot or cold working.

**Formulation** (Section 158.153) means (1) The process of mixing, blending, or dilution of one or more active ingredients with one or more other active or inert ingredients, without an intended chemical reaction, to obtain a manufacturing use product or an end use product, or (2) The repackaging of any registered product.

**Formulation of pesticide products** (Section 455.10) means the process of mixing, blending or diluting one or more pesticide active ingredients (PAIs) with one or more active or inert ingredients, without an intended chemical reaction to obtain a manufacturing use product or an end use product.

**Fossil fuel** (Section 72.2) means natural gas, petroleum, coal, or any form of solid, liquid, or gaseous fuel derived from such material.

**Fossil fuel** (Section 96.2) means natural gas, petroleum, coal, or any form of solid, liquid, or gaseous fuel derived from such material.

**Fossil fuel** (Section 97.2) means natural gas, petroleum, coal, or any form of solid, liquid, or gaseous fuel derived from such material.

**Fossil fuel fired** (Section 97.2) means, with regard to a unit: (1) For units that commenced operation before January 1, 1996, the combination of fossil fuel, alone or in combination with any other fuel, where fossil fuel actually combusted comprises more than 50 percent of the annual heat input on a Btu basis during 1995, or, if a unit had no heat input in 1995, during the last year of operation of the unit prior to 1995; (2) For units that commenced operation on or after January 1, 1996 and before January 1, 1997, the combination of fossil fuel, alone or in combination with any other fuel, where fossil fuel actually combusted comprises more than 50 percent of the annual heat input on a Btu basis during 1996; or (3) For units that commence operation on or after January 1, 1997: (i) The combination of fossil fuel, alone or in combination with any other fuel, where fossil fuel actually combusted comprises more than 50 percent of the annual heat input on a Btu basis during any year; or (ii) The combination of fossil fuel, alone or in combination with any other fuel, where fossil fuel is projected to comprise more than 50 percent of the annual heat input on a Btu basis during any year, provided that the unit shall be "fossil fuel-fired" as of the date, during such year, on which the unit begins combusting fossil fuel.

**Fossil fuel-fired** (Section 72.2) means the combustion of fossil fuel or any derivative of fossil fuel, alone or in combination with any other fuel, independent of the percentage of fossil fuel consumed in any calendar year (expressed in mmBtu).

**Fossil fuel-fired** (Section 96.2) means, with regard to a unit: (1) The combustion of fossil fuel, alone or in combination with any other fuel, where fossil fuel actually combusted comprises more than 50 percent of the annual heat input on a Btu basis during any year starting in 1995 or, if a unit had no heat input starting in 1995, during the last year of operation of the unit prior to 1995; or (2) The combustion of fossil fuel, alone or in combination with any other fuel, where fossil fuel is projected to comprise more than 50 percent of the annual heat input on a Btu basis during any year; provided that the unit shall be "fossil fuel-fired" as of the date, during such year, on which the unit begins combusting fossil fuel.

**Fossil fuel-fired steam generator** (Section 51.100) means a furnace or boiler used in the process of burning fossil fuel for the primary purpose of producing steam by heat transfer.

**Four-Wheel-Drive General Utility Vehicle** (Section 600.022-85) means a four-wheel-drive, general purpose automobile capable of off-highway operation that has a wheelbase not more than 110 inches and that has a body shape similar to a 1977 Jeep CJ-5 or CJ-7, or the 1977 Toyota Land Cruiser, as defined by the Secretary of Transportation at 49 CFR 553.4.

**Fractionation operation** (Section 264.1031) means a distillation operation or method used to separate a mixture of several volatile components of different boiling points in successive stages, each stage removing from the mixture some proportion of one of the components.

**Free available chlorine** (Section 423.11) shall mean the value obtained using the amperometric titration method for free available chlorine described in Standard Methods for the Examination of Water and Wastewater, page 112 ($13^{th}$ edition).

**Free liquids** (Section 260.10) means liquids which readily separate from the solid portion of a waste under ambient temperature and pressure.

**Free moisture** (Section 240.101) means liquid that will drain freely by gravity from solid materials.

**Free product** (Section 280.12) refers to a regulated substance that is present as a non-aqueous phase liquid (e.g., liquid not dissolved in water.)

**Free stall barn** (Section 412.11) shall mean specialized facilities wherein producing cows are permitted free movement between resting and feeding areas.

**Freeboard** (Section 260.10) means the vertical distance between the top of a tank or surface impoundment dike, and the surface of the waste contained therein.

**Freshly manufactured locomotive** (Section 92.2) means a locomotive which is powered by a freshly manufactured engine, and which contains fewer than 25 percent previously used parts (weighted by the dollar value of the parts).

**Freshly manufactured locomotive engine** (Section 92.2) means a new locomotive engine which has not been remanufactured.

**Freshwater** (Section 147.2902) "Underground source of drinking water."

**Friable when referring to material in a school building** (Section 763.83) means that the material, when dry, may be crumbled, pulverized, or reduced to powder by hand pressure, and includes previously nonfriable material after such previously nonfriable material becomes damaged to the extent that when dry it may be crumbled, pulverized, or reduced to powder by hand pressure.

**Friction surface** (Section 745.63) means an interior or exterior surface that is subject to abrasion or friction, including, but not limited to, certain window, floor, and stair surfaces.

**ft** (Section 401.11) means foot (feet).

**FTA** (Section 93.101) means the Federal Transit Administration of DOT.

**FTP** (Section 85.1502) The Federal Test Procedure at part 86.

**Fuel** (Section 600.022-85) means (i) gasoline and diesel fuel for gasoline- or diesel-powered automobiles or (ii) electrical energy for electrically powered automobiles.

**Fuel** (Section 79.2) means any material which is capable of releasing energy or power by combustion or other chemical or physical reaction.

**Fuel blending** (Section 437.2) means the process of combining waste, wastewater, or used material for the purpose of regenerating a fuel for reuse. However, fuel blending may be loosely applied to any process where recovered hydrocarbons are combined as a fuel product where some pretreatment operations generate wastewater.

**Fuel Economy** (Section 600.022-85) means (i) the average number of miles traveled by an automobile or group of automobiles per gallon of gasoline or diesel fuel consumed as computed in Sec. 600.113 or Sec. 600.207 or (ii) the equivalent petroleum-based fuel economy for an electrically powered automobile as determined by the Secretary of Energy.

**Fuel economy** (Section 610.11) means the average number of miles traveled by an automobile per gallon of gasoline (or equivalent amount of other fuel) consumed, as determined by the Administrator in accordance with procedures established under subpart D or F.

**Fuel Economy Data Vehicle** (Section 600.022-85) means a vehicle used for the purpose of determining fuel economy which is not a certification vehicle.

**Fuel evaporative emissions** (Section 86.082-2) means vaporized fuel emitted into the atmosphere from the fuel system of a motor vehicle.

**Fuel flowmeter QA operating quarter** (Section 72.2) means a unit operating quarter in which the unit combusts the fuel measured by the fuel flowmeter for at least 168 unit operating hours (as defined in this section) or more.

**Fuel flowmeter system** (Section 72.2) means an excepted monitoring system (as defined in this section) which provides a continuous record of the flow rate of fuel oil or gaseous fuel, in accordance with appendix D to part 75 of this chapter. A fuel flowmeter system consists of one or more fuel flowmeter components, all necessary auxiliary components (e.g., transmitters, transducers, etc.), and a data acquisition and handling system (DAHS).

**Fuel gas** (Section 65.2) means gases that are combusted to derive useful work or heat.

**Fuel gas system** (Section 65.2) means the offsite and onsite piping and flow and pressure control system that gathers gaseous stream(s) generated by onsite operations, may blend them with other sources of gas, and transports the gaseous stream for use as fuel gas in combustion devices or in-process combustion equipment, such as furnaces and gas turbines, either singly or in combination.

**Fuel manufacturer** (Section 79.2) means any person who, for sale or introduction into commerce, produces, manufactures, or imports a fuel or causes or directs the alteration of the chemical composition of a bulk fuel, or the mixture of chemical compounds in a bulk fuel, by adding to it an additive, except: A party (other than a fuel refiner or importer) who adds a quantity of additive(s) amounting to less than 1.0 percent by volume of the resultant additive(s)/fuel mixture is not thereby considered a fuel manufacturer. (2) A party (other than a fuel refiner or importer) who adds an oxygenate compound to fuel in any otherwise allowable amount is not thereby considered a fuel manufacturer.

**Fuel oil** (Section 72.2) means any petroleum-based fuel (including diesel fuel or petroleum derivatives such as oil tar) as defined by the American Society for Testing and Materials in ASTM D396-90a, "Standard Specification for Fuel Oils" (incorporated by reference in Sec. 72.13), and any recycled or blended petroleum products or petroleum by-products used as a fuel whether in a liquid, solid or gaseous state; provided that for purposes of the monitoring requirements of part 75 of this chapter, "fuel oil" shall be limited to the petroleum-based fuels for which applicable ASTM methods are specified in Appendices D, E, or F of part 75 of this chapter.

**Fuel supply agreement** (Section 72.2) means a legally binding agreement between a new IPP or a firm associated with a new IPP and a fuel supplier that establishes the terms and conditions under which the fuel supplier commits to provide fuel to be delivered to the new IPP.

**Fuel system** (Section 86.082-2) means the combination of fuel tank(s), fuel pump, fuel lines, and carburetor or fuel injection components, and includes all fuel system vents and fuel evaporative emission control system components.

**Fuel system** (Section 86.1803-01) means the combination of fuel tank(s), fuel pump, fuel lines, and carburetor or fuel injection components, and includes all fuel system vents and fuel evaporative emission control system components.

**Fuel system** (Section 90.3) means all components involved in the transport, metering, and mixture of the fuel from the fuel tankto the combustion chamber(s) including the following: fuel tank, fuel tank cap, fuel pump, fuel lines, oil injection metering system, carburetor or fuel injection components, and all fuel system vents.

**Fuel system** (Section 91.3) means all components involved in the transport, metering, and mixture of the fuel from the fuel tank to the combustion chamber(s) including the following: Fuel tank, fuel tank cap, fuel pump, fuel lines, oil injection metering system, carburetor or fuel injection components, and all fuel system vents.

**Fuel system** (Section 92.2) means the combination of fuel tank(s), fuel pump(s), fuel lines and filters, pressure regulator(s), and fuel injection components (or pressure regulator(s) and carburetor(s) if fuel injection is not employed), fuel system vents, and any other component involved in the delivery of fuel to the engine.

**Fuel system** (Section 94.2) means the combination of fuel tank(s), fuel pump(s), fuel lines and filters, pressure regulator(s), and fuel injection components, fuel system vents, and any other component involved in the delivery of fuel to the engine.

**Fuel usage time** (Section 72.2) means the portion of a clock hour during which a unit combusts a particular type of fuel. The fuel usage time, in hours, is expressed as a decimal fraction, with valid values ranging from 0.00 to 1.00.

**Fuel venting emissions** (Section 87.1) means raw fuel, exclusive of hydrocarbons in the exhaust emissions, discharged from aircraft gas turbine engines during all normal ground and flight operations.

**Fuel-fired heater** (Section 86.1702-99) means a fuel burning device which creates heat for the purpose of warming the passenger compartment of a vehicle but does not contribute to the propulsion of the vehicle.

**Fugitive dust, mist or vapor** (Section 129.2) means dust, mist or vapor containing a toxic pollutant regulated under this part which is emitted from any source other than through a stack.

**Fugitive emissions** (Section 57.103) means any air pollutants emitted to the atmosphere other than from a stack.

**Fugitive emissions** (Section 63.2) means those emissions from a stationary source that could not reasonably pass through a stack, chimney, vent, or other functionally equivalent opening. Under section 112 of the Act, all fugitive emissions are to be considered in determining whether a stationary source is a major source.

**Fugitive emissions** (Section 70.2) are those emissions which could not reasonably pass through a stack, chimney, vent, or other functionally-equivalent opening.

**Fugitive emissions** (Section 71.2) are those emissions which could not reasonably pass through a stack, chimney, vent, or other functionally-equivalent opening.

**Full load governed speed** (Section 89.2) is the maximum full load speed as specified by the manufacturer in the sales and service literature and certification application. This speed is the highest engine speed with an advertised power greater than zero.

**Full-time employee** (Section 372.3) means 2,000 hours per year of full-time equivalent employment. A facility would calculate the number of full-time employees by totaling the hours worked during the calendar year by all employees, including contract employees, and dividing that total by 2,000 hours.

**Full-time fellow** (Section 46.120) An individual enrolled in an academic educational program directly related to pollution abatement and control, and taking a minimum of 30 credit hours or an academic workload otherwise defined by the institution as a full-time curriculum for a school year. The fellow need not be pursuing a degree.

**Fume scrubber** (Section 420.121) means wet air pollution control devices used to remove and clean fumes originating from hot coating operations.

**Fume scrubber** (Section 420.91) means those pollution control devices used to remove and clean fumes originating in pickling operations.

**Fumigant** (Section 156.203) means any pesticide product that is a vapor or gas or forms a vapor or gas on application and whose method of pesticidal action is through the gaseous state.

**Fumigant** (Section 170.3) means any pesticide product that is a vapor or gas, or forms a vapor or gas on application, and whose method of pesticidal action is through the gaseous state.

**Functional space** (Section 763.83) means a room, group of rooms, or homogeneous area (including crawl spaces or the space between a dropped ceiling and the floor or roof deck above), such as classroom(s), a cafeteria, gymnasium, hallway(s), designated by a person accredited to prepare management plans, design abatement projects, or conduct response actions.

**Functionally equivalent component** (Section 270.2) means a component which performs the same function or measurement and which meets or exceeds the performance specifications of another component.

**Functions** (Section 164.2) The Administrator may delegate any or part of his authority to act in a given case under subparts B and C of this part to a judicial officer. The Administrator can separately delegate his authority to rule on interlocutory orders and motions, and may also delegate his authority to make findings of fact and draw conclusions of law in a particular proceeding, providing that this delegation shall not preclude the Judicial Officer from referring any motion or case to the Administrator when the Judicial Officer determines such referral to be appropriate. The Administrator, in deciding a case himself, may consult with and assign the preliminary drafting of conclusions of law and findings of fact to any judicial officer.

**Fund** (Section 307.14) means the Hazardous Substance Superfund established by section 9507 of the Internal Revenue Code of 1986.

**Fund or Superfund** (Section 305.3) means the Hazardous Substance Superfund established by section 9507 of the Internal Revenue Code of 1986.

**Fund or Trust Fund** (Section 300.5) means the Hazardous Substance Superfund established by section 9507 of the Internal Revenue Code of 1986.

**Funding period** (Section 35.102) The period of time specified in the grant agreement during which the recipient may expend or obligate funds for the purposes set forth in the agreement.

**Furnace** (Section 240.101) means the chambers of the combustion train where drying, ignition, and combustion of waste material and evolved gases occur.

**Furnace pull** (Section 426.121) shall mean that amount of glass drawn from the glass furnace or furnaces.

**Furnace pull** (Section 426.81) shall mean that amount of glass drawn from the glass furnace or furnaces.

**Future year subaccount** (Section 72.2) means a subaccount in an Allowance Tracking System account, established by the Administrator pursuant to Sec. 73.31 of this chapter, in which allowances are held for one of the 30 years following the later of 1995 or a current calendar year following 1995.

**FWPCA** (Section 220.2) means the Federal Water Pollution Control Act, as amended (33 U.S.C. 1251);

**FWS** (Section 233.2) means the U.S. Fish and Wildlife Service.

# G

**g** (Section 401.11) means gram(s).

**GAC10** (Section 141.2) means granular activated carbon filter beds with an empty-bed contact time of 10 minutes based on average daily flow and a carbon reactivation frequency of every 180 days.

**gal** (Section 401.11) means gallon(s).

**Galvanized basis material** (Section 465.02) means zinc coated steel, galvalum, brass and other copper base strip which is processed in coil coating.

**Galvanizing** (Section 420.121) means coating steel products with zinc by the hot dip process including the immersion of the steel product in a molten bath of zinc metal, and the related operations preceding and subsequent to the immersion phase.

**Garbage** (Section 435.41) means all kinds of victual, domestic, and operational waste, excluding fresh fish and parts thereof, generated during the normal operation of coastal oil and gas facility and liable to be disposed of continuously or periodically, except dishwater, graywater, and those substances that are defined or listed in other Annexes to MARPOL 73/78. A copy of MARPOL may be inspected at EPA's Water Docket; 1200 Pennsylvania Ave., NW., Washington, DC 20460.

**Garden hose** (Section 247.3) means a flexible tubing that conducts water to a specific location;

**Gas manufacturer's intermediate standard (GMIS)** (Section 72.2) means a compressed gas calibration standard that has been assayed and certified by direct comparison to a standard reference material (SRM), an SRM-equivalent PRM, a NIST/EPA-approved certified reference material (CRM), or a NIST traceable reference material (NTRM), in accordance with section 2.1.2.1 of the "EPA Traceability Protocol for Assay and Certification of Gaseous Calibration Standards," September 1997, EPA-600/R-97/121.

**Gas well** (Section 435.61) shall mean any well which produces natural gas in a ratio to the petroleum liquids produced greater than 15,000 cubic feet of gas per 1 barrel (42 gallons) of petroleum liquids.

**Gaseous fuel** (Section 72.2) means a material that is in the gaseous state at standard atmospheric temperature and pressure conditions and that is combusted to produce heat.

**Gaseous fuel** (Section 86.1803-01) means natural gas or liquefied petroleum gas.

**Gaseous fuel** (Section 92.2) means a fuel which is a gas at standard temperature and pressure. This includes both natural gas and liquefied petroleum gas.

**Gaseous fuels** (Section 86.1702-99) means liquefied petroleum gas, compressed natural gas, or liquefied natural gas fuels for use in motor vehicles.

**Gas-fired** (Section 72.2) means: (1) For all purposes under the Acid Rain Program, except for part 75 of this chapter, the combustion of: (i) Natural gas or other gaseous fuel (including coal-derived gaseous fuel), for at least 90.0 percent of the unit's average annual heat input during the previous three calendar years and for at least 85.0 percent of the annual heat input in each of those calendar years.

**Gasoline** (Section 80.2) means any fuel sold in any State \1\ for use in motor vehicles and motor vehicle engines, and commonly or commercially known or sold as gasoline.

**Gasoline blending stock, blendstock, or component** (Section 80.2) means any liquid compound which is blended with other liquid compounds to produce gasoline.

**Gathering lines** (Section 280.12) means any pipeline, equipment, facility, or building used in the transportation of oil or gas during oil or gas production or gathering operations.

**Gear oils** (Section 247.3) means petroleum-based oils used for lubricating machinery gears;

**General account** (Section 72.2) means an Allowance Tracking System account that is not a unit account.

**General account** (Section 96.2) means a NOX Allowance Tracking System account, established under subpart F of this part, that is not a compliance account or an overdraft account.

**General account** (Section 97.2) means a NOX Allowance Tracking System account, established under subpart F of this part, that is not a compliance account or an overdraft account.

**General commercial use** (Section 725.3) means use for commercial purposes other than research and development.

**General Counsel** (Section 1603.3) means the Board's principal legal officer, or a CSB attorney serving as Acting General Counsel.

**General environment** (Section 190.02) means the total terrestrial, atmospheric and aquatic environments outside sites upon which any operation which is part of a nuclear fuel cycle is conducted.

**General environment** (Section 191.02) means the total terrestrial, atmospheric, and aquatic environments outside sites within which any activity, operation, or process associated with the management and storage of spent nuclear fuel or radioactive waste is conducted.

**General environment** (Section 197.2) means everywhere outside the Yucca Mountain site, the Nellis Air Force Range, and the Nevada Test Site.

**General permit (NPDES and 404)** (Section 124.2) means an NPDES or 404 "permit" authorizing a category of discharges or activities under the CWA within a geographical area. For NPDES, a general permit means a permit issued under Sec. 122.28. For 404, a general permit means a permit issued under Sec. 233.37.

**General permit** (Section 122.2) means an NPDES "permit" issued under Sec. 122.28 authorizing a category of discharges under the CWA within a geographical area.

**General permit** (Section 232.2) means a permit authorizing a category of discharges of dredged or fill material under the Act. General permits are permits for categories of discharge which are similar in nature, will cause only minimal adverse environmental effects when performed separately, and will have only minimal cumulative adverse effect on the environment.

**General permit** (Section 70.2) means a part 70 permit that meets the requirements of Sec. 70.6(d).

**General permit** (Section 71.2) means a part 71 permit that meets the requirements of Sec. 71.6(d).

**General processing** (Section 410.31) shall mean the internal subdivision of the low water use processing subcategory for facilities described in Sec. 410.30 that do not qualify under the water jet weaving subdivision.

**General purpose unit of local government** (Section 310.3) means the governing body of a county, parish, municipality, city, town, township, Federally-recognized Indian tribe or similar governing body. This term does not include special purpose districts.

**Generation** (Section 243.101) means the act or process of producing solid waste.

**Generation** (Section 246.101) means the act or process of producing solid waste.

**Generator** (Section 144.3) means any person, by site location, whose act or process produces hazardous waste identified or listed in 40 CFR part 261.

**Generator** (Section 146.3) means any person, by site location, whose act or process produces hazardous waste identified or listed in 40 CFR part 261.

**Generator** (Section 260.10) means any person, by site, whose act or process produces hazardous waste identified or listed in part 261 of this chapter or whose act first causes a hazardous waste to become subject to regulation.

**Generator** (Section 270.2) means any person, by site location, whose act, or process produces "hazardous waste" identified or listed in 40 CFR part 261.

**Generator** (Section 273.9) means any person, by site, whose act or process produces hazardous waste identified or listed in part 261 of this chapter or whose act first causes a hazardous waste to become subject to regulation.

**Generator** (Section 72.2) means a device that produces electricity and was or would have been required to be reported as a generating unit pursuant to the United States Department of Energy Form 860 (1990 edition).

**Generator** (Section 96.2) means a device that produces electricity.

**Generator** (Section 97.2) means a device that produces electricity.

**Generator of PCB waste** (Section 761.13) means any person whose act or process produces PCBs that are regulated for disposal under subpart D of this part, or whose act first causes PCBs or PCB Items to become subject to the disposal requirements of subpart D of this part, or who has physical control over the PCBs when a decision is made that the use of the PCBs has been terminated and therefore is subject to the disposal requirements of subpart D of this part. Unless another provision of this part specifically requires a site-specific meaning, "generator of PCB waste" includes all of the sites of PCB waste generation owned or operated by the person who generates PCB waste.

**Generator Output capacity** (Section 72.2) means the full-load continuous rating of a generator under specific conditions as designed by the manufacturer.

**Genetic material necessary for the production** (Section 152.3) means both: Genetic material that encodes a substance or leads to the production of a substance, and regulatory regions. It does not include noncoding, nonexpressed nucleotide sequences.

**Genetic material necessary for the production** (Section 174.3) means both: Genetic material that encodes a substance or leads to the production of a substance; and regulatory regions. It does not include noncoding, nonexpressed nucleotide sequences.

**Genome** (Section 174.3) means the sum of the heritable genetic material in the plant, including genetic material in the nucleus and organelles.

**Genome** (Section 725.3) means the sum total of chromosomal and extrachromosomal genetic material of an isolate and any descendants derived under pure culture conditions from that isolate.

**Genus** (Section 132.2) mean acute value (GMAV) is the geometric mean of the SMAVs for the genus.

**Genus mean chronic value (GMCV)** (Section 132.2) is the geometric mean of the SMCVs for the genus.

**Gigawatt-year** (Section 190.02) refers to the quantity of electrical energy produced at the busbar of a generating station. A gigawatt is equal to one billion watts. A gigawatt-year is equivalent to the amount of energy output represented by an average electric power level of one gigawatt sustained for one year.

**Good engineering practice (GEP) stack height** (Section 51.100) means the greater of: (1) 65 meters, measured from the ground-level elevation at the base of the stack: (2)(i) For stacks in existence on January 12, 1979, and for which the owner or operator had obtained all applicable permits or approvals required under 40 CFR parts 51 and 52. $Hg = 2.5H$, provided the owner or operator produces evidence that this equation was actually relied on in establishing an emission limitation: (ii) For all other stacks, $Hg = H + 1.5L$ where: Hg = good engineering practice stack height, measured from the ground-level elevation at the base of the stack, H = height of nearby structure(s) measured from the ground-level elevation at the base of the stack. L = lesser dimension, height or projected width, of nearby structure(s) provided that the EPA, State or local control agency may require the use of a field study or fluid model to verify GEP stack height for the source: or (3) The height demonstrated by a fluid model or a field study approved by the EPA State or local control agency, which ensures that the emissions from a stack do not result in excessive concentrations of any air pollutant as a result of atmospheric downwash, wakes, or eddy effects created by the source itself, nearby structures or nearby terrain features. Nearby as used in Sec. 51.100(ii) of this part is defined for a specific structure or terrain feature and (1) For purposes of applying the formulae provided in Sec. 51.100(ii)(2) means that distance up to five times the lesser of the height or the width dimension of a structure, but not greater than 0.8 km (\1/2\ mile), and (2) For conducting demonstrations under Sec. 51.100(ii)(3) means not greater than 0.8 km (\1/2\ mile), except that the portion of a terrain feature may be considered to be nearby which falls within a distance of up to 10 times the maximum height (Ht) of the feature, not to exceed 2 miles if such feature achieves a height (Ht) 0.8 km from the stack that is at least 40 percent of the GEP stack height determined by the formulae provided in Sec. 51.100(ii)(2)(ii) of this part or 26 meters, whichever is greater, as measured from the ground-level elevation at the base of the stack. The height of the structure or terrain feature is measured from the ground-level elevation at the base of the stack.

**Good engineering practice** (Section 92.102) means those methods and practices which the Administrator determines to be consistent with scientific and engineering principles.

**Grab sample** (Section 471.02) is a single sample which is collected at a time and place most representative of total discharge.

**Grape juice canning** (Section 407.61) shall mean the processing of grape juice into the following products and product styles: Canned and frozen, fresh and stored, natural grape juice for the manufacture of juices, drinks, concentrates, jams, jellies, and other related finished products but not wine or other spirits. In terms of raw material processed 1000 kg (1000 lb) of grapes are equivalent to 834 liters (100 gallons) of grape juice.

**Grape pressing** (Section 407.61) shall mean the washing and subsequent handling including pressing, heating, and filtration of natural juice from all varieties of grapes for the purpose of manufacturing juice, drink, concentrate, and jelly but not wine or other spirits. In terms of raw material processed 1000 kg (1000 lb) of grapes are equivalent to 834 liters (100 gallons) of grape juice.

**Grate siftings** (Section 240.101) means the materials that fall from the solid waste fuel bed through the grate openings.

**Gray Iron** (Section 464.31) A cast iron that gives a gray fracture due to the presence of flake graphite.

**Great Lakes** (Section 132.2) means Lake Ontario, Lake Erie, Lake Huron (including Lake St. Clair), Lake Michigan, and Lake Superior; and the connecting channels (Saint Mary's River, Saint Clair River, Detroit River, Niagara River, and Saint Lawrence River to the Canadian Border).

**Great Lakes States and Great Lakes Tribes, or Great Lakes States and Tribes** (Section 132.2) means the States of Illinois, Indiana, Michigan, Minnesota, New York, Ohio, Pennsylvania, and Wisconsin, and any Indian Tribe as defined in this part which is located in whole or in part within the drainage basin of the Great Lakes, and for which EPA has approved water quality standards under section 303 of the Clean Water Act or which EPA has authorized to administer an NPDES program under section 402 of the Clean Water Act.

**Great Lakes System** (Section 132.2) means all the streams, rivers, lakes and other bodies of water within the drainage basin of the Great Lakes within the United States.

**Green engine factor** (Section 92.2) means a factor that is applied to emission measurements from a locomotive or locomotive engine that has had little or no service accumulation. The green engine factor adjusts emission measurements to be equivalent to emission measurements from a locomotive or locomotive engine that has had approximately 300 hours of use.

**Green Engine Factor** (Section 94.2) means a factor that is applied to emission measurements from an engine that has had little or no service accumulation. The Green Engine Factor adjusts emission measurements to be equivalent to emission measurements from an engine that has had approximately 300 hours of use.

**Greenhouse** (Section 170.3) means any operation engaged in the production of agricultural plants inside any structure or space that is enclosed with nonporous covering and that is of sufficient size to permit worker entry. This term includes, but is not limited to, polyhouses, mushroom houses, rhubarb houses, and similar structures. It does not include such structures as malls, atriums, conservatories, arboretums, or office buildings where agricultural plants are present primarily for aesthetic or climatic modification.

**Grinding** (Section 471.02) is the process of removing stock from a workpiece by the use of a tool consisting of abrasive grains held by a rigid or semi-rigid grinder. Grinding includes surface finishing, sanding, and slicing.

**Gross alpha particle activity** (Section 141.2) means the total radioactivity due to alpha particle emission as inferred from measurements on a dry sample.

**Gross beta particle activity** (Section 141.2) means the total radioactivity due to beta particle emission as inferred from measurements on a dry sample.

**Gross calorific value** (Section 240.101) means heat liberated when waste is burned completely and the products of combustion are cooled to the initial temperature of the waste. Usually expressed in British thermal units per pound.

**Gross cane** (Section 409.41) shall mean that amount of crop material as harvested, including field trash and other extraneous material.

**Gross cane** (Section 409.61) shall mean that amount of crop material as harvested, including field trash and other extraneous material.

**Gross Combination Weight Rating (GCWR)** (Section 202.10) means the value specified by the manufacturer as the loaded weight of a combination vehicle.

**Gross Combination Weight Rating (GCWR)** (Section 205.51) means the value specified by the manufacturer as the loaded weight of a combination vehicle.

**Gross power** (Section 89.2) means the power measured at the crankshaft or its equivalent, the engine being equipped only with the standard accessories (such as oil pumps, coolant pumps, and so forth) necessary for its operation on the test bed. **Alternators** (Section 89.2) must be used, if necessary, to run the engine. Fans, air conditioners, and other accessories may be used at the discretion of the manufacturer, but no power adjustments for these accessories may be made.

**Gross power** (Section 90.3) means the power measured at the crankshaft or its equivalent, the engine being equipped only with the standard accessories (such as oil pumps, coolant pumps, and so forth) necessary for its operation on the test bed.

**Gross power** (Section 91.3) means the power measured at the crankshaft or its equivalent (for outboards, the power may be measured at the propeller shaft), the engine being equipped only with the standard accessories (such as oil pumps, coolant pumps, and so forth) necessary for its operation on the test bed.

**Gross production of fiberboard products** (Section 429.11) means the air dry weight of hardboard or insulation board following formation of the mat and prior to trimming and finishing operations.

**Gross vehicle weight** (Section 86.082-2) means the manufacturer's gross weight rating for the individual vehicle.

**Gross vehicle weight** (Section 86.1803-01) means the manufacturer's gross weight rating for the individual vehicle.

**Gross Vehicle Weight Rating (GVWR)** (Section 202.10) means the value specified by the manufacturer as the loaded weight of a single vehicle.

**Gross Vehicle Weight Rating (GVWR)** (Section 205.51) means the value specified by the manufacturer as the loaded weight of a single vehicle.

**Gross vehicle weight rating (GVWR)** (Section 86.082-2) means the value specified by the manufacturer as the maximum design loaded weight of a single vehicle.

Gross vehicle weight rating (GVWR) means the value specified by the manufacturer as the maximum design loaded weight of a single vehicle.

**Gross Vehicle Weight Rating** (Section 600.022-85) means the manufacturer's gross weight rating for the individual vehicle.

**Ground water** (Section 144.3) means water below the land surface in a zone of saturation.

**Ground water** (Section 146.3) means water below the land surface in a zone of saturation.

**Ground water** (Section 147.2902) water below the land surface in a zone of saturation.

**Ground water** (Section 191.12) means water below the land surface in a zone of saturation.

**Ground water** (Section 258.2) means water below the land surface in a zone of saturation.

**Ground water** (Section 260.10) means water below the land surface in a zone of saturation.

**Ground water** (Section 270.2) means water below the land surface in a zone of saturation.

**Ground water as defined by section 101(12) of CERCLA** (Section 300.5) means water in a saturated zone or stratum beneath the surface of land or water.

**Ground water under the direct influence of surface water (GWUDI)** (Section 141.2) means any water beneath the surface of the ground with significant occurrence of insects or other macroorganisms, algae, or large-diameter pathogens such as Giardia lamblia or Cryptosporidium, or significant and relatively rapid shifts in water characteristics such as turbidity, temperature, conductivity, or pH which closely correlate to climatological or surface water conditions. Direct influence must be determined for individual sources in accordance with criteria established by the State. The State determination of direct influence may be based on site-specific measurements of water quality and/or documentation of well construction characteristics and geology with field evaluation.

**Groundwater** (Section 192) means water below the ground surface in a zone of saturation.

**Groundwater infiltration in Sec. 440.131** (Section 440.132) means that water which enters the treatment facility as a result of the interception of natural springs, aquifers, or run-off which percolates into the ground and seeps into the treatment facility's tailings pond or wastewater holding facility and that cannot be diverted by ditching or grouting the tailings pond or wastewater holding facility.

**Group 1 boiler** (Section 76.2) means a tangentially fired boiler or a dry bottom wall-fired boiler (other than a unit applying cell burner technology).

**Group 1 mixtures** (Section 455.10) means any product whose only pesticidal active ingredient(s) is: a common food/food constituent or non-toxic household item; or is a substance that is generally recognized as safe (GRAS) by the Food and Drug Administration (21 CFR 170.30, 182, 184, and 186) in accordance with good manufacturing practices, as defined by 21 CFR part 182; or is exempt from FIFRA under 40 CFR 152.25.

**Group 1 process vent** (Section 65.2) means a process vent for which the flow rate is greater than or equal to 0.011 standard cubic meter per minute (0.39 cubic feet per minute); th total concentration is greater than or equal to the appropriate value in table 1 of subpart D of this part, and the total resource effectiveness index value, calculated according to Sec. 65.64(h) is less than or equal to 1.0.

**Group 2 boiler** (Section 76.2) means a wet bottom wall-fired boiler, a cyclone boiler, a boiler applying cell burner technology, a vertically fired boiler, an arch-fired boiler, or any other type of utility boiler (such as a fluidized bed or stoker boiler) that is not a Group 1 boiler.

**Group 2 mixtures** (Section 455.10) means those chemicals listed in Table 9 to this part 455.

**Group 2A process vent** (Section 65.2) means a process vent that is not Group 1 or Group 2B for which monitoring and recordkeeping are required to demonstrate a total resource effectiveness index value greater than 1.0.

**Group 2B process vent** (Section 65.2) means a process vent that is not Group 1 or Group 2A for which monitoring and recordkeeping are not required to demonstrate a total resource effectiveness index value greater than 4.0, or which is exempt from control requirements due to the vent stream's flow rate, regulated material concentration, or total resource effectiveness index value.

**Guest instructor** (Section 745.223) means an individual designated by the training program manager or principal instructor to provide instruction specific to the lecture, hands-on activities, or work practice components of a course.

**Guidance** (Section 239.2) means policy memorandum, an application for approval under this Part, or other technical or policy documents that supplement state laws and regulations. These documents provide direction with regard to how state agencies should interpret their permit program requirements and must be consistent with state laws and regulations.

**Guidelines** (Section 766.3) means the Midwest Research Institute (MRI) publication Guidelines for the Determination of Polyhalogenated Dioxins and Dibenzofurans in Commercial Products, EPA contract No. 68-02-3938; MRI Project No. 8201-A(41), 1985.

# H

**ha** (Section 401.11) means hectare(s).

**Hair pulp** (Section 425.02) means the removal of hair by chemical dissolution.

**Hair save** (Section 425.02) means the physical or mechanical removal of hair which has not been chemically dissolved, and either selling the hair as a by-product or disposing of it as a solid waste.

**Haloacetic acids (five) (HAA5)** (Section 141.2) mean the sum of the concentrations in milligrams per liter of the haloacetic acid compounds (monochloroacetic acid, dichloroacetic acid, trichloroacetic acid, monobromoacetic acid, and dibromoacetic acid), rounded to two significant figures after addition.

**Halogen** (Section 141.2) means one of the chemical elements chlorine, bromine or iodine.

**Halogenated organic compounds or HOCs** (Section 268.2) means those compounds having a carbon-halogen bond which are listed under appendix III to this part.

**Halogenated vent stream or halogenated stream** (Section 65.2) means, for purposes of this part, a vent stream determined to be halogenated by the procedures specified in Sec. 65.85© for transfer racks and in Sec. 65.64(g) for process vents, as applicable.

**Halogens and hydrogen halides** (Section 65.2) means hydrogen chloride (HCl), chlorine (Cl2), hydrogen bromide (HBr), bromine (Br2), and hydrogen fluoride (HF).

**Ham processor** (Section 432.81) shall mean an operation which manufactures hams alone or in combination with other finished products at rates greater than 2730 kg (6000 lb) per day.

**Hand labor** (Section 170.3) means any agricultural activity performed by hand or with hand tools that causes a worker to have substantial contact with surfaces (such as plants, plant parts, or soil) that may contain pesticide residues. These activities include, but are not limited to, harvesting, detasseling, thinning, weeding, topping, planting, sucker removal, pruning, disbudding, roguing, and packing produce into containers in the field. Hand labor does not include operating, moving, or repairing irrigation or watering equipment or performing the tasks of crop advisors.

**Handheld equipment engine** (Section 90.3) means a nonroad engine that meets the requirements specified in Sec. 90.103(a)(2)(I) through (v).

**Handler** (Section 170.3) means any person, including a self-employed person: (1) Who is employed for any type of compensation by an agricultural establishment or commercial pesticide handling establishment to which subpart C of this part applies and who is: (i) Mixing, loading, transferring, or applying pesticides. (ii) Disposing of pesticides or pesticide containers. (iii) Handling opened containers of pesticides. (iv) Acting as a flagger. (v) Cleaning, adjusting, handling, or repairing the parts of mixing, loading, or application equipment that may contain pesticide residues. (vi) Assisting with the application of pesticides. (vii) Entering a greenhouse or other enclosed area after the application and before the inhalation exposure level listed in the labeling has been reached or one of the ventilation criteria established by this part ( Sec. 170.110©(3)) or in the labeling has been met: (A) To operate ventilation equipment. (B) To adjust or remove coverings used in fumigation. (C) To monitor air levels. (viii) Entering a treated area outdoors after application of any soil fumigant to adjust or remove soil coverings such as tarpaulins. (ix) Performing tasks as a crop advisor: (A) During any pesticide application. (B) Before the inhalation exposure level listed in the labeling has been reached or one of the ventilation criteria established by this part (Sec. 170.110©(3))

or in the labeling has been met. (C) During any restricted-entry interval. (2) The term does not include any person who is only handling pesticide containers that have been emptied or cleaned according to pesticide product labeling instructions or, in the absence of such instructions, have been subjected to triple-rinsing or its equivalent.

**Handler employer** (Section 170.3) means any person who is self-employed as a handler or who employs any handler, for any type of compensation.

**Hands-on skills assessment** (Section 745.223) means an evaluation which tests the trainees' ability to satisfactorily perform the work practices and procedures identified in Sec. 745.225(d), as well as any other skill taught in a training course.

**Hang-up** (Section 86.082-2) refers to the process of hydrocarbon molecules being adsorbed, condensed, or by any other method removed from the sample flow prior to reaching the instrument detector. It also refers to any subsequent desorption of the molecules into the sample flow when they are assumed to be absent.

**Hang-up** (Section 86.1803-01) refers to the process of hydrocarbon molecules being adsorbed, condensed, or by any other method removed from the sample flow prior to reaching the instrument detector. It also refers to any subsequent desorption of the molecules into the sample flow when they are assumed to be absent.

**Hang-up** (Section 92.102) refers to the process of hydrocarbon molecules being adsorbed, condensed, or by any other method removed from the sample flow prior to reaching the instrument detector. It also refers to any subsequent desorption of the molecules into the sample flow when they are assumed to be absent.

**HAP** (Section 63.601) metals mean those metals and their compounds (in particulate or volatile form) that are included on the list of hazardous air pollutants in section 112 of the Clean Air Act. HAP metals include, but are not limited to: antimony, arsenic, beryllium, cadmium, chromium, lead, manganese, nickel, and selenium expressed as particulate matter as measured by the methods and procedures in this subpart or an approved alternative method. For the purposes of this subpart, HAP metals are expressed as particulate matter as measured by 40 CFR part 60, appendix A, Method 5.

**Hardboard** (Section 429.11) means a panel manufactured from interfelted ligno-cellulosic fibers consolidated under heat and pressure to a density of 0.5 g/cu cm (31 lb/cu ft) or greater.

**Hardener** (Section 59.101) means a coating component specifically designed to promote a faster cure of an enamel finish.

**Hard-piping** (Section 265.1081) means pipe or tubing that is manufactured and properly installed in accordance with relevant standards and good engineering practices.

**Hard-piping** (Section 65.2) means pipe or tubing that is manufactured and installed using good engineering judgment and standards, such as ASME B31.3, Process Piping (available from the American Society of Mechanical Engineers, PO Box 2900, Fairfield, NJ 07007-2900).

**Hatchback** (Section 600.022-85) means a passenger automobile where the conventional luggage compartment, i.e., trunk, is replaced by a cargo area which is open to the passenger compartment and accessed vertically by a rear door which encompasses the rear window.

**Hazard** (Section 171.2) means a probability that a given pesticide will have an adverse effect on man or the environment in a given situation, the relative likelihood of danger or ill effect being dependent on a number of interrelated factors present at any given time.

**Hazard category** (Section 370.2) means any of the following: (1) Immediate (acute) health hazard, including highly toxic, toxic, irritant, sensitizer, corrosive, (as defined under Sec. 1910.1200 of Title 29 of the Code of Federal Regulations) and other hazardous chemicals that cause an adverse effect to a target organ and which effect usually occurs rapidly as a result of short term exposure and is of short duration; (2) Delayed (chronic) health hazard, including carcinogens (as defined under Sec. 1910.1200 of Title 29 of the Code of Federal

Regulations) and other hazardous chemicals that cause an adverse effect to a target organ and which effect generally occurs as a result of long term exposure and is of long duration; (3) Fire hazard, including flammable, combustible liquid, pyrophoric, and oxidizer (as defined under Sec. 1910.1200 of Title 29 of the Code of Federal Regulations); (4) Sudden release of pressure, including explosive and compressed gas (as defined under Sec. 1910.1200 of Title 29 of the Code of Federal Regulations); and (5) Reactive, including unstable reactive, organic peroxide, and water reactive (as defined under Sec. 1910.1200 of Title 29 of the Code of Federal Regulations).

**Hazard Ranking System (HRS)** (Section 300.5) means the method used by EPA to evaluate the relative potential of hazardous substance releases to cause health or safety problems, or ecological or environmental damage.

**Hazardous air pollutant** (Section 63.2) means any air pollutant listed in or pursuant to section 112(b) of the Act.

**Hazardous chemical** (Section 355.20) means any hazardous chemical as defined under Sec. 1910.1200© of Title 29 of the Code of Federal Regulations, except that such term does not include the following substances: (1) Any food, food additive, color additive, drug, or cosmetic regulated by the Food and Drug Administration. (2) Any substance present as a solid in any manufactured item to the extent exposure to the substance does not occur under normal conditions of use. (3) Any substance to the extent it is used for personal, family, or household purposes, or is present in the same form and concentration as a product packaged for distribution and use by the general public. (4) Any substance to the extent it is used in a research laboratory or a hospital or other medical facility under the direct supervision of a technically qualified individual. (5) Any substance to the extent it is used in routine agricultural operations or is a fertilizer held for sale by a retailer to the ultimate customer.

**Hazardous chemical** (Section 370.2) means any hazardous chemical as defined under Sec. 1910.1200© of Title 29 of the Code of Federal Regulations, except that such term does not include the following substances: (1) Any food, food additive, color additive, drug, or cosmetic regulated by the Food and Drug Administration. (2) Any substance present as a solid in any manufactured item to the extent exposure to the substance does not occur under normal conditions of use. (3) Any substance to the extent it is used for personal, family, or household purposes, or is present in the same form and concentration as a product packaged for distribution and use by the general public. (4) Any substance to the extent it is used in a research laboratory or a hospital or other medical facility under the direct supervision of a technically qualified individual. (5) Any substance to the extent it is used in routine agricultural operations or is a fertilizer held for sale by a retailer to the ultimate customer.

**Hazardous constituent or constituents** (Section 268.2) means those constituents listed in appendix VIII to part 261 of this chapter.

**Hazardous debris** (Section 268.2) means debris that contains a hazardous waste listed in subpart D of part 261 of this chapter, or that exhibits a characteristic of hazardous waste identified in subpart C of part 261 of this chapter. Any deliberate mixing of prohibited hazardous waste with debris that changes its treatment classification (i.e., from waste to hazardous debris) is not allowed under the dilution prohibition in Sec. 268.3.

**Hazardous substance** (Section 122.2) means any substance designated under 40 CFR part 116 pursuant to section 311 of CWA.

**Hazardous substance** (Section 300.5) as defined by section 101(14) of CERCLA, means: Any substance designated pursuant to section 311(b)(2)(A) of the CWA; any element, compound, mixture, solution, or substance designated pursuant to section 102 of CERCLA; any hazardous waste having the characteristics identified under or listed pursuant to section 3001 of the Solid Waste Disposal Act (but not including any waste the regulation of which under the Solid Waste Disposal Act (42 U.S.C. 6901 et seq.) has been suspended by Act of Congress); any toxic pollutant listed under section

307(a) of the CWA; any hazardous air pollutant listed under section 112 of the Clean Air Act (42 U.S.C. 7521 et seq.); and any imminently hazardous chemical substance or mixture with respect to which the EPA Administrator has taken action pursuant to section 7 of the Toxic Substances Control Act (15 U.S.C. 2601 et seq.). The term does not include petroleum, including crude oil or any fraction thereof which is not otherwise specifically listed or designated as a hazardous substance in the first sentence of this paragraph, and the term does not include natural gas, natural gas liquids, liquified natural gas, or synthetic gas usable for fuel (or mixtures of natural gas and such synthetic gas).

**Hazardous substance** (Section 302.3) means any substance designated pursuant to 40 CFR part 302; Hazardous waste shall have the meaning provided in 40 CFR 261.3;

**Hazardous substance** (Section 310.3) (1) Hazardous substance, as defined by section 101(14) of CERCLA, means: (i) Any substance designated pursuant to section 311(b)(2)(A) of the Federal Water Pollutión Control Act (Pub. L. 101-380, 33 U.S.C. 1251 et seq.); (ii) Any element, compound, mixture, solution, or substance designated pursuant to section 102 of CERCLA; (iii) Any hazardous waste having the characteristics identified under or listed pursuant to section 3001 of the Solid Waste Disposal Act (Pub. L. 89-272, 42 U.S.C.3259 et seq.) (but not including any waste the regulation of which under the Solid Waste Disposal Act has been suspended by Act of Congress); (iv) Any toxic pollutant listed under section 307(a) of the Federal Water Pollution Control Act (Pub. L. 101-380, 33 U.S.C. 1251 et seq.); (v) Any hazardous air pollutant listed under section 112 of the Clean Air Act (42 U.S.C. 7401-7642); and (vi) Any imminently hazardous chemical substance or mixture with respect to which the Administrator has taken action pursuant to section 7 of the Toxic Substances Control Act (Pub. L. 94-469, 15 U.S.C. 2601-2629). (2) The term does not include petroleum, including crude oil or any fraction thereof that is not otherwise specifically listed or designated as a hazardous substance under paragraphs (f)(1)(i) through (f)(1)(vi) of this section, and the term does not include natural gas, natural gas liquids, liquefied natural gas, or synthetic gas usable for fuel (or mixtures of natural gas and such synthetic gas).

**Hazardous substance as defined by section 101(14) of CERCLA** (Section 307.14) means: (1) Any substance designated pursuant to section 311(b)(2)(A) of the Federal Water Pollution Control Act (33 U.S.C. 1251 et seq.); (2) Any element, compound, mixture, solution, or substance designated pursuant to section 102 of CERCLA; (3) Any hazardous waste having the characteristics identified under or listed pursuant to section 3001 of the Solid Waste Disposal Act (42 U.S.C. 6801 et seq.) (but not including any waste the regulation of which under the Solid Waste Disposal Act has been suspended by Act of Congress); (4) Any toxic pollutant listed under section 307(a) of the Federal Water Pollution Control Act; (5) Any hazardous air pollutant listed under section 112 of the Clean Air Act (42 U.S.C. 7401 et seq.); and (6) Any imminently hazardous chemical substance or mixture with respect to which the Administrator of EPA (Administrator) has taken action pursuant to section 7 of the Toxic Substances Control Act (15 U.S.C. 2601 et seq.). The term does not include petroleum, including crude oil or any fraction thereof which is not otherwise specifically listed or designated as a hazardous substance under paragraphs (1) through (6) of this definition, and the term does not include natural gas, natural gas liquids, liquefied natural gas, or synthetic gas usable for fuel (or mixtures of natural gas and such synthetic gas).

**Hazardous substance UST system** (Section 280.12) means an underground storage tank system that contains a hazardous substance defined in section 101(14) of the Comprehensive Environmental Response, Compensation and Liability Act of 1980 (but not including any substance regulated as a hazardous waste under subtitle C) or any mixture of such substances and petroleum, and which is not a petroleum UST system.

**Hazardous substances** (Section 373.4) means that group of substances defined as hazardous under CERCLA 101(14), and that appear at 40 CFR 302.4.

**Hazardous waste** (Section 144.3) means a hazardous waste as defined in 40 CFR 261.3.

**Hazardous waste** (Section 146.3) means a hazardous waste as defined in 40 CFR 261.3.

**Hazardous waste** (Section 240.101) means any waste or combination of wastes which pose a substantial present or potential hazard to human health or living organisms because such wastes are nondegradable or persistent in nature or because they can be biologically magnified, or because they can be lethal, or because they may otherwise cause or tend to cause detrimental cumulative effects.

**Hazardous waste** (Section 243.101) means a waste or combination of wastes of a solid, liquid, contained gaseous, or semisolid form which may cause, or contribute to, an increase in mortality or an increase in serious irreversible, or incapacitating reversible illness, taking into account the toxicity of such waste, its persistence and degradability in nature, its potential for accumulation or concentration in tissue, and other factors that may otherwise cause or contribute to adverse acute or chronic effects on the health of persons or other organisms.

**Hazardous waste** (Section 260.10) means a hazardous waste as defined in Sec. 261.3 of this chapter.

**Hazardous waste** (Section 270.2) means a hazardous waste as defined in 40 CFR 261.3.

**Hazardous waste** (Section 62.1201) is defined in Sec. 261.3 of this chapter.

**Hazardous waste** (Section 745.223) means any waste as defined in 40 CFR 261.3.

**Hazardous waste burning cement kiln** (Section 62.1201) means a rotary kiln and any associated preheater or precalciner devices that produce clinker by heating limestone and other materials for subsequent production of cement for use in commerce, and that burns hazardous waste at any time.

**Hazardous waste combustor** (Section 62.1201) means a hazardous waste incinerator, hazardous waste burning cement kiln, or hazardous waste burning lightweight aggregate kiln.

**Hazardous waste constituent** (Section 260.10) means a constituent that caused the Administrator to list the hazardous waste in part 261, subpart D, of this chapter, or a constituent listed in table 1 of Sec. 261.24 of this chapter.

**Hazardous waste incinerator** (Section 62.1201) means a device defined as an incinerator in Sec. 260.10 of this chapter and that burns hazardous waste at any time. For purposes of this subpart, the hazardous waste incinerator includes all associated firing systems and air pollution control devices, as well as the combustion chamber equipment.

**Hazardous waste lightweight aggregate kiln** (Section 62.1201) means a rotary kiln that produces clinker by heating materials such as slate, shale and clay for subsequent production of lightweight aggregate used in commerce, and that burns hazardous waste at any time.

**Hazardous waste management facility ("HWM facility")** (Section 144.3) means all contiguous land, and structures, other appurtenances, and improvements on the land used for treating, storing, or disposing of hazardous waste. A facility may consist of several treatment, storage, or disposal operational units (for example, one or more landfills, surface impoundments, or combination of them).

**Hazardous Waste Management facility ("HWM facility")** (Section 146.3) means all contiguous land, and structures, other appurtenances, and improvements on the land used for treating, storing, or disposing of hazardous waste. A facility may consist of several treatment, storage, or disposal operational units (for example, one or more landfills, surface impoundments, or combination of them).

**Hazardous Waste Management facility (HWM facility)** (Section 270.2) means all contiguous land, and structures, other appurtenances, and improvements on the land, used for treating, storing, or disposing of hazardous waste. A facility may consist of several treatment, storage, or disposal operational units (for example, one or more landfills, surface impoundments, or combinations of them).

**Hazardous waste management unit** (Section 260.10) is a contiguous area of land on or in which hazardous waste is placed, or the largest area in which there is significant likelihood of mixing hazardous waste constituents in the same area. Examples of hazardous waste management units include a surface impoundment, a waste pile, a land treatment area, a landfill cell, an incinerator, a tank and its associated piping and underlying containment system and a container storage area. A container alone does not constitute a unit; the unit includes containers and the land or pad upon which they are placed.

**Hazardous waste management unit shutdown** (Section 264.1031) means a work practice or operational procedure that stops operation of a hazardous waste management unit or part of a hazardous waste management unit. An unscheduled work practice or operational procedure that stops operation of a hazardous waste management unit or part of a hazardous waste management unit for less than 24 hours is not a hazardous waste management unit shutdown. The use of spare equipment and technically feasible bypassing of equipment without stopping operation are not hazardous waste management unit shutdowns.

**Hazardous waste residence time** (Section 62.1201) means the time elapsed from cutoff of the flow of hazardous waste into the combustor (including, for example, the time required for liquids to flow from the cutoff valve into the combustor) until solid, liquid, and gaseous materials from the hazardous waste (excluding residues that may adhere to combustion chamber surfaces and excluding waste-derived recycled materials such as cement kiln dust and internally recycled metals) exit the combustion chamber. For combustors with multiple firing systems whereby the residence time may vary for the firing systems, the hazardous waste residence time for purposes of complying with this subpart means the longest residence time for any firing system in use at the time of the waste cutoff.

**HC+NOX** (Section 90.3) means total hydrocarbons plus oxides of nitrogen.

**HDD or 2,3,7,8-HDD** (Section 766.3) means any of the dibenzo-p-dioxins totally chlorinated or totally brominated at the following positions on the molecular structure: 2,3,7,8; 1,2,3,7,8; 1,2,3,4,7,8; 1,2,3,6,7,8; 1,2,3,7,8,9; and 1,2,3,4,7,8,9.

**HDF or 2,3,7,8-HDF** (Section 766.3) means any of the dibenzofurans totally chlorinated or totally brominated at the following positions on the molecular structure: 2,3,7,8; 1,2,3,7,8; 2,3,4,7,8; 1,2,3,4,7,8; 1,2,3,6,7,8; 1,2,3,7,8,9; 2,3,4,6,7,8; 1,2,3,4,6,7,8; and 1,2,3,4,7,8,9.

**Headband** (Section 211.203) The component of hearing protective device which applies force to, and holds in place on the head, the component which is intended to acoustically seal the ear canal.

**Health and safety study of a microorganism or health and safety study** (Section 725.3) means any study of any effect of a microorganism or microbial mixture on health or the environment or on both, including underlying data and epidemiological studies, studies of occupational exposure to a microorganism or microbial mixture, toxicological, clinical, and ecological, or other studies of a microorganism or microbial mixture, and any test performed under the Act. Microorganism identity is always part of a health and safety study of a microorganism. (1) It is intended that the term "health and safety study of a microorganism" be interpreted broadly. Not only is information which arises as a result of a formal, disciplined study included, but other information relating to the effects of a microorganism or microbial mixture on health or the environment is also included. Any data that bear on the effects of a microorganism on health or the environment would be included. (2) Examples include: (i) Tests for ecological or other environmental effects on invertebrates, fish, or other animals, and plants, including: Acute toxicity tests, chronic toxicity tests, critical life stage tests, behavioral tests, algal growth tests, seed germination tests, plant growth or damage tests, microbial function tests, bioconcentration or bioaccumulation tests, and model ecosystem (microcosm) studies. (ii) Long- and short-term tests of mutagenicity, carcinogenicity, or teratogenicity; dermatoxicity; cumulative, additive, and synergistic effects; and acute, subchronic, and chronic effects. (iii) Assessments of human and environmental exposure, including workplace exposure, and impacts of a particular microorganism or microbial mixture on the environment, including surveys, tests, and studies of: Survival and transport in air, water, and soil; ability to exchange genetic material with other microorganisms, ability to colonize human or animal guts, and ability to colonize plants. (iv) Monitoring data, when they have been aggregated and analyzed to measure the exposure of humans or the environment to a microorganism. (v) Any assessments of risk to health and the environment resulting from the manufacture, processing, distribution in commerce, use, or disposal of the microorganism.

**Health and safety study or study** (Section 716.3) means any study of any effect of a chemical substance or mixture on health or the environment or on both, including underlying data and epidemiological studies, studies of occupational exposure to a chemical substance or mixture, toxicological, clinical, and ecological or other studies of a chemical substance or mixture, and any test performed under TSCA. (1) It is intended that the term health and safety study be interpreted broadly. Not only is information which arises as a result of a formal, disciplined study included, but other information relating to the effects of a chemical substance or mixture on health or the environment is also included. Any data that bear on the effects of a chemical substance on health or the environment would be included. Chemical identity is part of, or underlying data to, a health and safety study. (2) Examples are: (i) Long- and short-term tests of mutagenicity, carcinogenicity, or teratogenicity; data on behavioral disorders; dermatoxicity; pharmacological effects; mammalian absorption, distribution, metabolism, and excretion; cumulative, additive, and synergistic effects; and acute, subchronic, and chronic effects. (ii) Tests for ecological or other environmental effects on invertebrates, fish, or other animals, and plants, including: Acute toxicity tests, chronic toxicity tests, critical life-stage tests, behavioral tests, algal growth tests, seed germination tests, plant growth or damage tests, microbial function tests, bioconcentration or bioaccumulation tests, and model ecosystem (microcosm) studies. (iii) Assessments of human and environmental exposure, including workplace exposure, and impacts of a particular chemical substance or mixture on the environment, including surveys, tests, and studies of: Biological, photochemical, and chemical degradation; structure/activity relationships; air, water, and soil transport; biomagnification and bioconcentration; and chemical and physical properties, e.g., boiling point, vapor pressure, evaporation rates from soil and water, octanol/ water partition coefficient, and water solubility. (iv) Monitoring data, when they have been aggregated and analyzed to measure the exposure of humans or the nvironment to a chemical substance or mixture.

**Health and safety study or study** (Section 720.3) means any study of any effect of a chemical substance or mixture on health or the environment or on both, including underlying data and epidemiological studies, studies of occupational exposure to a chemical substance or mixture, toxicological, clinical, and ecological, or other studies of a chemical substance or mixture, and any test performed under the Act. Chemical identity is always part of a health and safety study. (1) Not only is information which arises as a result of a formal, disciplined study included, but other information relating to the effects of a chemical substance or mixture on health or the environment is also included. Any data that bear on the effects of a chemical substance on health or the environment would be included. (2) Examples include: (i) Long- and short-term tests of mutagenicity, carcinogenicity, or teratogenicity; data on behavioral disorders; dermatoxicity; pharmacological effects; mammalian absorption, distribution, metabolism, and excretion; cumulative, additive, and synergistic effects; acute, subchronic, and chronic effects; and structure/activity analyses. (ii) Tests for ecological or other environmental effects on invertebrates, fish, or other animals, and plants, including: Acute toxicity tests, chronic toxicity tests, critical life stage tests, behavioral tests, algal growth tests, seed germination tests, plant growth or damage tests, microbial function tests, bioconcentration or bioaccumulation tests, and model ecosystem (microcosm) studies. (iii) Assessments of human and environmental exposure, including workplace exposure, and impacts of a particular chemical substance or mixture on the environment, including surveys, tests, and studies of: Biological, photochemical, and chemical degradation; air, water, and soil transport; biomagnification and bioconcentration; and chemical and physical properties, e.g., boiling point, vapor pressure, evaporation rates from soil and water, octanol/water partition coefficient, and water solubility. (iv) Monitoring data, when they have been aggregated and analyzed to measure the exposure of humans or the environment to a chemical substance or mixture. (v) Any assessments of risk to health and the environment resulting from the manufacture, processing, distribution in commerce, use, or disposal of the chemical substance.

**Hearing** (Section 164.2) means a public hearing which is conducted pursuant to the provisions of chapter 5, subchapter II of title 5 of the United States Code and the regulations of this part.

**Hearing** (Section 305.3) means a hearing on the record open to the public and conducted under this part.

**Hearing Clerk** (Section 104.2) means the Hearing Clerk, U.S. Environmental Protection Agency, 401 M Street SW., Washington, DC 20460.

**Hearing Clerk** (Section 164.2) means the Hearing Clerk, Environmental Protection Agency, Washington, DC 20460.

**Hearing clerk** (Section 209.3) means the hearing clerk of the Environmental Protection Agency.

**Hearing Clerk** (Section 305.3) means the Hearing Clerk, A-110, United States Environmental Protection Agency, 1200 Pennsylvania Ave., NW, Washington, DC 20460.

**Hearing clerk** (Section 72.2) means an EPA employee designated by the Administrator to establish a repository for all books, records, documents, and other materials relating to proceedings under part 78 of this chapter.

**Hearing Protective Device** (Section 211.203) Any device or material, capable of being worn on the head or in the ear canal, that is sold wholly or in part on the basis of its ability to reduce the level of sound entering the ear. This includes devices of which hearing protection may not be the primary function, but which are nonetheless sold partially as providing hearing protection to the user. This term is used interchangeably with the terms, "hearing protector" and "device."

**Heat input** (Section 72.2) means the product (expressed in mmBtu/time) of the gross calorific value of the fuel (expressed in Btu/lb) and the fuel feed rate into the combustion device (expressed in mass of fuel/time) and does not include the heat derived from preheated combustion air, recirculated flue gases, or exhaust from other sources.

**Heat input** (Section 96.2) means the product (in mmBtu/time) of the gross calorific value of the fuel (in Btu/lb) and the fuel feed rate into a combustion device (in mass of fuel/time), as measured, recorded, and reported to the Administrator by the NOX authorized account representative and as determined by the Administrator in accordance with subpart H of this part, and does not include the heat derived from preheated combustion air, recirculated flue gases, or exhaust from other sources.

**Heat input** (Section 97.2) means, with regard to a specified period to time, the product (in mmBtu/time) of the gross calorific value of the fuel (in Btu/lb) divided by 1,000,000 Btu/mmBtu and multiplied by the fuel feed rate into a combustion device (in lb of fuel/time), as measured, recorded, and reported to the Administrator by the NOX authorized account representative and as determined by the Administrator in accordance with subpart H of this part. Heat input does not include the heat derived from preheated combustion air, recirculated flue gases, or exhaust from other sources.

**Heat input rate** (Section 72.2) means the product (expressed in mmBtu/hr) of the gross calorific value of the fuel (expressed in mmBtu/mass of fuel) and the fuel feed rate into the combustion device (expressed in mass of fuel/hr) and does not include the heat derived from preheated combustion air, recirculated flue gases, or exhaust from other sources.

**Heat input rate** (Section 97.2) means the amount of heat input (in mmBtu) divided by unit operating time (in hr) or, with regard to a specific fuel, the amount of heat input attributed to the fuel (in mmBtu) divided by the unit operating time (in hr) during which the unit combusts the fuel.

**Heat treatment** (Section 467.02) is the application of heat of specified temperature and duration to change the physical properties of the metal.

**Heat treatment** (Section 468.02) shall mean the application or removal of heat to a workpiece to change the physical properties of the metal.

**Heat treatment** (Section 471.02) is the application of heat of specified temperature and duration to change the physical properties of the metal.

**Heating degree day** (Section 86.1803-01) means the number of degrees per day the daily average temperature is below 65 degrees Fahrenheit. The daily average temperature is the mean of the maximum and minimum temperature for a 24-hour period. The annual heating degree day value is derived by summing the daily heating degree days over a calendar year period.

**Heating oil** (Section 280.12) means petroleum that is No. 1, No. 2, No. 4--light, No. 4--heavy, No. 5--light, No. 5--heavy, and No. 6 technical grades of fuel oil: other residual fuel oils (including Navy Special Fuel Oil and Bunker C): and other fuels when used as substitutes for one of these fuel oils. Heating oil is typically used in the operation of heating equipment, boilers, or furnaces.

**Heavy light-duty truck** (Section 86.094-2) means any light-duty truck rated greater than 6000 lbs GVWR.

**Heavy light-duty truck** (Section 86.1803-01) means any light-duty truck rated greater than 6000 lbs GVWR. The LDT3 and LDT4 classifications comprise the heavy light-duty truck category.

**Heavy Light-Duty Truck** (Section 88.102-94) means any light-duty truck rated greater than 6000 lbs. GVWR.

**Heavy metal** (Section 191.12) means all uranium, plutonium, or thorium placed into a nuclear reactor.

**Heavy-duty engine** (Section 86.082-2) means any engine which the engine manufacturer could reasonably expect to be used for motive power in a heavy-duty vehicle.

**Heavy-duty engine** (Section 86.1803-01) means any engine which the engine manufacturer could reasonably expect to be used for motive power in a heavy-duty vehicle.

**Heavy-duty vehicle** (Section 86.082-2) means any motor vehicle rated at more than 8,500 pounds GVWR or that has a vehicle curb weight of more than 6,000 pounds or that has a basic vehicle frontal area in excess of 45 square feet.

**Heavy-duty vehicle** (Section 86.1803-01) means any motor vehicle rated at more than 8,500 pounds GVWR or that has a vehicle curb

**Heel** (Section 442.2) means any material remaining in a tank following unloading, delivery, or discharge of the transported cargo. Heels may also be referred to as container residue, residual materials or residuals.

**Heel** (Section 82.3) means the amount of a controlled substance that remains in a container after it is discharged or off-loaded (that is no more than ten percent of the volume of the container) and that the person owning or operating the container certifies the residual amount will remain in the container and be included in a future shipment, or be recovered for transformation, destruction or a non-emissive purpose.

**Hexavalent chromium (or chromium VI)** (Section 420.02) means the value obtained by the method specified in 40 CFR 136.3.

**Hide** (Section 425.02) means any animal pelt or skin as received by a tannery as raw material to be processed.

**High altitude** (Section 86.082-2) means any elevation over 1,219 meters (4,000 feet).

**High altitude** (Section 86.1803-01) means any elevation over 1,219 meters (4,000 feet).

**High occupancy area** (Section 761.13) means any area where PCB remediation waste has been disposed of on-site and where occupancy for any individual not wearing dermal and respiratory protection for a calendar year is: 840 hours or more (an average of 16.8 hours or more per week) for non-porous surfaces and 335 hours or more (an average of 6.7 hours or more per week) for bulk PCB remediation waste. Examples could include a residence, school, day care center, sleeping quarters, a single or multiple occupancy 40 hours per week work station, a school class room, a cafeteria in an industrial facility, a control room, and a work station at an assembly line.

**High temperature metals recovery** (Section 437.2) means a metals recovery process in which solid forms of metal-containing materials are processed with a heat-based pyrometallurgical technology to produce a metal product.

**High-altitude** (Section 92.2) means relating to an altitude greater than 4000 feet (1220 meters) and less than 7000 feet (2135 meters), or equivalent observed barometric test conditions of 25.7 to 22.7 inch Hg (88.5 to 78.1 kilopascals).

**High-altitude conditions** (Section 86.082-2) means a test altitude of 1,620 meters (5,315 feet), plus or minus 100 meters (328 feet), or equivalent observed barometric test conditions of 83.3<SUP>plus-minus</SUP>1 kilopascals.

**High-altitude conditions** (Section 86.1803-01) means a test altitude of 1,620 meters (5,315 feet), plus or minus 100 meters (328 feet), or equivalent observed barometric test conditions of 83.3 kPa (24.2 inches Hg) plus or minus 1 kPa (0.30 Hg).

**High-altitude reference point** (Section 86.082-2) means an elevation of 1,620 meters (5,315 feet) plus or minus 100 meters (328 feet), or equivalent observed barometric test conditions of 83.3 kPa (24.2 inches Hg), plus or minus 1 kPa (0.30 Hg).

**High-concentration PCBs** (Section 761.123) means PCBs that contain 500 ppm or greater PCBs, or those materials which EPA requires to be assumed to contain 500 ppm or greater PCBs in the absence of testing.

**High-contact industrial surface** (Section 761.123) means a surface in an industrial setting which is repeatedly touched, often for relatively long periods of time. Manned machinery and control panels are examples of high-contact industrial surfaces. High-contact industrial surfaces are generally of impervious solid material. Examples of low-contact industrial surfaces include ceilings, walls, floors, roofs, roadways and sidewalks in the industrial area, utility poles, unmanned machinery, concrete pads beneath electrical equipment, curbing, exterior structural building components, indoor vaults, and pipes.

**High-contact residential/commercial surface** (Section 761.123) means a surface in a residential/commercial area which is repeatedly touched, often for relatively long periods of time. Doors, wall areas below 6 feet in height, uncovered flooring, windowsills, fencing, bannisters, stairs, automobiles, and children's play areas such as outdoor patios and sidewalks are examples of high-contact residential/commercial surfaces. Examples of low-contact residential/commercial surfaces include interior ceilings, interior wall areas above 6 feet in height, roofs, asphalt roadways, concrete roadways, wooden utility poles, unmanned machinery, concrete pads beneath electrical equipment, curbing, exterior structural building components (e.g., aluminum/vinyl siding, cinder block, asphalt tiles), and pipes.

**High-efficiency particulate air (HEPA)** (Section 763.83) refers to a filtering system capable of trapping and retaining at least 99.97 percent of all monodispersed particles 0.3 [mu]m in diameter or larger.

**High-grade paper** (Section 246.101) means letterhead, dry copy papers, miscellaneous business forms, stationery, typing paper, tablet sheets, and computer printout paper and cards, commonly sold as "white ledger," "computer printout" and "tab card" grade by the wastepaper industry.

**High-level radioactive waste** (Section 191.02) as used in this part, means high-level radioactive waste as defined in the Nuclear Waste Policy Act of 1982 (Pub. L. 97-425).

**High-level radioactive waste** (Section 197.2) means: (1) The highly radioactive material resulting from the reprocessing of spent nuclear fuel, including liquid waste produced directly in reprocessing and any solid material derived from such liquid waste that contains fission products in sufficient concentrations; and (2) Other highly radioactive material that the Commission, consistent with existing law, determines by rule requires permanent isolation.

**High-level radioactive waste** (Section 227.30) means the aqueous waste resulting from the operation of the first cycle solvent extraction system, or equivalent, and the concentrated waste from subsequent extraction cycles, or equivalent, in a facility for reprocessing irradiated reactor fuels or irradiated fuel from nuclear power reactors.

**High-processing packinghouse** (Section 432.41) shall mean a packinghouse which processes both animals slaughtered at the site and additional carcasses from outside sources.

**High-throughput transfer racks** (Section 65.2) means those transfer racks that transfer greater than or equal to a total of 11.8 million liters per year (3.12 million gallons per year) of liquid containing regulated material.

**Highway** (Section 202.10) means the streets, roads, and public ways in any State.

**Highway** (Section 205.2) means the streets, roads, and public ways in any State.

**Highway Fuel Economy** (Section 600.022-85) means the fuel economy determined by operating a vehicle (or vehicles) over the driving schedule in the Federal highway fuel economy test procedure.

**Highway Fuel Economy Test** (Section 610.11) means the test procedure described in Sec. 600.111(b).

**Highway project** (Section 93.101) is an undertaking to implement or modify a highway facility or highway-related program. Such an undertaking consists of all required phases necessary for implementation. For analytical purposes, it must be defined sufficiently to: (1) Connect logical termini and be of sufficient length to address environmental matters on a broad scope; (2) Have independent utility or significance, i.e., be usable and be a reasonable expenditure even if no additional transportation improvements in the area are made; and (3) Not restrict consideration of alternatives for other reasonably foreseeable transportation improvements.

**Holder** (Section 280.200) is a person who, upon the effective date of this regulation or in the future, maintains indicia of ownership (as defined in Sec. 280.200©) primarily to protect a security interest (as defined in Sec. 280.200(f)(1)) in a petroleum UST or UST system or facility or property on which a petroleum UST or UST system is located. A holder includes the initial holder (such as a loan originator); any subsequent holder (such as a successor-in-interest or subsequent purchaser of the security interest on the secondary market); a guarantor of an obligation, surety, or any other person who holds ownership indicia primarily to protect a security interest; or a receiver or other person who acts on behalf or for the benefit of a holder.

**Holocene time** (Section 503.21) is the most recent epoch of the Quaternary period, extending from the end of the Pleistocene epoch to the present.

**Holtwood unit 17, Hunlock unit 6, and Sunbury units 1A, 1B, 2A, and 2B.** (Section 76.2) This definition shall exclude dry bottom turbo fired boilers.

**Homogeneous area** (Section 763.83) means an area of surfacing material, thermal system insulation material, or miscellaneous material that is uniform in color and texture.

**Homolog** (Section 766.3) means a group of isomers that have the same degree of halogenation. For example, the homologous class of tetrachlorodibenzo-p-dioxins consists of all dibenzo-p-dioxins containing four chlorine atoms. When the homologous classes discussed in this part are referred to, the following abbreviations for the prefix denoting the number of halogens are used: tetra-, T (4 atoms) penta-, Pe (5 atoms) hexa-, Hx (6 atoms) hepta-, Hp (7 atoms) HRGC means high resolution gas chromatography.

**Horizon year** (Section 93.101) is a year for which the transportation plan describes the envisioned transportation system according to Sec. 93.106.

**Hospitalization** (Section 159.153) means admission for treatment to a hospital, clinic or other health care facility. Treatment as an out-patient is not considered to be hospitalization.

**Host** (Section 171.2) means any plant or animal on or in which another lives for nourishment, development, or protection.

**Hot forming** (Section 420.71) means those steel operations in which solidified, heated steel is shaped by rolls.

**Hot pressing** (Section 471.02) is forming a powder metallurgy compact at a temperature high enough to effect concurrent sintering.

**Hot soak emissions** (Section 86.096-2) means evaporative emissions after termination of engine operation.

**Hot strip and sheet mill** (Section 420.71) means those steel hot forming operations that produce flat hot-rolled products other than plates.

**Hot water seal** (Section 467.02) is a heated water bath (heated to approximately 180 [deg]F) used to seal the surface coating on formed aluminum which has been anodized and coated. In establishing an effluent allowance for this operation, the hot water seal shall be classified as a cleaning or etching rinse.

**Hot well** (Section 264.1031) means a container for collecting condensate as in a steam condenser serving a vacuum-jet or steam-jet ejector.

**Hot work** (Section 68.3) means work involving electric or gas welding, cutting, brazing, or similar flame or spark-producing operations.

**Hotel power** (Section 92.2) means the power provided by an engine on a locomotive to operate equipment on passenger cars of a train; e.g., heating and air conditioning, lights, etc.

**Hot-soak emissions and Hot-soak losses** (Section 86.1803-01) means evaporative emissions after termination of engine operation.

**Hot-soak losses** (Section 86.082-2) means evaporative emissions after termination of engine operation.

**Hot-soak losses** (Section 86.096-2) means hot soak emissions.

**Hot-spot analysis** (Section 93.101) is an estimation of likely future localized CO and PM10 pollutant concentrations and a comparison of those concentrations to the national ambient air quality standards. Hot-spot analysis assesses impacts on a scale smaller than the entire nonattainment or maintenance area, including, for example, congested roadway intersections and highways or transit terminals, and uses an air quality dispersion model to determine the effects of emissions on air quality.

**Hour before and after** (Section 72.2) means, for purposes of the missing data substitution procedures of part 75 of this chapter, the quality-assured hourly SO2 or CO2 concentration, hourly flow rate, or hourly NOX emission rate recorded by a certified monitor during the unit operating hour immediately before and the unit operating hour immediately after a missing data period.

**Hour before and hour after** (Section 72.2) means, for purposes of the missing data substitution procedures of part 75 of this chapter, the quality-assured hourly SO2 or CO2 concentration, hourly flow rate, hourly NOX concentration, hourly moisture, hourly O2 concentration, or hourly NOX emission rate (as applicable) recorded by a certified monitor during the unit or stack operating hour immediately before and the unit or stack operating hour immediately after a missing data period.

**Housed lot** (Section 412.11) shall mean totally roofed buildings which may be open or completely enclosed on the sides wherein animals or poultry are housed over solid concrete or dirt floors, slotted (partially open) floors over pits or manure collection areas in pens, stalls or cages, with or without bedding materials and mechanical ventilation. For the purposes hereof, the term "housed lot" is synonymous with the terms "slotted floor" buildings (swine, beef), "barn" (dairy cattle) or "stable" (horses), "houses" (turkeys, chickens), which are terms widely used in the industry.

**Household "do-it-yourselfer" used oil** (Section 279.1) means oil that is derived from households, such as used oil generated by individuals who generate used oil through the maintenance of their personal vehicles.

**Household "do-it-yourselfer" used oil generator** (Section 279.1) means an individual who generates household "do-it-yourselfer" used oil.

**Household waste** (Section 258.2) means any solid waste (including garbage, trash, and sanitary waste in septic tanks) derived from households (including single and multiple residences, hotels and motels, bunkhouses, ranger stations, crew quarters, campgrounds, picnic grounds, and day-use recreation areas).

**Housing for the elderly** (Section 745.103) means retirement communities or similar types of housing reserved for households composed of one or more persons 62 years of age or more at the time of initial occupancy.

**hp** (Section 401.11) means horsepower.

**HRMS** (Section 766.3) means high resolution mass spectrometry.

**HUD** (Section 745.103) means the U.S. Department of Housing and Urban Development.

**Human cancer criterion (HCC)** (Section 132.2) is a Human Cancer Value (HCV) for a pollutant that meets the minimum data requirements for Tier I specified in appendix C of this part.

**Human cancer value (HCV)** (Section 132.2) is the maximum ambient water concentration of a substance at which a lifetime of exposure from either: drinking the water, consuming fish from the water, and water-related recreation activities; or consuming fish from the water, and water-related recreation activities, will represent a plausible upper-bound risk of contracting cancer of one in 100,000 using the exposure assumptions specified in the Methodologies for the Development of Human Health Criteria and Values in appendix C of this part.

**Human noncancer criterion (HNC)** (Section 132.2) is a Human Noncancer Value (HNV) for a pollutant that meets the minimum data requirements for Tier I specified in appendix C of this part.

**Human noncancer value (HNV)** (Section 132.2) is the maximum ambient water concentration of a substance at which adverse noncancer effects are not likely to occur in the human population from lifetime exposure via either: drinking the water, consuming fish from the water, and water-related recreation activities; or consuming fish from the water, and water-related recreation activities using the Methodologies for the Development of Human Health Criteria and Values in appendix C of this part.

**HWM facility** (Section 144.3) means "Hazardous Waste Management facility"

**HWM facility** (Section 146.3) means "Hazardous Waste Management facility."

**HWM facility** (Section 270.2) means Hazardous Waste Management facility.

**Hybrid electric vehicle (HEV)** (Section 86.1702-99) means any vehicle which is included in the definition of a "series hybrid electric vehicle," a "parallel hybrid electric vehicle," or a "battery assisted combustion engine vehicle."

**Hybrid generation facility** (Section 72.2) means a plant that generates electrical energy derived from a combination of qualified renewable energy (wind, solar, biomass, or geothermal) and one or more other energy resources.

**Hydraulic barking** (Section 429.11) means a wood processing operation that removes bark from wood by the use of water under a pressure of 6.8 atm (100 psia) or greater.

**Hydraulic fluids** (Section 247.3) means petroleum-based hydraulic fluids;

**Hydraulic lift tank** (Section 280.12) means a tank holding hydraulic fluid for a closed-loop mechanical system that uses compressed air or hydraulic fluid to operate lifts, elevators, and other similar devices.

**Hydraulic mulch** (Section 247.3) means a mulch that is a cellulose-based (paper or wood) protective covering that is mixed with water and applied through mechanical spraying in order to aid the germination of seeds and to prevent soil erosion;

**Hydrochloric acid pickling** (Section 420.91) means those operations in which steel products are immersed in hydrochloric acid solutions to chemically remove oxides and scale, and those rinsing operations associated with such immersions.

**Hydroseeding** (Section 247.3) means the process of spraying seeds mixed with water through a mechanical sprayer (hydroseeder). Hydraulic mulch, fertilizer, a tacking agent, or a wetting agent can also be added to the water/seed mix for enhanced performance;

**Hydrotesting** (Section 471.02) is the testing of piping or tubing by filling with water and pressurizing to test for integrity.

# I

**I** (Section 105.2) means the applicant for an award.

**Identification number** (Section 89.2) means a specification (for example, model number/serial number combination) which allows a particular nonroad engine to be distinguished from other similar engines.

**Identification number** (Section 91.3) means a unique specification (for example, model number/serial number combination) which allows a particular marine SI engine to be distinguished from other similar engines.

**Identification number** (Section 94.2) means a specification (for example, model number/serial number combination) which allows a particular engine to be distinguished from other similar engines.

**Identity** (Section 721.3) means any chemical or common name used to identify a chemical substance or a mixture containing that substance.

**Idle** (Section 201.1) means that condition where all engines capable of providing motive power to the locomotive are set at the lowest operating throttle position; and where all auxiliary non-motive power engines are not operating.

**Idle speed** (Section 92.2) means that speed, expressed as the number of revolutions of the crankshaft per unit of time (e.g., rpm), at which the engine is set to operate when not under load for purposes of propelling the locomotive.

**Immediate family** (Section 170.3) includes only spouse, children, stepchildren, foster children, parents, stepparents, foster parents, brothers, and sisters.

**Immediate use** (Section 721.3) A chemical substance is for the "immediate use" of a person if it is under the control of, and used only by, the person who transferred it from a labeled container and will only be used by that person within the work shift in which it is transferred from the labeled container.

**Impact surface** (Section 745.63) means an interior or exterior surface that is subject to damage by repeated sudden force such as certain parts of door frames.

**Impact-resistant coating** (Section 59.101) means a coating designed to resist chipping caused by road debris.

**Impaired waterbody** (Section 130.2) Any waterbody of the United States that does not attain and maintain water quality standards (as defined in 40 CFR part 131) throughout the waterbody due to an individual pollutant, multiple pollutants, or other causes of pollution, including any waterbody for which biological information indicates that it does not attain and maintain water quality standards. Where a waterbody receives a thermal discharge from one or more point sources, impaired means that the waterbody does not have or maintain a balanced indigenous population of shellfish, fish, and wildlife.

**Impervious Chemical protective clothing** (Section 721.3) is "impervious" to a chemical substance if the substance causes no chemical or mechanical degradation, permeation, or penetration of the chemical protective clothing under the conditions of, and the duration of, exposure.

**Impervious solid surfaces** (Section 761.123) means solid surfaces which are nonporous and thus unlikely to absorb spilled PCBs within the short period of time required for cleanup of spills under this policy. Impervious solid surfaces include, but are not limited to, metals, glass, aluminum siding, and enameled or laminated surfaces.

**Implementation** (Section 256.06) means putting the plan into practice by carrying out planned activities, including compliance and enforcement activities, or ensuring such activities are carried out.

**Implementing agency** (Section 191.12) means: (1) The Commission for facilities licensed by the Commission; (2) The Agency for those implementation responsibilities for the Waste Isolation Pilot Plant, under this part, given to the Agency by the Waste Isolation Pilot Plant Land Withdrawal Act (Pub. L. 102-579, 106 Stat. 4777) which, for the purposes of this part, are: (i) Determinations by the Agency that the Waste Isolation Pilot Plant is in compliance with subpart A of this part; (ii) Issuance of criteria for the certifications of compliance with subparts B and C of this part of the Waste Isolation Pilot Plant's compliance with subparts B and C of this part; (iii) Certifications of compliance with subparts B and C of this part of the Waste Isolation Pilot Plant's compliance with subparts B and C of this part; (iv) If the initial certification is made, periodic recertification of the Waste Isolation Pilot Plant's continued compliance with subparts B and C of this part; (v) Review and comment on performance assessment reports of the Waste Isolation Pilot Plant; and (vi) Concurrence by the Agency with the Department's determination under Sec. 191.02(i) that certain wastes do not need the degree of isolation required by subparts B and C of this part; and (3) The Department of Energy for any other disposal facility and all other implementation responsibilities for the Waste Isolation Pilot Plant, under this part, not given to the Agency.

**Implementing agency** (Section 239.2) means the state and/or local agency(ies) responsible for carrying out an approved state permit program.

**Implementing agency** (Section 280.12) means EPA, or, in the case of a state with a program approved under section 9004 (or pursuant to a memorandum of agreement with EPA), the designated state or local agency responsible for carrying out an approved UST program.

**Implementing agency** (Section 68.3) means the state or local agency that obtains delegation for an accidental release prevention program under subpart E, 40 CFR part 63. The implementing agency may, but is not required to, be the state or local air permitting agency. If no state or local agency is granted delegation, EPA will be the implementing agency for that state.

**Import** (Section 372.3) means to cause a chemical to be imported into the customs territory of the United States. For purposes of this definition, to cause means to intend that the chemical be imported and to control the identity of the imported chemical and the amount to be imported.

**Import** (Section 704.3) means to import for commercial purposes.

**Import** (Section 716.3) means to import for commercial purposes.

**Import** (Section 763.163) means to bring into the customs territory of the United States, except for: (1) Shipment through the customs territory of the United States for export without any use, processing, or disposal within the customs territory of the United States; or (2) entering the customs territory of the United States as a component of a product during normal personal or business activities involving use of the product.

**Import** (Section 82.3) means to land on, bring into, or introduce into, or attempt to land on, bring into, or introduce into any place subject to the jurisdiction of the United States whether or not such landing, bringing, or introduction constitutes an importation within the meaning of the customs laws of the United States, with the following exemptions: (1) Off-loading used or excess controlled substances or controlled products from a ship during servicing, (2) Bringing controlled substances into the U.S. from Mexico where the controlled substance had been admitted into Mexico in bond and was of U.S. origin, and (3) Bringing a controlled product into the U.S. when transported in a consignment of personal or household effects or in a similar non-commercial situation normally exempted from U.S. Customs attention.

**Import for commercial purposes** (Section 704.3) means to import with the purpose of obtaining an immediate or eventual commercial advantage for the importer, and includes the importation of any amount of a chemical substance or mixture. If a chemical substance or mixture containing impurities is imported for commercial purposes, then those impurities also are imported for commercial purposes.

**Import for commercial purposes** (Section 716.3) means to import with the purpose of obtaining an immediate or eventual commercial advantage for the importer, and includes the importation of any amount of a chemical substance or mixture. If a chemical substance or mixture containing impurities is imported for commercial purposes, then those impurities are also imported for commercial purposes.

**Import in bulk form** (Section 704.3) means to import a chemical substance (other than as part of a mixture or article) in any quantity, in cans, bottles, drums, barrels, packages, tanks, bags, or other containers, if the chemical substance is intended to be removed from the container and the substance has an end use or commercial purpose separate from the container.

**Import in bulk form (Section 712.3)** means to import a chemical substance (other than as part of a mixture or article) in any quantity, in cans, bottles, drums, barrels, packages, tanks, bags, or other containers used for purposes of transportation or containment, if the chemical substance has an end use or commercial purpose separate from the container.

**Importer** (Section 704.3) means (1) any person who imports any chemical substance or any chemical substance as part of a mixture or article into the customs territory of the United States, and includes: (i) The person primarily liable for the payment of any duties on the merchandise, or (ii) An authorized agent acting on his behalf (as defined in 19 CFR 1.11). (2) Importer also includes, as appropriate: (i) The consignee. (ii) The importer of record. (iii) The actual owner if an actual owner's declaration and superseding bond have been filed in accordance with 19 CFR 141.20. (iv) The transferee, if the right to draw merchandise in a bonded warehouse has been transferred in accordance with subpart C of 19 CFR part 144. (3) For the purposes of this definition, the customs territory of the United States consists of the 50 States, Puerto Rico, and the District of Columbia.

**Importer** (section 710.2) means any person who imports any chemical substance or any chemical substance as part of a mixture or article into the customs territory of the U.S. and includes: (1) The person primarily liable for the payment of any duties on the merchandise, or (2) An authorized agent acting on his behalf (as defined in 19 CFR 1.11).

**Importer** (Section 712.3) means anyone who imports a chemical substance, including a chemical substance as part of a mixture or article, into the customs territory of the U.S. and includes the person liable for the payment of any duties on the merchandise, or an authorized agent on his behalf. Importer also includes, as appropriate: (1) The consignee. (2) The importer of record. (3) The actual owner if an actual owner's declaration and superseding bond has been filed in accordance with 19 CFR 141.20. (4) The transferee, if the right to withdraw merchandise in a bonded warehouse has been transferred in accordance with subpart C of 19 CFR part 144. For the purposes of this definition, the customs territory of the U.S. consists of the 50 states, Puerto Rico, and the District of Columbia.

**Importer** (Section 716.3) means any person who imports a chemical substance, including a chemical substance as a part of a mixture or article, into the customs territory of the United States and includes the person primarily liable for the payment of any duties on the merchandise or an authorized agent acting on his behalf (as defined in 19 CFR 1.11). Importer also includes, as appropriate: (1) The consignee. (2) The importer of record. (3) The actual owner, if an actual owner's declaration and superseding bond has been filed in accordance with 19 CFR 141.20. (4) The transferee, if the right to draw merchandise in a bonded warehouse has been transferred in accordance with subpart C of 19 CFR part 144. For the purpose of this definition, the customs territory of the United States consists of the 50 States, Puerto Rico, and the District of Columbia.

**Importer** (Section 720.3) means any person who imports a chemical substance, including a chemical substance as part of a mixture or article, into the customs territory of the

United States. "Importer" includes the person primarily liable for the payment of any duties on the merchandise or an authorized agent acting on his or her behalf. The term also includes, as appropriate: (1) The consignee. (2) The importer of record. (3) The actual owner if an actual owner's declaration and superseding bond has been filed in accordance with 19 CFR 141.20; or (4) The transferee, if the right to draw merchandise in a bonded warehouse has been transferred in accordance with subpart C of 19 CFR part 144. (See "principal importer.")

**Importer** (Section 761.13) means any person defined as an "importer" at Sec. 720.3(l) of this chapter who imports PCBs or PCB Items and is under the jurisdiction of the United States.

**Importer** (Section 763.163) means anyone who imports a chemical substance, including a chemical substance as part of a mixture or article, into the customs territory of the United States. Importer includes the person primarily liable for the payment of any duties on the merchandise or an authorized agent acting on his or her behalf. The term includes as appropriate: (1) The consignee. (2) The importer of record. (3) The actual owner if an actual owner's declaration and superseding bond has been filed in accordance with 19 CFR 141.20. (4) The transferee, if the right to withdraw merchandise in a bonded warehouse has been transferred in accordance with subpart C of 19 CFR part 144.

**Importer** (Section 80.2) means a person who imports gasoline, gasoline blending stocks or components, or diesel fuel from a foreign country into the United States (including the Commonwealth of Puerto Rico, the Virgin Islands, Guam, American Samoa, and the Northern Mariana Islands).

**Importer** (Section 82.3) means the importer of record listed on U.S. Customs Service forms for imported controlled substances, used controlled substances or controlled products.

**Importer** (Section 92.2) means an entity or person who imports locomotives or locomotive engines from a foreign country into the United States (including the Commonwealth of Puerto Rico, the Virgin Islands, Guam, American Samoa, and the Northern Mariana Islands).

**Importer** (Section 94.2) means an entity or person who imports engines from a foreign country into the United States (including its territories).

**Importing country** (Section 262.81) means any designated OECD country in Sec. 262.58(a) (1) to which a transfrontier movement of wastes is planned or takes place for the purpose of submitting the wastes to recovery operations therein.

**Impoundments** (Section 421.11) the term within the impoundment for purposes of calculating the volume of process wastewater which may be discharged, shall mean the surface area within the impoundment at the maximum capacity plus the area of the inside and outside slopes of the impoundment dam and the surface area between the outside edge of the impoundment dam and seepage ditches upon which rain falls and is returned to the impoundment. For the purpose of such calculations, the surface area allowance for external appurtenances to the impoundment shall not be more than 30 percent of the water surface area within the impoundment dam at maximum capacity.

**Impregnation** (Section 471.02) is the process of filling pores of a formed powder part, usually with a liquid such as a lubricant, or mixing particles of a nonmetallic substance in a matrix of metal powder.

**Improved sinkhole** (Section 144.3) means a naturally occurring karst depression or other natural crevice found in volcanic terrain and other geologic settings which have been modified by man for the purpose of directing and emplacing fluids into the subsurface.

**Improved sinkhole** (Section 146.3) means a naturally occurring karst depression or other natural crevice found in volcanic terrain and other geologic settings which have been modified by man for the purpose of directing and emplacing fluids into the subsurface.

**Impulsive Noise** (Section 211.203) An acoustic event characterized by very short rise time and duration.

**Impurity** (Section 158.153) means any substance (or group of structurally similar substances if specified by the Agency) in a pesticide product other than an active ingredient or an inert ingredient, including unreacted starting materials, side reaction products, contaminants, and degradation products.

**Impurity** (Section 704.3) means a chemical substance which is unintentionally present with another chemical substance.

**Impurity** (section 710.2) means a chemical substance which is unintentionally present with another chemical substance.

**Impurity** (Section 712.3) means a chemical substance unintentionally present with another chemical substance or mixture.

**Impurity** (Section 716.3)means a chemical substance which is unintentionally present with another chemical substance.

**Impurity** (Section 720.3) means a chemical substance which is unintentionally present with another chemical substance.

**Impurity** (Section 761.13) means a chemical substance which is unintentionally present with another chemical substance.

**Impurity** (Section 79.2) means any chemical element present in an additive that is not included in the chemical formula or identified in the breakdown by element in the chemical composition of such additive.

**Impurity** (Section 790.3) means a chemical substance which is uninintentionally present with another chemical substance.

**Impurity** (Section 791.3) refers to a chemical substance unintentionally present with another chemical substance or mixture.

**Impurity associated with an active ingredient** (Section 158.153) means: (1) Any impurity present in the technical grade of active ingredient; and (2) Any impurity which forms in the pesticide product through reactions between the active ingredient and any other component of the product or packaging of the product.

**In a living plant** (Section 152.3) means inside the living plant, on the surface of the living plant, or as an exudate from the living plant.

**In a living plant** (Section 174.3) means inside the living plant, on the surface of the living plant, or as an exudate from the living plant.

**In food/medical service** (Section 65.2) means that a piece of equipment in regulated material service contacts a process stream used to manufacture a Food and Drug Administration-regulated product where leakage of a barrier fluid into the process stream would cause any of the following: (1) A dilution of product quality so that the product would not meet written specifications; (2) An exothermic reaction that is a safety hazard; (3) The intended reaction to be slowed down or stopped; or (4) An undesired side reaction to occur.

**In gas/vapor service** (Section 264.1031) means that the piece of equipment contains or contacts a hazardous waste stream that is in the gaseous state at operating conditions.

**In gas/vapor service** (Section 65.2) means that a piece of equipment in regulated material service contains a gas or vapor when in operation.

**In heavy liquid service** (Section 264.1031) means that the piece of equipment is not in gas/vapor service or in light liquid service.

**In heavy liquid service** (Section 65.2) means that a piece of equipment in regulated material service is not in gas/vapor service or in light liquid service.

**In light liquid service** (Section 264.1031) means that the piece of equipment contains or contacts a waste stream where the vapor pressure of one or more of the organic components in the stream is greater than 0.3 kilopascals (kPa) at 20 [deg]C, the total concentration of the pure organic components having a vapor pressure greater than 0.3 kilopascals (kPa) at 20 [deg]C is equal to or greater than 20 percent by weight, and the fluid is a liquid at operating conditions.

**In light liquid service** (Section 65.2) means that a piece of equipment in regulated material service contains a liquid that meets the following conditions: (1) The vapor pressure of one or more of the organic compounds is greater than 0.3 kilopascals at 20 [deg]C (0.04 pounds per square inch at 68 [deg]F); (2) The total concentration of the pure organic compound constituents having a vapor pressure greater than 0.3 kilopascals at 20 [deg]C (0.04 pounds per square inch at 68 [deg]F) is equal to or greater than 20 percent by weight of the total process stream; and (3) The fluid is a liquid at operating conditions. (Note: Vapor pressures may be determined by standard reference texts or American Society for Testing and Materials (ASTM) D-2879, available from American Society for Testing and Materials, 1916 Race Street, Philadelphia, Pennsylvania 19103; or University Microfilms International, 300 North Zeeb Road, Ann Arbor, Michigan 48106.)

**In light material service** (Section 265.1081) means the container is used to manage a material for which both of the following conditions apply: The vapor pressure of one or more of the organic constituents in the material is greater than 0.3 kilopascals (kPa) at 20 [deg] C; and the total concentration of the pure organic constituents having a vapor pressure greater than 0.3 kPa at 20 [deg] C is equal to or greater than 20 percent by weight.

**In liquid service** (Section 65.2) means that a piece of equipment in regulated material service is not in gas/vapor service.

**In operation** (Section 260.10) refers to a facility which is treating, storing, or disposing of hazardous waste.

**In operation** (Section 270.2) means a facility which is treating, storing, or disposing of hazardous waste.

**In or Near Commercial Buildings** (Section 761.13) means within the interior of, on the roof of, attached to the exterior wall of, in the parking area serving, or within 30 meters of a non-industrial non-substation building. Commercial buildings are typically accessible to both members of the general public and employees, and include: (1) Public assembly properties, (2) educational properties, (3) institutional properties, (4) residential properties, (5) stores, (6) office buildings, and (7) transportation centers (e.g., airport terminal buildings, subway stations, bus stations, or train stations).

**In regulated material service** (Section 65.2) means, for the purposes of the equipment leak provisions of subpart F of this part, equipment which meets the definition of "in volatile organic compound service," "in volatile hazardous air pollutant service," "in benzene service," "in vinyl chloride service," or "in organic hazardous air pollutant service" as defined in the referencing subpart.

**In situ sampling systems** (Section 264.1031) means nonextractive samplers or in-line samplers.

**In the Hands of the Manufacturer** (Section 86.602-84) means that vehicles are still in the possession of the manufacturer and have not had their bills of lading transferred to another person for the purpose of transporting.

**In vacuum service** (Section 264.1031) means that equipment is operating at an internal pressure that is at least 5 kPa below ambient pressure.

**In vacuum service** (Section 65.2) means that equipment is operating at an internal pressure that is at least 5 kilopascals (0.7 pounds per square inch) below ambient pressure.

**in.** (Section 401.11) means inch.

**Inability** (Section 169.1) The term "inability" means the incapacity of any person to maintain, furnish or permit access to any records under this Act and regulations, where such incapacity arises out of causes beyond the control and without the fault or negligence of such person. Such causes may include, but are not restricted to acts of God or of the public enemy, fires, floods, epidemics, quarantine restrictions, strikes, and unusually severe weather, but in every case, the failure must be beyond the control and without the fault or negligence of said person.

**Inactivation** (Section 725.3) means that living microorganisms are rendered nonviable.

**Inactive facility** (Section 256.06) means a facility which no longer receives solid waste.

**Inactive portion** (Section 260.10) means that portion of a facility which is not operated after the effective date of part 261 of this chapter. (See also "active portion" and "closed portion".)

**Inactive range** (Section 266.201) means a military range that is not currently being used, but that is still under military control and considered by the military to be a potential range area, and that has not been put to a new use that is incompatible with range activities.

**Inboard engine** (Section 91.3) means a four stroke marine SI engine that is designed such that the propeller shaft penetrates the hull of the marine vessel while the engine and the remainder of the drive unit is internal to the hull of the marine vessel.

**Incidental fallback** (Section 232.2) is the redeposit of small volumes of dredged material that is incidental to excavation activity in waters of the United States when such material falls back to substantially the same place as the initial removal. Examples of incidental fallback include soil that is disturbed when dirt is shoveled and the back-spill that comes off a bucket when such small volume of soil or dirt falls into substantially the same place from which it was initially removed.

**Incineration** (Section 240.101) means the controlled process which combustible solid, liquid, or gaseous wastes are burned and changed into noncombustible gases.

**Incinerator** (Section 240.101) means a facility consisting of one or more furnaces in which wastes are burned.

**Incinerator** (Section 260.10) means any enclosed device that: (1) Uses controlled flame combustion and neither meets the criteria for classification as a boiler, sludge dryer, or carbon regeneration unit, nor is listed as an industrial furnace; or (2) Meets the definition of infrared incinerator or plasma arc incinerator.

**Incinerator** (Section 65.2) means an enclosed combustion device that is used for destroying organic compounds. Auxiliary fuel may be used to heat waste gas to combustion temperatures. Any energy recovery section present is not physically formed into one manufactured or assembled unit with the combustion section; rather, the energy recovery section is a separate section following the combustion section and the two are joined by ducts or connections carrying flue gas. This energy recovery section limitation does not apply to an energy recovery section used solely to preheat the incoming vent stream or combustion air.

**Incinerator** (Section 761.13) means an engineered device using controlled flame combustion to thermally degrade PCBs and PCB Items. Examples of devices used for incineration include rotary kilns, liquid injection incinerators, cement kilns, and high temperature boilers.

**Incompatible waste** (Section 260.10) means a hazardous waste which is unsuitable for: (1) Placement in a particular device or facility because it may cause corrosion or decay of containment materials (e.g., container inner liners or tank walls); or (2) Commingling with another waste or material under uncontrolled conditions because the commingling might produce heat or pressure, fire or explosion, violent reaction, toxic dusts, mists, fumes, or gases, or flammable fumes or gases. (See part 265, appendix V, of this chapter for examples.)

**Incomplete gasoline-fueled heavy-duty vehicle** (Section 86.085-2) means any gasoline-fueled heavy-duty vehicle which does not have the primary load-carrying device, or passenger compartment, or engine compartment or fuel system attached.

**Incomplete heavy-duty vehicle** (Section 86.1803-01) means any heavy-duty vehicle which does not have the primary load carrying device or container attached.

**Incomplete truck** (Section 86.082-2) means any truck which does not have the primary load carrying device or container attached.

**Incomplete truck** (Section 86.1803-01) means any truck which does not have the primary load carrying device or container attached.

**Increase the frequency or severity** (Section 93.101) means to cause a location or region to exceed a standard more often or to cause a violation at a greater concentration than previously existed and/or would otherwise exist during the future period in question, if the project were not implemented.

**Increase the frequency or severity of any existing violation of any standard in any area** (Section 93.152) means to cause a nonattainment area to exceed a standard more often or to cause a violation at a greater concentration than previously existed and/or would otherwise exist during the future period in question, if the project were not implemented.

**Increments of progress** (Section 51.100) means steps toward compliance which will be taken by a specific source, including: (1) Date of submittal of the source's final control plan to the appropriate air pollution control agency; (2) Date by which contracts for emission control systems or process modifications will be awarded; or date by which orders will be issued for the purchase of component parts to accomplish emission control or process modification; (3) Date of initiation of on-site construction or installation of emission control equipment or process change; (4) Date by which on-site construction or installation of emission control equipment or process modification is to be completed; and (5) Date by which final compliance is to be achieved.

**Independent auditor** (Section 72.2) means a professional engineer who is not an employee or agent of the source being audited.

**Independent commercial importer (ICI)** (Section 85.1502) An importer who is not an original equipment manufacturer (OEM) (see definition below) or does not have a contractual agreement with an OEM to act as its authorized representative for the distribution of motor vehicles or motor vehicle engines in the U.S. market.

**Independent commercial importer (ICI)** (Section 89.602) An importer who is not an original engine manufacturer (OEM) (see definition below), but is the entity in whose name a certificate of conformity for a class of nonroad engines has been issued.

**Independent laboratory** (Section 610.11) means a test facility operated independently of any motor vehicle, motor vehicle engine, or retrofit device manufacturer capable of performing retrofit device evaluation tests. Additionally, the laboratory shall have no financial interests in the outcome of these tests other than a fee charged for each test performed.

**Independent Power Production Facility (IPP)** (Section 72.2) means a source that: (1) Is nonrecourse project financed, as defined by the Secretary of Energy at 10 CFR part 715; (2) Is used for the generation of electricity, eighty percent or more of which is sold at wholesale; and (3) Is a new unit required to hold allowances under Title IV of the Clean Air Act; but only if direct public utility ownership of the equipment comprising the facility does not exceed 50 percent.

**Independent printed circuit board manufacturer** (Section 433.11) shall mean a facility which manufacturers printed circuit boards principally for sale to other companies.

**Indian country** (Section 122.2) means: (1) All land within the limits of any Indian reservation under the jurisdiction of the United States Government, notwithstanding the issuance of any patent, and, including rights-of-way running through the reservation; (2) · All dependent Indian communities with the borders of the United States whether within the originally or subsequently acquired territory thereof, and whether within or without the limits of a state; and (3) All Indian allotments, the Indian titles to which have not been extinguished, including rights-of-way running through the same.

**Indian Country** (Section 350.1) means Indian country as defined in 18 U.S.C. 1151. That section defines Indian country as: (a) All land within the limits of any Indian reservation under the jurisdiction of the United States government, notwithstanding the issuance of any patent, and including rights-of-way running through the reservation; (b) All dependent Indian communities within the borders of the United States whether within the original or subsequently acquired territory thereof, and whether within or without the limits of a State; and (c) All Indian allotments, the Indian titles to which have not been extinguished, including rights-of-way running through the same.

**Indian Country** (Section 355.20) means Indian country as defined in 18 U.S.C. 1151. That section defines Indian country as: (a) All land within the limits of any Indian reservation under the jurisdiction of the United States government, notwithstanding the issuance of any patent, and including rights-of-way running through the reservation; (b) All dependent Indian communities within the borders of the United States whether within the original or subsequently acquired territory thereof, and whether within or without the limits of a State; and (c) All Indian allotments, the Indian titles to which have not been extinguished, including rights-of-way running through the same.

**Indian Country** (Section 370.2) means Indian country as defined in 18 U.S.C. 1151. That section defines Indian country as: (a) All land within the limits of any Indian reservation under the jurisdiction of the United States government, notwithstanding the issuance of any patent, and including rights-of-way running through the reservation; (b) All dependent Indian communities within the border of the United States whether within the original or subsequently acquired territory thereof, and whether within or without the limits of a State; and (c) All Indian allotments, the Indian titles to which have not been extinguished, including rights-of-way running through the same.

**Indian Country** (Section 372.3) means Indian country as defined in 18 U.S.C. 1151. That section defines Indian country as: (a) All land within the limits of any Indian reservation under the jurisdiction of the United States government, notwithstanding the issuance of any patent, and including rights-of-way running through the reservation; (b) All dependent Indian communities within the borders of the United States whether within the original or subsequently acquired territory thereof, and whether within or without the limits of a State; and (c) All Indian allotments, the Indian titles to which have not been extinguished, including rights-of-way running through the same.

**Indian country** (Section 50.1) is as defined in 18 U.S.C. 1151.

**Indian country** (Section 71.2) means: (1) All land within the limits of any Indian reservation under the jurisdiction of the United States government, notwithstanding the issuance of any patent, and including rights-of-way running through the reservation; (2) All dependent Indian communities within the borders of the United States whether within the original or subsequently acquired territory thereof, and whether within or without the limits of a State; and (3) All Indian allotments, the Indian titles to which have not been extinguished, including rights-of-way running through the same.

**Indian Governing Body** (Section 58.1) means the governing body of any tribe, band, or group of Indians subject to the jurisdiction of the United States and recognized by the United States as possessing power of self-government.

**Indian lands** (Section 144.3) means "Indian country" as defined in 18 U.S.C. 1151. That section defines Indian country as: (a) All land within the limits of any Indian reservation under the jurisdiction of the United States government, notwithstanding the issuance of any patent, and, including rights-of-way running through the reservation; (b) All dependent Indian communities within the borders of the United States whether within the original or subsequently acquired territory thereof, and whether within or without the limits of a State; and (c) All Indian allotments, the Indian titles to which have not been extinguished, including rights-of-way running through the same.

**Indian lands or Indian country** (Section 258.2) means: (1) All land within the limits of any Indian reservation under the jurisdiction of the United States Government, notwithstanding the issuance of any patent, and including rights-of-way running throughout the reservation; (2) All dependent Indian communities within the borders of the United States whether within the original or subsequently acquired territory thereof, and whether within or without the limits of the State; and (3) All Indian allotments, the Indian titles to which have not been extinguished, including rights of way running through the same.

**Indian Reservation** (Section 58.1) means any Federally recognized reservation established by treaty, agreement, executive order, or act of Congress.

**Indian Tribe** (Section 122.2) means any Indian Tribe, band, group, or community recognized by the Secretary of the Interior and exercising governmental authority over a Federal Indian reservation.

**Indian Tribe** (Section 124.2) means (in the case of UIC) any Indian Tribe having a federally recognized governing body carrying out substantial governmental duties and powers over a defined area. For the NPDES program, the term "Indian Tribe" means any Indian Tribe, band, group, or community recognized by the Secretary of the Interior and exercising governmental authority over a Federal Indian reservation.

**Indian Tribe** (Section 142.2) means any Indian Tribe having a Federally recognized governing body carrying out substantial governmental duties and powers over a defined area.

**Indian Tribe** (Section 144.3) means any Indian Tribe having a Federally recognized governing body carrying out substantial governmental duties and powers over a defined area.

**Indian Tribe** (Section 146.3) means any Indian Tribe having a Federally recognized governing body carrying out substantial governmental duties and powers over a defined area.

**Indian tribe** (Section 192) means any tribe, band, clan, group, pueblo, or community of Indians recognized as eligible for services provided by the Secretary of the Interior to Indians.

**Indian Tribe** (Section 232.2) means any Indian Tribe, band, group, or community recognized by the Secretary of the Interior and exercising governmental authority over a Federal Indian reservation.

**Indian Tribe** (Section 233.2) means any Indian Tribe, band, group, or community recognized by the Secretary of the Interior and exercising governmental authority over a Federal Indian reservation.

**Indian tribe** (Section 300.5) as defined by section 101(36) of CERCLA, means any Indian tribe, band, nation, or other organized group or community, including any Alaska Native village but not including any Alaska Native regional or village corporation, which is recognized as eligible for the special programs and services provided by the United States to Indians because of their status as Indians. "Indian tribe," as defined by OPA section 1001, means any Indian tribe, band, nation, or other organized group or community, but not including any Alaska Native regional or village corporation, which is recognized as eligible for the special programs and services provided by the United States to Indians because of their status as Indians and has governmental authority over lands belonging to or controlled by the tribe.

**Indian tribe** (Section 350.1) means those tribes federally recognized by the Secretary of the Interior.

**Indian tribe** (Section 355.20) means those tribes federally recognized by the Secretary of the Interior.

**Indian tribe** (Section 370.2) means those tribes federally recognized by the Secretary of the Interior.

**Indian tribe** (Section 372.3) means those tribes federally recognized by the Secretary of the Interior.

**Indian Tribe** (Section 501.2) means any Indian Tribe, band, group, or community recognized by the Secretary of the Interior and exercising governmental authority over a Federal Indian reservation.

**Indian Tribe Consortium or Tribal Consortium** (Section 49.2) means a group of two or more Indian tribes.

**Indian Tribe or Tribe** (Section 131.3) means any Indian Tribe, band, group, or community recognized by the Secretary of the Interior and exercising governmental authority over a Federal Indian reservation.

**Indian Tribe or Tribe** (Section 132.2) means any Indian Tribe, band, group, or community recognized by the Secretary of the Interior and exercising governmental authority over a Federal Indian reservation.

**Indian Tribe or Tribe** (Section 258.2) means any Indian tribe, band, nation, or community recognized by the Secretary of the Interior and exercising substantial governmental duties and powers on Indian lands.

**Indian tribe or tribe** (Section 49.2) means any Indian tribe, band, nation, or other organized group or community, including any Alaska Native village, which is federally recognized as eligible for the special programs and services provided by the United States to Indians because of their status as Indians.

**Indian Tribe or Tribe** (Section 71.2) means any Indian Tribe, band, nation, or other organized group or community, including any Alaskan native village, which is federally recognized as eligible for the special programs and services provided by the United States to Indians because of their status as Indians.

**Indicia of ownership** (Section 280.200) means evidence of a secured interest, evidence of an interest in a security interest, or evidence of an interest in real or personal property securing a loan or other obligation, including any legal or equitable title or deed to real or personal property acquired through or incident to foreclosure. Evidence of such interests include, but are not limited to, mortgages, deeds of trust, liens, surety bonds and guarantees of obligations, title held pursuant to a lease financing transaction in which the lessor does not select initially the leased property (hereinafter "lease financing transaction"), and legal or equitable title obtained pursuant to foreclosure. Evidence of such interests also includes assignments, pledges, or other rights to or other forms of encumbrance against property that are held primarily to protect a security interest. A person is not required to hold title or a security interest in order to maintain indicia of ownership.

**Indirect ammonia recovery system** (Section 420.11) means those systems which recover ammonium hydroxide as a by-product from coke oven gases and waste ammonia liquors.

**Indirect Discharge or Discharge** (Section 403.3) means the introduction of pollutants into a POTW from any non-domestic source regulated under section 307(b), (c) or (d) of the Act.

**Indirect discharger** (Section 122.2) means a nondomestic discharger introducing "pollutants" to a "publicly owned treatment works."

**Indirect Dischargers** (Section 455.41): Pollution prevention (P2) allowable discharge (excluding interior wastewater sources, leak and spill clean-up water, and floor wash) for this subpart means the quantity of/concentrations of pollutants in PFPR process wastewaters that remain after a facility has demonstrated that it is using the specified practices of the Pollution Prevention Alternative as listed in Table 8 to this part 455.

**Indirect emissions** (Section 93.152) means those emissions of a criteria pollutant or its precursors that: (1) Are caused by the Federal action, but may occur later in time and/or may be further removed in distance from the action itself but are still reasonably foreseeable; and (2) The Federal agency can practicably control and will maintain control over due to a continuing program responsibility of the Federal agency.

**Indirect information** (Section 86.094-2) means any information that is not specifically contained in the service literature, but is contained in items such as tools or equipment provided to franchised dealers (or others). Intermediary means any individual or entity, other than an original equipment manufacturer, which provides service or equipment to automotive technicians.

**Indirect information** (Section 86.1803-01) means any information that is not specifically contained in the service literature, but is contained in items such as tools or equipment provided to franchised dealers (or others).

**Individual** (Section 1516.2) means a citizen of the United States or an alien lawfully admitted for permanent residence.

**Individual** (Section 303.11) natural person, not a corporation or other legal entity nor an association of persons.

**Individual control strategy** (Section 122.2) is defined at 40 CFR 123.46©.

**Individual generation site** (Section 260.10) means the contiguous site at or on which one or more hazardous wastes are generated. An individual generation site, such as a large manufacturing plant, may have one or more sources of hazardous waste but is considered a single or individual generation site if the site or property is contiguous.

**Individual proficiency/RMP exam** (Section 195.2) means the exam which evaluates individuals who provide radon measurement services in a residential environment. Listed participant in an individual or organization who has met all the requirements for listing in the RMP and RCP programs.

**Industrial building** (Section 761.13) means a building directly used in manufacturing or technically productive enterprises. Industrial buildings are not generally or typically accessible to other than workers. Industrial buildings include buildings used directly in the production of power, the manufacture of products, the mining of raw materials, and the storage of textiles, petroleum products, wood and paper products, chemicals, plastics, and metals.

**Industrial drums** (Section 247.3) are cylindrical containers used for shipping and storing liquid or solid materials.

**Industrial furnace** (Section 260.10) means any of the following enclosed devices that are integral components of manufacturing processes and that use thermal treatment to accomplish recovery of materials or energy: (1) Cement kilns, (2) Lime kilns, (3) Aggregate kilns, (4) Phosphate kilns, (5) Coke ovens, (6) Blast furnaces, (7) Smelting, melting and refining furnaces (including pyrometallurgical devices such as cupolas, reverberator furnaces, sintering machine, roasters, and foundry furnaces), (8) Titanium dioxide chloride process oxidation reactors, (9) Methane reforming furnaces, (10) Pulping liquor recovery furnaces, (11) Combustion devices used in the recovery of sulfur values from spent sulfuric acid, (12) Halogen acid furnaces (HAFs) for the production of acid from halogenated hazardous waste generated by chemical production facilities where the furnace is located on the site of a chemical production facility, the acid product has a halogen acid content of at least 3%, the acid product is used in a manufacturing process, and, except for hazardous waste burned as fuel, hazardous waste fed to the furnace has a minimum halogen content of 20% as-generated, (13) Such other devices as the Administrator may, after notice and comment, add to this list on the basis of one or more of the following factors: (i) The design and use of the device primarily to accomplish recovery of material products; (ii) The use of the device to burn or reduce raw materials to make a material product; (iii) The use of the device to burn or reduce secondary materials as effective substitutes for raw materials, in processes using raw materials as principal feedstocks; (iv) The use of the device to burn or reduce secondary materials as ingredients in an industrial process to make a material product; (v) The use of the device in common industrial practice to produce a material product; and (vi) Other factors, as appropriate.

**Industrial furnace** (Section 372.3) means any of the following enclosed devices that are integral components of manufacturing processes and that use thermal treatment to accomplish recovery of materials or energy: (1) Cement kilns. (2) Lime k kilns. (3) Aggregate kilns. (4) Phosphate kilns. (5) Coke ovens. (6) Blast furnaces. (7) Smelting, melting and refining furnaces (including pyrometallurgical devices such as cupolas, reverberator furnaces, sintering machine, roasters, and foundry furnaces). (8) Titanium dioxide chloride process oxidation reactors. (9) Methane reforming furnaces. (10) Pulping liquor recovery furnaces. (11) Combustion devices used in the recovery of sulfur values from spent sulfuric acid. (12) Halogen acid furnaces (HAFs) for the production of acid from halogenated hazardous waste generated by chemical production facilities where the furnace is located on the site of a chemical production facility, the acid product has a halogen acid content of at least 3%, the acid product is used in a manufacturing process, and, except for hazardous waste burned as fuel, hazardous waste fed to the furnace has a minimum halogen content of 20% as-generated. (13) Such other devices as the Administrator may, after notice and comment, add to this list on the basis of one or more of the following factors: (i) The design and use of the device primarily to accomplish recovery of material products; (ii) The use of the device to burn or reduce raw materials to make a material product; (iii) The use of the device to burn or reduce secondary materials as effective substitutes for raw materials, in processes using raw materials as principal feedstocks; (iv) The use of the device to burn or reduce secondary materials as ingredients in an industrial process to make a material product; (v) The use of the device in common industrial practice to produce a material product; and (vi) Other factors, as appropriate.

**Industrial organization** (Section 105.2) means any company, corporation, association, partnership, firm, university, not-for-profit organization, or wastewater treatment facility, as well as a Federal, State or Tribal government wastewater treatment facility, or U.S. military command to the extent such government and other organizations operate in an "industrial" capacity in the treatment of wastes or abatement of pollution.

**Industrial solid waste** (Section 243.101) means the solid waste generated by industrial processes and manufacturing.

**Industrial solid waste** (Section 246.101) means the solid waste generated by industrial processes and manufacturing.

**Industrial solid waste** (Section 258.2) means solid waste generated by manufacturing or industrial processes that is not a hazardous waste regulated under subtitle C of RCRA. Such waste may include, but is not limited to, waste resulting from the following manufacturing processes: Electric power generation; fertilizer/agricultural chemicals; food and related products/by-products; inorganic chemicals; iron and steel manufacturing; leather and leather products; nonferrous metals manufacturing/foundries; organic chemicals; plastics and resins manufacturing; pulp and paper industry; rubber and miscellaneous plastic products; stone, glass, clay, and concrete products; textile manufacturing; transportation equipment; and water treatment. This term does not include mining waste or oil and gas waste.

**Industrial User or User** (Section 403.3) means a source of Indirect Discharge.

**Inert ingredient** (Section 152.3) means any substance (or group of structurally similar substances if designated by the Agency), other than an active ingredient, which is intentionally included in a pesticide product, except as provided by Sec. 174.3 of this chapter.

**Inert ingredient** (Section 158.153) means any substance (or group of structurally similar substances if designated by the Agency), other than an active ingredient, which is intentionally included in a pesticide product.

**Inert ingredient** (Section 174.3) means any substance, such as a selectable marker, other than the active ingredient, where the substance is used to confirm or ensure the presence of the active ingredient, and includes the genetic material necessary for the production of the substance, provided that genetic material is intentionally introduced into a living plant in addition to the active ingredient.

**Inertia weight** (Section 86.1803-01) class means the class, which is a group of equivalent test weights, into which a vehicle is grouped based on its test weight basis in accordance with the provisions of this Part 86.

**Inertia Weight Class** (Section 600.022-85) means the class, which is a group of test weights, into which a vehicle is grouped based on its loaded vehicle weight in accordance with the provisions of part 86.

**Inertia weight class** (Section 86.082-2) means the class, which is a group of test weights, into which a vehicle is grouped based on its loaded vehicle weight in accordance with the provisions of part 86.

**Infectious waste** (Section 240.101) means: (1) Equipment, instruments, utensils, and fomites of a disposable nature from the rooms of patients who are suspected to have or have been diagnosed as having a communicable disease and must, therefore, be isolated as required by public health agencies; (2) laboratory wastes such as pathological specimens (e.g., all tissues, specimens of blood elements, excreta, and secretions obtained from patients or laboratory animals) and disposable fomites (any substance that may harbor or transmit pathogenic organisms) attendant thereto; (3) surgical operating room pathologic specimens and disposable fomites attendant thereto and similar disposable materials from outpatient areas and emergency rooms.

**Infectious waste** (Section 243.101) means: (1) Equipment, instruments, utensils, and formites of a disposable nature from the rooms of patients who are suspected to have or have been diagnosed as having a communicable disease and must, therefore, be isolated as required by public health agencies; (2) laboratory wastes, such as pathological specimens (e.g., all tissues, specimens of blood elements, excreta, and secretions obtained from patients or laboratory animals) and disposable fomites (any substance that may harbor or transmit pathogenic organisms) attendant thereto; (3) surgical operating room pathologic specimens and disposable fomites attendant thereto, and similar disposable materials from outpatient areas and emergency rooms.

**Infectious waste** (Section 246.101) means: (1) Equipment, instruments, utensils, and fomites (any substance that may harbor or transmit pathogenic organisms) of a disposable nature from the rooms of patients who are suspected to have or have been diagnosed as having a communicable disease and must, therefore, be isolated as required by public health agencies; (2) laboratory wastes, such as pathological specimens (e.g. all tissues, specimens of blood elements, excreta, and secretions obtained from patients or laboratory animals) and disposable fomites attendant thereto; (3) surgical operating room pathologic specimens and disposable fomites attendant thereto and similar disposable materials from outpatient areas and emergency rooms.

**Infrared incinerator** (Section 260.10) means any enclosed device that uses electric powered resistance heaters as a source of radiant heat followed by an afterburner using controlled flame combustion and which is not listed as an industrial furnace.

**Inground tank** (Section 260.10) means a device meeting the definition of "tank" in Sec. 260.10 whereby a portion of the tank wall is situated to any degree within the ground, thereby preventing visual inspection of that external surface area of the tank that is in the ground.

**Inherent process equipment** (Section 64.1) means equipment that is necessary for the proper or safe functioning of the process, or material recovery equipment that the owner or operator documents is installed and operated primarily for purposes other than compliance with air pollution regulations. Equipment that must be operated at an efficiency higher than that achieved during normal process operations in order to comply with the applicable emission limitation or standard is not inherent process equipment. For the purposes of this part, inherent process equipment is not considered a control device.

**Inherently Low-Emission Vehicle** (Section 88.302-93) means any LDV or LDT conforming to the applicable Inherently Low-Emission Vehicle standard, or any HDV with an engine conforming to the applicable Inherently Low-Emission Vehicle standard. No dual-fuel or flexible-fuel vehicles shall be considered Inherently Low-Emission Vehicles unless they are certified to the applicable standard(s) on all fuel types for which they are designed to operate.

**Inherently waste-like materials** (Section 261.2) The following materials are solid wastes when they are recycled in any manner: (1) Hazardous Waste Nos. F020, F021 (unless used as an ingredient to make a product at the site of generation), F022, F023, F026, and F028. (2) Secondary materials fed to a halogen acid furnace that exhibit a characteristic of a hazardous waste or are listed as a hazardous waste as defined in subparts C or D of this part, except for brominated material that meets the following criteria: (i) The material must contain a bromine concentration of at least 45%; and (ii) The material must contain less than a total of 1% of toxic organic compounds listed in appendix VIII; and (iii) The material is processed continually on-site in the halogen acid furnace via direct conveyance (hard piping). (3) The Administrator will use the following criteria to add wastes to that list: (i) (A) The materials are ordinarily disposed of, burned, or incinerated; or (B) The materials contain toxic constituents listed in appendix VIII of part 261 and these constituents are not ordinarily found in raw materials or products for which the materials substitute (or are found in raw materials or products in smaller concentrations) and are not used or reused during the recycling process; and (ii) The material may pose a substantial hazard to human health and the environment when recycled. (e) Materials that are not solid waste when recycled. (1) Materials are not solid wastes when they can be shown to be recycled by being: (i) Used or reused as ingredients in an industrial process to make a product, provided the materials are not being reclaimed; or (ii) Used or reused as effective substitutes for commercial products; or (iii) Returned to the original process from which they are generated, without first being reclaimed or land disposed. The material must be returned as a substitute for feedstock materials. In cases where the original process to which the material is returned is a secondary process, the materials must be managed such that there is no placement on the land. In cases where the materials are generated and reclaimed within the primary mineral processing industry, the conditions of the exclusion found at Sec. 261.4(a) (17) apply rather than this paragraph. (2) The following materials are solid wastes, even if the recycling involves use, reuse, or return to the original process (described in paragraphs (e)(1) (i) through (iii) of this section): (i) Materials used in a manner constituting disposal, or used to produce products that are applied to the land; or (ii) Materials burned for energy recovery, used to produce a fuel, or contained in fuels; or (iii) Materials accumulated speculatively; or (iv) Materials listed in paragraphs (d)(1) and (d)(2) of this section. (f) Documentation of claims that materials are not solid wastes or are conditionally exempt from regulation. Respondents in actions to enforce regulations implementing subtitle C of RCRA who raise a claim that a certain material is not a solid waste, or is conditionally exempt from regulation, must demonstrate that there is a known market or disposition for the material, and that they meet the terms of the exclusion or exemption. In doing so, they must provide appropriate documentation (such as contracts showing that a second person uses the material as an ingredient in a production process) to demonstrate that the material is not a waste, or is exempt from regulation. In addition, owners or operators of facilities claiming that they actually are recycling materials must show that they have the necessary equipment to do so.

**Initial Certification Statement for this subpart** (Section 455.41) means a written submission to the appropriate permitting authority, e.g., the local Control Authority (the POTW) or NPDES permit writer which must be signed by the responsible corporate officer as defined in 40 CFR 403.12(l) or 40 CFR 122.22 and which: Lists and describes those product families, process lines and/or process units for which the PFPR facility is implementing the Pollution Prevention Alternative ("P2 Alternative"); (2) Describes the PFPR facility specific practices for each product family/process line/process unit which are to be practiced as part of the P2 Alternative; (3) Describes any justification allowing modification to the practices listed in Table 8 to this part 455; and (4) Lists the treatment system being used to obtain a P2 allowable discharge (as defined in 455.41).

**Initial compliance period** (Section 141.2) means the first full three-year compliance period which begins at least 18 months after promulgation, except for contaminants listed at Sec. 141.61(a) (19)-(21), (c) (19)-(33), and Sec. 141.62(b) (11)-(15), initial compliance period means the first full three-year compliance period after promulgation for systems with 150 or more service connections (January 1993-December 1995), and first full three-year compliance period after the effective date of the regulation (January 1996-December 1998) for systems having fewer than 150 service connections.

**Initial comprehensive performance test** (Section 62.1201) means the comprehensive performance test that is used as the basis for initially demonstrating compliance with the standards.

**Initial Decision** (Section 164.2) means the decision of the Administrative Law Judge supported by findings of fact and conclusions regarding all material issues of law, fact, or discretion, as well as reasons therefore. Such decision shall become the final decision and order of the Administrator without further proceedings unless an appeal there from is taken or the Administrator orders review thereof as herein provided.

**Initial startup** (Section 65.2) means, for new or reconstructed sources, the first time the source begins production. For additions or changes not defined as a new source by an applicable referencing subpart, initial startup means the first time additional or changed equipment is put into operation. Initial startup does not include operation solely for testing equipment. Initial startup does not include subsequent startup (as defined in this section) of process units following malfunctions or process unit shutdowns. Except for equipment leaks, initial startup also does not include subsequent startups (as defined in this section) of process units following changes in product for flexible operation units or following recharging of equipment in batch operation.

**Injection interval** (Section 148.2) means that part of the injection zone in which the well is screened, or in which the waste is otherwise directly emplaced.

**Injection well** (Section 144.3) means a "well" into which "fluids" are being injected.

**Injection well** (Section 146.3) means a "well" into which "fluids" are being injected.

**Injection well** (Section 147.2902) a well into which fluids are being injected.

**Injection well** (Section 260.10) means a well into which fluids are injected. (See also "underground injection".)

**Injection well** (Section 270.2) means a well into which fluids are being injected.

**Injection zone** (Section 144.3) means a geological "formation" group of formations, or part of a formation receiving fluids through a "well."

**Injection zone** (Section 146.3) means a geological "formation", group of formations, or part of a formation receiving fluids through a well.

**Injection zone** (Section 147.2902) a geological formation, group of formations, or part of a formation receiving fluids through a well.

**Injury** (Section 112.2) means a measurable adverse change, either long- or short-term, in the chemical or physical quality or the viability of a natural resource resulting either directly or indirectly from exposure to a discharge of oil, or exposure to a product of reactions resulting from a discharge of oil.

**Injury** (Section 68.3) means any effect on a human that results either from direct exposure to toxic concentrations; radiant heat; or overpressures from accidental releases or from the direct consequences of a vapor cloud explosion (such as flying glass, debris, and other projectiles) from an accidental release and that requires medical treatment or hospitalization.

**Inland waters, for the purposes of classifying the size of discharges** (Section 300.5) means those waters of the United States in the inland zone, waters of the Great Lakes, and specified ports and harbors on inland rivers.

**Inland zone** (Section 300.5) means the environment inland of the coastal zone excluding the Great Lakes and specified ports and harbors on inland rivers. The term inland zone delineates an area of federal responsibility for response action. Precise boundaries are determined by EPA/USCG agreements and identified in federal regional contingency plans.

**In-line kiln raw mill** (Section 62.1201) means a hazardous waste burning cement kiln design whereby kiln gas is ducted through the raw material mill for portions of time to facilitate drying and heating of the raw material.

**Inner liner** (Section 260.10) means a continuous layer of material placed inside a tank or container which protects the construction materials of the tank or container from the contained waste or reagents used to treat the waste.

**Inorganic metal-bearing waste** (Section 268.2) is one for which EPA has established treatment standards for metal hazardous constituents, and which does not otherwise contain significant organic or cyanide content as described in Sec. 268.3©(1), and is specifically listed in appendix XI of this part.

**Inorganic wastewater treatment chemicals** (Section 455.10) means inorganic chemicals that are commonly used in wastewater treatment systems to aid in the removal of pollutants through physical/chemical technologies such as chemical precipitation, flocculation, neutralization, chemical oxidation, hydrolysis and/or adsorption.

**In-plant monitoring point** (Section 439.1) means a location within a plant, where an individual process effluent can be exclusively monitored before it is diluted or mixed with other process wastewaters enroute to the end-of-pipe.

**In-process control technology** (Section 467.02) is the conservation of chemicals and water throughout the production operations to reduce the amount of wastewater to be discharged.

**In-process control technology** (Section 471.02) is the conservation of chemicals and water throughout the production operations to reduce the amount of wastewater to be discharged.

**In-situ leach methods** (Section 440.132) means the processes involving the purposeful introduction of suitable leaching solutions into a uranium ore body to dissolve the valuable minerals in place and the purposeful leaching of uranium ore in a static or semistatic condition either by gravity through an open pile, or by flooding a confined ore pile. It does not include the natural dissolution of uranium by ground waters, the incidental leaching of uranium by mine drainage, nor the rehabilitation of aquifiers and the monitoring of these aquifiers.

**In-situ sampling systems** (Section 65.2) means nonextractive samplers or in-line samplers.

**Inspect and qualify** (Section 92.2) means to determine that a previously used component or system meets all applicable criteria listed for the component or system in a certificate of conformity for remanufacturing (e.g., determine that the component or system is functionally equivalent to one that has not been used previously).

**Inspection** (Section 745.103) means: (1) A surface-by-surface investigation to determine the presence of lead-based paint as provided in section 302© of the Lead-Based Paint Poisoning and Prevention Act [42 U.S.C. 4822], and (2) The provision of a report explaining the results of the investigation.

**Inspection** (Section 745.223) means a surface-by-surface investigation to determine the presence of lead-based paint and the provision of a report explaining the results of the investigation.

**Inspection Criteria** (Section 204.51) means the rejection and acceptance numbers associated with a particular sampling plan.

**Inspection Criteria** (Section 205.51) means the rejection and acceptance numbers associated with a particular sampling plan.

**Inspection criteria** (Section 86.1002-2001) means the pass and fail numbers associated with a particular sampling plan.

**Inspection Criteria** (Section 86.1002-84) means the pass and fail numbers associated with a particular sampling plan.

**Inspection criteria** (Section 86.1002-97) means the pass and fail numbers associated with a particular sampling plan.

**Inspection Criteria** (Section 86.602-84) means the pass and fail numbers associated with a particular sampling plan.

**Inspection criteria** (Section 89.502) means the pass and fail numbers associated with a particular sampling plan.

**Installation** (Section 279.1) will be considered to have commenced if the owner or operator has obtained all federal, state, and local approvals or permits necessary to begin installation of the tank and if either (1) A continuous on-site installation program has begun, or (2) The owner or operator has entered into contractual obligations— which cannot be canceled or modified without substantial loss—for installation of the tank to be completed within a reasonable time.

**Installation inspector** (Section 260.10) means a person who, by reason of his knowledge of the physical sciences and the principles of engineering, acquired by a professional education and related practical experience, is qualified to supervise the installation of tank systems.

**Installer** (Section 92.2) means an individual or entity which assembles remanufactured locomotives or locomotive engines.

**Instantaneous monitoring** (Section 62.1201) means continuously sampling, detecting, and recording the regulated parameter without use of an averaging period.

**Institutional Biosafety Committee** (Section 725.3) means the committees described in the NIH Guidelines in section IV.B.2.

**Institutional solid waste** (Section 243.101) means solid wastes generated by educational, health care, correctional, and other institutional facilities.

**Institutional solid waste** (Section 246.101) means solid wastes generated by educational, health care, correctional and other institutional facilities.

**Institutional use** (Section 152.3) means any application of a pesticide in or around any property or facility that functions to provide a service to the general public or to public or private organizations, including but not limited to: (1) Hospitals and nursing homes. (2) Schools other than preschools and day care facilities. (3) Museums and libraries. (4) Sports facilities. (5) Office buildings.

**Instrumentation system** (Section 65.2) means a group of equipment components used to condition and convey a sample of the process fluid to analyzers and instruments for the purpose of determining process operating conditions (for example, composition, pressure, flow). Valves and connectors are the predominant type of equipment used in instrumentation systems; however, other types of equipment may also be included in these systems. Only valves nominally 0.5 inches and smaller in diameter and connectors nominally 0.75 inches and smaller in diameter are considered instrumentation systems for the purposes of subpart F of this part.

**Insulation board** (Section 429.11) means a panel manufactured from interfelted ligno-cellulosic fibers consolidated to a density of less than 0.5 g/cu cm (less than 31 lb/cu ft).

**Integrated facility** (Section 413.02) is defined as a facility that performs electroplating as only one of several operations necessary for manufacture of a product at a single physical location and has significant quantities of process wastewater from non-electroplating manufacturing operations. In addition, to qualify as an "integrated facility" one or more plant electroplating process wastewater lines must be combined prior to or at the point of treatment (or proposed treatment) with one or more plant sewers carrying process wastewater from non-electroplating manufacturing operations.

**Integrated refueling emission control system** (Section 86.098-2) means a system where vapors resulting from refueling are stored in a common vapor storage unit(s) with other evaporative emissions of the vehicle and are purged through a common purge system.

**Integrated refueling emission control system** (Section 86.1803-01) means a system where vapors resulting from refueling are stored in a common vapor storage unit(s) with other evaporative emissions of the vehicle and are purged through a common purge system.

**Integrated system** (Section 158.153) means a process for producing a pesticide product that: (1) Contains any active ingredient derived from a source that is not an EPA-registered product; or (2) Contains any active ingredient that was produced or acquired in a manner that does not permit its inspection by the Agency under FIFRA sec. 9(a) prior to its use in the process.

**Interested person** (Section 304.12) means the Administrator, any EPA employee, any party to the proceeding, any potentially responsible party associated with the facility concerned, any person who filed written comments in the proceeding, any participant or intervenor in the proceeding, all officers, directors, employees, consultants, and agents of any party, and any attorney of record for any of the foregoing persons.

**Interested person** (Section 72.2) means any person who submitted written comments or testified at a public hearing on the draft permit or other matter subject to notice and comment under the Acid Rain Program or any person who submitted his or her name to the Administrator or the permitting authority, as appropriate, to be placed on a list of persons interested in such matter. The Administrator or the permitting authority may update the list of interested persons from time to time by requesting additional written indication of continued interest from the persons listed and may delete from the list the name of any person failing to respond as requested.

**Interference** (Section 403.3) means a Discharge which, alone or in conjunction with a discharge or discharges from other sources, both: (1) Inhibits or disrupts the POTW, its treatment processes or operations, or its sludge processes, use or disposal; and (2) Therefore is a cause of a violation of any requirement of the POTW's NPDES permit (including an increase in the magnitude or duration of a violation) or of the prevention of sewage sludge use or disposal in compliance with the following statutory provisions and regulations or permits issued thereunder (or more stringent State or local regulations): Section 405 of the Clean Water Act, the Solid Waste Disposal Act (SWDA) (including title II, more commonly referred to as the Resource Conservation and Recovery Act (RCRA), and including State regulations contained in any State sludge management plan prepared pursuant to subtitle D of the SWDA), the Clean Air Act, the Toxic Substances Control Act, and the Marine Protection, Research and Sanctuaries Act.

**Interference** (Section 425.02) means the discharge of sulfides in quantities which can result in human health hazards and/or risks to human life, and an inhibition or disruption of POTW as defined in 40 CFR 403.3(i).

**Intergeneric microorganism** (Section 725.3) means a microorganism that is formed by the deliberate combination of genetic material originally isolated from organisms of different taxonomic genera. (1) The term "intergeneric microorganism" includes a microorganism which contains a mobile genetic element which was first identified in a microorganism in a genus different from the recipient microorganism. (2) The term "intergeneric microorganism" does not include a microorganism which contains introduced genetic material consisting of only well-characterized, non-coding regulatory regions from another genus.

**Interim approval** (Section 281.12) means the approval received by a state program that meets the requirements in Sec. 281.11© (1) and (2) for the time period defined in Sec. 281.11© (3).

**Interim authorization** (Section 270.2) means approval by EPA of a State hazardous waste program which has met the requirements of section 3006(g) (2) of RCRA and applicable requirements of part 271, subpart B.

**Interim certification** (Section 745.223) means the status of an individual who has successfully completed the appropriate training course in a discipline from an accredited training program, as defined by this section, but has not yet received formal certification in that discipline from EPA pursuant to Sec. 745.226. Interim certifications expire 6 months after the completion of the training course, and is equivalent to a certificate for the 6-month period.

**Interim controls** (Section 745.223) means a set of measures designed to temporarily reduce human exposure or likely exposure to lead-based paint hazards, including specialized cleaning, repairs, maintenance, painting, temporary containment, ongoing monitoring of lead-based paint hazards or potential hazards, and the establishment and operation of management and resident education programs.

**Interim non-Tier 2 vehicle, interim non-Tier 2 LDV/LLDT, interim non-Tier 2 HLDT/MDPV, or interim vehicle** (Section 86.1803-01) refer to 2004 or later model year light-duty vehicles, light-duty trucks or MDPVs, or a specific combination thereof, not certified to Tier 2 FTP exhaust emission standards during the Tier 2 phase-in period. Model year 2004 HLDTs belonging to test groups whose model year commences before December 21, 2003, are not interim non-Tier 2 HLDTs unless their manufacturer chooses to comply with the interim requirements applicable to HLDTs for all of its 2004 model year HLDTs as permitted in this subpart. Similarly 2004 model year heavy-duty vehicles whose model year commences before December 21, 2003, are not interim non-Tier 2 MDPVs unless their manufacturer chooses to comply with the interim requirements applicable to MDPVs for all of its 2004 model year MDPVs as permitted in this subpart. The terms interim non-Tier 2 vehicle, interim non-Tier 2 LDV, interim non-Tier 2 LDT, interim non-Tier 2 HLDT, interim non-Tier 2 MDPV, etc. have the same meaning without the words "non-Tier 2".

**Interior wastewater sources** (Section 455.10) means wastewater that is generated from cleaning or rinsing the interior of pesticide formulating, packaging or repackaging equipment; or from rinsing the interior of raw material drums, shipping containers or bulk storage tanks; or cooling water that comes in direct contact with pesticide active ingredients (PAIs) during the formulating, packaging or repackaging process.

**Interior window sill** (Section 745.63) means the portion of the horizontal window ledge that protrudes into the interior of the room.

**Intermediary** (Section 86.1803-01) means any individual or entity, other than a manufacturer, which provides service or equipment to automotive technicians.

**Intermediate** (Section 704.3) means any chemical substance that is consumed, in whole or in part, in chemical reactions used for the intentional manufacture of other chemical substances or mixtures, or that is intentionally present for the purpose of altering the rates of such chemical reactions.

**Intermediate** (section 710.2) means any chemical substance: (1) Which is intentionally removed from the equipment in which it is manufactured, and (2) which either is consumed in whole or in part in chemical reaction(s) used for the intentional manufacture of other chemical substance(s) or mixture(s), or is intentionally present for the purpose of altering the rate of such chemical reaction(s). Note: The equipment in which it was manufactured includes the reaction vessel in which the chemical substance was manufactured and other equipment which is strictly ancillary to the reaction vessel, and any other equipment through which the chemical substance may flow during a continuous flow process, but does not include tanks or other vessels in which the chemical substance is stored after its manufacture.

**Intermediate** (Section 712.3) means any chemical substance that is consumed, in whole or in part, in chemical reactions used for the intentional manufacture of other chemical substances or mixtures, or that is intentionally present for the purpose of altering the rates of such chemical reactions. (See also paragraph (j) of this section.)

**Intermediate** (Section 720.3) means any chemical substance that is consumed, in whole or in part, in chemical reactions used for the intentional manufacture of another chemical substance(s) or mixture(s), or that is intentionally present for the purpose of altering the rates of such chemical reactions.

**Intermediate bulk container ("IBC" or "Tote")** (Section 442.2) means a completely enclosed storage vessel used to hold liquid, solid, or gaseous commodities or cargos which are in direct contact with the container interior. IBCs may be loaded onto flat beds for either truck or rail transport, or onto ship decks for water transport. IBCs are portable containers with 450 liters (119 gallons) to 3000 liters (793 gallons) capacity. IBCs are also commonly referred to as totes or tote bins.

**Intermediate change to monitoring** (Section 65.2) means a modification to federally required monitoring involving "proven technology" (generally accepted by the scientific community as equivalent or better) that is applied on a site-specific basis and that may have the potential to decrease the stringency of the associated emission limitation or standard. Though site-specific, an intermediate change may set a national precedent for a source category and may ultimately result in a revision to the federally required monitoring. Examples of intermediate changes to monitoring include, but are not limited to: (1) Use of a continuous monitoring system (CEMS) in lieu of a parameter monitoring approach; (2) Decreased frequency for non-continuous parameter monitoring or physical inspections; (3) Changes to quality control requirements for parameter monitoring; and (4) Use of an electronic data reduction system in lieu of manual data reduction.

**Intermediate change to test method** (Section 65.2) means a within-method modification to a federally enforceable test method involving "proven technology" (generally accepted by the scientific community as equivalent or better) that is applied on a site-specific basis and that may have the potential to decrease the stringency of the associated emission limitation or standard. Though site-specific, an intermediate change may set a national precedent for a source category and may ultimately result in a revision to the federally enforceable test method. In order to be approved, an intermediate change must be validated according to EPA Method 301 (40 CFR part 63, appendix A) to demonstrate that it provides equal or improved accuracy or precision. Examples of intermediate changes to a test method include, but are not limited to: (1) Modifications to a test method's sampling procedure including substitution of sampling equipment that has been demonstrated for a particular sample matrix; and use of a different impinger absorbing solution; (2) Changes in sample recovery procedures and analytical techniques, such as changes to sample holding times and use of a different analytical finish with proven capability for the analyte of interest; and (3) "Combining" a federally required method with another proven method for application to processes emitting multiple pollutants.

**Intermediate premanufacture notice or intermediate PMN** (Section 700.43) means any PMN submitted to EPA for a chemical substance which is an intermediate (as "intermediate" is defined in Sec. 720.3 of this chapter) in the production of a final product, provided that the PMN for the intermediate is submitted to EPA at the same time as, and together with, the PMN for the final product and that the PMN for the intermediate identifies the final product and describes the chemical reactions leading from the intermediate to the final product. If PMNs are submitted to EPA at the same time for several intermediates used in the production of a final product, each of those is an intermediate PMN if they all identify the final product and every other associated intermediate PMN and are submitted to EPA at the same time as, and together with, the PMN for the final product.

**Intermediate speed** (Section 86.082-2) means peak torque speed if peak torque speed occurs between 60 and 75 percent of rated speed. If the peak torque speed is less than 60 percent of rated speed, intermediate speed means 60 percent of rated speed. If the peak torque speed is greater than 75 percent of rated speed, intermediate speed means 75 percent of rated speed.

**Intermediate speed** (Section 89.2) means peak torque speed if peak torque speed occurs from 60 to 75 percent of rated speed. If peak torque speed is less than 60 percent of rated speed, intermediate speed means 60 percent of rated speed. If peak torque speed is greater than 75 percent of rated speed, intermediate speed means 75 percent of rated speed.

**Intermediate Speed** (Section 94.2) means peak torque speed if peak torque speed occurs from 60 to 75 percent of maximum test speed. If peak torque speed is less than 60 percent of maximum test speed, intermediate speed means 60 percent of maximum test speed. If peak torque speed is greater than 75 percent of maximum test speed, intermediate speed means 75 percent of maximum test speed.

**Intermediate Temperature Cold Testing** (Section 86.094-2) means testing done pursuant to the driving cycle and testing conditions contained in 40 CFR part 86, subpart C, at temperatures between 25 deg.F (-4 deg.C) and 68 deg.F (20 deg.C).

**Intermediate temperature cold testing** (Section 86.1803-01) means testing done pursuant to the driving cycle and testing conditions contained in subpart C of this part, at temperatures between 25 deg.F (-4 deg. C) and 68 deg. F (20 deg. C).

**Intermittent control system (ICS)** (Section 51.100) means a dispersion technique which varies the rate at which pollutants are emitted to the atmosphere according to meteorological conditions and/or ambient concentrations of the pollutant, in order to prevent ground-level concentrations in excess of applicable ambient air quality standards. Such a dispersion technique is an ICS whether used alone, used with other dispersion techniques, or used as .a supplement to continuous emission controls (i.e., used as a supplemental control system).

**Intermittent operations** (Section 471.02) means the industrial users does not have a continuous operation.

**Intermodal tank container** (Section 442.2) means a completely enclosed storage vessel used to hold liquid, solid, or gaseous commodities or cargos which come in direct contact with the tank interior. Intermodal tank containers may be loaded onto flat beds for either truck or rail transport, or onto ship decks for water transport. Containers larger than 3000 liters capacity are considered intermodal tank containers. Containers smaller than 3000 liters capacity are considered IBCs.

**Internal floating roof** (Section 265.1081) means a cover that rests or floats on the material surface (but not necessarily in complete contact with it) inside a tank that has a fixed roof.

**Internal floating roof or IFR** (Section 65.2) means a pontoon-type (noncontact) or double-deck-type (contact) roof that is designed to rest or float on the stored liquid surface inside a storage vessel that has a fixed roof.

**International shipment** (Section 260.10) means the transportation of hazardous waste into or out of the jurisdiction of the United States.

**International System of Units** (Section 191.12) is the version of the metric system which has been established by the International Bureau of Weights and Measures and is administered in the United States by the National Institute of Standards and Technology. The abbreviation for this system is "SI."

**Interstate agency** (Section 122.2) means an agency of two or more States established by or under an agreement or compact approved by the Congress, or any other agency of two or more States having substantial powers or duties pertaining to the control of pollution as determined and approved by the Administrator under the CWA and regulations.

**Interstate agency** (Section 124.2) means an agency of two or more States established by or under an agreement or compact approved by the Congress, or any other agency of two or more States having substantial powers or duties pertaining to the control of pollution as determined and approved by the Administrator under the "appropriate Act and regulations."

**Interstate Agency** (Section 142.2) means an agency of two or more States established by or under an agreement or compact approved by the Congress, or any other agency of two or more States or Indian Tribes having substantial powers or duties pertaining to the control of pollution as determined and approved by the Administrator.

**Interstate Agency** (Section 144.3) means an agency of two or more States established by or under an agreement or compact approved by the Congress, or any other agency of two or more States or Indian Tribes having substantial powers or duties pertaining to the control of pollution as determined and approved by the Administrator under the "appropriate Act and regulations."

**Interstate agency** (Section 233.2) means an agency of two or more States established by or under an agreement or compact approved by the Congress, or any other agency of two or more States having substantial powers or duties pertaining to the control of pollution.

**Interstate Commerce** (Section 201.1) means the commerce between any place in a State and any place in another State, or between places in the same State through another State, whether such commerce moves wholly by rail or partly by rail and partly by motor vehicle, express, or water. This definition of "interstate commerce" for purposes of this regulation is similar to the definition of "interstate commerce" in section 203(a) of the Interstate Commerce Act (49 U.S.C. 303(a)).

**Interstate commerce** (Section 202.10) means the commerce between any place in a State and any place in another State or between places in the same State through another State, whether such commerce moves wholly by motor vehicle or partly by motor vehicle and partly by rail, express, water or air. This definition of "interstate commerce" for purposes of these regulations is the same as the definition of "interstate commerce" in section 203(a) of the Interstate Commerce Act. [49 U.S.C. 303(a)]

**Intervener** (Section 209.3) means a person who files a motion to be made a party under Sec. 209.15 or Sec. 209.16, and whose motion is approved.

**Introduced genetic material** (Section 725.3) means genetic material that is added to, and remains as a component of, the genome of the recipient.

**In-use aircraft gas turbine engine** (Section 87.1) means an aircraft gas turbine engine which is in service.

**In-Use Compliance Period for purposes of in-use testing** (Section 85.1402) means a period of 150,000 miles.

**In-use vehicle** (Section 86.1803-01) means a customer owned and operated vehicle which is not under the control of the manufacturer, dealerships or their agents. Leased vehicles will be considered in-use vehicles for the purpose of this subpart if the vehicles meet the criteria specified in Sec. 86.1845-01.

**In-use verification program (IUVP)** (Section 86.1803-01) means the testing program conducted by manufacturers which gathers in-use emission data in accordance with Sec. 86.1848-01.

**Inventory** (Section 720.3) means the list of chemical substances manufactured or processed in the United States that EPA compiled and keeps current under section 8(b) of the Act.

**Inventory form** (Section 370.2) means the Tier I and Tier II emergency and hazardous chemical inventory forms set forth in subpart D of this part.

**Inventory of open dumps** (Section 256.06) means the inventory required under section 4005(b) and is defined as the list published by EPA of those disposal facilities which do not meet the criteria.

**Investor-owned utility** (Section 72.2) means a utility that is organized as a tax-paying for-profit business.

**Iron and steel** (Section 420.11) means those by-product cokemaking operations other than merchant cokemaking operations.

**Iron blast furnace** (Section 420.31) means all blast furnaces except ferromanganese blast furnaces.

**ISO 9001-registered facility** (Section 53.1) means a manufacturing facility that is either: (1) An International Organization for Standardization (ISO) 9001-registered manufacturing facility, registered to the ISO 9001 standard (by the Registrar Accreditation Board (RAB) of the American Society for Quality Control (ASQC) in the United States), with registration maintained continuously. (2) A facility that can be demonstrated, on the basis of information submitted to the EPA, to be operated according to an EPA-approved and periodically audited quality system which meets, to the extent appropriate, the same general requirements as an ISO 9001-registered facility for the design and manufacture of designated reference and equivalent method samplers and monitors.

**ISO-certified auditor** (Section 53.1) means an auditor who is either certified by the Registrar Accreditation Board (in the United States) as being qualified to audit quality systems using the requirements of recognized standards such as ISO 9001, or who, based on information submitted to the EPA, meets the same general requirements as provided for ISO-certified auditors.

**Isokinetic sampling** (Section 60.2) means sampling in which the linear velocity of the gas entering the sampling nozzle is equal to that of the undisturbed gas stream at the sample point.

**Issuance of a part 70 permit** (Section 60.2) will occur, if the State is the permitting authority, in accordance with the requirements of part 70 of this chapter and the applicable, approved State permit program. When the EPA is the permitting authority, issuance of a Title V permit occurs immediately after the EPA takes final action on the final permit.

**Issuance of a part 70 permit** (Section 61.02) will occur, if the State is the permitting authority, in accordance with the requirements of part 70 of this chapter and the applicable, approved State permit program. When the EPA is the permitting authority, issuance of a title V permit occurs immediately after the EPA takes final action on the final permit.

**Issuance of a part 70 permit** (Section 63.2) will occur, if the State is the permitting authority, in accordance with the requirements of part 70 of this chapter and the applicable, approved State permit program. When the EPA is the permitting authority, issuance of a title V permit occurs immediately after the EPA takes final action on the final permit.

**Issuance of an NSO** (Section 57.103) means the final transmittal of the NSO pursuant to Sec. 57.107 (a) by an issuing agency (other than EPA) to EPA for approval, or the publication of an NSO issued by EPA in the Federal Register.

**Issuing agency, unless otherwise specifically indicated,** (Section 57.103) means the State or local air pollution control agency to which a smelter's owner has applied for an NSO, or which has issued the NSO, or EPA, when the NSO application has been made to EPA. Any showings or demonstrations required to be made under this part to the issuing agency, when not EPA, are subject to independent determinations by EPA.

# J

**Jams and jellies** (Section 407.81) shall include the production of jams, jellies and preserves defined as follows: The combination of fruit and fruit concentrate, sugar, pectin, and other additives in an acidic medium resulting in a gelatinized and thickened finished product.

**Job shop** (Section 433.11) shall mean a facility which owns not more than 50% (annual area basis) of the materials undergoing metal finishing.

**Joint sponsor** (Section 790.3) means a person who sponsors testing pursuant to section 4(b)(3)(A) of the Act.

**Joint sponsorship** (Section 790.3) means the sponsorship of testing by two or more persons in accordance with section 4(b)(3)(A) of the Act.

**Joint submitters** (Section 700.43) means two or more persons who submit a section 5 notice together.

**Judicial Officer** (Section 164.2) means an officer or employee of the Agency designated as a judicial officer, pursuant to these rules, who shall meet the qualifications and perform functions as herein provided.

# K

**k cu m** (Section 401.11) means 1000 cubic meter(s).

**Kappa number** (Section 430.21) is a measure of the lignin content in unbleached pulp, determined after pulping and prior to bleaching.

**kg** (Section 401.11) means kilogram(s).

**Kilowatthour saved or savings** (Section 72.2) means the net savings in electricity use (expressed in Kwh) that result directly from a utility's energy conservation measures or programs.

**kkg** (Section 401.11) means 1000 kilogram(s).

**Known human effects** (Section 717.3) means a commonly recognized human health effect of a particular substance or mixture as described either in: (i) Scientific articles or publications abstracted in standard reference sources. (ii) The firm's product labeling or material safety data sheets (MSDS). (2) However, an effect is not a "known human effect" if it: (i) Was a significantly more severe toxic effect than previously described. (ii) Was a manifestation of a toxic effect after a significantly shorter exposure period or lower exposure level than described. (iii) Was a manifestation of a toxic effect by an exposure route different from that described.

**Known to or reasonably ascertainable** (Section 704.3) by means all information in a person's possession or control, plus all information that a reasonable person similarly situated might be expected to possess, control, or know.

**Known to or reasonably ascertainable** (Section 712.3) by means all information in a person's possession or control, plus all information that a reasonable person similarly situated might be expected to possess, control, or know, or could obtain without unreasonable burden.

**Known to or reasonably ascertainable** (Section 720.3) by means all information in a person's possession or control, plus all information that a reasonable person similarly situated might be expected to possess, control, or know.

**kw** (Section 401.11) means kilowatt(s).

**kwh** (Section 401.11) means kilowatt hour(s).

# L

**l** (Section 401.11) means liter(s).

**Label** (Section 211.203) That item, as described in this regulation, which is inscribed on, affixed to or appended to a product, its packaging, or both for the purpose of giving noise reduction effectiveness information appropriate to the product.

**Label** (Section 59.101) means any written, printed, or graphic matter affixed to or appearing upon any automobile refinish coating or coating component container or package for purposes of identifying or giving information on the product, use of the product, or contents of the container or package.

**Label** (Section 600.022-85) means a sticker that contains fuel economy information and is affixed to new automobiles in accordance with subpart D of this part.

**Laboratory** (Section 761.13) means a facility that analyzes samples for PCBs and is unaffiliated with any entity whose activities involve PCBs.

**Lacquer** (Section 59.101) means a thermoplastic coating which dries primarily by solvent evaporation, and which is resoluble in its original solvent.

**Laminated paperboard** (Section 247.3) means board made from one or more plies of kraft paper bonded together, with or without facers, that is used for decorative, structural, or insulating purposes;

**Lamp** (Section 260.10) also referred to as "universal waste lamp", is defined as the bulb or tube portion of an electric lighting device. A lamp is specifically designed to produce radiant energy, most often in the ultraviolet, visible, and infra-red regions of the electromagnetic spectrum. Examples of common universal waste electric lamps include, but are not limited to, fluorescent, high intensity discharge, neon, mercury vapor, high pressure sodium, and metal halide lamps.

**Lamp, also referred to as "universal waste lamp"** (Section 273.9) is defined as the bulb or tube portion of an electric lighting device. A lamp is specifically designed to produce radiant energy, most often in the ultraviolet, visible, and infra-red regions of the electromagnetic spectrum. Examples of common universal waste electric lamps include, but are not limited to, fluorescent, high intensity discharge, neon, mercury vapor, high pressure sodium, and metal halide lamps.

**Land** (Section 192.11) means any surface or subsurface land that is not part of a disposal site and is not covered by an occupiable building.

**Land application** (Section 503.11) is the spraying or spreading of sewage sludge onto the land surface; the injection of sewage sludge below the land surface; or the incorporation of sewage sludge into the soil so that the sewage sludge can either condition the soil or fertilize crops or vegetation grown in the soil.

**Land application unit** (Section 272.2) means an area where wastes are applied onto or incorporated into the soil surface (excluding manure spreading operations) for agricultural purposes or for treatment and disposal.

**Land disposal** (Section 268.2) means placement in or on the land, except in a corrective action management unit or staging pile, and includes, but is not limited to, placement in a landfill, surface impoundment, waste pile, injection well, land treatment facility, salt dome formation, salt bed formation, underground mine or cave, or placement in a concrete vault, or bunker intended for disposal purposes.

**Land treatment facility** (Section 260.10) means a facility or part of a facility at which hazardous waste is applied onto or incorporated into the soil surface; such facilities are disposal facilities if the waste will remain after closure.

**Land with a high potential for public exposure** (Section 503.31) is land that the public uses frequently. This includes, but is not limited to, a public contact site and a reclamation site located in a populated area (e.g, a construction site located in a city).

**Land with a low potential for public exposure** (Section 503.31) is land that the public uses infrequently. This includes, but is not limited to, agricultural land, forest, and a reclamation site located in an unpopulated area (e.g., a strip mine located in a rural area).

**Landfill** (Section 260.10) means a disposal facility or part of a facility where hazardous waste is placed in or on land and which is not a pile, a land treatment facility, a surface impoundment, an underground injection well, a salt dome formation, a salt bed formation, an underground mine, a cave, or a corrective action management unit.

**Landfill** (Section 272.2) means an area of land or an excavation in which wastes are placed for permanent disposal, and that is not a land application unit, surface impoundment, injection well, or waste pile.

**Landfill cell** (Section 260.10) means a discrete volume of a hazardous waste landfill which uses a liner to provide isolation of wastes from adjacent cells or wastes. Examples of landfill cells are trenches and pits.

**Landfill directly associated with an industrial or commercial operation** (Section 445.2) means: (1) A landfill located on the same site as industrial or commercial operations; and (2) A landfill not located on the same site as the industrial or commercial operations (off-site), but "wholly-owned" by the industrial or commercial facility and primarily dedicated to receiving waste from the related industrial or commercial facility.

**Landfill unit** (Section 445.2) means an area of land or an excavation in which wastes are placed for permanent disposal, that is not a land application or land treatment unit, surface impoundment, underground injection well, waste pile, salt dome formation, a salt bed formation, an underground mine or a cave as these terms are defined in 40 CFR 257.2, 258.2 and 264.10.

**Landfill wastewater** (Section 445.2) means all wastewater associated with, or produced by, landfilling activities except for sanitary wastewater, non-contaminated storm water, contaminated ground water, and wastewater from recovery pumping wells. Landfill wastewater includes, but is not limited to, leachate, gas collection condensate, drained free liquids, laboratory derived wastewater, contaminated storm water and contact washwater from washing truck, equipment, and railcar exteriors and surface areas which have come in direct contact with solid waste at the landfill facility.

**Lapse** (Section 93.101) means that the conformity determination for a transportation plan or TIP has expired, and thus there is no currently conforming transportation plan and TIP.

**Large** (Section 407.61) shall mean a point source that processes a total annual raw material production of fruits, vegetables, specialties and other products that exceeds 9,080 kkg (10,000 tons) per year.

**Large** (Section 407.71) shall mean a point source that processes a total annual raw material production of fruits, vegetables, specialties and other products that exceeds 9,080 kkg (10,000 tons) per year.

**Large** (Section 407.81) shall mean a point source that processes a total annual raw material production of fruits, vegetables, specialties and other products that exceeds 9,080 kkg (10,000 tons) per year.

**Large Quantity Handler of Universal Waste** (Section 273.9) means a universal waste handler (as defined in this section) who accumulates 5,000 kilograms or more total of universal waste (batteries, pesticides, thermostats, or lamps, calculated collectively) at any time. This designation as a large quantity handler of universal waste is retained through the end of the calendar year in which 5,000 kilograms or more total of universal waste is accumulated.

**Large water system** (Section 141.2) for the purpose of subpart I of this part only, means a water system that serves more than 50,000 persons.

**Large-sized plants** (Section 428.71) shall mean plants which process more than

10,430 kg/day (23,000 lbs/day) of raw materials.

**Lateral expansion** (Section 258.2) means a horizontal expansion of the waste boundaries of an existing MSWLF unit.

**Latex paint** (Section 247.3) means a water-based decorative or protective covering having a latex binder;

**Law enforcement vehicle** (Section 88.302-94) means any vehicle which is primarily operated by a civilian or military police officer or sheriff, or by personnel of the Federal Bureau of Investigation, the Drug Enforcement Administration, or other agencies of the federal government, or by state highway patrols, municipal law enforcement, or other similar law enforcement agencies, and which is used for the purpose of law enforcement activities including, but not limited to, chase, apprehension, surveillance, or patrol of people engaged in or potentially engaged in unlawful activities. For federal law enforcement vehicles, the definition contained in Executive Order 12759, Section 11:

**Lawn edging** (Section 247.3) means a barrier used between lawns and landscaped areas or garden beds to prevent grass roots or weeds from spreading to the landscaped areas;

**lb** (Section 401.11) means pound(s).

**LC50** (Section 116.3) means that concentration of material which is lethal to one-half of the test population of aquatic animals upon continuous exposure for 96 hours or less.

**LC50** (Section 132.2) is a statistically or graphically estimated concentration that is expected to be lethal to 50 percent of a group of organisms under specified conditions.

**LDV/T** (Section 86.1803-01) means light-duty vehicles and light-duty trucks collectively, without regard to category.

**Leachate** (Section 258.2) means a liquid that has passed through or emerged from solid waste and contains soluble, suspended, or miscible materials removed from such waste.

**Leachate** (Section 260.10) means any liquid, including any suspended components in the liquid that has percolated through or drained from hazardous waste.

**Leachate** (Section 272.2) means liquid that has passed through or emerged from solid waste and contains soluble, suspended or miscible materials removed from such wastes.

**Leachate collection system** (Section 503.21) is a system or device installed immediately above a liner that is designed, constructed, maintained, and operated to collect and remove leachate from a sewage sludge unit.

**Lead** (Section 415.671) shall mean the total lead present in the process wastewater stream exiting the wastewater treatment system.

**Lead** (Section 416.61) shall mean total lead.

**Lead** (Section 420.02) means total lead and is determined by the method specified in 40 CFR 136.3.

**Lead additive** (Section 80.2) means any substance containing lead or lead compounds.

**Lead additive manufacturer** (Section 80.2) means any person who produces a lead additive or sells a lead additive under his own name.

**Lead administrative trustee** (Section 300.5) means a natural resource trustee who is designated on an incident-by-incident basis for the purpose of preassessment and damage assessment and chosen by the other trustees whose natural resources are affected by the incident. The lead administrative trustee facilitates effective and efficient communication during response operations between the OSC and the other natural resource trustees conducting activities associated with damage assessment, and is responsible for applying to the OSC for access to response operations resources on behalf of all trustees for initiation of a damage assessment.

**Lead agency** (Section 300.5) means the agency that provides the OSC/RPM to plan and implement response actions under the NCP. EPA, the USCG, another federal agency, or a state (or political subdivision of a state) operating pursuant to a contract or cooperative agreement executed pursuant to section 104(d)(1) of CERCLA, or designated pursuant to a Superfund Memorandum of Agreement (SMOA) entered into pursuant to subpart F of the NCP or other agreements may be the lead agency for a response action. In the case of a release of a hazardous substance, pollutant, or contaminant, where the release is on, or the sole source of the release is from, any facility or vessel under the jurisdiction, custody, or control of Department of Defense (DOD) or Department of Energy (DOE), then DOD or DOE will be the lead agency. Where the release is on, or the sole source of the release is from, any facility or vessel under the jurisdiction, custody, or control of a federal agency other than EPA, the USCG, DOD, or DOE, then that agency will be the lead agency for remedial actions and removal actions other than emergencies. The federal agency maintains its lead agency responsibilities whether the remedy is selected by the federal agency for non-NPL sites or by EPA and the federal agency or by EPA alone under CERCLA section 120. The lead agency will consult with the support agency, if one exists, throughout the response process.

**Lead service line** (Section 141.2) means a service line made of lead which connects the water main to the building inlet and any lead pigtail, gooseneck or other fitting which is connected to such lead line.

**Lead state agency** (Section 239.2) means the state agency which has the legal authority and oversight responsibilities to implement the permit program or other system of prior approval and conditions to ensure that facilities regulated under section 4010© of Subtitle D of RCRA comply with the requirements of the approved state permit program and/or has been designated as lead agency.

**Lead-based paint** (Section 745.103) means paint or other surface coatings that contain lead equal to or in excess of 1.0 milligram per square centimeter or 0.5 percent by weight.

**Lead-based paint** (Section 745.223) means paint or other surface coatings that contain lead equal to or in excess of 1.0 milligrams per square centimeter or more than 0.5 percent by weight.

**Lead-based paint activities** (Section 745.223) means, in the case of target housing and child-occupied facilities, inspection, risk assessment, and abatement, as defined in this subpart.

**Lead-based paint free housing** (Section 745.103) means target housing that has been found to be free of paint or other surface coatings that contain lead equal to or in excess of 1.0 milligram per square centimeter or 0.5 percent by weight.

**Lead-based paint hazard** (Section 745.103) means any condition that causes exposure to lead from lead-contaminated dust, lead-contaminated soil, or lead-contaminated paint that is deteriorated or present in accessible surfaces, friction surfaces, or impact surfaces that would result in adverse human health effects as established by the appropriate Federal agency.

**Lead-based paint hazard** (Section 745.223) means any condition that causes exposure to lead from lead-contaminated dust, lead-contaminated soil, or lead-contaminated paint that is deteriorated or present in accessible surfaces, friction surfaces, or impact surfaces that would result in adverse human health effects as identified by the Administrator pursuant to TSCA section 403.

**Lead-based paint hazard** (Section 745.63) means hazardous lead-based paint, dust-lead hazard or soil-lead hazard as identified in Sec. 745.65.

**Lead-hazard screen** (Section 745.223) is a limited risk assessment activity that involves limited paint and dust sampling as described in Sec. 745.227©.

**Leak or leaking** (Section 761.13) means any instance in which a PCB Article, PCB Container, or PCB Equipment has any PCBs on any portion of its external surface.

**Leak-detection system** (Section 260.10) means a system capable of detecting the failure of either the primary or secondary containment structure or the presence of a release of hazardous waste or accumulated liquid in the secondary containment structure. Such a system must employ operational controls (e.g., daily visual inspections for releases into the secondary containment system of aboveground tanks) or consist of an interstitial monitoring device designed to detect continuously and automatically the failure of the primary or secondary containment structure or the presence of a release of hazardous waste into the secondary containment structure.

**Least-cost plan or least-cost planning process** (Section 72.2) means an energy conservation and electric power planning methodology meeting the requirements of Sec. 73.82(a)(4) of this chapter.

**Legal defense cost** (Section 280.92) is any expense that an owner or operator or provider of financial assurance incurs in defending against claims or actions brought, (1) By EPA or a state to require corrective action or to recover the costs of corrective action; (2) By or on behalf of a third party for bodily injury or property damage caused by an accidental release; or (3) By any person to enforce the terms of a financial assurance mechanism.

**Legionella** (Section 141.2) means a genus of bacteria, some species of which have caused a type of pneumonia called Legionnaires Disease.

**LEPC** (Section 1400.2) means a Local Emergency Planning Committee created under the Emergency Planning and Community Right-to-Know Act, 42 U.S.C. 11001 et seq.

**Lessee** (Section 745.103) means any entity that enters into an agreement to lease, rent, or sublease target housing, including but not limited to individuals, partnerships, corporations, trusts, government agencies, housing agencies, Indian tribes, and nonprofit organizations.

**Lessor** (Section 745.103) means any entity that offers target housing for lease, rent, or sublease, including but not limited to individuals, partnerships, corporations, trusts, government agencies, housing agencies, Indian tribes, and nonprofit organizations.

**Level of quantitation or LOQ** (Section 766.3) means the lowest concentration at which HDDs/HDFs can be reproducibly measured in a specific chemical substance within specified confidence limits, as described in this part.

**License or permit** (Section 121.1) means any license or permit granted by an agency of the Federal Government to conduct any activity which may result in any discharge into the navigable waters of the United States.

**Licensed site** (Section 192.31) means the area contained within the boundary of a location under the control of persons generating or storing uranium byproduct materials under a license issued pursuant to section 84 of the Act. For purposes of this subpart, "licensed site" is equivalent to "regulated unit" in subpart F of part 264 of this chapter.

**Licensing or permitting agency** (Section 121.1) means any agency of the Federal Government to which application is made for a license or permit.

**Life-of-the-unit** (Section 96.2) firm power contractual arrangement means a unit participation power sales agreement under which a utility or industrial customer reserves, or is entitled to receive, a specified amount or percentage of nameplate capacity and associated energy from any specified unit and pays its proportional amount of such unit's total costs, pursuant to a contract: (1) For the life of the unit; (2) For a cumulative term of no less than 30 years, including contracts that permit an election for early termination; or (3) For a period equal to or greater than 25 years or 70 percent of the economic useful life of the unit determined as of the time the unit is built, with option rights to purchase or release some portion of the nameplate capacity and associated energy generated by the unit at the end of the period.

**Life-of-the-unit, firm power contractual arrangement** (Section 72.2) means a unit participation power sales agreement under which a utility or industrial customer reserves, or is entitled to receive, a specified amount or percentage of nameplate capacity and associated energy generated by any specified generating unit and pays its proportional amount of such unit's total costs, pursuant to a contract: (1) For the life of the unit; (2) For a cumulative term of no less than 30 years, including contracts that permit an election for early termination; or (3) For a period equal to or greater than 25 years or 70 percent of the economic useful life of the unit determined as of the time the unit was built, with option rights to purchase or release some portion of the nameplate capacity and associated energy generated by the unit at the end of the period.

**Life-of-the-unit, firm power contractual arrangement** (Section 97.2) means a unit participation power sales agreement under which a utility or industrial customer reserves, or is entitled to receive, a specified amount or percentage of nameplate capacity and associated energy from any specified unit and pays its proportional amount of such unit's total costs, pursuant to a contract: (1) For the life of the unit; (2) For a cumulative term of no less than 30 years, including contracts that permit an election for early termination; or (3) For a period equal to or greater than 25 years or 70 percent of the economic useful life of the unit determined as of the time the unit is built, with option rights to purchase or release some portion of the nameplate capacity and associated energy generated by the unit at the end of the period.

**Light light-duty truck** (Section 86.094-2) means any light-duty truck rated up through 6000 lbs GVWR.

**Light light-duty truck** (Section 86.1803-01) means any light-duty truck rated up through 6000 lbs GVWR. The LDT1 and LDT2 classifications compose the light light-duty truck category.

**Light Light-Duty Truck** (Section 88.102-94) means any light-duty truck rated through 6000 lbs GVWR.

**Light-duty truck** (Section 86.082-2) means any motor vehicle rated at 8,500 pounds GVWR or less which as a vehicle curb weight of 6,000 pounds or less and which has a basic vehicle frontal area of 45 square feet or less, which is: (1) Designed primarily for purposes of transportation of property or is a derivation of such a vehicle, or (2) Designed primarily for transportation of persons and has a capacity of more than 12 persons, or (3) Available with special features enabling off-street or off-highway operation and use.

**Light-duty truck** (Section 86.1803-01) means any motor vehicle rated at 8,500 pounds GVWR or less which has a curb weight of 6,000 pounds or less and which has a basic vehicle frontal area of 45 square feet or less, which is: (1) Designed primarily for purposes of transportation of property or is a derivation of such a vehicle; or (2) Designed primarily for transportation of persons and has a capacity of more than 12 persons; or (3) Available with special features enabling off-street or off-highway operation and use.

**Light-duty truck 1 (LDT1)** (Section 86.1803-01) means any light light-duty truck up through 3750 lbs loaded vehicle weight.

**Light-duty truck 1** (Section 86.094-2) means any light light-duty truck up through 3750 lbs loaded vehicle weight.

**Light-duty truck 2 (LDT2)** (Section 86.1803-01) means any light light-duty truck greater than 3750 lbs loaded vehicle weight.

**Light-duty truck 2** (Section 86.094-2) means any light light-duty truck greater than 3750 lbs loaded vehicle weight.

**Light-duty truck 3 (LDT3)** (Section 86.1803-01) means any heavy light-duty truck up through 5750 lbs adjusted loaded vehicle weight.

**Light-duty truck 3** (Section 86.094-2) means any heavy light-duty truck up through 5750 lbs adjusted loaded vehicle weight.

**Light-duty truck 4 (LDT4)** (Section 86.1803-01) means any heavy light-duty truck greater than 5750 lbs adjusted loaded vehicle weight.

**Light-duty truck 4** (Section 86.094-2) means any heavy light-duty truck greater than 5750 lbs adjusted loaded vehicle weight.

**Light-duty vehicle** (Section 86.082-2) means a passenger car or passenger car derivative capable of seating 12 passengers or less.

**Light-duty vehicle** (Section 86.1803-01) means a passenger car or passenger car derivative capable of seating 12 passengers or less.

**Lima beans** (Section 407.71) shall mean the processing of lima beans into the following product styles: Canned and frozen, green and white, all varieties and sizes.

**Limited use groundwater** (Section 192.11) means groundwater that is not a current or potential source of drinking water because (1) the concentration of total dissolved solids is in excess of 10,000 mg/l, or (2) widespread, ambient contamination not due to activities involving residual radioactive materials from a designated processing site exists that cannot be cleaned up using treatment methods reasonably employed in public water systems, or (3) the quantity of water reasonably available for sustained continuous use is less than 150 gallons per day. The parameters for determining the quantity of water reasonably available shall be determined by the Secretary with the concurrence of the Commission.

**Liner** (Section 260.10) means a continuous layer of natural or man-made materials, beneath or on the sides of a surface impoundment, landfill, or landfill cell, which restricts the downward or lateral escape of hazardous waste, hazardous waste constituents, or leachate.

**Liner** (Section 503.21) is soil or synthetic material that has a hydraulic conductivity of 1x10-7 centimeters per second or less.

**Liquefied petroleum gas** (Section 80.2) means a liquid hydrocarbon fuel that is stored under pressure and is composed primarily of species that are gases at atmospheric conditions (temperature = 25 deg.C and pressure = 1 atm), excluding natural gas.

**Liquefied petroleum gas** (Section 86.094-2) means a liquid hydrocarbon fuel that is stored under pressure and is composed primarily of species that are gases at atmospheric conditions (temperature = 25 deg.C and pressure = 1 atm), excluding natural gas.

**Liquefied petroleum gas** (Section 86.1803-01) means a liquid hydrocarbon fuel that is stored under pressure and is composed primarily of species that are gases at atmospheric conditions (temperature = 25 deg. C and pressure = 1 atm), excluding natural gas.

**Liquefied petroleum gas** (Section 92.2) means the commercial product marketed as liquefied petroleum gas or propane.

**Liquid PCBs** (Section 761.13) means a homogenous flowable material containing PCBs and no more than 0.5 percent by weight non-dissolved material.

**Liquid trap** (Section 280.12) means sumps, well cellars, and other traps used in association with oil and gas production, gathering, and extraction operations (including gas production plants), for the purpose of collecting oil, water, and other liquids. These liquid traps may temporarily collect liquids for subsequent disposition or reinjection into a production or pipeline stream, or may collect and separate liquids from a gas stream.

**Liquid-mounted seal** (Section 265.1081) means a foam or liquid-filled primary seal mounted in contact with the hazardous waste between the tank wall and the floating roof continuously around the circumference of the tank.

**Liquid-mounted seal** (Section 65.2) means a foam-or liquid-filled continuous seal mounted in contact with the stored liquid.

**Liquids dripping** (Section 65.2) means any visible leakage from a seal including dripping, spraying, misting, clouding, and ice formation. Indications of liquids dripping include puddling or new stains that are indicative of an existing evaporated drip.

**Listed mixture** (Section 716.3) means any mixture listed in Sec. 716.120.

**Lithology** (Section 146.3) means the description of rocks on the basis of their physical and chemical characteristics.

**Lithology** (Section 147.2902) the description of rocks on the basis of their physical and chemical characteristics.

**Lithosphere** (Section 191.12) means the solid part of the Earth below the surface, including any ground water contained within it.

**Living area** (Section 745.223) means any area of a residential dwelling used by one or more children age 6 and under, including, but not limited to, living rooms, kitchen areas, dens, play rooms, and children's bedrooms.

**Living plant** (Section 152.3) means a plant, plant organ, or plant part that is alive, viable, or dormant. Examples of plant parts include, but are not limited to, seeds, fruits, leaves, roots, stems, flowers, and pollen.

**Living plant** (Section 174.3) means a plant, plant organ, or plant part that is alive, viable, or dormant. Examples of plant parts include, but are not limited to, seeds, fruits, leaves, roots, stems, flowers, and pollen.

**Load allocation (LA)** (Section 130.2) The portion of a receiving water's loading capacity that is attributed either to one of its existing or future nonpoint sources of pollution or to natural background sources. Load allocations are best estimates of the loading, which may range from reasonably accurate estimates to gross allotments, depending on the availability of data and appropriate techniques for predicting the loading. Wherever possible, natural and nonpoint source loads should be distinguished.

**Load allocation (LA)** (Section 132.2) is the portion of a receiving water's loading capacity that is attributed either to one of its existing or future nonpoint sources or to natural background sources, as more fully defined at 40 CFR 130.2(g). Nonpoint sources include: in-place contaminants, direct wet and dry deposition, groundwater inflow, and overland runoff.

**Load allocation** (Section 130.2) The portion of a TMDL's pollutant load allocated to a nonpoint source, storm water source for which a National Pollutant Discharge Elimination System (NPDES) permit is not required, atmospheric deposition, ground water, or background source of pollutants.

**Load Cell** (Section 201.1) means a device external to the locomotive, of high electrical resistance, used in locomotive testing to simulate engine loading while the locomotive is stationary. (Electrical energy produced by the diesel generator is dissipated in the load cell resistors instead of the traction motors).

**Load or loading** (Section 130.2) An amount of matter or thermal energy that is introduced into a receiving water; to introduce matter or thermal energy into a receiving water. Loading may be either man-caused (pollutant loading) or natural (natural background loading).

**Load or loading** (Section 130.2) An amount of matter or thermal energy that is introduced into a receiving water; to introduce matter or thermal energy into a receiving water. Loading of pollutants may be either man-caused or natural (natural background loading).

**Loaded vehicle weight** (Section 86.082-2) means the vehicle curb weight plus 300 pounds.

**Loaded vehicle weight** (Section 86.1803-01) means the vehicle's curb weight plus 300 pounds.

**Loaded Vehicle Weight** (Section 88.102-94) is defined as the curb weight plus 300 lbs.

**Loading** (Section 745.63) means the quantity of a specific substance present per unit of surface area, such as the amount of lead in micrograms contained in the dust collected from a certain surface area divided by the surface area in square feet or square meters.

**Loading capacity** (Section 130.2) The greatest amount of loading that a water can receive without violating water quality standards.

**Loading capacity** (Section 132.2) is the greatest amount of loading that a water can receive without violating water quality standards.

**Loading cycle** (Section 65.2) means the time period from the beginning of filling a tank truck or railcar until flow to the control device ceases as determined by the flow indicator.

**Local agency** (Section 51.100) means any local government agency other than the State agency, which is charged with responsibility for carrying out a portion of the plan.

**Local agency** (Section 58.1) means any local government agency, other than the State agency, which is charged with the responsibility for carrying out a portion of the plan.

**Local air quality modeling analysis** (Section 93.152) means an assessment of localized impacts on a scale smaller than the entire nonattainment or maintenance area, including, for example, congested roadway intersections and highways or transit terminals, which uses an air quality dispersion model to determine the effects of emissions on air quality.

**Local education agency** (Section 47.105) means any education agency as defined in section 198 of the Elementary and Secondary Education Act of 1965 (20 U.S.C. 3381) and shall include any tribal education agency, as defined in Sec. 47.105(f).

**Local education agency** (Section 763.83) means: (1) Any local educational agency as defined in section 198 of the Elementary and Secondary Education Act of 1965 (20 U.S.C. 3381). (2) The owner of any nonpublic, nonprofit elementary, or secondary school building. (3) The governing authority of any school operated under the defense dependent's education system provided for under the Defense Dependents' Education Act of 1978 (20 U.S.C. 921, et seq.).

**Local emergency planning committee or committee** (Section 350.1) means the local emergency planning committee appointed by the emergency response commission.

**Local emergency response plan** (Section 310.3) means the emergency plan prepared by the Local Emergency Planning Committee (LEPC) as required by section 303 of the Emergency Planning and Community Right-to-Know Act of 1986 (EPCRA or SARA Title III).

**Local government** (Section 280.92) shall have the meaning given this term by applicable state law and includes Indian tribes. The term is generally intended to include: (1) Counties, municipalities, townships, separately chartered and operated special districts (including local government public transit systems and redevelopment authorities), and independent school districts authorized as governmental bodies by state charter or constitution; and (2) Special districts and independent school districts established by counties, municipalities, townships, and other general purpose governments to provide essential services.

**Local government** (Section 745.223) means a county, city, town, borough, parish, district, association, or other public body (including an agency comprised of two or more of the foregoing entities) created under State law.

**Location** (Section 88.302-94) means any building, structure, facility, or installation which; is owned or operated by a person, or is under the control of a person; is located on one or more contiguous properties and contains or could contain a fueling pump or pumps for the use of the vehicles owned or controlled by that person.

**Locomotive** (Section 201.1) means for the purpose of this regulation, a self-propelled vehicle designed for and used on railroad tracks in the transport or rail cars, including self-propelled rail passenger vehicles.

**Locomotive** (Section 85.1602) The definition of locomotive specified in 40 CFR 92.2 applies to this subpart.

**Locomotive** (Section 92.2) means a self-propelled piece of on-track equipment designed for moving or propelling cars that are designed to carry freight, passengers or other equipment, but which itself is not designed or intended to carry freight, passengers (other than those operating the locomotive) or other equipment. Other equipment which is designed for operation both on highways and rails; specialized railroad equipment for maintenance, construction, post accident recovery of equipment, and repairs; and other similar equipment; and vehicles propelled by engines with rated horsepower of less than 750 kW (1006 hp) are not locomotives (see 40 CFR Parts 86 and 89 for this equipment).

**Locomotive engine** (Section 92.2) means an engine incorporated into a locomotive or intended for incorporation into a locomotive.

**Locomotive Load Cell Test Stand** (Section 201.1) means the load cell Sec. 201.1(o) and associated structure, equipment, trackage and locomotive being tested.

**London Amendments** (Section 82.3) means the Montreal Protocol, as amended at the Second Meeting of the Parties to the Montreal Protocol in London in 1990.

**Loose-fill insulation** (Section 247.3) means insulation in granular, nodular, fibrous, powdery, or similar form, designed to be installed by pouring, blowing or hand placement;

**Low altitude** (Section 86.082-2) means any elevation equal to or less than 1,219 meters (4,000 feet).

**Low altitude** (Section 86.1602) means any elevation less than or equal to 1,219 meters (4,000 feet).

**Low altitude** (Section 86.1803-01) means any elevation equal to or less than 1,219 meters (4,000 feet).

**Low altitude conditions** (Section 86.082-2) means a test altitude less than 549 meters (1,800 feet).

**Low altitude conditions** (Section 86.1803-01) means a test altitude less than 549 meters (1,800 feet).

**Low emission vehicle (LEV)** (Section 86.1702-99) means any vehicle certified to the low emission vehicle standards specified in this subpart.

**Low hour engine** (Section 92.2) means an engine during the interval between the time that normal assembly operations and adjustments are completed and the time that 300 additional operating hours have been accumulated (including hours accumulated during emission testing if performed).

**Low hour engine** (Section 94.2) means an engine during the interval between the time that normal assembly operations and adjustments are completed and the time that 300 additional operating hours have been accumulated (including hours of operation accumulated during emission testing, if performed).

**Low idle speed** (Section 92.2) means a speed which is less than normal idle speed, expressed as the number of revolutions of the crankshaft per unit of time, at which an engine can be set when not under load for purposes of propelling the locomotive.

**Low mass emissions unit** (Section 72.2) means an affected unit that is a gas-fired or oil-fired unit, burns only natural gas or fuel oil and qualifies under Sec. 75.19 of this chapter.

**Low mass emissions unit** (Section 72.2) means an affected unit that is "gas-fired" or "oil-fired" (as defined in this section), and that qualifies to use the low mass emissions excepted methodology in Sec. 75.19 of this chapter.

**Low mileage locomotive** (Section 92.2) means a locomotive during the interval between the time that normal assembly operations and adjustments are completed and the time that either 10,000 miles of locomotive operation or 300 additional operating hours have been accumulated (including emission testing if performed).

**Low NOX burners and low NOX burner technology** (Section 76.2) means commercially available combustion modification NOX controls that minimize NOX formation by introducing coal and its associated combustion air into a boiler such that initial combustion occurs in a manner that promotes rapid coal devolatilization in a fuel-rich (i.e., oxygen deficient) environment and introduces additional air to achieve a final fuel-lean (i.e., oxygen rich) environment to complete the combustion process. This definition shall include the staging of any portion of the combustion air using air nozzles or registers located inside any waterwall hole that includes a burner. This definition shall exclude the staging of any portion of the combustion air using air nozzles or ports located outside any waterwall hole that includes a burner (commonly referred to as NOX ports or separated overfire air ports).

**Low occupancy area** (Section 761.13) means any area where PCB remediation waste has been disposed of on-site and where occupancy for any individual not wearing dermal and respiratory protection for a calendar year is: less than 840 hours (an average of 16.8 hours per week) for non-porous surfaces and less than 335 hours (an average of 6.7 hours per week) for bulk PCB remediation waste. Examples could include an electrical substation or a location in an industrial facility where a worker spends small amounts of time per week (such as an unoccupied area outside a building, an electrical equipment vault, or in the non-office space in a warehouse where occupancy is transitory).

**Low volume manufacturer, for a particular model year** (Section 86.1702-99) means any vehicle manufacturer that: Is considered a "small volume manufacturer" by the State of California according to the State of California regulatory definition of "small volume manufacturer", contained in the California Regulatory Requirements Applicable to the National Low

**Low volume waste sources** (Section 423.11) means, taken collectively as if from one source, wastewater from all sources except those for which specific limitations are otherwise established in this part. Low volume wastes sources include, but are not limited to: wastewaters from wet scrubber air pollution control systems, ion exchange water treatment system, water treatment evaporator blowdown, laboratory and sampling streams, boiler blowdown, floor drains, cooling tower basin cleaning wastes, and recirculating house service water systems. Sanitary and air conditioning wastes are not included.

**Low-concentration PCBs** (Section 761.123) means PCBs that are tested and found to contain less than 500 ppm PCBs, or those PCB-containing materials which EPA requires to be assumed to be at concentrations below 500 ppm (i.e., untested mineral oil dielectric fluid).

**Low-Emission Vehicle** (Section 88.102-94) means any light-duty vehicle or light-duty truck conforming to the applicable Low-Emission Vehicle standard, or any heavy-duty vehicle with an engine conforming to the applicable Low-Emission Vehicle standard.

**Lower explosive limit for methane gas** (Section 503.21) is the lowest percentage of methane gas in air, by volume, that propagates a flame at 25 degrees Celsius and atmospheric pressure.

**Lowest observed adverse effect level (LOAEL)** (Section 132.2) is the lowest tested dose or concentration of a substance which resulted in an observed adverse effect in exposed test organisms when all higher doses or concentrations resulted in the same or more severe effects.

**Low-gloss coating** (Section 59.101) means a coating which exhibits a gloss reading less than or equal to 25 on a 60 deg. glossmeter.

**Low-Noise-Emission Product Determination** (Section 203.1) means the Administrator's determination whether or not a product, for which a properly filed application has been received, meets the low-noise-emission product criterion.

**Low-throughput transfer racks** (Section 65.2) means those transfer racks that transfer less than a total of 11.8 million liters per year (3.12 million gallons per year) of liquid containing regulated material.

**Luminescent materials** (Section 469.41) shall mean materials that emit light upon excitation by such energy sources as photons, electrons, applied voltage, chemical reactions or mechanical energy and which are specifically used as coatings in fluorescent lamps and cathode ray tubes. Luminescent materials include, but are not limited to, calcium halophosphate, yttrium oxide, zinc sulfide, and zinc-cadmium sulfide.

**LWK (live weight killed)** (Section 432.11) shall mean the total weight of the total number of animals slaughtered during the time to which the effluent limitations apply; i.e., during any one day or any period of thirty consecutive days.

**LWK (live weight killed)** (Section 432.21) shall mean the total weight of the total number of animals slaughtered during the time to which the effluent limitations apply; i.e., during any one day or any period of thirty consecutive days.

**LWK (live weight killed)** (Section 432.41) shall mean the total weight of the total number of animals slaughtered during the time to which the effluent limitations apply; i.e., during any one day or any period of thirty consecutive days.

# M

**m** (Section 401.11) means meter(s).

**M10** (Section 435.11) means those offshore facilities continuously manned by ten (10) or more persons.

**M10** (Section 435.41) means those offshore facilities continuously manned by ten (10) or more persons.

**M9IM** (Section 435.11) means those offshore facilities continuously manned by nine (9) or fewer persons or only intermittently manned by any number of persons.

**M9IM** (Section 435.41) means those offshore facilities continuously manned by nine (9) or fewer persons or only intermittently manned by any number of persons.

**MACT** (Section 62.1201) means maximum achievable control technology.

**Mail or serve by mail** (Section 72.2) means to submit or serve by means other than personal service.

**Maintain** (Section 1516.2) means maintain, collect, use or disseminate.

**Maintenance** (Section 280.12) means the normal operational upkeep to prevent an underground storage tank system from releasing product.

**Maintenance area** (Section 93.101) means any geographic region of the United States previously designated nonattainment pursuant to the CAA Amendments of 1990 and subsequently redesignated to attainment subject to the requirement to develop a maintenance plan under section 175A of the CAA, as amended.

**Maintenance area** (Section 93.152) means an area with a maintenance plan approved under section 175A of the Act.

**Maintenance Event** (Section 85.1402) means a single maintenance activity for which the engine is removed from service. Once the engine is returned to service, the maintenance event is considered done.

**Maintenance instructions** (Section 204.2) means those instructions for maintenance, use, and repair, which the Administrator is authorized to require pursuant to section 6© (1) of the Act.

**Maintenance instructions or instructions** (Section 205.2) means those instructions for maintenance, use, and repair, which the Administrator is authorized to require pursuant to section 6©(1) of the Act.

**Maintenance plan** (Section 93.101) means an implementation plan under section 175A of the CAA, as amended.

**Maintenance plan** (Section 93.152) means a revision to the applicable SIP, meeting the requirements of section 175A of the Act.

**Major change** (Section 68.3) means introduction of a new process, process equipment, or regulated substance, an alteration of process chemistry that results in any change to safe operating limits, or other alteration that introduces a new hazard.

**Major change to monitoring** (Section 65.2) means a modification to federally required monitoring that uses "unproven technology or procedures" (not generally accepted by the scientific community) or is an entirely new method (sometimes necessary when the required monitoring is unsuitable). A major change to monitoring may be site-specific or may apply to one or more source categories and will almost always set a national precedent. Examples of major changes to monitoring include, but are not limited to: (1) Use of a new monitoring approach developed to apply to a control technology not contemplated in the applicable regulation in this part; (2) Use of a predictive emission monitoring system (PEMS) in place of a required continuous emission monitoring system (CEMS); (3) Use of alternative calibration procedures that do not involve calibration gases or test cells; (4) Use of an analytical technology that differs from that specified by a performance specification; (5) Decreased monitoring frequency for a continuous emission monitoring system, continuous opacity monitoring system, predictive emission monitoring system, or continuous parameter monitoring system; (6) Decreased monitoring frequency for a leak detection and repair program; and (7) Use of alternative averaging times for reporting purposes.

**Major change to test method** (Section 65.2) means a modification to a federally enforceable test method that uses "unproven technology or procedures" (not generally accepted by the scientific community) or is an entirely new method (sometimes necessary when the required test method is unsuitable). A major change to a test method may be site-specific or may apply to one or more source categories and will almost always set a national precedent. In order to be approved, a major change must be validated according to EPA Method 301 (40 CFR part 63, appendix A) Examples of major changes to a test method include, but are not limited to: (1) Use of an unproven analytical finish; (2) Use of a method developed to fill a test method gap; (3) Use of a new test method developed to apply to a control technology not contemplated in the applicable regulation in this part; and (4) Combining two or more sampling/analytical methods (at least one unproven) into one for application to processes emitting multiple pollutants.

**Major Cylinder Component** (Section 85.1402) means piston assembly, cylinder liner, connecting rod, or piston ring set.

**Major disaster** (Section 109.2) means any hurricane, tornado, storm, flood, high water, wind-driven water, tidal wave, earthquake, drought, fire, or other catastrophe in any part of the United States which, in the determination of the President, is or threatens to become of sufficient severity and magnitude to warrant disaster assistance by the Federal Government to supplement the efforts and available resources of States and local governments and relief organizations in alleviating the damage, loss, hardship, or suffering caused thereby.

**Major facility** (Section 122.2) means any NPDES "facility or activity" classified as such by the Regional Administrator, or, in the case of "approved State programs," the Regional Administrator in conjunction with the State Director.

**Major facility** (Section 124.2) means any RCRA, UIC, NPDES, or 404 "facility or activity" classified as such by the Regional Administrator, or, in the case of "approved State programs," the Regional Administrator in conjunction with the State Director.

**Major facility** (Section 144.3) means any UIC "facility or activity" classified as such by the Regional Administrator, or, in the case of approved State programs, the Regional Administrator in conjunction with the State Director.

Major facility means any facility or activity classified as such by the Regional Administrator, or, in the case of approved State programs, the Regional Administrator in conjunction with the State Director.

**Major source** (Section 63.2) means any stationary source or group of stationary sources located within a contiguous area and under common control that emits or has the potential to emit considering controls, in the aggregate, 10 tons per year or more of any hazardous air pollutant or 25 tons per year or more of any combination of hazardous air pollutants, unless the Administrator establishes a lesser quantity, or in the case of radionuclides, different criteria from those specified in this sentence.

**Major source** (Section 64.1) shall have the same meaning as provided under part 70 or 71 of this chapter.

**Major source** (Section 70.2) means any stationary source (or any group of stationary sources that are located on one or more contiguous or adjacent properties, and are under common control of the same person (or persons under common control)) belonging to a single major industrial grouping and that are described in paragraph (1), (2), or (3) of this definition. For the purposes of defining "major source," a stationary source or group of stationary sources shall be considered part of a single industrial grouping if all of the pollutant emitting activities at such source or group of sources on contiguous or adjacent properties belong to the same Major Group (i.e., all have the same two-digit code) as described in the Standard Industrial Classification Manual, 1987. (1)A major source under section 112 of the Act, which is defined as: (i)For pollutants other than radionuclides, any stationary source or group of stationary sources located within a contiguous area and under common control that emits or has the potential to emit, in the aggregate, 10 tons per year (tpy) or more of any hazardous air pollutant which has been listed pursuant to section 112(b) of the Act, 25 tpy or more of any combination of such hazardous air pollutants, or such lesser quantity as the Administrator may establish by rule. Notwithstanding the preceding sentence, emissions from any oil or gas exploration or production well (with its associated equipment) and emissions from any pipeline compressor or pump station shall not be aggregated with emissions from other similar units, whether or not such units are in a contiguous area or under common control, to determine whether such units or stations are major sources; or (ii) For radionuclides, "major source" shall have the meaning specified by the Administrator by rule. (2) A major stationary source of air pollutants, as defined in section 302 of the Act, that directly emits or has the potential to emit, 100 tpy or more of any air pollutant (including any major source of fugitive emissions of any such pollutant, as determined by rule by the Administrator). The fugitive emissions of a stationary source shall not be considered in determining whether it is a major stationary source for the purposes of section 302(j) of the Act, unless the source belongs to one of the following categories of stationary source: (i) Coal cleaning plants (with thermal dryers); (ii) Kraft pulp mills; (iii) Portland cement plants; (iv) Primary zinc smelters; (v) Iron and steel mills; (vi) Primary aluminum ore reduction plants; (vii) Primary copper smelters; (viii) Municipal incinerators capable of charging more than 250 tons of refuse per day; (ix) Hydrofluoric, sulfuric, or nitric acid plants; (x) Petroleum refineries; (xi) Lime plants; (xii) Phosphate rock processing plants; (xiii) Coke oven batteries; (xiv) Sulfur recovery plants; (xv) Carbon black plants (furnace process); (xvi) Primary lead smelters; (xvii) Fuel conversion plants; (xviii) Sintering plants; (xix) Secondary metal production plants; (xx) Chemical process plants; (xxi) Fossil-fuel boilers (or combination thereof) totaling more than 250 million British thermal units per hour heat input; (xxii) Petroleum storage and transfer units with a total storage capacity exceeding 300,000 barrels; (xxiii) Taconite ore processing plants; (xxiv) Glass fiber processing plants; (xxv) Charcoal production plants; (xxvi) Fossil-fuel-fired steam electric plants of more than 250 million British thermal units per hour heat input; or (xxvii) Any other stationary source category, which as of August 7, 1980 is being regulated under section 111 or 112 of the Act. (3) A major stationary source as defined in part D of title I of the Act, including: (i) For ozone nonattainment areas, sources with the potential to emit 100 tpy or more of volatile organic compounds or oxides of nitrogen in areas classified as "marginal" or "moderate," 50 tpy or more in areas classified as "serious," 25 tpy or more in areas classified as "severe," and 10 tpy or more in areas classified as "extreme"; except that the references in this paragraph to 100, 50, 25 and 10 tpy of nitrogen oxides shall not apply with respect to any source for which the Administrator has made a finding, under section 182(f) (1) or (2) of the Act, that requirements under section 182(f) of the Act do not apply; (ii) For ozone transport regions established pursuant to section 184 of the Act, sources with the potential to emit 50 tpy or more of volatile organic compounds; (iii) For carbon monoxide nonattainment areas: (A) That are classified as "serious," and (B) in which stationary sources contribute significantly to carbon monoxide levels as determined under rules issued by the Administrator, sources with the potential to emit 50 tpy or more of carbon monoxide; and (iv) For particulate matter (PM-10) nonattainment areas classified as "serious," sources with the potential to emit 70 tpy or more of PM-10.

**Major source** (Section 71.2) means any stationary source (or any group of stationary sources that are located on one or more contiguous or adjacent properties, and are under common control of the same person (or persons under common control)), belonging to a single major industrial grouping and that are described in paragraph (1), (2), or (3) of this definition. For the purposes of defining "major source," a stationary source or group of stationary sources shall be considered part of a single industrial grouping if all of the pollutant emitting activities at such source or group of sources on contiguous or adjacent properties belong to the same Major Group (i.e., all have the same two-digit code) as described in the Standard Industrial Classification Manual, 1987. (1) A major source under section 112 of the Act, which is defined as: (i) For pollutants other than radionuclides, any stationary source or group of stationary sources located within a contiguous area and under common control that emits or has the potential to emit, in the aggregate, 10 tpy or more of any hazardous air pollutant which has been listed pursuant to section 112(b) of the Act, 25 tpy or more of any combination of such hazardous air pollutants, or such lesser quantity as the Administrator may establish by rule. Notwithstanding the preceding sentence, emissions from any oil or gas exploration or production well (with its associated equipment) and emissions from any pipeline compressor or pump station shall not be aggregated with emissions from other similar units, whether or not such units are in a contiguous area or under common control, to determine whether such units or stations are major sources; or (ii) For radionuclides, "major source" shall have the meaning specified by the Administrator by rule. (2) A major stationary source of air pollutants or any group of stationary sources as defined in section 302 of the Act, that directly emits, or has the potential to emit, 100 tpy or more of any air pollutant (including any major source of fugitive emissions of any such pollutant, as determined by rule by the Administrator). The fugitive emissions of a stationary source shall not be considered in determining whether it is a major stationary source for the purposes of section 302(j) of the Act, unless the source belongs to one of the following categories of stationary source: (i) Coal cleaning plants (with thermal dryers); (ii) Kraft pulp mills; (iii) Portland cement plants; (iv) Primary zinc smelters; \(v) Iron and steel mills; (vi) Primary aluminum ore reduction plants; (vii) Primary copper smelters; (viii) Municipal incinerators capable of charging more than 250 tons of refuse per day; (ix) Hydrofluoric, sulfuric, or nitric acid plants; (x) Petroleum refineries; (xi) Lime plants; (xii) Phosphate rock processing plants; (xiii) Coke oven batteries; (xiv) Sulfur recovery plants; (xv) Carbon black plants (furnace process); (xvi) Primary lead smelters; (xvii) Fuel conversion plants; (xviii) Sintering plants; (xix) Secondary metal production plants; (xx) Chemical process plants; (xxi) Fossil-fuel boilers (or combination thereof) totaling more than 250 million British thermal units per hour heat input; (xxii) Petroleum storage and transfer units with a total storage capacity exceeding 300,000 barrels; (xxiii) Taconite ore processing plants; (xxiv) Glass fiber processing plants; (xxv) Charcoal production plants; (xxvi) Fossil-fuel-fired steam electric plants of more than 250 million British thermal units per hour heat input; or (xxvii) Any other stationary source category which, as of August 7, 1980, is being regulated under section 111 or 112 of the Act. (3) A major stationary source as defined in part D of title I of the Act, including: (i) For ozone nonattainment areas, sources with the potential to emit 100 tpy or more of volatile organic compounds or oxides of nitrogen in areas classified as "marginal" or "moderate," 50 tpy or more in areas classified as "serious"; 25 tpy or more in areas classified as "severe," and 10 tpy or more in areas classified as "extreme"; except that the references in this paragraph (3)(i) to 100, 50, 25, and 10 tpy of nitrogen oxides shall not apply with respect to any source for which the Administrator has made a finding, under section 182(f) (1) or (2) of the Act, that requirements under section 182(f) of the Act do not apply; (ii) For ozone transport regions established pursuant to section 184 of the Act, sources with the potential to emit 50 tpy or more of volatile organic compounds; (iii) For carbon monoxide nonattainment areas: (A) That are classified as "serious," and (B) in which stationary sources contribute significantly to carbon monoxide levels as determined under rules issued by the Administrator, sources with the potential to emit 50 tpy or more of carbon monoxide; and (iv) For particulate matter (PM-10) nonattainment areas classified as "serious," sources with the potential to emit 70 tpy or more of PM-10.

**Major stationary source** (Section 66.3) means any stationary facility or source of air pollutants which directly emits, or has the potential to emit, one hundred tons per year or more of any air pollutant regulated by EPA under the Clean Air Act.

**Make available for use** (Section 171.2) means to distribute, sell, ship, deliver for shipment, or receive and (having so received) deliver, to any person. However, the term excludes transactions solely between persons who are pesticide producers, registrants, wholesalers, or retail sellers, acting only in those capacities. Limited. The following definitions apply only to dealers, dealerships and transactions in States or on Indian Reservations where EPA conducts a Federal Pesticide Applicator Certification Program.

**Malfunction** (Section 264.1031) means any sudden failure of a control device or a hazardous waste management unit or failure of a hazardous waste management unit to operate in a normal or usual manner, so that organic emissions are increased.

**Malfunction** (Section 265.1081) means any sudden, infrequent, and not reasonably preventable failure of air pollution control equipment, process equipment, or a process to operate in a normal or usual manner. Failures that are caused in part by poor maintenance or careless operation are not malfunctions.

**Malfunction** (Section 57.103) means any unanticipated and unavoidable failure of air pollution control equipment or process equipment or of a process to operate in a normal or usual manner. Failures that are caused entirely or in part by poor design, poor maintenance, careless operation, or any other preventable upset condition or preventable equipment breakdown shall not be considered malfunctions. A malfunction exists only for the minimum time necessary to implement corrective measures.

**Malfunction** (Section 60.2) means any sudden, infrequent, and not reasonably preventable failure of air pollution control equipment, process equipment, or a process to operate in a normal or usual manner. Failures that are caused in part by poor maintenance or careless operation are not malfunctions.

**Malfunction** (Section 63.2) means any sudden, infrequent, and not reasonably preventable failure of air pollution control and monitoring equipment, process equipment, or a process to operate in a normal or usual manner. Failures that are caused in part by poor maintenance or careless operation are not malfunctions.

**Malfunction** (Section 65.2) means any sudden, infrequent, and not reasonably preventable failure of air pollution control equipment, monitoring equipment, process equipment, or a process to operate in a normal or usual manner. Failures that are caused in part by poor maintenance or careless operation are not malfunctions. Malfunctions that do not affect a regulated source or compliance with this part are not malfunctions for purposes of this part.

**Malfunction** (Section 86.082-2) means not operating according to specifications (e.g., those specifications listed in the application for certification).

**Malfunction** (Section 86.1803-01) means not operating according to specifications (e.g., those specifications listed in the certification application).

**Malfunction** (Section 92.2) means a condition in which the operation of a component in a locomotive or locomotive engine occurs in a manner other than that specified by the certifying manufacturer or remanufacturer (e.g., as specified in the application for certification); or the operation of the locomotive or locomotive engine in that condition.

**Malfunction** (Section 94.2) means a condition in which the operation of a component in an engine occurs in a manner other than that specified by the certifying manufacturer (e.g., as specified in the application for certification); or the operation of an engine in that condition.

**Malleable Iron** (Section 464.31) A cast iron made by a prolonged anneal of white cast iron in which decarburization or graphitization, or both, take place to eliminate some or all of the cementite.

Graphite is present in the form of temper carbon.

**Management** (Section 191.02) means any activity, operation, or process (except for transportation) conducted to prepare spent nuclear fuel or radioactive waste for storage or disposal, or the activities associated with placing such fuel or waste in a disposal system.

**Management authority** (Section 228.2) means the EPA organizational entity assigned responsibility for implementing the management functions identified in Sec. 228.3.

**Management measures** (Section 130.2) Best practical and economically achievable measures to control the addition of pollutants to waters of the United States through the application of nonpoint pollution control practices, technologies, processes, siting criteria, operating methods, best management practices, or other alternatives.

**Management of migration** (Section 300.5) means actions that are taken to minimize and mitigate the migration of hazardous substances or pollutants or contaminants and the effects of such migration. Measures may include, but are not limited to, management of a plume of contamination, restoration of a drinking water aquifer, or surface water restoration.

**Management or hazardous waste management** (Section 260.10) means the systematic control of the collection, source separation, storage, transportation, processing, treatment, recovery, and disposal of hazardous waste.

**Management systems review** (Section 194.2) means the qualitative assessment of a data collection operation or organization(s) to establish whether the prevailing quality management structure, policies, practices, and procedures are adequate to ensure that the type and quality of data needed are obtained.

**Manifest** (Section 144.3) means the shipping document originated and signed by the "generator", which contains the information required by subpart B of 40 CFR part 262.

**Manifest** (Section 260.10) means the shipping document EPA form 8700-22 and, if necessary, EPA form 8700-22A, originated and signed by the generator in accordance with the instructions included in the appendix to part 262.

**Manifest** (Section 270.2) means the shipping document originated and signed by the generator which contains the information required by subpart B of 40 CFR part 262.

**Manifest** (Section 761.13) means the shipping document EPA form 8700-22 and any continuation sheet attached to EPA form 8700-22, originated and signed by the generator of PCB waste in accordance with the instructions included with the form and subpart K of this part.

**Manifest document number** (Section 260.10) means the U.S. EPA twelve digit identification number assigned to the generator plus a unique five digit document number assigned to the Manifest by the generator for recording and reporting purposes.

**Man-made beta particle and photon emitters** (Section 141.2) means all radionuclides emitting beta particles and/or photons listed in Maximum Permissible Body Burdens and Maximum Permissible Concentration of Radionuclides in Air or Water for Occupational Exposure, NBS Handbook 69, except the daughter products of thorium-232, uranium-235 and uranium-238.

**Manned Control Center** (Section 761.13) means an electrical power distribution control room where the operating conditions of a PCB Transformer are continuously monitored during the normal hours of operation (of the facility), and, where the duty engineers, electricians, or other trained personnel have the capability to deenergize a PCB Transformer completely within 1 minute of the receipt of a signal indicating abnormal operating conditions such as an overtemperature condition or overpressure condition in a PCB Transformer.

**Manual** (Section 66.3) means the Noncompliance Penalties Instruction Manual which accompanies these regulations. This Manual appears as appendix B to these regulations.

**Manual method** (Section 53.1) means a method for measuring concentrations of an ambient air pollutant in which sample collection, analysis, or measurement, or some combination thereof, is performed manually. A method for PM10 or PM2.5 which utilizes a sampler that requires manual preparation, loading, and weighing of filter samples is considered a manual method even though the sampler may be capable of automatically collecting a series of sequential samples.

**Manual-grade strapping** (Section 247.3) refers to straps of material used with transport packaging to hold products in place on pallets or in other methods of commercial, bulk shipment. Strapping can also prevent tampering and pilferage during shipping.

**Manufacture** (Section 372.3) means to produce, prepare, import, or compound a toxic chemical. Manufacture also applies to a toxic chemical that is produced coincidentally during the manufacture, processing, use, or disposal of another chemical or mixture of chemicals, including a toxic chemical that is separated from that other chemical or mixture of chemicals as a byproduct, and a toxic chemical that remains in that other chemical or mixture of chemicals as an impurity.

**Manufacture** (Section 704.3) means to manufacture for commercial purposes.

**Manufacture** (section 710.2) means to produce or manufacture in the United States or import into the customs territory of the United States.

**Manufacture** (Section 712.3) for commercial purposes means to import, produce, or manufacture with the purpose of obtaining an immediate or eventual commercial advantage for the manufacturer and includes, among other things, such "manufacture" of any amount of a chemical substance or mixture: (1) For commercial distribution, including for test marketing. (2) For use by the manufacturer, including use for product research and development, or as an intermediate. Manufacture for commercial purposes also applies to substances that are produced coincidentally during the manufacture, processing, use, or disposal of another substance or mixture, including byproducts and coproducts that are separated from that other substance or mixture, and impurities that remain in that substance or mixture. Byproducts and impurities may not in themselves have commercial value. They are nonetheless produced for the purpose of obtaining a commercial advantage since they are part of the manufacture of a chemical produced for a commercial purpose.

**Manufacture** (Section 716.3) means to manufacture for commercial purposes.

**Manufacture** (Section 720.3) means to produce or manufacture in the United States or import into the customs territory of the United States.

**Manufacture** (Section 761.13) means to produce, manufacture, or import into the customs territory of the United States.

**Manufacture** (Section 763.163) means to produce or manufacture in the United States.

**Manufacture for commercial purposes** (Section 704.3) means: (1) To import, produce, or manufacture with the purpose of obtaining an immediate or eventual commercial advantage for the manufacturer, and includes among other things, such "manufacture" of any amount of a chemical substance or mixture: (i) For commercial distribution, including for test marketing. (ii) For use by the manufacturer, including use for product research and development, or as an intermediate. (2) Manufacture for commercial purposes also applies to substances that are produced coincidentally during the manufacture, processing, use, or disposal of another substance or mixture, including both byproducts that are separated from that other substance or mixture and impurities that remain in that substance or mixture. Such byproducts and impurities may, or may not, in themselves have commercial value. They are nonetheless produced for the purpose of obtaining a commercial advantage since they are part of the manufacture of a chemical product for a commercial purpose.

**Manufacture for commercial purposes** (Section 717.3) means to import, produce, or manufacture with the purpose of obtaining an immediate or eventual commercial advantage for the manufacturer, and includes, among other things, such "manufacture" of any amount of a chemical substance or mixture: (i) For distribution in commerce, including for test marketing. (ii) For use by the manufacturer, including use for product research and development, or as an intermediate. (2) Manufacture for commercial purposes also applies to substances that are produced coincidentally during the manufacture, processing, use, or disposal of another substance or mixture, including both byproducts that are separated from that other substances or mixture and impurities that remain in that substance or mixture. Such byproducts and impurities may, or may not, in themselves have commercial value. They are nonetheless produced for the purpose of obtaining a commercial advantage since they are part of the manufacture of a chemical product for a commercial purpose.

**Manufacture for commercial purposes means** (Section 716.3) : (1) To produce, with the purpose of obtaining an immediate or eventual commercial advantage for the manufacturer, and includes among other things such "manufacture" of any amount of a chemical substance or mixture: (i) For commercial distribution, including for test marketing. (ii) For use by the manufacturer, including use for product research and development, or as an intermediate. (2) Manufacture for commercial purposes also applies to substances that are produced coincidentally during the manufacture, processing, use, or disposal of another substance or mixture, including byproducts and impurities. Such byproducts and impurities may, or may not, in themselves have commercial value. They are nonetheless produced for the purpose of obtaining a commercial advantage since they are part of the manufacture of a chemical product for a commercial purpose.

**Manufacture of electronic crystals** (Section 469.22) means the growing of crystals and/or the production of crystal wafers for use in the manufacture of electronic devices.

**Manufacture of semiconductors** (Section 469.12) means those processes, beginning with the use of crystal wafers, which lead to or are associated with the manufacture of semiconductor devices.

**Manufacture or import "for commercial purposes"** (section 710.2) means to manufacture or import: (1) For distribution in commerce, including for test marketing purposes, or (2) For use by the manufacturer, including for use as an intermediate.

**Manufacture or import for commercial purposes** (Section 720.3) means: (1) To import, produce, or manufacture with the purpose of obtaining an immediate or eventual commercial advantage for the manufacturer or importer, and includes, among other things, "manufacture" of any amount of a chemical substance or mixture: (i) For commercial distribution, including for test marketing. (ii) For use by the manufacturer, including use for product research and development or as an intermediate. (2) The term also applies to substances that are produced coincidentally during the manufacture, processing, use, or disposal of another substance or mixture, including byproducts that are separated from that other substance or mixture and impurities that remain in that substance or mixture. Byproducts and impurities without separate commercial value are nonetheless produced for the purpose of obtaining a commercial advantage, since they are part of the manufacture of a chemical substance for commercial purposes.

**Manufacture or process** (Section 717.3) means to manufacture or process for commercial purposes.

**Manufacture solely for export** (Section 720.3) means to manufacture or import for commercial purposes a chemical substance solely for export from the United States under the following restrictions on activities in the United States: (1) Distribution in commerce is limited to purposes of export or processing solely for export as defined in Sec. 721.3 of this chapter. (2) The manufacturer or importer, and any person to whom the substance is distributed for purposes of export or processing solely for export (as defined in Sec. 721.3 of this chapter), may not use the substance except in small quantities solely for research and development in accordance with Sec. 720.36.

**Manufacture, import, or process for commercial purposes** (Section 725.3) means: (1) To import, produce, manufacture, or process with the purpose of obtaining an immediate or eventual commercial advantage for the manufacturer, importer, or processor, and includes, among other things, "manufacture" or "processing" of any amount of a microorganism or microbial mixture: (i) For commercial distribution, including for test marketing. (ii) For use by the manufacturer, including use for product research and development or as an intermediate. (2) The term also applies to substances that are produced coincidentally during the manufacture, processing, use, or disposal of another microorganism or microbial mixture, including byproducts that are separated from that other microorganism or microbial mixture and impurities that remain in that microorganism or microbial mixture. Byproducts and impurities without separate commercial value are nonetheless produced for the purpose of obtaining a commercial advantage, since they are part of the manufacture or processing of a microorganism for commercial purposes.

**Manufacturer** (Section 129.2) means any establishment engaged in the mechanical or chemical transformation of materials or substances into new products including but not limited to the blending of materials such as pesticidal products, resins, or liquors.

**Manufacturer** (Section 205.2) means any person engaged in the manufacturing or assembling of new products, or the importing of new products for resale, or who acts for and is controlled by any such person in connection with the distribution of such products.

**Manufacturer** (Section 211.203) As stated in the Act "means any person engaged in the manufacturing or assembling of new products, or the importing of new products for resale, or who acts for, and is controlled by, any such person in connection with the distribution of such products."

**Manufacturer** (Section 610.11) means a person or company which is engaged in the business of producing or assembling, and which has primary control over the design specifications, of a retrofit device for which a fuel economy improvement claim is made.

**Manufacturer** (Section 704.3) means a person who imports, produces, or manufactures a chemical substance. A person who extracts a component chemical substance from a previously existing chemical substance or a complex combination of substances is a manufacturer of that component chemical substance.

**Manufacturer** (Section 716.3) means a person who produces or manufactures a chemical substance. A person who extracts a component chemical substance from a previously existing chemical substance or a complex combination of substances is a manufacturer of that component chemical substance.

**Manufacturer** (Section 720.3) means a person who imports, produces, or manufactures a chemical substance. A person who extracts a component chemical substance from a previously existing chemical substance or a complex combination of substances is a manufacturer of that component chemical substance. A person who contracts with a manufacturer to manufacture or produce a chemical substance is also a manufacturer if (1) the manufacturer manufactures or produces the substance exclusively for that person, and (2) that person specifies the identity of the substance and controls the total amount produced and the basic technology for the plant process.

**Manufacturer** (Section 763.163) means a person who produces or manufactures in the United States.

**Manufacturer** (Section 85.1902) shall be given the meaning ascribed to it by section 214 of the Act.

**Manufacturer** (Section 92.2) means an individual or entity engaged in the manufacturing or assembling of freshly manufactured locomotives or freshly manufactured locomotive engines; or the importing of locomotives or locomotive engines originally manufactured on or after January 1, 1973 and not remanufactured.

(See Secs. 92.1© and 92.209 for applicability of this term.)

**Manufacturer** (Section 94.2) means any person engaged in the manufacturing or assembling of new engines or importing such engines for resale, or who acts for and is under the control of any such person in connection with the distribution of such engines. The term manufacturer includes post-manufacturer marinizers, but does not include any dealer with respect to new engines received by such person in commerce.

**Manufacturer parts** (Section 86.1602) are parts produced or sold by the manufacturer of the motor vehicle or motor vehicle engine.

**Manufacturer-owned engine** (Section 94.2) means an uncertified marine engine that is owned and controlled by a manufacturer, is used for product development, and is not sold or leased.

**Manufacturer-owned marine engine** (Section 91.1002) means an uncertified marine SI engine owned and controlled by a marine SI engine manufacturer and used in a manner not involving lease or sale by itself or in a marine vessel or piece of equipment employed from year to year in the ordinary course of business for product development, production method assessment, or market promotion purposes.

**Manufacturer-owned nonroad engine** (Section 90.902) means an uncertified nonroad engine owned and controlled by a nonroad engine manufacturer and used in a manner not involving lease or sale by itself or in a vehicle employed from year to year in the ordinary course of business for product development, production method assessment, and market promotion purposes.

**Manufacturing process** (Section 761.13) means all of a series of unit operations operating at a site, resulting in the production of a product.

**Manufacturing stream** (Section 721.3) means all reasonably anticipated transfer, flow, or disposal of a chemical substance, regardless of physical state or concentration, through all intended operations of manufacture, including the cleaning of equipment.

**Manufacturing use product** (Section 152.3) means any pesticide product that is not an end-use product.

**Manufacturing use product** (Section 158.153) means any pesticide product other than an end use product. A product may consist of the technical grade of active ingredient only, or may contain inert ingredients, such as stabilizers or solvents.

**Manufacturing-use product** (Section 162.151) means any pesticide product other than a product to be labeled with directions for end use. This term includes any product intended for use as a pesticide after re-formulation or repackaging.

**Marine diesel engine** (Section 89.2) means a compression-ignition engine that is intended to be installed on a vessel.

**Marine engine** (Section 94.2) means an engine that is installed or intended to be installed on a marine vessel. This definition does not include portable auxiliary engines for which the fueling, cooling and exhaust systems are not integral parts of the vessel.

**Marine engine manufacturer** (Section 91.3) means any person engaged in the manufacturing or assembling of new marine SI engines or the importing of such engines for resale, or who acts for and is under the control of any such person in connection with the distribution of such engines. A marine SI engine manufacturer does not include any dealer with respect to new marine SI engines received by such person in commerce.

**Marine generated waste** (Section 437.2) means any waste, wastewater, and/or used material generated as part of the normal maintenance and operation of a ship, boat, or barge operating on inland, coastal, or open waters, or while berthed.

**Marine Pollution Control Device, abbreviated MPCD** (Section 1700.3) means any equipment or management practice installed or used on an Armed Forces vessel that is designed to receive, retain, treat, control, or discharge a discharge incidental to the normal operation of a vessel, and that is determined by the

Administrator and Secretary to be the most effective equipment or management practice to reduce the environmental impacts of the discharge consistent with the considerations in Clean Water Act section 312(n)(2)(B).

**Marine sanitation device** (Section 140.1) includes any equipment for installation onboard a vessel and which is designed to receive, retain, treat, or discharge sewage and any process to treat such sewage;

**Marine spark-ignition engine** (Section 91.3) means any engine used to propel a marine vessel, which utilizes the spark-ignition combustion cycle.

**Marine vessel** (Section 91.3) means every description of watercraft or another artificial contrivance used, or capable of being used, as a means of transportation on water, as defined in 1 U.S.C. 3 (1992).

**Marine vessel** (Section 94.2) has the meaning specified in the General Provisions of the United States Code, 1 U.S.C. 3.

**Marine vessel manufacturer** (Section 91.3) means any person engaged in the manufacturing or assembling of new marine vessels or importing such marine vessels for resale, or who acts for and is under the control of any such person in connection with the distribution of such vehicles. A marine vessel manufacturer does not include any dealer with respect to new marine vessels received by such person in commerce.

**Mark** (Section 761.13) means the descriptive name, instructions, cautions, or other information applied to PCBs and PCB Items, or other objects subject to these regulations.

**Marked** (Section 761.13) means the marking of PCB Items and PCB storage areas and transport vehicles by means of applying a legible mark by painting, fixation of an adhesive label, or by any other method that meets the requirements of these regulations.

**Market/Marketers** (Section 761.13) means the processing or distributing in commerce, or the person who processes or distributes in commerce, used oil fuels to burners or other marketers, and may include the generator of the fuel if it markets the fuel directly to the burner.

**MARPOL 73/78** (Section 110.1) means the International Convention for the Prevention of Pollution from Ships, 1973, as modified by the Protocol of 1978 relating thereto, Annex I, which regulates pollution from oil and which entered into force on October 2, 1983;

**Mass of pollutant that can be discharged** (Section 463.2) is the pollutant mass calculated by multiplying the pollutant concentration times the average process water usage flow rate.

**Master Inventory File** (section 710.2) means EPA's comprehensive list of chemical substances which constitute the Chemical Substances Inventory compiled under section 8(b) of the Act. It includes substances reported under subpart A of this part and substances reported under part 720 of this chapter for which a Notice of Commencement of Manufacture or Import has been received under Sec. 720.120 of this chapter.

**Material** (Section 220.2) means matter of any kind or description, including, but not limited to, dredged material, solid waste, incinerator residue, garbage, sewage, sewage sludge, munitions, radiological, chemical, and biological warfare agents, radioactive materials, chemicals, biological and laboratory waste, wreck or discarded equipment, rock, sand, excavation debris, industrial, municipal, agricultural, and other waste, but such term does not mean sewage from vessels within the meaning of section 312 of the FWPCA. Oil within the meaning of section 311 of the FWPCA shall constitute "material" for purposes of this subchapter H only to the extent that it is taken on board a vessel or aircraft for the primary purpose of dumping.

**Material Safety Data Sheet or MSDS** (Section 370.2) means the sheet required to be developed under Sec. 1910.1200(g) of Title 29 of the Code of Federal Regulations.

**Mats** (Section 247.3) are temporary or semipermanent protective floor coverings used for numerous applications, including home and office carpet protection, car and truck floor board protection, traction on slippery surfaces, cushion from floor hardness, and reduction of injury risk during athletic events.

**Maximum as applied to BAT effluent limitations and NSPS for drilling fluids and drill cuttings** (Section 435.11) means the maximum concentration allowed as measured in any single sample of the barite for determination of cadmium and mercury content.

**Maximum as applied to BAT effluent limitations and NSPS for drilling fluids and drill cuttings** (Section 435.41) means the maximum concentration allowed as measured in any single sample of the barite for determination of cadmium and mercury content.

**Maximum contaminant level (MCL)** (Section 159.153) means the maximum permissible level, established by EPA, for a contaminant in water which is delivered to any user of a public water system.

**Maximum contaminant level** (Section 141.2) means the maximum permissable level of a contaminant in water which is delivered to any user of a public water system.

**Maximum contaminant level** (Section 142.2) means the maximum permissible level of a contaminant in water which is delivered to the free flowing outlet of the ultimate user of a public water system; except in the case of turbidity where the maximum permissible level is measured at the point of entry to the distribution system. Contaminants added to the water under circumstances controlled by the user, except for those resulting from corrosion of piping and plumbing caused by water quality are excluded from this definition.

**Maximum contaminant level goal or MCLG** (Section 141.2) means the maximum level of a contaminant in drinking water at which no known or anticipated adverse effect on the health of persons would occur, and which allows an adequate margin of safety. Maximum contaminant level goals are nonenforceable health goals.

**Maximum Continuous Steam Flow at 100% of Load** (Section 76.2) means the maximum capacity of a boiler as reported in item 3 (Maximum Continuous Steam Flow at 100% Load in thousand pounds per hour), Section C ( design parameters), Part III (boiler information) of the Department of Energy's Form EIA-767 for 1995.

**Maximum daily discharge limitation** (Section 122.2) means the highest allowable "daily discharge."

**Maximum design heat input** (Section 96.2) means the ability of a unit to combust a stated maximum amount of fuel per hour on a steady state basis, as determined by the physical design and physical characteristics of the unit.

**Maximum design heat input** (Section 97.2) means the ability of a unit to combust a stated maximum amount of fuel per hour (in mmBtu/hr) on a steady state basis, as determined by the physical design and physical characteristics of the unit.

**Maximum extent practicable** (Section 112.2) means the limitations used to determine oil spill planning resources and response times for on-water recovery, shoreline protection, and cleanup for worst case discharges from onshore non-transportation-related facilities in adverse weather. It considers the planned capability to respond to a worst case discharge in adverse weather, as contained in a response plan that meets the requirements in Sec. 112.20 or in a specific plan approved by the Regional Administrator.

**Maximum for any one day** (Section 435.41) as applied to BPT, BCT and BAT effluent limitations and NSPS for oil and grease in produced water means the maximum concentration allowed as measured by the average of four grab samples collected over a 24-hour period that are analyzed separately. Alternatively, for BAT and NSPS the maximum concentration allowed may be determined on the basis of physical composition of the four grab samples prior to a single analysis.

**Maximum for any one day and average of daily values for thirty consecutive days** (Section 407.61) shall be based on the daily average mass of material processed during the peak thirty consecutive day production period.

**Maximum for any one day and average of daily values for thirty consecutive days** (Section 407.71) shall be based on the daily average mass of raw material processed during the peak thirty consecutive day production period.

**Maximum for any one day and average of daily values for thirty consecutive days** (Section 407.81) shall be based on the daily average mass of final product produced during the peak thirty consecutive day production period.

**Maximum for any one day as applied to BPT, BCT and BAT effluent limitations and NSPS for oil and grease in produced water** (Section 435.11) means the maximum concentration allowed as measured by the average of four grab samples collected over a 24-hour period that are analyzed separately. Alternatively, for BAT and NSPS the maximum concentration allowed may be determined on the basis of physical composition of the four grab samples prior to a single analysis.

**Maximum organic vapor pressure** (Section 265.1081) means the sum of the individual organic constituent partial pressures exerted by the material contained in a tank, at the maximum vapor pressure-causing conditions (i.e., temperature, agitation, pH effects of combining wastes, etc.) reasonably expected to occur in the tank. For the purpose of this subpart, maximum organic vapor pressure is determined using the procedures specified in Sec. 265.1084© of this subpart.

**Maximum potential hourly heat input** (Section 72.2) means an hourly heat input used for reporting purposes when a unit lacks certified monitors to report heat input. If the unit intends to use appendix D of part 75 of this chapter to report heat input, this value should be calculated, in accordance with part 75 of this chapter, using the maximum fuel flow rate and the maximum gross calorific value. If the unit intends to use a flow monitor and a diluent gas monitor, this value should be reported, in accordance with part 75 of this chapter, using the maximum potential flow rate and either the maximum carbon dioxide concentration (in percent CO2) or the minimum oxygen concentration (in percent O2).

**Maximum potential hourly heat input** (Section 96.2) means an hourly heat input used for reporting purposes when a unit lacks certified monitors to report heat input. If the unit intends to use appendix D of part 75 of this chapter to report heat input, this value should be calculated, in accordance with part 75 of this chapter, using the maximum fuel flow rate and the maximum gross calorific value. If the unit intends to use a flow monitor and a diluent gas monitor, this value should be reported, in accordance with part 75 of this chapter, using the maximum potential flowrate and either the maximum carbon dioxide concentration (in percent CO2) or the minimum oxygen concentration (in percent O2).

**Maximum potential hourly heat input** (Section 97.2) means an hourly heat input (in mmBtu/hr) used for reporting purposes when a unit lacks certified monitors to report heat input. If the unit intends to use appendix D of part 75 of this chapter to report heat input, this value should be calculated, in accordance with part 75 of this chapter, using the maximum fuel flow rate and the maximum gross calorific value. If the unit intends to use a flow monitor and a diluent gas monitor, this value should be reported, in accordance with part 75 of this chapter, using the maximum potential flowrate and either the maximum carbon dioxide concentration (in percent CO2) or the minimum oxygen concentration (in percent O2).

**Maximum potential NOX emission rate** (Section 72.2) means the emission rate of nitrogen oxides (in lb/mmBtu) calculated in accordance with section 3 of appendix F of part 75 of this chapter, using the maximum potential nitrogen oxides concentration as defined in section 2 of appendix A of part 75 of this chapter, and either the maximum oxygen concentration (in percent O2) or the minimum carbon dioxide concentration (in percent CO2) under all operating conditions of the unit except for unit start-up, shutdown, and upsets.

**Maximum potential NOX emission rate** (Section 96.2) means the emission rate of nitrogen oxides (in lb/mmBtu) calculated in accordance with section 3 of appendix F of part 75 of this chapter, using the maximum potential nitrogen oxides concentration as defined in section 2 of appendix A of part 75 of this chapter, and either the maximum oxygen concentration (in percent O2) or the minimum carbon dioxide concentration (in percent CO2), under all operating conditions of the unit except for unit start up, shutdown, and upsets.

**Maximum potential NOX emission rate** (Section 97.2) means the emission rate of nitrogen oxides (in lb/mmBtu) calculated in accordance with section 3 of appendix F of part 75 of this chapter, using the maximum potential concentration of NOX under section 2 of appendix A of part 75 of this chapter, and either the maximum oxygen concentration (in percent O2) or the minimum carbon dioxide concentration (in percent CO2), under all operating conditions of the unit except for unit start up, shutdown, and upsets.

**Maximum potential NOX emission rate or MER** (Section 72.2) means the emission rate of nitrogen oxides (in lb/mmBtu) calculated in accordance with section 3 of appendix F to part 75 of this chapter, using the maximum potential nitrogen oxides concentration (MPC), as defined in section 2.1.2.1 of appendix A to part 75 of this chapter, and either the maximum oxygen concentration (in percent O2) or the minimum carbon dioxide concentration (in percent CO2) under all operating conditions of the unit except for unit start-up, shutdown, and upsets. The diluent cap value, as defined in this section, may be used in lieu of the maximum O2 or minimum CO2 concentration to calculate the MER. As a second alternative, when the NOX MPC is determined from emission test results or from historical CEM data, as described in section 2.1.2.1 of appendix A to part 75 of this chapter, quality-assured diluent gas (i.e., O2 or CO2) data recorded concurrently with the MPC may be used to calculate the MER. For the purposes of Secs. 75.4(f), 75.19(b)(3), and 75.33(c)(7) in part 75 of this chapter and section 2.5 in appendix E to part 75 of this chapter, the MER is specific to the type of fuel combusted in the unit.

**Maximum production capacity** (Section 57.103) means either the maximum demonstrated rate at which a smelter has produced its principal metallic final product under the process equipment configuration and operating procedures prevailing on or before August 7, 1977, or a rate which the smelter is able to demonstrate by calculation is attainable with process equipment existing on August 7, 1977. The rate may be expressed as a concentrate feed rate to the smelter.

**Maximum Rated Capacity** (Section 204.51) means that the portable air compressor, operating at the design full speed with the compressor on load, delivers its rated cfm output and pressure, as defined by the manufacturer.

**Maximum rated horsepower** (Section 86.082-2) means the maximum brake horsepower output of an engine as stated by the manufacturer in his sales and service literature and his application for certification under Sec. 86.082-21.

**Maximum rated horsepower** (Section 92.2) means the maximum brake horsepower output of an engine.

**Maximum rated hourly heat input** (Section 72.2) means a unit-specific maximum hourly heat input (mmBtu) which is the higher of the manufacturer's maximum rated hourly heat input or the highest observed hourly heat input.

**Maximum rated hourly heat input** (Section 96.2) means a unit-specific maximum hourly heat input (mmBtu) which is the higher of the manufacturer's maximum rated hourly heat input or the highest observed hourly heat input.

**Maximum rated hourly heat input** (Section 97.2) means a unit specific maximum hourly heat input (in mmBtu/hr) which is the higher of the manufacturer's maximum rated hourly heat input or the highest observed hourly heat input.

**Maximum rated hourly heat input rate** (Section 72.2) means a unit-specific maximum hourly heat input rate (mmBtu/hr) which is the higher of the manufacturer's maximum rated hourly heat input rate or the highest observed hourly heat input rate.

**Maximum rated RPM** (Section 205.151) means the engine speed measured in revolutions per minute (RPM) at which peak net brake power (SAE J-245) is developed for motorcycles of a given configuration.

**Maximum rated torque** (Section 86.082-2) means the maximum torque produced by an engine as stated by the manufacturer in his sales and service literature and his application for certification under Sec. 86.082-21.

**Maximum residual disinfectant level (MRDL)** (Section 141.2) means a level of a disinfectant added for water treatment that may not be exceeded at the consumer's tap without an unacceptable possibility of adverse health effects. For chlorine and chloramines, a PWS is in compliance with the MRDL when the running annual average of monthly averages of samples taken in the distribution system, computed quarterly, is less than or equal to the MRDL. For chlorine dioxide, a PWS is in compliance with the MRDL when daily samples are taken at the entrance to the distribution system and no two consecutive daily samples exceed the MRDL. MRDLs are enforceable in the same manner as maximum contaminant levels under Section 1412 of the Safe Drinking Water Act. There is convincing evidence that addition of a disinfectant is necessary for control of waterborne microbial contaminants. Notwithstanding the MRDLs listed in Sec. 141.65, operators may increase residual disinfectant levels of chlorine or chloramines (but not chlorine dioxide) in the distribution system to a level and for a time necessary to protect public health to address specific microbiological contamination problems caused by circumstances such as distribution line breaks, storm runoff events, source water contamination, or cross-connections.

**Maximum residual disinfectant level goal (MRDLG)** (Section 141.2) means the maximum level of a disinfectant added for water treatment at which no known or anticipated adverse effect on the health of persons would occur, and which allows an adequate margin of safety. MRDLGs are nonenforceable health goals and do not reflect the benefit of the addition of the chemical for control of waterborne microbial contaminants.

**Maximum Sound Level** (Section 201.1) means the greatest A-weighted sound level in decibels measured during the designated time interval or during the event, with either fast meter response Sec. 201.1(l) or slow meter response Sec. 201.1(ii) as specified. It is abbreviated as Lmax.

**Maximum Test Power** (Section 94.2) means: (1) For Category 1 engines, the power output observed at the maximum test speed with the maximum fueling rate possible. (2) For Category 2 engines, 90 percent of the power output observed at the maximum test speed with the maximum fueling rate possible.

**Maximum Test Torque** (Section 94.2) means the torque output observed at the test speed with the maximum fueling rate possible at that speed.

**Maximum Total Trihalomethane Potential (MTP)** (Section 141.2) means the maximum concentration of total trihalomethanes produced in a given water containing a disinfectant residual after 7 days at a temperature of 25 deg.C or above.

**Maximum weighted mass ratio averaged over all NAF well sections for BAT effluent limitations and NSPS for base fluid retained on cuttings** (Section 435.11) means the weighted average base fluid retention for all NAF well sections as determined by the API Recommended Practice 13B-2, using the methods and averaging calculations presented in Appendix 7 of subpart A of this part.

**Mayonnaise and salad dressings** (Section 407.81) shall be defined as the emulsified and non-emulsified semisolid food prepared from the combining of edible vegetable oil with acidifying, and egg yolk containing ingredients, or gum and starch combinations to which certain colorings, spices, and flavorings have been added.

**Mbbl** (Section 419.11) means one thousand barrels (one barrel is equivalent to 42 gallons).

**Measurement method** (Section 195.2) is a means of measuring radon gas or radon decay products encompassing similar measurement devices, sampling techniques, or analysis procedures.

**Measurement Period** (Section 201.1) means a continuous period of time during which noise of railroad yard operations is assessed, the beginning and finishing times of which may be selected after completion of the measurements.

**Meat cutter** (Section 432.61) shall mean an operation which fabricates, cuts, or otherwise produces fresh meat cuts and related finished products from livestock carcasses, at rates greater than 2730 kg (6000 lb) per day.

**Mechanical accessory horsepower** (Section 92.2) means the sum of mechanical horsepower generated by an engine to supply accessories. Mechanical accessory horsepower does not include power supplied to the main alternator or generator, power used to circulate engine coolant or engine lubricant, or power used to supply fuel to the engine.

**Mechanical integrity** (Section 68.3) means the process of ensuring that process equipment is fabricated from the proper materials of construction and is properly installed, maintained, and replaced to prevent failures and accidental releases.

**Mechanism** (Section 56.1) means an administrative procedure, guideline, manual, or written statement.

**Medical treatment** (Section 68.3) means treatment, other than first aid, administered by a physician or Registered professional personnel under standing orders from a physician.

**Medium** (Section 407.61) shall mean a point source that processes a total annual raw material production of fruits, vegetables, specialties and other products that is between 1,816 kkg (2,000 tons) per year and 9,080 kkg (10,000 tons) per year.

**Medium** (Section 407.71) shall mean a point source that processes a total annual raw material production of fruits, vegetables, specialties and other products that is between 1,816 kkg (2,000 tons) per year and 9,080 kkg (10,000 tons) per year.

**Medium** (Section 407.81) shall mean a point source that processes a total annual raw material production of fruits, vegetables, specialties and other products that is between 1,816 kkg (2,000 tons) per year and 9,080 kkg (10,000 tons) per year.

**Medium-duty passenger vehicle (MDPV)** (Section 86.1803-01) means any heavy-duty vehicle (as defined in this subpart) with a gross vehicle weight rating (GVWR) of less than 10,000 pounds that is designed primarily for the transportation of persons. The MDPV definition does not include any vehicle which: (1) Is an "incomplete truck" as defined in this subpart; or (2) Has a seating capacity of more than 12 persons; or (3) Is designed for more than 9 persons in seating rearward of the driver's seat; or (4) Is equipped with an open cargo area (for example, a pick-up truck box or bed) of 72.0 inches in interior length or more. A covered box not readily accessible from the passenger compartment will be considered an open cargo area for purposes of this definition.

**Medium-size water system** (Section 141.2) for the purpose of subpart I of this part only, means a water system that serves greater than 3,300 and less than or equal to 50,000 persons.

**Medium-sized plants** (Section 428.61) shall mean plants which process between 3,720 kg/day (8,200 lbs/day) and 10,430 kg/day (23,000 lbs/day) of raw materials.

**Meeting** (Section 1517.2) means the deliberations of at least two Council members where such deliberations determine or result in the joint conduct or disposition of official collegial Council business, but does not include deliberations to take actions to open or close a meeting under Secs. 1517.4 and 1517.5 or to release or withhold information under Secs. 1517.4 and 1517.7. "Meeting" shall not be construed to prevent Council members from considering individually Council business that is circulated to them sequentially in writing.

**Meeting** (Section 1603.3) means the deliberations of at least a quorum of Members where such deliberations determine or result in the joint conduct or disposition of official CSB business, and includes conference telephone calls or other exchanges otherwise coming within the definition. A meeting does not include: (1) Notation voting or similar consideration of business, whether by circulation of material to the Members individually in writing or by a polling of the Members individually by telephone. Action by at least a quorum of Members to: (i) Open or to close a meeting or to release or to withhold information pursuant to Sec. 1603.7; (ii) Set an agenda for a proposed meeting(s); (iii) Call a meeting on less than seven days' notice as permitted by Sec. 1603.9(b); or (iv) Change the subject matter or the determination to open or to close a publicly announced meeting under Sec. 1603.10(b). (3) A session attended by at least a quorum of Members for the purpose of having the Board's staff or expert consultants to the Board brief or otherwise provide information to the Board concerning any matters within the purview of the Board under its authorizing statute, provided that the Board does not engage in deliberations that determine or result in the joint conduct or disposition of official CSB business on such matters. (4) A session attended by at least a quorum of Members for the purpose of having the Environmental Protection Agency or Occupational Safety and Health Administration (including contractors of those agencies) or other persons or organizations brief or otherwise provide information to the Board concerning any matters within the purview of the Board under its authorizing statute, provided that the Board does not engage in deliberations that determine or result in the joint conduct or disposition of official CSB business on such matters. (5) A gathering of Members for the purpose of holding informal preliminary discussions or exchange of views which do not effectively predetermine official action.

**Melt** (Section 409.21) shall mean that amount of raw material (raw sugar) contained within aqueous solution at the beginning of the process for production of refined cane sugar.

**Member** (Section 1603.3) means an individual duly appointed and confirmed to the collegial body known as the Board.

**Member of the public** (Section 190.02) means any individual that can receive a radiation dose in the general environment, whether he may or may not also be exposed to radiation in an occupation associated with a nuclear fuel cycle. However, an individual is not considered a member of the public during any period in which he is engaged in carrying out any operation which is part of a nuclear fuel cycle.

**Member of the public** (Section 191.02) means any individual except during the time when that individual is a worker engaged in any activity, operation, or process that is covered by the Atomic Energy Act of 1954, as amended.

**Member of the public** (Section 197.2) means anyone who is not a radiation worker for purposes of worker protection.

**Member of the public or person** (Section 1400.2) means an individual.

**Merchant** (Section 420.11) means those by-product cokemaking operations which provide more than fifty percent of the coke produced to operations, industries, or processes other than iron making blast furnaces associated with steel production.

**Mercury** (Section 416.61) shall mean the total mercury present in the process wastewater stream exiting the mercury treatment system.

**Metal cleaning waste** (Section 423.11) means any wastewater resulting from cleaning [with or without chemical cleaning compounds] any metal process equipment including, but not limited to, boiler tube cleaning, boiler fireside cleaning, and air preheater cleaning.

**Metal powder production operations** (Section 471.02) are mechanical process operations which convert metal to a finely divided form.

**Metal preparation** (Section 466.02) means any and all of the metal processing steps preparatory to applying the enamel slip. Usually this includes cleaning, pickling and applying a nickel flash or chemical coating.

**Metal-bearing wastes** (Section 437.2) means wastes and/or used materials from manufacturing or processing facilities or other commercial operations that contain significant quantities of metal pollutants, but not significant quantities of oil and grease (generally less than 100 mg/L). Examples of these wastes are spent electroplating baths and sludges, metal-finishing rinse water and sludges, chromate wastes, blow-down water and sludges from air pollution control, spent anodizing solutions, incineration air pollution control wastewaters, waste liquid mercury, cyanide containing wastes greater than 136 mg/L, and waste acids and bases with or without metals.

**Metallic shoe seal** (Section 265.1081) means a continuous seal that is constructed of metal sheets which are held vertically against the wall of the tank by springs, weighted levers, or other mechanisms and is connected to the floating roof by braces or other means. A flexible coated fabric (envelope) spans the annular space between the metal sheet and the floating roof.

**Metallic shoe seal or mechanical shoe seal** (Section 65.2) means metal sheets that are held vertically against the wall of the storage vessel by springs, weighted levers, or other mechanisms and connected to the floating roof by braces or other means. A flexible coated fabric (envelope) spans the annular space between the metal sheet and the floating roof.

**Metallo-organic active ingredients** (Section 455.31) means carbon containing active ingredients containing one or more metallic atoms in the structure.

**Metalworking fluid** (Section 721.3) means a liquid of any viscosity or color containing intentionally added water and used in metal machining operations for the purpose of cooling, lubricating, or rust inhibition.

**Meteorological measurements** (Section 58.1) means measurements of wind speed, wind direction, barometric pressure, temperature, relative humidity, and solar radiation.

**Methane reactivity adjustment factor** (Section 86.1702-99) means a factor applied to the mass of methane emissions from natural gas fueled vehicles for the purpose of determining the gasoline equivalent ozone-forming potential of the methane emissions.

**Methanol** (Section 80.2) means any fuel sold for use in motor vehicles and commonly known or commercially sold as methanol or MXX, where XX is the percent methanol (CH3OH) by volume.

**Methanol** (Section 86.1702-99) means any fuel for motor vehicles and motor vehicle engines that is composed of either commercially available or chemically pure methanol (CH3OH) and gasoline as specified in Sec. 86.1771 (Fuel Specifications). The required fuel blend is based on the type of methanol-fueled vehicle being certified and the particular aspect of the certification procedure being conducted.

**Methanol** (Section 92.2) means a fuel that contains at least 50 percent methanol (methyl alcohol, (CH3OH)) by volume.

**Methanol vehicle** (Section 86.1702-99) means any motor vehicle that is engineered and designed to be operated using methanol as a fuel.

**Methanol-fueled** (Section 86.090-2) means any motor vehicle or motor vehicle engine that is engineered and designed to be operated using methanol fuel (i.e., a fuel that contains at least 50 percent methanol (CH<INF>3</INF>OH) by volume) as fuel. Flexible fuel vehicles are methanol-fueled vehicles.

**Methanol-fueled vehicle** (Section 86.1803-01) means any motor vehicle or motor vehicle engine that is engineered and designed to be operated using methanol fuel (i.e., a fuel that contains at least 50 percent methanol (CH3OH) by volume) as fuel.

**Method 1654A** (Section 435.11) refers to Method 1654, Revision A, entitled "PAH Content of Oil by HPLC/UV," December 1992, which is published in Methods for the Determination of Diesel, Mineral, and Crude Oils in Offshore Oil and Gas Industry Discharges, EPA-821-R-92-008. This incorporation by reference was approved by the Director of the Federal Register in accordance with 5 U.S.C. 552(a) and 1 CFR part 51. Copies may be obtained from the National Technical Information Service, Springfield, VA 22161, 703-605-6000. Copies may be inspected at the Office of the Federal Register, 800 North Capitol Street, NW., Suite 700, Washington, DC. A copy may also be inspected at EPA's Water Docket, 1200 Pennsylvania Ave., NW., Washington, DC 20460.

**Method detection level** (Section 132.2) is the minimum concentration of an analyte (substance) that can be measured and reported with a 99 percent confidence that the analyte concentration is greater than zero as determined by the procedure set forth in appendix B of 40 CFR part 136.

**Method of aspiration** (Section 94.2) means the method whereby air for fuel combustion enters the engine (e.g., naturally aspirated or turbocharged).

Method of aspiration means the method whereby air for fuel combustion enters the engine (e.g., natural or turbocharged).

**Metropolitan planning organization (MPO)** (Section 93.101) is that organization designated as being responsible, together with the State, for conducting the continuing, cooperative, and comprehensive planning process under 23 U.S.C. 134 and 49 U.S.C. 5303. It is the forum for cooperative transportation decision-making.

**Metropolitan Planning Organization (MPO)** (Section 93.152) is that organization designated as being responsible, together with the State, for conducting the continuing, cooperative, and comprehensive planning process under 23 U.S.C. 134 and 49 U.S.C. 1607.

**Metropolitan Statistical Area (MSA)** (Section 58.1) as designated by the most recent decennial U.S. Census of Population Report.

**mg (Section 401.11)** means milligram(s).

**mg/l** (Section 133.101) Milligrams per liter.

**Mg/L** (Section 439.1) means milligrams per liter or parts per million (ppm).

**Mgal** (Section 419.11) means one thousand gallons.

**Microbial commercial activity notice or MCAN** (Section 700.43) means any notice for microorganisms submitted to EPA pursuant to section 5(a)(1) of the Act in accordance with subpart D of part 725 of this chapter.

**Microbial commercial activity notice or MCAN** (Section 725.3) means a notice for microorganisms submitted to EPA pursuant to section 5(a)(1) of the Act in accordance with subpart D of this part.

**Microbial mixture** (Section 725.3) means any combination of microorganisms or microorganisms and other chemical substances, if the combination does not occur in nature and is not an article.

**Microorganism** (Section 725.3) means an organism classified, using the 5-kingdom classification system of Whittacker, in the kingdoms Monera (or Procaryotae), Protista, Fungi, and the Chlorophyta and the Rhodophyta of the Plantae, and a virus or virus-like particle.

**Microorganisms** (Section 455.10) means registered pesticide active ingredients that are biological control agents listed in 40 CFR 152.20(a)(3) including Eucaryotes (protozoa, algae, fungi), Procaryotes (bacteria), and Viruses.

**Mid-yard** (Section 745.63) means an area of a residential yard approximately midway between the dripline of a residential building and the nearest property boundary or between the driplines of a residential building and another building on the same property.

**Milestone** (Section 192.31) means an enforceable date by which action, or the occurrence of an event, is required for purposes of achieving compliance with the 20 pCi/m\2\-s flux standard.

**Milestone** (Section 93.101) has the meaning given in sections 182(g) (1) and 189© of the CAA. A milestone consists of an emissions level and the date on which it is required to be achieved.

**Milestone** (Section 93.152) has the meaning given in sections 182(g)(1) and 189©(1) of the Act.

**Military** (Section 266.201) means the Department of Defense (DOD), the Armed Services, Coast Guard, National Guard, Department of Energy (DOE), or other parties under contract or acting as an agent for the foregoing, who handle military munitions.

**Military engine** (Section 86.082-2) means any engine manufactured solely for the Department of Defense to meet military specifications.

**Military munition is not a solid waste when** (Section 266.202) : (1) Used for its intended purpose, including: (i) Use in training military personnel or explosives and munitions emergency response specialists (including training in proper destruction of unused propellant or other munitions); or (ii) Use in research, development, testing, and evaluation of military munitions, weapons, or weapon systems; or (iii) Recovery, collection, and on-range destruction of unexploded ordnance and munitions fragments during range clearance activities at active or inactive ranges. However, "use for intended purpose" does not include the on-range disposal or burial of unexploded ordnance and contaminants when the burial is not a result of product use. (2) An unused munition, or component thereof, is being repaired, reused, recycled, reclaimed, disassembled, reconfigured, or otherwise subjected to materials recovery activities, unless such activities involve use constituting disposal as defined in 40 CFR 261.2©(1), or burning for energy recovery as defined in 40 CFR 261.2©(2).

**Military munitions** (Section 260.10) means all ammunition products and components produced or used by or for the U.S. Department of Defense or the U.S. Armed Services for national defense and security, including military munitions under the control of the Department of Defense, the U.S. Coast Guard, the U.S. Department of Energy (DOE), and National Guard personnel. The term military munitions includes: confined gaseous, liquid, and solid propellants, explosives, pyrotechnics, chemical and riot control agents, smokes, and incendiaries used by DOD components, including bulk explosives and chemical warfare agents, chemical munitions, rockets, guided and ballistic missiles, bombs, warheads, mortar rounds, artillery ammunition, small arms ammunition, grenades, mines, torpedoes, depth charges, cluster munitions and dispensers, demolition charges, and devices and components thereof. Military munitions do not include wholly inert items, improvised explosive devices, and nuclear weapons, nuclear devices, and nuclear components thereof. However, the term does include non-nuclear components of nuclear devices, managed under DOE's nuclear weapons program after all required sanitization operations under the Atomic Energy Act of 1954, as amended, have been completed.

**Military munitions** (Section 266.201) is as defined in 40 CFR 260.10.

**Military range** (Section 266.201) means designated land and water areas set aside, managed, and used to conduct research on, develop, test, and evaluate military munitions and explosives, other ordnance, or weapon systems, or to train military personnel in their use and handling. Ranges include firing lines and positions, maneuver areas, firing lanes, test pads, detonation pads, impact areas, and buffer zones with restricted access and exclusionary areas.

**Milking center** (Section 412.11) shall mean a separate milking area with storage and cooling facilities adjacent to a free stall barn or cowyard dairy operation.

**Milkroom** (Section 412.11) shall mean milk storage and cooling rooms normally used for stall barn dairies.

**Mill** (Section 440.132) is a preparation facility within which the metal ore is cleaned, concentrated, or otherwise processed before it is shipped to the customer, refiner, smelter, or manufacturer. A mill includes all ancillary operations and structures necessary to clean, concentrate, or otherwise process metal ore, such as ore and gangue storage areas and loading facilities.

**Milling** (Section 471.02) is the mechanical treatment of a nonferrous metal to produce powder, or to coat one component of a powder mixture with another.

**Mine** (Section 436.21) shall mean an area of land, surface or underground, actively mined for the production of crushed and broken stone from natural deposits.

**Mine** (Section 438.181) shall mean an area of land, surface or underground, actively used for or resulting from the extraction of a mineral from natural deposits.

**Mine** (Section 440.132) is an active mining area, including all land and property placed under, or above the surface of such land, used in or resulting from the work of extracting metal ore or minerals from their natural deposits by any means or method, including secondary recovery of metal ore from refuse or other storage piles, wastes, or rock dumps and mill tailings derived from the mining, cleaning, or concentration of metal ores.

**Mine dewatering** (Section 436.21) shall mean any water that is impounded or that collects in the mine and is pumped, drained or otherwise removed from the mine through the efforts of the mine operator. However, if a mine is also used for treatment of process generated waste water, discharges of commingled water from the facilities shall be deemed discharges of process generated waste water.

**Mine drainage** (Section 434.11) means any drainage, and any water pumped or siphoned, from an active mining area or a post-mining area.

**Mine drainage** (Section 436.381) shall mean any water drained, pumped or siphoned from a mine.

**Mine drainage** (Section 440.132) means any water drained, pumped, or siphoned from a mine.

**Mineral fiber insulation** (Section 247.3) means insulation (rock wool or fiberglass) which is composed principally of fibers manufactured from rock, slag or glass, with or without binders;

**Mineral Oil PCB Transformer** (Section 761.13) means any transformer originally designed to contain mineral oil as the dielectric fluid and which has been tested and found to contain 500 ppm or greater PCBs.

**Minimum** (Section 435.41) as applied to BAT effluent limitations and NSPS for drilling fluids and drill cuttings means the minimum 96-hour LC50 value allowed as measured in any single sample of the discharged waste stream. Minimum as applied to BPT and BCT effluent limitations and NSPS for sanitary wastes means the minimum concentration value allowed as measured in any single sample of the discharged waste stream.

**Minimum as applied to BAT effluent limitations and NSPS for drilling fluids and drill cuttings** (Section 435.11) means the minimum 96-hour LC50 value allowed as measured in any single sample of the discharged waste stream. Minimum as applied to BPT and BCT effluent limitations and NSPS for sanitary wastes means the minimum concentration value allowed as measured in any single sample of the discharged waste stream.

**Minimum Level (ML)** (Section 132.2) is the concentration at which the entire analytical system must give a recognizable signal and acceptable calibration point. The ML is the concentration in a sample that is equivalent to the concentration of the lowest calibration standard analyzed by a specific analytical procedure, assuming that all the method-specified sample weights, volumes and processing steps have been followed.

**Minimum level (ML)** (Section 430.0) The level at which the analytical system gives recognizable signals and an acceptable calibration point. The following minimum levels apply to pollutants in this part: (Seed the regulation under this part).

**Minimum level** (Section 439.1) means the level at which an analytical system gives recognizable signals and an acceptable calibration point.

**Mining overburden returned to the mine site** (Section 260.10) means any material overlying an economic mineral deposit which is removed to gain access to that deposit and is then used for reclamation of a surface mine.

**Mining wastes** (Section 243.101) means residues which result from the extraction of raw materials from the earth.

**Mining wastes** (Section 246.101) means residues which result from the extraction of raw materials from the earth.

**Minor change to monitoring** (Section 65.2) means: (1) A modification to federally required monitoring that: (i) Does not decrease the stringency of the compliance and enforcement measures of the relevant standard; (ii) Has no national significance (e.g., does not affect implementation of the applicable regulation in this part for other affected sources, does not set a national precedent, and individually does not result in a revision to the monitoring requirements); and (iii) Is site-specific, made to reflect or accommodate the operational characteristics, physical constraints, or safety concerns of an affected source. (2) Examples of minor changes to monitoring include, but are not limited to: (i) Modifications to a sampling procedure, such as use of an improved sample conditioning system to reduce maintenance requirements; (ii) Increased monitoring frequency; and (iii) Modification of the environmental shelter to moderate temperature fluctuation and thus protect the analytical instrumentation.

**Minor change to test method** (Section 65.2) means: (1) A modification to a federally enforceable test method that: (i) Does not decrease the stringency of the emission limitation or standard; (ii) Has no national significance (e.g., does not affect implementation of the applicable regulation in this part for other affected sources, does not set a national precedent, and individually does not result in a revision to the test method); and (iii) Is site-specific, made to reflect or accommodate the operational characteristics, physical constraints, or safety concerns of an affected source. (2) Examples of minor changes to a test method include, but are not limited to: (i) Field adjustments in a test method's sampling procedure, such as a modified sampling traverse or location to avoid interference from an obstruction in the stack, increasing the sampling time or volume, use of additional impingers for a high moisture situation, accepting particulate emission results for a test run that was conducted with a lower than specified temperature, substitution of a material in the sampling train that has been demonstrated to be more inert for the sample matrix; and (ii) Changes in recovery and analytical techniques such as a change in quality control/quality assurance requirements needed to adjust for analysis of a certain sample matrix.

**Miscellaneous ACM** (Section 763.83) means miscellaneous material that is ACM in a school building.

**Miscellaneous material** (Section 763.83) means interior building material on structural components, structural members or fixtures, such as floor and ceiling tiles, and does not include surfacing material or thermal system insulation.

**Miscellaneous oil spill control agent** (Section 300.5) is any product, other than a dispersant, sinking agent, surface washing agent, surface collecting agent, bioremediation agent, burning agent, or sorbent that can be used to enhance oil spill cleanup, removal, treatment, or mitigation.

**Miscellaneous unit** (Section 260.10) means a hazardous waste management unit where hazardous waste is treated, stored, or disposed of and that is not a container, tank, surface impoundment, pile, land treatment unit, landfill, incinerator, boiler, industrial furnace, underground injection well with appropriate technical standards under part 146 of this chapter, containment building, corrective action management unit, unit eligible for a research, development, and demonstration permit under 40 CFR 270.65, or staging pile.

**Miscellaneous waste stream** (Section 468.02) shall mean the following additional waste streams related to forming copper: hydrotesting, sawing, surface milling, and maintenance.

**Miscellaneous wastewater streams** (Section 461.2) shall mean the combined wastewater streams from the process operations listed below for each subcategory. If a plant has one of these streams then the plant receives the entire miscellaneous waste stream allowance. (1) Cadmium subcategory. Cell wash, electrolyte preparation, floor and equipment wash, and employee wash. (2) Lead subcategory. Floor wash, wet air pollution control, battery repair, laboratory, hand wash, and respirator wash. (3) Lithium subcategory. Floor and equipment wash, cell testing, and lithium scrap disposal. (4) Zinc subcategory. Cell wash, electrolyte preparation, employee wash, reject cell handling, floor and equipment wash.

**Miscellaneous wastewater streams** (Section 461.2) shall mean the combined wastewater streams from the process operations listed below for each subcategory. If a plant has one of these streams then the plant receives the entire miscellaneous waste stream allowance. (1) Cadmium subcategory. Cell wash, electrolyte preparation, floor and equipment wash, and employee wash. (2) Lead subcategory. Floor wash, wet air pollution control, battery repair, laboratory, hand wash, and respirator wash. (3) Lithium subcategory. Floor and equipment wash, cell testing, and lithium scrap disposal. (4) Zinc subcategory. Cell wash, electrolyte preparation, employee wash, reject cell handling, floor and equipment wash.

**Mischmetal** (Section 421.271) refers to a rare earth metal alloy comprised of the natural mixture of rare earths to about 94-99 percent. The balance of that alloy includes traces of other elements and one to two percent iron.

**Missing data period** (Section 72.2) means the total number of consecutive hours during which any component part of a certified CEMS or approved alternative monitoring system is not providing quality-assured data, regardless of the reason.

**Missing data period** (Section 72.2) means the total number of consecutive hours during which any certified CEMS or approved alternative monitoring system is not providing quality-assured data, regardless of the reason.

**Mitigation or mitigation system** (Section 68.3) means specific activities, technologies, or equipment designed or deployed to capture or control substances upon loss of containment to minimize exposure of the public or the environment. Passive mitigation means equipment, devices, or technologies that function without human, mechanical, or other energy input. Active mitigation means equipment, devices, or technologies that need human, mechanical, or other energy input to function.

**Mixed fertilizer** (Section 418.71) shall mean a mixture of wet and/or dry straight fertilizer materials, mixed fertilizer materials, fillers and additives prepared through chemical reaction to a given formulation.

**Mixing instructions** (Section 59.101) means the coating or coating component manufacturer's or importer's specification of the quantities of coating components for mixing a coating.

**Mixing zone** (Section 230.3) means a limited volume of water serving as a zone of initial dilution in the immediate vicinity of a discharge point where receiving water quality may not meet quality standards or other requirements otherwise applicable to the receiving water. The mixing zone should be considered as a place where wastes and water mix and not as a place where effluents are treated.

**Mixing, compounding, and formulating operations** (Section 439.41) means processes that put pharmaceutical products in dosage forms.

**Mixture** (Section 116.3) means any combination of two or more elements and/or compounds in solid, liquid, or gaseous form except where such substances have undergone a chemical reaction so as to become inseparable by physical means.

**Mixture** (Section 355.20) means a heterogeneous association of substances where the various individual substances retain their identities and can usually be separated by mechanical means. Includes solutions or compounds but does not include alloys or amalgams.

**Mixture** (Section 372.3) means any combination of two or more chemicals, if the combination is not, in whole or in part, the result of a chemical reaction. However, if the combination was produced by a chemical reaction but could have been produced without a chemical reaction, it is also treated as a mixture. A mixture also includes any combination which consists of a chemical and associated impurities.

**Mixture** (section 710.2) means any combination of two or more chemical substances if the combination does not occur in nature and is not, in whole or in part, the result of a chemical reaction; except that "mixture" does include: (1) Any combination which occurs, in whole or in part, as a result of a chemical reaction if the combination could have been manufactured for commercial purposes without a chemical reaction at the time the chemical substances comprising the combination were combined and if, after the effective date or premanufacture notification requirements, none of the chemical substances comprising the combination is a new chemical substance, and (2) Hydrates of a chemical substance or hydrated ions formed by association of a chemical substance with water.

**Mixture** (Section 712.3) means any combination of two or more chemical substances if the combination does not occur in nature and is not, in whole or in part, the result of a chemical reaction; except that mixture does include (1) any combination which occurs, in whole or in part, as a result of a chemical reaction if the combination could have been manufactured for commercial purposes without a chemical reaction at the time the chemical substances comprising the combination were combined, and if all of the chemical substances comprising the combination are included in the EPA, TSCA Chemical Substance Inventory after the effective date of the premanufacture notification requirement under 40 CFR part 720, and (2) hydrates of a chemical substance or hydrated ions formed by association of a chemical substance with water. The term mixture includes alloys, inorganic glasses, ceramics, frits, and cements, including Portland cement.

**Mixture** (Section 720.3) means any combination of two or more chemical substances if the combination does not occur in nature and is not, in whole or in part, the result of a chemical reaction; except "mixture" does include (1) any combination which occurs, in whole or in part, as a result of a chemical reaction if the combination could have been manufactured for commercial purposes without a chemical reaction at the time the chemical substances comprising the combination were combined, and if all of the chemical substances comprising the combination are not new chemical substances, and (2) hydrates of a chemical substance or hydrated ions formed by association of a chemical substance with water, so long as the nonhydrated form is itself not a new chemical substance.

**Mixture** (Section 761.13) means any combination of two or more chemical substances if the combination does not occur in nature and is not, in whole or in part, the result of a chemical reaction; except that such term does include any combination which occurs, in whole or in part, as a result of a chemical reaction if none of the chemical substances comprising the combination is a new chemical substance and if the combination could have been manufactured for commercial purposes without a chemical reaction at the time the chemical substances comprising the combination were combined.

**ml/l** (Section 434.11) means milliliters per liter.

**mm** (Section 401.11) means millimeter(s).

**Mobile equipment** (Section 59.101) means any equipment that is physically capable of being driven or drawn upon a highway including, but not limited to, the following types of equipment: construction vehicles (such as mobile cranes, bulldozers, concrete mixers); farming equipment (wheel tractor, plow, pesticide sprayer); hauling equipment (truck trailers, utility bodies, camper shells); and miscellaneous equipment (street cleaners, golf carts).

**Mobile genetic element or MGE** (Section 725.3) means an element of genetic material that has the ability to move genetic material within and between organisms. "Mobile genetic elements" include all plasmids, viruses, transposons, insertion sequences, and other classes of elements with these general properties.

**Mobile source** (Section 117.1) means any vehicle, rolling stock, or other means of transportation which contains or carries a reportable quantity of a hazardous substance.

**MOD Director** (Section 85.1402) means Director of Manufacturers Operations Division, Office of Mobile Sources—Office of Air and Radiation of the Environmental Protection Agency. Office Director means the Director for the Office of Mobile Sources- Office of Air and Radiation of the Environmental Protection Agency or an authorized representative of the Office Director.

**MOD Director** (Section 85.2102) means Director of Manufacturers Operations Division, Office of Mobile Sources—Office of Air and Radiation of the Environmental Protection Agency.

**Model** (Section 86.082-2) means a specific combination of car line, body style, and drivetrain configuration.

**Model** (Section 86.1803-01) means a specific combination of car line, body style, and drivetrain configuration.

**Model specific code** (Section 205.151) means the designation used for labeling purposes in Secs. 205.158 and 205.169 for identifying the motorcycle manufacturer, class, and "advertised engine displacement," respectively.

**Model Type** (Section 600.022-85) means a unique combination of car line, basic engine, and transmission class.

**Model type** (Section 86.082-2) means a unique combination of car line, basic engine, and transmission class.

**Model type** (Section 86.1803-01) means a unique combination of car line, basic engine, and transmission class.

**Model year** (Section 85.2302) means the manufacturer's annual production period (as determined under Sec. 85.2304) which includes January 1 of such calendar year, provided, that if the manufacturer has no annual production period, the term "model year" shall mean the calendar year.

**Model year (MY)** (Section 89.2) means the manufacturer's annual new model production period which includes January 1 of the calendar year, ends no later than December 31 of the calendar year, and does not begin earlier than January 2 of the previous calendar year. Where a manufacturer has no annual new model production period, model year means calendar year.

**Model year (MY)** (Section 90.3) means the manufacturer's annual new model production period which includes January 1 of the calendar year, ends no later than December 31 of the calendar year, and does not begin earlier than January 2 of the previous calendar year. Where a manufacturer has no annual new model production period, model year means calendar year.

**Model year** (Section 204.51) means the manufacturer's annual production period which includes January 1 of such calendar year; Provided, that if the manufacturer has no annual production period, the term "model year" shall mean the calendar year.

**Model year** (Section 205.151) means the manufacturer's annual production period, which includes January 1 of any calendar year, or if the manufacturer has no annual production period, the term "model year" shall mean the calendar year.

**Model year** (Section 205.51) means the manufacturer's annual production period which includes January 1 of such calendar year: Provided, that if the manufacturer has no annual production period, the term "model year" shall mean the calendar year.

**Model Year** (Section 600.022-85) means the manufacturer's annual production period (as determined by the Administrator) which includes January 1 of such calendar year. If a manufacturer has no annual production period, the term model year means the calendar year.

**Model year** (Section 85.1502) The manufacturer's annual production period (as determined by the Administrator) which includes January 1 of such calendar year; Provided, That if the manufacturer has no annual production period, the term "model year" shall mean the calendar year in which a vehicle is modified. A certificate holder shall be deemed to have produced a vehicle or engine when the certificate holder has modified the nonconforming vehicle or engine.

**Model Year** (Section 85.2102) means the manufacturer's annual production period (as determined by the Office Director) which includes January 1 of such calendar year; however, if the manufacturer has no annual production period, the term "model year" shall mean the calendar year.

**Model Year** (Section 85.502) means the manufacturer's annual production period (as determined by the Administrator) which includes January 1 of such calendar year:

Provided, That if the manufacturer has no annual production period, the term model year shall mean the calendar year.

**Model year** (Section 86.082-2) means the manufacturer's annual production period (as determined by the Administrator) which includes January 1 of such calendar year: Provided, That if the manufacturer has no annual production period, the term model year shall mean the calendar year.

**Model year** (Section 86.1803-01) means the manufacturer's annual production period (as determined by the Administrator) which includes January 1 of such calendar year: Provided that if the manufacturer has no annual production period, the term "model year" shall mean the calendar year.

**Model year** (Section 86.402-78) means the manufacturer's annual production period (as determined by the Administrator) which includes January first of such calendar year. If the manufacturer has no annual production period, the term model year shall mean the calendar year. Motorcycle means any motor vehicle with a headlight, taillight, and stoplight and having: Two wheels, or Three wheels and a curb mass less than or equal to 680 kilograms (1499 pounds). Oxides of nitrogen means the sum of the nitric oxide and nitrogen dioxide contained in a gas sample as if the nitric oxide were in the form of nitrogen dioxide.

**Model year** (Section 91.3) means the manufacturer's annual new model production period which includes January 1 of the calendar year for which the model year is named, ends no later than December 31 of the calendar year, and does not begin earlier than January 2 of the previous calendar year. Where a manufacturer has no annual new model production period, model year means the calendar year.

**Model year** (Section 92.2) means a calendar year; except where the Administrator determines a different production period which includes January 1 of such calendar year.

**Model year** (Section 94.2) means the manufacturer's annual new model production period which includes January 1 of the calendar year, ends no later than December 31 of the calendar year, and does not begin earlier than January 2 of the previous calendar year. Where a manufacturer has no annual new model production period, model year means calendar year.

**Model year for imported engines** (Section 89.602) The manufacturer's annual production period (as determined by the Administrator) which includes January 1 of the calendar year; provided, that if the manufacturer has no annual production period, the term "model year" means the calendar year in which a nonroad engine is modified. An independent commercial importer (ICI) is deemed to have produced a nonroad engine when the ICI has modified (including labeling) the nonconforming nonroad engine to meet applicable emission requirements.

**Model year, as it applies to the clean fuel vehicle fleet purchase requirements** (Section 88.302-94) means September 1 through August 31.

**Modification** (Section 55.2) shall have the meaning given in the applicable requirements incorporated into Secs. 55.13 and 55.14 of this part, except that for two years following the date of promulgation of this part the definition given in section 111(a) of the Act shall apply for the purpose of determining the required date of compliance with this part, as set forth in Sec. 55.3 of this part.

**Modification** (Section 60.2) means any physical change in, or change in the method of operation of, an existing facility which increases the amount of any air pollutant (to which a standard applies) emitted into the atmosphere by that facility or which results in the emission of any air pollutant (to which a standard applies) into the atmosphere not previously emitted.

**Modification** (Section 194.2) means action(s) taken by the Administrator that alters the terms or conditions of certification pursuant to section 8(d) (1) of the WIPP LWA. Modification of any certification shall comply with this part and part 191 of this chapter.

**Monitor** (Section 58.1) is a generic term for an instrument, sampler, analyzer, or other device that measures or assists in the measurement of atmospheric air pollutants and which is acceptable for use in ambient air surveillance under the provisions of appendix C to this part, including both point and open path analyzers that have been designated as either reference or equivalent methods under part 53 of this chapter and air samplers that are specified as part of a manual method that has been designated as a reference or equivalent method under part 53 of this chapter. Monitoring path for an open path analyzer is the actual path in space between two geographical locations over which the pollutant concentration is measured and averaged. Monitoring path length of an open path analyzer is the length of the monitoring path in the atmosphere over which the average pollutant concentration measurement (path-averaged concentration) is determined. See also, optical measurement path length.

**Monitor accuracy** (Section 72.2) means the closeness of the measurement made by a CEMS or by one of its component parts to the reference value of the emissions or volumetric flow being measured, expressed as the difference between the measurement and the reference value.

**Monitor operating hour** (Section 72.2) means any unit operating hour or portion thereof over which a CEMS, or other monitoring system approved by the Administrator under part 75 of this chapter is operating, regardless of the number of measurements (i.e., data points) collected during the hour or portion of an hour.

**Monitoring** (Section 63.2) means the collection and use of measurement data or other information to control the operation of a process or pollution control device or to verify a work practice standard relative to assuring compliance with applicable requirements. Monitoring is composed of four elements: (1) Indicator(s) of performance—the parameter or parameters you measure or observe for deonstrating proper operation of the pollution control measures or compliance with the applicable emissions limitation or standard. Indicators of performance may include direct or predicted emissions measurements (including opacity), operational parametric values that correspond to process or control device (and capture system) efficiencies or emissions rates, and recorded findings of inspection of work practice activities, materials tracking, or design characteristics. Indicators may be expressed as a single maximum or minimum value, a function of process variables (for example, within a range of pressure drops), a particular operational or work practice status (for example, a damper position, completion of a waste recovery task, materials tracking), or an interdependency between two or among more than two variables. (2) Measurement techniques—the means by which you gather and record information of or about the indicators of performance. The components of the measurement technique include the detector type, location and installation specifications, inspection procedures, and quality assurance and quality control measures. Examples of measurement techniques include continuous emission monitoring systems, continuous opacity monitoring systems, continuous parametric monitoring systems, and manual inspections that include making records of process conditions or work practices. (3) Monitoring frequency—the number of times you obtain and record monitoring data over a specified time interval. Examples of monitoring frequencies include at least four points equally spaced for each hour for continuous emissions or parametric monitoring systems, at least every 10 seconds for continuous opacity monitoring systems, and at least once per operating day (or week, month, etc.) for work practice or design inspections. (4) Averaging time—the period over which you average and use data to verify proper operation of the pollution control approach or compliance with the emissions limitation or standard. Examples of averaging time include a 3-hour average in units of the emissions limitation, a 30-day rolling average emissions value, a daily average of a control device operational parametric range, and an instantaneous alarm.

**Monitoring** (Section 64.1) means any form of collecting data on a routine basis to determine or otherwise assess compliance with emission limitations or standards. Recordkeeping may be considered monitoring where such records are used to determine or assess compliance with an emission limitation or standard (such as records of raw material content and usage, or records documenting compliance with work practice requirements). The conduct of compliance method tests, such as the procedures in appendix A to part 60 of this chapter, on a routine periodic basis may be considered monitoring (or as a supplement to other monitoring), provided that requirements to conduct such tests on a one-time basis or at such times as a regulatory authority may require on a non-regular basis are not considered monitoring requirements for purposes of this paragraph. Monitoring may include one or more than one of the following data collection techniques, where appropriate for a particular circumstance: (1) Continuous emission or opacity monitoring systems. (2) Continuous process, capture system, control device or other relevant parameter monitoring systems or procedures, including a predictive emission monitoring system. (3) Emission estimation and calculation procedures (e.g., mass balance or stoichiometric calculations). (4) Maintenance and analysis of records of fuel or raw materials usage. (5) Recording results of a program or protocol to conduct specific operation and maintenance proedures. (6) Verification of emissions, process parameters, capture system parameters, or control device parameters using portable or in situ measurement devices. (7) Visible emission observations. (8) Any other form of measuring, recording, or verifying on a routine basis emissions, process parameters, capture system parameters, control device parameters or other factors relevant to assessing compliance with emission limitations or standards.

**Monitoring device** (Section 60.2) means the total equipment, required under the monitoring of operations sections in applicable subparts, used to measure and record (if applicable) process parameters.

**Monitoring Planning Area (MPA)** (Section 58.1) means a contiguous geographic area with established, well defined boundaries, such as a metropolitan statistical area, county or State, having a common area that is used for planning monitoring locations for PM2.5. MPAs may cross State boundaries, such as the Philadelphia PA-NJ MSA, and be further subdivided into community monitoring zones. MPAs are generally oriented toward areas with populations greater than 200,000, but for convenience, those portions of a State that are not associated with MSAs can be considered as a single MPA. MPAs must be defined, where applicable, in a State PM monitoring network description.

**Monitoring system** (Section 61.02) means any system, required under the monitoring sections in applicable subparts, used to sample and condition (if applicable), to analyze, and to provide a record of emissions or process parameters.

**Monitoring system** (Section 96.2) means any monitoring system that meets the requirements of subpart H of this part, including a continuous emissions monitoring system, an excepted monitoring system, or an alternative monitoring system.

**Monitoring system** (Section 97.2) means any monitoring system that meets the requirements of subpart H of this part, including a continuous emissions monitoring system, an excepted monitoring system, or an alternative monitoring system.

**Monovent** (Section 62.1201) means an exhaust configuration of a building or emission control device (e.g. positive pressure fabric filter) that extends the length of the structure and has a width very small in relation to its length (i.e., length to width ratio is typically greater than 5:1). The exhaust may be an open vent with or without a roof, louvered vents, or a combination of such features.

**Monthly average** (Section 425.02) means the arithmetic average of eight (8) individual data points from effluent sampling and analysis during any calendar month.

**Monthly average** (Section 503.11) is the arithmetic mean of all measurements taken during the month.

**Montreal Anniversary amendments** (Section 82.3) means the Montreal Protocol, as amended at the Ninth Meeting of the Parties to the Montreal Protocol in Montreal in 1997.

**Montreal Protocol (1987)** (Section 82.3) means the Montreal Protocol, as originally adopted by the Parties in 1987.

**Montreal Protocol** (Section 82.3) means the Montreal Protocol on Substances that Deplete the Ozone Layer, a protocol to the Vienna Convention for the Protection of the Ozone Layer, including adjustments adopted by the Parties thereto and amendments that have entered into force.

**Most stringent federally enforceable emissions limitation** (Section 72.2) means the most stringent emissions limitation for a given pollutant applicable to the unit, which has been approved by the Administrator under the Act, whether in a State implementation plan approved pursuant to title I of the Act, a new source performance standard, or otherwise. To determine the most stringent emissions limitation for sulfur dioxide, each limitation shall be converted to lbs/mmBtu, using the appropriate conversion factors in appendix B of this part; provided that for determining the most stringent emissions limitation for sulfur dioxide for 1985, each limitation shall also be annualized, using the appropriate annualization factors in appendix A of this part.

**Most stringent State or Federal NOX emissions limitation** (Section 96.2) means, with regard to a NOX Budget opt-in source, the lowest NOX emissions limitation (in terms of lb/mmBtu) that is applicable to the unit under State or Federal law, regardless of the averaging period to which the emissions limitation applies.

**Most stringent State or Federal NOX emissions limitation** (Section 97.2) means, with regard to a NOX Budget opt-in unit, the lowest NOX emissions limitation (in lb/mmBtu) that is applicable to the unit under State or Federal law, regardless of the averaging period to which the emissions limitation applies.

**Motor carrier** (Section 202.10) means a common carrier by motor vehicle, a contract carrier by motor vehicle, or a private carrier of property by motor vehicle as those terms are defined by paragraphs (14), (15), and (17) of section 203(a) of the Interstate Commerce Act [49 U.S.C. 303(a)].

**Motor Controller** (Section 600.022-85) means an electronic or electro-mechanical device to convert energy stored in an energy storage device into a form suitable to power the traction motor.

**Motor fuel** (Section 280.12) means petroleum or a petroleum-based substance that is motor gasoline, aviation gasoline, No. 1 or No. 2 diesel fuel, or any grade of gasohol, and is typically used in the operation of a motor engine.

**Motor vehicle** (Section 202.10) means any vehicle, machine, tractor, trailer, or semi trailer propelled or drawn by mechanical power and used upon the highways in the transportation of passengers or property, or any combination thereof, but does not include any vehicle, locomotive, or car operated exclusively on a rail or rails.

**Motor vehicle diesel fuel** (Section 80.2) means any diesel fuel, or any distillate product, that is used, intended for use, or made available for use, (z) Aromatic content is the aromatic hydrocarbon content in volume percent as determined by ASTM standard test method D 1319-88, entitled "Standard Test Method for Hydrocarbon Types in Liquid Petroleum Products by Fluorescent Indicator Adsorption". ASTM test method D 1319-88 is incorporated by reference. This incorporation by reference was approved by the Director of the Federal Register in accordance with 5 U.S.C. 552(a) and 1 CFR part 51. A copy may be obtained from the American Society for Testing and Materials, 1916 Race Street, Philadelphia, PA 19103. A copy may be inspected at the Air Docket Section (A-130), room M-1500, U.S. Environmental Protection Agency, Docket No. A-86-03, 401 M Street SW., Washington, DC 20460 or at the Office of the Federal Register, 800 North Capitol Street, NW., suite 700, Washington, DC.

**Motor vehicle diesel fuel additive** (Section 80.2) means any substance not composed solely of carbon and/or hydrogen, or of diesel blendstocks, that is added, intended for adding, used, or offered for use in motor vehicle diesel fuel subsequent to the production of diesel fuel by processing crude oil from refinery processing units, or in diesel motor vehicle fuel systems.

**Motor vehicle emissions budget** (Section 93.101) is that portion of the total allowable emissions defined in the submitted or approved control strategy implementation plan revision or maintenance plan for a certain date for the purpose of meeting reasonable further progress milestones or demonstrating attainment or maintenance of the NAAQS, for any criteria pollutant or its precursors, allocated to highway and transit vehicle use and emissions.

**Motor vehicles held for lease or rental to the general public** (Section 88.302-94) means a vehicle that is owned or controlled primarily for the purpose of short-term rental or extended-term leasing (with or without maintenance), without a driver, pursuant to a contract.

**Motorcycle** (Section 205.151) means any motor vehicle, other than a tractor, that: (i) Has two or three wheels; (ii) Has a curb mass less than or equal to 680 kg (1499 lb); and (iii) Is capable, with an 80 kg (176 lb) driver, of achieving a maximum speed of at least 24 km/h (15 mph) over a level paved surface.

**Motorcycle** (Section 86-402-98) means any motor vehicle with a headlight, taillight, and stoplight and having: Two wheels, or Three wheels and a curb mass less than or equal to 793 kilograms (1749 pounds).

**Motorcycle noise level** (Section 205.151) means the A-weighted noise level of a motorcycle as measured by the acceleration test procedure.

**Movement** (Section 260.10) means that hazardous waste transported to a facility in an individual vehicle.

**Mpc** (Section 427.71) shall mean 1000 pieces of floor tile.

**MSBu** (Section 406.11) shall mean 1000 standard bushels.

**MSBu** (Section 406.41) shall mean 1,000 standard bushels.

**MSDS** (Section 721.3) means material safety data sheet, the written listing of data for the chemical substance as required under Sec. 721.72©.

**MTEC** (Section 62.1201) means maximum theoretical emissions concentration of metals or HCl/Cl, expressed as [mu]g/dscm, and is calculated by dividing the feedrate by the gas flowrate.

**MTEC** (Section 62.1201) means maximum theoretical emissions concentration.

**Muffler** (Section 202.10) means a device for abating the sound of escaping gases of an internal combustion engine.

**Multi-colored topcoat** (Section 59.101) means a topcoat that exhibits more than one color, is packaged in a single container, and camouflages surface defects on areas of heavy use, such as cargo beds and other surfaces of trucks and other utility vehicles.

**Multi-family dwelling** (Section 745.223) means a structure that contains more than one separate residential dwelling unit, which is used or occupied, or intended to be used or occupied, in whole or in part, as the home or residence of one or more persons.

**Multi-family housing** (Section 745.83) means a housing property consisting of more than four dwelling units.

**Multi-fuel** (Section 86.094-2) means capable of operating on two or more different fuel types, either separately or simultaneously.

**Multi-fuel** (Section 86.1803-01) means capable of operating on two or more different fuel types, either separately or simultaneously.

**Multi-header generator** (Section 72.2) means a generator served by ductwork from more than one unit.

**Multi-header unit** (Section 72.2) means a unit with ductwork serving more than one generator.

**Multiple Ferrous Melting Furnace Scrubber Configuration** (Section 464.31) A multiple ferrous melting furnace scrubber configuration is a configuration where two or more discrete wet scrubbing devices are employed in series in a single melting furnace exhaust gas stream. The ferrous melting furnace scrubber mass allowance shall be given to each discrete wet scrubbing device that has an associated wastewater discharge in a multiple ferrous melting furnace scrubber configuration. The mass allowance for each discrete wet scrubber shall be identical and based on the air flow of the exhaust gas stream that passes through the multiple scrubber configuration.

**Multiple stack configuration** (Section 72.2) refers to an exhaust configuration in which the flue gases from a particular unit discharge to the atmosphere through two or more stacks. The term also refers to a unit for which emissions are monitored in two or more ducts leading to the exhaust stack, in lieu of monitoring at the stack.

**Multiple stands** (Section 420.101) means those recirculation or direct application cold rolling mills which include more than one stand of work rolls.

**Multiple wastestream CWT facility** (Section 437.2) means a CWT facility which accepts waste in more than one CWT subcategory (metals, oils, or organics) and combines any portion of these different subcategory wastes at any point prior to the compliance discharge sampling location.

**Municipal separate storm sewer system** (Section 122.2) is defined at Sec. 122.26 (b)(4) and (b)(7).

**Municipal solid waste landfill (MSWLF)** (Section 272.2) unit means a discrete area of land or an excavation that receives household waste, and that is not a land application unit, surface impoundment, injection well, or waste pile, as those terms are defined in this section. A MSWLF unit also may receive other types of RCRA Subtitle D wastes, such as commercial solid waste, nonhazardous sludge, and industrial solid waste. Such a landfill may be publicly or privately owned. An MSWLF unit may be a new MSWLF unit, an existing MSWLF unit or a lateral expansion.

**Municipal solid waste landfill unit** (Section 258.2) means a discrete area of land or an excavation that receives household waste, and that is not a land application unit, surface impoundment, injection well, or waste pile, as those terms are defined under Sec. 257.2. A MSWLF unit also may receive other types of RCRA Subtitle D wastes, such as commercial solid waste, nonhazardous sludge, conditionally exempt small quantity generator waste and industrial solid waste. Such a landfill may be publicly or privately owned. A MSWLF unit may be a new MSWLF unit, an existing MSWLF unit or a lateral expansion.

**Municipal solid wastes** (Section 240.101) means normally, residential and commercial solid wastes generated within a community.

**Municipal solid wastes** (Section 761.13) means garbage, refuse, sludges, wastes, and other discarded materials resulting from residential and non-industrial operations and activities, such as household activities, office functions, and commercial housekeeping wastes.

**Municipality** (Section 122.2) means a city, town, borough, county, parish, district, association, or other public body created by or under State law and having jurisdiction over disposal of sewage, industrial wastes, or other wastes, or an Indian tribe or an authorized Indian tribal organization, or a designated and approved management agency under section 208 of CWA.

**Municipality** (Section 142.2) means a city, town, or other public body created by or pursuant to State law, or an Indian Tribe which does not meet the requirements of subpart H of this part.

**Municipality** (Section 501.2) means a city, town, borough, county, parish, district, association, or other public body (including an intermunicipal agency of two or more of the foregoing entities) created under State law (or an Indian tribe or an authorized Indian tribal organization), or a designated and approved management agency under section 208 of the Clean Water Act. This definition includes a special district created under State law such as a water district, sewer district, sanitary district, utility district, drainage district, or similar entity, or an integrated waste management facility as defined in section 201(e) of the CWA, as amended, that has as one of its principal responsibilities the treatment, transport, or disposal of sewage sludge.

**Mushrooms** (Section 407.71) shall mean the processing of mushrooms into the following product styles: Canned, frozen, dehydrated, all varieties, shapes and sizes.

**Mw** (Section 401.11) means megawatt(s).

**Mwh** (Section 401.11) means megawatt hour(s).

**Mwh** (Section 424.11) shall mean megawatt hour(s) of electrical energy consumed in the smelting process (furnace power consumption).

# N

**N/V** (Section 86.1803-01) means the ratio of engine speed in revolutions per minute (rpm) to vehicle speed in miles per hour in the top transmission gear. At the manufacturer's option, either the 1:1 transmission gear ratio or the lowest numerical gear ratio available in the transmission will be used to determine N/V.

**NAAQS and National Ambient Air Quality Standards,** (Section 57.103) unless otherwise specified, refer only to the National Primary and Secondary Ambient Air Quality Standards for sulfur dioxide.

**NAICS** (Section 68.3) means North American Industry Classification System.

**Nameplate capacity** (Section 72.2) means the maximum electrical generating output (expressed in MWe) that a generator can sustain over a specified period of time when not restricted by seasonal or other deratings, as listed in the NADB under the data field "NAMECAP" if the generator is listed in the NADB or as measured in accordance with the United States Department of Energy standards if the generator is not listed in the NADB.

**Nameplate capacity** (Section 96.2) means the maximum electrical generating output (in MWe) that a generator can sustain over a specified period of time when not restricted by seasonal or other deratings as measured in accordance with the United States Department of Energy standards.

**Nameplate capacity** (Section 97.2) means the maximum electrical generating output (in MWe) that a generator can sustain over a specified period of time when not restricted by seasonal or other deratings as measured in accordance with the United States Department of Energy standards.

**NAMS** (Section 58.1) means National Air Monitoring Station(s). Collectively the NAMS are a subset of the SLAMS ambient air quality monitoring network.

**Naphthalene (or priority pollutant No. 55)** (Section 420.02) means the value obtained by the standard method Number 610 specified in 44 FR 69464, 69571 (December 3, 1979).

**National Allowance Data Base or NADB** (Section 72.2) means the data base established by the Administrator under section 402(4)© of the Act.

**National ambient air quality standards (NAAQS)** (Section 93.101) are those standards established pursuant to section 109 of the CAA.

**National ambient air quality standards (NAAQS)** (Section 93.152) are those standards established pursuant to section 109 of the Act and include standards for carbon monoxide (CO), lead (Pb), nitrogen dioxide (NO2), ozone, particulate matter (PM-10), and sulfur dioxide (SO2).

**National Contingency Plan** (Section 310.3) means the National Oil and Hazardous Substances Pollution Contingency Plan (40 CFR part 300).

**National Contingency Plan or NCP** (Section 304.12) means the National Oil and Hazardous Substances Pollution Contingency Plan, developed under section 311©(2) of the Federal Water Pollution Control Act, 33 U.S.C. 1251, et seq., as amended, revised periodically pursuant to section 105 of CERCLA, 42 U.S.C. 9605, and published at 40 CFR part 300.

**National Contingency Plan or NCP** (Section 305.3) means the National Oil and Hazardous Substances Pollution Contingency Plan developed under section 311© of the Clean Water Act and revised pursuant to section 105 of CERCLA (40 CFR part 300).

**National Contingency Plan, or NCP** (Section 307.14) means the National Oil and Hazardous Substances Pollution Contingency Plan (40 CFR part 300) developed under section 311© of the

Federal Water Pollution Control Act and revised pursuant to section 105 of CERCLA.

**National Panel of Environmental Arbitrators or Panel** (Section 304.12) means a panel of environmental arbitrators selected and maintained by the Association to arbitrate cost recovery claims under this part.

**National Pollutant Discharge Elimination System (NPDES)** (Section 122.2) means the national program for issuing, modifying, revoking and reissuing, terminating, monitoring and enforcing permits, and imposing and enforcing pretreatment requirements, under sections 307, 402, 318, and 405 of CWA. The term includes an "approved program."

**National Pollutant Discharge Elimination System (NPDES)** (Section 136.2) means the national system for the issuance of permits under section 402 of the Act and includes any State or interstate program which has been approved by the Administrator, in whole or in part, pursuant to section 402 of the Act.

**National Pollutant Discharge Elimination System** (Section 270.2) means the national program for issuing, modifying, revoking and reissuing, terminating, monitoring and enforcing permits, and imposing and enforcing pretreatment requirements, under sections 307, 402, 318, and 405 of the CWA. The term includes an approved program.

**National Pollution Funds Center (NPFC)** (Section 300.5) means the entity established by the Secretary of Transportation whose function is the administration of the Oil Spill Liability Trust Fund (OSLTF). Among the NPFC's duties are: providing appropriate access to the OSLTF for federal agencies and states for removal actions and for federal trustees to initiate the assessment of natural resource damages; providing appropriate access to the OSLTF for claims; and coordinating cost recovery efforts.

**National Pretreatment Standard or Pretreatment Standard** (Section 117.1) means any regulation containing pollutant discharge limits promulgated by the EPA in accordance with section 307 (b) and (c) of the Act, which applies to industrial users of a publicly owned treatment works. It further means any State or local pretreatment requirement applicable to a discharge and which is incorporated into a permit issued to a publicly owned treatment works under section 402 of the Act.

**National Pretreatment Standard, Pretreatment Standard, or Standard** (Section 403.3) means any regulation containing pollutant discharge limits promulgated by the EPA in accordance with section 307 (b) and (c) of the Act, which applies to Industrial Users. This term includes prohibitive discharge limits established pursuant to Sec. 403.5.

**National primary drinking water regulation** (Section 142.2) means any primary drinking water regulation contained in part 141 of this chapter.

**National Priorities List (NPL)** (Section 300.5) means the list, compiled by EPA pursuant to CERCLA section 105, of uncontrolled hazardous substance releases in the United States that are priorities for long-term remedial evaluation and response.

**National program guidance** (Section 35.102) Guidance issued by EPA's National Program Managers for establishing and maintaining effective environmental programs. This guidance establishes national goals, objectives, and priorities as well as the core performance measures and other information to be used in monitoring progress. The guidance may also set out specific environmental strategies, criteria for evaluating programs, and other elements of program implementation.

**National Response Center** (Section 310.3) means the national communications center located in Washington, DC, that receives and relays notice of oil discharge or releases of hazardous substances to appropriate Federal officials.

**National response system (NRS)** (Section 300.5) is the mechanism for coordinating response actions by all levels of government in support of the OSC/RPM. The NRS is composed of the NRT, RRTs, OSC/RPM, Area Committees, and Special Teams and related support entities. The NRS is capable of expanding or contracting to accommodate the response effort required by the size or complexity of the discharge or release.

**National security exemption** (Section 204.2) means an exemption from the prohibitions of section 10(a) (1), (2), (3), and (5) of the Act, which may be granted under section 10(b)(1) of the Act for the purpose of national security.

**National security exemption** (Section 205.2) means an exemption from the prohibitions of section 10(a) (1), (2), (3), and (5) of the Act, which may be granted under section 10(b)(1) of the Act for the purpose of national security.

**National security exemption** (Section 211.102) means an exemption from the prohibitions of section 10(a) (3) and (5) of the Act, which may be granted under section 10(b) (1) of the Act in cases involving national security.

**National security exemption** (Section 85.1702) means an exemption which may be granted under section 203(b)(1) of the Act for the purpose of national security.

**National security exemption** (Section 90.902) means an exemption granted under Sec. 90.1004(b) for the purpose of national security.

**National security exemption** (Section 91.1002) means an exemption which may be granted under Sec. 91.1104 (b) for the purpose of national security.

**National standard** (Section 51.100) means either a primary or secondary standard.

**National Strike Force (NSF)** (Section 300.5) is a special team established by the USCG, including the three USCG Strike Teams, the Public Information Assist Team (PIAT), and the National Strike Force Coordination Center. The NSF is available to assist OSCs/RPMs in their preparedness and response duties.

**National Strike Force Coordination Center (NSFCC), authorized as the National Response Unit by CWA sections 311(a)(23) and (j)(2)** (Section 300.5) means the entity established by the Secretary of the department in which the USCG is operating at Elizabeth City, North Carolina with responsibilities that include administration of the USCG Strike Teams, maintenance of response equipment inventories and logistic networks, and conducting a national exercise program.

**Nations complying with, but not joining, the Protocol** (Section 82.3) means any nation listed in Appendix C, Annex 2, to this subpart.

**Natural gas** (Section 72.2) means a naturally occurring fluid mixture of hydrocarbons (e.g., methane, ethane, or propane) produced in geological formations beneath the Earth's surface that maintains a gaseous state at standard atmospheric temperature and pressure under ordinary conditions. Natural gas contains 1.0 grain or less of hydrogen sulfide per 100 standard cubic feet and the hydrogen sulfide constitutes more than 50% (by weight) of the total sulfur in the gas fuel. Additionally, natural gas must meet either be composed of at least 70% methane by volume or have a gross calorific value between 950 and 1100 Btu per standard cubic foot. Natural gas does not include the following gaseous fuels: landfill gas, digester gas, refinery gas, sour gas, blast furnace gas, coal-derived gas, producer gas, coke oven gas, or any gaseous fuel produced in a process which might result in highly variable sulfur content or heating value.

**Natural gas** (Section 72.2) means Natural gas contains 20.0 grains or less of total sulfur per 100 standard cubic feet.

**Natural gas** (Section 80.2) means a fuel whose primary constituent is methane.

**Natural gas** (Section 86.094-2) means a fuel whose primary constituent is methane. Non-Methane Hydrocarbon Equivalent means the sum of the carbon mass emissions of non-oxygenated non-methane hydrocarbons, methanol, formaldehyde, or other organic compounds that are separately measured, expressed as gasoline-fueled vehicle hydrocarbons. In the case of exhaust emissions, the hydrogen-to-carbon ratio of the equivalent hydrocarbon is 1.85:1. In the case of diurnal and hot soak emissions, the hydrogen-to-carbon ratios of the equivalent hydrocarbons are 2.33:1 and 2.2:1, respectively.

**Natural gas** (Section 86.1702-99) means either compressed natural gas or liquefied natural gas.

**Natural gas** (Section 86.1803-01) means a fuel whose primary constituent is methane.

**Natural gas** (Section 92.2) means the commercial product marketed as natural gas whose primary constituent is methane.

**Natural gas pipeline system** (Section 761.13) means natural gas gathering facilities, natural gas pipe, natural gas compressors, natural gas storage facilities, and natural gas pipeline appurtenances (including instrumentation and vessels directly in contact with transported natural gas such as valves, regulators, drips, filter separators, etc., but not including air compressors).

**Natural gas processing plant (gas plant)** (Section 68.3) means any processing site engaged in the extraction of natural gas liquids from field gas, fractionation of mixed natural gas liquids to natural gas products, or both, classified as North American Industrial Classification System (NAICS) code 211112 (previously Standard Industrial Classification (SIC) code 1321).

**Natural gas vehicle** (Section 86.1702-99) means any motor vehicle that is engineered and designed to be operated using either compressed natural gas or liquefied natural gas.

**Natural resources** (Section 300.5) means land, fish, wildlife, biota, air, water, ground water, drinking water supplies, and other such resources belonging to, managed by, held in trust by, appertaining to, or otherwise controlled by the United States (including the resources of the exclusive economic zone defined by the Magnuson Fishery Conservation and Management Act of 1976), any state or local government, any foreign government, any Indian tribe, or, if such resources are subject to a trust restriction on alienation, any member of an Indian tribe.

**Navigable waters** (Section 110.1) means the waters of the United States, including the territorial seas. The term includes: (a) All waters that are currently used, were used in the past, or may be susceptible to use in interstate or foreign commerce, including all waters that are subject to the ebb and flow of the tide; (b) Interstate waters, including interstate wetlands; (c) All other waters such as intrastate lakes, rivers, streams (including intermittent streams), mudflats, sandflats, and wetlands, the use, degradation, or destruction of which would affect or could affect interstate or foreign commerce including any such waters: (1) That are or could be used by interstate or foreign travelers for recreational or other purposes; (2) From which fish or shellfish are or could be taken and sold in interstate or foreign commerce; (3) That are used or could be used for industrial purposes by industries in interstate commerce; (d) All impoundments of waters otherwise defined as navigable waters under this section; (e) Tributaries of waters identified in paragraphs (a) through (d) of this section, including adjacent wetlands; and (f) Wetlands adjacent to waters identified in paragraphs (a) through (e) of this section: Provided, That waste treatment systems (other than cooling ponds meeting the criteria of this paragraph) are not waters of the United States; Navigable waters do not include prior converted cropland. Notwithstanding the determination of an area's status as prior converted cropland by any other federal agency, for the purposes of the Clean Water Act, the final authority regarding Clean Water Act jurisdiction remains with EPA.

**Navigable waters** (Section 116.3) is defined in section 502(7) of the Act to mean "waters of the United States, including the territorial seas," and includes, but is not limited to: (1) All waters which are presently used, or were used in the past, or may be susceptible to use as a means to transport interstate or foreign commerce, including all waters which are subject to the ebb and flow of the tide, and including adjacent wetlands; the term wetlands as used in this regulation shall include those areas that are inundated or saturated by surface or ground water at a frequency and duration sufficient to support, and that under normal circumstances do support, a prevelance of vegetation typically adapted for life in saturated soil conditions. Wetlands generally include swamps, marshes, bogs and similar areas; the term adjacent means bordering, contiguous or neighboring; (2) Tributaries of navigable waters of the United States, including adjacent wetlands; (3) Interstate waters, including wetlands; and (4) All other waters of the United States such as intrastate lakes, rivers, streams, mudflats, sandflats and wetlands, the use, degradation or destruction of which affect interstate commerce including, but not limited to: (i) Intrastate lakes, rivers, streams, and wetlands which are utilized by interstate travelers for recreational or other purposes; and (ii) Intrastate lakes, rivers, streams, and wetlands from which fish or shellfish are or could be taken and sold in interstate commerce; and (iii) Intrastate lakes, rivers, streams, and wetlands which are utilized for industrial purposes by industries in interstate commerce. Navigable waters do not include prior converted cropland. Notwithstanding the determination of an area's status as prior converted cropland by any other federal agency, for the purposes of the Clean Water Act, the final authority regarding Clean Water Act jurisdiction remains with EPA.

**Navigable waters** (Section 117.1) means "waters of the United States, including the territorial seas." This term includes: (1) All waters which are currently used, were used in the past, or may be susceptible to use in interstate or foreign commerce, including all waters which are subject to the ebb and flow of the tide; (2) Interstate waters, including interstate wetlands; (3) All other waters such as intrastate lakes, rivers, streams, (including intermittent streams), mudflats, sandflats, and wetlands, the use, degradation or destruction of which would affect or could affect interstate or foreign commerce including any such waters: (i) Which are or could be used by interstate or foreign travelers for recreational or other purposes; (ii) From which fish or shellfish are or could be taken and sold in interstate or foreign commerce; (iii) Which are used or could be used for industrial purposes by industries in interstate commerce; (4) All impoundments of waters otherwise defined as navigable waters under this paragraph; (5) Tributaries of waters identified in paragraphs (i) (1) through (4) of this section, including adjacent wetlands; and (6) Wetlands adjacent to waters identified in paragraphs (i) (1) through (5) of this section ("Wetlands" means those areas that are inundated or saturated by surface or ground water at a frequency and duration sufficient to support, and that under normal circumstances do support, a prevalence of vegetation typically adapted for life in saturated soil conditions. Wetlands generally included playa lakes, swamps, marshes, bogs, and similar areas such as sloughs, prairie potholes, wet meadows, prairie river overflows, mudflats, and natural ponds): Provided, That waste treatment systems (other than cooling ponds meeting the criteria of this paragraph) are not waters of the United States. Navigable waters do not include prior converted cropland. Notwithstanding the determination of an area's status as prior converted cropland by any other federal agency, for the purposes of the Clean Water Act, the final authority regarding Clean Water Act jurisdiction remains with EPA.

**Navigable waters** (Section 300.5) as defined by 40 CFR 110.1, means the waters of the United States, including the territorial seas. The term includes: (1) All waters that are currently used, were used in the past, or may be susceptible to use in interstate or foreign commerce, including all waters that are subject to the ebb and flow of the tide; (2) Interstate waters, including interstate wetlands; (3) All other waters such as intrastate lakes, rivers, streams (including intermittent streams), mudflats, sandflats, and wetlands, the use, degradation, or destruction of which would affect or could affect interstate or foreign commerce including any such waters; (i) That are or could be used by interstate or foreign travelers for recreational or other purposes; (ii) From which fish or shellfish are or could be taken and sold in interstate or foreign commerce; (iii) That are used or could be used for industrial purposes by industries in interstate commerce; (4) All impoundments of waters otherwise defined as navigable waters under this section; (5) Tributaries of waters identified in paragraphs (a) through (d) of this definition, including adjacent wetlands; and (6) Wetlands adjacent to waters identified in paragraphs (a) through (e) of this definition: Provided, that waste treatment systems (other than cooling ponds meeting the criteria of this paragraph) are not waters of the United States. (7) Waters of the United States do not include prior converted cropland. Notwithstanding the determination of an area's status as prior converted cropland by any other federal agency, for the purposes of the Clean Water Act, the final authority regarding Clean Water Act jurisdiction remains with EPA.

**Navigable waters** (Section 401.11) includes: All navigable waters of the United States; tributaries of navigable waters of the United States; interstate waters; intrastate lakes, rivers, and streams which are utilized by interstate travelers for recreational or other purposes; intrastate lakes, rivers, and streams from which fish or shellfish are taken and sold in interstate commerce; and intrastate lakes, rivers, and streams which are utilized for industrial purposes by industries in interstate commerce. Navigable waters do not include prior converted cropland. Notwithstanding the determination of an area's status as prior converted cropland by any other federal agency, for the purposes of the Clean Water Act, the final authority regarding Clean Water Act jurisdiction remains with EPA.

**Navigable waters or navigable waters of the United States** (Section 302.3) means waters of the United States, including the territorial seas;

**Near the first service connection** (Section 141.2) means at one of the 20 percent of all service connections in the entire system that are nearest the water supply treatment facility, as measured by water transport time within the distribution system.

**Nearest Onshore Area (NOA)** (Section 55.2) means, with respect to any existing or proposed OCS source, the onshore area that is geographically closest to that source.

**Neat oil** (Section 467.02) is a pure oil with no or few impurities added. In aluminum forming its use is mostly as a lubricant.

**Neat oil** (Section 471.02) is a pure oil with no or few impurities added. In nonferrous metals forming, its use is mostly as a lubricant.

**Neat soap** (Section 417.11) shall mean the solution of completely saponified and purified soap containing about 20-30 percent water which is ready for final formulation into a finished product.

**Neat soap** (Section 417.31) shall mean the solution of completely saponified and purified soap containing about 20-30 percent water which is ready for final formulation into a finished product.

**Necessary costs** (Section 307.14) means "necessary response costs" as required by section 111(a)(2) of CERCLA for Fund reimbursement of a preauthorized response action. Necessary response costs are costs determined to be: (1) Required (based upon the site-specific circumstances); (2) Reasonable (nature and amount do not exceed that estimated or which would be incurred by a prudent person); (3) Allocable (incurred specifically for the site at issue); and (4) Otherwise allowable (consistent with the limitations and exclusions under the appropriate Federal cost principles). See OMB Circular A-122 (non-profit organizations); OMB Circular A-87 (States and political subdivisions); and 48 CFR part 31, subparts 31.1 and 31.2 (profit-making organizations).

**Negligible residue** (Section 180.1) means any amount of a pesticide chemical remaining in or on a raw agricultural commodity or group of raw agricultural commodities that would result in a daily intake regarded as toxicologically insignificant on the basis of scientific judgment of adequate safety data. Ordinarily this will add to the diet an amount which will be less than ½,000$^{th}$ of the amount that has been demonstrated to have no effect from feeding studies on the most sensitive animal species tested. Such toxicity studies shall usually include at least 90-day feeding studies in two species of mammals.

**NEPA** (Section 93.101) means the National Environmental Policy Act of 1969, as amended (42 U.S.C. 4321 et seq.).

**NEPA process completion** (Section 93.101) for the purposes of this subpart, with respect to FHWA or FTA, means the point at which there is a specific action to make a determination that a project is categorically excluded, to make a Finding of No Significant Impact, or to issue a record of decision on a Final Environmental Impact Statement under NEPA.

**NERC region** (Section 72.2) means the North American Electric Reliability Council region or, if any, subregion.

**Net** (Section 409.21) shall mean the addition of pollutants.

**Net cane** (Section 409.61) shall mean that amount of "gross cane" less the weight of extraneous material.

**Net income neutrality** (Section 72.2) means, in the case of energy conservation measures undertaken by an investor-owned utility whose rates are regulated by a State utility regulatory authority, rates and charges established by the State utility regulatory authority that ensure that the net income earned by the utility on its State-jurisdictional equity investment will be no lower as a consequence of its expenditures on cost-effective qualified energy conservation measures and any associated lost sales than it would have been had the utility not made such expenditures, or that the State utility regulatory authority has implemented a ratemaking approach designed to meet this objective.

**Neutralization** (Section 420.91) means those acid pickling operations that do not include acid recovery or acid regeneration processes.

**New** (Section 85.1602) means a domestic or imported nonroad vehicle or nonroad engine the equitable or legal title to which has never been transferred to an ultimate purchaser. Where the equitable or legal title to an engine or vehicle is not transferred to an ultimate purchaser until after the engine or vehicle is placed into service, then the engine or vehicle will no longer be new after it is placed into service. A nonroad engine or vehicle is placed into service when it is used for its functional purposes. The term ultimate purchaser means, with respect to any new nonroad vehicle or new nonroad engine, the first person who in good faith purchases such new nonroad vehicle or new nonroad engine for purposes other than resale. This definition of new shall not apply to locomotives or engines used in locomotives.

**New** (Section 89.2) for purposes of this part, means a nonroad engine, nonroad vehicle, or nonroad equipment the equitable or legal title to which has never been transferred to an ultimate purchaser. Where the equitable or legal title to the engine, vehicle, or equipment is not transferred to an ultimate purchaser until after the engine, vehicle, or equipment is placed into service, then the engine, vehicle, or equipment will no longer be new after it is placed into service. A nonroad engine, vehicle, or equipment is placed into service when it is used for its functional purposes. With respect to imported nonroad engines, nonroad vehicles, or nonroad equipment, the term new means an engine, vehicle, or piece of equipment that is not covered by a certificate of conformity issued under this part at the time of importation, and that is manufactured after the effective date of a regulation issued under this part which is applicable to such engine, vehicle, or equipment (or which would be applicable to such engine, vehicle, or equipment had it been manufactured for importation into the United States).

**New** (Section 90.3) for the purposes of this part, means a nonroad engine or nonroad vehicle the equitable or legal title to which has never been transferred to an ultimate purchaser. Where the equitable or legal title to the engine or vehicle is not transferred to an ultimate purchaser until after the engine or vehicle is placed into service, then the engine or vehicle will no longer be new after it is placed into service. A nonroad engine or vehicle is placed into service when it is used for its functional purposes. With respect to imported nonroad engines or nonroad vehicles, the term "new" means an engine or vehicle that is not covered by a certificate of conformity issued under this part at the time of importation, and that is manufactured after the effective date of a regulation issued under this part which is applicable to such engine or vehicle (or which would be applicable to such engine or vehicle had it been manufactured for importation into the United States).

**New** (Section 91.3) for purposes of this part, means a nonroad engine, nonroad vehicle, or nonroad equipment the equitable or legal title to which has never been transferred to an ultimate purchaser. Where the equitable or legal title to the engine, vehicle or equipment is not transferred to an ultimate purchaser until after the engine, vehicle, or equipment is placed into service, then the engine, vehicle, or equipment will no longer be new after it is placed into service. A nonroad engine, vehicle, or equipment is placed into service when it is used for its functional purposes. With respect to imported nonroad engines, nonroad vehicles, or nonroad equipment, the term "new" means an engine, vehicle, or piece of equipment that is not covered by a certificate of conformity issued under this part at the time of importation, and that is manufactured after the effective date of a regulation issued under this part which is applicable to such engine, vehicle, or equipment, or which would be applicable to such engine, vehicle, or equipment had it been manufactured for importation into the United States.

**New affected source** (Section 63.2) means the collection of equipment, activities, or both within a single contiguous area and under common control that is included in a section 112© source category or subcategory that is subject to a section 112(d) or other relevant standard for new sources. This definition of "new affected source," and the criteria to be utilized in implementing it, shall apply to each section 112(d) standard for which the initial proposed rule is signed by the Administrator after June 30, 2002. Each relevant standard will define the term "new affected source," which will be the same as the "affected source" unless a different collection is warranted based on consideration of factors including: (1) Emission reduction impacts of controlling individual sources versus groups of sources; (2) Cost effectiveness of controlling individual equipment; (3) Flexibility to accommodate common control strategies; (4) Cost/benefits of emissions averaging; (5) Incentives for pollution prevention; (6) Feasibility and cost of controlling processes that share common equipment (e.g., product recovery devices); (7) Feasibility and cost of monitoring; and (8) Other relevant factors.

**New aircraft turbine engine** (Section 87.1) means an aircraft gas turbine engine which has never been in service.

**New chemical** (Section 166.3) means an active ingredient not contained in any currently registered pesticide.

**New chemical substance** (section 710.2) means any chemical substance which is not included in the inventory compiled and published under subsection 8(b) of the Act.

**New chemical substance** (Section 720.3) means any chemical substance which is not included on the Inventory.

**New Class I engine family** (Section 90.3) means any group of engines that employ a design that is different from engine families that the engine manufacturer has previously certified, and does not include any engine family certified on the basis of carryover data or any engine family that differs from another engine family solely as a result of a running change.

**New Class II Wells** (Section 147.2902) wells constructed or converted after the effective date of this program, or which are under construction on the effective date of this program.

**New covered fleet vehicle** (Section 88.302-94) means a vehicle that has not been previously controlled by the current purchaser, regardless of the model year, except as follows: Vehicles that were manufactured before the start of the fleet program for such vehicle's weight class, vehicles transferred due to the purchase of a company not previously controlled by the purchaser or due to a consolidation of business operations, vehicles transferred as part of an employee transfer, or vehicles transferred for seasonal requirements (i.e., for less than 120 days) are not considered new. States are permitted to discontinue the use of the fourth exception for fleet operators who abuse the discretion afforded them. This definition of new covered fleet vehicle is distinct from the definition of new vehicle as it applies to manufacturer certification, including the certification of vehicles to the clean fuel standards.

**New discharger** (Section 122.2) means any building, structure, facility, or installation: (a) From which there is or may be a "discharge of pollutants;" (b) That did not commence the "discharge of pollutants" at a particular "site" prior to August 13, 1979; ©

Which is not a "new source;" and (d) Which has never received a finally effective NDPES permit for discharges at that "site." This definition includes an "indirect discharger" which commences discharging into "waters of the United States" after August 13, 1979. It also includes any existing mobile point source (other than an offshore or coastal oil and gas exploratory drilling rig or a coastal oil and gas developmental drilling rig) such as a seafood processing rig, seafood processing vessel, or aggregate plant, that begins discharging at a "site" for which it does not have a permit; and any offshore or coastal mobile oil and gas exploratory drilling rig or coastal mobile oil and gas developmental drilling rig that commences the discharge of pollutants after August 13, 1979, at a "site" under EPA's permitting jurisdiction for which it is not covered by an individual or general permit and which is located in an area determined by the Regional Administrator in the issuance of a final permit to be an area or biological concern. In determining whether an area is an area of biological concern, the Regional Administrator shall consider the factors specified in 40 CFR 125.122(a) (1) through (10). An offshore or coastal mobile exploratory drilling rig or coastal mobile developmental drilling rig will be considered a "new discharger" only for the duration of its discharge in an area of biological concern.

**New engine used in a locomotive** (Section 85.1602) means new locomotive engine, as defined in 40 CFR 92.2.

**New Great Lakes discharger** (Section 132.2) is any building, structure, facility, or installation from which there is or may be a "discharge of pollutants" (as defined in 40 CFR 122.2) to the Great Lakes System, the construction of which commenced after March 23, 1997.

**New hazardous waste management facility or new facility** (Section 260.10) means a facility which began operation, or for which construction commenced after October 21, 1976. (See also "Existing hazardous waste management facility".)

**New HWM facility** (Section 270.2) means a Hazardous Waste Management facility which began operation or for which construction commenced after November 19, 1980.

**New independent power production facility or new IPP** (Section 72.2) means a unit that: (1) Commences commercial operation on or after November 15, 1990; (2) Is nonrecourse project-financed, as defined in 10 CFR part 715; (3) Sells 80% of electricity generated at wholesale; and (4) Does not sell electricity to any affiliate or, if it does, demonstrates it cannot obtain the required allowances from such an affiliate.

**New injection wells** (Section 144.3) means an "injection well" which began injection after a UIC program for the State applicable to the well is approved or prescribed.

**New locomotive** (Section 85.1602) The definition of new locomotive specified in 40 CFR 92.2 applies to this subpart.

**New locomotive or new locomotive engine** (Section 92.2) means: (1) (i) A locomotive or locomotive engine the equitable or legal title to which has never been transferred to an ultimate purchaser; or (ii) A locomotive or locomotive engine which has been remanufactured, but has not been placed back into service. (2) Where the equitable or legal title to a locomotive or locomotive engine is not transferred prior to its being placed into service, the locomotive or locomotive engine ceases to be new when it is placed into service. (3) With respect to imported locomotives or locomotive engines, the term "new locomotive" or "new locomotive engine" means a locomotive or locomotive engine that is not covered by a certificate of conformity under this part at the time of importation, and that was manufactured or remanufactured after the effective date of the emission standards in this part which is applicable to such locomotive or engine (or which would be applicable to such locomotive or engine had it been manufactured or remanufactured for importation into the United States). (4) Notwithstanding paragraphs (1) through (3) of this definition, locomotives and locomotive engines which were originally manufactured before January 1, 1973 and which have not been upgraded are not new. (5) Notwithstanding paragraphs (1) through (3) of this definition, locomotives and locomotive engines which are owned by a small railroad and which have never been remanufactured into a certified configuration are not new.

**New marine engine** (Section 94.2) means: (1) (i) A marine engine, the equitable or legal title to which has never been transferred to an ultimate purchaser; (ii) A marine engine installed on a vessel, the equitable or legal title to such vessel has never been transferred to an ultimate purchaser; or (iii) A marine engine that has not been placed into service on a vessel. (2) Where the equitable or legal title to an engine or vessel is not transferred to an ultimate purchaser prior to its being placed into service, the engine ceases to be new after it is placed into service. (3) With respect to imported engines, the term "new marine engine" means an engine that is not covered by a certificate of conformity under this part at the time of importation, and that was manufactured after the starting date of the emission standards in this part which are applicable to such engine (or which would be applicable to such engine had it been manufactured for importation into the United States).

**New microorganism** (Section 725.3) means a microorganism not included on the Inventory.

**New MSWLF unit** (Section 258.2) means any municipal solid waste landfill unit that has not received waste prior to October 9, 1993, or prior to October 9, 1997 if the MSWLF unit meets the conditions of Sec. 258.1(f)(1).

**New product** (Section 162.151) means a pesticide product which is not a federally registered product.

**New product** (Section 205.2) means (i) a product the equitable or legal title of which has never been transferred to an ultimate purchaser, or (ii) a product which is imported or offered for importation into the United States and which is manufactured after the effective date of a regulation under section 6 or 8 which would have been applicable to such product had it been manufactured in the United States.

**New source** (Section 122.2) means any building, structure, facility, or installation from which there is or may be a "discharge of pollutants," the construction of which commenced: (a) After promulgation of standards of performance under section 306 of CWA which are applicable to such source, or (b) After proposal of standards of performance in accordance with section 306 of CWA which are applicable to such source, but only if the standards are promulgated in accordance with section 306 within 120 days of their proposal.

**New source** (Section 129.2) means any source discharging a toxic pollutant, the construction of which is commenced after proposal of an effluent standard or prohibition applicable to such source if such effluent standard or prohibition is thereafter promulgated in accordance with section 307.

**New source** (Section 401.11) means any building, structure, facility or installation from which there is or may be the discharge of pollutants, the construction of which is commenced after the publication of proposed regulations prescribing a standard of performance under section 306 of the Act which will be applicable to such source if such standard is thereafter promulgated in accordance with section 306 of the Act.

**New Source** (Section 403.3) means any building, structure, facility or installation from which there is or may be a Discharge of pollutants, the construction of which commenced after the publication of proposed Pretreatment Standards under section 307© of the Act which will be applicable to such source if such Standards are thereafter promulgated in accordance with that section, provided that: (i) The building, structure, facility or installation is constructed at a site at which no other source is located; or (ii) The building, structure, facility or installation totally replaces the process or production equipment that causes the discharge of pollutants at an existing source; or (iii) The production or wastewater generating processes of the building, structure, facility or installation are substantially independent of an existing source at the same site. In determining whether these are substantially independent, factors such as the extent to which the new facility is integrated with the existing plant, and the extent to which the new facility is engaged in the same general type of activity as the existing source should be considered. (2) Construction on a site at which an existing source is located results in a modification rather than a new source if the construction does not create a new building, structure, facility or installation meeting the criteria of paragraphs (k)(1)(ii), or (k)(1)(iii) of this section but otherwise alters, replaces, or adds to existing process or production equipment. (3) Construction of a new source as defined under this paragraph has commenced if the owner or operator has: (i) Begun, or caused to begin as part of a continuous onsite construction program: (A) Any placement, assembly, or installation of facilities or equipment; or (B) Significant site preparation work including clearing, excavation, or removal of existing buildings, structures, or facilities which is necessary for the placement, assembly, or installation of new source facilities or equipment; or (ii) Entered into a binding contractual obligation for the purchase of facilities or equipment which are intended to be used in its operation within a reasonable time. Options to purchase or contracts which can be terminated or modified without substantial loss, and contracts for feasibility, engineering, and design studies do not constitute a contractual obligation under this paragraph.

**New source** (Section 430.0) (1) Notwithstanding the criteria codified at 40 CFR 122.29(b)(1), a source subject to subpart B or E of this part is a "new source" if it meets the definition of "new source" at 40 CFR 122.2 and: (i) It is constructed at a site at which no other source is located; or (ii) It totally replaces the process or production equipment that causes the discharge of pollutants at an existing source, including the total replacement of a fiber line that causes the discharge of pollutants at an existing source, except as provided in paragraph (j)(2) of this section; or (iii) Its processes are substantially independent of an existing source at the same site. In determining whether these processes are substantially independent, the Director shall consider such factors as the extent to which the new facility is integrated with the existing plant; and the extent to which the new facility is engaged in the same general type of activity as the existing source. (2) The following are examples of changes made by mills subject to subparts B or E of this part that alone do not cause an existing mill to become a "new source": (i) Upgrades of existing pulping operations; (ii) Upgrades or replacement of pulp screening and washing operations; (iii) Installation of extended cooking and/or oxygen delignification systems or other post-digester, pre-bleaching delignification systems; (iv) Bleach plant modifications including changes in methods or amounts of chemical applications, new chemical applications, installation of new bleaching towers to facilitate replacement of sodium or calcium hypochlorite, and installation of new pulp washing systems; or (v) Total replacement of process or production equipment that causes the discharge of pollutants at an existing source (including a replacement fiber line), but only if such replacement is performed for the purpose of achieving limitations that have been included in the discharger's NPDES permit pursuant to Sec. 430.24(b).

**New source** (Section 435.11) means any facility or activity of this subcategory that meets the definition of "new source" under 40 CFR 122.2 and meets the criteria for determination of new sources under 40 CFR 122.29(b) applied consistently with all of the following definitions:(1) Water area as used in "site" in 40 CFR 122.29 and 122.2 means the water area and water body floor beneath any exploratory, development, or production facility where such facility is conducting its exploratory, development or production activities. (ii) Significant site preparation work as used in 40 CFR 122.29 means the process of surveying, clearing or preparing an area of the water body floor for the purpose of constructing or placing a development or production facility on or over the site. (2) "New Source" does not include facilities covered by an existing NPDES permit immediately prior to the effective date of these guidelines pending EPA issuance of a new source NPDES permit.

**New source** (Section 435.41) means any facility or activity of this subcategory that meets the definition of "new source" under 40 CFR 122.2 and meets the criteria for determination of new sources under 40 CFR

122.29(b) applied consistently with all of the following definitions: (i) Water area as used in "site" in 40 CFR 122.29 and 122.2 means the water area and water body floor beneath any exploratory, development, or production facility where such facility is conducting its exploratory, development or production activities. (ii) Significant site preparation work as used in 40 CFR 122.29 means the process of surveying, clearing or preparing an area of the water body floor for the purpose of constructing or placing a development or production facility on or over the site. (2) "New Source" does not include facilities covered by an existing NPDES permit immediately prior to the effective date of these guidelines pending EPA issuance of a new source NPDES permit.

**New source** (Section 61.02) means any stationary source, the construction or modification of which is commenced after the publication in the Federal Register of proposed national emission standards for hazardous air pollutants which will be applicable to such source.

**New source** (Section 62.1201) means any affected source the construction or reconstruction of which is commenced after April 19, 1996.

**New source** (Section 63.2) means any affected source the construction or reconstruction of which is commenced after the Administrator first proposes a relevant emission standard under this part establishing an emission standard applicable to such source.

**New source coal mine** (Section 434.11) means a coal mine (excluding coal preparation plants and coal preparation plant associated areas) including an abandoned mine which is being re-mined. (i) The construction of which is commenced after May 4, 1984; or (ii) Which is determined by the EPA Regional Administrator to constitute a "major alteration". In making this determination, the Regional Administrator shall take into account whether one or more of the following events resulting in a new, altered or increased discharge of pollutants has occurred after May 4, 1984 in connection with the mine for which the NPDES permit is being considered: (A) Extraction of a coal seam not previously extracted by that mine; (B) Discharge into a drainage area not previously affected by wastewater discharge from the mine; (C) Extensive new surface disruption at the mining operation; (D) A construction of a new shaft, slope, or drift; and (E) Such other factors as the Regional Administrator deems relevant. (2) No provision in this part shall be deemed to affect the classification as a new source of a facility which was classified as a new source coal mine under previous EPA regulations, but would not be classified as a new source under this section, as modified. Nor shall any provision in this part be deemed to affect the standards applicable to such facilities, except as provided in Sec. 434.65 of this chapter.

**New source or new OCS source** (Section 55.2) shall have the meaning given in the applicable requirements of Secs. 55.13 and 55.14 of this part, except that for two years following the date of promulgation of this part, the definition given in Sec. 55.3 of this part shall apply for the purpose of determining the required date of compliance with this part.

**New source remining operation** (Section 434.70) means a remining operation at a coal mine where mining first commences after February 22, 2002 and subsequently becomes an abandoned mine.

**New tank** (Section 279.1) means a tank that will be used to store or process used oil and for which installation has commenced after the effective date of the authorized used oil program for the State in which the tank is located.

**New tank system** (Section 280.12) means a tank system that will be used to contain an accumulation of regulated substances and for which installation has commenced after December 22, 1988. (See also "Existing Tank System.")

**New tank system or new tank component** (Section 260.10) means a tank system or component that will be used for the storage or treatment of hazardous waste and for which installation has commenced after July 14, 1986; except, however, for purposes of Sec. 264.193(g) (2) and Sec. 265.193(g) (2), a new tank system is one for which construction commences after July 14, 1986. (See also "existing tank system.")

**New unit** (Section 72.2) means a unit that commences commercial operation on or after November 15, 1990, including any such unit that serves a generator with a nameplate capacity of 25 MWe or less or that is a simple combustion turbine.

**New use, when used with respect to a product containing a particular active ingredient** (Section 152.3) means: (1) Any proposed use pattern that would require the establishment of, the increase in, or the exemption from the requirement of, a tolerance or food additive regulation under section 408 or 409 of the Federal Food, Drug and Cosmetic Act; (2) Any aquatic, terrestrial, outdoor, or forestry use pattern, if no product containing the active ingredient is currently registered for that use pattern; or (3) Any additional use pattern that would result in a significant increase in the level of exposure, or a change in the route of exposure, to the active ingredient of man or other organisms.

**New uses of asbestos** (Section 763.163) means commercial uses of asbestos not identified in Sec. 763.165 the manufacture, importation or processing of which would be initiated for the first time after August 25, 1989.

**New vessel** (Section 140.1) refers to any vessel on which construction was initiated on or after January 30, 1975;

**New vessel** (Section 94.2) means: (1)(i) A vessel, the equitable or legal title to which has never been transferred to an ultimate purchaser; or (ii) A vessel that has been modified such that the value of the modifications exceeds 50 percent of the value of the modified vessel. (2) Where the equitable or legal title to a vessel is not transferred to an ultimate purchaser prior to its being placed into service, the vessel ceases to be new when it is placed into service.

**NFPA** (Section 68.3) means the National Fire Protection Association.

**NIC** (Section 62.1201) means notification of intent to comply.

**Nickel** (Section 415.361) shall mean the total nickel present in the process wastewater stream exiting the wastewater treatment system.

**Nickel** (Section 415.471) shall mean the total nickel present in the process wastewater stream exiting the wastewater treatment system.

**Nickel** (Section 415.651) shall mean the total nickel present in the process wastewater stream exiting the wastewater treatment system.

**Nickel** (Section 420.02) means total nickel and is determined by the method specified in 40 CFR 136.3.

**NIH Guidelines** (Section 725.3) means the National Institutes of Health (NIH) "Guidelines for Research Involving Recombinant DNA Molecules" (July 5, 1994).

**Ninetieth (90$^{th}$) percentile** (Section 72.2) means a value that would divide an ordered set of increasing values so that at least 90 percent are less than or equal to the value and at least 10 percent are greater than or equal to the value.

**Ninety-fifth (95$^{th}$) percentile** (Section 72.2) means a value that would divide an ordered set of increasing values so that at least 95 percent of the set are less than or equal to the value and at least 5 percent are greater than or equal to the value.

**NIOSH** (Section 721.3) means the National Institute for Occupational Safety and Health of the U.S. Department of Health and Human Services.

**NIST traceable reference material (NTRM)** (Section 72.2) means a calibration gas mixture tested by and certified by the National Institutes of Standards and Technologies (NIST) to have a certain specified concentration of gases. NTRMs may have different concentrations from those of standard reference materials.

**NIST/EPA-approved certified reference material or NIST/EPA-approved CRM** (Section 72.2) means a calibration gas mixture that has been approved by EPA and the National Institutes of Standards and Technologies (NIST) as having specific known chemical or physical property values certified by a technically valid procedure as evidenced by a certificate or other documentation issued by a certifying standard-setting body.

**Nitric acid plant** (Section 51.100) means any facility producing nitric acid 30 to 70 percent in strength by either the pressure or atmospheric pressure process.

**Nitrification capability** (Section 439.1) means the capability of a POTW treatment system to oxidize ammonia or ammonium salts initially to nitrites (via Nitrosomonas bacteria) and subsequently to nitrates (via Nitrobacter bacteria). Criteria for determining the nitrification capability of a POTW treatment system are: bioassays confirming the presence of nitrifying bacteria; and analyses of the nitrogen balance demonstrating a reduction in the concentration of ammonia or ammonium salts and an increase in the concentrations of nitrites and nitrates.

**Nitrogen oxides** (Section 60.2) means all oxides of nitrogen except nitrous oxide, as measured by test methods set forth in this part.

**NMFS** (Section 233.2) means the National Marine Fisheries Service.

**NMHC+NOX** (Section 90.3) means nonmethane hydrocarbons plus oxides of nitrogen.

**No detectable organic emissions** (Section 265.1081) means no escape of organics to the atmosphere as determined using the procedure specified in Sec. 265.1084 (d) of this subpart.

**No discharge of free oil** (Section 435.11) means that waste streams may not be discharged that contain free oil as evidenced by the monitoring method specified for that particular stream, e.g., deck drainage or miscellaneous discharges cannot be discharged when they would cause a film or sheen upon or discoloration of the surface of the receiving water; drilling fluids or cuttings may not be discharged when they fail the static sheen test defined in Appendix 1 of subpart A of this part.

**No discharge of free oil** (Section 435.41) means that waste streams may not be discharged that contain free oil as evidenced by the monitoring method specified for that particular stream, e.g., deck drainage or miscellaneous discharges cannot be discharged when they would cause a film or sheen upon or discoloration of the surface of the receiving water; drilling fluids or cuttings may not be discharged when they fail the static sheen test defined in Appendix 1 of subpart A of this part.

**No effective pesticides** (Section 166.3) are available under the Act that have labeled uses registered for control of the pest under the conditions of the emergency; and (2) No economically or environmentally feasible alternative practices which provide adequate control are available; and (3) The situation: (i) Involves the introduction or dissemination of a pest new to or not theretofore known to be widely prevalent or distributed within or throughout the United States and its territories; or (ii) Will present significant risks to human health; or (iii) Will present significant risks to threatened or endangered species, beneficial organisms, or the environment; or (iv) Will cause significant economic loss due to: (A) An outbreak or an expected outbreak of a pest; or (B) A change in plant growth or development caused by unusual environmental conditions where such change can be rectified by the use of a pesticide(s).

**No observed adverse effect level (NOAEL)** (Section 132.2) is the highest tested dose or concentration of a substance which resulted in no observed adverse effect in exposed test organisms where higher doses or concentrations resulted in an adverse effect.

**No observed effect concentration (NOEC)** (Section 132.2) is the highest concentration of toxicant to which organisms are exposed in a full life-cycle or partial life-cycle (short-term) test, that causes no observable adverse effects on the test organisms (i.e., the highest concentration of toxicant in which the values for the observed responses are not statistically significantly different from the controls).

**NO2** (Section 58.1) means nitrogen dioxide. NO means nitrogen oxide. NOX means oxides of nitrogen and is defined as the sum of the concentrations of NO2 and NO.

**No-discharge zone** (Section 1700.3) means an area of specified waters established pursuant to this regulation into which one or more specified discharges incidental to the normal operation of Armed Forces vessels, whether treated or untreated, are prohibited.

**Noise control system** (Section 205.151) means any vehicle part, component or system, the purpose of which includes control or the reduction of noise emitted from a vehicle, including all exhaust system components.

**Noise Control System** (Section 205.51) includes any vehicle part, component or system the primary purpose of which is to control or cause the reduction of noise emitted from a vehicle.

**Noise emission standard** (Section 205.151) means the noise levels in Sec. 205.152 or Sec. 205.166.

**Noise emission test** (Section 204.51) means a test conducted pursuant to the measurement methodology specified in Sec. 204.54.

**Noise emission test** (Section 205.151) means a test conducted pursuant to a measurement methodology specified in this subpart.

**Noise emission test** (Section 205.51) means a test conducted pursuant to the measurement methodology specified in this subpart.

**Noise Reduction Rating (NRR)** (Section 211.203) A single number noise reduction factor in decibels, determined by an empirically derived technique which takes into account performance variation of protectors in noise reducing effectiveness due to differing noise spectra, fit variability and the mean attenuation of test stimuli at the one-third octave band test frequencies.

**Nominal concentration** (Section 158.153) means the amount of an ingredient which is expected to be present in a typical sample of a pesticide product at the time the product is produced, expressed as a percentage by weight.

**Nominal fuel tank capacity** (Section 86.082-2) means the volume of the fuel tank(s), specified by the manufacturer to the nearest tenth of a U.S. gallon, which may be filled with fuel from the fuel tank filler inlet.

**Nominal fuel tank capacity** (Section 86.1803-01) means the volume of the fuel tank(s), specified by the manufacturer to the nearest tenth of a U.S. gallon, which may be filled with fuel from the fuel tank filler inlet.

**Nominee** (Section 105.2) means a candidate recommended by the State or Tribe or EPA for consideration for a CWA Recognition Award.

**Non-aqueous drilling fluid** (Section 435.11) means the continuous phase and suspending medium for solids is a water-immiscible fluid, such as oleaginous materials (e.g., mineral oil, enhanced mineral oil, paraffinic oil, C16-C18 internal olefins, and C8-C16 fatty acid/2-ethylhexyl esters).

**Nonattainment area** (Section 93.101) means any geographic region of the United States which has been designated as nonattainment under section 107 of the CAA for any pollutant for which a national ambient air quality standard exists.

**Nonattainment area** (Section 93.152) means an area designated as nonattainment under section 107 of the Act and described in 40 CFR part 81.

**Nonautomated monitoring and recording system** (Section 65.2) means manual reading of values measured by monitoring instruments and manual transcription of those values to create a record. Nonautomated systems do not include strip charts nor circular charts.

**Non-coding regulatory region** (Section 725.3) means a segment of introduced genetic material for which: (1) The regulatory region and any inserted flanking nucleotides do not code for protein, peptide, or functional ribonucleic acid molecules. (2) The regulatory region solely controls the activity of other regions that code for protein or peptide molecules or act as recognition sites for the initiation of nucleic acid or protein synthesis.

**Noncoding, nonexpressed nucleotide sequences** (Section 152.3) means the nucleotide sequences are not transcribed and are not involved in gene expression. Examples of noncoding, nonexpressed nucleotide sequences include, but are not limited to, linkers, adapters, homopolymers, and sequences of restriction enzyme recognition sites.

**Noncoding, nonexpressed nucleotide sequences** (Section 174.3) means the nucleotide sequences are not transcribed and are not involved in gene expression. Examples of noncoding, nonexpressed nucleotide sequences include, but are not limited to, linkers, adapters, homopolymers, and sequences of restriction enzyme recognition sites.

**Noncommercial education broadcasting entities** (Section 47.105) means any noncommercial educational broadcasting station (and/or its legal nonprofit affiliates) as defined and licensed by the Federal Communications Commission.

**Noncommercial purposes** (Section 280.12) with respect to motor fuel means not for resale.

**Non-commercial scientific institution** (Section 1601.3) refers to an institution that is not operated on a commercial basis as that term is used above in defining commercial-use requester, and which is operated solely for the purpose of conducting scientific research the results of which are not intended to promote any particular product or industry.

**Non-community water system** (Section 141.2) means a public water system that is not a community water system. A non-community water system is either a "transient non-community water system (TWS)" or a "non-transient non-community water system (NTNCWS)."

**Nonconforming locomotive or nonconforming locomotive engine** (Section 92.2) means a locomotive or locomotive engine which is not covered by a certificate of conformity prior to importation or being offered for importation (or for which such coverage has not been adequately demonstrated to EPA); or a locomotive or locomotive engine which was originally covered by a certificate of conformity, but which is not in a certified configuration, or otherwise does not comply with the conditions of that certificate of conformity. (Note: Domestic locomotives and locomotive engines which are not covered by a certificate of conformity prior to their introduction into U.S. commerce are considered to be noncomplying locomotives and locomotive engines.)

**Nonconforming marine engine** (Section 91.702) A marine SI engine which is not covered by a certificate of conformity prior to importation or being offered for importation (or for which such coverage has not been adequately demonstrated to EPA). Also, a marine SI engine which was originally covered by a certificate of conformity, but subsequently altered or modified such that it is no longer in a certified configuration.

**Nonconforming marine engine** (Section 94.2) means a marine engine which is not covered by a certificate of conformity prior to importation or being offered for importation (or for which such coverage has not been adequately demonstrated to EPA); or a marine engine which was originally covered by a certificate of conformity, but which is not in a certified configuration, or otherwise does not comply with the conditions of that certificate of conformity. Note: This definition does not include domestic marine engines which are not covered by a certificate of conformity prior to their introduction into U.S. commerce; such engines are considered to be "noncomplying marine engines."

**Nonconforming nonroad engine** (Section 89.602) A nonroad engine which is not covered by a certificate of conformity prior to final or conditional admission (or for which such coverage has not been adequately demonstrated to EPA) and which has not been finally admitted into the United States under the provisions of Sec. 89.605 or Sec. 89.609.

**Nonconforming vehicle or engine** (Section 85.1502) A motor vehicle or motor vehicle engine which is not covered by a certificate of conformity prior to final or conditional importation and which has not been finally admitted into the United States under the provisions of Sec. 85.1505, Sec. 85.1509 or the applicable provisions of Sec. 85.1512. Excluded from this definition are vehicles admitted under provisions of Sec. 85.1512 covering EPA approved manufacturer and U.S. Government Agency catalyst and O2 sensor control programs.

**Noncontact cooling water pollutants** (Section 401.11) means pollutants present in noncontact cooling waters.

**Non-contaminated storm water** (Section 437.2) means storm water which does not come in direct contact with CWT wastes, the waste handling and treatment areas, or other CWT wastewater that is defined in paragraph (d) of this section.

**Non-contaminated storm water** (Section 445.2) means storm water which does not come in direct contact with landfill wastes, the waste handling and treatment areas, or landfill wastewater that is defined in paragraph (f) of this section. Non-contaminated storm water includes storm water which flows off the cap, cover, intermediate cover, daily cover, and/or final cover of the landfill.

**Non-continuous discharger** (Section 430.0) (1) Except as provided in paragraph (k)(2) of this section, a non-continuous discharger is a mill which is prohibited by the NPDES authority from discharging pollutants during specific periods of time for reasons other than treatment plant upset control, such periods being at least 24 hours in duration. A mill shall not be deemed a non-continuous discharger unless its permit, in addition to setting forth the prohibition described above, requires compliance with the effluent limitations established for non-continuous dischargers and also requires compliance with maximum day and average of 30 consecutive days effluent limitations. Such maximum day and average of 30 consecutive days effluent limitations for non-continuous dischargers shall be established by the NPDES authority in the form of concentrations which reflect wastewater treatment levels that are representative of the application of the best practicable control technology currently available, the best conventional pollutant control technology, or new source performance standards in lieu of the maximum day and average of 30 consecutive days effluent limitations for conventional pollutants set forth in each subpart. (2) A mill is a non-continuous discharger for the purposes of determining applicable effluent limitations under subpart B or E of this part (other than conventional limits for existing sources) if, for reasons other than treatment plant upset control (e.g., protecting receiving water quality), the mill is prohibited by the NPDES authority from discharging pollutants during specific periods of time or if it is required to release its discharge on a variable flow or pollutant loading rate basis.

**Non-continuous discharger** (Section 464.02) is a plant which does not discharge pollutants during specific periods of time for reasons other than treatment plant upset, such periods being at least 24 hours in duration. A typical example of a non-continuous discharger is a plant where wastewaters are routinely stored for periods in excess of 24 hours to be treated on a batch basis. For non-continuous discharging direct discharging plants, NPDES permit authorities shall apply the mass-based annual average effluent limitations or standards and the concentration-based maximum day and maximum for monthly average effluent limitations or standards established in the regulations. POTWs may elect to establish concentration-based standards for non-continuous discharges to POTWs. They may do so by establishing concentration-based pretreatment standards equivalent to the mass-based standards provided in Secs. 464.15, 464.16, 464.25, 464.26, 464.35, 464.36, 464.45, and 464.46 of the regulations. Equivalent concentration standards may be established by following the procedures outlined in Sec. 464.03(b).

**Noncontract cooling water** (Section 401.11) means water used for cooling which does not come into direct contact with any raw material, intermediate product, waste product or finished product.

**Non-conventional pollutants** (Section 439.1) means parameters that are neither conventional pollutants (40 CFR 401.16), nor "toxic" pollutants (40 CFR 401.15).

**Non-Core Stable Standards** (Section 86.1702-99) means the standards and other requirements listed in Sec. 86.1707(d)(9)(i)(G) through (L).

**Noncorrugating medium furnish subdivision mills** (Section 430.101) are mills where recycled corrugating medium is not used in the production of paperboard.

**Non-detect (ND)** (Section 439.1) means a concentration value below the minimum level that can be reliably measured by the analytical method.

**Non-emission-related maintenance** (Section 86.088-2) means that maintenance which does not substantially affect emissions and which does not have a lasting effect on the emissions deterioration of the vehicle or engine during normal in-use operation once the maintenance is performed.

**Non-emission-related maintenance** (Section 86.1803-01) means that maintenance which does not substantially affect emissions and which does not have a lasting effect on the emissions deterioration of the vehicle or engine during normal in-use operation once the maintenance is performed.

**Non-enclosed process** (Section 721.3) means any equipment system (such as an open-top reactor, storage tank, or mixing vessel) in which a chemical substance is manufactured, processed, or otherwise used where significant direct contact of the bulk chemical substance and the workplace air may occur.

**Nonferrous metal** (Section 471.02) is any pure metal other than iron or any metal alloy for which a metal other than iron is its major constituent in percent by weight.

**Non-food uses** (Section 180.2003) are those uses that are not likely to yield residues in food or feed crops, meat, milk, poultry or eggs.

**Nonfriable** (Section 763.83) means material in a school building which when dry may not be crumbled, pulverized, or reduced to powder by hand pressure.

**Nonimpervious solid surfaces** (Section 761.123) means solid surfaces which are porous and are more likely to absorb spilled PCBs prior to completion of the cleanup requirements prescribed in this policy. Nonimpervious solid surfaces include, but are not limited to, wood, concrete, asphalt, and plasterboard.

**Non-industrial use** (Section 721.3) means use other than at a facility where chemical substances or mixtures are manufactured, imported, or processed.

**Non-integrated refueling emission control system** (Section 86.098-2) means a system where fuel vapors from refueling are stored in a vapor storage unit assigned solely to the function of storing refueling vapors. Refueling emissions means evaporative emissions that emanate from a motor vehicle fuel tank(s) during a refueling operation. Refueling emissions canister(s) means any vapor storage unit(s) that is exposed to the vapors generated during refueling. Resting losses means evaporative emissions that may occur continuously, that are not diurnal emissions, hot soak emissions, refueling emissions, running losses, or spitback emissions.

**Non-integrated refueling emission control system** (Section 86.1803-01) means a system where fuel vapors from refueling are stored in a vapor storage unit assigned solely to the function of storing refueling vapors.

**Non-isolated intermediate** (Section 704.3) means any intermediate that is not intentionally removed from the equipment in which it is manufactured, including the reaction vessel in which it is manufactured, equipment which is ancillary to the reaction vessel, and any equipment through which the substance passes during a continuous flow process, but not including tanks or other vessels in which the substance is stored after its manufacture. Mechanical or gravity transfer through a closed system is not considered to be intentional removal, but storage or transfer to shipping containers "isolates" the substance by removing it from process equipment in which it is manufactured.

**Nonisolated intermediate** (section 710.2) means any intermediate that is not intentionally removed from the equipment in which it is manufactured, including the reaction vessel in which it is manufactured, equipment which is ancillary to the reaction vessel, and any equipment through which the substance passes during a continuous flow process, but not including tanks or other vessels in which the substance is stored after its manufacture. Note: Any chemical substances manufactured, imported or processed in quantities less than 1,000 pounds annually shall be presumed to be manufactured, imported or processed for research and development purposes. No person may report for the inventory any chemical substance in such quantities unless that person can certify, that the substance was not manufactured, imported, or processed solely in small quantities for research and development, as defined in this section.

**Non-isolated intermediate** (Section 712.3) means any intermediate that is not intentionally removed from the equipment in which it is manufactured, including the reaction vessel in which it is manufactured, equipment which is ancillary to the reaction vessel, and any equipment through which the substance passes during a continuous flow process, but not including tanks or other vessels in which the substance is stored after its manufacture. (See also paragraph (f) of this section.)

**Nonisolated intermediate** (Section 720.3) means any intermediate that is not intentionally removed from the equipment in which it is manufactured, including the reaction vessel in which it is manufactured, equipment which is ancillary to the reaction vessel, and any equipment through which the chemical substance passes during a continuous flow process, but not including tanks or other vessels in which the substance is stored after its manufacture.

**Non-liquid PCBs** (Section 761.13) means materials containing PCBs that by visual inspection do not flow at room temperature (25 [deg]C or 77 [deg]F) or from which no liquid passes when a 100 g or 100 ml representative sample is placed in a mesh number 60 [plusmn] 5 percent paint filter and allowed to drain at room temperature for 5 minutes.

**Non-locomotive-specific engine** (Section 92.2) means an engine that is sold for and used in non-locomotive applications more than for locomotive applications.

**Non-Methane Hydrocarbon Equivalent** (Section 86.1803-01) means the sum of the carbon mass emissions of non-oxygenated non-methane hydrocarbons, methanol, formaldehyde, or other organic compounds that are separately measured, expressed as gasoline-fueled vehicle hydrocarbons. In the case of exhaust emissions, the hydrogen-to-carbon ratio of the equivalent hydrocarbon is 1.85:1. In the case of diurnal and hot soak emissions, the hydrogen-to-carbon ratios of the equivalent hydrocarbons are 2.33:1 and 2.2:1, respectively.

**Non-methane Hydrocarbon Equivalent** (Section 88.102-94) means the sum of the carbon mass emissions of non-oxygenated non-methane hydrocarbons plus the carbon mass emissions of alcohols, aldehydes, or other organic compounds which are separately measured in accordance with the applicable test procedures of 40 CFR part 86, expressed as gasoline-fueled vehicle non-methane hydrocarbons. In the case of exhaust emissions, the hydrogen-to-carbon ratio of the equivalent hydrocarbon is 1.85:1. In the case of diurnal and hot soak emissions, the hydrogen-to-carbon ratios of the equivalent hydrocarbons are 2.33:1 and 2.2:1 respectively.

**Non-methane Organic Gas** (Section 88.102-94) is defined as in section 241(3) Clean Air Act as amended (42 U.S.C. 7581(3)).

**Non-methane organic gases (NMOG)** (Section 86.1702-99) means the sum of oxygenated and non-oxygenated hydrocarbons contained in a gas sample as measured in accordance with Chapter 5 of the California Regulatory Requirements Applicable to the National Low Emission Vehicle Program (October, 1996). These requirements are incorporated by reference (see Sec. 86.1).

**Non-methane organic gases (NMOG)** (Section 86.1803-01) means the sum of oxygenated and non-oxygenated hydrocarbons contained in a gas sample as measured in accordance with the California Non-Methane Organic Gas Test Procedures.

These requirements are incorporated by reference (see Sec. 86.1)

**Non-oxygenated hydrocarbon** (Section 86.090-2) means organic emissions measured by a flame ionization detector, excluding methanol.

**Non-oxygenated hydrocarbon** (Section 86.1803-01) means organic emissions measured by a flame ionization detector, excluding methanol.

**Nonpassenger Automobile** (Section 600.022-85) means an automobile that is not a passenger automobile, as defined by the Secretary of Transportation at 49 CFR 523.5.

**Non-PCB Transformer** (Section 761.13) means any transformer that contains less than 50 ppm PCB; except that any transformer that has been converted from a PCB Transformer or a PCB-Contaminated Transformer cannot be classified as a non-PCB Transformer until reclassification has occurred, in accordance with the requirements of Sec. 761.30(a)(2)(v).

**Nonperishable raw agricultural commodity** (Section 180.1) means any raw agricultural commodity not subject to rapid decay or deterioration that would render it unfit for consumption. Examples are cocoa beans, coffee beans, field-dried beans, field-dried peas, grains, and nuts. Not included are eggs, milk, meat, poultry, fresh fruits, and vegetables such as onions, parsnips, potatoes, and carrots.

**Non-petroleum** (Section 112.2) oil means oil of any kind that is not petroleum-based, including but not limited to: Fats, oils, and greases of animal, fish, or marine mammal origin; and vegetable oils, including oils from seeds, nuts, fruits, and kernels.

**Non-plug-in combustion controls** (Section 76.2) means the replacement, in a cell burner boiler, of the portions of the waterwalls containing the cell burners by new portions of the waterwalls containing low NOX burners or low NOX burners with overfire air.

**Non-porous surface** (Section 761.13) means a smooth, unpainted solid surface that limits penetration of liquid containing PCBs beyond the immediate surface. Examples are: smooth uncorroded metal; natural gas pipe with a thin porous coating originally applied to inhibit corrosion; smooth glass; smooth glazed ceramics; impermeable polished building stone such as marble or granite; and high density plastics, such as polycarbonates and melamines, that do not absorb organic solvents.

**Nonprocess waste water** (Section 428.11) Water used only for tread cooling.

**Nonprofit** (Section 745.223) means an entity which has demonstrated to any branch of the Federal Government or to a State, municipal, tribal or territorial government, that no part of its net earnings inure to the benefit of any private shareholder or individual.

**Non-regeneration emission test** (Section 86.1702-99) means a complete emission test which does not include a regeneration.

**Nonrepairable** (Section 65.2) means that it is technically infeasible to repair a piece of equipment from which a leak has been detected without a process unit shutdown.

**Nonrestricted access areas** (Section 761.123) means any area other than restricted access, outdoor electrical substations, and other restricted access locations, as defined in this section. In addition to residential/ commercial areas, these areas include unrestricted access rural areas (areas of low density development and population where access is uncontrolled by either man-made barriers or naturally occurring barriers, such as rough terrain, mountains, or cliffs).

**Nonroad engine** (Section 85.1602) means: (1) Except as discussed in paragraph (2) of this definition, a nonroad engine is any internal combustion engine: (i) In or on a piece of equipment that is self-propelled or serves a dual purpose by both propelling itself and performing another function (such as garden tractors, off-highway mobile cranes and bulldozers); or (ii) In or on a piece of equipment that is intended to be propelled while performing its function (such as lawnmowers and string trimmers); or (iii) That, by itself or in or on a piece of equipment, is portable or transportable, meaning designed to be and capable of being carried or moved from one location to another. Indicia of transportability include, but are not limited to, wheels, skids, carrying handles, dolly, trailer, or platform. (2) An internal combustion engine is not a nonroad engine if: (i) The engine is used to propel a motor vehicle or a vehicle used solely for competition, or is subject to standards promulgated under section 202 of the Act; or (ii) The engine is regulated by a federal New Source Performance Standard promulgated under section 111 of the Act; or (iii) The engine otherwise included in paragraph (1)(iii) of this definition remains or will remain at a location for more than 12 consecutive months or a shorter period of time for an engine located at a seasonal source. A location is any single site at a building, structure, facility, or installation. Any engine (or engines) that replaces an engine at a location and that is intended to perform the same or similar function as the engine replaced will be included in calculating the consecutive time period. An engine located at a seasonal source is an engine that remains at a seasonal source during the full annual operating period of the seasonal source. A seasonal source is a stationary source that remains in a single location on a permanent basis (i.e., at least two years) and that operates at that single location approximately three (or more) each year. This paragraph does not apply to an engine after the engine is removed from the location.

**Nonroad engine** (Section 89.2) means: (1) Except as discussed in paragraph (2) of this definition, a nonroad engine is any internal combustion engine: (i) In or on a piece of equipment that is self-propelled or serves a dual purpose by both propelling itself and performing another function (such as garden tractors, off-highway mobile cranes and bulldozers); or (ii) In or on a piece of equipment that is intended to be propelled while performing its function (such as lawnmowers and string trimmers); or (iii) That, by itself or in or on a piece of equipment, is portable or transportable, meaning designed to be and capable of being carried or moved from one location to another. Indicia of transportability include, but are not limited to, wheels, skids, carrying handles, dolly, trailer, or platform. (2) An internal combustion engine is not a nonroad engine if: (i) the engine is used to propel a motor vehicle or a vehicle used solely for competition, or is subject to standards promulgated under section 202 of the Act; or (ii) the engine is regulated by a federal New Source Performance Standard promulgated under section 111 of the Act; or (iii) the engine otherwise included in paragraph (1)(iii) of this definition remains or will remain at a location for more than 12 consecutive months or a shorter period of time for an engine located at a seasonal source. A location is any single site at a building, structure, facility, or installation. Any engine (or engines) that replaces an engine at a location and that is intended to perform the same or similar function as the engine replaced will be included in calculating the consecutive time period. An engine located at a seasonal source is an engine that remains at a seasonal source during the full annual operating period of the seasonal source. A seasonal source is a stationary source that remains in a single location on a permanent basis (i.e., at least two years) and that operates at that single location approximately three months (or more) each year. This paragraph does not apply to an engine after the engine is removed from the location.

**Nonroad engine** (Section 91.3) has the meaning as defined in 40 CFR 89.2.

**Nonroad engine means (Section 90.3)** (1) Except as discussed in paragraph (2) of this definition, any internal combustion engine: (i) In or on a piece of equipment that is self-propelled or serves a dual purpose by both propelling itself and performing another function (such as garden tractors, off-highway mobile cranes, and bulldozers); or (ii) In or on a piece of equipment that is intended to be propelled while performing its function (such as lawnmowers and string trimmers); or (iii) That, by itself or in or on a piece of equipment, is portable or transportable, meaning designed to be and capable of being carried or moved from one location to another. Indicia of transportability include, but are not limited to, wheels, skids, carrying handles, dolly, trailer, or platform. (2) An internal combustion engine is not a nonroad engine if: (i) The engine is used to propel a motor vehicle or a vehicle used solely for competition, or is subject to standards promulgated under section 202 of the Act; or (ii)The engine is regulated by a federal New Source Performance Standard promulgated under section 111 of the Act; or (iii) The engine otherwise included in paragraph (1)(iii) of this definition remains or will remain at a location for more than 12 consecutive months or a shorter period of time for an engine located at a seasonal source. A location is any site at a building, structure, facility, or installation. Any engine (or engines) that replaces an engine at a location and that is intended to perform the same or similar function as the engine replaced will be included in calculating the consecutive time period. An engine located at a seasonal source is an engine that remains at a seasonal source during the full annual operating period of the seasonal source. A seasonal source is a stationary source that remains in a single location on a permanent basis (i.e., at least two years) and that operates at that single location approximately three months (or more) each year. This paragraph does not apply to an engine after the engine is removed from the location.

**Nonroad equipment** (Section 89.2) means equipment that is powered by nonroad engines.

**Nonroad equipment** (Section 91.3) has the meaning as defined in 40 CFR 89.2.

**Nonroad vehicle** (Section 89.2) means a vehicle that is powered by a nonroad engine as defined in this section and that is not a motor vehicle or a vehicle used solely for competition.

**Nonroad vehicle** (Section 90.3) means a vehicle that is powered by a nonroad engine as defined in this section and that is not a motor vehicle or a vehicle used solely for competition. Nonroad vehicle also includes equipment that is powered by nonroad engines.

**Nonroad vehicle** (Section 91.3) has the meaning as defined in 40 CFR 89.2.

**Nonroad vehicle manufacturer** (Section 90.3) means any person engaged in the manufacturing or assembling of new nonroad vehicles or importing such vehicles for resale, or who acts for and is under the control of any such person in connection with the distribution of such vehicles. A nonroad vehicle manufacturer does not include any dealer with respect to new nonroad vehicles received by such person in commerce.

**Nonroad vehicle or nonroad equipment manufacturer** (Section 89.2) means any person engaged in the manufacturing or assembling of new nonroad vehicles or equipment r importing such vehicles or equipment for resale, or who acts for and is under the control of any such person in connection with the distribution of such vehicles or equipment. A nonroad vehicle or equipment manufacturer does not include any dealer with respect to new nonroad vehicles or equipment received by such person in commerce. A nonroad vehicle or equipment manufacturer does not include any person engaged in the manufacturing or assembling of new nonroad vehicles or equipment who does not install an engine as part of that manufacturing or assembling process. All nonroad vehicle or equipment manufacturing entities that are under the control of the same person are considered to be a single nonroad vehicle or nonroad equipment manufacturer.

**Non-target organism** (Section 159.153) means any organism for which pesticidal control was either not intended or not legally permitted by application of a pesticide.

**Non-target organism** (Section 171.2) means a plant or animal other than the one against which the pesticide is applied.

**Non-title V permit** (Section 96.2) means a federally enforceable permit administered by the permitting authority pursuant to the CAA and regulatory authority under the CAA, other than title V of the CAA and part 70 or 71 of this chapter.

**Non-title V permit** (Section 97.2) means a federally enforceable permit administered by the permitting authority pursuant to the Clean Air Act and regulatory authority under the Clean Air Act, other than title V of the Clean Air Act and part 70 or 71 of this chapter.

**Non-transient non-community water system or NTNCWS** (Section 141.2) means a public water system that is not a community water system and that regularly serves at least 25 of the same persons over 6 months per year.

**Nonwastewaters** (Section 268.2) are wastes that do not meet the criteria for wastewaters in paragraph (f) of this section.

**Normal ambient value** (Section 228.2) means that concentration of a chemical species reasonably anticipated to be present in the water column, sediments, or biota in the absence of disposal activities at the disposal site in question.

**Normal idle** (Section 92.2) means relating to the idle throttle-notch position for locomotives that have one throttle-notch position, or the highest the idle throttle-notch position for locomotives that have two throttle-notch positions.

**Normal operation of a spray tower** (Section 417.151) shall mean operation utilizing formulations that present limitted air quality problems from stack gases and associated need for extensive wet scrubbing, and without more than 6 turnarounds in a 30 consecutive day period, thus permitting essentially complete recycle of waste water.

**Northeast Trading Region (NTR)** (Section 86.1702-99) means the region comprised of the states that meet the conditions specified under Sec. 86.1705(d).

**Not-for-profit organization** (Section 47.105) means an organization, association, or institution described in section 501©(3) of the Internal Revenue Code of 1986, which is exempt from taxation pursuant to the provisions of section 501(a) of such Code.

**Notifier** (Section 262.81) means the person under the jurisdiction of the exporting country who has, or will have at the time the planned transfrontier movement commences, possession or other forms of legal control of the wastes and who proposes their transfrontier movement for the ultimate purpose of submitting them to recovery operations. When the United States (U.S.) is the exporting country, notifier is interpreted to mean a person domiciled in the U.S.

**NOX allowance** (Section 96.2) means an authorization by the permitting authority or the Administrator under the NOX Budget Trading Program to emit up to one ton of nitrogen oxides during the control period of the specified year or of any year thereafter.

**NOX allowance** (Section 97.2) means a limited authorization by the Administrator under the NOX Budget Trading Program to emit up to one ton of nitrogen oxides during the control period of the specified year or of any year thereafter, except as provided under Sec. 97.54(f). No provision of the NOX Budget Trading Program, the NOX Budget permit application, the NOX Budget permit, or an exemption under Sec. 97.4(b) or Sec. 97.5 and no provision of law shall be construed to limit the authority of the United States to terminate or limit such authorization, which does not constitute a property right. For purposes of all sections of this part except Sec. 97.41, Sec. 97.42, Sec. 97.43, or Sec. 97.88, "NOX allowance" also includes an authorization to emit up to one ton of nitrogen oxides during the control period of the specified year or of any year thereafter by the permitting authority or the Administrator in accordance with a State NOX Budget Trading Program established, and approved and administered by the Administrator, pursuant to Sec. 51.121 of this chapter.

**NOX allowance deduction or deduct NOX allowances** (Section 96.2) means the permanent withdrawal of NOX allowances by the Administrator from a NOX Allowance Tracking System compliance account or overdraft account to account for the number of tons of NOX emissions from a NOX Budget unit for a control period, determined in accordance with subpart H of this part, or for any other allowance surrender obligation under this part.

**NOX allowance deduction or deduct NOX allowances** (Section 97.2) means the permanent withdrawal of NOX allowances by the Administrator from a NOX Allowance Tracking System compliance account or overdraft account to account for the number of tons of NOX emissions from a NOX Budget unit for a control period, determined in accordance with subparts H and F of this part, or for any other NOX allowance withdrawal requirement under this part.

**NOX Allowance Tracking System** (Section 96.2) means the system by which the Administrator records allocations, deductions, and transfers of NOX allowances under the NOX Budget Trading Program.

**NOX Allowance Tracking System** (Section 97.2) means the system by which the Administrator records allocations, deductions, and transfers of NOX allowances under the NOX Budget Trading Program.

**NOX Allowance Tracking System account** (Section 96.2) means an account in the NOX Allowance Tracking System established by the Administrator for purposes of recording the allocation, holding, transferring, or deducting of NOX allowances.

**NOX Allowance Tracking System account** (Section 97.2) means an account in the NOX Allowance Tracking System established by the Administrator for purposes of recording the allocation, holding, transferring, or deducting of NOX allowances.

**NOX allowance transfer deadline** (Section 96.2) means midnight of November 30 or, if November 30 is not a business day, midnight of the first business day thereafter and is the deadline by which NOX allowances may be submitted for recordation in a NOX Budget unit's compliance account, or the overdraft account of the source where the unit is located, in order to meet the unit's NOX Budget emissions limitation for the control period immediately preceding such deadline.

**NOX allowance transfer deadline** (Section 97.2) means midnight of November 30 or, if November 30 is not a business day, midnight of the first business day thereafter and is the deadline by which NOX allowances must be submitted for recordation in a NOX Budget unit's compliance account, or the overdraft account of the source where the unit is located, in order to meet the unit's NOX Budget emissions limitation for the control period immediately preceding such deadline.

**NOX allowances held or hold NOX allowances** (Section 96.2) means the NOX allowances recorded by the Administrator, or submitted to the Administrator for recordation, in accordance with subparts F and G of this part, in a NOX Allowance Tracking System account.

**NOX allowances held or hold NOX allowances** (Section 97.2) means the NOX allowances recorded by the Administrator, or submitted to the Administrator for recordation, in accordance with subparts F and G of this part, in a NOX Allowance Tracking System account.

**NOX authorized account representative** (Section 96.2) means, for a NOX Budget source or NOX Budget unit at the source, the natural person who is authorized by the owners and operators of the source and all NOX Budget units at the source, in accordance with subpart B of this part, to represent and legally bind each owner and operator in matters pertaining to the NOX Budget Trading Program or, for a general account, the natural person who is authorized, in accordance with subpart F of this part, to transfer or otherwise dispose of NOX allowances held in the general account.

**NOX authorized account representative** (Section 97.2) means, for a NOX Budget source or NOX Budget unit at the source, the natural person who is authorized by the owners and operators of the source and all NOX Budget units at the source, in accordance with subpart B of this part, to represent and legally bind each owner and operator in matters pertaining to the NOX Budget Trading Program or, for a general account, the natural person who is authorized, in accordance with subpart F of this part, to transfer or otherwise dispose of NOX allowances held in the general account.

**NOX Budget emissions limitation** (Section 96.2) means, for a NOX Budget unit, the tonnage equivalent of the NOX allowances available for compliance deduction for the unit and for a control period under Sec. 96.54(a) and (b), adjusted by any deductions of such NOX allowances to account for actual utilization under Sec. 96.42(e) for the control period or to account for excess emissions for a prior control period under Sec. 96.54(d) or to account for withdrawal from the NOX Budget Program, or for a change in regulatory status, for a NOX

Budget opt-in source under Sec. 96.86 or Sec. 96.87.

**NOX Budget emissions limitation** (Section 97.2) means, for a NOX Budget unit, the tonnage equivalent of the NOX allowances available for compliance deduction for the unit under Sec. 97.54(a), (b), (e), and (f) in a control period adjusted by deductions of such NOX allowances to account for actual heat input under Sec. 97.42(e) for the control period or to account for excess emissions for a prior control period under Sec. 97.54(d) or to account for withdrawal from the NOX Budget Trading Program, or for a change in regulatory status, of a NOX Budget opt-in unit under Sec. 97.86 or Sec. 97.87.

**NOX Budget opt-in permit** (Section 96.2) means a NOX Budget permit covering a NOX Budget opt-in source.

**NOX Budget opt-in permit** (Section 97.2) means a NOX Budget permit covering a NOX Budget opt-in unit.

**NOX Budget opt-in source** (Section 96.2) means a unit that has been elected to become a NOX Budget unit under the NOX Budget Trading Program and whose NOX Budget opt-in permit has been issued and is in effect under subpart I of this part.

**NOX Budget opt-in unit** (Section 97.2) means a unit that has been elected to become a NOX Budget unit under the NOX Budget Trading Program and whose NOX Budget opt-in permit has been issued and is in effect under subpart I of this part.

**NOX Budget permit** (Section 96.2) means the legally binding and federally enforceable written document, or portion of such document, issued by the permitting authority under this part, including any permit revisions, specifying the NOX Budget Trading Program requirements applicable to a NOX Budget source, to each NOX Budget unit at the NOX Budget source, and to the owners and operators and the NOX authorized account representative of the NOX Budget source and each NOX Budget unit.

**NOX Budget permit** (Section 97.2) means the legally binding and federally enforceable written document, or portion of such document, issued by the permitting authority under this part, including any permit revisions, specifying the NOX Budget Trading Program requirements applicable to a NOX Budget source, to each NOX Budget unit at the NOX Budget source, and to the owners and operators and the NOX authorized account representative of the NOX Budget source and each NOX Budget unit.

**NOX Budget source** (Section 96.2) means a source that includes one or more NOX Budget units.

**NOX Budget source** (Section 97.2) means a source that includes one or more NOX Budget units.

**NOX Budget Trading Program** (Section 96.2) means a multi-state nitrogen oxides air pollution control and emission reduction program established in accordance with this part and pursuant to Sec. 51.121 of this chapter, as a means of mitigating the interstate transport of ozone and nitrogen oxides, an ozone precursor.

**NOX Budget Trading Program** (Section 97.2) means a multistate nitrogen oxides air pollution control and emission reduction program established by the Administrator in accordance with this part and pursuant to Sec. 52.34 of this chapter, as a means of mitigating the interstate transport of ozone and nitrogen oxides, an ozone precursor.

**NOX Budget unit** (Section 96.2) means a unit that is subject to the NOX Budget Trading Program emissions limitation under Sec. 96.4 or Sec. 96.80.

**NOX Budget unit** (Section 97.2) means a unit that is subject to the NOX Budget Trading Program emissions limitation under Sec. 97.4(a) or Sec. 97.80.

**NPDES** (Section 110.1) means National Pollutant Discharge Elimination System;

**NPDES** (Section 122.2) means "National Pollutant Discharge Elimination System."

**NPDES** (Section 133.101) National Pollutant Discharge Elimination System.

**NPDES** (Section 270.2) means National Pollutant Discharge Elimination System.

**NPDES Permit or Permit** (Section 403.3) means a permit issued to a POTW pursuant to section 402 of the Act.

**NPDES State** (Section 403.3) means a State (as defined in 40 CFR 122.2) or Interstate water pollution control agency with an NPDES permit program approved pursuant to section 402(b) of the Act.

**NPL** (Section 307.14) means the National Priorities List established pursuant to section 105 of CERCLA and 40 CFR 300.425, which consists of uncontrolled hazardous substance facilities in the United States that need to be addressed under CERCLA authorities. Only NPL sites are eligible for Fund-financed remedial action.

**NSPS** (Section 467.02) means new source performance standards under section 306 of the Act.

**NTIS** (Section 761.13) means the National Technical Information Service, U.S. Department of Commerce, 5285 Port Royal Rd., Springfield, VA 22161.

**Nuclear fuel cycle** (Section 190.02) means the operations defined to be associated with the production of electrical power for public use by any fuel cycle through utilization of nuclear energy.

**Nucleic acids** (Section 174.3) means ribosides or deoxyribosides of adenine, thymine, guanine, cytosine, and uracil; polymers of the deoxyribose-5'-monophosphates of thymine, cytosine, guanine, and adenine linked by successive 3'-5' phosphodiester bonds (also known as deoxyribonucleic acid); and polymers of the ribose-5'-monophosphates of uracil, cytosine, guanine, and adenine linked by successive 3'-5' phosphodiester bonds (also known as ribonucleic acid). The term does not apply to nucleic acid analogues (e.g., dideoxycytidine), or polymers containing nucleic acid analogues.

**Nursery** (Section 170.3) means any operation engaged in the outdoor production of any agricultural plant to produce cut flowers and ferns or plants that will be used in their entirety in another location. Such plants include, but are not limited to, flowering and foliage plants or trees; tree seedlings; live Christmas trees; vegetable, fruit, and ornamental transplants; and turfgrass produced for sod.

**NWPA** (Section 191.02) means the Nuclear Waste Policy Act of 1982 (Pub. L. 97-425).

# O

**O3** (Section 58.1) means ozone.

**Objective Evidence of an emission related repair** (Section 85.2102) means all diagnostic information and data, the actual parts replaced during repair, and any other information directly used to support a warranty claim, or to support denial of such a claim.

**Occurrence** (Section 280.92) means an accident, including continuous or repeated exposure to conditions, which results in a release from an underground storage tank. Note: This definition is intended to assist in the understanding of these regulations and is not intended either to limit the meaning of "occurrence" in a way that conflicts with standard insurance usage or to prevent the use of other standard insurance terms in place of "occurrence."

**Ocean or ocean waters** (Section 220.2) means those waters of the open seas lying seaward of the baseline from which the territorial sea is measured, as provided for in the Convention on the Territorial Sea and the Contiguous Zone (15 UST 1606; TIAS 5639); this definition includes the waters of the territorial sea, the contiguous zone and the oceans as defined in section 502 of the FWPCA.

**Ocean/sea tanker** (Section 442.2) means a self or non-self-propelled vessel constructed or adapted to transport liquid, solid or gaseous commodities or cargos in bulk in cargo spaces (or tanks) through oceans and seas, where the commodity or cargo carried comes in direct contact with the tank interior. There are no maximum or minimum vessel or tank volumes.

**OCS source** (Section 55.2) means any equipment, activity, or facility which: (1) Emits or has the potential to emit any air pollutant; (2) Is regulated or authorized under the Outer Continental Shelf Lands Act ("OCSLA") (43 U.S.C. Sec. 1331 et seq.); and (3) Is located on the OCS or in or on waters above the OCS. This definition shall include vessels only when they are: (1) Permanently or temporarily attached to the seabed and erected thereon and used for the purpose of exploring, developing or producing resources there from, within the meaning of section 4(a) (1) of OCSLA (43 U.S.C. Sec. 1331 et seq.); or (2) Physically attached to an OCS facility, in which case only the stationary sources aspects of the vessels will be regulated.

**Octave Band Attenuation** (Section 211.203) The amount of sound reduction determined according to the measurement procedure of Sec. 211.206 for one-third octave bands of noise.

**OECD area** (Section 262.81) means all land or marine areas under the national jurisdiction of any designated OECD member country in Sec. 262.58. When the regulations refer to shipments to or from an OECD country, this means OECD area.

**Office** (Section 164.2) There may be designated for the Agency one or more judicial officers, one of whom may be Chief Judicial Officer. As work requires, there may be a judicial officer designated to act for the purpose of a particular case. All prior designations of judicial officer shall stay in force until further notice.

**Office Director** (Section 85.2102) means the Director for the Office of Mobile Sources—Office of Air and Radiation of the Environmental Protection Agency or other authorized representative of the Office Director.

**Office Director-Approved Emission Test or Emission Short Test** (Section 85.2102) means any test prescribed under 40 CFR 85.2201 et seq., and meeting all of the requirements thereunder.

**Official use** (Section 1400.2) means an action of a Federal, State, or local government agency or an entity described in paragraph (e)(3) of this section intended to carry out a function relevant to preventing, planning for, or responding to accidental releases.

**Off-kg (off-lb)** (Section 471.02) means the mass of metal or metal alloy removed from a forming operation at the end of a process cycle for transfer to a different machine or process.

**Off-kilogram (off-pound)** (Section 467.02) shall mean the mass of aluminum or aluminum alloy removed from a forming or ancillary operation at the end of a process cycle for transfer to a different machine or process.

**Off-kilogram (off-pound)** (Section 468.02) shall mean the mass of copper or copper alloy removed from a forming or ancillary operation at the end of a process cycle for transfer to a different machine or process.

**Off-road motorcycle** (Section 205.151) means any motorcycle that is not a street motorcycle or competition motorcycle.

**Offset plan** (Section 72.2) means a plan pursuant to part 77 of this chapter for offsetting excess emissions of sulfur dioxide that have occurred at an affected unit in any calendar year.

**Offshore facility** (Section 112.2) means any facility of any kind located in, on, or under any of the navigable waters of the United States, which is not a transportation-related facility.

**Offshore facility** (Section 116.3) means any facility of any kind located in, on, or under, any of the navigable waters of the United States, and any facility of any kind which is subject to the jurisdiction of the United States and is located in, on, or under any other waters, other than a vessel or a public vessel;

**Offshore facility** (Section 302.3) means any facility of any kind located in, on, or under, any of the navigable waters of the United States, and any facility of any kind which is subject to the jurisdiction of the United States and is located in, on, or under any other waters, other than a vessel or a public vessel;

**Offshore facility as defined by section 101(17) of CERCLA and section 311(a)(11) of the CWA** (Section 300.5) means any facility of any kind located in, on, or under any of the navigable waters of the United States, and any facility of any kind which is subject to the jurisdiction of the United States and is located in, on, or under any other waters, other than a vessel or a public vessel.

**Off-site** (Section 270.2) means any site which is not on-site.

**Off-site** (Section 437.2) means outside the boundaries of a facility.

**Off-site** (Section 444.11) means outside the boundaries of a facility.

**Off-site** (Section 445.2) means outside the boundaries of a facility.

**Offsite** (Section 68.3) means areas beyond the property boundary of the stationary source, and 'areas within the property boundary to which the public has routine and unrestricted access during or outside business hours.

**Off-site consequence analysis (OCA) data elements** (Section 1400.2) means the results of the off-site consequence analysis conducted by a stationary source pursuant to 40 CFR part 68, subpart B, when presented in a format different than sections 2 through 5 of a risk management plan or anyAdministrator-created electronic database.

**Off-site consequence analysis (OCA) information** (Section 1400.2) means sections 2 through 5 of a risk management plan (consisting of an evaluation of one or more worst-case release scenarios or alternative release scenarios) for an identified facility and any electronic database created by the Administrator from those sections.

**Off-site consequence analysis (OCA) rankings** (Section 1400.2) means any statewide or national rankings of identified stationary sources derived from OCA information.

**Oil** (Section 109.2) means oil of any kind or in any form, including, but not limited to, petroleum, fuel oil, sludge, oil refuse, and oil mixed with wastes other than dredged spoil.

**Oil** (Section 112.2) means oil of any kind or in any form, including, but not limited to petroleum, fuel oil, sludge, oil refuse and oil mixed with wastes other than dredged spoil.

**Oil** (Section 113.3) means oil of any kind or in any form, including but no limited to, petroleum, fuel oil, sludge, oil refuse, and oil mixed with wastes other than dredged spoil.

**Oil** (Section 426.121) shall mean those components of a waste water amenable to measurement by the technique or techniques described in the most recent addition of Standard Methods for the analysis of grease in polluted waters, waste waters, and effluents, such as Standard Methods, 13th Edition, 2nd Printing, page 407.

**Oil** (Section 426.81) shall mean those components of a waste water amenable to measurement by the technique or techniques described in the most recent addition of Standard Methods for the analysis of grease in polluted waters, waste waters, and effluents, such as Standard Methods, 13th Edition, 2nd Printing, page 407.

**Oil and grease (or O&G)** (Section 420.02) means the value obtained by the method specified in 40 CFR 136.3.

**Oil and grease** (Section 408.11) shall mean those components of a waste water amenable to measurement by the method described in Methods for Chemical Analysis of Water and Wastes, 1971, Environmental Protection Agency, Analytical Quality Control Laboratory, page 217.

**Oil and grease** (Section 410.11) shall mean total recoverable oil and grease as measured by the procedure listed in 40 CFR part 136.

**Oil and grease** (Section 432.11) shall mean those components of process waste water amenable to measurement by the method described in Methods for Chemical Analysis of Water and Wastes, 1971, EPA, Analytical Quality Control Laboratory, page 217.

**Oil and grease** (Section 432.21) shall mean those components of process waste water amenable to measurement by the method described in Methods for Chemical Analysis of Water and Wastes, 1971, EPA, Analytical Quality Control Laboratory, page 217.

**Oil and grease** (Section 432.41) shall mean those components of process waste water amenable to measurement by the method described in Methods for Chemical Analysis of Water and Wastes, 1971, EPA, Analytical Quality Control Laboratory, page 217.

**Oil as defined by section 311(a)(1) of the CWA** (Section 300.5) means oil of any kind or in any form, including, but not limited to, petroleum, fuel oil, sludge, oil refuse, and oil mixed with wastes other than dredged spoil. Oil, as defined by section 1001 of the OPA means oil of any kind or in any form, including, but not limited to, petroleum, fuel oil, sludge, oil refuse, and oil mixed with wastes other than dredged spoil, but does not include petroleum, including crude oil or any fraction thereof, which is specifically listed or designated as a hazardous substance under subparagraphs (A) through (F) of section 101(14) of the Comprehensive Environmental Response, Compensation, and Liability Act (42 U.S.C. 9601) and which is subject to the provisions of that Act.

**Oil Spill Liability Trust Fund (OSLTF)** (Section 300.5) means the fund established under section 9509 of the Internal Revenue Code of 1986 (26 U.S.C. 9509).

**Oil Spill Removal Organization** (Section 112.2) means an entity that provides oil spill response resources, and includes any for-profit or not-for-profit contractor, cooperative, or in-house response resources that have been established in a geographic area to provide required response resources.

**Oil-based** (Section 435.11) means the continuous phase of the drilling fluid consists of diesel oil, mineral oil, or some other oil, but contains no synthetic material or enhanced mineral oil.

**Oil-based** (Section 435.41) means the continuous phase of the drilling fluid consists of diesel oil, mineral oil, or some other oil, but contains no synthetic material or enhanced mineral oil.

**Oil-fired** (Section 72.2) means: (1) For all purposes under the Acid Rain Program, except part 75 of this chapter, the combustion of: (i) Fuel oil for more than 10.0 percent of the average annual heat input during the previous three calendar years or for more than 15.0 percent of the annual heat input during any one of those calendar years; and (ii) Any solid, liquid or gaseous fuel (including coal-derived gaseous fuel), other than coal or any other coal-derived solid or liquid fuel, for the remaining heat input, if any. (2) For purposes of part 75 of this chapter, combustion of only fuel oil and gaseous fuels, provided that the unit involved does not meet the definition of gas-fired.

**Oily absorbent recycling** (Section 437.2) means the process of recycling oil-soaked or contaminated disposable rags, paper, or pads for the purpose of regenerating a fuel for reuse.

**Oily wastes** (Section 437.2) means wastes and/or used materials that contain oil and grease (generally at or in excess of 100 mg/L) from manufacturing or processing facilities or other commercial operations. Examples of these wastes are used oils, oil-water emulsions or mixtures, lubricants, coolants, contaminated groundwater clean-up from petroleum sources, used petroleum products, oil spill clean-up, bilge water, rinse/wash waters from petroleum sources, interceptor wastes, off-specification fuels, underground storage tank remediation waste, and tank clean out from petroleum or oily sources.

**Olives** (Section 407.61) shall mean the processing of olives into the following product styles: Canned all varieties, fresh and stored, green ripe, black ripe, Spanish, Sicilian, and any other styles to which spices, acids, and flavorings may have been added.

**On ground tank** (Section 260.10) means a device meeting the definition of "tank" in Sec. 260.10 and that is situated in such a way that the bottom of the tank is on the same level as the adjacent surrounding surface so that the external tank bottom cannot be visually inspected.

**On site** (Section 761.13) means within the boundaries of a contiguous property unit.

**On the premises where stored with respect to heating oil** (Section 280.12) means UST systems located on the same property where the stored heating oil is used.

**Once through cooling water** (Section 423.11) means water passed through the main cooling condensers in one or two passes for the purpose of removing waste heat.

**Once-through cooling water** (Section 419.11) shall mean those waters discharged that are used for the purpose of heat removal and that do not come into direct contact with any raw material, intermediate, or finished product.

**One-hour period** (Section 60.2) means any 60-minute period commencing on the hour.

**One-hour period** (Section 65.2)means the 60-minute period commencing on the hour.

**One-hour period, unless otherwise defined in an applicable subpart,** (Section 63.2) means any 60-minute period commencing on the hour.

**One-minute average** (Section 62.1201) means the average of detector responses calculated at least every 60 seconds from responses obtained at least every 15 seconds.

**On-scene coordinator (OSC)** (Section 300.5) means the federal official predesignated by EPA or the USCG to coordinate and direct responses under subpart D, or the government official designated by the lead agency to coordinate and direct removal actions under subpart E of the NCP.

**On-Scene Coordinator** (Section 113.3) is the single Federal representative designated pursuant to the National Oil and Hazardous Substances Pollution Contingency Plan and identified in approved Regional Oil and Hazardous Substances Pollution Contingency Plans.

**Onshore** (Section 435.51) shall mean all land areas landward of the territorial seas as defined in 40 CFR 125.1(gg).

**Onshore area** (Section 55.2) means a coastal area designated as an attainment, nonattainment, or unclassifiable area by EPA in accordance with section 107 of the Act. If the boundaries of an area designated pursuant to section 107 of the Act do not coincide with the boundaries of a single onshore air pollution control agency, then onshore area shall mean a coastal area defined by the jurisdictional boundaries of an air pollution control agency.

**Onshore facility** (Section 112.2) means any facility of any kind located in, on, or under any land within the United States, other than submerged lands, which is not a transportation-related facility.

**Onshore facility** (Section 116.3) means any facility (including, but not limited to, motor vehicles and rolling stock) of any kind located in, on, or under, any land within the United States other than submerged land;

**Onshore facility** (Section 302.3) means any facility (including, but not limited to, motor vehicles and rolling stock) of any kind located in, on, or under, any land or non-navigable waters within the United States;

**Onshore facility as defined by section 101(18) of CERCLA** (Section 300.5) means any facility (including, but not limited to, motor vehicles and rolling stock) of any kind located in, on, or under any land or non-navigable waters within the United States; and, as defined by section 311(a)(10) of the CWA, means any facility (including, but not limited to, motor vehicles and rolling stock) of any kind located in, on, or under any land within the United States other than submerged land.

**Onshore Oil Storage Facility** (Section 113.3) means any facility (excluding motor vehicles and rolling stock) of any kind located in, on, or under, any land within the United States, other than submerged land.

**On-site** (Section 260.10) means the same or geographically contiguous property which may be divided by public or private right-of-way, provided the entrance and exit between the properties is at a cross-roads intersection, and access is by crossing as opposed to going along, the right-of-way. Non-contiguous properties owned by the same person but connected by a right-of-way which he controls and to which the public does not have access, is also considered on-site property.

**On-site** (Section 270.2) means on the same or geographically contiguous property which may be divided by public or private right(s)-of-way, provided the entrance and exit between the properties is at a cross-roads intersection, and access is by crossing as opposed to going along, the right(s)-of-way. Non-contiguous properties owned by the same person but connected by a right-of-way which the person controls and to which the public does not have access, is also considered on-site property.

**On-site** (Section 273.9) means the same or geographically contiguous property which may be divided by public or private right-of-way, provided that the entrance and exit between the properties is at a cross-roads intersection, and access is by crossing as opposed to going along the right of way. Non-contiguous properties owned by the same person but connected by a right-of-way which he controls and to which the public does not have access, are also considered on-site property.

**On-site** (Section 300.5) means the areal extent of contamination and all suitable areas in very close proximity to the contamination necessary for implementation of the response action.

**On-site** (Section 437.2) means within the boundaries of a facility. A facility may encompass land areas that are bisected by public thoroughfares but are under the control of a common owner.

**On-site** (Section 442.2) means within the contiguous and non-contiguous established boundaries of a facility.

**On-site** (Section 444.11) means within the boundaries of a facility.

**On-site** (Section 445.2) means within the boundaries of a facility.

**On-site Compliance Paperwork** (Section 455.41) for this subpart means data or information maintained in the offices of the PFPR facility which supports the initial and periodic certification statements as follows: (1) Lists and describes those product families, process lines and/or process units for which the facility is implementing the P2 Alternative; (2) Describes the facility specific practices for each product family/process line/process unit which are to be practiced as part of the P2 Alternative; (3) Describes any justification allowing modification to the practices listed in Table

8 to this part 455; (4) Includes a written discussion demonstrating that the treatment system being used contains the appropriate pollution control technologies (or equivalent systems/pesticide manufacturing systems) for removing the PAIs which may be found in the wastewater; (5) Establishes a method for demonstrating to the permitting/control authority that the treatment system is well operated and maintained; and (6) Includes a discussion of the rationale for choosing the method of demonstration.

**Onsite or on-site** (Section 65.2) means, with respect to records required to be maintained by this part, that the records are stored at a location within a plant site that encompasses the regulated source. Onsite includes, but is not limited to, storage at the regulated source to which the records pertain, or storage in central files elsewhere at the plant site.

**Opacity** (Section 60.2) means the degree to which emissions reduce the transmission of light and obscure the view of an object in the background.

**Opacity** (Section 63.2) means the degree to which emissions reduce the transmission of light and obscure the view of an object in the background. For continuous opacity monitoring systems, opacity means the fraction of incident light that is attenuated by an optical medium.

**Opacity** (Section 72.2) means the degree to which emissions reduce the transmission of light and obscure the view of an object in the background.

**Opacity** (Section 86.082-2) means the fraction of a beam of light, expressed in percent, which fails to penetrate a plume of smoke.

**Opacity** (Section 89.2) means the fraction of a beam of light, expressed in percent, which fails to penetrate a plume of smoke.

**Opacity** (Section 92.2) means the fraction of a beam of light, expressed in percent, which fails to penetrate a plume of smoke as measured and calculated under the provisions of subpart B of this part.

**Open burning** (Section 240.101) means burning of solid wastes in the open, such as in an open dump.

**Open burning** (Section 258.2) means the combustion of solid waste without: (1) Control of combustion air to maintain adequate temperature for efficient combustion, (2) Containment of the combustion reaction in an enclosed device to provide sufficient residence time and mixing for complete combustion, and (3) Control of the emission of the combustion products.

**Open burning** (Section 260.10) means the combustion of any material without the following characteristics: (1) Control of combustion air to maintain adequate temperature for efficient combustion, (2) Containment of the combustion-reaction in an enclosed device to provide sufficient residence time and mixing for complete combustion, and (3) Control of emission of the gaseous combustion products. (See also "incineration" and "thermal treatment".)

**Open burning** (Section 761.13) means the combustion of any PCB regulated for disposal, in a manner not approved or otherwise allowed under subpart D of this part, and without any of the following: (1) Control of combustion air to maintain adequate temperature for efficient combustion. (2) Containment of the combustion reaction in an enclosed device to provide sufficient residence time and mixing for complete combustion. (3) Control of emission of the gaseous combustion products.

**Open combustion** (Section 420.41) means those basic oxygen furnace steelmaking wet air cleaning systems which are designed to allow excess air to enter the air pollution control system for the purpose of combusting the carbon monoxide in furnace gases.

**Open dump** (Section 240.101) means a land disposal site at which solid wastes are disposed of in a manner that does not protect the environment, are susceptible to open burning, and are exposed to the elements, vectors, and scavengers.

**Open dump** (Section 272.2) means a facility for the disposal of solid waste which does not comply with this part.

**Open hearth furnace steelmaking** (Section 420.41) means the production of steel from molten iron, steel scrap, fluxes, and various combinations thereof, in refractory lined fuel-fired furnaces equipped with regenerative chambers to recover heat from the flue and combustion gases.

**Open lot** (Section 412.11) shall mean pens or similar confinement areas with dirt, or concrete (or paved or hard) surfaces wherein animals or poultry are substantially or entirely exposed to the outside environment except for possible small portions affording some protection by windbreaks, small shed-type shade areas. For the purposes hereof the term "open lot" is synonymous with the terms "cowyard" (dairy cattle), "pasture lot" (swine), and "dirt lot" (swine, sheep or turkeys), "dry lot" (swine, cattle, sheep, or turkeys) which are terms widely used in the industry.

**Open path analyzer** (Section 58.1) is an automated analytical method that measures the average atmospheric pollutant concentration in situ along one or more monitoring paths having a monitoring path length of 5 meters or more and that has been designated as a reference or equivalent method under the provisions of part 53 of this chapter.

**Open site** (Section 202.10) means an area that is essentially free of large sound-reflecting objects, such as barriers, walls, board fences, signboards, parked vehicles, bridges, or buildings.

**Open waters of the Great Lakes (OWGLs)** (Section 132.2) means all of the waters within Lake Erie, Lake Huron (including Lake St. Clair), Lake Michigan, Lake Ontario, and Lake Superior lakeward from a line drawn across the mouth of tributaries to the Lakes, including all waters enclosed by constructed breakwaters, but not including the connecting channels.

**Open-ended valve or line** (Section 264.1031) means any valve, except pressure relief valves, having one side of the valve seat in contact with hazardous waste and one side open to the atmosphere, either directly or through open piping.

**Open-ended valve or line** (Section 65.2) means any valve except relief valves having one side of the valve seat in contact with process fluid and one side open to the atmosphere, either directly or through open piping.

**Operable unit** (Section 300.5) means a discrete action that comprises an incremental step toward comprehensively addressing site problems. This discrete portion of a remedial response manages migration, or eliminates or mitigates a release, threat of a release, or pathway of exposure. The cleanup of a site can be divided into a number of operable units, depending on the complexity of the problems associated with the site.

**Operable unit** (Section 307.14) means a discrete action that comprises an incremental step toward comprehensively addressing site problems. This discrete portion of a remedial response manages migration, or eliminates or mitigates a release, threat of release, or pathway of exposure. The cleanup of a site can be divided into a number of operable units, depending on the complexity of the problems associated with the site. Operable units may address geographical portions of a site, specific site problems, or initial phases of an action, or may consist of any set of actions performed over time or any actions that are concurrent but located in different parts of a site. Operable units will not impede implementation of subsequent actions, including final action at the site.

**Operable units** (Section 300.5) may address geographical portions of a site, specific site problems, or initial phases of an action, or may consist of any set of actions performed over time or any actions that are concurrent but located in different parts of a site.

**Operated by the same producer, when used with respect to two establishments** (Section 152.3) means that each such establishment is either owned by, or leased for operation by and under the control of, the same person. The term does not include establishments owned or operated by different persons, regardless of contractual agreement between such persons.

**Operating** (Section 72.2) when referring to a combustion or process source seeking entry into the Opt-in Program, means that the source had documented consumption of fuel input for more than 876 hours in the 6 months immediately preceding the submission of a combustion source's opt-in application under Sec. 74.16(a) of this chapter.

**Operating** (Section 96.2) means, with regard to a unit under Secs. 96.22(d)(2) and 96.80, having documented heat input for more than 876 hours in the 6 months immediately preceding the submission of an application for an initial NOX Budget permit under Sec. 96.83(a).

**Operating** (Section 97.2) means, with regard to a unit under Secs. 97.22(d)(2) and 97.80, having documented heat input for more than 876 hours in the 6 months immediately preceding the submission of an application for an initial NOX Budget permit under Sec. 97.83(a). The unit's documented heat input will be determined in accordance with part 75 of this chapter if the unit was otherwise subject to the requirements of part 75 of this chapter during that 6-month period or will be based on the best available data reported to the Administrator for the unit if the unit was not otherwise subject to the requirements of part 75 of this chapter during that 6-month period.

**Operating hours** (Section 89.2) means: (1) For engine storage areas or facilities, all times during which personnel other than custodial personnel are at work in the vicinity of the storage area or facility and have access to it. (2) For all other areas or facilities, all times during which an assembly line is in operation or all times during which testing, maintenance, service accumulation, production or compilation of records, or any other procedure or activity related to certification testing, to translation of designs from the test stage to the production stage, or to engine manufacture or assembly is being carried out in a facility.

**Operating hours** (Section 90.3) means: (1) For engine storage areas or facilities, all times during which personnel other than custodial personnel are at work in the vicinity of the storage area or facility and have access to it. (2) For all other areas or facilities, all times during which an assembly line is in operation or all times during which testing, maintenance, service accumulation, production or compilation of records, or any other procedure or activity related to certification testing, to translation of designs from the test stage to the production stage, or to engine manufacture or assembly is being carried out in a facility.

**Operating hours** (Section 91.3) means: (1) For engine storage areas or facilities, all times during which personnel other than custodial personnel are at work in the vicinity of the storage area or facility and have access to it. (2) For all other areas or facilities, all times during which an assembly line is in operation or all times during which testing, maintenance, service accumulation, production or compilation of records, or any other procedure or activity related to certification testing, to translation of designs from the test stage to the production stage, or to engine manufacture or assembly is being carried out in a facility.

**Operating period** (Section 76.2) means a period of time of not less than three consecutive months and that occurs not more than one month prior to applying for an alternative emission limitation demonstration period under Sec. 76.10, during which the owner or operator of an affected unit that cannot meet the applicable emission limitation: (1) Operates the installed NOX emission controls in accordance with primary vendor specifications and procedures, with the unit operating under normal conditions; and (2) records and reports quality-assured continuous emission monitoring (CEM) and unit operating data according to the methods and procedures in part 75 of this chapter.

**Operating permit** (Section 72.2) means a permit issued under part 70 of this chapter and any other regulations implementing title V of the Act.

**Operating record** (Section 62.1201) means a documentation retained at the facility for ready inspection by authorized officials of all information required by the standards to document and maintain compliance with the applicable regulations, including data and information, reports, notifications, and communications with regulatory officials.

**Operating requirements** (Section 62.1201) means operating terms or conditions, limits, or operating parameter limits developed under this subpart that ensure compliance with the emission standards.

**Operation** (Section 280.200) means, for purposes of this subpart, the use, storage, filling, or dispensing of petroleum contained in an UST or UST system.

**Operation** (Section 413.11) shall mean any step in the electroplating process in which a metal is electrodeposited on a basis material and which is followed by a rinse; this includes the related operations of alkaline cleaning, acid pickle, stripping, and coloring when each operation is followed by a rinse.

**Operation and maintenance (O&M)** (Section 300.5) means measures required to maintain the effectiveness of response actions.

**Operational** (Section 192.31) means that a uranium mill tailings pile or impoundment is being used for the continued placement of uranium byproduct material or is in standby status for such placement. A tailings pile or impoundment is operational from the day that uranium byproduct material is first placed in the pile or impoundment until the day final closure begins.

**Operational life** (Section 280.12) refers to the period beginning when installation of the tank system has commenced until the time the tank system is properly closed under Subpart G.

**Operations and maintenance program** (Section 763.83) means a program of work practices to maintain friable ACBM in good condition, ensure clean up of asbestos fibers previously released, and prevent further release by minimizing and controlling friable ACBM disturbance or damage.

**Operator** (Section 256.06) includes facility owners and operators.

**Operator** (Section 258.2) means the person(s) responsible for the overall operation of a facility or part of a facility.

**Operator** (Section 260.10) means the person responsible for the overall operation of a facility.

**Operator** (Section 280.12) means any person in control of, or having responsibility for, the daily operation of the UST system.

**Operator** (Section 610.11) means any person who installs, services or maintains a retrofit device in an automobile or who operates an automobile with a retrofit device installed.

**Operator** (Section 85.1402) means transit authority, state, city department, or private or public entity controlling the use of one or more urban buses.

**Operator** (Section 96.2) means any person who operates, controls, or supervises a NOX Budget unit, a NOX Budget source, or unit for which an application for a NOX Budget opt-in permit under Sec. 96.83 is submitted and not denied or withdrawn and shall include, but not be limited to, any holding company, utility system, or plant manager of such a unit or source.

**Operator** (Section 97.2) means any person who operates, controls, or supervises a NOX Budget unit, a NOX Budget source, or a unit for which an application for a NOX Budget opt-in permit under Sec. 97.83 is submitted and not denied or withdrawn and shall include, but not be limited to, any holding company, utility system, or plant manager of such a unit or source.

**OPPTS** (Section 179.3) means the Agency's Office of Prevention, Pesticides and Toxic Substances.

**Opt in or opt into** (Section 72.2) means to elect to become an affected unit under the Acid Rain Program through the issuance of the final effective opt-in permit under Sec. 74.14 of this chapter.

**Optical measurement path length** (Section 58.1) is the actual length of the optical beam over which measurement of the pollutant is determined. The path-integrated pollutant concentration measured by the analyzer is divided by the optical measurement path length to determine the path-averaged concentration. Generally, the optical measurement path length is: (1) Equal to the monitoring path length for a (bistatic) system having a transmitter and a receiver at opposite ends of the monitoring path; (2) Equal to twice the monitoring path length for a (monostatic) system having a transmitter and receiver at one end of the monitoring path and a mirror or retroreflector at the other end; or (3) Equal to some multiple of the monitoring path length for more complex systems having multiple passes of the measurement beam through the monitoring path.

**Optimal corrosion control treatment** (Section 141.2) for the purpose of subpart I of this part only, means the corrosion control treatment that minimizes the lead and copper concentrations at users' taps while insuring that the treatment does not cause the water system to violate any national primary drinking water regulations.

**Opt-in** (Section 96.2) means to be elected to become a NOX Budget unit under the NOX Budget Trading Program through a final, effective NOX Budget opt-in permit under subpart I of this part.

**Opt-in** (Section 97.2) means to be elected to become a NOX Budget unit under the NOX Budget Trading Program through a final, effective NOX Budget opt-in permit under subpart I of this part.

**Opt-in area**. (Section 80.2) An area which becomes a covered area under Sec. 80.70 pursuant to section 211(k)(6) of the Clean Air Act.

**Opt-in permit** (Section 72.2) means the legally binding written document that is contained within the Acid Rain permit and sets forth the requirements under part 74 of this chapter for a combustion source or a process source that opts into the Acid Rain Program.

**Opt-in source** (Section 72.2) means a combustion source or process source that has elected to become an affected unit under the Acid Rain Program and whose opt-in permit has been issued and is in effect.

**Option** (Section 86.082-2) means any available equipment or feature not standard equipment on a model.

**Option, in the context of a vehicle design feature** (Section 86.1803-01) means any available equipment or feature not standard equipment on a model.

**Order** (Section 108.2) means any order issued by the Administrator under section

**Organ** (Section 190.02) means any human organ exclusive of the dermis, the epidermis, or the cornea.

**Organic material non-methane hydrocarbon equivalent (or OMNMHCE) for methanol-fueled vehicles** (Section 86.1702-99) means the sum of the carbon mass contribution of non-oxygenated hydrocarbons (excluding methane), methanol, and formaldehyde as contained in a gas sample, expressed as gasoline-fueled hydrocarbons. For ethanol-fueled vehicles, organic material non-methane hydrocarbon equivalent (OMNMHCE) means the sum of carbon mass contribution of non-oxygenated hydrocarbons (excluding methane), methanol, ethanol, formaldehyde and acetaldehyde as contained in a gas sample, expressed as gasoline-fueled hydrocarbons.

**Organic monitoring device** (Section 65.2) means a device used to indicate the concentration level of organic compounds based on a detection principle such as infrared, photo ionization, or thermal conductivity.

**Organic wastes** (Section 437.2) means wastes and/or used materials that contain organic pollutants, but not a significant quantity of oil and grease (generally less than 100 mg/L) from manufacturing or processing facilities or other commercial operations. Examples of these wastes are landfill leachate, contaminated groundwater clean-up from non-petroleum sources, solvent-bearing wastes, off-specification organic product, still bottoms, byproduct glycols, wastewater from paint washes, wastewater from adhesives and/or epoxies, wastewater from chemical product operations, and tank clean-out from organic, non-petroleum sources.

**Organization** (Section 195.2) is any individual, sole proprietorship, partnership, business, company, corporation, college or university, government agency (includes Federal, State and local government entities), laboratory, or institution.

**Original data submitter** (Section 152.83) means the person who possesses all rights to exclusive use or compensation under FIFRA section 3©(1)(D) in a study originally submitted in support of an application for registration, amended registration, reregistration, or experimental use permit, or to maintain an existing registration in effect. The term includes the person who originally submitted the study, any person to whom the rights under FIFRA section 3© (1) (D) have been transferred, or the authorized representative of a group of joint data developers.

**Original Engine Configuration** (Section 85.1402) means the engine configuration at time of initial sale.

**Original engine manufacturer (OEM)** (Section 89.602) The entity which originally manufactured the nonroad engine.

**Original engine manufacturer (OEM)** (Section 91.702) The entity which originally manufactured the marine engine.

**Original equipment manufacturer (OEM)** (Section 85.1502) The entity which originally manufactured the motor vehicle or motor vehicle engine prior to conditional importation.

**Original Equipment Manufacturer (OEM)** (Section 86.1803-01) means the manufacturer responsible for the design and production of a vehicle or component. This manufacturer will be fully knowledgeable of any production changes made to the design of the vehicle or component and shall be able to track the individual vehicles or component with regard to such production changes.

**Original Equipment Part** (Section 85.1402) means a part present in or on an engine at the time an urban bus is originally sold to the ultimate purchaser.

**Original Equipment Part** (Section 85.2102) means a part present in or on a vehicle at the time the vehicle is sold to the ultimate purchaser, except for components installed by a dealer which are not manufactured by the vehicle manufacturer or are not installed at the direction of the vehicle manufacturer.

**Original manufacture** (Section 92.2) means the event of freshly manufacturing a locomotive or locomotive engine. The date of original manufacture is the date of final assembly; except as provided in Sec. 92.11. Where a locomotive or locomotive engine is manufactured under Sec. 92.11, the date of original manufacture is the date on which the final assembly of locomotive or locomotive engine was originally scheduled.

**Original production (OP) year** (Section 85.1502) The calendar year in which the motor vehicle or motor vehicle engine was originally produced by the OEM.

**Original production (OP) year** (Section 89.602) The calendar year in which the nonroad engine was originally produced by the OEM.

**Original production (OP) years old** (Section 85.1502) The age of a vehicle as determined by subtracting the original production year of the vehicle from the calendar year of importation.

**Original production (OP) years old** (Section 89.602) The age of a nonroad engine as determined by subtracting the original production year of the nonroad engine from the calendar year of importation.

**Original remanufacture** (Section 92.2) means the first remanufacturing of a locomotive or locomotive engine at which the locomotive or locomotive engines is subject to the emission standards of this part.

**Ornamental** (Section 171.2) means trees, shrubs, and other plantings in and around habitations generally, but not necessarily located in urban and suburban areas, including residences, parks, streets, retail outlets, industrial and institutional buildings.

**OSHA** (Section 68.3) means the U.S. Occupational Safety and Health Administration. Owner or operator means any person who owns, leases, operates, controls, or supervises a stationary source.

**Other coatings** (Section 420.121) means coating steel products with metals other than zinc or terne metal by the hot dip process including the immersion of the steel product in a molten bath of metal, and the related operations preceding the subsequent to the immersion phase.

**Other container** (Section 503.11) is either an open or closed receptacle. This includes, but is not limited to, a bucket, a box, a carton, and a vehicle or trailer with a load capacity of one metric ton or less.

**Other restricted access (nonsubstation) locations** (Section 761.123) means areas other than electrical substations that are at least 0.1 kilometer (km) from a residential/commercial area and limited by man-made barriers (e.g., fences and walls) to substantially limited by naturally occurring barriers such as mountains, cliffs, or rough terrain. These areas generally include industrial facilities and extremely remote rural locations. (Areas where access is restricted but are less than 0.1 km from a residential/commercial area are considered to be residential/ commercial areas.)

**Other significant evidence** (Section 154.3) means factually significant information that relates to the uses of the pesticide and their adverse risk to man or to the environment but does not include evidence based only on misuse of the pesticide unless such misuse is widespread and commonly recognized practice.

**Otherwise subject to the jurisdiction of the United States** (Section 116.3) means subject to the jurisdiction of the United States by virtue of United States citizenship, United States vessel documentation or numbering, or as provided for by international agreement to which the United States is a party. A discharge in connection with activities under the Outer Continental Shelf Lands Act or the Deepwater Port Act of 1974, or which may affect natural resources belonging to, appertaining to, or under the exclusive management authority of the United States (including resources under the Fishery Conservation and Management Act of 1976), means: (1) A discharge into any waters beyond the contiguous zone from any vessel or onshore or offshore facility, which vessel or facility is subject to or is engaged in activities under the Outer Continental Shelf Lands Act or the Deepwater Port Act of 1974, and (2) any discharge into any waters beyond the contiguous zone which contain, cover, or support any natural resource belonging to, appertaining to, or under the exclusive management authority of the United States (including resources under the Fishery Conservation and Management Act of 1976).

**Otherwise use** (Section 372.3) means any use of a toxic chemical, including a toxic chemical contained in a mixture or other trade name product or waste, that is not covered by the terms "manufacture" or "process." Otherwise use of a toxic chemical does not include disposal, stabilization (without subsequent distribution in commerce), or treatment for destruction unless: (1) The toxic chemical that was disposed, stabilized, or treated for destruction was received from off-site for the purposes of further waste management; or (2) The toxic chemical that was disposed, stabilized, or treated for destruction was manufactured as a result of waste management activities on materials received from off-site for the purposes of further waste management activities. Relabeling or redistributing of the toxic chemical where no repackaging of the toxic chemical occurs does not constitute otherwise use or processing of the toxic chemical.

**Otto-cycle** (Section 86.090-2) means type of engine with operating characteristics significantly similar to the theoretical Otto combustion cycle. The use of a throttle during normal operation is indicative of an Otto-cycle engine.

**Otto-cycle** (Section 86.1803-01) means type of engine with operating characteristics significantly similar to the theoretical Otto combustion cycle. The use of a throttle during normal operation is indicative of an Otto-cycle engine.

**Outboard engine** (Section 91.3) is a marine SI engine that, when properly mounted on a marine vessel in the position to operate, houses the engine and drive unit external to the hull of the marine vessel.

**Outcome** (Section 35.102) The environmental result, effect, or consequence that will occur from carrying out an environmental program or activity that is related to an environmental or programmatic goal or objective. Outcomes must be quantitative, and they may not necessarily be achievable during a grant funding period. See "output."

**Outdoor electrical substations** (Section 761.123) means outdoor, fenced-off, and restricted access areas used in the transmission and/or distribution of electrical power Outdoor electrical substations restrict public access by being fenced or walled off as defined under Sec. 761.30(l)(1)(ii). For purposes of this TSCA policy, outdoor electrical substations are defined as being located at least 0.1 km from a residential/commercial area. Outdoor fenced-off and restricted access areas used in the transmission and/or distribution of electrical power which are located less than 0.1. km from a residential/commercial area are considered to be residential/commercial areas.

**Outdoor use** (Section 152.161) means any pesticide application that occurs outside enclosed manmade structures or the consequences of which extend beyond enclosed manmade structures, including, but not limited to, pulp and paper mill water treatments and industrial cooling water treatments.

**Outer continental shelf** (Section 55.2) shall have the meaning provided by section 2 of the OCSLA (43 U.S.C. Sec. 1331 et seq.).

**Out-of-control period** (Section 72.2) means any period: (1) Beginning with the hour corresponding to the completion of a daily calibration error, linearity check, or quality assurance audit that indicates that the instrument is not measuring and recording within the applicable performance specifications; and (2) Ending with the hour corresponding to the completion of an additional calibration error, linearity check, or quality assurance audit following corrective action that demonstrates that the instrument is measuring and recording within the applicable performance specifications.

**Output** (Section 35.102) An environmental activity or effort and associated work products related to an environmental goal or objective that will be produced or provided over a period of time or by a specified date. Outputs may be quantitative or qualitative but must be measurable during a grant funding period. See "outcome."

**Overburden** (Section 372.3) means the unconsolidated material that overlies a deposit of useful materials or ores. It does not include any portion of ore or waste rock.

**Overdraft account** (Section 96.2) means the NOX Allowance Tracking System account, established by the Administrator under subpart F of this part, for each NOX Budget source where there are two or more NOX Budget units.

**Overdraft account** (Section 97.2) means the NOX Allowance Tracking System account, established by the Administrator under subpart F of this part, for each NOX Budget source where there are two or more NOX Budget units.

**Overfill release** (Section 280.12) is a release that occurs when a tank is filled beyond its capacity, resulting in a discharge of the regulated substance to the environment.

**Overhead valve engine** (Section 90.3) means an otto-cycle, four stroke engine in which the intake and exhaust valves are located above the combustion chamber within the cylinder head. Such engines are sometimes referred to as "valve-in-head" engines.

**Oversubscription payment deadline** (Section 72.2) means 30 calendar days prior to the allowance transfer deadline.

**Over-the-Head Position** (Section 211.203) The mode of use of a device with a headband, in which the headband is worn such that it passes over the user's head. This is contrast to the behind-the-head and under-the-chin positions.

**Own or control** (Section 704.3) means ownership of 50 percent or more of a company's voting stock or other equity rights, or the power to control the management and policies of that company. A company may own or control one or more sites. A company may be owned or controlled by a foreign or domestic parent company.

**Owned or controlled by the parent company** (Section 712.3) means the parent owns or controls 50 percent or more of the other company's voting stock or other equity rights, or has the power to control the management and policies of the other company.

**Owned or operated, leased or otherwise controlled by such person** (Section 88.302-94) means either of the following: (1) Such person holds the beneficial title to such vehicle; or (2) Such person uses the vehicle for transportation purposes pursuant to a contract or similar arrangement, the term of such contract or similar arrangement is for a period of 120 days or more, and such person has control over the vehicle pursuant to the definition of control of this section.

**Owner** (Section 170.3) means any person who has a present possessory interest (fee, leasehold, rental, or other) in an agricultural establishment covered by this part. A person who has both leased such agricultural establishment to another person and granted that same person the right and full authority to manage and govern the use of such agricultural establishment is not an owner for purposes of this part.

**Owner** (Section 258.2) means the person(s) who owns a facility or part of a facility.

**Owner** (Section 260.10) means the person who owns a facility or part of a facility.

**Owner** (Section 280.12) means: (a) In the case of an UST system in use on November 8, 1984, or brought into use after that date, any person who owns an UST system used for storage, use, or dispensing of regulated substances; and (b) In the case of any UST system in use before November 8, 1984, but no longer in use on that date, any person who owned such UST immediately before the discontinuation of its use.

**Owner** (Section 72.2) means any of the following persons: (1) Any holder of any portion of the legal or equitable title in an affected unit or in a combustion source or process source; or (2) Any holder of a leasehold interest in an affected unit or in a combustion source or process source; or (3) Any purchaser of power from an affected unit or from a combustion source or process source under a life-of-the-unit, firm power contractual arrangement as the term is defined herein and used in section 408(i) of the Act. However, unless expressly provided for in a leasehold agreement, owner shall not include a passive lessor, or a person who has an equitable interest through such lessor, whose rental payments are not based, either directly or indirectly, upon the revenues or income from the affected unit; or (4) With respect to any Allowance Tracking System general account, any person identified in the submission required by Sec. 73.31© of this chapter that is subject to the binding agreement for the authorized account representative to represent that person's ownership interest with respect to allowances.

**Owner** (Section 745.103) means any entity that has legal title to target housing, including but not limited to individuals, partnerships, corporations, trusts, government agencies, housing agencies, Indian tribes, and nonprofit organizations, except where a mortgagee holds legal title to property serving as collateral for a mortgage loan, in which case the owner would be the mortgagor.

**Owner** (Section 85.2102) means the original purchaser or any subsequent purchaser of a vehicle.

**Owner** (Section 96.2) means any of the following persons: (1) Any holder of any portion of the legal or equitable title in a NOX Budget unit or in a unit for which an application for a NOX Budget opt-in permit under Sec. 96.83 is submitted and not denied or withdrawn; or (2) Any holder of a leasehold interest in a NOX Budget unit or in a unit for which an application for a NOX Budget opt-in permit under Sec. 96.83 is submitted and not denied or withdrawn; or (3) Any purchaser of power from a NOX Budget unit or from a unit for which an application for a NOX Budget opt-in permit under Sec. 96.83 is submitted and not denied or withdrawn under a life-of-the-unit, firm power contractual arrangement. However, unless expressly provided for in a leasehold agreement, owner shall not include a passive lessor, or a person who has an equitable interest through such lessor, whose rental payments are not based, either directly or indirectly, upon the revenues or income from the NOX Budget unit or the unit for which an application for a NOX Budget opt-in permit under Sec. 96.83 is submitted and not denied or withdrawn; or (4) With respect to any general account, any person who has an ownership interest with respect to the NOX allowances held in the general account and who is subject to the binding agreement for the NOX authorized account representative to represent that person's ownership interest with respect to NOX allowances.

**Owner** (Section 97.2) means any of the following persons: (1) Any holder of any portion of the legal or equitable title in a NOX Budget unit or in a unit for which an application for a NOX Budget opt-in permit under Sec. 97.83 is submitted and not denied or withdrawn; or (2) Any holder of a leasehold interest in a NOX Budget unit or in a unit for which an application for a NOX Budget opt-in permit under Sec. 97.83 is submitted and not denied or withdrawn; or (3) Any purchaser of power from a NOX Budget unit or from a unit for which an application for a NOX Budget opt-in permit under Sec. 97.83 is submitted and not denied or withdrawn under a life-of-the-unit, firm power contractual arrangement. However, unless expressly provided for in a leasehold agreement, owner shall not include a passive lessor, or a person who has an equitable interest through such lessor, whose rental payments are not based, either directly or indirectly, upon the revenues or income from the NOX Budget unit or the unit for which an application for a NOX Budget opt-in permit under Sec. 97.83 is submitted and not denied or withdrawn; or (4) With respect to any general account, any person who has an ownership interest with respect to the NOX allowances held in the general account and who is subject to the binding agreement for the NOX authorized account representative to represent that person's ownership interest with respect to NOX allowances.

**Owner or operator** (Section 112.2) means any person owning or operating an onshore facility or an offshore facility, and in the case of any abandoned offshore facility, the person who owned or operated such facility immediately prior to such abandonment.

**Owner or operator** (Section 122.2) means the owner or operator of any "facility or activity" subject to regulation under the NPDES program.

**Owner or operator** (Section 124.2) means owner or operator of any "facility or activity" subject to regulation under the RCRA, UIC, NPDES, or 404 programs.

**Owner or operator** (Section 129.2) means any person who owns, leases, operates, controls, or supervises a source as defined above.

**Owner or operator** (Section 144.3) means the owner or operator of any "facility or activity" subject to regulation under the UIC program.

**Owner or operator** (Section 146.3) means the owner or operator of any facility or activity subject to regulation under the RCRA, UIC, NPDES, or 404 programs.

**Owner or operator** (Section 232.2) means the owner or operator of any activity subject to regulation under the 404 program.

**Owner or operator** (Section 270.2) means the owner or operator of any facility or activity subject to regulation under RCRA.

**Owner or operator** (Section 280.92) when the owner or operator are separate parties, refers to the party that is obtaining or has obtained financial assurances.

**Owner or operator** (Section 51.100) means any person who owns, leases, operates, controls, or supervises a facility, building, structure, or installation which directly or indirectly result or may result in emissions of any air pollutant for which a national standard is in effect.

**Owner or operator** (Section 60.2) means any person who owns, leases, operates, controls, or supervises an affected facility or a stationary source of which an affected facility is a part.

**Owner or operator** (Section 61.02) means any person who owns, leases, operates, controls, or supervises a stationary source.

**Owner or operator** (Section 63.2) means any person who owns, leases, operates, controls, or supervises a stationary source.

**Owner or operator** (Section 64.1) means any person who owns, leases, operates, controls or supervises a stationary source subject to this part.

**Owner or operator** (Section 65.2) means any person who owns, leases, operates, controls, or supervises a regulated source or a stationary source of which a regulated source is a part.

**Owner or operator** (Section 66.3) means any person who owns, leases, operates or supervises a facility, building, structure or installation which emits or has the potential to emit any air pollutant regulated by EPA under the Act.

**Owner or operator** (Section 72.2) means any person who is an owner or who operates, controls, or supervises an affected unit, affected source, combustion source, or process source and shall include, but not be limited to, any holding company, utility system, or plant manager of an affected unit, affected source, combustion source, or process source.

**Owner/operator** (Section 147.2902) the owner or operator of any facility or activity subject to regulation under the Osage UIC program.

**Owner's Manual** (Section 85.2102) means the instruction booklet normally provided to the purchaser of a vehicle.

**Oxides of nitrogen** (Section 86.082-2) means the sum of the nitric oxide and nitrogen dioxide contained in a gas sample as if the nitric oxide were in the form of nitrogen dioxide.

**Oxides of nitrogen** (Section 86.1803-01) means the sum of the nitric oxide and nitrogen dioxide contained in a gas sample as if the nitric oxide were in the form of nitrogen dioxide.

**Oxides of nitrogen** (Section 92.2) means nitric oxide and nitrogen dioxide. Oxides of nitrogen are expressed quantitatively as if the nitric oxide were in the form of nitrogen dioxide (oxides of nitrogen are assumed to have a molecular weight equivalent to nitrogen dioxide).

**Oxides of nitrogen** (Section 94.2) means nitric oxide and nitrogen dioxide. Oxides of nitrogen are expressed quantitatively as if the nitric oxide were in the form of nitrogen dioxide (oxides of nitrogen are assumed to have a molecular weight equivalent to nitrogen dioxide).

**Oxygenate** (Section 80.2) means any substance which, when added to gasoline, increases the oxygen content of that gasoline. Lawful use of any of the substances or any combination of these substances requires that they be "substantially similar" under section 211(f)(1) of the Clean Air Act, or be permitted under a waiver granted by the Administrator under the authority of section 211(f)(4) of the Clean Air Act.

**Oxygenate blender** (Section 80.2) means any person who owns, leases, operates, controls, or supervises an oxygenate blending facility, or who owns or controls the blendstock or gasoline used or the gasoline produced at an oxygenate blending facility.

**Oxygenate blending facility** (Section 80.2) means any facility (including a truck) at which oxygenate is added to gasoline or blendstock, and at which the quality or quantity of gasoline is not altered in any other manner except for the addition of deposit control additives.

**Oxygenate compound** (Section 79.2) means an oxygen-containing, ashless organic compound, such as an alcohol or ether, which may be used as a fuel or fuel additive.

**Oxygenated gasoline** (Section 80.2) means gasoline which contains a measurable amount of oxygenate.

**Ozone deterioration factor** (Section 86.1702-99) means a factor applied to the mass of NMOG emissions from TLEVs, LEVs, or ULEVs which accounts for changes in the ozone-forming potential of the NMOG emissions from a vehicle as it accumulates mileage.

**Ozone nonattainment area** (Section 72.2) means an area designated as a nonattainment area for ozone under subpart C of part 81 of this chapter.

**Ozone season** (Section 72.2) means the period of time beginning May 1 of a year and ending on September 30 of the same year, inclusive.

**Ozone Transport Commission States or OTC States** (Section 86.1702-99) means the States of Connecticut, Delaware, Maine,

Maryland, Massachusetts, New Hampshire, New Jersey, New York, Pennsylvania, Rhode Island, Vermont and Virginia, and the District of Columbia.

**Ozone transport region** (Section 72.2) means the ozone transport region designated under Section 184 of the Act.

# P

**Package** (Section 211.203) The container in which a hearing protective device is presented for purchase or use. The package in some cases may be the same as the carrying case.

**Package or packaging** (Section 152.3) means the immediate container or wrapping, including any attached closure(s), in which the pesticide is contained for distribution, sale, consumption, use, or storage. The term does not include any shipping or bulk container used for transporting or delivering the pesticide unless it is the only such package.

**Package or packaging** (Section 157.21) means the immediate container or wrapping, including any attached closure(s), in which the pesticide is contained for distribution, sale, consumption, use or storage. The term does not include any shipping or bulk container used for transporting or delivering the pesticide unless it is the only such package.

**Packaging of pesticide products** (Section 455.10) means enclosing or placing a formulated pesticide product into a marketable container.

**Packer** (Section 146.3) means a device lowered into a well to produce a fluid-tight seal.

**Packer** (Section 147.2902) a device lowered into a well to produce a fluid-tight seal within the casing.

**Packinghouse** (Section 432.41) shall mean a plant that both slaughters animals and subsequently processes carcasses into cured, smoked, canned or other prepared meat products.

**PAH (as phenanthrene)** (Section 435.11) means polynuclear aromatic hydrocarbons reported as phenanthrene.

**Paint in poor condition** (Section 745.223) means more than 10 square feet of deteriorated paint on exterior components with large surface areas; or more than 2 square feet of deteriorated paint on interior components with large surface areas (e.g., walls, ceilings, floors, doors); or more than 10 percent of the total surface area of the component is deteriorated on interior or exterior components with small surface areas (window sills, baseboards, soffits, trim).

**Pallet** (Section 247.3) means a portable platform for storing or moving cargo or freight;

**Pamphlet** (Section 745.83) means the EPA pamphlet developed under section 406(a) of TSCA for use in complying with this and other rulemakings under Title IV of TSCA and the Residential Lead-Based Paint Hazard Reduction Act, or any State or Tribal pamphlet approved by EPA pursuant to 40 CFR 745.326 that is developed for the same purpose. This includes reproductions of the pamphlet when copied in full and without revision or deletion of material from the pamphlet (except for the addition or revision of State or local sources of information).

**PAMS** (Section 58.1) means Photochemical Assessment Monitoring Stations.

**Paper** (Section 247.3) means one of two broad subdivisions of paper products, the other being paperboard. Paper is generally lighter in basis weight, thinner, and more flexible than paperboard. Sheets 0.012 inch or less in thickness are generally classified as paper. Its primary uses are for printing, writing, wrapping, and sanitary purposes. However, in this guideline, the term paper is also used as a generic term that includes both paper and paperboard.

**Paper product** (Section 247.3) means any item manufactured from paper or paperboard. The term paper product is used in this guideline to distinguish such items as boxes, doilies, and paper towels from printing and writing papers.

**Parallel hybrid electric vehicle** (Section 86.1702-99) means any vehicle which allows power to be delivered to the driven wheels by either a combustion engine and/or by a battery powered electric motor.

**Parameters** (Section 444.11) are defined as Parameters at 40 CFR 136.2 in Table 1B, which also cites the approved methods of analysis. (1) Arsenic means total arsenic, Parameter 6. (2) Cadmium means total cadmium, Parameter 12. (3) Chromium means total chromium, Parameter 19. (4) Copper means total copper, Parameter 22. (5) Lead means total lead, Parameter 32. (6) Mercury means total mercury, Parameter 35. (7) pH means hydrogen ion, Parameter 28. (8) Silver means total silver, Parameter 62. (9) Titanium means total titanium, Parameter 72. (10) TSS means total suspended solids, Parameter 55. (11) Zinc means total zinc, Parameter 75.

**Parent company** (Section 704.3) is a company that owns or controls another company.

**Park benches and picnic tables** (Section 247.3) are recreational furniture found in parks, outdoor recreational facilities, and the grounds of office buildings and other facilities.

**Parking stop** (Section 247.3) means a barrier used to mark parking spaces and keep parked vehicles from rolling beyond a designated parking area;

**Part 70 or 71 permit** (Section 64.1) shall have the same meaning as provided under part 70 or 71 of this chapter, provided that it shall also refer to a permit issued, renewed, amended, revised, or modified under any federal permit program promulgated under title V of the Act.

**Part 70 or 71 permit application** (Section 64.1) shall mean an application (including any supplement to a previously submitted application) that is submitted by the owner or operator in order to obtain a part 70 or 71 permit.

**Part 70 permit** (Section 60.2) means any permit issued , renewed, or revised pursuant to part 70 of this chapter.

**Part 70 permit** (Section 61.02) means any permit issued, renewed, or revised pursuant to part 70 of this chapter.

**Part 70 permit** (Section 65.2) means any permit issued, renewed, or revised pursuant to part 70 of this chapter.

**Part 70 permit** (Section 71.2) means any permit or group of permits covering a part 70 source that has been issued, renewed, amended or revised pursuant to 40 CFR part 70.

**Part 70 permit or permit (unless the context suggests otherwise)** (Section 70.2) means any permit or group of permits covering a part 70 source that is issued, renewed, amended, or revised pursuant to this part.

**Part 70 program or State program** (Section 70.2) means a program approved by the Administrator under this part.

**Part 70 program or State program** (Section 71.2) means a program approved by the Administrator under 40 CFR part 70.

**Part 70 source** (Section 70.2) means any source subject to the permitting requirements of this part, as provided in Sec. Sec. 70.3(a) and 70.3(b) of this part.

**Part 70 source** (Section 71.2) means any source subject to the permitting requirements of 40 CFR part 70, as provided in Secs. 70.3(a) and 70.3(b).

**Part 71 permit, or permit (unless the context suggests otherwise)** (Section 71.2) means any permit or group of permits covering a part 71 source that has been issued, renewed, amended or revised pursuant to this part.

**Part 71 program** (Section 71.2) means a Federal operating permits program under this part.

**Part 71 source** (Section 71.2) means any source subject to the permitting requirements of this part, as provided in Secs. 71.3(a) and 71.3(b).

**Partial closure** (Section 260.10) means the closure of a hazardous waste management unit in accordance with the applicable closure requirements of parts 264 and 265 of this chapter at a facility that contains other active hazardous waste management units. For example, partial closure may include the closure of a tank (including its associated piping and underlying containment systems), landfill cell, surface impoundment, waste pile, or other hazardous waste management unit, while other units of the same facility continue to operate.

**Partially-Covered Fleet** (Section 88.302-93) pertains to a vehicle fleet in a covered area which contains both covered fleet vehicles and non-covered fleet vehicles, i.e., exempt from covered fleet purchase requirements.

**Participant** (Section 172.1) means any person acting as a representative of the permittee and responsible for making available for use, or supervising the use or evaluation of, an experimental use pesticide to be applied at a specific application site.

**Participant** (Section 195.2) is an individual or organization engaged in radon measurement and/or mitigation activities or in offering radon measurement and/or mitigation services to consumers and others, whose proficiency program application EPA has accepted.

**Participating PRP** (Section 304.12) is any potentially responsible party who has agreed, pursuant to Sec. 304.21 of this part, to submit one or more issues arising in an EPA claim for resolution pursuant to the procedures established by this part.

**Particulate matter** (Section 51.100) means any airborne finely divided solid or liquid material with an aerodynamic diameter smaller than 100 micrometers.

**Particulate matter** (Section 60.2) means any finely divided solid or liquid material, other than uncombined water, as measured by the reference methods specified under each applicable subpart, or an equivalent or alternative method.

**Particulate matter emissions** (Section 51.100)means all finely divided solid or liquid material, other than uncombined water, emitted to the ambient air as measured by applicable reference methods, or an equivalent or alternative method, specified in this chapter, or by a test method specified in an approved State implementation plan.

**Particulate matter monitoring network description, required by Sec. 58.20(f),** (Section 58.1) means a detailed plan, prepared by control agencies and submitted to EPA for approval that describes their PM2.5 and PM10 air quality surveillance network.

**Parts per million, carbon or ppm C** (Section 92.102) means the concentration of an organic compound in a gas expressed as parts per million (by volume or by moles) multiplied by the number of carbon atoms in a molecule of that compound.

**Part-time fellow** (Section 46.120) An individual enrolled in an academic educational program directly related to pollution abatement and control and taking at least 6 credit hours but less than 30 credit hours per school year, or an academic workload otherwise defined by the institution as less than a full-time curriculum. The fellow need not be pursuing a degree.

**Party** (Section 104.2) means the Environmental Protection Agency as the proponent of an effluent standard or standards, and any person who files an objection pursuant to Sec. 104.3 hereof.

**Party** (Section 108.2) means an employee filing a request under Sec. 108.3, any employee similarly situated, the employer of any such employee, and the Regional Administrator or his designee.

**Party** (Section 164.2) means any person, group, organization, or Federal agency or department that participates in a hearing.

**Party** (Section 209.3) means the Environmental Protection Agency, the respondent(s) and any interveners.

**Party** (Section 304.12) means EPA and any person who has agreed, pursuant to Sec. 304.21 of this part, to submit one or more issues arising in an EPA claim for resolution pursuant to the procedures established by this part, and any person who has been granted leave to intervene pursuant to Sec. 304.24(a) of this part.

**Party** (Section 305.3) means EPA or any person that participates in a proceeding under this part as a Requestor.

**Party** (Section 307.14) means EPA or a claimant.

**Party** (Section 791.3) refers to a person subject to a section 4 test rule, who: (1) Seeks reimbursement from another person under these rules, or (2) From whom reimbursement is sought under these rules.

**Party** (Section 82.3) means any foreign state that is listed in appendix C to this subpart (pursuant to instruments of ratification, acceptance, or approval deposited with the Depositary of the United Nations Secretariat), as having ratified the specified control measure in effect under the Montreal Protocol. Thus, for purposes of the trade bans specified in Sec. 82.4(k)(2) pursuant to the London Amendments, only those foreign states that are listed in appendix C to this subpart as having ratified both the 1987 Montreal Protocol and the London Amendments shall be deemed to be Parties.

**Party to the proceeding** (Section 173.2) shall mean the State or the Agency's Office of Enforcement.

**Pass Through** (Section 403.3) means a Discharge which exits the POTW into waters of the United States in quantities or concentrations which, alone or in conjunction with a discharge or discharges from other sources, is a cause of a violation of any requirement of the POTW's NPDES permit (including an increase in the magnitude or duration of a violation).

**Passenger** (Section 94.2) has the meaning given by 46 U.S.C. 2101(21). This generally means that a passenger is a person that pays to be on the vessel.

**Passenger Automobile** (Section 600.022-85) means any automobile which the Secretary determines is manufactured primarily for use in the transportation of no more than 10 individuals.

**Passenger locomotive** (Section 92.2) means a locomotive designed and constructed for the primary purpose of propelling passenger trains, and providing power to the passenger cars of the train for such functions as heating, lighting and air conditioning.

**Passive institutional control** (Section 191.12) means: (1) Permanent markers placed at a disposal site, (2) public records and archives, (3) government ownership and regulations regarding land or resource use, and (4) other methods of preserving knowledge about the location, design, and contents of a disposal system.

**Past year** (Section 167.3) means the calendar year immediately prior to that in which the report is submitted.

**Pasture** (Section 503.11) is land on which animals feed directly on feed crops such as legumes, grasses, grain stubble, or stover.

**Pathogenic organisms** (Section 503.31) are disease-causing organisms. These include, but are not limited to, certain bacteria, protozoa, viruses, and viable helminth ova.

**Pb** (Section 58.1) means lead.

**PCB and PCBs** (Section 761.13) means any chemical substance that is limited to the biphenyl molecule that has been chlorinated to varying degrees or any combination of substances which contains such substance. Refer to Sec. 761.1(b) for applicable concentrations of PCBs. PCB and PCBs as contained in PCB items are defined in Sec. 761.3. For any purposes under this part, inadvertently generated non-Aroclor PCBs are defined as the total PCBs calculated following division of the quantity of monochlorinated biphenyls by 50 and dichlorinated biphenyls by 5.

**PCB Article** (Section 761.13) means any manufactured article, other than a PCB Container, that contains PCBs and whose surface(s) has been in direct contact with PCBs. "PCB Article" includes capacitors, transformers, electric motors, pumps, pipes and any other manufactured item (1) which is formed to a specific shape or design during manufacture, (2) which has end use function(s) dependent in whole or in part upon its shape or design during end use, and (3) which has either no change of chemical composition during its end use or only those changes of composition which have no commercial purpose separate from that of the PCB Article.

**PCB Article Container** (Section 761.13) means any package, can, bottle, bag, barrel, drum, tank, or other device used to contain PCB Articles or PCB Equipment, and whose surface(s) has not been in direct contact with PCBs.

**PCB bulk product waste** (Section 761.13) means waste derived from manufactured products containing PCBs in a non-liquid state, at any concentration where the concentration at the time of designation for disposal was [ge]50 ppm PCBs. PCB bulk product waste does not include PCBs or PCB Items regulated for disposal under Sec. 761.60(a) through (c), Sec. 761.61, Sec. 761.63, or Sec. 761.64. PCB bulk product waste includes, but is not limited to: (1) Non-liquid bulk wastes or debris from the demolition of buildings and other man-made structures manufactured, coated, or serviced with PCBs. PCB bulk product waste does not include debris from the demolition of buildings or other man-made structures that is contaminated by spills from regulated PCBs which have not been disposed of, decontaminated, or otherwise cleaned up in accordance with subpart D of this part. (2) PCB-containing wastes from the shredding of automobiles, household appliances, or industrial appliances. (3) Plastics (such as plastic insulation from wire or cable; radio, television and computer casings; vehicle parts; or furniture laminates); preformed or molded rubber parts and components; applied dried paints, varnishes, waxes or other similar coatings or sealants; caulking; adhesives; paper; Galbestos; sound deadening or other types of insulation; and felt or fabric products such as gaskets. (4) Fluorescent light ballasts containing PCBs in the potting material.

**PCB Capacitor** (Section 761.13) means any capacitor that contains [ge]500 ppm PCB. Concentration assumptions applicable to capacitors appear under Sec. 761.2.

**PCB Container** (Section 761.13) means any package, can, bottle, bag, barrel, drum, tank, or other device that contains PCBs or PCB Articles and whose surface(s) has been in direct contact with PCBs.

**PCB Equipment** (Section 761.13) means any manufactured item, other than a PCB Container or a PCB Article Container, which contains a PCB Article or other PCB Equipment, and includes microwave ovens, electronic equipment, and fluorescent light ballasts and fixtures.

**PCB field screening test** (Section 761.13) means a portable analytical device or kit which measures PCBs. PCB field screening tests usually report less than or greater than a specific numerical PCB concentration. These tests normally build in a safety factor which increases the probability of a false positive report and decreases the probability of a false negative report. PCB field screening tests do not usually provide: an identity record generated by an instrument; a quantitative comparison record from calibration standards; any identification of PCBs; and/or any indication or identification of interferences with the measurement of the PCBs. PCB field screening test technologies include, but are not limited to, total chlorine colorimetric tests, total chlorine x-ray fluorescence tests, total chlorine microcoulometric tests, and rapid immunoassay tests.

**PCB household waste** (Section 761.13) means PCB waste that is generated by residents on the premises of a temporary or permanent residence for individuals (including individually owned or rented units of a multi-unit construction), and that is composed primarily of materials found in wastes generated by consumers in their homes. PCB household waste includes unwanted or discarded non-commercial vehicles (prior to shredding), household items, and appliances or appliance parts and wastes generated on the premises of a residence for individuals as a result of routine household maintenance by or on behalf of the resident. Bulk or commingled liquid PCB wastes at concentrations of [ge]50 ppm, demolition and renovation wastes, and industrial or heavy duty equipment with PCBs are not household wastes.

**PCB Item** (Section 761.13) means any PCB Article, PCB Article Container, PCB Container, PCB Equipment, or anything that deliberately or unintentionally contains or has as a part of it any PCB or PCBs.

**PCB remediation waste** (Section 761.13) means waste containing PCBs as a result of a spill, release, or other unauthorized disposal, at the following concentrations: Materials disposed of prior to April 18, 1978, that are currently at concentrations [ge] 50 ppm PCBs, regardless of the concentration of the original spill; materials which are currently at any volume or concentration where the original source was [ge] 500 ppm PCBs beginning on April 18, 1978, or [ge] 50 ppm PCBs beginning on July 2, 1979; and materials which are currently at any concentration if the PCBs are spilled or released from a source not authorized for use under this part. PCB remediation waste means soil, rags, and other debris generated as a result of any PCB spill cleanup, including, but not limited to: (1) Environmental media containing PCBs, such as soil and gravel; dredged materials, such as sediments, settled sediment fines, and aqueous decantate from sediment. (2) Sewage sludge containing <50 ppm PCBs and not in use according to Sec. 761.20(a)(4); PCB sewage sludge; commercial or industrial sludge contaminated as the result of a spill of PCBs including sludges located in or removed from any pollution control device; aqueous decantate from an industrial sludge. (3) Buildings and other man-made structures (such as concrete floors, wood floors, or walls contaminated from a leaking PCB or PCB-Contaminated Transformer), porous surfaces, and non-porous surfaces.

**PCB sewage sludge** (Section 761.13) means sewage sludge as defined in 40 CFR 503.9(w) which contains [ge]50 ppm PCBs, as measured on a dry weight basis.

**PCB Transformer** (Section 761.13) means any transformer that contains [ge]500 ppm PCBs. For PCB concentration assumptions applicable to transformers containing 1.36 kilograms (3 lbs.) or more of fluid other than mineral oil, see Sec. 761.2. For provisions permitting reclassification of electrical equipment, including PCB Transformers, containing [ge]500 ppm PCBs to PCB-Contaminated Electrical Equipment, see Sec. 761.30(a) and (h).

**PCB waste(s)** (Section 761.13) means those PCBs and PCB Items that are subject to the disposal requirements of subpart D of this part.

**PCB/radioactive waste** (Section 761.13) means PCBs regulated for disposal under subpart D of this part that also contain source, special nuclear, or byproduct material subject to regulation under the Atomic Energy Act of 1954, as amended, or naturally-occurring or accelerator-produced radioactive material.

**PCB-Contaminated** (Section 761.13) means a non-liquid material containing PCBs at concentrations [ge]50 ppm but <500 ppm; a liquid material containing PCBs at concentrations [ge]50 ppm but <500 ppm or where insufficient liquid material is available for analysis, a non-porous surface having a surface concentration 10 [mu]g/100 cm2 but <100 [mu]g/100 cm2, measured by a standard wipe test as defined in Sec. 761.123.

**PCB-Contaminated Electrical Equipment** (Section 761.13) means any electrical equipment including, but not limited to, transformers (including those used in railway locomotives and self-propelled cars), capacitors, circuit breakers, reclosers, voltage regulators, switches (including sectionalizers and motor starters), electromagnets, and cable, that contains PCBs at concentrations of [ge]50 ppm and <500 ppm in the contaminating fluid. In the absence of liquids, electrical equipment is PCB-Contaminated if it has PCBs at 10 [mu]g/100 cm2 and <100 [mu]g/100 cm2 as measured by a standard wipe test (as defined in Sec. 761.123) of a non-porous surface.

**PCBs** (Section 761.123) means polychlorinated biphenyls as defined under Sec. 761.3. As specified under Sec. 761.1(b), no requirements may be avoided through dilution of the PCB concentration.

**Peaches** (Section 407.61) shall mean the processing of peaches into the following product styles: Canned or frozen, all varieties, peeled, pitted and unpitted, whole, halves, sliced, diced, and any other cuts, nectar, and concentrate but not dehydrated.

**Peak torque speed** (Section 86.082-2) means the speed at which an engine develops maximum torque.

**Peaking unit** (Section 72.2) means: (1) A unit that has: (i) An average capacity factor of no more than 10.0 percent during the previous three calendar years and (ii) A capacity factor of no more than 20.0 percent in each of those calendar years. (2) For purposes of part 75 of this chapter, a unit may initially qualify as a peaking unit if the designated representative demonstrates to the satisfaction of the administrator that the requirements of paragraph (1) of this definition are met, or will in the future be met, through one of the following submissions: (i) For a unit for which a monitoring plan has not been submitted under Sec. 75.62, the designated representative submits either: (A) Capacity factor data for the unit for the three calendar years immediately preceding the date of initial submission of the monitoring plan for the unit under Sec. 75.62; or (B) If a unit does not have capacity factor data for one or more of the three calendar years immediately preceding the date of initial submission of the monitoring plan for the unit under Sec. 75.62, all available capacity factor data, beginning with the date on which the unit commenced commercial operation; and projected capacity factor data. (ii) For a unit for which a monitoring plan has already been submitted under Sec. 75.62, that has not qualified as a peaking unit under paragraph (2)(i) of this definition, and where capacity factor changes, the designated representative submits either: (A) Three calendar years of data following the change in the unit's capacity factor showing an average capacity factor of no more than 10.0 percent during the three previous calendar years and a capacity factor of no more than 20.0 percent in each of those calendar years; or (B) One calendar year of data following the change in the unit's capacity factor showing a capacity factor of no more than 10.0 percent and a statement that this changed pattern of operation resulting in a capacity factor less than 10.0 percent is considered permanent and is projected to continue for the foreseeable future. (3) For purposes of part 75 of this chapter, a unit that initially qualifies as a peaking unit must meet the criteria in paragraph (1) of this definition each year in order to continue to qualify as a peaking unit. If such a unit fails to meet such criteria for a given year, the unit no longer qualifies as a peaking unit starting January 1 of the year after the year for which the criteria are not met. If a unit failing to meet the criteria in paragraph (1) of this definition initially qualified as a peaking unit under paragraph (2) of this definition, the unit may qualify as a peaking unit for a subsequent year only if the designated representative submits the data specified in paragraph (2)(ii)(A) of this definition.

**Peaking unit** (Section 72.2) means: (4) A unit required to comply with the provisions of subpart H of part 75 of this chapter, under a State or Federal NOX mass emissions reduction program, may, pursuant to Sec. 75.74©(11) in part 75 of this chapter, qualify as a peaking unit on an ozone season basis rather than an annual basis, if the owner or operator reports NOX mass emissions and heat input data only during the ozone season.

**Pears** (Section 407.61) shall mean the processing of pears into the following product styles: Canned, peeled, halved, sliced, diced, and any other cuts, nectar and concentrate but not dehydrated.

**Peas** (Section 407.71) shall mean the processing of peas into the following product styles: Canned and frozen, all varieties and sizes, whole.

**Percent elongation at break** (Section 238.20) means the percent increase in length of the plastic material caused by a tensile load. Percent elongation at break shall be calculated by dividing the extension at the moment of rupture of the specimen by the initial gage length of the specimen and multiplying by 100.

**Percent load** (Section 86.082-2) means the fraction of the maximum available torque at a specified engine speed.

**Percent monitor data availability** (Section 97.2) means, for purposes of Sec. 97.43 (a)(1) and Sec. 94.84(b), total unit operating hours for which quality-assured data were recorded under subpart H of this part in a control period, divided by 3,672 hours per control period, and multiplied by 100%.

**Percent removal** (Section 133.101) A percentage expression of the removal efficiency across a treatment plant for a given pollutant parameter, as determined from the 30-day average values of the raw wastewater influent pollutant concentrations to the facility and the 30-day average values of the effluent pollutant concentrations for a given time period.

**Perfected** (Section 307.14) means the point at which EPA determines that the written demand for a sum certain (i.e., claim) has the documentation necessary to substantiate the appropriateness of the amounts claimed; i.e., the claim is technically complete.

**Performance assessment** (Section 191.12) means an analysis that: (1) Identifies the processes and events that might affect the disposal system; (2) examines the effects of these processes and events on the performance of the disposal system; and (3) estimates the cumulative releases of radionuclides, considering the associated uncertainties, caused by all significant processes and events. These estimates shall be incorporated into an overall probability distribution of cumulative release to the extent practicable.

**Performance audit** (Section 63.2) means a procedure to analyze blind samples, the content of which is known by the Administrator, simultaneously with the analysis of performance test samples in order to provide a measure of test data quality.

**Performance evaluation** (Section 63.2) means the conduct of relative accuracy testing, calibration error testing, and other measurements used in validating the continuous monitoring system data.

**Performance evaluation sample** (Section 141.2) means a reference sample provided to a laboratory for the purpose of demonstrating that the laboratory can successfully analyze the sample within limits of performance specified by the Agency. The true value of the concentration of the reference material is unknown to the laboratory at the time of the analysis.

**Performance Partnership Agreement** (Section 35.102) A negotiated agreement signed by the EPA Regional Administrator and an appropriate official of a State agency and designated as a Performance Partnership Agreement. Such agreements typically set out jointly developed goals, objectives, and priorities; the strategies to be used in meeting them; the roles and responsibilities of the State and EPA; and the measures to be used in assessing progress. A Performance Partnership Agreement may be used as all or part of a work plan for a grant if it meets the requirements for a work plan set out in Sec. 35.107.

**Performance Partnership Grant** (Section 35.102) A single grant combining funds from more than one environmental program. A Performance Partnership Grant may provide for administrative savings or programmatic flexibility to direct grant resources where they are most needed to address public health and environmental priorities (see also Sec. 35.130). Each Performance Partnership Grant has a single, integrated budget and recipients do not need to account for grant funds in accordance with the funds' original environmental program sources.

**Performance test** (Section 63.2) means the collection of data resulting from the execution of a test method (usually three emission test runs) used to demonstrate compliance with a relevant emission standard as specified in the performance test section of the relevant standard.

**Performance test** (Section 65.2) means the collection of data resulting from the execution of a test method (usually three emission test runs) used to demonstrate compliance with a relevant emission standard as specified in the performance test section of the relevant standard.

**Performance-based organic decontamination fluid (PODF)** (Section 761.13) means kerosene, diesel fuel, terpene hydrocarbons, and terpene hydrocarbon/alcohol mixtures.

**Periodic Certification Statement** (Section 455.41) for this subpart means a written submission to the appropriate permitting authority, e.g., the local Control Authority (the POTW) or NPDES permit writer, which states that the P2 Alternative is being implemented in the manner set forth in the control mechanism (for indirect dischargers) or NPDES permit (for direct dischargers) or that a justification allowing modification of the practices listed in Table 8 to this part 455 has been implemented resulting in a change in the pollution prevention practices conducted at the facility. The Periodic Certification Statement must be signed by the responsible corporate officer as defined in 40 CFR 403.12(l) or 40 CFR 122.22.

**Periodically regenerating trap oxidizer system** (Section 86.1702-99) means a trap oxidizer system that utilizes, during normal driving conditions for cleaning the trap, an automated regeneration mode which can be easily detected.

**Periodically regenerating trap oxidizer system** (Section 86.1803-01) means a trap oxidizer that utilizes, during normal driving conditions, an automated regeneration mode for cleaning the trap, the operation of which can be easily detected.

**Perlite composite board** (Section 247.3) means insulation board composed of expanded perlite and fibers formed into rigid, flat, rectangular units with a suitable sizing material incorporated in the product. It may have on one or both surfaces a facing or coating to prevent excessive hot bitumen strike-in during roofing installation;

**Permanent Radon Barrier** (Section 192.31) means the final radon barrier constructed to achieve compliance with, including attainment of, the limit on releases of radon-222 in Sec. 192.32(b) (1) (ii).

**Permanently covered soil** (Section 745.223) means soil which has been separated from human contact by the placement of a barrier consisting of solid, relatively impermeable materials, such as pavement or concrete. Grass, mulch, and other landscaping materials are not considered permanent covering.

**Permit** (Section 122.2) means an authorization, license, or equivalent control document issued by EPA or an "approved State" to implement the requirements of this part and parts 123 and 124. "Permit" includes an NPDES "general permit" (Sec. 122.28). Permit does not include any permit which has not yet been the subject of final agency action, such as a "draft permit" or a "proposed permit."

**Permit** (Section 124.2) means an authorization, license, or equivalent control document issued by EPA or an "approved State" to implement the requirements of this part and parts 122, 123, 144, 145, 233, 270, and 271. "Permit" includes RCRA "permit by rule" (Sec. 270.60), UIC area permit (Sec. 144.33), NPDES or 404 "general permit" (Secs. 270.61, 144.34, and 233.38). Permit does not include RCRA interim status (Sec. 270.70), UIC authorization by rule (Sec. 144.21), or any permit which has not yet been the subject of final agency action, such as a "draft permit" or a "proposed permit."

**Permit** (Section 129.2) means a permit for the discharge of pollutants into navigable waters under the National Pollutant Discharge Elimination System established by section 402 of the Act and implemented in regulations in 40 CFR parts 124 and 125.

**Permit** (Section 144.3) means an authorization, license, or equivalent control document issued by EPA or an approved State to implement the requirements of this part, parts 145, 146 and 124. "Permit" includes an area permit (Sec. 144.33) and an emergency permit (Sec. 144.34). Permit does not include UIC authorization by rule (Sec. 144.21), or any permit which has not yet been the subject of final agency action, such as a "draft permit."

**Permit** (Section 146.3) means an authorization, license, or equivalent control document issued by EPA or an "approved State" to implement the requirements of this part and parts 124, 144, and 145. Permit does not include RCRA interim status (Sec. 122.23), UIC authorization by rule (Secs. 144.21 to 144.26 and 144.15), or any permit which has not yet been the subject of final agency action, such as a "draft permit" or a "proposed permit."

**Permit** (Section 147.2902) an authorization issued by EPA to implement UIC program requirements. Permit does not include the UIC authorization by rule or any permit which has not yet been the subject of final Agency action.

**Permit** (Section 232.2) means a written authorization issued by an approved State to implement the requirements of part 233, or by the Corps under 33 CFR parts 320-330. When used in these regulations, "permit" includes "general permit" as well as individual permit.

**Permit** (Section 256.06) is an entitlement to commence and continue operation of a facility as long as both procedural and performance standards are met. The term "permit" includes any functional equivalent such as a registration or license.

**Permit** (Section 270.2) means an authorization, license, or equivalent control document issued by EPA or an approved State to implement the requirements of this part and parts 271 and 124. Permit includes permit by rule (Sec. 270.60), and emergency permit (Sec. 270.61). Permit does not include RCRA interim status (subpart G of this part), or any permit which has not yet been the subject of final agency action, such as a draft permit or a proposed permit.

**Permit** (Section 501.2) means an authorization, license, or equivalent control document issued by EPA or an "approved State program" to implement the requirements of this part.

**Permit documents** (Section 239.2) means permit applications, draft and final permits, or other documents that include applicable design and management conditions in accordance with the Subtitle D federal revised criteria, found at 40 CFR part 257, subpart B and 40 CFR part 258, and the technical and administrative information used to explain the basis of permit conditions.

**Permit modification** (Section 63.2) means a change to a title V permit as defined in regulations codified in this chapter to implement title V of the Act (42 U.S.C. 7661).

**Permit modification** (Section 70.2) means a revision to a part 70 permit that meets the requirements of Sec. 70.7 (e) of this part.

**Permit modification** (Section 71.2) means a revision to a part 71 permit that meets the requirements of Sec. 71.7(e).

**Permit or prior approval and conditions** (Section 239.2) means any authorization, license, or equivalent control document issued under the authority of the state regulating the location, design, operation, ground-water monitoring, closure, post-closure care, corrective action, and financial assurance of Subtitle D regulated facilities.

**Permit program** (Section 60.2) means a comprehensive State operating permit system established pursuant to title V of the Act (42 U.S.C. 7661) and regulations codified in part 70 of this chapter and applicable State regulations, or a comprehensive Federal operating permit system established pursuant to title V of the Act and regulations codified in this chapter.

**Permit program** (Section 61.02) means a comprehensive State operating permit system established pursuant to title V of the Act (42 U.S.C. 7661) and regulations codified in part 70 of this chapter and applicable State regulations, or a comprehensive Federal operating permit system established pursuant to title V of the Act and regulations codified in this chapter.

**Permit program** (Section 63.2) means a comprehensive State operating permit system established pursuant to title V of the Act (42 U.S.C. 7661) and regulations codified in part 70 of this chapter and applicable State regulations, or a comprehensive Federal operating permit system established pursuant to title V of the Act and regulations codified in this chapter.

**Permit program** (Section 65.2) means a comprehensive State operating permit system established pursuant to title V of the Act (42 U.S.C. 7661) and regulations codified in part 70 of this chapter and applicable State regulations, or a comprehensive Federal operating permit system established pursuant to title V of the Act and regulations codified in part 71 of this chapter.

**Permit program costs** (Section 70.2) means all reasonable (direct and indirect) costs required to develop and administer a permit program, as set forth in Sec. 70.9(b) of this part (whether such costs are incurred by the permitting authority or other State or local agencies that do not issue permits directly, but that support permit issuance or administration).

**Permit program costs** (Section 71.2) means all reasonable (direct and indirect) costs required to administer an operating permits program, as set forth in Sec. 71.9(b).

**Permit revision** (Section 63.2) means any permit modification or administrative permit amendment to a title V permit as defined in regulations codified in this chapter to implement title V of the Act (42 U.S.C. 7661).

**Permit revision** (Section 70.2) means any permit modification or administrative permit amendment.

**Permit revision** (Section 71.2) means any permit modification or administrative permit amendment.

**Permit revision** (Section 72.2) means a permit modification, fast track modification, administrative permit amendment, or automatic permit amendment, as provided in subpart H of this part.

**Permit-by-rule** (Section 270.2) means a provision of these regulations stating that a facility or activity is deemed to have a RCRA permit if it meets the requirements of the provision.

**Permittee** (Section 172.1) means any applicant to whom an experimental use permit has been granted.

**Permitting authority** (Section 230.3) means the District Engineer of the U.S. Army Corps of Engineers or such other individual as may be designated by the Secretary of the Army to issue or deny permits under section 404 of the Act; or the State Director of a permit program approved by EPA under section 404(g) and section 404(h) or his delegated representative.

**Permitting authority** (Section 60.2) means: (1) The State air pollution control agency, local agency, other State agency, or other agency authorized by the Administrator to carry out a permit program under part 70 of this chapter; or (2) The Administrator, in the case of EPA-implemented permit programs under title V of the Act (42 U.S.C. 7661).

**Permitting authority** (Section 61.02) means: (1) The State air pollution control agency, local agency, other State agency, or other agency authorized by the Administrator to carry out a permit program under part 70 of this chapter; or (2) The Administrator, in the case of EPA-implemented permit programs under title V of the Act (42 U.S.C. 7661).

**Permitting authority** (Section 63.2)means: (1) The State air pollution control agency, local agency, other State agency, or other agency authorized by the Administrator to carry out a permit program under part 70 of this chapter; or (2) The Administrator, in the case of EPA-implemented permit programs under title V of the Act (42 U.S.C. 7661).

**Permitting authority** (Section 64.1) shall have the same meaning as provided under part 70 or 71 of this chapter.

**Permitting authority** (Section 65.2) means one of the following: (1) The State air pollution control agency, local agency, other State agency, or other agency authorized by the Administrator to carry out a permit program under part 70 of this chapter; or (2) The Administrator, in the case of EPA-implemented permit programs under title V of the Act (42 U.S.C. 7661) and part 71 of this chapter.

**Permitting authority** (Section 70.2) means either of the following: (1) The Administrator, in the case of EPA-implemented programs; or (2) The State air pollution control agency, local agency, other State agency, or other agency authorized by the Administrator to carry out a permit program under this part.

**Permitting authority** (Section 71.2) means one of the following: (1) The Administrator, in the case of EPA-implemented programs; (2) A delegate agency authorized by the Administrator to carry out a Federal permit program under this part; or (3) The State air pollution control agency, local agency, other State agency, Indian Tribe, or other agency authorized by the Administrator to carry out a permit program under 40 CFR part 70.

**Permitting authority** (Section 72.2) means either: (1) When the Administrator is responsible for administering Acid Rain permits under subpart G of this part, the Administrator or a delegatee agency authorized by the Administrator; or (2) The State air pollution control agency, local agency, other State agency, or other agency authorized by the Administrator to administer Acid Rain permits under subpart G of this part and part 70 of this chapter.

**Permitting authority** (Section 96.2) means the State air pollution control agency, local agency, other State agency, or other agency authorized by the Administrator to issue or revise permits to meet the requirements of the NOX Budget Trading Program in accordance with subpart C of this part.

**Permitting authority** (Section 97.2) means the State air pollution control agency, local agency, other State agency, or other agency authorized by the Administrator to issue or revise permits to meet the requirements of the NOX Budget Trading Program in accordance with subpart C of this part.

**Person** (Section 104.2) means an individual, corporation, partnership, association, state, municipality or other political subdivision of a state, or any interstate body.

**Person** (Section 112.2) includes an individual, firm, corporation, association, and a partnership.

**Person** (Section 122.2) means an individual, association, partnership, corporation, municipality, State or Federal agency, or an agent or employee thereof.

**Person** (Section 124.2) means an individual, association, partnership, corporation, municipality, State, Federal, or Tribal agency, or an agency or employee thereof.

**Person** (Section 141.2) means an individual; corporation; company; association; partnership; municipality; or State, Federal, or tribal agency.

**Person** (Section 154.3) means an applicant, registrant, manufacturer, pesticide user, environmental group, labor union, or other individual or group of individuals interested in pesticide regulation.

**Person** (Section 205.2) means an individual, corporation, partnership, or association, and except as provided in sections 11(e) and 12(a) of the Act includes any officer, employee, department, agency or instrumentality of the United States, a State or any political subdivision of a State.

**Person** (Section 231.2) means an individual, corporation, partnership, association, Federal agency, state, municipality, or commission, or political subdivision of a state, or any interstate body.

**Person** (Section 247.3) means an individual, trust, firm, joint stock company, corporation (including a government corporation), partnership, association, Federal agency, State, municipality, commission, political subdivision of a State, or any interstate body;

**Person** (Section 280.12) means an individual, trust, firm, joint stock company, Federal agency, corporation, state, municipality, commission, political subdivision of a state, or any interstate body. "Person" also includes a consortium, a joint venture, a commercial entity, and the United States Government.

**Person** (Section 700.43) means a manufacturer, importer, or processor.

**Person** (Section 704.3) includes any individual, firm, company, corporation, joint venture, partnership, sole proprietorship, association, or any other business entity; any State or political subdivision thereof; any municipality; any interstate body; and any department, agency, or instrumentality of the Federal Government.

**Person** (Section 72.2) includes an individual, corporation, partnership, association, State, municipality, political subdivision of a State, any agency, department, or instrumentality of the United States, and any officer, agent, or employee thereof.

**Person** (Section 82.3) means any individual or legal entity, including an individual, corporation, partnership, association, state, municipality, political subdivision of a state, Indian tribe; any agency, department, or instrumentality of the United States; and any officer, agent, or employee thereof.

**Person** (Section 88.302-94) includes an individual, corporation, partnership, association, State, municipality, political subdivision of a State, and any agency, department, or instrumentality of the United States and any officer, agent, or employee thereof.

**Person as defined by section 101(21) of CERCLA** (Section 300.5) means an individual, firm, corporation, association, partnership, consortium, joint venture, commercial entity, United States government, state, municipality, commission, political subdivision of a state, or any interstate body. As defined by section 1001 of the OPA, "person" means an individual, corporation, partnership, association, state, municipality, commission, or political subdivision of a state, or any interstate body.

**Person as defined by section 101(21) of CERCLA** (Section 307.14) means an individual, firm, corporation, association, partnership, consortium, joint venture, commercial entity, United States Government, State, municipality, commission, political subdivision of a State, or any interstate body.

**Personal protective equipment** (Section 721.3) means any chemical protective clothing or device placed on the body to prevent contact with, and exposure to, an identified chemical substance or substances in the work area. Examples include, but are not limited to, chemical protective clothing, aprons, hoods, chemical goggles, face splash shields, or equivalent eye protection, and various types of respirators. Barrier creams are not included in this definition.

**Personal watercraft engine (PWC)** (Section 91.3) is a marine SI engine that does not meet the definition of outboard engine, inboard engine or sterndrive engine, except that the Administrator in his or her discretion may classify a PWC as an inboard or sterndrive engine if it is comparable in technology and emissions to an inboard or sterndrive engine.

**Personnel or facility personnel** (Section 260.10) means all persons who work, at, or oversee the operations of, a hazardous waste facility, and whose actions or failure to act may result in noncompliance with the requirements of part 264 or 265 of this chapter.

**Pest means**: (1) Any insect, rodent, nematode, fungus, weed, or (2) any other form of terrestrial or aquatic plant or animal life or virus, bacteria, or other micro-organism (except virusus, bacteria, or other micro-organisms on or in living man or other living animals) which the Administrator declares to be a pest under section 25©(1) of Pub. L. 94-140, Federal Insecticide, Fungicide and Rodenticide Act.

**Pest problem** (Section 162.151) means (1) a pest infestation and its consequences, or (2) any condition for which the use of plant regulators, defoliants, or desiccants would be appropriate.

**Pesticidal product** (Section 167.3) means a pesticide, active ingredient, or device.

**Pesticidal product report** (Section 167.3) means information showing the types and amounts of pesticidal products which were: (1) Produced in the past calendar year; (2) produced in the current calendar year; and, (3) sold or distributed in the past calendar year. For active ingredients, the pesticidal product report must include information on the types and amounts of an active ingredient for which there is actual or constructive knowledge of its use or intended use as a pesticide. This pesticidal product report also pertains to those products produced for export only which must also be reported. A positive or a negative annual report is required in order to maintain registration for the establishment.

**Pesticidal substance** (Section 174.3) means a substance that is intended to be produced and used in a living plant, or in the produce thereof, for a pesticidal purpose, during any part of a plant's life cycle (e.g., in the embryo, seed, seedling, mature plant).

**Pesticidal substance, when referring to a plant-incorporated protectant only** (Section 152.3) means a substance that is intended to be produced and used in a living plant, or in the produce thereof, for a pesticidal purpose during any part of a plant's life cycle (e.g., in the embryo, seed, seedling, mature plant).

**Pesticide** (Section 152.3) means any substance or mixture of substances intended for preventing, destroying, repelling, or mitigating any pest, or intended for use as a plant regulator, defoliant, or desiccant, other than any article that: (1) Is a new animal drug under FFDCA sec. 201(w), or (2) Is an animal drug that has been determined by regulation of the Secretary of Health and Human Services not to be a new animal drug, or (3) Is an animal feed under FFDCA sec. 201(x) that bears or contains any substances described by paragraph (s) (1) or (2) of this section.

**Pesticide** (Section 159.153) means a pesticide product which is or was registered by EPA, and each active ingredient, inert ingredient, impurity, metabolite, contaminant or degradate contained in, or derived from, such pesticide product.

**Pesticide** (Section 260.10) means any substance or mixture of substances intended for preventing, destroying, repelling, or mitigating any pest, or intended for use as a plant regulator, defoliant, or desiccant, other than any article that: (1) Is a new animal drug under FFDCA section 201(w), or (2) Is an animal drug that has been determined by regulation of the Secretary of Health and Human Services not to be a new animal drug, or (3) Is an animal feed under FFDCA section 201(x) that bears or contains any substances described by paragraph (1) or (2) of this definition.

**Pesticide** (Section 273.9) means any substance or mixture of substances intended for preventing, destroying, repelling, or mitigating any pest, or intended for use as a plant regulator, defoliant, or desiccant, other than any article that: (a) Is a new animal drug under FFDCA section 201(w), or (b) Is an animal drug that has been determined by regulation of the Secretary of Health and Human Services not to be a new animal drug, or (c) Is an animal feed under FFDCA section 201(x) that bears or contains any substances described by paragraph (a) or (b) of this section.

**Pesticide** (Section 455.10) means any substance or mixture of substances intended for preventing, destroying, repelling, or mitigating any pest.

**Pesticide** (section 710.2) shall have the meaning contained in the Federal Insecticide, Fungicide, and Rodenticide Act, 7 U.S.C. 136 et seq., and the regulations issued thereunder.

**Pesticide** (Section 720.3) has the meaning contained in the Federal Insecticide, Fungicide, and Rodenticide Act, 7 U.S.C. 136 et seq. and the regulations issued under it.

**Pesticide chemical** (Section 177.3) means any substance which alone, or in chemical combination with or in formulation with one or more other substances, is a "pesticide" within the meaning of FIFRA and which is used in the production, storage, or transportation of any raw agricultural commodity or processed food. The term includes any substance that is an active ingredient, intentionally-added inert ingredient, or impurity of such a "pesticide."

**Pesticide chemical** (Section 180.1) means any substance that is a pesticide within the meaning of the Federal Insecticide, Fungicide, and Rodenticide Act, including all active and inert ingredients of such pesticide.

**Pesticide chemical and raw agricultural commodity** (Section 163.2) shall have the same meanings as they have in paragraphs (q) and ®, respectively, of section 201 of the Act.

**Pesticide chemical residue** (Section 180.1) means a residue on or in a raw agricultural commodity or processed food of: (1) A Pesticide chemical; or (2) Any other added substance that is present on or in the commodity or food primarily as a result of the metabolism or other degradation of a pesticide chemical.

**Pesticide chemicals** (Section 455.10) means the sum of all active ingredients manufactured at each facility covered by this part.

**Pesticide product** (Section 152.3) means a pesticide in the particular form (including composition, packaging, and labeling) in which the pesticide is, or is intended to be, distributed or sold. The term includes any physical apparatus used to deliver or apply the pesticide if distributed or sold with the pesticide.

**Pesticide residue** (Section 177.3) means a residue of a pesticide chemical or of any metabolite or degradation product of a pesticide chemical.

**Pesticide use** (Section 154.3) means a use of a pesticide (described in terms of the application site and other applicable identifying factors) that is included in the labeling of a pesticide product which is registered, or for which an application for registration is pending, and the terms and conditions (or proposed terms and conditions) of registration for the use.

**Petition** (Section 163.2) means a petition filed with the Administrator, Environmental Protection Agency pursuant to section 408(d) (1) of the Act.

**Petitioner** (Section 164.2) means any person adversely affected by a notice of the Administrator who requests a public hearing.

**Petitioner** (Section 350.1) is any person who submits a petition under this regulation requesting disclosure of a chemical identity claimed as trade secret.

**Petrochemical operations** (Section 419.31) shall mean the production of second-generation petrochemicals (i.e., alcohols, ketones, cumene, styrene, etc.) or first generation petrochemicals and isomerization products (i.e. BTX, olefins, cyclohexane, etc.) when 15 percent or more of refinery production is as first-generation petrochemicals and isomerization products.

**Petroleum cargos** (Section 442.2) mean products of the fractionation or straight distillation of crude oil, redistillation of unfinished petroleum derivatives, cracking, or other refining processes. For purposes of this rule, petroleum cargos also include products obtained from the refining or processing of natural gas and coal. For purposes of this rule, specific examples of petroleum products include but are not limited to: asphalt; benzene; coal tar; crude oil; cutting oil; ethyl benzene; diesel fuel; fuel additives; fuel oils; gasoline; greases; heavy, medium, and light oils; hydraulic fluids, jet fuel; kerosene; liquid petroleum gases (LPG) including butane and propane; lubrication oils; mineral spirits; naphtha; olefin, paraffin, and other waxes; tall oil; tar; toluene; xylene; and waste oil.

**Petroleum fuel** (Section 86.094-2) means liquid fuels normally derived from crude oil, excluding liquefied petroleum gas. Gasoline and diesel fuel are petroleum fuels.

**Petroleum fuel** (Section 86.1803-01) means liquid fuels normally derived from crude oil, excluding liquefied petroleum gas. Gasoline and diesel fuel are petroleum fuels.

**Petroleum fuel** (Section 92.2) means a fuel primarily derived from crude oil (e.g., gasoline or diesel fuel).

**Petroleum marketing** (Section 280.200) means the distribution, transfer, or sale of petroleum or petroleum products for wholesale or retail purposes.

**Petroleum marketing facilities** (Section 280.92) include all facilities at which petroleum is produced or refined and all facilities from which petroleum is sold or transferred to other petroleum marketers or to the public.

**Petroleum marketing firms** (Section 280.92) are all firms owning petroleum marketing facilities. Firms owning other types of facilities with USTs as well as petroleum marketing facilities are considered to be petroleum marketing firms.

**Petroleum oil** (Section 112.2) means petroleum in any form, including but not limited to crude oil, fuel oil, mineral oil, sludge, oil refuse, and refined products.

**Petroleum production** (Section 280.200) means the production of crude oil or other forms of petroleum (as defined in Sec. 280.12) as well as the production of petroleum products from purchased materials.

**Petroleum refining** (Section 280.200) means the cracking, distillation, separation, conversion, upgrading, and finishing of refined petroleum or petroleum products.

**Petroleum refining facility** (Section 279.1) means an establishment primarily engaged in producing gasoline, kerosine, distillate fuel oils, residual fuel oils, and lubricants, through fractionation, straight distillation of crude oil, redistillation of unfinished petroleum derivatives, cracking or other processes (i.e., facilities classified as SIC 2911).

**Petroleum refining process unit** (Section 68.3) means a process unit used in an establishment primarily engaged in petroleum refining as defined in NAICS code 32411 for petroleum refining (formerly SIC code 2911) and used for the following: Producing transportation fuels (such as gasoline, diesel fuels, and jet fuels), heating fuels (such as kerosene, fuel gas distillate, and fuel oils), or lubricants; Separating petroleum; or Separating, cracking, reacting, or reforming intermediate petroleum streams. Examples of such units include, but are not limited to, petroleum based solvent units, alkylation units, catalytic hydrotreating, catalytic hydrorefining, catalytic hydrocracking, catalytic reforming, catalytic cracking, crude distillation, lube oil processing, hydrogen production, isomerization, polymerization, thermal processes, and blending, sweetening, and treating processes. Petroleum refining process units include sulfur plants.

**Petroleum UST system** (Section 280.12) means an underground storage tank system that contains petroleum or a mixture of petroleum with de minimis quantities of other regulated substances. Such systems include those containing motor fuels, jet fuels, distillate fuel oils, residual fuel oils, lubricants, petroleum solvents, and used oils.

**PFPR/Manufacturer** (Section 455.10) means a pesticide formulating, packaging and repackaging facility that also performs pesticide manufacturing on-site and commingles their PFPR process wastewaters and pesticide manufacturing process wastewaters.

**pH** (Section 420.02) means the value obtained by the standard method specified in 40 CFR 136.3.

**pH** (Section 503.31) means the logarithm of the reciprocal of the hydrogen ion concentration measured at 25 [deg]Centigrade or measured at another temperature and then converted to an equivalent value at 25 [deg]Centigrade.

**Phase 1 engine** (Section 90.3) means any handheld or nonhandheld engine, that was produced under a certificate of conformity issued under the regulations in this part to the standard levels defined for Phase 1.

**Phase 2 engine** (Section 90.3) means any handheld and nonhandheld engine that was produced under a certificate of conformity under the regulations in this part to the standards defined for Phase 2 engines.

**Phase I** (Section 72.2) means the Acid Rain Program period beginning January 1, 1995 and ending December 31, 1999.

**Phase I unit** (Section 72.2) means any affected unit, except an affected unit under part 74 of this chapter, that is subject to an Acid Rain emissions reduction requirement or Acid Rain emissions limitation beginning in Phase I; or any unit exempt under Sec. 72.8 that, but for such exemption, would be subject to an Acid Rain emissions reduction requirement or Acid Rain emissions limitation beginning in Phase I.

**Phase II** (Section 72.2) means the Acid Rain Program period beginning January 1, 2000, and continuing into the future thereafter.

**Phase II unit** (Section 72.2) means any affected unit, except an affected unit under part 74 of this chapter, that is subject to an Acid Rain emissions reduction requirement or Acid Rain emissions limitation during Phase II only.

**Phenolic insulation** (Section 247.3) means insulation made with phenolic plastics which are plastics based on resins made by the condensation of phenols, such as phenol or cresol, with aldehydes;

**Phenols 4AAP (or phenolic compounds)** (Section 420.02) means the value obtained by the method specified in 40 CFR 136.3.

**Phosphate rock calciner** (Section 63.601) means the equipment used to remove moisture and organic matter from phosphate rock through direct or indirect heating.

**Phosphate rock dryer** (Section 63.601) means the equipment used to reduce the moisture content of phosphate rock through direct or indirect heating.

**Phosphate rock feed** (Section 63.601) means all material entering any phosphate rock dryer or phosphate rock calciner including moisture and extraneous material as well as the following ore materials: fluorapatite, hydroxylapatite, chlorapatite, and carbonateapatite.

**Physical chemical treatment system** (Section 420.11) means those full scale coke plant wastewater treatment systems incorporating full scale granular activated carbon adsorption units which were in operation prior to January 7, 1981, the date of proposal of this regulation.

**Physical construction** (Section 270.2) means excavation, movement of earth, erection of forms or structures, or similar activity to prepare an HWM facility to accept hazardous waste.

**Pickles processed** (Section 407.61) shall mean the processing of pickles, cucumbers and other vegetables, all varieties, sizes and types, made after fermentation and storage.

**Pickles, fresh** (Section 407.61) shall mean the processing of fresh cucumbers and other vegetables, all varieties, all sizes from whole to relish, all styles, cured after packing.

**Pickles, salt stations** (Section 407.61) shall mean the handling and subsequent preserving of cucumbers and other vegetables at salting stations or tankyards, by salt and other chemical additions necessary to achieve proper fermentation for the packing of processed pickle products. Limitations include allowances for the discharge of spent brine, tank wash, tank soak, and cucumber wash waters. At locations where both salt station and process pack operations (Sec. 407.61(o)) occur, additive allowances shall be made for both of these sources in formulation of effluent limitations. The effluent limitations are to be calculated based upon the total annual weight (1000 lb, kkg) of raw product processed at each of the salt station and process pack operations. Allowances for contaminated stormwater runoff should be considered in NPDES permit formulation on a case-by-case basis.

**Pickling bath** (Section 468.02) shall mean any chemical bath (other than alkaline cleaning) through which a workpiece is processed.

**Pickling fume scrubber** (Section 468.02) shall mean the process of using an air pollution control device to remove particulates and fumes from air above a pickling bath by entraining the pollutants in water.

**Pickling rinse** (Section 468.02) shall mean a rinse, other than an alkaline cleaning rinse, through which a workpiece is processed. A rinse consisting of a series of rinse tanks is considered as a single rinse.

**Pickling rinse for forged parts** (Section 468.02) shall mean a rinse, other than an alkaline cleaning rinse, through which forged parts are processed. A rinse consisting of a series of rinse tanks is considered as a single rinse.

**Pickup Truck** (Section 600.022-85) means a nonpassenger automobile which has a passenger compartment and an open cargo bed.

**Picocurie (pCi)** (Section 141.2) means the quantity of radioactive material producing 2.22 nuclear transformations per minute.

**Pieces** (Section 427.71) shall mean floor tile measured in the standard size of 12"x 12" x \3/32\".

**Pile** (Section 260.10) means any non-containerized accumulation of solid, nonflowing hazardous waste that is used for treatment or storage and that is not a containment building.

**Pilot-scale operation** (Section 439.1) means processing equipment being operated at an intermediate stage between laboratory-scale and full-scale operation for the purpose of developing a new product or manufacturing process.

**Pineapples** (Section 407.61) shall mean the processing of pineapple into the following product styles: Canned, peeled, sliced, chunk, tidbit, diced, crushed, and any other related piece size, juice and concentrate. It also specifically includes the on-site production of by-products such as alcohol, sugar or animal feed.

**Pipe and tube mill** (Section 420.71) means those steel hot forming operations that produce butt welded or seamless tubular steel products.

**Pipe or Piping** (Section 280.12) means a hollow cylinder or tubular conduit that is constructed of non-earthen materials.

**Pipe, tube and other** (Section 420.91) means those acid pickling operations that pickle pipes, tubes or any steel product other than those included in paragraphs (k), (l) and (m) of this section.

**Pipeline** (Section 437.2) means an open or closed conduit used for the conveyance of material. A pipeline includes a channel, pipe, tube, trench, or ditch, or fixed delivery system.

**Pipeline facilities (including gathering lines)** (Section 280.12) are new and existing pipe rights-of-way and any associated equipment, facilities, or buildings.

**Pipeline natural gas** (Section 72.2) means a naturally occurring fluid mixture of hydrocarbons (e.g., methane, ethane, or propane) produced in geological formations beneath the Earth's surface that maintains a gaseous state at standard atmospheric temperature and pressure under ordinary conditions, and which is provided by a supplier through a pipeline. Pipeline natural gas contains 0.5 grains or less of total sulfur per 100 standard cubic feet. Additionally, pipeline natural gas must either be composed of at least 70 percent methane by volume or have a gross calorific value between 950 and 1100 Btu per standard cubic foot.

**Pipeline natural gas** (Section 72.2) means natural gas, as defined in this section, that is provided by a supplier through a pipeline and that contains 0.3 grains or less of hydrogen sulfide per 100 standard cubic feet and the hydrogen sulfide in content of the gas constitutes at least 50% (by weight) of the total sulfur in the fuel.

**Plan** (Section 51.100) means an implementation plan approved or promulgated under section 110 of 172 of the Act.

**Plan** (Section 58.1) means an implementation plan, approved or promulgated pursuant to section 110 of the Clean Air Act.

**Planning** (Section 256.06) includes identifying problems, defining objectives, collecting information, analyzing alternatives and determining necessary activities and courses of action.

**Planning target** (Section 35.102) The amount of funds that the Regional Administrator suggests a grant applicant consider in developing its application, including the work plan, for an environmental program.

**Plans** (Section 240.101) means reports and drawings, including a narrative operating description, prepared to describe the facility and its proposed operation.

**Plant** (Section 174.3) for plant-incorporated protectants, means an organism classified using the 5-kingdom classification system of Whittaker in the kingdom Plantae. This includes, but is not limited to, bryophytes such as mosses, pteridophytes such as ferns, gymnosperms such as conifers, and angiosperms such as most major crop plants.

**Plant** (Section 82.3) means one or more facilities at the same location owned by or under common control of the same person.

**Plant site** (Section 65.2) means all contiguous or adjoining property that is under common control, including properties that are separated only by a road or other public right-of-way. Common control includes properties that are owned, leased, or operated by the same entity, parent entity, subsidiary, or any combination thereof.

**Plant-incorporated protectant** (Section 152.3) means a pesticidal substance that is intended to be produced and used in a living plant, or in the produce thereof, and the genetic material necessary for production of such a pesticidal substance. It also includes any inert ingredient contained in the plant, or produce thereof.

**Plant-incorporated protectant** (Section 174.3) means a pesticidal substance that is intended to be produced and used in a living plant, or in the produce thereof, and the genetic material necessary for production of such a pesticidal substance. It also includes any inert ingredient contained in the plant, or produce thereof.

**Plasma arc incinerator** (Section 260.10) means any enclosed device using a high intensity electrical discharge or arc as a source of heat followed by an afterburner using controlled flame combustion and which is not listed as an industrial furnace.

**Plastic fencing** (Section 247.3) means a barrier with an open-weave pattern that can be used to control drifting snow or sand by restricting the force of wind and to provide a warning or barrier in construction and other areas;

**Plastic lumber landscaping timbers and posts** (Section 247.3) are used to enhance the appearance of and control erosion in parks, highways, housing developments, urban plazas, zoos, and the exteriors of office buildings, military facilities, schools, and other public use areas.

**Plastic material** (Section 463.2) is a synthetic organic polymer (i.e., a thermoset polymer, a thermoplastic polymer, or a combination of a natural polymer and a thermoset or thermoplastic polymer) that is solid in its final form and that was shaped by flow. The material can be either a homogeneous polymer or a polymer combined with fillers, plasticizers, pigments, stabilizers, or other additives.

**Plastics molding and forming** (Section 463.2) is a manufacturing process in which plastic materials are blended, molded, formed, or otherwise processed into intermediate or final products.

**Plate mill** (Section 420.71) means those steel hot forming operations that produce flat hot-rolled products which are (1) between 8 and 48 inches wide and over 0.23 inches thick; or (2) greater than 48 inches wide and over 0.18 inches thick.

**Plate soak** (Section 461.2) shall mean the process operation of soaking or reacting lead subcategory battery plates, that are more than 2.5 mm (0.100 in) thick, in sulfuric acid.

**Plate soak** (Section 461.2) shall mean the process operation of soaking or reacting lead subcategory battery plates, that are more than 2.5 mm (0.100 in) thick, in sulfuric acid.

**Play area** (Section 745.63) means an area of frequent soil contact by children of less than 6 years of age as indicated by, but not limited to, such factors including the following: the presence of play equipment (e.g., sandboxes, swing sets, and sliding boards), toys, or other children's possessions, observations of play patterns, or information provided by parents, residents, care givers, or property owners.

**Playground equipment** (Section 247.3) includes many components, like slides, merry-go-rounds, hand rails, etc., and is found in parks, schools, child care facilities, institutions, multiple family dwellings, restaurants, resort and recreational developments, and other public use areas.

**Plugging** (Section 144.3) means the act or process of stopping the flow of water, oil or gas into or out of a formation through a borehole or well penetrating that formation.

**Plugging** (Section 146.3) means the act or process of stopping the flow of water, oil or gas into or out of a formation through a borehole or well penetrating that formation.

**Plugging** (Section 147.2902) the act or process of stopping the flow of water, oil or gas into or out of a formation through a borehole or well penetrating that formation.

**Plugging record** (Section 146.3) means a systematic listing of permanent or temporary abandonment of water, oil, gas, test, exploration and waste injection wells, and may contain a well log, description of amounts and types of plugging material used, the method employed for plugging, a description of formations which are sealed and a graphic log of the well showing formation location, formation thickness, and location of plugging structures.

**Plug-in combustion controls** (Section 76.2) means the replacement, in a cell burner boiler, of existing cell burners by low NOX burners or low NOX burners with overfire air.

**Plums** (Section 407.61) shall mean the processing of plums into the following product styles: Canned and frozen, pitted and unpitted, peeled and unpeeled, blanched and unblanched, whole, halved, and other piece size.

**PM10** (Section 51.100) means particulate matter with an aerodynamic diameter less than or equal to a nominal 10 micrometers as measured by a reference method based on appendix J of part 50 of this chapter and designated in accordance with part 53 of this chapter or by an equivalent method designated in accordance with part 53 of this chapter.

**PM10** (Section 58.1)means particulate matter with an aerodynamic diameter less than or equal to a nominal 10 micrometers as measured by a reference method based on appendix J of part 50 of this chapter and designated in accordance with part 53 of this chapter or by an equivalent method designated in accordance with part 53 of this chapter.

**PM10 emissions** (Section 51.100) means finely divided solid or liquid material, with an aerodynamic diameter less than or equal to a nominal 10 micrometers emitted to the ambient air as measured by an applicable reference method, or an equivalent or alternative method, specified in this chapter or by a test method specified in an approved State implementation plan.

**PM10 sampler** (Section 53.1) means a device, associated with a manual method for measuring PM10, designed to collect PM10 from an ambient air sample, but lacking the ability to automatically analyze or measure the collected sample to determine the mass concentrations of PM10 in the sampled air.

**PM2.5** (Section 58.1) means particulate matter with an aerodynamic diameter less than or equal to a nominal 2.5 micrometers as measured by a reference method based on 40 CFR part 50, Appendix L, and designated in accordance with part 53 of this chapter or by an equivalent method designated in accordance with part 53 of this chapter.

**PM2.5 sampler** (Section 53.1) means a device, associated with a manual method for measuring PM2.5, designed to collect PM2.5 from an ambient air sample, but lacking the ability to automatically analyze or measure the collected sample to determine the mass concentrations of PM2.5 in the sampled air.

**Point analyzer** (Section 58.1) is an automated analytical method that measures pollutant concentration in an ambient air sample extracted from the atmosphere at a specific inlet probe point and that has been designated as a reference or equivalent method in accordance with part 53 of this chapter.

**Point of disinfectant application** (Section 141.2) is the point where the disinfectant is applied and water downstream of that point is not subject to recontamination by surface water runoff.

**Point of first sale** (Section 86.1702-99) is the location where the completed light-duty vehicle or light-duty truck is purchased, also known as the final product purchase location. The point of first sale may be a retail customer, dealer, distributor, fleet operator, broker, secondary manufacturer, or any other entity which comprises the point of first sale. In cases where the end user purchases the completed vehicle directly from the manufacturer, the end user is the point of first sale.

**Point of first sale** (Section 86.1803-01) means the location where the completed vehicle is first purchased. This term is synonymous with final product purchase location. The point of first sale may be a retail customer, dealer, distributor, fleet operator, broker, secondary manufacturer, or any other entity which purchases a vehicle from a manufacturer. In cases where the end user purchases the completed vehicle directly from the manufacturer, the end user is the point of first sale.

**Point of injection** (Section 144.3) means the last accessible sampling point prior to waste fluids being released into the subsurface environment through a Class V injection well. For example, the point of injection of a Class V septic system might be the distribution box—the last accessible sampling point before the waste fluids drain into the underlying soils. For a dry well, it is likely to be the well bore itself.

**Point of injection for Class V wells** (Section 146.3) means the last accessible sampling point prior to waste fluids being released into the subsurface environment through a Class V injection well. For example, the point of injection of a Class V septic system might be the distribution box—the last accessible sampling point before the waste fluids drain into the underlying soils. For a dry well, it is likely to be the well bore itself.

**Point of waste origination** (Section 265.1081) means as follows: (1) When the facility owner or operator is the generator of the hazardous waste, the point of waste origination means the point where a solid waste produced by a system, process, or waste management unit is determined to be a hazardous waste as defined in 40 CFR part 261. Note: In this case, this term is being used in a manner similar to the use of the term "point of generation" in air standards established for waste management operations under authority of the Clean Air Act in 40 CFR parts 60, 61, and 63.] (2) When the facility owner and operator are not the generator of the hazardous waste, point of waste origination means the point where the owner or operator accepts delivery or takes possession of the hazardous waste.

**Point of waste treatment** (Section 265.1081) means the point where a hazardous waste to be treated in accordance with Sec. 265.1083©(2) of this subpart exits the treatment process. Any waste determination shall be made before the waste is conveyed, handled, or otherwise managed in a manner that allows the waste to volatilize to the atmosphere.

**Point source** (Section 122.2) means any discernible, confined, and discrete conveyance, including but not limited to, any pipe, ditch, channel, tunnel, conduit, well, discrete fissure, container, rolling stock, concentrated animal feeding operation, landfill leachate collection system, vessel or other floating craft from which pollutants are or may be discharged. This term does not include return flows from irrigated agriculture or agricultural storm water runoff. (See Sec. 122.3).

**Point source** (Section 260.10) means any discernible, confined, and discrete conveyance, including, but not limited to any pipe, ditch, channel, tunnel, conduit, well, discrete fissure, container, rolling stock, concentrated animal feeding operation, or vessel or other floating craft, from which pollutants are or may be discharged. This term does not include return flows from irrigated agriculture.

**Point source** (Section 401.11) means any discernible, confined and discrete conveyance, including but not limited to any pipe, ditch, channel, tunnel, conduit, well, discrete fissure, container, rolling stock, concentrated animal feeding operation, or vessel or other floating craft, from which pollutants are or may be discharged.

**Point source** (Section 51.100) means the following: (1) For particulate matter, sulfur oxides, carbon monoxide, volatile organic compounds (VOC) and nitrogen dioxide— (i) Any stationary source the actual emissions of which are in excess of 90.7 metric tons (100 tons) per year of the pollutant in a region containing an area whose 1980 urban place population, as defined by the U.S. Bureau of the Census, was equal to or greater than 1 million. (ii) Any stationary source the actual emissions of which are in excess of 22.7 metric tons (25 tons) per year of the pollutant in a region containing an area whose 1980 urban place population, as defined by the U.S. Bureau of the Census, was less than 1 million; or(2) For lead or lead compounds measured as elemental lead, any stationary source that actually emits a total of 4.5 metric tons (5 tons) per year or more.

**Point-of-entry treatment device (POE)** (Section 141.2) is a treatment device applied to the drinking water entering a house or building for the purpose of reducing contaminants in the drinking water distributed throughout the house or building.

**Point-of-use treatment device (POU)** (Section 141.2) is a treatment device applied to a single tap used for the purpose of reducing contaminants in drinking water at that one tap.

**Political subdivision** (Section 307.14) means any general purpose unit of a local or State government.

**Political subdivision of State** (Section 105.2) means a municipality, city, town, borough, county, parish, district, association, or other public body (including an intermunicipal agency of two or more of the foregoing entities) created by or pursuant to State law.

**Pollutant** (Section 122.2) means dredged spoil, solid waste, incinerator residue, filter backwash, sewage, garbage, sewage sludge, munitions, chemical wastes, biological materials, radioactive materials (except those regulated under the Atomic Energy Act of 1954, as amended (42 U.S.C. 2011 et seq.)), heat, wrecked or discarded equipment, rock, sand, cellar dirt and industrial, municipal, and agricultural waste discharged into water. It does not mean: (a) Sewage from vessels; or (b) Water, gas, or other material which is injected into a well to facilitate production of oil or gas, or water derived in association with oil and gas production and disposed of in a well, if the well used either to facilitate production or for disposal purposes is approved by authority of the State in which the well is located, and if the State determines that the injection or disposal will not result in the degradation of ground or surface water resources. Note: Radioactive materials covered by the Atomic Energy Act are those encompassed in its definition of source, byproduct, or special nuclear materials. Examples of materials not covered include radium and accelerator-produced isotopes. See Train v. Colorado Public Interest Research Group, Inc., 426 U.S. 1 (1976).

**Pollutant** (Section 130.2) Dredged spoil, solid waste, incinerator residue, sewage, garbage, sewage sludge, munitions, chemical wastes, biological materials, radioactive materials (except those regulated under Atomic Energy Act of 1954, as amended (42 U.S.C. 2011 et seq.)), heat, wrecked or discarded equipment, rock, sand, cellar dirt, and industrial, municipal, and agricultural waste discharged into water. This term does not mean: "sewage from vessels" within the meaning of section 312 of the Clean Water Act; or water, gas, or other material that is injected into a well to facilitate production of oil or gas, or water derived in association with oil or gas production and disposed of in a well, if the well used either to facilitate production or for disposal purposes is approved by authority of the State in which the well is located, and if the State determines that such injection or disposal will not result in the degradation of ground or surface water resources. (See Clean Water Act section 502(6).)

**Pollutant** (Section 230.3) means dredged spoil, solid waste, incinerator residue, sewage, garbage, sewage sludge, munitions, chemical wastes, biological materials, radioactive materials not covered by the Atomic Energy Act, heat, wrecked or discarded equipment, rock, sand, cellar dirt, and industrial, municipal, and agricultural waste discharged into water. The legislative history of the Act reflects that "radioactive materials" as included within the definition of "pollutant" in section 502 of the Act means only radioactive materials which are not encompassed in the definition of source, byproduct, or special nuclear materials as defined by the Atomic Energy Act of 1954, as amended, and regulated under the Atomic Energy Act. Examples of radioactive materials not covered by the Atomic Energy Act and, therefore, included within the term "pollutant", are radium and accelerator produced isotopes. See Train v. Colorado Public Interest Research Group, Inc., 426 U.S. 1 (1976).

**Pollutant** (Section 401.11) means dredged spoil, solid waste, incinerator residue, sewage, garbage, sewage sludge, munitions, chemical wastes, biological materials, radioactive materials, heat, wrecked or discarded equipment, rock, sand, cellar dirt and industrial, municipal and agricultural waste discharged into water. It does not mean (1) sewage from vessels or (2) water, gas or other material which is injected into a well to facilitate production of oil or gas, or water derived in association with oil or gas production and disposed of in a well, if the well, used either to facilitate production or for disposal purposes, is approved by authority of the State in which the well is located, and if such State determines that such injection or disposal will not result in degradation of ground or surface water resources.

**Pollutant concentration monitor** (Section 72.2) means that component of the continuous emission monitoring system that measures the concentration of a pollutant in a unit's flue gas.

**Pollutant or contaminant** (Section 310.3) as defined by section 104(a)(2) of CERCLA, includes, but is not limited to, any element, substance, compound, or mixture, including disease-causing agents, which after release into the environment and upon exposure, ingestion, inhalation, or assimilation into any organism, either directly from the environment or indirectly by ingestion through food chains, will or may reasonably be anticipated to cause death, disease, behavioral abnormalities, cancer, genetic mutation, physiological malfunctions (including malfunctions in reproduction) or physical deformations, in such organisms or their offspring. The term does not include petroleum, including crude oil and any fraction thereof that is not otherwise specifically listed or designated as a hazardous substance under section 101(14)(A) through (F) of CERCLA, nor does it include natural gas, liquefied natural gas, or synthetic gas of pipeline quality (or mixtures of natural gas and such synthetic gas).

**Pollutant or contaminant as defined by section 101(33) of CERCLA** (Section 300.5) shall include, but not be limited to, any element, substance, compound, or mixture, including disease-causing agents, which after release into the environment and upon exposure, ingestion, inhalation, or assimilation into any organism, either directly from the environment or indirectly by ingestion through food chains, will or may reasonably be anticipated to cause death, disease, behavioral abnormalities, cancer, genetic mutation, physiological malfunctions (including malfunctions in reproduction) or physical deformations, in such organisms or their offspring. The term does not include petroleum, including crude oil or any fraction thereof which is not otherwise specifically listed or designated as a hazardous substance under section 101(14) (A) through (F) of CERCLA, nor does it include natural gas, liquified natural gas, or synthetic gas of pipeline quality (or mixtures of natural gas and such synthetic gas). For purposes of the NCP, the term pollutant or contaminant means any pollutant or contaminant that may present an imminent and substantial danger to public health or welfare of the United States.

**Pollutant or Contaminant as defined by section 101(33) of CERCLA** (Section 307.14) includes, but is not limited to, any element, substance, compound, or mixture, including disease-causing agents, which after release into the environment and upon exposure, ingestion, inhalation, or assimilation into any organism, either directly from the environment or indirectly by ingestion through food chains, will or may reasonably be anticipated to cause death, disease, behavioral abnormalities, cancer, genetic mutation, physiological malfunctions (including malfunctions in reproduction) or physical deformations in such organisms or their offspring. The term does not include petroleum, including crude oil and any fraction thereof which is not otherwise specifically listed or designated as a hazardous substance under section 101(14)(A) through (F) of the Act, nor does it include natural gas, liquefied natural gas, or synthetic gas of pipeline quality (or mixtures of natural gas and such synthetic gas).

**Pollutant-specific emissions unit** (Section 64.1) means an emissions unit considered separately with respect to each regulated air pollutant.

**Pollution** (Section 130.2) The man-made or man-induced alteration of the chemical, physical, biological, and radiological integrity of water.

**Pollution** (Section 130.2) The man-made or man-induced alteration of the chemical, physical, biological, and radiological integrity of water. (See Clean Water Act section 502(19).)

**Pollution** (Section 230.3) means the man-made or man-induced alteration of the chemical, physical, biological or radiological integrity of an aquatic ecosystem.

**Pollution** (Section 401.11) means the man-made or man induced alteration of the chemical, physical, biological and radiological integrity of water.

**Pollution abatement area** (Section 434.70) means the part of the permit area that is causing or contributing to the baseline pollution load of pre-existing discharges. The pollution abatement area must include, to the extent practicable, areas adjacent to and nearby the remining operation that also must be affected to reduce the pollution load of the pre-existing discharges and may include the immediate location of the pre-existing discharges.

**Pollution prevention (P2) allowable discharge for interior wastewater sources, leak and spill cleanup water, and floor wash for this subpart** (Section 455.41) means the quantity of/concentrations of pollutants in PFPR process wastewaters that remain after a facility has demonstrated that it is using the specified practices of the Pollution Prevention Alternative as listed in Table 8 to this part 455 and that have been pretreated using appropriate pollution control technologies, as defined in Sec.

455.10(g), or a pesticide manufacturer's treatment system, or an equivalent system, used individually, or in any combination to achieve a sufficient level of pollutant reduction. Pretreatment requirements may be modified or waived by the Control Authority (POTW) to the extent that removal credits have been granted by the POTW in accordance with 40 CFR 403.7, provided the granting of such credits does not result in pass through or interference as defined in 40 CFR 403.3 and complies with the provisions of 40 CFR 403.5. The facility must demonstrate that the appropriate pollution control technology is properly maintained and operated.

**Pollution Prevention Allowable Discharge for this subpart** (Section 442.2) means the quantity of/concentrations of pollutants in wastewaters being discharged to publicly owned treatment works after a facility has demonstrated compliance with the Pollutant Management Plan provisions in Secs. 442.15(b), 442.16(b), 442.25(b), or 442.26(b) of this part.

**Polybrominated dibenzofurans** (Section 766.3) refers to any member of a class of dibenzofurans with two to eight bromine substituents.

**Polybrominated dibenzo-p-dioxin or PBDD** (Section 766.3) means to any member of a class of dibenzo-p-dioxins with two to eight bromine substituents.

**Polychlorinated biphenyls or PCBs** (Section 268.2) are halogenated organic compounds defined in accordance with 40 CFR 761.3.

**Polychlorinated dibenzofuran** (Section 766.3) means any member of a class of dibenzofurans with two to eight chlorine substituents.

**Polychlorinated dibenzo-p-dioxin or PCDD** (Section 766.3) means any member of a class of dibenzo-p-dioxins with two to eight chlorine substituents.

**Polyhalogenated dibenzofuran or PHDF** (Section 766.3) means any member of a class of dibenzofurans containing two to eight chlorine, bromine, or a combination of chlorine and bromine substituents.

**Polyhalogenated dibenzo-p-dioxin or PHDD** (Section 766.3) means any member of a class of dibenzo-p-dioxins containing two to eight chlorine substituents or two to eight bromine substituents.

**Polyisocyanurate insulation** (Section 247.3) means insulation produced principally by the polymerization of polymeric polyisocyanates, usually in the presence of polyhydroxyl compounds with the addition of cell stabilizers, blowing agents, and appropriate catalyst to produce a polyisocyanurate chemical structure;

**Polymerizing monomer** (Section 65.2) means, for the purposes of this part, a compound which may form polymer buildup in pump mechanical seals resulting in rapid mechanical seal failure.

**Polystyrene insulation** (Section 247.3) means an organic foam composed principally of polymerized styrene resin processed to form a homogenous rigid mass of cells;

**Polyurethane insulation** (Section 247.3) means insulation composed principally of the catalyzed reaction product of polyisocyanates and polyhydroxyl compounds, processed usually with a blowing agent to form a rigid foam having a predominantly closed cell structure;

**Pond water surface area** (Section 421.11) for the purpose of calculating the volume of waste water shall mean the area within the impoundment for rainfall and the actual water surface area for evaporation.

**Pond water surface area** (Section 421.61) when used for the purpose of calculating the volume of wastewater which may be discharged shall mean the water surface area of the pond created by the impoundment for storage of process wastewater at normal operating level. This surface shall in no case be less than one-third of the surface area of the maximum amount of water which could be contained by the impoundment. The normal operating level shall be the average level of the pond during the preceding calendar month.

**Pool chemicals** (Section 455.10) means pesticide products that are intended to disinfect or sanitize, reducing or mitigating growth or development of microbiological organisms including bacteria, algae, fungi or viruses in the water of swimming pools, hot tubs, spas or other such areas, in the household and/or institutional environment, as provided in the directions for use on the product label.

**Population** (Section 68.3) means the public.

**Population of CCDFs** (Section 194.2) means all possible complementary, cumulative distribution functions (CCDFs) that can be generated from all disposal system parameter values used in performance assessments.

**Population of estimates** (Section 194.2) means all possible estimates of radiation doses and radionuclide concentrations that can be generated from all disposal system parameter values used in compliance assessments.

**Population-oriented monitoring (or sites)** (Section 58.1) applies to residential areas, commercial areas, recreational areas, industrial areas, and other areas where a substantial number of people may spend a significant fraction of their day.

**Porcelain enameling** (Section 466.02) means the entire process of applying a fused vitreous enamel coating to a metal basis material. Usually this includes metal preparation and coating operations.

**Porous surface** (Section 761.13) means any surface that allows PCBs to penetrate or pass into itself including, but not limited to, paint or coating on metal; corroded metal; fibrous glass or glass wool; unglazed ceramics; ceramics with a porous glaze; porous building stone such as sandstone, travertine, limestone, or coral rock; low-density plastics such as styrofoam and low-density polyethylene; coated (varnished or painted) or uncoated wood; concrete or cement; plaster; plasterboard; wallboard; rubber; fiberboard; chipboard; asphalt; or tar paper. For purposes of cleaning and disposing of PCB remediation waste, porous surfaces have different requirements than non-porous surfaces.

**Portable air compressor or compressor** (Section 204.51) means any wheel, skid, truck, or railroad car mounted, but not self-propelled, equipment designed to activate pneumatic tools. This consists of an air compressor (air end), and a reciprocating rotary or turbine engine rigidly connected in permanent alignment and mounted on a common frame. Also included are all cooling, lubricating, regulating, starting, and fuel systems, and all equipment necessary to constitute a complete, self-contained unit with a rated capacity of 75 cfm or greater which delivers air at pressures greater than 50 psig, but does not include any pneumatic tools themselves.

**Posing an exposure risk to food or feed** (Section 761.13) means being in any location where human food or animal feed products could be exposed to PCBs released from a PCB Item. A PCB Item poses an exposure risk to food or feed if PCBs released in any way from the PCB Item have a potential pathway to human food or animal feed. EPA considers human food or animal feed to include items regulated by the U.S. Department of Agriculture or the Food and Drug Administration as human food or animal feed; this includes direct additives. Food or feed is excluded from this definition if it is used or stored in private homes.

**Positive test result** (Section 766.3) means: (1) Any resolvable gas chromatographic peak for any 2,3,7,8-HDD or HDF which exceeds the LOQ listed under Sec. 766.27 for that congener, or (2) exceeds LOQs approved by EPA under Sec. 766.28.

**Possession or control** (Section 704.3) means in the possession or control of any person, or of any subsidiary, partnership in which the person is a general partner, parent company, or any company or partnership which the parent company owns or controls, if the subsidiary, parent company, or other company or partnership is associated with the person in the research, development, test marketing, or commercial marketing of the substance in question. Information is in the possession or control of a person if it is: (1) In the person's own files including files maintained by employees of the person in the course of their employment.(2) In commercially available data bases to which the person has purchased access. (3) Maintained in the files in the course of employment by other agents of the person who are associated with research, development, test marketing, or commercial marketing of the chemical substance in question.

**Possession or control** (Section 720.3) means in possession or control of the submitter, or of any subsidiary, partnership in which the submitter is a general partner, parent company, or any company or partnership which the parent company owns or controls, if the subsidiary, parent company, or other company or partnership is associated with the submitter in the research, development, test marketing, or commercial marketing of the chemical substance in question. (A parent company owns or controls another company if the parent owns or controls 50 percent or more of the other company's voting stock. A parent company owns or controls any partnership in which it is a general partner). Information is included within this definition if it is: (1) In files maintained by submitter's employees who are: (i) Associated with research, development, test marketing, or commercial marketing of the chemical substance in question. (ii) Reasonably likely to have such data. (2) Maintained in the files of other agents of the submitter who are associated with research, development, test marketing, or commercial marketing of the chemical substance in question in the course of their employment as such agents.

**Postconsumer material** (Section 247.3) means a material or finished product that has served its intended use and has been diverted or recovered from waste destined for disposal, having completed its life as a consumer item. Postconsumer material is a part of the broader category of recovered materials.

**Postconsumer recovered paper** (Section 247.3) means: (1) Paper, paperboard and fibrous wastes from retail stores, office buildings, homes and so forth, after they have passed through their end-usage as a consumer item including: Used corrugated boxes; old newspapers; old magazines; mixed waste paper; tabulating cards and used cordage; and (2) All paper, paperboard and fibrous wastes that enter and are collected from municipal solid waste;

**Post-consumer waste (PCW) (Section 246.101)** means a material or product that has served its intended use and has been discarded for disposal or recovery after passing through the hands of a final consumer.

**Post-disposal period** (Section 192) means the period of time beginning immediately after the disposal period and ending at termination of the monitoring period established under Sec. 192.03.

**Post-manufacture marinizer** (Section 89.2) means a person who produces a marine diesel engine by substantially modifying a certified or uncertified complete or partially complete engine, and is not controlled by the manufacturer of the base engine or by an entity that also controls the manufacturer of the base engine. For the purpose of this definition, "substantially modify" means changing an engine in a way that could change engine emission characteristics.

**Post-manufacture marinizer** (Section 94.2) means an entity that produces a marine engine by modifying a non-marine engine, whether certified or uncertified, complete or partially complete, where such entity is not controlled by the manufacturer of the base engine or by an entity that also controls the manufacturer of the base engine. In addition, vessel manufacturers that substantially modify marine engines are post-manufacture marinizers. For the purpose of this definition, "substantially modify" means changing an engine in a way that could change engine emission characteristics.

**Post-mining area** (Section 434.11) means: (1) A reclamation area or (2) The underground workings of an underground coal mine after the extraction, removal, or recovery of coal from its natural deposit has ceased and prior to bond release.

**Post-removal site control** (Section 300.5) means those activities that are necessary to sustain the integrity of a Fund-financed removal action following its conclusion. Post-removal site control may be a removal or remedial action under CERCLA. The term includes, without being limited to, activities such as relighting gas flares, replacing filters, and collecting leachate.

**Potatoes** (Section 407.71) shall mean the processing of sweet potatoes into the following product styles: Canned, peeled, solid, syrup, and vacuum packed. The following white potato product styles are also included: Canned, peeled, white, all varieties, whole and sliced.

**Potential damage** (Section 763.83) means circumstances in which: (1) Friable ACBM is in an area regularly used by building occupants, including maintenance personnel, in the course of their normal activities. (2) There are indications that there is a reasonable likelihood that the material or its covering will become damaged, deteriorated, or delaminated due to factors such as changes in building use, changes in operations and maintenance practices, changes in occupancy, or recurrent damage.

**Potential electrical output capacity** (Section 72.2) means the MWe capacity rating for the units which shall be equal to 33 percent of the maximum design heat input capacity of the steam generating unit, as calculated according to appendix D of part 72.

**Potential electrical output capacity** (Section 97.2) means 33 percent of a unit's maximum design heat input.

**Potential emissions** (Section 55.2) means the maximum emissions of a pollutant from an OCS source operating at its design capacity. Any physical or operational limitation on the capacity of a source to emit a pollutant, including air pollution control equipment and restrictions on hours of operation or on the type or amount of material combusted, stored, or processed, shall be treated as a limit on the design capacity of the source if the limitation is federally enforceable. Pursuant to section 328 of the Act, emissions from vessels servicing or associated with an OCS source shall be considered direct emissions from such a source while at the source, and while enroute to or from the source when within 25 miles of the source, and shall be included in the "potential to emit" for an OCS source. This definition does not alter or affect the use of this term for any other purposes under Secs. 55.13 or 55.14 of this part, except that vessel emissions must be included in the "potential to emit" as used in Secs. 55.13 and 55.14 of this part.

**Potential significant damage** (Section 763.83) means circumstances in which: (1) Friable ACBM is in an area regularly used by building occupants, including maintenance personnel, in the course of their normal activities. (2) There are indications that there is a reasonable likelihood that the material or its covering will become significantly damaged, deteriorated, or delaminated due to factors such as changes in building use, changes in operations and maintenance practices, changes in occupancy, or recurrent damage. (3) The material is subject to major or continuing disturbance, due to factors including, but not limited to, accessibility or, under certain circumstances, vibration or air erosion.

**Potential to emit** (Section 63.2) means the maximum capacity of a stationary source to emit a pollutant under its physical and operational design. Any physical or operational limitation on the capacity of the stationary source to emit a pollutant, including air pollution control equipment and restrictions on hours of operation or on the type or amount of material combusted, stored, or processed, shall be treated as part of its design if the limitation or the effect it would have on emissions is federally enforceable.

**Potential to emit** (Section 64.1) shall have the same meaning as provided under part 70 or 71 of this chapter, provided that it shall be applied with respect to an "emissions unit" as defined under this part in addition to a "stationary source" as provided under part 70 or 71 of this chapter.

**Potential to emit** (Section 66.3) means the capability at maximum design capacity to emit a pollutant after the application of air pollution control equipment. Annual potential shall be based on the larger of the maximum annual rated capacity of the stationary source assuming continuous operation, or on a projection of actual annual emissions. Enforceable permit conditions on the type of materials combusted or processed may be used in determining the annual potential. Fugitive emissions, to the extent quantifiable, will be considered in determining annual potential for those stationary sources whose fugitive emissions are regulated by the applicable state implementation plan.

**Potential to emit** (Section 70.2) means the maximum capacity of a stationary source to emit any air pollutant under its physical and operational design. Any physical or operational limitation on the capacity of a source to emit an air pollutant, including air pollution control equipment and restrictions on hours of operation or on the type or amount of material combusted, stored, or processed, shall be treated as part of its design if the limitation is enforceable by the Administrator. This term does not alter or affect the use of this term for any other purposes under the Act, or the term "capacity factor" as used in title IV of the Act or the regulations promulgated thereunder.

**Potentially responsible party (PRP)** (Section 310.3) means any person who may be liable under section 107 of CERCLA for a release or threatened release of hazardous substances or pollutants or contaminants.

**Potentially responsible party or PRP** (Section 304.12) means any person who may be liable pursuant to section 107(a) of CERCLA, 42 U.S.C. 9607(a), for response costs incurred and to be incurred by the United States not inconsistent with NCP.

**POTW** (Section 122.2) is defined at Sec. 403.3 of this chapter.

**POTW** (Section 270.2) means publicly owned treatment works.

**POTW** (Section 430.0) Publicly owned treatment works as defined at 40 CFR 403.3(o).

**POTW** (Section 439.1) means publicly owned treatment works (40 CFR 403.3).

**POTW** (Section 444.11) means a publicly owned treatment works.

**POTW** (Section 464.02) shall mean "publicly owned treatment works."

**POTW** (Section 501.2) means a publicly owned treatment works.

**POTW Treatment Plant** (Section 403.3) means that portion of the POTW which is designed to provide treatment (including recycling and reclamation) of municipal sewage and industrial waste.

**Powder forming** (Section 471.02) includes forming and compressing powder into a fully dense finished shape, and is usually done within closed dies.

**Powder or dry solid form** (Section 721.3) means a state where all or part of the substance would have the potential to become fine, loose, solid particles.

**Power assembly** (Section 92.2) means the components of an engine in which combustion of fuel occurs, and consists of the cylinder, piston and piston rings, valves and ports for admission of charge air and discharge of exhaust gases, fuel injection components and controls, cylinder head and associated components.

**Power distribution system** (Section 72.2) means the portion of an electricity grid owned or operated by a utility that is dedicated to delivering electric energy to customers.

**Power purchase commitment** (Section 72.2) means a commitment or obligation of a utility to purchase electric power from a facility pursuant to: (1) A power sales agreement; (2) A state regulatory authority order requiring a utility to: (i) Enter into a power sales agreement with the facility; (ii) Purchase from the facility; or (iii) Enter into arbitration concerning the facility for the purpose of establishing terms and conditions of the utility's purchase of power; (3) A letter of intent or similar instrument committing to purchase power (actual electrical output or generator output capacity) from the source at a previously offered or lower price and a power sales agreement applicable to the source is executed within the time frame established by the terms of the letter of intent but no later than November 15, 1993 or, where the letter of intent does not specify a time frame, a power sale agreement applicable to the source is executed on or before November 15, 1993; or (4) A utility competitive bid solicitation that has resulted in the selection of the qualifying facility or independent power production facility as the winning bidder.

**Power sales agreement** (Section 72.2) is a legally binding agreement between a QF, IPP, new IPP, or firm associated with such facility and a regulated electric utility that establishes the terms and conditions for the sale of power from the facility to the utility.

**Power setting** (Section 87.1) means the power or thrust output of an engine in terms of kilonewtons thrust for turbojet and turbofan engines and shaft power in terms of kilowatts for turboprop engines.

**Practicable** (Section 157.21) when used with respect to child-resistant packaging, means that the packaging can be mass produced and can be used in assembly line production.

**Practicable** (Section 230.3) means available and capable of being done after taking into consideration cost, existing technology, and logistics in light of overall project purposes.

**Practicable** (Section 247.3) means capable of being used consistent with: Performance in accordance with applicable specifications, availability at a reasonable price, availability within a reasonable period of time, and maintenance of a satisfactory level of competition;

**Practical knowledge** (Section 171.2) means the possession of pertinent facts and comprehension together with the ability to use them in dealing with specific problems and situations.

**Practice** (Section 272.2) means the act of disposal of solid waste.

**Preauthorization** (Section 305.3) means EPA's prior approval to submit a claim against the Fund for necessary response costs incurred as a result of carrying out the NCP.

**Preauthorization** (Section 307.14) means EPA's prior approval to submit a claim against the Fund for necessary response costs incurred as a result of carrying out the NCP. The process of preauthorization consists of three steps: (1) EPA's receipt of the application for preauthorization; (2) EPA's review and analysis of the application; and (3) EPA's issuance of the Preauthorization Decision Document, which sets forth the terms and conditions for reimbursement.

**Preauthorized response actions** (Section 307.14) are response actions approved through the preauthorization process.

**Pre-certification vehicle** (Section 85.1702) means an uncertified vehicle which a manufacturer employs in fleets from year to year in the ordinary course of business for product development, production method assessment, and market promotion purposes, but in a manner not involving lease or sale.

**Pre-certification vehicle engine** (Section 85.1702) means an uncertified heavy-duty engine owned by a manufacturer and used in a manner not involving lease or sale in a vehicle employed from year to year in the ordinary course of business for product development, production method assessment and market promotion purposes.

**Precious** (Section 466.02) metal means gold, silver, or platinum group metals and the principal alloys of those metals.

**Precious metals** (Section 421.261) shall mean gold, platinum, palladium, rhodium, iridium, osmium, and ruthenium.

**Precious metals** (Section 468.02) shall mean gold, platinum, palladium and silver and their alloys. Any alloy containing 30 or greater percent by weight of precious metals is considered a precious metal.

**Precious metals** (Section 471.02) include gold, platinum, palladium, and silver and their alloys. Any alloy containing 30 or greater percent by weight of precious metals is considered a precious metal alloy.

**Precision** (Section 86.082-2) means the standard deviation of replicated measurements.

**Precision** (Section 86.1803-01) means the standard deviation of replicated measurements.

**Precision** (Section 92.102) means the standard deviation of replicated measurements, or one-half of the readability, whichever is greater; except where explicitly noted otherwise.

**Preconditioning** (Section 610.11) means the operation of an automobile through one (1) EPA Urban Dynamometer Driving Schedule, described in 40 CFR part 86.

**Precursor** (Section 766.3) means a chemical substance which is not contaminated due to the process conditions under which it is manufactured, but because of its molecular structure, and under favorable process conditions, it may cause or aid the formation of HDDs/HDFs in other chemicals in which it is used as a feedstock or intermediate.

**Precursors of a criteria pollutant** (Section 93.152) are: (1) For ozone, nitrogen oxides (NOx), unless an area is exempted from NOx requirements under section 182(f) of the Act, and volatile organic compounds (VOC); and (2) For PM-10, those pollutants described in the PM-10 nonattainment area applicable SIP as significant contributors to the PM-10 levels.

**Predictive emission monitoring system (PEMS)** (Section 64.1) means a system that uses process and other parameters as inputs to a computer program or other data reduction system to produce values in terms of the applicable emission limitation or standard.

**Pre-existing discharge** (Section 434.70) means any discharge resulting from mining activities that have been abandoned prior to the time of a remining permit application. This term shall include a pre-existing discharge that is relocated as a result of the implementation of best management practices (BMPs) contained in the Pollution Abatement Plan.

**Preheater tower combustion gas monitoring location** (Section 62.1201) means a location within the preheater tower of a dry process cement kiln downstream (in terms of gas flow) of all hazardous waste firing locations and where a representative sample of combustion gas to measure combustion efficiency can be monitored.

**Preliminary analysis** (Section 610.11) means the engineering analysis performed by EPA prior to testing prescribed by the Administrator based on data and information submitted by a manufacturer or available from other sources.

**Preliminary assessment (PA) under CERCLA** (Section 300.5) means review of existing information and an off-site reconnaissance, if appropriate, to determine if a release may require additional investigation or action. A PA may include an on-site reconnaissance, if appropriate.

**Premanufacture notice or PMN** (Section 700.43) means any notice submitted to EPA pursuant to section 5(a)(1)(A) of the Act in accordance with part 720 of this chapter or Sec. 723.250 of this chapter.

**Prerinse/presteam** (Section 442.2) means a rinse, typically with hot or cold water, performed at the beginning of the cleaning sequence to remove residual material from the tank interior.

**Present in the same form and concentration as a product packaged for distribution and use by the general public** (Section 370.2) means a substance packaged in a similar manner and present in the same concentration as the substance when packaged for use by the general public, whether or not it is intended for distribution to the general public or used for the same purpose as when it is packaged for use by the general public.

**Presentation of credentials** (Section 89.2) means the display of the document designating a person as an EPA enforcement officer or EPA authorized representative.

**Presentation of credentials** (Section 90.3) means the display of the document designating a person as an EPA enforcement officer or EPA authorized representative.

**Presentation of credentials** (Section 91.3) means the display of the document designating a person as an EPA enforcement officer or EPA authorized representative.

**Presentation of credentials** (Section 94.2) means the display of the document designating a person as an EPA enforcement officer.

**Preshipment applications, with respect to class I, Group VI controlled substances,** (Section 82.3) are those non-quarantine applications applied within 21 days prior to export to meet the official requirements of the importing country or existing official requirements of the exporting country. Official requirements are those which are performed by, or authorized by, a national plant, animal, environmental, health or stored product authority.

**Presiding Officer** (Section 104.2) means the Chief Administrative Law Judge of the Agency or a person designated by the Chief Administrative Law Judge or by the Administrator to preside at a hearing under this part, in accordance with Sec. 104.6 hereof.

**Presiding Officer** (Section 164.2) means any person designated by the Administrator to conduct an expedited hearing.

**Presiding Officer** (Section 173.2) means an attorney appointed by the Administrator to conduct the rescission proceeding. The Presiding Officer shall be an employee or representative of the Agency and shall not have had prior direct connection with the specific proceeding except in circumstances where subsequent hearings are in order.

**Presiding Officer** (Section 305.3) means the Administrative Law Judge designated by the Chief Administrative Law Judge, or the Chief Administrative Law Judge himself, in the absence of such designation, to conduct a hearing pursuant to this part.

**Presiding Officer** (Section 72.2) means an Administrative Law Judge appointed under 5 U.S.C. 3105 and designated to preside at a hearing in an appeal under part 78 of this chapter or an EPA lawyer designated to preside at any such hearing under Sec. 78.6(b)(3)(ii) of this chapter.

**Presolve wash** (Section 442.2) means the use of diesel, kerosene, gasoline, or any other type of fuel or solvent as a tank interior cleaning solution.

**Pressure** (Section 146.3) means the total load or force per unit area acting on a surface.

**Pressure** (Section 147.2902) the total load or force per unit area acting on a surface.

**Pressure release** (Section 264.1031) means the emission of materials resulting from the system pressure being greater than the set pressure of the pressure relief device.

**Pressure release** (Section 65.2) means the emission of materials resulting from the system pressure being greater than the set pressure of the relief device. This release can be one release or a series of releases over a short time period.

**Pressure relief device or valve** (Section 65.2) means a device used to prevent operating pressures from exceeding the maximum allowable working pressure of the process equipment. A common pressure relief device is a spring-loaded pressure relief valve. Devices that are actuated either by a pressure of less than or equal to 2.5 pounds per square inch gauge or by a vacuum are not pressure relief devices.

**Pretreatment** (Section 403.3) means the reduction of the amount of pollutants, the elimination of pollutants, or the alteration of the nature of pollutant properties in wastewater prior to or in lieu of discharging or otherwise introducing such pollutants into a POTW. The reduction or alteration may be obtained by physical, chemical or biological processes, process changes or by other means, except as prohibited by Sec. 403.6(d). Appropriate pretreatment technology includes control equipment, such as equalization tanks or facilities, for protection against surges or slug loadings that might interfere with or otherwise be incompatible with the POTW. However, where wastewater from a regulated process is mixed in an equalization facility with unregulated wastewater or with wastewater from another regulated process, the effluent from the equalization facility must meet an adjusted pretreatment limit calculated in accordance with Sec. 403.6(e).

**Pretreatment control authority** (Section 414.10) means: (1) The POTW if the POTW's submission for its pretreatment program has been approved in accordance with the requirements of 40 CFR 403.11, or (2) The Approval Authority if the submission has not been approved.

**Pretreatment requirements** (Section 403.3) means any substantive or procedural requirement related to Pretreatment, other than a National Pretreatment Standard, imposed on an Industrial User.

**Pretreatment wash primer** (Section 59.101) means a primer that contains a minimum of 0.5 percent acid, by weight, that is applied directly to bare metal surfaces to provide corrosion resistance and to promote adhesion of subsequent coatings.

**Preventive measures** (Section 763.83) means actions taken to reduce disturbance of ACBM or otherwise eliminate the reasonable likelihood of the material's becoming damaged or significantly damaged.

**Previously certified gasoline, or PCG,** (Section 80.2) means gasoline or RBOB that previously has been included in a batch for purposes of complying with the standards in Subparts D, E, H, and J of this part, as appropriate.

**Primarily to protect a security interest** (Section 280.200) as used in this subpart, does not include indicia of ownership held primarily for investment purposes, nor ownership indicia held primarily for purposes other than as protection for a security interest. A holder may have other, secondary reasons for maintaining indicia of ownership, but the primary reason why any ownership indicia are held must be as protection for a security interest.

**Primarily to protect a security interest** (Section 280.200) means that the holder's indicia of ownership are held primarily for the purpose of securing payment or performance of an obligation.

**Primarily used** (Section 85.1602) means used 51 percent or more.

**Primary enforcement responsibility** (Section 142.2) means the primary responsibility for administration and enforcement of primary drinking water regulations and related requirements applicable to public water systems within a State.

**Primary Exporter** (Section 262.51) means any person who is required to originate the manifest for a shipment of hazardous waste in accordance with 40 CFR part 262, subpart B, or equivalent State provision, which specifies a treatment, storage, or disposal facility in a receiving country as the facility to which the hazardous waste will be sent and any intermediary arranging for the export.

**Primary fuel** (Section 65.2) means the fuel that provides the principal heat input to the device. To be considered primary, the fuel must be able to sustain operation without the addition of other fuels.

**Primary fuel** (Section 92.2) means that type of fuel (e.g., diesel fuel) that is consumed in the greatest quantity (mass basis) when the locomotive or locomotive engine is operated in use.

**Primary fuel** (Section 94.2) means that type of fuel (e.g., petroleum distillate diesel fuel) that is expected to be consumed in the greatest quantity (volume basis) when the engine is operated in use.

**Primary fuel or primary fuel supply** (Section 72.2) means the main fuel type (expressed in mmBtu) consumed by an affected unit for the applicable calendar year.

**Primary industry category** (Section 122.2) means any industry category listed in the NRDC settlement agreement (Natural Resources Defense Council et al. v. Train, 8 E.R.C. 2120 (D.D.C. 1976), modified 12 E.R.C. 1833 (D.D.C. 1979)); also listed in appendix A of part 122. Privately owned treatment works means any device or system which is (a) used to treat wastes from any facility whose operator is not the operator of the treatment works and (b) not a "POTW."

**Primary intended service class** (Section 86.085-2) means: (a) The primary service application group for which a heavy-duty diesel engine is designed and marketed, as determined by the manufacturer. The primary intended service classes are designated as light, medium, and heavy heavy-duty diesel engines. The determination is based on factors such as vehicle GVW, vehicle usage and operating patterns, other vehicle design characteristics, engine horsepower, and other engine design and operating characteristics.(1) Light heavy-duty diesel engines usually are non-sleeved and not designed for rebuild; their rated horsepower generally ranges from 70 to 170. Vehicle body types in this group might include any heavy-duty vehicle built for a light-duty truck chassis, van trucks, multi-stop vans, recreational vehicles, and some single axle straight trucks. Typical applications would include personal transportation, light-load commercial hauling and delivery, passenger service, agriculture, and construction. The GVWR of these vehicles is normally less than 19,500 lbs. (2) Medium heavy-duty diesel engines may be sleeved or non-sleeved and may be designed for rebuild. Rated horsepower generally ranges from 170 to 250. Vehicle body types in this group would typically include school buses, tandem axle straight trucks, city tractors, and a variety of special purpose vehicles such as small dump trucks, and trash compactor trucks. Typical applications would include commercial short haul and intra-city delivery and pickup. Engines in this group are normally used in vehicles whose GVWR varies from 19,500-33,000 lbs. (3) Heavy heavy-duty diesel engines are sleeved and designed for multiple rebuilds. Their rated horsepower generally exceeds 250. Vehicles in this group are normally tractors, trucks, and buses used in inter-city, long-haul applications. These vehicles normally exceed 33,000 lbs GVWR.

**Primary intended service class** (Section 86.090-2) means: (a) The primary service application group for which a heavy-duty diesel engine is designed and marketed, as determined by the manufacturer. The primary intended service classes are designated as light, medium, and heavy heavy-duty diesel engines. The determination is based on factors such as vehicle GVW, vehicle usage and operating patterns, other vehicle design characteristics, engine horsepower, and other engine design and operating characteristics. (1) Light heavy-duty diesel engines usually are non-sleeved and not designed for rebuild; their rated horsepower generally ranges from 70 to 170. Vehicle body types in this group might include any heavy-duty vehicle built for a light-duty truck chassis, van trucks, multi-stop vans, recreational vehicles, and some single axle straight trucks. Typical applications would include personal transportation, light-load commercial hauling and delivery, passenger service, agriculture, and construction. The GVWR of these vehicles is normally less than 19,500 lbs. (2) Medium heavy-duty diesel engines may be sleeved or non-sleeved and may be designed for rebuild. Rated horsepower generally ranges from 170 to 250. Vehicle body types in this group would typically include school buses, tandem axle straight trucks, city tractors, and a variety of special purpose vehicles such as small dump trucks, and trash compactor trucks. Typical applications would include commercial short haul and intra-city delivery and pickup. Engines in this group are normally used in vehicles whose GVWR varies from 19,500-33,000 lbs. (3) Heavy heavy-duty diesel engines are sleeved and designed for multiple rebuilds. Their rated horsepower generally exceeds 250. Vehicles in this group are normally tractors, trucks, and buses used in inter-city, long-haul applications. These vehicles normally exceed 33,000 lbs. GVWR.

**Primary measurement services (primary)** (Section 195.2) refers to radon measurement services using a specific device which services include the capability to read and/or analyze the results generated from the device.

**Primary metal cast** (Section 464.31) shall mean the metal that is poured in the greatest quantity at an individual plant.

**Primary Metropolitan Statistical Area (PMSA)** (Section 58.1) is a separate component of a consolidated metropolitan statistical area. For the purposes of this part, PMSA is used interchangeably with MSA.

**Primary mill** (Section 420.71) means those steel hot forming operations that reduce ingots to blooms or slabs by passing the ingots between rotating steel rolls. The first hot forming operation performed on solidified steel after it is removed from the ingot molds is carried out on a "primary mill".

**Primary Panel** (Section 211.203) The surface that is considered to be the front surface or that surface which is intended for initial viewing at the point of ultimate sale or the point of distribution for use.

**Primary standard** (Section 51.100) means a national primary ambient air quality standard promulgated pursuant to section 109 of the Act.

**Primary vendor** (Section 76.2) means the vendor of the NOX emission control system who has primary responsibility for providing the equipment, service, and technical expertise necessary for detailed design, installation, and operation of the controls, including process data, mechanical drawings, operating manuals, or any combination thereof.

**Primer** (Section 59.101) means any coating applied prior to the application of a topcoat for the purpose of corrosion resistance and/or adhesion.

**Primer-sealer** (Section 59.101) means any coating applied prior to the application of a topcoat for the purpose of corrosion resistance, adhesion of the topcoat, and/or color uniformity and to promote the ability of an undercoat to resist penetration by the topcoat.

**Primer-surfacer** (Section 59.101) means any coating applied prior to the application of a topcoat for the purpose of filling surface imperfections in the substrate, corrosion resistance, and/or adhesion of the topcoat.

**Principal importer** (Section 720.3) means the first importer who, knowing that a new chemical substance will be imported rather than manufactured domestically, specifies the identity of the chemical substance and the total amount to be imported. Only persons who are incorporated, licensed, or doing business in the United States may be principal importers.

**Principal importer** (Section 721.3) means the first importer who, knowing that a chemical substance will be imported for a significant new use rather than manufactured in the United States, specifies the chemical substance and the amount to be imported. Only persons who are incorporated, licensed, or doing business in the United States may be principal importers.

**Principal instructor** (Section 745.223) means the individual who has the primary responsibility for organizing and teaching a particular course.

**Principal place of business** (Section 171.2) means the principal location, either residence or office, in the State in which an individual, partnership, or corporation applies pesticides. Limited. The following definitions apply only to dealers, dealerships and transactions in States or on Indian Reservations where EPA conducts a Federal Pesticide Applicator Certification Program.

**Principal sponsor** (Section 790.3) means an individual sponsor or the joint sponsor who assumes primary responsibility for the direction of a study and for oral and written communication with EPA.

**Printer ribbon** (Section 247.3) means a nylon fabric designed to hold ink and used in dot matrix and other types of impact printers;

**Priority pollutants** (Section 414.10) means the toxic pollutants listed in 40 CFR 401.15.

**Priority Pollutants** (Section 455.10) means the toxic pollutants listed in 40 CFR part 423, appendix A.

**Private applicator** (Section 171.2) means a certified applicator who uses or supervises the use of any pesticide which is classified for restricted use for purposes of producing any agricultural commodity on property owned or rented by him or his employer or (if applied without compensation other than trading of personal services between producers of agricultural commodities) on the property of another person.

**Private carrier of property by motor vehicle** (Section 202.10) means any person not included in terms "common carrier by motor vehicle" or "contract carrier by motor vehicle", who or which transports in interstate or foreign commerce by motor vehicle property of which such person is the owner, lessee, or bailee, when such transportation is for sale, lease, rent or bailment, or in furtherance of any commercial enterprise.

**Probationary calibration error test** (Section 72.2) means an on-line calibration error test performed in accordance with section 2.1.1 of appendix B to part 75 of this chapter that is used to initiate a conditionally valid data period.

**Probe** (Section 58.1) is the actual inlet where an air sample is extracted from the atmosphere for delivery to a sampler or point analyzer for pollutant analysis.

**Proceeding** (Section 305.3) means the entire process of review of a claim conducted pursuant to this part that is initiated by a Request for a Hearing. A hearing is part of a proceeding.

**Process** (Section 372.3) means the preparation of a toxic chemical, after its manufacture, for distribution in commerce: (1) In the same form or physical state as, or in a different form or physical state from, that in which it was received by the person so preparing such substance, or (2) As part of an article containing the toxic chemical. Process also applies to the processing of a toxic chemical contained in a mixture or trade name product.

**Process** (Section 68.3) means any activity involving a regulated substance including any use, storage, manufacturing, handling, or on-site movement of such substances, or combination of these activities. For the purposes of this definition, any group of vessels that are interconnected, or separate vessels that are located such that a regulated substance could be involved in a potential release, shall be considered a single process.

**Process** (Section 704.3) means to process for commercial purposes.

**Process** (section 710.2) means the preparation of a chemical substance or mixture, after its manufacture, for distribution in commerce (1) in the same form or physical state as, or in a different form or physical state from, that in which it was received by the person so preparing such substance or mixture, or (2) as part of a mixture or article containing the chemical substance or mixture.

**Process** (Section 716.3) means to process for commercial purposes.

**Process** (Section 720.3) means the preparation of a chemical substance or mixture, after its manufacture, for distribution in commerce (1) in the same form or physical state as, or in a different form or physical state from, that in which it was received by the person so preparing such substance or mixture, or (2) as part of a mixture or article containing the chemical substance or mixture.

**Process** (Section 761.13) means the preparation of a chemical substance or mixture, after its manufacture, for distribution in commerce: (1) In the same form or physical state as, or in a different form or physical state from, that in which it was received by the person so preparing such substance or mixture, or (2) As part of an article containing the chemical substance or mixture.

**Process** (Section 763.163) has the same meaning as in section 3 of the Act TSCA).

**Process for "commercial purposes"** (section 710.2) means to process (1) for distribution in commerce, including for test marketing purposes, or (2) for use as an intermediate.

**Process for commercial purposes** (Section 704.3) means the preparation of a chemical substance or mixture after its manufacture for distribution in commerce with the purpose of obtaining an immediate or eventual commercial advantage for the processor. Processing of any amount of a chemical substance or mixture is included in this definition. If a chemical substance or mixture containing impurities is processed for commercial purposes, then the impurities also are processed for commercial purposes.

**Process for commercial purposes** (Section 712.3) means the preparation of a chemical substance or mixture, after its manufacture, for distribution in commerce with the purpose of obtaining an immediate or eventual commercial advantage for the processor. Processing of any amount of a chemical substance or mixture is included. If a chemical or mixture containing impurities is processed for commercial purposes, then those impurities are also processed for commercial purposes.

**Process for commercial purposes** (Section 716.3) means the preparation of a chemical substance or mixture, after its manufacture, for distribution in commerce with the purpose of obtaining an immediate or eventual commercial advantage for the processor. Processing of any amount of a chemical substance or mixture is included. If a chemical substance or mixture containing impurities is processed for commercial purposes, then those impurities are also processed for commercial purposes.

**Process for commercial purposes** (Section 717.3) means the preparation of a chemical substance or mixture, after its manufacture, for distribution in commerce with the purpose of obtaining an immediate or eventual commercial advantage for the processor. Processing of any amount of a chemical substance or mixture is included. If a chemical substance or mixture containing impurities is processed for commercial purposes, then those impurities are also processed for commercial purposes.

**Process for commercial purposes** (Section 721.3) means the preparation of a chemical substance or mixture containing the chemical substance, after manufacture of the substance, for distribution in commerce with the purpose of obtaining an immediate or eventual commercial advantage for the processor. Processing of any amount of a chemical substance or mixture containing the chemical substance is included in this definition. If a chemical substance or mixture containing impurities is processed for commercial purposes, the impurities also are processed for commercial purposes.

**Process generated waste water** (Section 412.11) shall mean water directly or indirectly used in the operation of a feedlot for any or all of the following: Spillage or overflow from animal or poultry watering systems; washing, cleaning or flushing pens, barns, manure pits or other feedlot facilities; direct contact swimming, washing or spray cooling of animals; and dust control.

**Process generated waste water** (Section 412.21) shall mean water directly or indirectly used in the operation of a feedlot for any or all of the following: Spillage or overflow from animal or poultry watering systems; washing, cleaning or flushing pens, barns, manure pits or other feedlot facilities; direct contact swimming, washing or spray cooling of animals; and dust control.

**Process generated waste water** (Section 436.21) shall mean any waste water used in the slurry transport of mined material, air emissions control, or processing exclusive of mining. The term shall also include any other water which becomes commingled with such waste water in a pit, pond, lagoon, mine, or other facility used for treatment of such waste water.

**Process generated waste water** (Section 438.181) shall mean any waste water used in the slurry transport of mined material, air emissions control, or processing exclusive of mining. The term shall also include any other water which becomes commingled with such waste water in a pit, pond lagoon, mine, or other facility used for settling or treatment of such waste water.

**Process heater** (Section 264.1031) means a device that transfers heat liberated by burning fuel to fluids contained in tubes, including all fluids except water that are heated to produce steam.

**Process heater** (Section 65.2) means an enclosed combustion device that transfers heat liberated by burning fuel directly to process streams or to heat transfer liquids other than water. A process heater may, as a secondary function, heat water in unfired heat recovery sections.

**Process solely for export** (Section 721.3) means to process for commercial purposes solely for export from the United States under the following restrictions on activity in the United States: Processing must be performed at sites under the control of the processor; distribution in commerce is limited to purposes of export; and the processor may not use the chemical substance except in small quantities solely for research and development.

**Process stream** (Section 721.3) means all reasonably anticipated transfer, flow, or disposal of a chemical substance, regardless of physical state or concentration, through all intended operations of processing, including the cleaning of equipment.

**Process unit** (Section 65.2) means the equipment specified in the definitions of process unit or chemical manufacturing process unit in the applicable referencing subpart. If the referencing subpart does not define process unit, then, for the purposes of this part, process unit means the equipment assembled and connected by pipes or ducts to process raw materials and to manufacture an intended product.

**Process unit shutdown** (Section 65.2) means a work practice or operational procedure that stops production from a process unit or part of a process unit during which it is technically feasible to clear process material from a process unit or part of a process unit consistent with safety constraints and during which repairs can be effected. An unscheduled work practice or operational procedure that stops production from a process unit or part of a process unit for less than 24 hours is not a process unit shutdown. An unscheduled work practice or operational procedure that would stop production from a process unit or part of a process unit for a shorter period of time than would be required to clear the process unit or part of the process unit of materials and start up the unit, and would result in greater emissions than delay of repair of leaking components until the next scheduled process unit shutdown is not a process unit shutdown. The use of spare equipment and technically feasible bypassing of equipment without stopping production are not process unit shutdowns.

**Process vent** (Section 264.1031) means any open-ended pipe or stack that is vented to the atmosphere either directly, through a vacuum-producing system, or through a tank (e.g., distillate receiver, condenser, bottoms receiver, surge control tank, separator tank, or hot well) associated with hazardous waste distillation, fractionation, thin-film evaporation, solvent extraction, or air or steam stripping operations.

**Process vent** (Section 65.2) means a process vent or vent stream as they are defined in the referencing subpart.

**Process waste water** (Section 117.1) means any water which, during manufacturing or processing, comes into direct contact with or results from the production or use of any raw material, intermediate product, finished product, byproduct, or waste product.

**Process waste water** (Section 401.11) means any water which, during manufacturing or processing, comes into direct contact with or results from the production or use of any raw material, intermediate product, finished product, by-product, or waste product.

**Process waste water** (Section 412.11) shall mean any process generated waste water and any precipitation (rain or snow) which comes into contact with any manure, litter or bedding, or any other raw material or intermediate or final material or product used in or resulting from the production of animals or poultry or direct products (e.g. milk, eggs).

**Process waste water** (Section 412.21) shall mean any process generated waste water and any precipitation (rain or snow) which comes into contact with any manure, litter or bedding, or any other raw material or intermediate or final material or product used in or resulting from the production of animal or poultry or direct products (e.g. milk, eggs).

**Process waste water** (Section 422.41) means any water which, during manufacturing or processing, comes into direct contact with or results from the production or use of any raw material, intermediate product, finished product, by-product, or waste product. The term "process waste water" does not include contaminated nonprocess waste water, as defined below.

**Process waste water** (Section 422.51) means any water which, during manufacturing or processing, comes into direct contact with or results from the production or use of any raw material, intermediate product, finished product, by-product, or waste product. The term "process waste water" does not include contaminated non-process waste water, as defined below.

**Process waste water** (Section 428.11) shall mean, in the case of tire and inner tube plants constructed before 1959, discharges from the following: Soapstone solution applications; steam cleaning operations; air pollution control equipment; unroofed process oil unloading areas; mold cleaning operations; latex applications; and air compressor receivers. Discharges from other areas of such plants shall not be classified as process waste water for the purposes of this section. Except as provided in paragraphs © and (e) of this section, the term process waste water shall have the meaning set forth in Sec. 401.11(q) of this chapter.

**Process waste water** (Section 458.11) shall mean waters which result from baghouse operations or thermal quench operations.

**Process waste water** (Section 458.21) shall mean waters which result from baghouse operations or thermal quench operations.

**Process waste water pollutants** (Section 401.11) means pollutants present in process waste water.

**Process wastes** (Section 129.2) means any designated toxic pollutant, whether in wastewater or otherwise present, which is inherent to or unavoidably resulting from any manufacturing process, including that which comes into direct contact with or results from the production or use of any raw material, intermediate product, finished product, by-product or waste product and is discharged into the navigable waters.

**Process wastewater** (Section 122.2) means any water which, during manufacturing or processing, comes into direct contact with or results from the production or use of any raw material, intermediate product, finished product, byproduct, or waste product.

**Process wastewater** (Section 415.241) means any water which, during manufacturing or processing, comes into direct contact with or results from the production or use of any raw material, intermediate product, finished product, by-product, or waste product. The term "process wastewater" does not include contaminated nonprocess wastewater, as defined below.

**Process wastewater** (Section 415.311) means any water which, during manufacturing or processing, comes into direct contact with or results from the production or use of any raw material, intermediate product, finished product, by-product, or waste product. The term "process wastewater" does not include contaminated nonprocess wastewater, as defined below.

**Process wastewater** (Section 415.331) means any water which, during manufacturing or processing, comes into direct contact with or results from the production or use of any raw material, intermediate product, finished product, by-product, or waste product. The term "process wastewater" does not include contaminated nonprocess wastewater, as defined below.

**Process wastewater** (Section 415.381) means any water which, during manufacturing or processing, comes into direct contact with or results from the production or use of any raw material, intermediate product, finished product, by-product, or waste product. The term "process wastewater" does not include contaminated nonprocess wastewater, as defined below.

**Process wastewater** (Section 415.401) means any water which, during manufacturing or processing, comes into direct contact with or results from the production or use of any raw material, intermediate product, finished product, by-product, or waste product. The term "process wastewater" does not include contaminated non-process wastewater, as defined below.

**Process wastewater** (Section 415.431) means any water which, during manufacturing or processing, comes into direct contact with or results from the production or use of any raw material, intermediate product, finished product, by-product, or waste product. The term "process wastewater" does not include contaminated non-process wastewater, as defined below.

**Process wastewater** (Section 415.441) means any water which, during manufacturing or processing, comes into direct contact with or results from the production or use of any raw material, intermediate product, finished product, by-product, or waste product. The term "process wastewater" does not include contaminated non-process wastewater, as defined below.

**Process wastewater** (Section 415.531) means any water which, during manufacturing or processing, comes into direct contact with or results from the production or use of any raw material, intermediate product, finished product, by-product, or waste product. The term "process wastewater" does not include contaminated non-process wastewater, as defined below.

**Process wastewater** (Section 415.551) means any water which, during manufacturing or processing, comes into direct contact with or results from the production or use of any raw material, intermediate product, finished product, by-product, or waste product. The term "process wastewater" does not include contaminated non-process wastewater, as defined below.

**Process wastewater** (Section 415.91) means any water which, during manufacturing or processing, comes into direct contact with or results from the production or use of any raw material, intermediate product, finished product, by-product, or waste product. The term "process wastewater" does not include contaminated non-process wastewater, as defined below.

**Process wastewater** (Section 418.11) means any water which, during manufacturing or processing, comes into direct contact with or results from the production or use of any raw material, intermediate product, finished product, by-product, or waste product. The term "process wastewater" does not include contaminated non-process wastewater, as defined below.

**Process wastewater** (Section 429.11) specifically excludes noncontact cooling water, material storage yard runoff (either raw material or processed wood storage), and boiler blowdown. For the dry process hardboard, veneer, finishing, particleboard, and sawmills and planing mills subcategories, fire control water is excluded from the definition.

**Process wastewater** (Section 430.0) For subparts B and E only, process wastewater is any water that, during manufacturing or processing, comes into direct contact with or results from the production or use of any raw material, intermediate product, finished product, byproduct, or waste product. For purposes of subparts B and E of this part, process wastewater includes boiler blowdown; wastewaters from water treatment and other utility operations; blowdowns from high rate (e.g., greater than 98 percent) recycled non-contact cooling water systems to the extent they are mixed and co-treated with other process wastewaters; wastewater, including leachates, from landfills owned by pulp and paper mills subject to subpart B or E of this part if the wastewater is commingled with wastewater from the mill's manufacturing or processing facility; and storm waters from the immediate process areas to the extent they are mixed and co-treated with other process wastewaters. For purposes of this part, contaminated groundwaters from on-site or off- site groundwater remediation projects are not process wastewater.

**Process wastewater** (Section 439.1) as defined at 40 CFR 122.2 and for the purposes of this part, does not include the following: (1) Trimethyl silanol, any active anti-microbial materials, process wastewater from imperfect fermentation batches, and process area spills. Discharges containing such materials are not subject to the limitations and standards of this part. (2) Non-contact cooling water, utility wastewaters, general site surface runoff, groundwater (e.g., contaminated groundwaters from on-site or off-site groundwater remediation projects), and other non-process water generated on site. Discharges of such waters and wastewaters are not subject to the limitations and standards of this part.

**Process wastewater** (Section 443.21) shall mean any water which, during the manufacturing process, comes into direct contact with any raw material, intermediate product, by-product, or product used in or resulting from the production of paving asphalt concrete.

**Process wastewater** (Section 455.41) for this subpart, means all wastewater associated with pesticide formulating, packaging and repackaging except for sanitary water, non-contact cooling water and those wastewaters excluded from the applicability of the rule in Sec. 455.40.

**Process wastewater** (Section 455.61) for this subpart, means all wastewater except for sanitary water and those wastewaters excluded from the applicability of the rule in Sec. 455.60.

**Process wastewater pollutants** (Section 415.241) means pollutants present in process wastewater.

**Process wastewater pollutants** (Section 415.311) means pollutants present in process wastewater.

**Process wastewater pollutants** (Section 415.331) means pollutants present in process wastewater.

**Process wastewater pollutants** (Section 415.381) means pollutants present in process wastewater.

**Process wastewater pollutants** (Section 415.401) means pollutants present in process wastewater.

**Process wastewater pollutants** (Section 415.431) means pollutants present in process wastewater.

**Process wastewater pollutants** (Section 415.441) means pollutants present in process wastewater.

**Process wastewater pollutants** (Section 415.531) means pollutants present in process wastewater.

**Process wastewater pollutants** (Section 415.551) means pollutants present in process wastewater.

**Process wastewater pollutants** (Section 415.91) means pollutants present in process wastewater.

**Process wastewater pollutants** (Section 443.11) shall mean any pollutants present in the process wastewaters and rainwater runoff.

**Process wastewater pollutants** (Section 443.21) shall mean any pollutants present in the process wastewater.

**Process water** (Section 463.2) is any raw, service, recycled, or reused water that contacts the plastic product or contacts shaping equipment surfaces such as molds and mandrels that are, or have been, in contact with the plastic product.

**Processing** (Section 279.1) means chemical or physical operations designed to produce from used oil, or to make used oil more amenable for production of, fuel oils, lubricants, or other used oil-derived product. Processing includes, but is not limited to: blending used oil with virgin petroleum products, blending used oils to meet the fuel specification, filtration, simple distillation, chemical or physical separation and re-refining.

**Processing site** (Section 192) means: (1) Any site, including the mill, designated by the Secretary under Section 102(a)(1) of the Act; and (2) Any other real property or improvement thereon which is in the vicinity of such site, and is determined by the Secretary, in consultation with the Commission, to be contaminated with residual radioactive materials derived from such site. \

**Processor** (Section 238.20) means the persons or entities that produce ring carriers ready for use as beverage carriers.

**Processor** (Section 704.3) means any person who processes a chemical substance or mixture.

**Processor** (section 710.2) means any person who processes a chemical substance or mixture.

**Processor** (Section 720.3) means any person who processes a chemical substance or mixture.

**Processor** (Section 763.163) has the same meaning as in section 3 of the Act (TSCA).

**Procurement item** (Section 247.3) means any device, good, substance, material, product, or other item, whether real or personal property, which is the subject of any purchase, barter, or other exchange made to procure such item;

**Procuring agency** (Section 247.3) means any Federal agency, or any State agency or agency of a political subdivision of a State, which is using appropriated Federal funds for such procurement, or any person contracting with any such agency with respect to work performed under such contract;

**Produce** (Section 167.3) means to manufacture, prepare, propagate, compound, or process any pesticide, including any pesticide produced pursuant to section 5 of the Act, any active ingredient or device, or to package, repackage, label, relabel, or otherwise change the container of any pesticide or device.

**Produce** (Section 92.2) means to manufacture or remanufacture. Where a certificate holder does not actually assemble the locomotives or locomotive engines that it manufactures or remanufactures, produce means to allow other entities to assemble locomotives or locomotive engines under the certificate holder's certificate.

**Produce thereof, when referring to plant-incorporated protectants only** (Section 152.3) means a product of a living plant containing a plant-incorporated protectant, where the pesticidal substance is intended to serve a pesticidal purpose after the product has been separated from the living plant. Examples of such products include, but are not limited to, agricultural produce, grains, and lumber. Products such as raw agricultural commodities bearing pesticide chemical residues are not "produce thereof" when the residues are not intended to serve a pesticidal purpose in the produce.

**Produce thereof, when used with respect to plants containing plant- incorporated protectants only** (Section 174.3) means a product of a living plant containing a plant-incorporated protectant, where the pesticidal substance is intended to serve a pesticidal purpose after the product has been separated from the living plant. Examples of such products include, but are not limited to, agricultural produce, grains, and lumber. Products such as raw agricultural commodities bearing pesticide chemical residues are not "produce thereof" when the residues are not intended to serve a pesticidal purpose in the produce.

**Produced sand** (Section 435.11) means the slurried particles used in hydraulic fracturing, the accumulated formation sands and scales particles generated during production. Produced sand also includes desander discharge from the produced water waste stream, and blowdown of the water phase from the produced water treating system.

**Produced sand** (Section 435.41) means the slurried particles used in hydraulic fracturing, the accumulated formation sands and scales particles generated during production. Produced sand also includes desander discharge from the produced water waste stream, and blowdown of the water phase from the produced water treating system.

**Produced water** (Section 435.11) means the water (brine) brought up from the hydrocarbon-bearing strata during the extraction of oil and gas, and can include formation water, injection water, and any chemicals added downhole or during the oil/water separation process.

**Produced water** (Section 435.41) means the water (brine) brought up from the hydrocarbon-bearing strata during the extraction of oil and gas, and can include formation water, injection water, and any chemicals added downhole or during the oil/water separation process.

**Produced water** (Section 68.3) means water extracted from the earth from an oil or natural gas production well, or that is separated from oil or natural gas after extraction.

**Producer** (Section 167.3) means any person, as defined by the Act, who produces any pesticide, active ingredient, or device (including packaging, repackaging, labeling and relabeling).

**Producer** (Section 169.1) The term "producer" means the person, as defined by the Act, who produces or imports any pesticide or device or active ingredient used in producing a pesticide.

**Product** (Section 415.201) shall mean sodium sulfite.

**Product** (Section 415.541) means sodium bisulfite.

**Product** (Section 203.1) means any manufactured article or goods or component thereof; except that such term does not include—(i) Any aircraft, aircraft engine, propeller or appliance, as such terms are defined in section 101 of the Federal Aviation Act of 1958; or (ii)(a) Any military weapons or equipment which are designed for combat use; (b) any rockets or equipment which are designed for research, experimental or developmental work to be performed by the National Aeronautics and Space Administration; or © to the extent provided by regulations of the Administrator, any other machinery or equipment designed for use in experimental work done by or for the Federal Government.

**Product** (Section 204.2) means any construction equipment for which regulations have been promulgated under this part and includes "test product."

**Product** (Section 205.2) means any transportation equipment for which regulations have been promulgated under this part and includes "test product."

**Product** (Section 211.102) means any noise-producing or noise-reducing product for which regulations have been promulgated under part 211; the term includes "test product".

**Product** (Section 410.61) shall mean the final carpet produced or processed including the primary backing but excluding the secondary backing.

**Product** (Section 415.161) shall mean sodium chloride.

**Product** (Section 415.171) shall mean sodium dichromate.

**Product** (Section 415.221) shall mean titanium dioxide.

**Product** (Section 415.231) means aluminum fluoride produced by the dry process in which partially dehydrated alumina hydrate is reacted with hydrofluoric acid gas.

**Product** (Section 415.241) shall mean ammonium chloride.

**Product** (Section 415.281) shall mean boric acid.

**Product** (Section 415.331) shall mean carbon monoxide plus hydrogen.

**Product** (Section 415.341) means chrome pigments.

**Product** (Section 415.361) shall mean copper salts.

**Product** (Section 415.41) shall mean calcium chloride.

**Product** (Section 415.421) means hydrogen cyanide.

**Product** (Section 415.451) shall mean lithium carbonate.

**Product** (Section 415.471) shall mean nickel salts.

**Product** (Section 415.511) shall mean potassium iodide.

**Product** (Section 415.531) shall mean silver nitrate.

**Product** (Section 415.641) shall mean cadmium pigment or cadmium salt.

**Product** (Section 415.651) shall mean cobalt salts.

**Product** (Section 415.661) shall mean sodium chlorate.

**Product** (Section 415.671) shall mean zinc chloride.

**Product** (Section 415.91) shall mean hydrogen peroxide as a one hundred percent hydrogen peroxide solution.

**Product** (Section 416.61) shall mean chlorine.

**Product** (Section 418.31) shall mean the 100 percent urea content of the material manufactured.

**Product** (Section 418.41) shall mean the 100 percent ammonium nitrate content of the material manufactured.

**Product** (Section 421.11) shall mean alumina.

**Product** (Section 421.21) shall mean hot aluminum metal.

**Product** (Section 421.31) shall mean hot aluminum metal.

**Product** (Section 421.51) means electrolytically refined copper.

**Product** (Section 421.81) shall mean zinc metal.

**Product** (Section 421.91) means 100 percent equivalent sulfuric acid, $H_2SO_4$ capacity.

**Product** (Section 439.11) means pharmaceutical products derived from fermentation processes.

**Product** (Section 439.21) means any substance manufactured by an extraction process, including blood fractions, vaccines, serums, animal bile derivatives, endocrine products and medicinal products such as alkaloids that are isolated from botanical drugs and herbs.

**Product** (Section 439.31) means any pharmaceutical product manufactured by chemical synthesis.

**Product** (Section 439.41) means any pharmaceutical product manufactured by blending, mixing, compounding, and formulating pharmaceutical ingredients. The term includes pharmaceutical preparations for both human and veterinary use, such as ampules, tablets, capsules, vials, ointments, medicinal powders, solutions, and suspensions.

**Product** (Section 454.11) shall mean char and charcoal briquets.

**Product** (Section 454.21) shall mean gum rosin and turpentine.

**Product** (Section 454.31) shall mean products from wood rosin, turpentine and pine oil.

**Product** (Section 454.41) shall mean tall oil rosin, pitch and fatty acids.

**Product** (Section 454.51) shall mean essential oils.

**Product** (Section 454.61) shall mean rosin-based derivatives.

**Product** (Section 457.11) shall mean dynamite, nitroglycerin, cyclotrimethylene trinitramine (RDX), cyclotetramethylene tetranitramine (HMX), and trinitrotoluene (TNT).

**Product** (Section 457.31) shall mean products from plants which blend explosives and market a final product, and plants that fill shells and blasting caps. Examples of such installations would be plants manufacturing ammonium nitrate and fuel oil (ANFO), nitrocarbonitrate (NCN), slurries, water gels, and shells.

**Product** (Section 458.11) shall mean carbon black manufactured by the furnace process.

**Product** (Section 458.21) shall mean carbon black manufactured by the thermal process.

**Product** (Section 458.31) shall mean carbon black manufactured by the channel process.

**Product** (Section 459.11) shall mean articles developed or printed by photographic processes, such as paper prints, slides, negatives, enlargements, movie film and other sensitized materials.

**Product** (Section 460.11) shall mean service resulting from the hospital activity in terms of 1,000 occupied beds.

**Product frosted** (Section 426.121) shall mean that portion of the furnace pull associated with the fraction of finished incandescent lamp envelopes which is frosted; this quantity shall be calculated by multiplying furnace pull by the fraction of finished incandescent lamp envelopes which is frosted.

**Product or pesticide product** (Section 162.151) means a pesticide offered for distribution and use, and includes any labeled container and any supplemental labeling.

**Product stewardship** (Section 437.2) means a manufacturer's treatment or recovery of its own unused products, shipping and storage containers with product residues, off-specification products, and does not include spent or used materials from use of its products.

**Product testing** (Section 471.02) includes operations such as dye penetrant testing, hydrotesting, and ultrasonic testing.

**Production** (Section 430.0) (1) For all limitations and standards specified in this part except those pertaining to AOX and chloroform: Production shall be defined as the annual off-the-machine production (including off-the-machine coating where applicable) divided by the number of operating days during that year. Paper and paperboard production shall be measured at the off-the-machine moisture content, except for subpart C of this part (as it pertains to pulp and paperboard production at unbleached kraft mills including linerboard or bag paper and other mixed products, and to pulp and paperboard production using the unbleached kraft neutral sulfite semi-chemical (cross recovery) process), and subparts F and J of this part (as they pertain to paperboard production from wastepaper from noncorrugating medium furnish or from corrugating medium furnish) where paper and paperboard production shall be measured in air-dry-tons (10% moisture content). Market pulp shall be measured in air-dry tons (10% moisture). Production shall be determined for each mill based upon past production practices, present trends, or committed growth. (2) For AOX and chloroform limitations and standards specified in subparts B and E of this part: Production shall be defined as the annual unbleached pulp production entering the first stage of the bleach plant divided by the number of operating days during that year. Unbleached pulp production shall be measured in air-dried-metric-tons (10% moisture) of brownstock pulp entering the bleach plant at the stage during which chlorine or chlorine-containing compounds are first applied to the pulp. In the case of bleach plants that use totally chlorine free bleaching processes, unbleached pulp production shall be measured in air-dried-metric tons (10% moisture) of brownstock pulp entering the first stage of the bleach plant from which wastewater is discharged. Production shall be determined for each mill based upon past production practices, present trends, or committed growth. (o) TCDD. 2,3,7,8-tetrachlorodibenzo-p-dioxin. (p) TCDF. 2,3,7,8-tetrachlorodibenzofuran.

**Production** (Section 82.3) means the manufacture of a controlled substance from any raw material or feedstock chemical, but does not include: (1) The manufacture of a controlled substance that is subsequently transformed; (2) The reuse or recycling of a controlled substance; (3) Amounts that are destroyed by the approved technologies; or (4) Amounts that are spilled or vented unintentionally.

**Production** (Section 86.1702-99) is the number of vehicles and/or trucks that a manufacturer produces in a given model year that are subject to the provisions of this subpart and are included in the same averaging set.

**Production allowances** (Section 82.3) means the privileges granted by this subpart to produce controlled substances; however, production allowances may be used to produce controlled substances only in conjunction with consumption allowances. A person's production allowances are the total of the allowances obtained under Secs. 82.7, 82.5 and 82.9, and as may be modified under Sec. 82.12 (transfer of allowances).

**Production area size** (Section 443.11) shall mean that area in which the oxidation towers, loading facilities, and all buildings that house product processes are located.

**Production changes** (Section 89.602) Those changes in nonroad engine configuration, equipment, or calibration which are made by an OEM or ICI in the course of nonroad engine production and required to be reported under Sec. 89.123.

**Production facility** (Section 435.11) means any fixed or mobile structure subject to this Subpart that is either engaged in well completion or used for active recovery of hydrocarbons from producing formations.

**Production facility** (Section 435.41) means any fixed or mobile structure subject to this subpart that is either engaged in well completion or used for active recovery of hydrocarbons from producing formations. It includes facilities that are engaged in hydrocarbon fluids separation even if located separately from wellheads.

**Production normalizing mass (/kkg)** (Section 467.02) for each core or ancillary operation is the mass (off-kkg or off-lb) processed through that operation.

**Production Volume** (Section 600.022-85) means, for a domestic manufacturer, the number of vehicle units domestically produced in a particular model year but not exported, and for a foreign manufacturer, means the number of vehicle units of a particular model imported into the United States.

**Production volume** (Section 704.3) means the quantity of a substance which is produced by a manufacturer, as measured in kilograms or pounds.

**Production weighted particulate average** (Section 86.090-2) means the manufacturer's production-weighted average particulate emission level, for certification purposes, of all of its diesel engine families included in the light-duty particulate averaging program. It is calculated at the end of the model year by multiplying each family particulate emission limit by its respective production, summing those terms, and dividing the sum by the total production of the effected families. Those vehicles produced for sale in California or at high altitude shall each be averaged separately from those produced for sale in any other area.

**Production-weighted average** (Section 86.085-2) means the manufacturer's production-weighted average particulate emission level, for certification purposes, of all of its diesel engine families included in the particulate averaging program. It is calculated at the end of the model year by multiplying each family particulate emission limit by its respective production, summing these terms, and dividing the sum by the total production of the effected families. Those vehicles produced for sale in California or at high altitude shall each be averaged separately from those produced for sale in any other area.

**Production-weighted average** (Section 86.087-2) means the manufacturer's production-weighted average particulate emission level, for certification purposes, of all of its diesel engine families included in the particulate averaging program. It is calculated at the end of the model year by multiplying each family particulate emission limit by its respective production, summing these terms, and dividing the sum by the total production of the affected families. Those vehicles produced for sale in California or at high altitude shall each be averaged separately from those produced for sale in any other area. Diesel light-duty trucks with a loaded vehicle weight equal to or greater than 3,751 lbs (LDDT2s) shall only be averaged with other diesel light-duty trucks with a loaded vehicle weight equal to or greater than 3,751 lbs produced by that manufacturer.

**Production-weighted NO<INF>X</INF> average** (Section 86.088-2) means the manufacturer's production-weighted average NO<INF>X</INF> emission level, for certification purposes, of all of its light-duty truck engine families included in the NO<INF>X</INF> averaging program. It is calculated at the end of the model year by multiplying each family NO<INF>X</INF> emission limit by its respective production, summing those terms, and dividing the sum by the total production of the effected families. Those vehicles produced for sale in California or at high altitude shall each be averaged separately from those produced for sale in any other area.

**Production-weighted particulate average** (Section 86.088-2) means the manufacturer's production-weighted average particulate emission level, for certification purposes, of all of its diesel engine families included in the particulate averaging program. It is calculated at the end of the model year by multiplying each family particulate emission limit by its respective production, summing those terms, and dividing the sum by the total production of the effected families. Those vehicles produced for sale in California or at high altitude shall each be averaged separately from those produced for sale in any other area. (Secs. 202, 203, 206, 207, 208, 301a, Clean Air Act, as amended; 42

**Program directive** (Section 56.1) means any formal written statement by the Administrator, the Deputy Administrator, the Assistant Administrator, a Staff Office Director, the General Counsel, a Deputy Assistant Administrator, an Associate General Counsel, or a division Director of an Operational Office that is intended to guide or direct Regional Offices in the implementation or enforcement of the provisions of the act.

**Prohibit specification** (Section 231.2) means to prevent the designation of an area as a present or future disposal site.

**Prohibited** (Section 129.2) means that the constituent shall be absent in any discharge subject to these standards, as determined by any analytical method.

**Project** (Section 144.3) means a group of wells in a single operation.

**Project** (Section 146.3) means a group of wells in a single operation.

**Project** (Section 93.101) means a highway project or transit project.

**Property damage** (Section 280.92) shall have the meaning given this term by applicable state law. This term shall not include those liabilities which, consistent with standard insurance industry practices, are excluded from coverage in liability insurance policies for property damage.

However, such exclusions for property damage shall not include corrective action associated with releases from tanks which are covered by the policy.

**Proportional sampling** (Section 60.2) means sampling at a rate that produces a constant ratio of sampling rate to stack gas flow rate.

**Propose to manufacture, import, or process** (Section 704.3) means that a person has made a firm management decision to commit financial resources for the manufacture, import, or processing of a specified chemical substance or mixture.

**Propose to manufacture, import, or process** (Section 716.3) means that a person has made a management decision to commit financial resources toward the manufacture, importation, or processing of a substance or mixture.

**Proposed Acid Rain permit or proposed permit** (Section 72.2) means, in the case of a State operating permit program, the version of an Acid Rain permit that the permitting authority submits to the Administrator after the public comment period, but prior to completion of the EPA permit review period, as provided for in part 70 of this chapter.

**Proposed permit** (Section 122.2) means a State NPDES "permit" prepared after the close of the public comment period (and, when applicable, any public hearing and administrative appeals) which is sent to EPA for review before final issuance by the State. A "proposed permit" is not a "draft permit."

**Proposed permit** (Section 70.2) means the version of a permit that the permitting authority proposes to issue and forwards to the Administrator for review in compliance with Sec. 70.8.

**Proposed permit** (Section 71.2) means the version of a permit that the delegate agency proposes to issue and forwards to the Administrator for review in compliance with Sec. 71.10(d).

**Propulsion engine** (Section 94.2) means an engine that moves a vessel through the water or directs the movement of a vessel.

**Propulsion marine diesel engine** (Section 89.2) means a marine diesel engine that is intended to move a vessel through the water or direct the movement of a vessel.

**Protective equipment** (Section 171.2) means clothing or any other materials or devices that shield against unintended exposure to pesticides.

**Protective finding** (Section 93.101) means a determination by EPA that a submitted control strategy implementation plan revision contains adopted control measures or written commitments to adopt enforceable control measures that fully satisfy the emissions reductions requirements relevant to the statutory provision for which the implementation plan revision was submitted, such as reasonable further progress or attainment.

**Protocol** (Section 790.3) means the plan and procedures which are to be followed in conducting a test.

**Proven emission control systems** (Section 86.092-2) are emission control components or systems (and fuel metering systems) that have completed full durability testing evaluation over a vehicle's useful life in some other certified engine family, or have completed bench or road testing demonstrated to be equal or more severe than certification mileage accumulation requirements. Alternatively, proven components or systems are those that are determined by EPA to be of comparable functional quality and manufactured using comparable materials and production techniques as components or systems which have been durability demonstrated in some other certified engine family. In addition, the components or systems must be employed in an operating environment (e.g., temperature, exhaust flow, etc.,) similar to that experienced by the original or comparable components or systems in the original certified engine family.

**Proven emission control systems** (Section 86.1803-01) are emission control components or systems (and fuel metering systems) that have completed full durability testing evaluation over a vehicle's useful life in some other certified test group, or have completed bench or road testing demonstrated to be equal or more severe than certification mileage accumulation requirements. Alternatively, proven components or systems are those that are determined by EPA to be of comparable functional quality and manufactured using comparable materials and production techniques as components or systems which have been durability demonstrated in some other certified test group. In addition, the components or systems must be employed in an operating environment (e.g., temperature, exhaust flow, etc.,) similar to that experienced by the original or comparable components or systems in the original certified test group.

**Provide for in the phrase "the plan shall (should) provide for"** (Section 256.06) means explain, establish or set forth steps or courses of action.

**Provider of financial assurance** (Section 280.92) means an entity that provides financial assurance to an owner or operator of an underground storage tank through one of the mechanisms listed in Secs. 280.95-280.103, including a guarantor, insurer, risk retention group, surety, issuer of a letter of credit, issuer of a state-required mechanism, or a state.

**PSD station** (Section 58.1) means any station operated for the purpose of establishing the effect on air quality of the emissions from a proposed source for purposes of prevention of significant deterioration as required by Sec. 51.24 (n) of part 51 of this chapter.

**PSES** (Section 467.02) means pretreatment standards for existing sources, under section 307(b) of the Act.

**PSNS** (Section 467.02) means pretreatment standards for new sources, under section 307© of the Act.

**Public** (Section 68.3) means any person except employees or contractors at the stationary source.

**Public contact site** (Section 503.11) is land with a high potential for contact by the public. This includes, but is not limited to, public parks, ball fields, cemeteries, plant nurseries, turf farms, and golf courses.

**Public or Private Agricultural Research Agency or Educational Institution (Section 172.21)** means any organization engaged in research pertaining to the agricultural use of pesticides, or any educational institution engaged in pesticides research. Any research agency or educational institution whose principal function is to promote, or whose principal source of income is directly derived from, the sale or distribution of pesticides (or their active ingredients) does not come within the meaning of this term.

**Public participation** (Section 300.5) see the definition for community relations.

**Public receptor** (Section 68.3) means offsite residences, institutions (e.g., schools, hospitals), industrial, commercial, and office buildings, parks, or recreational areas inhabited or occupied by the public at any time without restriction by the stationary source where members of the public could be exposed to toxic concentrations, radiant heat, or overpressure, as a result of an accidental release.

**Public record** (Section 117.1) means the NPDES permit application or the NPDES permit itself and the materials comprising the administrative record for the permit decision specified in Sec. 124.18 of this chapter.

**Public service** (Section 445.2) means the provision of landfill waste disposal services to individual members of the general public; publicly-owned organizations (schools, universities, government agencies, municipalities) and not-for-profit organizations for which the landfill does not receive a fee or other remuneration.

**Public vessel** (Section 116.3) means a vessel owned or bareboat-chartered and operated by the United States, or a State or political subdivision thereof, or by a foreign nation, except when such vessel is engaged in commerce.

**Public vessel as defined by section 311(a)(4) of the CWA** (Section 300.5) means a vessel owned or bareboat-chartered and operated by the United States, or by a state or political subdivision thereof, or by a foreign nation, except when such vessel is engaged in commerce.

**Public water system** (Section 141.2) means a system for the provision to the public of water for human consumption through pipes or, after August 5, 1998, other constructed conveyances, if such system has at least fifteen service connections or regularly serves an average of at least twenty-five individuals daily at least 60 days out of the year. Such term includes: any collection, treatment, storage, and distribution facilities under control of the operator of such system and used primarily in connection with such system; and any collection or pretreatment storage facilities not under such control which are used primarily in connection with such system. Such term does not include any "special irrigation district." A public water system is either a "community water system" or a "noncommunity water system."

**Public water system** (Section 143.2) means a system for the provision to the public of piped water for human consumption, if such a system has at least fifteen service connections or regularly serves an average of at least twenty-five individuals daily at least 60 days out of the year. Such term includes (1) any collection, treatment, storage, and distribution facilities under control of the operator of such system and used primarily in connection with such system, and (2) any collection or pretreatment storage facilities not under such control which are used primarily in connection with such system. A public water system is either a "community water system" or a "non-community water system."

**Public water system** (Section 191.22) means a system for the provision to the public of piped water for human consumption, if such system has at least fifteen service connections or regularly serves at least twenty-five individuals. Such term includes: (1) Any collection, treatment, storage, and distribution facilities under control of the operator of such system and used primarily in connection with such system; and (2) Any collection or pretreatment storage facilities not under such control which are used primarily in connection with such system.

**Public water system or PWS** (Section 142.2) means a system for the provision to the public of water for human consumption through pipes or, after August 5, 1998, other constructed conveyances, if such system has at least fifteen service connections or regularly serves an average of at least twenty-five individuals daily at least 60 days out of the year. Such term includes:

**Publicly owned treatment works (POTW)** (Section 270.2) means any device or system used in the treatment (including recycling and reclamation) of municipal sewage or industrial wastes of a liquid nature which is owned by a State or municipality. This definition includes sewers, pipes, or other conveyances only if they convey wastewater to a POTW providing treatment.

**Publicly owned treatment works** (Section 122.2) is defined at 40 CFR 403.3.

**Publicly owned treatment works** (Section 501.2) means a treatment works treating domestic sewage that is owned by a municipality or State.

**Publicly Owned Treatment Works or POTW** (Section 117.1) means a treatment works as defined by section 212 of the Act, which is owned by a State or municipality (as defined by section 502(4) of the Act). This definition includes any sewers that convey wastewater to such a treatment works, but does not include pipes, sewers or other conveyances not connected to a facility providing treatment. The term also means the municipality as defined in section 502(4) of the Act, which has jurisdiction over the indirect discharges to and the discharges from such a treatment works.

**Publicly owned treatment works or POTW** (Section 260.10) means any device or system used in the treatment (including recycling and reclamation) of municipal sewage or industrial wastes of a liquid nature which is owned by a "State" or "municipality" (as defined by section 502(4) of the CWA). This definition includes sewers, pipes, or other conveyances only if they convey wastewater to a POTW providing treatment.

**Publicly Owned Treatment Works or POTW** (Section 403.3) means a treatment works as defined by section 212 of the Act, which is owned by a State or municipality (as defined by section

502(4) of the Act). This definition includes any devices and systems used in the storage, treatment, recycling and reclamation of municipal sewage or industrial wastes of a liquid nature. It also includes sewers, pipes and other conveyances only if they convey wastewater to a POTW Treatment Plant. The term also means the municipality as defined in section 502(4) of the Act, which has jurisdiction over the Indirect Discharges to and the discharges from such a treatment works.

**Purchaser** (Section 745.103) means an entity that enters into an agreement to purchase an interest in target housing, including but not limited to individuals, partnerships, corporations, trusts, government agencies, housing agencies, Indian tribes, and nonprofit organizations.

**Purchasing** (Section 247.3) means the act of and the function of responsibility for the acquisition of equipment, materials, supplies, and services, including: Buying, determining the need, selecting the supplier, arriving at a fair and reasonable price and terms and conditions, preparing the contract or purchase order, and follow-up;

**Purified phosphoric acid process line** 8 means any process line which uses a HAP as a solvent in the separation of impurities from the product acid for the purposes of rendering that product suitable for industrial, manufacturing or food grade uses.

# Q

**QA** (Section 766.3) means quality assurance.

**QA operating quarter** (Section 72.2) means a calendar quarter in which there are at least 168 unit operating hours (as defined in this section) or, for a common stack or bypass stack, a calendar quarter in which there are at least 168 stack operating hours (as defined in this section).

**QC** (Section 766.3) means quality control.

**Qualification** (Section 164.2) A judicial officer shall be a permanent or temporary employee or officer of the Agency who may perform other duties for the Agency. Such judicial officer shall not be employed by the Office of Hazardous Materials Control or have any connection with the preparation or presentation of evidence for a hearing.

**Qualified expert** (Section 159.153) means one who, by virtue of his or her knowledge, skill, experience, training, or education, could be qualified by a court as an expert to testify on issues related to the subject matter on which he or she renders a conclusion or opinion. Under Rule 702 of the Federal Rules of Evidence, a person may be qualified as an expert on a particular matter by virtue of "knowledge, skill, experience, training, or education." In general, EPA wants registrants to report information when a person has relevant expert credentials, e.g., a medical doctor giving a medical opinion, a plant pathologist giving an opinion on plant pathology, etc.

**Qualified Ground-Water Scientist** (Section 260.10) means a scientist or engineer who has received a baccalaureate or post-graduate degree in the natural sciences or engineering, and has sufficient training and experience in ground-water hydrology and related fields as may be demonstrated by state registration, professional certifications, or completion of accredited university courses that enable that individual to make sound professional judgments regarding ground-water monitoring and contaminant fate and transport.

**Qualified ground-water scientist** (Section 503.21) is an individual with a baccalaureate or post-graduate degree in the natural sciences or engineering who has sufficient training and experience in ground-water hydrology and related fields, as may be demonstrated by State registration, professional certification, or completion of accredited university programs, to make sound professional judgments regarding ground-water monitoring, pollutant fate and transport, and corrective action.

**Qualified incinerator** (Section 761.13) means one of the following: (1) An incinerator approved under the provisions of Sec. 761.70. Any level of PCB concentration can be destroyed in an incinerator approved under Sec. 761.70. (2) A high efficiency boiler which complies with the criteria of Sec. 761.71(a)(1), and for which the operator has given written notice to the appropriate EPA Regional Administrator in accordance with the notification requirements for the burning of mineral oil dielectric fluid under Sec. 761.71(a)(2). (3) An incinerator approved under section 3005© of the Resource Conservation and Recovery Act (42 U.S.C. 6925©) (RCRA). (4) Industrial furnaces and boilers which are identified in 40 CFR 260.10 and 40 CFR 279.61 (a)(1) and (2) when operating at their normal operating temperatures (this prohibits feeding fluids, above the level of detection, during either startup or shutdown operations).

**Qualified researcher** (Section 1400.2) means a researcher who receives OCA information pursuant to 42 U.S.C. 7412®(7)(H)(vii).

**Qualifying facility (QF)** (Section 72.2) means a "qualifying small power production facility" within the meaning of section 3(17) © of the Federal Power Act or a "qualifying cogeneration facility" within the meaning of section 3(18)(B) of the Federal Power Act.

**Qualifying Phase I technology** (Section 72.2) means a technological system of continuous emission reduction that is demonstrated to achieve a ninety (90) percent (or greater) reduction in emissions of sulfur dioxide from the emissions that would have resulted from the use of fossil fuels that were not subject to treatment prior to combustion, as provided in Sec. 72.42.

**Qualifying power purchase commitment** (Section 72.2) means a power purchase commitment in effect as of November 15, 1990 without regard to changes to that commitment so long as: (1) The identity of the electric output purchaser; or (2) The identity of the steam purchaser and the location of the facility, remain unchanged as of the date the facility commences commercial operation; and (3) The terms and conditions of the power purchase commitment are not changed in such a way as to allow the costs of compliance with the Acid Rain Program to be shifted to the purchaser.

**Qualifying repowering technology** (Section 72.2) means: (1) Replacement of an existing coal-fired boiler with one of the following clean coal technologies: Atmospheric or pressurized fluidized bed combustion, integrated gasification combined cycle, magnetohydrodynamics, direct and indirect coal-fired turbines, integrated gasification fuel cells, or as determined by the Administrator, in consultation with the Secretary of Energy, a derivative of one or more of these technologies, and any other technology capable of controlling multiple combustion emissions simultaneously with improved boiler or generation efficiency and with significantly greater waste reduction relative to the performance of technology in widespread commercial use as of the date of enactment of the Clean Air Act Amendments of 1990; or (2) Any oil- or gas-fired unit that has been awarded clean coal technology demonstration funding as of January 1, 1991, by the Department of Energy.

**Quality assurance** (Section 194.2) means those planned and systematic actions necessary to provide adequate confidence that the disposal system will comply with the disposal regulations set forth in part 191 of this chapter. Quality assurance includes quality control, which comprises those actions related to the physical characteristics of a material, structure, component, or system that provide a means to control the quality of the material, structure, component, or system to predetermined requirements.

**Quality assurance project plan (QAPP)** (Section 300.5) is a written document, associated with all remedial site sampling activities, which presents in specific terms the organization (where applicable), objectives, functional activities, and specific quality assurance (QA) and quality control (QC) activities designed to achieve the data quality objectives of a specific project(s) or continuing operation(s). The QAPP is prepared for each specific project or continuing operation (or group of similar projects or continuing operations). The QAPP will be prepared by the responsible program office, regional office, laboratory, contractor, recipient of an assistance agreement, or other organization. For an enforcement action, potentially responsible parties may prepare a QAPP subject to lead agency approval.

**Quality assurance unit** (Section 160.3) means any person or organizational element, except the study director, designated by testing facility management to perform the duties relating to quality assurance of the studies.

**Quality assurance unit** (Section 792.3) means any person or organizational element, except the study director, designated by testing facility management to perform the duties relating to quality assurance of the studies.

**Quality-assured monitor operating hour** (Section 72.2) means any unit operating hour or portion thereof over which a certified CEMS, or other monitoring system approved by the Administrator under part 75 of this chapter, is operating: (1) Within the performance specifications set forth in part 75, appendix A of this chapter and the quality assurance/quality control procedures set forth in part 75, appendix B of this chapter, without unscheduled maintenance, repair, or adjustment; and (2) In accordance with Sec. 75.10(d), (e), and (f) of this chapter.

**Quantifiable Level/Level of Detection** (Section 761.13) means 2 micrograms per gram from any resolvable gas chromatographic peak, i.e. 2 ppm.

**Quantification level** (Section 132.2) is a measurement of the concentration of a contaminant obtained by using a specified laboratory procedure calibrated at a specified concentration above the method detection level. It is considered the lowest concentration at which a particular contaminant can be quantitatively measured using a specified laboratory procedure for monitoring of the contaminant.

**Quantitative structure activity relationship (QSAR)** (Section 132.2) or structure activity relationship (SAR) is a mathematical relationship between a property (activity) of a chemical and a number of descriptors of the chemical. These descriptors are chemical or physical characteristics obtained experimentally or predicted from the structure of the chemical.

**Quarantine applications,** (Section 82.3) with respect to class I, Group VI controlled substances, are treatments to prevent the introduction, establishment and/or spread of quarantine pests (including diseases), or to ensure their official control, where: (1) Official control is that performed by, or authorized by, a national plant, animal or environmental protection or health authority; (2) Quarantine pests are pests of potential importance to the areas endangered thereby and not yet present there, or present but not widely distributed and being officially controlled.

# R

**Radiation** (Section 190.02) means any or all of the following: Alpha, beta, gamma, or X-rays; neutrons; and high-energy electrons, protons, or other atomic particles; but not sound or radio waves, nor visible, infrared, or ultraviolet light.

**Radioactive material** (Section 190.02) means any material which spontaneously emits radiation.

**Radioactive material** (Section 191.12) means matter composed of or containing radionuclides, with radiological half-lives greater than 20 years, subject to the Atomic Energy Act of 1954, as amended.

**Radioactive material** (Section 197.2) means matter composed of or containing radionuclides subject to the Atomic Energy Act of 1954, as amended (42 U.S.C. 2014 et seq.). Radioactive material includes, but is not limited to, high-level radioactive waste and spent nuclear fuel.

**Radioactive Waste** (Section 144.3) means any waste which contains radioactive material in concentrations which exceed those listed in 10 CFR part 20, appendix B, table II, column 2.

**Radioactive waste** (Section 146.3) means any waste which contains radioactive material in concentrations which exceed those listed in 10 CFR part 20, appendix B, table II column 2.

**Radioactive waste** (Section 191.02) as used in this part, means the high-level and transuranic radioactive waste covered by this part.

**Radon Contractor Proficiency (RCP) program** (Section 195.2) refers to EPA's program to evaluate radon mitigation contractors and the contractor's ability to communicate information to the public.

**Radon Measurement Proficiency (RMP) program** (Section 195.2) refers to EPA's program to evaluate organizations and individuals offering measurement services to consumers. It provides a means for organizations to demonstrate their proficiency in measuring radon and its decay products in indoor air.

**Radon mitigation contractor** (Section 195.2) means a contractor who provides radon mitigation services to the public.

**Rail Car** (Section 201.1) means a non-self-propelled vehicle designed for and used on railroad tracks.

**Rail Tank Car** (Section 442.2) means a completely enclosed storage vessel pulled by a locomotive that is used to transport liquid, solid, or gaseous commodities or cargos over railway access lines. A rail tank car storage vessel may have one or more storage compartments and the stored commodities or cargos come in direct contact with the tank interior. There are no maximum or minimum vessel or tank volumes.

**Railroad** (Section 201.1) means all the roads in use by any common carrier operating a railroad, whether owned or operated under a contract, agreement, or lease.

**Railroad** (Section 92.2) means a commercial entity that operates locomotives to transport passengers or freight.

**Railroad grade crossing surfaces** (Section 247.3) are materials placed between railroad tracks, and between the track and the road at highway and street railroad crossings, to enhance automobile and pedestrian safety.

**Raisins** (Section 407.61) shall mean the production of raisins from the following products: Dried grapes, all varieties, bleached and unbleached, which have been cleaned and washed prior to packaging.

**Random Incident Field** (Section 211.203) A sound field in which the angle of arrival of sound at a given point in space is random in time.

**Range land** (Section 503.11) is open land with indigenous vegetation.

**Range of concentration** (Section 79.2) means the highest concentration, the lowest concentration, and the average concentration of an additive in a fuel.

**Rare earth metals** (Section 421.271) refers to the elements scandium, yttrium, and lanthanum to lutetium, inclusive.

**Rated horsepower** (Section 92.2) means the maximum horsepower output of a locomotive engine in use.

**Rated output (rO)** (Section 87.1) means the maximum power/thrust available for takeoff at standard day conditions as approved for the engine by the Federal Aviation Administration, including reheat contribution where applicable, but excluding any contribution due to water injection.

**Rated pressure ratio (rPR)** (Section 87.1) means the ratio between the combustor inlet pressure and the engine inlet pressure achieved by an engine operating at rated output.

**Rated speed** (Section 86.082-2) means the speed at which the manufacturer specifies the maximum rated horsepower of an engine.

**Rated speed** (Section 89.2) is the maximum full load governed speed for governed engines and the speed of maximum horsepower for ungoverned engines.

**Raw agricultural commodities** (Section 180.1) include, among other things, fresh fruits, whether or not they have been washed and colored or otherwise treated in their unpeeled natural form; vegetables in their raw or natural state, whether or not they have been stripped of their outer leaves, waxed, prepared into fresh green salads, etc.; grains, nuts, eggs, raw milk, meats, and similar agricultural produce. It does not include foods that have been processed, fabricated, or manufactured by cooking, freezing, dehydrating, or milling. Where raw agricultural commodities bearing residues that have been exempted from the requirement of a tolerance, or which are within a tolerance permitted under section 408 are used, the processed foods will not be considered unsafe within the meaning of section 406 if: (1) The poisonous or deleterious pesticide residues have been removed to the extent possible in good manufacturing practice; and (2) The concentration of the pesticide in the preserved or processed food when ready to eat is not greater than the tolerance permitted on the raw agricultural commodity. (g) For the purpose of computing fees as required by Sec. 180.33, each group of related crops listed in Sec. 180.34(e) and each crop group or subgroup listed in Sec. 180.41 is counted as a single raw agricultural commodity in a petition or request for tolerances or exemption from the requirement of a tolerance. (h) Tolerances and exemptions established for pesticide chemicals in or on the general category of raw agricultural commodities listed in column A apply to the corresponding specific raw agricultural commodities listed in column B. However, a tolerance or exemption for a specific commodity in column B does not apply to the general category in column A. Unless otherwise specified, tolerances and exemptions established under the regulations in this part apply to residues from only preharvest application of the chemical. (j) Unless otherwise specified in this paragraph or in tolerance regulations prescribed in this part for specific pesticide chemicals, the raw agricultural commodity to be examined for pesticide residues shall consist of the whole raw agricultural commodity. (1) The raw agricultural commodity bananas, when examined for pesticide residues, shall not include any crown tissue or stalk. (2) Shell shall be removed and discarded from nuts before examination for pesticide residues. (3) Caps (hulls) shall be removed and discarded from strawberries before examination for pesticide residues. (4) Stems shall be removed and discarded from melons before examination for pesticide residues. (5) Roots, stems, and outer sheaths (or husks) shall be removed and discarded from garlic bulbs and dry bulb onions, and only the garlic cloves and onion bulbs shall be examined for pesticide residues. (6) Where a tolerance is established on a root vegetable including tops and/or with tops, and the tops and the roots are marketed together, they shall be analyzed separately and neither the pesticide residue on the roots nor the pesticide residue on the tops shall exceed the tolerance level, except that in the case of carrots, parsnips, and rutabagas, the tops shall be removed and discarded before analyzing roots for pesticide residues. (7) The crowns (leaves at the top of the fruit) shall be removed and discarded from pineapples before examination for pesticide residues. (8) The term lima beans mean the beans and the pod. (9) The term peanuts mean the peanut meat after removal of the hulls.

**Raw data** (Section 160.3) means any laboratory worksheets, records, memoranda, notes, or exact copies thereof, that are the result of original observations and activities of a study and are necessary for the reconstruction and evaluation of the report of that study. In the event that exact transcripts of raw data have been prepared (e.g., tapes which have been transcribed verbatim, dated, and verified accurate by signature), the exact copy or exact transcript may be substituted for the original source as raw data. "Raw data" may include photographs, microfilm or microfiche copies, computer printouts, magnetic media, including dictated observations, and recorded data from automated instruments.

**Raw data** (Section 792.3) means any laboratory worksheets, records, memoranda, notes, or exact copies thereof, that are the result of original observations and activities of a study and are necessary for the reconstruction and evaluation of the report of that study. In the event that exact transcripts of raw data have been prepared (e.g., tapes which have been transcribed verbatim, dated, and verified accurate by signature), the exact copy or exact transcript may be substituted for the original source as raw data. "Raw data" may include photographs, microfilm or microfiche copies, computer printouts, magnetic media, including dictated observations, and recorded data from automated instruments.

**Raw material** (Section 425.02) means the hides received by the tannery except for facilities covered by subpart D and subpart I where raw material means the hide or split in the condition in which it is first placed into a wet process.

**Raw material** (Section 428.101) shall mean all latex solids used in the manufacture of latex-dipped, latex-extruded, and latex-molded products.

**Raw material** (Section 428.11) shall mean all natural and synthetic rubber, carbon black, oils, chemical compounds, fabric and wire used in the manufacture of pneumatic tires and inner tubes or components thereof.

**Raw material** (Section 428.111) shall mean all latex solids used in the manufacture of latex foam.

**Raw material** (Section 428.51) shall mean all natural and synthetic rubber, carbon black, oils, chemical compounds, and fabric used in the manufacture of general molded, extruded, and fabricated rubber products.

**Raw material** (Section 432.101) or as abbreviated herein, "RM", shall mean the basic input materials to a renderer composed of animal and poultry trimmings, bones, meat scraps, dead animals, feathers and related usable by-products.

**Raw material equivalent** (Section 428.51) shall be equal to the raw material usage multiplied by the volume of air scrubbed via wet scrubbers divided by the total volume of air scrubbed.

**Raw material feed** (Section 62.1201) means the prepared and mixed materials, which include but are not limited to materials such as limestone, clay, shale, sand, iron ore, mill scale, cement kiln dust and fly ash that are fed to a cement or lightweight aggregate kiln. Raw material feed does not include the fuels used in the kiln to produce heat to form the clinker product.

**RCRA** (Section 124.2) means the Solid Waste Disposal Act as amended by the Resource Conservation and Recovery Act of 1976 (Pub. L. 94-580, as amended by Pub. L. 95-609, 42 U.S.C. 6901 et seq).

**RCRA** (Section 144.3) means the Solid Waste Disposal Act as amended by the Resource Conservation and Recovery Act of 1976 (Pub. L. 94-580, as amended by Pub. L. 95-609, Pub. L. 96-510, 42 U.S.C. 6901 et seq.).

**RCRA** (Section 146.3) means the Solid Waste Disposal Act as amended by the Resource Conservation and Recovery Act of 1976 (Pub. L. 94-580, as amended by Pub. L. 95-609, 42 U.S.C. 6901 et seq.).

**RCRA** (Section 270.2) means the Solid Waste Disposal Act as amended by the Resource Conservation and Recovery Act of 1976 (Pub. L. 94-580, as amended by Pub. L. 95-609 and Pub. L. 96-482, 42 U.S.C. 6901 et seq.)

**RCRA** (Section 761.13) means the Resource Conservation and Recovery Act (40 U.S.C. 6901 et seq.).

**RCRA approved test method** (Section 372.3) includes Test Method 9095 (Paint Filter Liquids Test) in "Test Methods for Evaluating Solid Waste, Physical/ Chemical Methods," EPA Publication No. SW-846, Third Edition, September 1986, as amended by Update I, November 15, 1992.

**Reactivation of a very clean coal-fired electric utility steam generating unit** (Section 60.2) means any physical change or change in the method of operation associated with the commencement of commercial operations by a coal-fired utility unit after a period of discontinued operation where the unit: (1) Has not been in operation for the two-year period prior to the enactment of the Clean Air Act Amendments of 1990, and the emissions from such unit continue to be carried in the permitting authority's emissions inventory at the time of enactment; (2) Was equipped prior to shut-down with a continuous system of emissions control that achieves a removal efficiency for sulfur dioxide of no less than 85 percent and a removal efficiency for particulates of no less than 98 percent; (3) Is equipped with low-NOx burners prior to the time of commencement of operations following reactivation; and (4) Is otherwise in compliance with the requirements of the Clean Air Act.

**Reactivity adjustment factor** (Section 86.1702-99) means a fraction applied to the mass of NMOG emission from a vehicle powered by a fuel other than conventional gasoline for the purpose of determining a gasoline-equivalent NMOG emission value. The reactivity adjustment factor is defined as the ozone-forming potential of the exhaust from a vehicle powered by a fuel other than conventional gasoline divided by the ozone-forming potential of conventional gasoline vehicle exhaust.

**Readability** (Section 92.102) means the smallest difference in measured values that can be detected. For example, the readability for a digital display with two decimal places would be 0.01.

**Real-Ear Protection at Threshold** (Section 211.203) The mean value in decibels of the occluded threshold of audibility (hearing protector in place) minus the open threshold of audibility (ears open and uncovered) for all listeners on all trials under otherwise identical test conditions.

**Reasonable assistance** (Section 204.2) means providing timely and unobstructed access to test products or products and records required by this part and opportunity for copying such records or testing such test products.

**Reasonable assistance** (Section 205.2) means providing timely and unobstructed access to test products or products and records required by this part, and opportunity for copying such records or testing such test products.

**Reasonable assurance** (Section 130.2) Reasonable assurance means a demonstration that TMDLs will be implemented through regulatory or voluntary actions, including management measures or other controls, by Federal, State or local governments, authorized Tribes, or individuals. (1) For point sources regulated under section 402 of the Clean Water Act, the demonstration of reasonable assurance must identify procedures that ensure that NPDES permits will be issued, reissued, or revised as expeditiously as practicable to implement applicable TMDL wasteload allocations for point sources. (2) For nonpoint sources, storm water sources for which an NPDES permit is not required, atmospheric deposition, ground water or background sources of a pollutant, the demonstration of reasonable assurance must show that management measures or other control actions to implement the load allocations contained in each TMDL meet the following four-part test: they specifically apply to the pollutant(s) and the waterbody for which the TMDL is being established; they will be implemented as expeditiously as practicable; they will be accomplished through reliable and effective delivery mechanisms; and they will be supported by adequate water quality funding.

**Reasonable Expense** (Section 85.2102) means any expense incurred due to repair of a warranty failure caused by a non-original equipment certified part, including, but not limited to, all charges in any expense categories that would be considered payable by the involved vehicle manufacturer to its authorized dealer under a similar warranty situation where an original equipment part was the cause of the failure. Included in "reasonable expense" are any additional costs incurred specifically due to the processing of a claim involving a certified aftermarket part or parts as covered in these regulations. The direct parts and labor expenses of carrying out repairs is immediately chargeable to the part manufacturer. All charges beyond the actual parts and labor repair expenses must be amortized over the number of claims and/or over a number of years in a manner that would be considered consistent with generally accepted accounting principles. These expense categories shall include but are not limited to the cost of labor, materials, record keeping, special handling, and billing as a result of replacement of a certified aftermarket part.

**Reasonably available control technology (RACT)** (Section 51.100) means devices, systems, process modifications, or other apparatus or techniques that are reasonably available taking into account: (1) The necessity of imposing such controls in order to attain and maintain a national ambient air quality standard; (2) The social, environmental, and economic impact of such controls; and (3) Alternative means of providing for attainment and maintenance of such standard. (This provision defines RACT for the purposes of Sec. 51.341(b) only.)

**Reasonably foreseeable emissions** (Section 93.152) are projected future indirect emissions that are identified at the time the conformity determination is made; the location of such emissions is known and the emissions are quantifiable, as described and documented by the Federal agency based on its own information and after reviewing any information presented to the Federal agency.

**Reburning** (Section 76.2) means reducing the coal and combustion air to the main burners and injecting a reburn fuel (such as gas or oil) to create a fuel-rich secondary combustion zone above the main burner zone and final combustion air to create a fuel-lean burnout zone. The formation of NOX is inhibited in the main burner zone due to the reduced combustion intensity, and NOX is destroyed in the fuel-rich secondary combustion zone by conversion to molecular nitrogen.

**Recall program** (Section 86.1803-01) means the program administered by the Agency under the authority of CAA section 207, and regulations in 40 CFR part 85.

**Receive or receipt of** (Section 72.2) means the date the Administrator or a permitting authority comes into possession of information or correspondence (whether sent in writing or by authorized electronic transmission), as indicated in an official correspondence log, or by a notation made on the information or correspondence, by the Administrator or the permitting authority in the regular course of business.

**Receive or receipt of** (Section 96.2) means, when referring to the permitting authority or the Administrator, to come into possession of a document, information, or correspondence (whether sent in writing or by authorized electronic transmission), as indicated in an official correspondence log, or by a notation made on the document, information, or correspondence, by the permitting authority or the Administrator in the regular course of business.

**Receive or receipt of** (Section 97.2) means, when referring to the permitting authority or the Administrator, to come into possession of a document, information, or correspondence (whether sent in writing or by authorized electronic transmission), as indicated in an official correspondence log, or by a notation made on the document, information, or correspondence, by the permitting authority or the Administrator in the regular course of business.

**Receiving country** (Section 262.51) means a foreign country to which a hazardous waste is sent for the purpose of treatment, storage or disposal (except short-term storage incidental to transportation).

**Receiving Property** (Section 201.1) means any residential or commercial property that receives the sound from railroad facility operations, but that is not owned or operated by a railroad; except that occupied residences located on property owned or controlled by the railroad are included in the definition of "receiving property." For purposes of this definition railroad crew sleeping quarters located on property owned or controlled by the railroad are not considered as residences. If, subsequent to the publication date of these regulations, the use of any property that is currently not applicable to this regulation changes, and it is newly classified as either residential or commercial, it is not receiving property until four years have elapsed from the date of the actual change in use.

**Receiving Property Measurement Location** (Section 201.1) means a location on receiving property that is on or beyond the railroad facility boundary and that meets the receiving property measurement location criteria of subpart C.

**Re-certification** (Section 194.2) means any action taken by the Administrator pursuant to section 8(f) of the WIPP LWA.

**Recharge** (Section 149.2) means a process, natural or artificial, by which water is added to the saturated zone of an aquifer.

**Recharge Area** (Section 149.2) means an area in which water reaches the zone of saturation (ground water) by surface infiltration; in addition, a major recharge area is an area where a major part of the recharge to an aquifer occurs.

**Recipient** (Section 721.3) means any person who purchases or otherwise obtains a chemical substance directly from a person who manufacturers, imports, or processes the substance.

**Recipient of funds designated under title 23 U.S.C. or the Federal Transit Laws** (Section 93.101) means any agency at any level of State, county, city, or regional government that routinely receives title 23 U.S.C. or Federal Transit Laws funds to construct FHWA/FTA projects, operate FHWA/FTA projects or equipment, purchase equipment, or undertake other services or operations via contracts or agreements. This definition does not include private landowners or developers, or contractors or entities that are only paid for services or products created by their own employees.

**Recipient plant** (Section 174.3) means the living plant in which the plant-incorporated protectant is intended to be produced and used.

**Recirculated cooling water** (Section 423.11) means water which is passed through the main condensers for the purpose of removing waste heat, passed through a cooling device for the purpose of removing such heat from the water and then passed again, except for blowdown, through the main condenser.

**Recirculation** (Section 420.101) means those cold rolling operations which include recirculation of rolling solutions at all mill stands.

**Reclamation area** (Section 434.11) means the surface area of a coal mine which has been returned to required contour and on which revegetation (specifically, seeding or planting) work has commenced.

**Reclamation site** (Section 503.11) is drastically disturbed land that is reclaimed using sewage sludge. This includes, but is not limited to, strip mines and construction sites.

**Recognized laboratory** (Section 745.223) means an environmental laboratory recognized by EPA pursuant to TSCA section 405(b) as being capable of performing an analysis for lead compounds in paint, soil, and dust.

**Recognized trader** (Section 262.81) means a person who, with appropriate authorization of concerned countries, acts in the role of principal to purchase and subsequently sell wastes; this person has legal control of such wastes from time of purchase to time of sale; such a person may act to arrange and facilitate transfrontier movements of wastes destined for recovery operations.

**Recombinant DNA** (Section 174.3) means the genetic material has been manipulated in vitro through the use of restriction endonucleases and/or other enzymes that aid in modifying genetic material, and subsequently introduced into the genome of the plant.

**Recommencing discharger** (Section 122.2) means a source which recommences discharge after terminating operations.

**Recommended Decision** (Section 164.2) means the recommended findings and conclusions of the Presiding Officer in an expedited hearing.

**Reconfigured emission-data vehicle** (Section 86.082-2) means an emission-data vehicle obtained by modifying a previously used emission-data vehicle to represent another emission-data vehicle.

**Reconfigured emission-data vehicle** (Section 86.1803-01) means an emission-data vehicle obtained by modifying a previously used emission-data vehicle to represent another emission-data vehicle.

**Reconstruction, unless otherwise defined in a relevant standard,** (Section 63.2) means the replacement of components of an affected or a previously nonaffected source to such an extent that: (1) The fixed capital cost of the new components exceeds 50 percent of the fixed capital cost that would be required to construct a comparable new source; and (2) It is technologically and economically feasible for the reconstructed source to meet the relevant standard(s) established by the Administrator (or a State) pursuant to section 112 of the Act. Upon reconstruction, an affected source, or a stationary source that becomes an affected source, is subject to relevant standards for new sources, including compliance dates, irrespective of any change in emissions of hazardous air pollutants from that source.

**Record** (Section 1516.2) means any item or collection or grouping of information about an individual that is maintained by the Council (including, but not limited to, his or her employment history, payroll information, and financial transactions), and that contains his or her name, or an identifying number, symbol, or other identifying particular assigned to the individual such as a social security number.

**Record** (Section 1601.3) includes any writing, drawing, map, recording, tape, film, photo, or other documentary material by which information is preserved.

**Recordation, record, or recorded** (Section 72.2) means, with regard to allowances, the transfer of allowances by the Administrator from one Allowance Tracking System account or subaccount to another.

**Recordation, record, or recorded** (Section 96.2) means, with regard to NOX allowances, the movement of NOX allowances by the Administrator from one NOX Allowance Tracking System account to another, for purposes of allocation, transfer, or deduction.

**Recordation, record, or recorded** (Section 97.2) means, with regard to NOX allowances, the movement of NOX allowances by the Administrator from one NOX Allowance Tracking System account to another, for purposes of allocation, transfer, or deduction.

**Recoverable resources** (Section 246.101) means materials that still have useful physical, chemical, or biological properties after serving their original purpose and can, therefore, be reused or recycled for the same or other purposes.

**Recovered materials** (Section 247.3) for purposes of purchasing paper and paper products, means waste material and byproducts that have been recovered or diverted from solid waste, but such term does not include those materials and byproducts generated from, and commonly reused within, an original manufacturing process. In the case of paper and paper products, the term recovered materials includes: (1) Postconsumer materials such as—(i) Paper, paperboard, and fibrous wastes from retail stores, office buildings, homes, and so forth, after they have passed through their end-usage as a consumer item, including: Used corrugated boxes; old newspapers; old magazines; mixed waste paper; tabulating cards; and used cordage; and (ii) All paper, paperboard, and fibrous wastes that enter and are collected from municipal solid waste, and (2) Manufacturing, forest residues, and other wastes such as—(i) Dry paper and paperboard waste generated after completion of the papermaking process (that is, those manufacturing operations up to and including the cutting and trimming of the paper machine reel in smaller rolls of rough sheets) including: Envelope cuttings, bindery trimmings, and other paper and paperboard waste, resulting from printing, cutting, forming, and other converting operations; bag, box, and carton manufacturing wastes; and butt rolls, mill wrappers, and rejected unused stock; and (ii) Finished paper and paperboard from obsolete inventories of paper and paperboard manufacturers, merchants, wholesalers, dealers, printers, converters, or others; (iii) Fibrous byproducts of harvesting, manufacturing, extractive, or wood-cutting processes, flax, straw, linters, bagasse, slash, and other forest residues; (iv) Wastes generated by the conversion of goods made from fibrous material (that is, waste rope from cordage manufacture, textile mill waste, and cuttings); and (v) Fibers recovered from waste water which otherwise would enter the waste stream.

**Recovered materials** (Section 247.3) means waste materials and byproducts which have been recovered or diverted from solid waste, but such term does not include those materials and byproducts generated from, and commonly reused within, an original manufacturing process;

**Recovery** (Section 246.101) means the process of obtaining materials or energy resources from solid waste.

**Recovery** (Section 437.2) means the recycling or processing of a waste, wastewater or used material such that the material, or a portion thereof, may be reused or converted to a raw material, intermediate, or product. Recovery does not include the re-use of treated or untreated wastewater in place of potable or pure water in industrial processes such as the use of secondary POTW effluents as non-contact cooling water, storm water in place of process water, or the re-use of spent chemicals in place of virgin treatment chemicals.

**Recovery device** (Section 65.2) means an individual unit of equipment capable of and normally used for the purpose of recovering chemicals for fuel value (i.e., net positive heating value), use, reuse, or for sale for fuel value, use, or reuse. Equipment capable of and used for the purpose of recovering chemicals, but not normally for use, reuse or sale, are not recovery devices but are control devices. Examples of equipment that may be recovery devices include absorbers, carbon adsorbers, condensers, oil-water separators or organic-water separators, or organic removal devices such as decanters, strippers, or thin-film evaporation units.

**Recovery facility** (Section 262.81) means an entity which, under applicable domestic law, is operating or is authorized to operate in the importing country to receive wastes and to perform recovery operations on them.

**Recovery operations** (Section 262.81) means activities leading to resource recovery, recycling, reclamation, direct re-use or alternative uses as listed in Table 2.B of the Annex of OECD Council Decision C(88)90(Final) of 27 May 1988, (available from the Environmental Protection Agency, RCRA Information Center (RIC), 1235 Jefferson-Davis Highway, first floor, Arlington, VA 22203 (Docket F-94-IEHF-FFFFF) and the Organization for Economic Co-operation and Development, Environment Directorate, 2 rue Andre Pascal, 75775 Paris Cedex 16, France) which include: R1 Use as a fuel (other than in direct incineration) or other means to generate energy R2 Solvent reclamation/regeneration, R3 Recycling/reclamation of organic substances which are not used as solvents, R4 Recycling/reclamation of metals and metal compounds, R5 Recycling/reclamation of other inorganic materials, R6 Regeneration of acids or bases, R7 Recovery of components used for pollution control, R8 Recovery of components from catalysts, R9 Used oil re-refining or other reuses of previously used oil, R10 Land treatment resulting in benefit to agriculture or ecological improvement, R11 Uses of residual materials obtained from any of the operations numbered R1-R10, R12 Exchange of wastes for submission to any of the operations numbered R1-R11, 6R13 Accumulation of material intended for any operation in Table 2.B

**Recreational marine engine** (Section 94.2) means a propulsion marine engine that is intended by the manufacturer to be installed on a recreational vessel, and which is permanently labeled as follows: "THIS ENGINE IS CATEGORIZED AS A RECREATIONAL ENGINE UNDER 40 CFR PART 94, AND IS NOT SUBJECT TO THE EMISSION STANDARDS OF THAT PART. INSTALLATION OF THIS ENGINE IN ANY NONRECREATIONAL VESSEL IS A VIOLATION OF FEDERAL LAW SUBJECT TO CIVIL PENALTY."

**Recreational vessel** (Section 94.2) means a vessel that is intended by the vessel manufacturer to be operated primarily for pleasure or leased, rented or chartered to another for the latter's pleasure. For this definition, the term "operated primarily for pleasure or leased, rented or chartered to another for the latter's pleasure" does not include the following vessels: (1) Vessels of less than 100 gross tons that carry more than 6 passengers (as defined in this section). (2) Vessels of 100 gross tons or more that carry one or more passengers (as defined in this section). (3) Vessels used solely for competition.

**Recycled material** (Section 246.101) means a material that is used in place of a primary, raw or virgin material in manufacturing a product.

**Recycled PCBs** (Section 761.13) means those PCBs which appear in the processing of paper products or asphalt roofing materials from PCB-contaminated raw materials. Processes which recycle PCBs must meet the following requirements: (1) There are no detectable concentrations of PCBs in asphalt roofing material products leaving the processing site. (2) The concentration of PCBs in paper products leaving any manufacturing site processing paper products, or in paper products imported into the United States, must have an annual average of less than 25 ppm with a 50 ppm maximum. (3) The release of PCBs at the point at which emissions are vented to ambient air must be less than 10 ppm. (4) The amount of Aroclor PCBs added to water discharged from an asphalt roofing processing site must at all times be less than 3 micrograms per liter ([mu]g/L) for total Aroclors (roughly 3 parts per billion (3 ppb)). Water discharges from the processing of paper products must at all times be less than 3 micrograms per liter ([mu]g/L) for total Aroclors (roughly 3 ppb), or comply with the equivalent mass-based limitation. (5) Disposal of any other process wastes at concentrations of 50 ppm or greater must be in accordance with subpart D of this part.

**Recycling** (Section 246.101) means the process by which recovered materials are transformed into new products.

**Reduced utilization** (Section 72.2) means a reduction, during any calendar year in Phase I, in the heat input (expressed in mmBtu for the calendar year) at a Phase I unit below the unit's baseline, where such reduction subjects the unit to the requirement to submit a reduced utilization plan under Sec. 72.43; or, in the case of an opt-in source, means a reduction in the average utilization, as specified in Sec. 74.44 of this chapter, of an opt-in source below the opt-in source's baseline.

**Reducer** (Section 59.101) means any solvent used to thin enamels.

**Reduction** (Section 745.103) means measures designed to reduce or eliminate human exposure to lead-based paint hazards through methods including interim controls and abatement.

**Reduction** (Section 745.223) means measures designed to reduce or eliminate human exposure to lead-based paint hazards through methods including interim controls and abatement.

**Reference** (Section 60.2) method means any method of sampling and analyzing for an air pollutant as specified in the applicable subpart.

**Reference method** (Section 50.1) means a method of sampling and analyzing the ambient air for an air pollutant that is specified as a reference method in an appendix to this part, or a method that has been designated as a reference method in accordance with part 53 of this chapter; it does not include a method for which a reference method designation has been cancelled in accordance with Sec. 53.11 or Sec. 53.16 of this chapter.

**Reference method** (Section 53.1) means a method of sampling and analyzing the ambient air for an air pollutant that is specified as a reference method in an appendix to part 50 of this chapter, or a method that has been designated as a reference method in accordance with this part; it does not include a method for which a reference method designation has been canceled in accordance with Sec. 53.11 or Sec. 53.16.

**Reference method** (Section 58.1) means a method of sampling and analyzing the ambient air for an air pollutant that will be specified as a reference method in an appendix to part 50 of this chapter, or a method that has been designated as a reference method in accordance with this part; it does not include a method for which a reference method designation has been canceled in accordance with Sec. 53.11 or Sec. 53.16 of this chapter.

**Reference method** (Section 61.02) means any method of sampling and analyzing for an air pollutant, as described in appendix B to this part.

**Reference method** (Section 65.2) means any method of sampling and analyzing for an air pollutant as specified in an applicable subpart, the appendices to 40 CFR part 60 or 63, or in appendix B of 40 CFR part 61.

**Reference method** (Section 72.2) means any direct test method of sampling and analyzing for an air pollutant as specified in part 60, appendix A of this chapter.

**Reference method** (Section 96.2) means any direct test method of sampling and analyzing for an air pollutant as specified in appendix A of part 60 of this chapter.

**Reference method** (Section 97.2) means any direct test method of sampling and analyzing for an air pollutant as specified in appendix A of part 60 of this chapter.

**Reference substance** (Section 160.3) means any chemical substance or mixture, or analytical standard, or material other than a test substance, feed, or water, that is administered to or used in analyzing the test system in the course of a study for the purposes of establishing a basis for comparison with the test substance for known chemical or biological measurements.

**Reference substance** (Section 792.3) means any chemical substance or mixture, or analytical standard, or material other than a test substance, feed, or water, that is administered to or used in analyzing the test system in the course of a study for the purposes of establishing a basis for comparison with the test substance for known chemical or biological measurements.

**Reference value or reference signal** (Section 72.2) means the known concentration of a calibration gas, the known value of an electronic calibration signal, or the known value of any other measurement standard approved by the Administrator, assumed to be the true value for the pollutant or diluent concentration or volumetric flow being measured.

**Referencing subpart** (Section 65.2) means 40 CFR part 60, subparts Ka, Kb, VV, DDD, III, NNN, and RRR; 40 CFR part 61, subparts V, Y, and BB; and 40 CFR part 63, subparts G and H.

**Refilling establishment** (Section 455.10) means an establishment where the activity of repackaging pesticide product into refillable containers occurs.

**Refiner** (Section 80.2) means any person who owns, leases, operates, controls, or supervises a refinery.

**Refinery** (Section 80.2) means any facility, including but not limited to, a plant, tanker truck, or vessel where gasoline or diesel fuel is produced, including any facility at which blendstocks are combined to produce gasoline or diesel fuel, or at which blendstock is added to gasoline or diesel fuel.

**Reformulated gasoline** (Section 80.2) means any gasoline whose formulation has been certified under Sec. 80.40, which meets each of the standards and requirements prescribed under Sec. 80.41, and which contains less than the maximum concentration of the marker specified in Sec. 80.82 that is allowed for reformulated gasoline under Sec. 80.82.

**Reformulated gasoline blendstock for oxygenate blending, or RBOB** (Section 80.2) means a petroleum product which, when blended with a specified type and percentage of oxygenate, meets the definition of reformulated gasoline, and to which the specified type and percentage of oxygenate is added other than by the refiner or importer of the RBOB at the refinery or import facility where the RBOB is produced or imported.

**Reformulated gasoline credit** (Section 80.2) means the unit of measure for the paper transfer of oxygen or benzene content resulting from reformulated gasoline which contains more than 2.1 weight percent of oxygen or less than 0.95 volume percent benzene.

**Refractory metals** (Section 471.02) includes the metals of columbium, tantalum, molybdenum, rhenium, tungsten and vanadium and their alloys.

**Refueling emissions** (Section 86.1803-01) means evaporative emissions that emanate from a motor vehicle fuel tank(s) during a refueling operation.

**Refueling emissions canister(s)** (Section 86.1803-01) means any vapor storage unit(s) that is exposed to the vapors generated during refueling.

**Regeneration** (Section 86.1702-99) means the process of oxidizing accumulated particulate matter. It may occur continually or periodically.

**Regeneration emission test** (Section 86.1702-99) means a complete emission test which includes a regeneration.

**Regeneration interval** (Section 86.1702-99) means the interval from the start of a regeneration to the start of the next regeneration.

**Region** (Section 51.100) means an area designated as an air quality control region (AQCR) under section 107© of the Act.

**Regional Administrator** (Section 124.2) means the Regional Administrator of the appropriate Regional Office of the Environmental Protection Agency or the authorized representative of the Regional Administrator.

**Regional Administrator** (Section 232.2) means the Regional Administrator of the appropriate Regional Office of the Environmental Protection Agency or the authorized representative of the Regional Administrator.

**Regional Administrator** (Section 239.2) means any one of the ten Regional Administrators of the U.S. Environmental Protection Agency or any authorized representative.

**Regional Administrator** (Section 260.10) means the Regional Administrator for the EPA Region in which the facility is located, or his designee.

**Regional Office** (Section 51.100) means one of the ten (10) EPA Regional Offices.

**Regional supplemental guidance** (Section 35.102) Guidance to environmental program applicants prepared by the Regional Administrator, based on the national program guidance and specific regional and applicant circumstances, for use in preparing a grant application.

**Regional water and/or wastewater projects** (Section 93.152) include construction, operation, and maintenance of water or wastewater conveyances, water or wastewater treatment facilities, and water storage reservoirs which affect a large portion of a nonattainment or maintenance area.

**Regionally significant action** (Section 93.152) means a Federal action for which the direct and indirect emissions of any pollutant represent 10 percent or more of a nonattainment or maintenance area's emission inventory for that pollutant.

**Regionally significant project** (Section 93.101) means a transportation project (other than an exempt project) that is on a facility which serves regional transportation needs (such as access to and from the area outside of the region, major activity centers in the region, major planned developments such as new retail malls, sports complexes, etc., or transportation terminals as well as most terminals themselves) and would normally be included in the modeling of a metropolitan area's transportation network, including at a minimum all principal arterial highways and all fixed guideway transit facilities that offer an alternative to regional highway travel.

**Registrant** (Section 159.153) includes any person who holds, or ever held, a registration for a pesticide product issued under FIFRA section 3 or 24©.

**Registrant** (Section 164.2) means any person who has registered a pesticide pursuant to the provisions of the Act.

**Registration Division** (Section 180.1) means the unit established within the Environmental Protection Agency charged with administration of the Pesticide Residue amendment to the Federal Food, Drug, and Cosmetic Act (section 408).

**Regraded area** (Section 434.80) means the surface area of a coal mine that has been returned to required contour.

**Regulated air pollutant** (Section 64.1) shall have the same meaning as provided under part 70 or 71 of this chapter.

**Regulated air pollutant** (Section 70.2) means the following: (1) Nitrogen oxides or any volatile organic compounds; (2)Any pollutant for which a national ambient air quality standard has been promulgated; (3) Any pollutant that is subject to any standard promulgated under section 111 of the Act; (4) Any Class I or II substance subject to a standard promulgated under or established by title VI of the Act; or (5) Any pollutant subject to a standard promulgated under section 112 or other requirements established under section 112 of the Act, including sections 112(g), (j), and ® of the Act, including the following: (i) Any pollutant subject to requirements under section 112(j) of the Act. If the Administrator fails to promulgate a standard by the date established pursuant to section 112(e) of the Act, any pollutant for which a subject source would be major shall be considered to be regulated on the date 18 months after the applicable date established pursuant to section 112(e) of the Act; and (ii) Any pollutant for which the requirements of section 112(g)(2) of the Act have been met, but only with respect to the individual source subject to section 112(g)(2) requirement.

**Regulated air pollutant** (Section 71.2) means the following: (1) Nitrogen oxides or any volatile organic compounds; (2) Any pollutant for which a national ambient air quality standard has been promulgated; (3) Any pollutant that is subject to any standard promulgated under section 111 of the Act; (4) Any Class I or II substance subject to a standard promulgated under or established by title VI of the Act; or (5) Any pollutant subject to a standard promulgated under section 112 of the Act or other requirements established under section 112 of the Act, including sections 112 (g), (j), and ® of the Act, including the following: (i) Any pollutant subject to requirements under section 112(j) of the Act. If the Administrator fails to promulgate a standard by the date established pursuant to section 112(e) of the Act, any pollutant for which a subject source would be major shall be considered to be regulated on the date 18 months after the applicable date established pursuant to section 112(e) of the Act; and (ii) Any pollutant for which the requirements of section 112(g)(2) of the Act have been met, but only with respect to the individual source subject to section 112(g)(2) requirements.

**Regulated chemical** (Section 707.63) means any chemical substance or mixture for which export notice is required under Sec. 707.60.

**Regulated material** (Section 65.2) means, for the purposes of this part, the material regulated by the specific referencing subpart, including volatile organic liquids (VOL), volatile organic compounds (VOC), organic hazardous air pollutants (HAP's), benzene, vinyl chloride, or other chemicals or groups of chemicals.

**Regulated pest** (Section 171.2) means a specific organism considered by a State or Federal agency to be a pest requiring regulatory restrictions, regulations, or control procedures in order to protect the host, man and/or his environment.

**Regulated pollutant (for fee calculation), which is used only for purposes of Sec. 71.9©** (Section 71.2) means any regulated air pollutant except the following: (1) Carbon monoxide; (2) Any pollutant that is a regulated air pollutant solely because it is a Class I or II substance subject to a standard promulgated under or established by title VI of the Act; or (3) Any pollutant that is a regulated air pollutant solely because it is subject to a standard or regulation under section 112 of the Act.

**Regulated pollutant** (Section 70.2) (for presumptive fee calculation), which is used only for purposes of Sec. 70.9(b)(2), means any "regulated air pollutant" except the following: (1) Carbon monoxide; (2) Any pollutant that is a regulated air pollutant solely because it is a Class I or II substance to a standard promulgated under or established by title VI of the Act; or (3) Any pollutant that is a regulated air pollutant solely because it is subject to a standard or regulation under section 112® of the Act.

**Regulated source** (Section 65.2) means, for the purposes of this part, the stationary source, the group of stationary sources, or the portion of a stationary source that is regulated by a relevant standard or other requirement established pursuant to this part, or 40 CFR part 60, 61, or 63.

**Regulated substance** (Section 280.12) means: (a) Any substance defined in section 101(14) of the Comprehensive Environmental Response, Compensation and Liability Act (CERCLA) of 1980 (but not including any substance regulated as a hazardous waste under subtitle C), and (b) Petroleum, including crude oil or any fraction thereof that is liquid at standard conditions of temperature and pressure (60 degrees Fahrenheit and 14.7 pounds per square inch absolute). The term "regulated substance" includes but is not limited to petroleum and petroleum-based substances comprised of a complex blend of hydrocarbons derived from crude oil though processes of separation, conversion, upgrading, and finishing, such as motor fuels, jet fuels, distillate fuel oils, residual fuel oils, lubricants, petroleum solvents, and used oils.

**Regulated substance** (Section 68.3) is any substance listed pursuant to section 112®(3) of the Clean Air Act as amended, in Sec. 68.130.

**Regulation promulgation schedule** (Section 63.2) means the schedule for the promulgation of emission standards under this part, established by the Administrator pursuant to section 112(e) of the Act and published in the Federal Register.

**Regulatory agency** (Section 190.02) means the government agency responsible for issuing regulations governing the use of sources of radiation or radioactive materials or emissions therefrom and carrying out inspection and enforcement activities to assure compliance with such regulations.

**Regulatory agency** (Section 192.31) means the U.S. Nuclear Regulatory Commission.

**Regulatory region** (Section 152.3) means genetic material that controls the expression of the genetic material that encodes a pesticidal substance or leads to the production of a pesticidal substance. Examples of regulatory regions include, but are not limited to, promoters, enhancers, and terminators.

**Regulatory region** (Section 174.3) means genetic material that controls the expression of the genetic material that encodes a pesticidal substance or leads to the production of a pesticidal substance. Examples of regulatory regions include, but are not limited to, promoters, enhancers, and terminators.

**Regulatory time frame** (Section 194.2) means the time period beginning at disposal and ending 10,000 years after disposal.

**Reimbursement period** (Section 766.3) means the period that begins when the data from the last test to be completed under this part for a specific chemical substance listed in Sec. 766.25 is submitted to EPA, and ends after an amount of time equal to that which had been required to develop that data or 5 years, whichever is later.

**Reimbursement period** (Section 790.3) refers to a period that begins when the data from the last non-duplicative test to be completed under a test rule are submitted to EPA and ends after an amount of time equal to that which had been required to develop data or after five years, whichever is later.

**Reimbursement period** (Section 791.3) refers to a period that begins when the data from the last non-duplicative test to be completed under a test rule is submitted to EPA and ends after an amount of time equal to that which had been required to develop that data or after 5 years, whichever is later.

**Rejection of a batch** (Section 204.51) means that the number of non-complying compressors in the batch sample is greater than or equal to the rejection number as determined by the appropriate sampling plan.

**Rejection of a batch** (Section 205.51) means the number of noncomplying vehicles in the batch sample is greater than or equal to the rejection number as determined by the appropriate sampling plan.

**Rejection of a batch sequence** (Section 204.51) means that the number of rejected batches in a sequence is greater than or equal to the sequence rejection number as determined by the appropriate sampling plan.

**Rejection of a Batch sequence** (Section 205.51) means that the number of rejected batches in a sequence is equal to or greater than the rejection number as determined by the appropriate sampling plan.

**Related local government agencies** (Section 1400.2) means local government agencies, such as police, fire, emergency management, and planning departments, that are involved in chemical emergency planning, prevention, or response.

**Related state government agencies** (Section 1400.2) means State government agencies, such as emergency management, environmental protection, health, and natural resources departments, that are involved in chemical emergency planning, prevention, or response.

**Relative accuracy** (Section 72.2) means a statistic designed to provide a measure of the systematic and random errors associated with data from continuous emission monitoring systems, and is expressed as the absolute mean difference between the pollutant concentration or volumetric flow measured by the pollutant concentration or flow monitor and the value determined by the applicable reference method(s) plus the 2.5 percent error confidence coefficient of a series of tests divided by the mean of the reference method tests in accordance with part 75 of this chapter.

**Release** (Section 280.12) means any spilling, leaking, emitting, discharging, escaping, leaching or disposing from an UST into ground water, surface water or subsurface soils.

**Release** (Section 302.3) means any spilling, leaking, pumping, pouring, emitting, emptying, discharging, injecting, escaping, leaching, dumping, or disposing into the environment, but excludes (1) any release which results in exposure to persons solely within a workplace, with respect to a claim which such persons may assert against the employer of such persons, (2) emissions from the engine exhaust of a motor vehicle, rolling stock, aircraft, vessel, or pipeline pumping station engine, (3) release of source, byproduct, or special nuclear material from a nuclear incident, as those terms are defined in the Atomic Energy Act of 1954, if such release is subject to requirements with respect to financial protection established by the Nuclear Regulatory Commission under section 170 of such Act, or for the purposes of section 104 of the Comprehensive Environmental Response, Compensation, and Liability Act or any other response action, any release of source, byproduct, or special nuclear material from any processing site designated under section 102(a)(1) or 302(a) of the Uranium Mill Tailings Radiation Control Act of 1978, and (4) the normal application of fertilizer;

**Release** (Section 355.20) means any spilling, leaking, pumping, pouring, emitting, emptying, discharging, injecting, escaping, leaching, dumping, or disposing into the environment (including the abandonment or discarding of barrels, containers, and other closed receptacles) of any hazardous chemical, extremely hazardous substance, or CERCLA hazardous substance.

**Release** (Section 372.3) means any spilling, leaking, pumping, pouring, emitting, emptying, discharging, injecting, escaping, leaching, dumping, or disposing into the environment (including the abandonment or discarding of barrels, containers, and other closed receptacles) of any toxic chemical.

**Release** (Section 373.4) is defined as specified by CERCLA 101(22).

**Release as defined by section 101(22) of CERCLA** (Section 300.5) means any spilling, leaking, pumping, pouring, emitting, emptying, discharging, injecting, escaping, leaching, dumping, or disposing into the environment (including the abandonment or discarding of barrels, containers, and other closed receptacles containing any hazardous substance or pollutant or contaminant), but excludes: Any release which results in exposure to persons solely within a workplace, with respect to a claim which such persons may assert against the employer of such persons; emissions from the engine exhaust of a motor vehicle, rolling stock, aircraft, vessel, or pipeline pumping station engine; release of source, byproduct, or special nuclear material from a nuclear incident, as those terms are defined in the Atomic Energy Act of 1954, if such release is subject to requirements with respect to financial protection established by the Nuclear Regulatory Commission under section 170 of such Act, or, for the purposes of section 104 of CERCLA or any other response action, any release of source, byproduct, or special nuclear material from any processing site designated under section 102(a)(1) or 302(a) of the Uranium Mill Tailings Radiation Control Act of 1978 (42 U.S.C. 7901 et seq.): and the normal application of fertilizer. For purposes of the NCP, release also means threat of release.

**Release detection** (Section 280.12) means determining whether a release of a regulated substance has occurred from the UST system into the environment or into the interstitial space between the UST system and its secondary barrier or secondary containment around it.

**Release, as defined by section 101(22) of CERCLA** (Section 310.3) means any spilling, leaking, pumping, pouring, emitting, emptying, discharging, injection, escaping, leaching, dumping, or disposing into the environment, but excludes: any release that results in exposure to persons solely within a workplace, with respect to a claim that such persons may assert against the employer of such persons; emissions from the engine exhaust of a motor vehicle, rolling stock, aircraft, vessel, or pipeline pumping station engine; release of source, by-product or special nuclear materials from a nuclear incident, as those terms are defined in the Atomic Energy Act of 1954 (42 U.S.C. 2011 et seq.), if such release is subject to requirements with respect to financial protection established by the Nuclear Regulatory Commission under section 170 of such act, or, for the purpose of section 104 of CERCLA or any other response action, any release of source, by-product, or special nuclear material from any processing site designated under section 122(a)(1) or 302(a) of the Uranium Mill Tailings Radiation Control Act of 1978 (Pub. L. 95-604, 42 U.S.C. 2014 et seq.); and the normal application of fertilizer. For purposes of this part, release also means the threat of release.

**Relevant and appropriate requirements** (Section 300.5) means those cleanup standards, standards of control, and other substantive requirements, criteria, or limitations promulgated under federal environmental or state environmental or facility siting laws that, while not "applicable" to a hazardous substance, pollutant, contaminant, remedial action, location, or other circumstance at a CERCLA site, address problems or situations sufficiently similar to those encountered at the CERCLA site that their use is well suited to the particular site. Only those state standards that are identified in a timely manner and are more stringent than federal requirements may be relevant and appropriate.

**Relevant standard** (Section 63.2) means: (1) An emission standard; (2) An alternative emission standard; (3) An alternative emission limitation; or (4) An equivalent emission limitation established pursuant to section 112 of the Act that applies to the collection of equipment, activities, or both regulated by such standard or limitation. A relevant standard may include or consist of a design, equipment, work practice, or operational requirement, or other measure, process, method, system, or technique (including prohibition of emissions) that the Administrator (or a State) establishes for new or existing sources to which such standard or limitation applies. Every relevant standard established pursuant to section 112 of the Act includes subpart A of this part, as provided by Sec. 63.1 (a)(4), and all applicable appendices of this part or of other parts of this chapter that are referenced in that standard.

**Relief device or valve** (Section 65.2) means a device or valve used only to release an unplanned, nonroutine discharge. A relief device or valve discharge can result from an operator error, a malfunction such as a power failure or equipment failure, or other unexpected cause that requires immediate venting of gas from process equipment in order to avoid safety hazards or equipment damage.

**Rem** (Section 141.2) means the unit of dose equivalent from ionizing radiation to the total body or any internal organ or organ system. A "millirem (mrem)" is 1/1000 of a rem.

**Remanufacture** (Section 92.2) means: (1) (i) To replace, or inspect and qualify, each and every power assembly of a locomotive or locomotive engine, whether during a single maintenance event or cumulatively within a five year period; or (ii) To upgrade a locomotive or locomotive engine; or (iii) To convert a locomotive or locomotive engine to enable it to operate using a fuel other than it was originally manufactured to use; or (iv) To install a remanufactured engine or a freshly manufactured engine into a previously used locomotive. (2) Remanufacture also means the act of remanufacturing.

**Remanufacture system or remanufacturing system** (Section 92.2) means all components (or specifications for components) and instructions necessary to remanufacture a locomotive or locomotive engine in accordance with applicable requirements of this part.

**Remanufactured locomotive** (Section 92.2) means either a locomotive which is powered by a remanufactured locomotive engine, or a repowered locomotive.

**Remanufactured locomotive engine** (Section 92.2) means a locomotive engine which has been remanufactured.

**Remanufacturer** (Section 92.2) means an individual or entity that is engaged in the manufacture or assembly of remanufactured locomotives or locomotive engines, (including: Entities that design or produce the emission-related parts used in remanufacturing; entities that install parts in an existing locomotive or locomotive engine to remanufacture it; and entities that own or operate the locomotive or locomotive engine and provide specifications as to how an engine is to be remanufactured (i.e., specifying who will perform the work, when the work is to be performed, what parts are to be used, or how to calibrate the adjustable parameters of the engine)); or an importer of remanufactured locomotives or locomotive engines. (See Secs. 92.1© and 92.209 for applicability of this term.)

**Remedial action** (Section 192) means any action performed under section 108 of the Act.

**Remedial Action Plan (RAP)** (Section 270.2) means a special form of RCRA permit that a facility owner or operator may obtain instead of a permit issued under Secs. 270.3 through 270.66, to authorize the treatment, storage or disposal of hazardous remediation waste (as defined in Sec. 260.10 of this chapter) at a remediation waste management site.

**Remedial design (RD)** (Section 300.5) means the technical analysis and procedures which follow the selection of remedy for a site and result in a detailed set of plans and specifications for implementation of the remedial action.

**Remedial investigation (RI)** (Section 300.5) is a process undertaken by the lead agency to determine the nature and extent of the problem presented by the release. The RI emphasizes data collection and site characterization, and is generally performed concurrently and in an interactive fashion with the feasibility study. The RI includes sampling and monitoring, as necessary, and includes the gathering of sufficient information to determine the necessity for remedial action and to support the evaluation of remedial alternatives.

**Remedial project manager (RPM)** (Section 300.5) means the official designated by the lead agency to coordinate, monitor, or direct remedial or other response actions under subpart E of the NCP.

**Remediation waste** (Section 260.10) means all solid and hazardous wastes, and all media (including ground water, surface water, soils, and sediments) and debris that are managed for implementing cleanup.

**Remediation waste management site** (Section 260.10) means a facility where an owner or operator is or will be treating, storing or disposing of hazardous remediation wastes. A remediation waste management site is not a facility that is subject to corrective action under 40 CFR 264.101, but is subject to corrective action requirements if the site is located in such a facility.

**Remedy or remedial action (RA)** (Section 300.5) means those actions consistent with permanent remedy taken instead of, or in addition to, removal action in the event of a release or threatened release of a hazardous substance into the environment, to prevent or minimize the release of hazardous substances so that they do not migrate to cause substantial danger to present or future public health or welfare or the environment. The term includes, but is not limited to, such actions at the location of the release as storage, confinement, perimeter protection using dikes, trenches, or ditches, clay cover, neutralization, cleanup of released hazardous substances and associated contaminated materials, recycling or reuse, diversion, destruction, segregation of reactive wastes, dredging or excavations, repair or replacement of leaking containers, collection of leachate and runoff, on-site treatment or incineration, provision of alternative water supplies, any monitoring reasonably required to assure that such actions protect the public health and welfare and the environment and, where appropriate, post-removal site control activities. The term includes the costs of permanent relocation of residents and businesses and community facilities (including the cost of providing "alternative land of equivalent value" to an Indian tribe pursuant to CERCLA section 126(b)) where EPA determines that, alone or in combination with other measures, such relocation is more cost-effective than, and environmentally preferable to, the transportation, storage, treatment, destruction, or secure disposition off-site of such hazardous substances, or may otherwise be necessary to protect the public health or welfare; the term includes off-site transport and off-site storage, treatment, destruction, or secure disposition of hazardous substances and associated contaminated materials. For the purpose of the NCP, the term also includes enforcement activities related thereto.

**Removal** (Section 763.83) means the taking out or the stripping of substantially all ACBM from a damaged area, a functional space, or a homogeneous area in a school building.

**Removal costs as defined by section 1001 of the OPA** (Section 300.5) means the costs of removal that are incurred after a discharge of oil has occurred, or in any case in which there is a substantial threat of a discharge of oil, the costs to prevent, minimize, or mitigate oil pollution from such an incident.

**Remove or removal** (Section 109.2) refers to the removal of the oil from the water and shorelines or the taking of such other actions as may be necessary to minimize or mitigate damage to the public health or welfare, including, but not limited to, fish, shellfish, wildlife, and public and private property, shorelines, and beaches.

**Remove or removal** (Section 113.3) means the removal of the oil from the water and shorelines or the taking of such other actions as the Federal On-Scene Coordinator may determine to be necessary to minimize or mitigate damage to the public health or welfare, including but not limited to, fish, shellfish, wildlife, and public and private property, shorelines, and beaches.

**Remove or removal** (Section 117.1) refers to removal of the oil or hazardous substances from the water and shoreline or the taking of such other actions as may be necessary to minimize or mitigate damage to the public health or welfare, including, but not limited to, fish, shellfish, wildlife, and public and private property, shorelines, and beaches.

**Remove or removal as defined by section 311(a)(8) of the CWA** (Section 300.5) refers to containment and removal of oil or hazardous substances from the water and shorelines or the taking of such other actions as may be necessary to minimize or mitigate damage to the public health or welfare of the United States (including, but not limited to, fish, shellfish, wildlife, public and private property, and shorelines and beaches) or to the environment. For the purpose of the NCP, the term also includes monitoring of action to remove a discharge. As defined by section 101(23) of CERCLA, remove or removal means the cleanup or removal of released hazardous substances from the environment; such actions as may be necessary taken in the event of the threat of release of hazardous substances into the environment; such actions as may be necessary to monitor, assess, and evaluate the release or threat of release of hazardous substances; the disposal of removed material; or the taking of such other actions as may be necessary to prevent, minimize, or mitigate damage to the public health or welfare of the United States or to the environment, which may otherwise result from a release or threat of release. The term includes, in addition, without being limited to, security fencing or other measures to limit access, provision of alternative water supplies, temporary evacuation and housing of threatened individuals not otherwise provided for, action taken under section 104(b) of CERCLA, post-removal site control, where appropriate, and any emergency assistance which may be provided under the Disaster Relief Act of 1974. For the purpose of the NCP, the term also includes enforcement activities related thereto.

**Renderer** (Section 432.101) shall mean an independent or off-site rendering operation, conducted separate from a slaughterhouse, packinghouse or poultry dressing or processing plant, which manufactures at rates greater than 75,000 pounds of raw material per day of meat meal, tankage, animal fats or oils, grease, and tallow, and may cure cattle hides, but excluding marine oils, fish meal, and fish oils.

**Renewal** (Section 70.2) means the process by which a permit is reissued at the end of its term.

**Renewal** (Section 71.2) means the process by which a permit is reissued at the end of its term.

**Renovation** (Section 745.83) means the modification of any existing structure, or portion thereof, that results in the disturbance of painted surfaces, unless that activity is performed as part of an abatement as defined by this part (40 CFR 745.223). The term renovation includes (but is not limited to): the removal or modification of painted surfaces or painted components (e.g., modification of painted doors, surface preparation activity (such as sanding, scraping, or other such activities that may generate paint dust)); the removal of large structures (e.g., walls, ceiling, large surface replastering, major re-plumbing); and window replacement.

**Renovator** (Section 745.83) means any person who performs for compensation a renovation.

**Repackaging of pesticide products** (Section 455.10) means the transfer of a pesticide formulation (or PAI) from one container to another without a change in composition of the formulation or the labeling content, for sale or distribution.

**Repair** (Section 280.12) means to restore a tank or UST system component that has caused a release of product from the UST system.

**Repair** (Section 763.83) means returning damaged ACBM to an undamaged condition or to an intact state so as to prevent fiber release.

**Repaired** (Section 264.1031) means that equipment is adjusted, or otherwise altered, to eliminate a leak.

**Repaired** (Section 65.2) means, for the purposes of subparts F and G of this part, that equipment meets the following conditions: (1) Is adjusted, or otherwise altered, to eliminate a leak as defined in the applicable section of this part; and (2) Unless otherwise specified in applicable provisions of this part, is monitored as specified in Sec. 65.104(b) and Sec. 65.143(c) to verify that emissions from the equipment are below the applicable leak definition.

**Repeat compliance period** (Section 141.2) means any subsequent compliance period after the initial compliance period.

**Replacement in kind** (Section 68.3) means a replacement that satisfies the design specifications.

**Replacement unit** (Section 260.10) means a landfill, surface impoundment, or waste pile unit (1) from which all or substantially all of the waste is removed, and (2) that is subsequently reused to treat, store, or dispose of hazardous waste. "Replacement unit" does not apply to a unit from which waste is removed during closure, if the subsequent reuse solely involves the disposal of waste from that unit and other closing units or corrective action areas at the facility, in accordance with an approved closure plan or EPA or State approved corrective action.

**Replacement unit** (Section 72.2) means an affected unit replacing the thermal energy provided by an opt-in source, where both the affected unit and the opt-in source are governed by a thermal energy plan.

**Reportable quantities** (Section 117.1) means quantities that may be harmful as set forth in Sec. 117.3, the discharge of which is a violation of section 311(b)(3) and requires notice as set forth in Sec. 117.21.

**Reportable quantity** (Section 302.3) means that quantity, as set forth in this part, the release of which requires notification pursuant to this part;

**Reportable quantity** (Section 355.20) means, for any CERCLA hazardous substance, the reportable quantity established in table 302.4 of 40 CFR part 302, for such substance, for any other substance, the reportable quantity is one pound.

**Reporter** (Section 1603.3) means a CSB employee designated by the General Counsel, under Sec. 1603.5©, to attend and prepare a written summary of all briefings described in paragraphs ©(3) and ©(4) of this section and all informal preliminary discussions described in paragraph ©(5) of this section.

**Repower** (Section 92.2) means replacement of the engine in a previously used locomotive with a freshly manufactured locomotive engine. Replacing a locomotive engine with a freshly manufactured locomotive engine in a locomotive that has a refurbished or reconditioned chassis such that less than 25 of the parts of the locomotive were previously used (as weighted by dollar value) is not repowering.

**Repowered locomotive** (Section 92.2) means a locomotive that has been repowered with a freshly manufactured engine.

**Repowering** (Section 60.2) means replacement of an existing coal-fired boiler with one of the following clean coal technologies: atmospheric or pressurized fluidized bed combustion, integrated gasification combined cycle, magnetohydrodynamics, direct and indirect coal-fired turbines, integrated gasification fuel cells, or as determined by the Administrator, in consultation with the Secretary of Energy, a derivative of one or more of these technologies, and any other technology capable of controlling multiple combustion emissions simultaneously with improved boiler or generation efficiency and with significantly greater waste reduction relative to the performance of technology in widespread commercial use as of November 15, 1990. Repowering shall also include any oil and/or gas-fired unit which has been awarded clean coal technology demonstration funding as of January 1, 1991, by the Department of Energy.

**Representative of the news media** (Section 1601.3) refers to any person actively gathering news for an entity that is organized and operated to publish or broadcast news to the public. The term news means information that is about current events or that would be of current interest to the public. For freelance journalists to be regarded as working for a news organization, they must demonstrate a solid basis for expecting publication through that organization. A publication contract would be the clearest proof, but components shall also look to the past publication record of a requester in making this determination.

**Representative sample** (Section 260.10) means a sample of a universe or whole (e.g., waste pile, lagoon, ground water) which can be expected to exhibit the average properties of the universe or whole.

**Request for a Hearing** (Section 305.3) means a written notice requesting an administrative hearing of the total or partial denial of a claim by the Claims Official. Such hearing shall be governed by this part.

**Requester** (Section 1601.3) means any person, including an individual, Indian tribe, partnership, corporation, association, or public or private organization other than a Federal agency, that requests access to records in the possession of the CSB.

**Requestor** (Section 305.3) is the party who files a Request for a Hearing.

**Requirements and standards** (Section 761.123) means: (1) "Requirements" as used in this policy refers to both the procedural responses and numerical decontamination levels set forth in this policy as constituting adequate cleanup of PCBs. (2) "Standards" refers to the numerical decontamination levels set forth in this policy.

**Re-refined oils** 8 means used oils from which the physical and chemical contaminants acquired through previous use have been removed through a refining process;

**Re-refining** (Section 437.2) means the processing of used oil using distillation, hydrotreating, and/or other treatment employing acid, caustic, solvent, clay and/or chemicals in order to produce high quality base stock for lubricants or other petroleum products.

**Re-refining distillation bottoms** (Section 279.1) means the heavy fraction produced by vacuum distillation of filtered and dehydrated used oil. The composition of still bottoms varies with column operation and feedstock.

**Research and development (R&D) for PCB disposal** (Section 761.13) means demonstrations for commercial PCB disposal approvals, pre-demonstration tests, tests of major modifications to previously approved PCB disposal technologies, treatability studies for PCB disposal technologies which have not been approved, development of new disposal technologies, and research on chemical transformation processes including, but not limited to, biodegradation.

**Research and development facility** (Section 63.601) means research or laboratory operations whose primary purpose is to conduct research and development into new processes and products, where the operations are under the close supervision of technically trained personnel, and where the facility is not engaged in the manufacture of products for commercial sale in commerce or other off-site distribution, except in a de minimis manner.

**Research gas material (RGM)** (Section 72.2) means a calibration gas mixture developed by agreement of a requestor and the National Institutes for Standards and Technologies (NIST) that NIST analyzes and certifies as "NIST traceable." RGMs may have concentrations different from those of standard reference materials.

**Research gas mixture (RGM)** (Section 72.2) means a calibration gas mixture developed by agreement of a requestor and NIST that NIST analyzes and certifies as "NIST traceable." RGMs may have concentrations different from those of standard reference materials.

**Research, development, and demonstration source** (Section 62.1201) means a source engaged in laboratory, pilot plant, or prototype demonstration operations: (1) Whose primary purpose is to conduct research, development, or short-term demonstration of an innovative and experimental hazardous waste treatment technology or process; and (2) Where the operations are under the close supervision of technically-trained personnel.

**Reseller** (Section 80.2) means any person who purchases gasoline or diesel fuel identified by the corporate, trade, or brand name of a refiner from such refiner or a distributor and resells or transfers it to retailers or wholesale purchaser-consumers displaying the refiner's brand, and whose assets or facilities are not substantially owned, leased, or controlled by such refiner.

**Reserved credits** (Section 92.302) means credits that have been generated but have not yet been reviewed by EPA or used to demonstrate compliance under the averaging provisions of this subpart.

**Reserved credits** (Section 94.302) means credits that have been generated but have not yet been reviewed by EPA or used to demonstrate compliance under the averaging provisions of this subpart.

**Residential building** (Section 745.63) means a building containing one or more residential dwellings.

**Residential dwelling** (Section 745.103) means: (1) A single-family dwelling, including attached structures such as porches and stoops; or (2) A single-family dwelling unit in a structure that contains more than one separate residential dwelling unit, and in which each such unit is used or occupied, or intended to be used or occupied, in whole or in part, as the residence of one or more persons.

**Residential dwelling** (Section 745.223) means (1) a detached single family dwelling unit, including attached structures such as porches and stoops; or (2) a single family dwelling unit in a structure that contains more than one separate residential dwelling unit, which is used or occupied, or intended to be used or occupied, in whole or in part, as the home or residence of one or more persons.

**Residential Property** (Section 201.1) means any property that is used for any of the purposes described in the following standard land use codes (ref. Standard Land Use Coding Manual. U.S. DOT/FHWA Washington, DC, reprinted March 1977): 1, Residential: 651, Medical and other Health Services; 68, Educational Services; 691, Religious Activities: and 711, Cultural Activities.

**Residential solid waste** (Section 243.101) means the wastes generated by the normal activities of households, including, but not limited to, food wastes, rubbish, ashes, and bulky wastes.

**Residential solid waste** (Section 246.101) means the wastes generated by the normal activities of households, including but not limited to, food wastes, rubbish, ashes, and bulky wastes.

**Residential tank** (Section 280.12) is a tank located on property used primarily for dwelling purposes.

**Residential use** (Section 152.3) means use of a pesticide directly: (1) On humans or pets, (2) In, on, or around any structure, vehicle, article, surface, or area associated with the household, including but not limited to areas such as non-agricultural outbuildings, non-commercial greenhouses, pleasure boats and recreational vehicles, or (3) In any preschool or day care facility.

**Residential use** (Section 157.21) means use of a pesticide or device: (1) Directly on humans or pets; (2) In, on, or around any structure, vehicle, article, surface or area associated with the household, including but not limited to areas such as non-agricultural outbuildings, non-commercial greenhouses, pleasure boats and recreational vehicles; or (3) In or around any preschool or day care facility.

**Residential/commercial areas** (Section 761.123) means those areas where people live or reside, or where people work in other than manufacturing or farming industries. Residential areas include housing and the property on which housing is located, as well as playgrounds, roadways, sidewalks, parks, and other similar areas within a residential community. Commercial areas are typically accessible to both members of the general public and employees and include public assembly properties, institutional properties, stores, office buildings, and transportation centers.

**Residual disinfectant concentration ("C" in CT calculations)** (Section 141.2) means the concentration of disinfectant measured in mg/l in a representative sample of water.

**Residual emissions** (Section 55.2) means the difference in emissions from an OCS source if it applies the control requirements(s) imposed pursuant to Sec. 55.13 or Sec. 55,14 of this part and emissions from that source if it applies a substitute control requirement pursuant to an exemption granted under Sec. 55.7 of this part.

**Residual radioactive material** (Section 192) means: (1) Waste (which the Secretary determines to be radioactive) in the form of tailings resulting from the processing of ores for the extraction of uranium and other valuable constituents of the ores; and (2) Other wastes (which the Secretary determines to be radioactive) at a processing site which relate to such processing, including any residual stock of unprocessed ores or low-grade materials.

**Residue** (Section 240.101) means all the solids that remain after completion of thermal processing, including bottom ash, fly ash, and grate siftings.

**Respond or response as defined by section 101(25) of CERCLA** (Section 300.5) means remove, removal, remedy, or remedial action, including enforcement activities related thereto.

**Respond or Response as defined by section 101(25) of CERCLA** (Section 307.14) means remove, removal, remedy, and remedial action, all such terms (including removal and remedial action) including enforcement activities related thereto.

**Respondent** (Section 164.2) means the Assistant Administrator of the Office of Hazardous Materials Control of the Agency.

**Respondent** (Section 209.3) means any person against whom a complaint has been issued under this subpart.

**Response action** (Section 304.12) means remove, removal, remedy and remedial action, as those terms are defined by section 101 of CERCLA, 42 U.S.C. 9601, including enforcement activities related thereto.

**Response action** (Section 763.83) means a method, including removal, encapsulation, enclosure, repair, operations and maintenance, that protects human health and the environment from friable ACBM.

**Response claim** (Section 307.14) means a preauthorized demand in writing for a sum certain for response costs referred to in section 111(a)(2) of CERCLA, including certain costs of actions referred to in section 122(b)(1) of CERCLA.

**Response costs** (Section 304.12) means all costs of removal or remedial action incurred and to be incurred by the United States at a facility pursuant to section 104 of CERCLA, 42 U.S.C. 9604, including, but not limited to, all costs of investigation and information gathering, planning and implementing a response action, administration, enforcement, litigation, interest and indirect costs.

**Responsible agency** (Section 240.101) means the organizational element that has the legal duty to ensure that owners, operators, or users of facilities comply with these guidelines.

**Responsible agency** (Section 243.101) means the organizational element that has the legal duty to ensure compliance with these guidelines.

**Responsible official** (Section 56.1) means the EPA Administrator or any EPA employee who is accountable to the Administrator for carrying out a power or duty delegated under section 301(a)(1) of the act, or is accountable in accordance with EPA's formal organization for a particular program or function as described in part 1 of this title.

**Responsible official** (Section 63.2) means one of the following: (1) For a corporation: A president, secretary, treasurer, or vice president of the corporation in charge of a principal business function, or any other person who performs similar policy or decision-making functions for the corporation, or a duly authorized representative of such person if the representative is responsible for the overall operation of one or more manufacturing, production, or operating facilities and either: (i) The facilities employ more than 250 persons or have gross annual sales or expenditures exceeding $25 million (in second quarter 1980 dollars); or (ii) The delegation of authority to such representative is approved in advance by the Administrator. (2) For a partnership or sole proprietorship: a general partner or the proprietor, respectively. (3) For a municipality, State, Federal, or other public agency: either a principal executive officer or ranking elected official. For the purposes of this part, a principal executive officer of a Federal agency includes the chief executive officer having responsibility for the overall operations of a principal geographic unit of the agency (e.g., a Regional Administrator of the EPA). (4) For affected sources (as defined in this part) applying for or subject to a title V permit: "responsible official" shall have the same meaning as defined in part 70 or Federal title V regulations in this chapter (42 U.S.C. 7661), whichever is applicable.

**Responsible official** (Section 70.2) means one of the following: (1) For a corporation: a president, secretary, treasurer, or vice-president of the corporation in charge of a principal business function, or any other person who performs similar policy or decision-making functions for the corporation, or a duly authorized representative of such person if the representative is responsible for the overall operation of one or more manufacturing, production, or operating facilities applying for or subject to a permit and either: (i) The facilities employ more than 250 persons or have gross annual sales or expenditures exceeding $25 million (in second quarter 1980 dollars); or (ii) The delegation of authority to such representatives is approved in advance by the permitting authority; (2) For a partnership or sole proprietorship: a general partner or the proprietor, respectively; (3) For a municipality, State, Federal, or other public agency: Either a principal executive officer or ranking elected official. For the purposes of this part, a principal executive officer of a Federal Agency includes the chief executive officer having responsibility for the overall operations of a principal geographic unit of the agency (e.g., a Regional Administrator of EPA); or (4) For affected sources: (i) The designated representative in so far as actions, standards, requirements, or prohibitions under title IV of the Act or the regulations promulgated thereunder are concerned; and (ii) The designated representative for any other purposes under part 70. Section 502(b) (10) changes are changes that contravene an express permit term. Such changes do not include changes that would violate applicable requirements or contravene federally enforceable permit terms and conditions that are monitoring (including test methods), recordkeeping, reporting, or compliance certification requirements.

**Responsible official** (Section 71.2) means one of the following: (1) For a corporation: a president, secretary, treasurer, or vice-president of the corporation in charge of a principal business function, or any other person who performs similar policy or decision-making functions for the corporation, or a duly authorized representative of such person if the representative is responsible for the overall operation of one or more manufacturing, production, or operating facilities applying for or subject to a permit and either: (i) the facilities employ more than 250 persons or have gross annual sales or expenditures exceeding $25 million (in second quarter 1980 dollars); or (ii) the delegation of authority to such representative is approved in advance by the permitting authority; (2) For a partnership or sole proprietorship: a general partner or the proprietor, respectively; (3) For a municipality, State, Federal, or other public agency: Either a principal executive officer or ranking elected official. For the purposes of this part, a principal executive officer of a Federal agency includes the chief executive officer having responsibility for the overall operations of a principal geographic unit of the agency (e.g., a Regional Administrator of EPA); or (4) For affected sources: (i) The designated representative insofar as actions, standards, requirements, or prohibitions under title IV of the Act or 40 CFR parts 72 through 78 are concerned; and (ii) The designated representative for any other purposes under part 71. Section 502(b)(10) changes are changes that contravene an express permit term. Such changes do not include changes that would violate applicable requirements or contravene federally enforceable permit terms and conditions that are monitoring (including test methods), recordkeeping, reporting, or compliance certification requirements.

**Responsible party** (Section 761.123) means the owner of the PCB equipment, facility, or other source of PCBs or his/her designated agent (e.g., a facility manager or foreman).

**Responsible party as defined by section 1001 of the OPA** (Section 300.5) means the following: (1) Vessels—In the case of a vessel, any person owning, operating, or demise chartering the vessel. (2) Onshore Facilities—In the case of an onshore facility (other than a pipeline), any person owning or operating the facility, except a federal agency, state, municipality, commission, or political subdivision of a state, or any interstate body, that as the owner transfers possession and right to use the property to another person by lease, assignment, or permit. (3) Offshore Facilities—In the case of an offshore facility (other than a pipeline or a deepwater port licensed under the Deepwater Port Act of 1974 (33 U.S.C. 1501 et seq.)), the lessee or permittee of the area in which the facility is located or the holder of a right of use and easement granted under applicable state law or the Outer Continental Shelf Lands Act (43 U.S.C. 1301-1356) for the area in which the facility is located (if the holder is a different person than the lessee or permittee), except a federal agency, state, municipality, commission, or political subdivision of a state, or any interstate body, that as owner transfers possession and right to use the property to another person by lease, assignment, or permit. (4) Deepwater Ports—In the case of a deepwater port licensed under the Deepwater Port Act of 1974 (33 U.S.C. 1501-1524), the licensee. (5) Pipelines—In the case of a pipeline, any person owning or operating the pipeline. (6) Abandonment—In the case of an abandoned vessel, onshore facility, deepwater port, pipeline, or offshore facility, the person who would have been responsible parties immediately prior to the abandonment of the vessel or facility.

**Resting losses** (Section 86.096-2) means evaporative emissions that may occur continuously, that are not diurnal emissions, hot soak emissions, running losses, or spitback emissions.

**Resting losses** (Section 86.1803-01) means evaporative emissions that may occur continuously, that are not diurnal emissions, hot soak emissions, refueling emissions, running losses, or spitback emissions.

**Restricted use pesticide** (Section 171.2) means a pesticide that is classified for restricted use under the provisions of section 3(d)(1)© of the Act.

**Restricted use pesticide retail dealer** (Section 171.2) means any person who makes available for use any restricted use pesticide, or who offers to make available for use any such pesticide. Limited. The following definitions apply only to dealers, dealerships and transactions in States or on Indian Reservations where EPA conducts a Federal Pesticide Applicator Certification Program.

**Restricted-entry interval** (Section 156.203) means the time after the end of a pesticide application during which entry to the treated area is restricted.

**Restricted-entry interval** (Section 170.3) means the time after the end of a pesticide application during which entry into the treated area is restricted.

**Restroom divider/partition** (Section 247.3) means a barrier used to provide privacy in public restroom facilities;

**Retail facility** (Section 68.3) means a stationary source at which more than one-half of the income is obtained from direct sales to end users or at which more than one-half of the fuel sold, by volume, is sold through a cylinder exchange program.

**Retail outlet** (Section 80.2) means any establishment at which gasoline, diesel fuel, methanol, natural gas or liquefied petroleum gas is sold or offered for sale for use in motor vehicles.

**Retailer** (Section 717.3) means a person who distributes in commerce a chemical substance, mixture, or article to ultimate purchasers who are not commercial entities.

**Retailer** (Section 80.2) means any person who owns, leases, operates, controls, or supervises a retail outlet.

**Retan-wet finish** (Section 425.02) means the final processing steps performed on a tanned hide including, but not limited to, the following wet processes: retan, bleach, color, and fatliquor.

**Retarder (Active)** (Section 201.1) means a device or system for decelerating rolling rail cars and controlling the degree of deceleration on a car by car basis.

**Retarder Sound** (Section 201.1) means a sound which is heard and identified by the observer as that of a retarder, and that causes a sound level meter indicator at fast meter response Sec. 201.1(l) to register an increase of at least ten decibels above the level observed immediately before hearing the sound.

**Retread tire** (Section 247.3) means a worn automobile, truck, or other motor vehicle tire whose tread has been replaced;

**Retrofill** (Section 761.13) means to remove PCB or PCB-contaminated dielectric fluid and to replace it with either PCB, PCB-contaminated, or non-PCB dielectric fluid.

**Retrofit** (Section 610.11) means the addition of a new item, modification or removal of an existing item of equipment beyond that of regular maintenance, on an automobile after its initial manufacture.

**Retrofit device or device** (Section 610.11) means: (i) Any component, equipment, or other device (except a flow measuring instrument or other driving aid, or lubricant or lubricant additive) which is designed to be installed in or on an automobile as an addition to, as a replacement for, or through alteration or modification of, any original component, or other devices; or (ii) Any fuel additive which is to be added to the fuel supply of an automobile by means other than fuel dispenser pumps; and (iii) Which any manufacturer, dealer, or distributor of such device represents will provide higher fuel economy than would have resulted with the automobile as originally equipped, as determined under rules of the Administrator.

**Retrofitted configuration** (Section 610.11) means the test configuration after adjustment of engine calibrations to the retrofit specifications and after all retrofit hardware has been installed.

**Reverberation Time** (Section 211.203) The time that would be required for the mean-square sound pressure level, originally in a steady state, to fall 60 dB after the source is stopped.

**Review** (Section 1601.3) refers to the process of examining a record, in response to a FOIA request, to determine whether any portion of that record may be withheld under one or more of the FOIA exemptions. It also includes the processing of any record for disclosure; for example, redacting information that is exempt from disclosure under the FOIA. Review does not include time spent resolving general legal or policy issues regarding the use of FOIA exemptions.

**Review Officer** (Section 305.3) means the EPA Administrator or his delegatee who is authorized to exercise all powers and duties prescribed or delegated under the Act or this part to him.

**Revocation** (Section 194.2) means any action taken by the Administrator to terminate the certification pursuant to section 8(d) (1) of the WIPP LWA.

**Rice** (Section 406.61) shall mean rice delivered to a plant before processing.

**Ring carrier** (Section 238.20) means any plastic ring carrier device that contains at least one hole greater than 1\3/4\ inches in diameter which is made, used, or designed for the purpose of packaging, transporting, or carrying multipackaged cans or bottles.

**Risk assessment** (Section 745.103) means an on-site investigation to determine and report the existence, nature, severity, and location of lead-based paint hazards in residential dwellings, including: (1) Information gathering regarding the age and history of the housing and occupancy by children under age 6; (2) Visual inspection; (3) Limited wipe sampling or other environmental sampling techniques; (4) Other activity as may be appropriate; and (5) Provision of a report explaining the results of the investigation.

**Risk assessment** (Section 745.223) means (1) an on-site investigation to determine the existence, nature, severity, and location of lead-based paint hazards, and (2) the provision of a report by the individual or the firm conducting the risk assessment, explaining the results of the investigation and options for reducing lead-based paint hazards.

**Risk associated dose (RAD)** (Section 132.2) is a dose of a known or presumed carcinogenic substance in (mg/kg)/day which, over a lifetime of exposure, is estimated to be associated with a plausible upper bound incremental cancer risk equal to one in 100,000.

**Risk management plan (RMP)** (Section 1400.2) means a risk management plan submitted to the Administrator by an owner or operator of a stationary source pursuant to 40 CFR part 68, subpart G.

**RMP** (Section 68.3) means the risk management plan required under subpart G of this part.

**Rock wool insulation** (Section 247.3) means insulation which is composed principally from fibers manufactured from slag or natural rock, with or without binders;

**Rod, wire and coil** (Section 420.91) means those acid pickling operations that pickle rod, wire or coiled rod and wire products.

**Roll bonding** (Section 471.02) is the process by which a permanent bond is created between two metals by rolling under high pressure in a bonding mill (co-rolling).

**Rollboard** (Section 763.163) means an asbestos-containing product made of paper that is produced in a continuous sheet, is flexible, and is rolled to achieve a desired thickness. Asbestos rollboard consists of two sheets of asbestos paper laminated together. Major applications of this product include: office partitioning; garage paneling; linings for stoves and electric switch boxes; and fire-proofing agent for security boxes, safes, and files.

**Rolling** (Section 467.02) is the reduction in thickness or diameter of a workpiece by passing it between lubricated steel rollers. There are two subcategories based on the rolling process. In the rolling with neat oils subcategory, pure or neat oils are used as lubricants for the rolling process. In the rolling with emulsions subcategory, emulsions are used as lubricants for the rolling process.

**Rolling** (Section 468.02) shall mean the reduction in the thickness or diameter of a workpiece by passing it between rollers.

**Rolling** (Section 471.02) is the reduction in thickness or diameter of a workpiece by passing it between lubricated steel rollers.

**Rolling average** (Section 62.1201) means the average of all one-minute averages over the averaging period.

**Room (Section 745.63)** means a separate part of the inside of a building, such as a bedroom, living room, dining room, kitchen, bathroom, laundry room, or utility room. To be considered a separate room, the room must be separated from adjoining rooms by built-in walls .or archways that extend at least 6 inches from an intersecting wall. Half walls or bookcases count as room separators if built-in. Movable or collapsible partitions or partitions consisting solely of shelves or cabinets are not considered built-in walls. A screened in porch that is used as a living area is a room.

**Round, rounded or rounding** (Section 86.1803-01) means, unless otherwise specified, that numbers will be rounded according to ASTM-E29-93a, which is incorporated by reference in this part pursuant to Sec. 86.1.

**Round, rounded or rounding** (Section 90.3) means, unless otherwise specified, that numbers will be rounded according to ASTM-E29-93a, which is incorporated by reference in this part pursuant to Sec. 90.7.

**Rounded** (Section 600.022-85) means a number shortened to the specific number of decimal places in accordance with the Round Off Method specified in ASTM E 29-67.

**Routed to a process or route to a process** (Section 65.2) means the emissions are conveyed to any enclosed portion of a process unit where the emissions are predominantly recycled and/or consumed in the same manner as a material that fulfills the same function in the process and/or transformed by chemical reaction into materials that are not regulated materials and/or incorporated into a product; and/or recovered.

**Routine maintenance area** (Section 763.83) means an area, such as a boiler room or mechanical room, that is not normally frequented by students and in which maintenance employees or contract workers regularly conduct maintenance activities.

**Routine use** (Section 1516.2) means with respect to the disclosure of a record, the use of such record for a purpose which is compatible with the purpose for which it was collected.

**Rubbish** (Section 243.101) means a general term for solid waste, excluding food wastes and ashes, taken from residences, commercial establishments, and institutions.

**Run** (Section 60.2) means the net period of time during which an emission sample is collected. Unless otherwise specified, a run may be either intermittent or continuous within the limits of good engineering practice.

**Run** (Section 61.02) means the net period of time during which an emission sample is collected. Unless otherwise specified, a run may be either intermittent or continuous within the limits of good engineering practice.

**Run** (Section 62.1201) means the net period of time during which an air emission sample is collected under a given set of operating conditions. Three or more runs constitutes a test. Unless otherwise specified, a run may be either intermittent or continuous.

**Run** (Section 63.2) means one of a series of emission or other measurements needed to determine emissions for a representative operating period or cycle as specified in this part.

**Run** (Section 65.2) means one of a series of emission or other measurements needed to determine emissions for a representative operating period or cycle as specified in this part. Unless otherwise specified, a run may be either intermittent or continuous within the limits of good engineering practice.

**Run average** (Section 62.1201) means the average of the one-minute average parameter values for a run.

**Running change** (Section 86.1803-01) means a change to a vehicle or addition of a model which occurs after certification but during vehicle production.

**Running changes** (Section 85.1502) Those changes in vehicle or engine configuration, equipment or calibration which are made by an OEM or ICI in the course of motor vehicle or motor vehicle engine production.

**Running loss** (Section 86.082-2) means fuel evaporative emissions resulting from an average trip in an urban area or the simulation of such a trip.

**Running losses** (Section 86.096-2) means evaporative emissions that occur during vehicle operation.

**Running losses** (Section 86.1803-01) means evaporative emissions that occur during vehicle operation.

**Run-off** (Section 258.2) means any rainwater, leachate, or other liquid that drains over land from any part of a facility.

**Run-off** (Section 260.10) means any rainwater, leachate, or other liquid that drains over land from any part of a facility.

**Runoff** (Section 419.11) shall mean the flow of storm water resulting from precipitation coming into contact with petroleum refinery property.

**Run-on** (Section 258.2) means any rainwater, leachate, or other liquid that drains over land onto any part of a facility.

**Run-on** (Section 260.10) means any rainwater, leachate, or other liquid that drains over land onto any part of a facility.

**Rupture of a PCB Transformer** (Section 761.13) means a violent or non-violent break in the integrity of a PCB Transformer caused by an overtemperature and/or overpressure condition that results in the release of PCBs.

# S

**Safety device** (Section 265.1081) means a closure device such as a pressure relief valve, frangible disc, fusible plug, or any other type of device which functions exclusively to prevent physical damage or permanent deformation to a unit or its air emission control equipment by venting gases or vapors directly to the atmosphere during unsafe conditions resulting from an unplanned, accidental, or emergency event. For the purpose of this subpart, a safety device is not used for routine venting of gases or vapors from the vapor headspace underneath a cover such as during filling of the unit or to adjust the pressure in this vapor headspace in response to normal daily diurnal ambient temperature fluctuations. A safety device is designed to remain in a closed position during normal operations and open only when the internal pressure, or another relevant parameter, exceeds the device threshold setting applicable to the air emission control equipment as determined by the owner or operator based on manufacturer recommendations, applicable regulations, fire protection and prevention codes, standard engineering codes and practices, or other requirements for the safe handling of flammable, ignitable, explosive, reactive, or hazardous materials.

**Safety margin** (Section 93.101) means the amount by which the total projected emissions from all sources of a given pollutant are less than the total emissions that would satisfy the applicable requirement for reasonable further progress, attainment, or maintenance.

**Sale for purposes other than resale** (Section 761.13) means sale of PCBs for purposes of disposal and for purposes of use, except where use involves sale for distribution in commerce. PCB Equipment which is first leased for purposes of use any time before July 1, 1979, will be considered sold for purposes other than resale.

**Sales** (Section 88.202-94) means vehicles that are produced, sold, and distributed (in accordance with normal business practices and applicable franchise agreements) in the State of California, including owners of covered fleets under subpart C of part 86 of this chapter. The manufacturer can choose at their option from one of the following three methods for determining sales: (i) Sales is defined as sales to the ultimate purchaser. (ii) Sales is defined as vehicle sales by a manufacturer to a dealer, distributor, fleet operator, broker, or any other entity which comprises the first point of sale. (iii) Sales is defined as equivalent to the production of vehicles for the state of California. This option can be revoked if it is determined that the production and actual sales numbers do not exhibit a functional equivalence per the language of Sec. 86.708-94(b)(1) of this chapter.

**Salt bath descaling, oxidizing** (Section 420.81) means the removal of scale from semi-finished steel products by the action of molten salt baths other than those containing sodium hydride.

**Salt bath descaling, reducing** (Section 420.81) means the removal of scale from semi-finished steel products by the action of molten salt baths containing sodium hydride.

**Sample system** (Section 87.1) means the system which provides for the transportation of the gaseous emission sample from the sample probe to the inlet of the instrumentation system.

**Sampling connection system** (Section 264.1031) means an assembly of equipment within a process or waste management unit used during periods of representative operation to take samples of the process or waste fluid. Equipment used to take non-routine grab samples is not considered a sampling connection system.

**Sampling connection system** (Section 65.2) means an assembly of equipment within a process unit used during periods of representative operation to take samples of the process fluid. Equipment used to take nonroutine grab samples is not considered a sampling connection system.

**Sanitary landfill** (Section 240.101) means a land disposal site employing an engineered method of disposing of solid wastes on land in a manner that minimizes environmental hazards by spreading the solid wastes in thin layers, compacting the solid wastes to the smallest practical volume, and applying and compacting cover material at the end of each operating day.

**Sanitary landfill** (Section 272.2) means a facility for the disposal of solid waste which complies with this part.

**Sanitary survey** (Section 141.2) means an onsite review of the water source, facilities, equipment, operation and maintenance of a public water system for the purpose of evaluating the adequacy of such source, facilities, equipment, operation and maintenance for producing and distributing safe drinking water.

**Sanitary survey** (Section 142.2) means an onsite review of the water source, facilities, equipment, operation and maintenance of a public water system for the purpose of evaluating the adequacy of such source, facilities, equipment, operation and maintenance for producing and distributing safe drinking water.

**Sanitary waste** (Section 144.3) means liquid or solid wastes originating solely from humans and human activities, such as wastes collected from toilets, showers, wash basins, sinks used for cleaning domestic areas, sinks used for food preparation, clothes washing operations, and sinks or washing machines where food and beverage serving dishes, glasses, and utensils are cleaned. Sources of these wastes may include single or multiple residences, hotels and motels, restaurants, bunkhouses, schools, ranger stations, crew quarters, guard stations, campgrounds, picnic grounds, day-use recreation areas, other commercial facilities, and industrial facilities provided the waste is not mixed with industrial waste.

**Sanitary waste** (Section 146.3) means liquid or solid wastes originating solely from humans and human activities, such as wastes collected from toilets, showers, wash basins, sinks used for cleaning domestic areas, sinks used for food preparation, clothes washing operations, and sinks or washing machines where food and beverage serving dishes, glasses, and utensils are cleaned. Sources of these wastes may include single or multiple residences, hotels and motels, restaurants, bunkhouses, schools, ranger stations, crew quarters, guard stations, campgrounds, picnic grounds, day-use recreation areas, other commercial facilities, and industrial facilities provided the waste is not mixed with industrial waste.

**Sanitary waste** (Section 435.11) means the human body waste discharged from toilets and urinals located within facilities subject to this Subpart.

**Sanitary waste** (Section 435.41) means the human body waste discharged from toilets and urinals located within facilities subject to this subpart.

**Sanitized** (Section 350.1) means a version of a document from which information claimed as trade secret or confidential has been omitted or withheld.

**Sanitizer products** (Section 455.10) means pesticide products that are intended to disinfect or sanitize, reducing or mitigating growth or development of microbiological organisms including bacteria, fungi or viruses on inanimate surfaces in the household, institutional, and/or commercial environment and whose labeled directions for use result in the product being discharged to Publicly Owned Treatment Works (POTWs). This definition shall also include sanitizer solutions as defined by 21 CFR 178.1010 and pool chemicals as defined in this section (455.10(q)). This definition does not include liquid chemical sterilants (including sporicidals) exempted by Sec. 455.40(f) or otherwise, industrial preservatives, and water treatment microbiocides other than pool chemicals.

**SARA** (Section 280.12) means the Superfund Amendments and Reauthorization Act of 1986.

**SARA** (Section 300.5) is the Superfund Amendments and Reauthorization Act of 1986. In addition to certain free-standing provisions of law, it includes amendments to CERCLA, the Solid Waste Disposal Act, and the Internal Revenue Code. Among the free-standing provisions of law is Title III of SARA, also known as the "Emergency Planning and Community Right-to-Know

**SAROAD site identification form** (Section 58.1) is one of the several forms in the SAROAD system. It is the form which provides a complete description of the site (and its surroundings) of an ambient air quality monitoring station.

**Satellite vehicle** (Section 243.101) means a small collection vehicle that transfers its load into a larger vehicle operating in conjunction with it.

**Saturated zone** (Section 258.2) means that part of the earth's crust in which all voids are filled with water.

**Saturated zone or zone of saturation** (Section 260.10) means that part of the earth's crust in which all voids are filled with water.

**Sauerkraut canning** (Section 407.71) shall mean the draining and subsequent filling and canning of fermented cabbage and juice.

**Sauerkraut cutting** (Section 407.71) shall mean the trimming, cutting, and subsequent preparatory handling of cabbage necessary for and including brining and fermentation, and subsequent tank soaking.

**Sausage and luncheon meat processor** (Section 432.71) shall mean an operation which cuts fresh meats, grinds, mixes, seasons, smokes or otherwise produces finished products such as sausage, bologna and luncheon meats at rates greater than 2730 kg (6000 lb) per day.

**Sawing** (Section 471.02) is cutting a workpiece with a band, blade, or circular disc having teeth.

**SC03** (Section 86.000-2) means the test cycle, described in Sec. 86.160-00 and listed in appendix I, paragraph (h), of this part, which is designed to represent driving immediately following startup.

**SC03** (Section 86.1803-01) means the test cycle, described in Sec. 86.160-00 and listed in Appendix I, paragraph (h), of this part, which is designed to represent driving immediately following startup.

**Scarfing** (Section 420.71) means those steel surface conditioning operations in which flames generated by the combustion of oxygen and fuel are used to remove surface metal imperfections from slabs, billets, or blooms.

**Scavenging** (Section 243.101) means the uncontrolled and unauthorized removal of materials at any point in the solid waste management system.

**SCF** (Section 464.02) shall means standard cubic feet.

**Schedule of compliance** (Section 122.2) means a schedule of remedial measures included in a "permit", including an enforceable sequence of interim requirements (for example, actions, operations, or milestone events) leading to compliance with the CWA and regulations.

**Schedule of compliance** (Section 124.2) means a schedule of remedial measures included in a "permit," including an enforceable sequence of interim requirements (for example, actions, operations, or milestone events) leading to compliance with the "appropriate Act and regulations."

**Schedule of compliance** (Section 144.3) means a schedule of remedial measures included in a "permit," including an enforceable sequence of interim requirements (for example, actions, operations, or milestone events) leading to compliance with the "appropriate Act and regulations."

**Schedule of compliance** (Section 270.2) means a schedule of remedial measures included in a permit, including an enforceable sequence of interim requirements (for example, actions, operations, or milestone events) leading to compliance with the Act and regulations.

**Schedule of compliance** (Section 72.2) means an enforceable sequence of actions, measures, or operations designed to achieve or maintain compliance, or correct non-compliance, with an applicable requirement of the Acid Rain Program, including any applicable Acid Rain permit requirement.

**Scheduled maintenance** (Section 57.103) means any periodic procedure, necessary to maintain the integrity or reliability of emissions control performance, which can be anticipated and scheduled in advance. In sulfuric acid plants, it includes among other items the screening or replacement of catalyst, the re-tubing of heat exchangers and the routine repair and cleaning of gas handling/cleaning equipment.

**Scheduled Maintenance** (Section 85.1402) means those maintenance events required by the equipment certifier in order to ensure that the retrofitted engine will maintain its emissions performance over the in-use compliance period.

**Scheduled maintenance** (Section 86.082-2) means any adjustment, repair, removal, disassembly, cleaning, or replacement of vehicle components or systems which is performed on a periodic basis to prevent part failure or vehicle (if the engine were installed in a vehicle) malfunction.

**Scheduled maintenance** (Section 86.1803-01) means any adjustment, repair, removal, disassembly, cleaning, or replacement of vehicle components or systems which is performed on a periodic basis to prevent part failure or vehicle (if the engine were installed in a vehicle) malfunction, or anticipated as necessary from inspection to correct an overt indication of vehicle malfunction or failure for which periodic maintenance is not appropriate.

**Scheduled maintenance** (Section 86.402-78) means any adjustment, repair, removal, disassembly, cleaning, or replacement of vehicle components or systems which is performed on a periodic basis to prevent part failure or vehicle malfunction, or anticipated as necessary to correct an overt indication of vehicle malfunction or failure for which periodic maintenance is not appropriate.

**Scheduled maintenance** (Section 90.3) means any adjustment, repair, removal, disassembly, cleaning, or replacement of components or systems required by the manufacturer to be performed on a periodic basis to prevent part failure or vehicle or engine malfunction, or those actions anticipated as necessary to correct an overt indication of malfunction or failure for which periodic maintenance is not appropriate.

**Scheduled maintenance** (Section 91.3) means any adjustment, repair, removal, disassembly, cleaning, or replacement of components or systems required by the manufacturer to be performed on a periodic basis to prevent part failure or marine vessel or engine malfunction, or those actions anticipated as necessary to correct an overt indication of malfunction or failure for which periodic maintenance is not appropriate.

**School** (Section 763.83) means any elementary or secondary school as defined in section 198 of the Elementary and Secondary Education Act of 1965 (20 U.S.C. 2854).

**School building** (Section 763.83) means: (1) Any structure suitable for use as a classroom, including a school facility such as a laboratory, library, school eating facility, or facility used for the preparation of food. (2) Any gymnasium or other facility which is specially designed for athletic or recreational activities for an academic course in physical education. (3) Any other facility used for the instruction or housing of students or for the administration of educational or research programs. (4) Any maintenance, storage, or utility facility, including any hallway, essential to the operation of any facility described in this definition of "school building" under paragraphs (1), (2), or (3). (5) Any portico or covered exterior hallway or walkway. (6) Any exterior portion of a mechanical system used to condition interior space.

**SDWA** (Section 124.2) means the Safe Drinking Water Act (Pub. L. 95-523, as amended by Pub. L. 95-1900; 42 U.S.C. 300f et seq).

**SDWA** (Section 144.3) means the Safe Drinking Water Act (Pub. L. 93-523, as amended; 42 U.S.C. 300f et seq.).

**SDWA** (Section 146.3) means the Safe Drinking Water Act (Pub. L. 95-523, as amended by Pub. L. 95-190, 42 U.S.C. 300(f) et seq.).

**SDWA** (Section 270.2) means the Safe Drinking Water Act (Pub. L. 95-523, as amended by Pub. L. 95-1900; 42 U.S.C. 3001 et seq.).

**Seafood** (Section 408.11) shall mean the raw material, including freshwater and saltwater fish and shellfish, to be processed, in the form in which it is received at the processing plant.

**Search** (Section 1601.3) refers to the time spent looking for material that is responsive to a request, including page-by-page or line-by-line identification of material within a document. The CSB shall ensure that searches are conducted in the most efficient and least expensive manner reasonably possible.

**Secondary fuel** (Section 65.2) means a fuel fired through a burner other than the primary fuel burner that provides supplementary heat in addition to the heat provided by the primary fuel.

**Secondary industry category** (Section 122.2) means any industry category which is not a "primary industry category."

**Secondary maximum contaminant levels** (Section 143.2) means SMCLs which apply to public water systems and which, in the judgement of the Administrator, are requisite to protect the public welfare. The SMCL means the maximum permissible level of a contaminant in water which is delivered to the free flowing outlet of the ultimate user of public water system. Contamimants added to the water under circumstances controlled by the user, except those resulting from corrosion of piping and plumbing caused by water quality, are excluded from this definition.

**Secondary radon measurement services (secondary)** (Section 195.2) refers to radon measurement services that do not include the reading or the ability to analyze the results of the measurement devices used. These services may include placement and retrieval of devices, reporting results, and/or consultation with consumers.

**Secondary standard** (Section 51.100) means a national secondary ambient air quality standard promulgated pursuant to section 109 of the Act.

**Secretary** (Section 122.2) means the Secretary of the Army, acting through the Chief of Engineers.

**Secretary** (Section 1700.3) means the Secretary of the United States Department of Defense or that person's authorized representative.

**Secretary** (Section 192) means the Secretary of Energy.

**Secretary** (Section 194.2) means the Secretary of Energy.

**Secretary** (Section 232.2) means the Secretary of the Army acting through the Chief of Engineers.

**Secretary** (Section 600.022-85) means the Secretary of Transportation or his authorized representative.

**Secretary** (Section 745.103) means the Secretary of Housing and Urban Development.

**Secretary** (Section 87.1) means the Secretary of Transportation and any other officer or employee of the Department of Transportation to whom the authority involved may be delegated.

**Secretary of Energy** (Section 600.022-85) means the Secretary of Energy or his authorized representative.

**Secretary of Energy** (Section 72.2) means the Secretary of the United States Department of Energy or the Secretary's duly authorized representative.

**Section 177 Program** (Section 86.1702-99) means State regulations or other laws, except ZEV Mandates, that apply to any of the following categories of motor vehicles: Passenger cars, light-duty trucks up through 6,000 pounds GVWR, and medium-duty vehicles from 6,001 to 14,000 pounds GVWR if designed to operate on gasoline, as these categories of motor vehicles are defined in the California Code of Regulations, Title 13, Division 3, Chapter 1, Article 1, Section 1900.

**Section 304(a) criteria** (Section 131.3) are developed by EPA under authority of section 304(a) of the Act based on the latest scientific information on the relationship that the effect of a constituent concentration has on particular aquatic species and/or human health. This information is issued periodically to the States as guidance for use in developing criteria.

Section 404 program or State 404 program or 404 means an "approved State program" to regulate the "discharge of dredged material" and the "discharge of fill material" under section 404 of the Clean Water Act in "State regulated waters."

**Section 5 notice** (Section 700.43) means any PMN, consolidated PMN, intermediate PMN, significant new use notice, exemption notice, exemption application, any MCAN or consolidated MCAN submitted under section 5 of the Act.

**Section mill** (Section 420.71) means those steel hot forming operations that produce a variety of finished and semi-finished steel products other than the products of those mills specified below in paragraphs (d), (e), (g), and (h) of this section.

**Security interest** (Section 280.200) means an interest in a petroleum UST or UST system or in the facility or property on which a petroleum UST or UST system is located, created or established for the purpose of securing a loan or other obligation. Security interests include but are not limited to mortgages, deeds of trusts, liens, and title pursuant to lease financing transactions. Security interests may also arise from transactions such as sale and leasebacks, conditional sales, installment sales, trust receipt transactions, certain assignments, factoring agreements, accounts receivable financing arrangements, and consignments, if the transaction creates or establishes an interest in an UST or UST system or in the facility or property on which the UST or UST system is located, for the purpose of securing a loan or other obligation.

**Sediment** (Section 434.80) means undissolved organic and inorganic material transported or deposited by water.

**Sediment toxicity** (Section 435.11) as applied to BAT effluent limitations and NSPS for drilling fluids and drill cuttings refers to the ASTM E 1367-92 method: "Standard Guide for Conducting 10-day Static Sediment Toxicity Tests with Marine and Estuarine Amphipods," 1992, with Leptocheirus plumulosus as the test organism and sediment preparation procedures specified in Appendix 3 of 40 CFR part 435, subpart A. This incorporation by reference was approved by the Director of the Federal Register in accordance with 5 U.S.C. 552(a) and 1 CFR part 51. Copies may be obtained from the American Society for Testing and Materials, 100 Barr Harbor Drive, West Conshohocken, PA, 19428. Copies may be inspected at the Office of the Federal Register, 800 North Capitol Street, NW., Suite 700, Washington, DC. A copy may also be inspected at EPA's Water Docket, 1200 Pennsylvania Ave., NW., Washington, DC 20460.

**Sediment yield** (Section 434.80) means the sum of the soil losses from a surface minus deposition in macro-topographic depressions, at the toe of the hillslope, along field boundaries, or in terraces and channels sculpted into the hillslope.

**Sedimentation** (Section 141.2) means a process for removal of solids before filtration by gravity or separation.

**Segments for the papergrade sulfite subcategory** (Section 430.51) are defined as follows: (1) The calcium-, magnesium-, or sodium-based sulfite pulp segment consists of papergrade sulfite mills where pulp and paper are produced using an acidic cooking liquor of calcium, magnesium, or sodium sulfite, unless those mills are specialty grade sulfite mills; (2) The ammonium-based sulfite pulp segment consists of papergrade sulfite mills where pulp and paper are produced using an acidic cooking liquor of ammonium sulfite, unless those mills are specialty grade sulfite mills; (3) The specialty grade sulfite pulp segment consists of those papergrade sulfite mills where a significant portion of production is characterized by pulp with a high percentage of alpha cellulose and high brightness sufficent to produce end products such as plastic molding compounds, saturating and laminating products, and photographic papers. The specialty grade segment also includes those mills where a major portion of production is 91 ISO brightness and above.

**Seismic impact zone** (Section 503.21) is an area that has a 10 percent or greater probability that the horizontal ground level acceleration of the rock in the area exceeds 0.10 gravity once in 250 years.

**Selective catalytic reduction** (Section 76.2) means a noncombustion control technology that destroys NOX by injecting a reducing agent (e.g., ammonia) into the flue gas that, in the presence of a catalyst (e.g., vanadium, titanium, or zeolite), converts NOX into molecular nitrogen and water.

**Selective noncatalytic reduction** (Section 76.2) means a noncombustion control technology that destroys NOX by injecting a reducing agent (e.g., ammonia, urea, or cyanuric acid) into the flue gas, downstream of the combustion zone that converts NOX to molecular nitrogen, water, and when urea or cyanuric acid are used, to carbon dioxide (CO2).

**Selenium** (Section 415.361) shall mean the total selenium present in the process wastewater stream exiting the wastewater treatment system.

**Selenium** (Section 415.641) shall mean the total selenium present in the process wastewater stream exiting the wastewater treatment system.

**Seller** (Section 745.103) means any entity that transfers legal title to target housing, in whole or in part, in return for consideration, including but not limited to individuals, partnerships, corporations, trusts, government agencies, housing agencies, Indian tribes, and nonprofit organizations. The term "seller" also includes: (1) An entity that transfers shares in a cooperatively owned project, in return for consideration; and (2) An entity that transfers its interest in a leasehold, in jurisdictions or circumstances where it is legally permissible to separate the fee title from the title to the improvement, in return for consideration.

**Seller** (Section 92.302) means the entity that provides credits during a trade or transfer.

**Seller** (Section 94.302) means the entity that provides credits during a trade.

**Semiconductors** (Section 469.12) means solid state electrical devices which perform functions such as information processing and display, power handling, and interconversion between light energy and electrical energy.

**Semi-wet** (Section 420.41) means those steelmaking air cleaning systems that use water for the sole purpose of conditioning the temperature and humidity of furnace gases such that the gases may be cleaned in dry air pollution control systems.

**Senior management official** (Section 350.1) means an official with management responsibility for the person or persons completing the report, or the manager of environmental programs for the facility or establishments, or for the corporation owning or operating the facility or establishments responsible for certifying similar reports under other environmental regulatory requirements.

**Senior management official** (Section 372.3) means an official with management responsibility for the person or persons completing the report, or the manager of environmental programs for the facility or establishments, or for the corporation owning or operating the facility or establishments responsible for certifying similar reports under other environmental regulatory requirements.

**Sensor** (Section 264.1031) means a device that measures a physical quantity or the change in a physical quantity, such as temperature, pressure, flow rate, pH, or liquid level.

**Sensor** (Section 65.2) means a device that measures a physical quantity or the change in a physical quantity, such as temperature, pressure, flow rate, pH, or liquid level.

**Separate collection** (Section 246.101) means collecting recyclable materials which have been separated at the point of generation and keeping those materials separate from other collected solid waste in separate compartments of a single collection vehicle or through the use of separate collection vehicles.

**Separator tank** (Section 264.1031) means a device used for separation of two immiscible liquids.

**Septage** (Section 122.2) means the liquid and solid material pumped from a septic tank, cesspool, or similar domestic sewage treatment system, or a holding tank when the system is cleaned or maintained.

**Septage** (Section 501.2) means the liquid and solid material pumped from a septic tank, cesspool, or similar domestic sewage treatment system, or a holding tank, when the system is cleaned or maintained.

**Septic system** (Section 144.3) means a "well" that is used to emplace sanitary waste below the surface and is typically comprised of a septic tank and subsurface fluid distribution system or disposal system.

**Septic system** (Section 146.3) means a "well" that is used to emplace sanitary waste below the surface and is typically comprised of a septic tank and subsurface fluid distribution system or disposal system.

**Septic tank** (Section 280.12) is a water-tight covered receptacle designed to receive or process, through liquid separation or biological digestion, the sewage discharged from a building sewer. The effluent from such receptacle is distributed for disposal through the soil and settled solids and scum from the tank are pumped out periodically and hauled to a treatment facility.

**Sequential samples for PM samplers** (Section 53.1) means two or more PM samples for sequential (but not necessarily contiguous) time periods that are collected automatically by the same sampler without the need for intervening operator service.

**SERC** (Section 1400.2) means a State Emergency Response Commission created under the EmergencyPlanning and Community Right-to-Know Act, 42 U.S.C. 11001 et seq.

**Serial number** (Section 205.151) means the identification number assigned by the manufacturer to a specific production unit.

**Serial number** (Section 72.2) means, when referring to allowances, the unique identification number assigned to each allowance by the Administrator, pursuant to Sec. 73.34(d) of this chapter.

**Serial number** (Section 96.2) means, when referring to NOX allowances, the unique identification number assigned to each NOX allowance by the Administrator, under Sec. 96.53©.

**Serial number** (Section 97.2) means, when referring to NOX allowances, the unique identification number assigned to each NOX allowance by the Administrator, under Sec. 97.53©.

**Series hybrid electric vehicle** (Section 86.1702-99) means any vehicle which allows power to be delivered to the driven wheels solely by a battery powered electric motor, but which also incorporates the use of a combustion engine to provide power to the battery and/or electric motor.

**Serious acute effects** (Section 721.3) means human injury or human disease processes that have a short latency period for development, result from short-term exposure to a chemical substance, or are a combination of these factors and which are likely to result in death or severe or prolonged incapacitation.

**Serious chronic effects** (Section 721.3) means human injury or human disease processes that have a long latency period for development, result from long-term exposure to a chemical substance, or are a combination of these factors and which are likely to result in death or severe or prolonged incapacitation.

**Service connection** (Section 141.2) as used in the definition of public water system, does not include a connection to a system that delivers water by a constructed conveyance other than a pipe if: (1) The water is used exclusively for purposes other than residential uses (consisting of drinking, bathing, and cooking, or other similar uses); (2) The State determines that alternative water to achieve the equivalent level of public health protection provided by the applicable national primary drinking water regulation is provided for residential or similar uses for drinking and cooking; or (3) The State determines that the water provided for residential or similar uses for drinking, cooking, and bathing is centrally treated or treated at the point of entry by the provider, a pass-through entity, or the user to achieve the equivalent level of protection provided by the applicable national primary drinking water regulations.

**Service connection** (Section 142.2) as used in the definition of public water system, does not include a connection to a system that delivers water by a constructed conveyance other than a pipe if: (1) The water is used exclusively for purposes other than residential uses (consisting of drinking, bathing, and cooking, or other similar uses); (2) The Administrator or the State exercising primary enforcement responsibility for public water systems, determines that alternative water to achieve the equivalent level of public health protection provided by the applicable national primary drinking water regulation is provided for residential or similar uses for drinking and cooking; or (3) The Administrator or the State exercising primary enforcement responsibility for public water systems, determines that the water provided for residential or similar uses for drinking, cooking, and bathing is centrally treated or treated at the point of entry by the provider, a pass-through entity, or the user to achieve the equivalent level of protection provided by the applicable national primary drinking water regulations.

**Service life** (Section 94.2) means the total life of an engine. Service life begins when the engine is originally manufactured and continues until the engine is permanently removed from service.

**Service life** (Section 92.2) means the total life of a locomotive or locomotive engine. Service life begins when the locomotive or locomotive engine is originally manufactured and continues until the locomotive or locomotive engine is permanently removed from service.

**Service line sample** (Section 141.2) means a one-liter sample of water collected in accordance with Sec. 141.86(b)(3), that has been standing for at least 6 hours in a service line.

**Set pressure** (Section 65.2) means, for the purposes of subparts F and G of this part, the pressure at which a properly operating pressure relief device begins to open to relieve atypical process system operating pressure.

**Settleable solids** (Section 434.11) is that matter measured by the volumetric method specified in Sec. 434.64.

**Sewage** (Section 140.1) means human body wastes and the wastes from toilets and other receptacles intended to receive or retain body wastes;

**Sewage from vessels** (Section 122.2) means human body wastes and the wastes from toilets and other receptacles intended to receive or retain body wastes that are discharged from vessels and regulated under section 312 of CWA, except that with respect to commercial vessels on the Great Lakes this term includes graywater. For the purposes of this definition, "graywater" means galley, bath, and shower water.

**Sewage Sludge** (Section 122.2) means any solid, semi-solid, or liquid residue removed during the treatment of municipal waste water or domestic sewage. Sewage sludge includes, but is not limited to, solids removed during primary, secondary, or advanced waste water treatment, scum, septage, portable toilet pumpings, type III marine sanitation device pumpings (33 CFR part 159), and sewage sludge products. Sewage sludge does not include grit or screenings, or ash generated during the incineration of sewage sludge.

**Sewage sludge** (Section 272.2) means solid, semi-solid, or liquid residue generated during the treatment of domestic sewage in a treatment works. Sewage sludge includes, but is not limited to, domestic septage; scum or solids removed in primary, secondary, or advanced wastewater treatment processes; and a material derived from sewage sludge. Sewage sludge does not include ash generated during the firing of sewage sludge in a sewage sludge incinerator or grit and screenings generated during preliminary treatment of domestic sewage in a treatment works.

**Sewage sludge** (Section 761.13) means sewage sludge as defined in Sec. 503.9(w) of this chapter that contains <50 ppm (on a dry weight basis) PCBs.

Sewage sludge means any solid, semi-solid, or liquid residue removed during the treatment of municipal waste water or domestic sewage. Sewage sludge includes, but is not limited to, solids removed during primary, secondary or advanced waste water treatment, scum, septage, portable toilet pumpings, Type III Marine Sanitation device pumpings (33 CFR part 159), and sewage sludge products. Sewage sludge does not include grit, screenings, or ash generated during the incineration of sewage sludge.

**Sewage sludge unit** (Section 503.21) is land on which only sewage sludge is placed for final disposal. This does not include land on which sewage sludge is either stored or treated. Land does not include waters of the United States, as defined in 40 CFR 122.2.

**Sewage sludge unit boundary** (Section 503.21) is the outermost perimeter of an active sewage sludge unit.

**Sewage sludge use or disposal practice** (Section 122.2) means the collection, storage, treatment, transportation, processing, monitoring, use, or disposal of sewage sludge.

**Sewage Treatment Works** (Section 220.2) means municipal or domestic waste treatment facilities of any type which are publicly owned or regulated to the extent that feasible compliance schedules are determined by the availability of funding provided by Federal, State, or local governments.

**Sexually compatible, when referring to plants** (Section 174.3) means a viable zygote is formed only through the union of two gametes through conventional breeding.

**Shaft power** (Section 87.1) means only the measured shaft power output of a turboprop engine.

**Shall** (Section 256.06) denotes requirements for the development and implementation of the State plan.

**Shallow drilling** (Section 194.2) means those drilling events in the Delaware Basin that do not reach a depth of 2,150 feet below the surface relative to where such drilling occurred.

**Sheen** (Section 110.1) means an iridescent appearance on the surface of water; Sludge means an aggregate of oil or oil and other matter of any kind in any form other than dredged spoil having a combined specific gravity equivalent to or greater than water; United States means the States, the District of Columbia, the Commonwealth of Puerto Rico, Guam, American Samoa, the Virgin Islands, and the Trust Territory of the Pacific Islands;

**Shift** (Section 204.51) means the regular production work period for one group of workers.

**Shift** (Section 205.51) means the regular production work period for one group of workers.

**Short-term test indicative of carcinogenic potential** (Section 721.3) means either any limited bioassay that measures tumor or preneoplastic induction, or any test indicative of interaction of a chemical substance with DNA (i.e., positive response in assays for gene mutation, chromosomal aberrations, DNA damage and repair, or cellular transformation).

**Short-term test indicative of the potential to cause a developmentally toxic effect** (Section 721.3) means either any in vivo preliminary development toxicity screen conducted in a mammalian species, or any in vitro developmental toxicity screen, including any test system other than the intact pregnant mammal, that has been extensively evaluated and judged reliable for its ability to predict the potential to cause developmentally toxic effects in intact systems across a broad range of chemicals or within a class of chemicals that includes the substance of concern.

**Shot casting** (Section 471.02) is the production of shot by pouring molten metal in finely divided streams to form spherical particles.

**Should** (Section 256.06) denotes recommendations for the development and implementation of the State plan.

**Shower divider/partition** (Section 247.3) means a water-proof barrier used to provide privacy in public shower facilities;

**Shutdown** (Section 60.2) means the cessation of operation of an affected facility for any purpose.

**Shutdown** (Section 63.2) means the cessation of operation of an affected source or portion of an affected source for any purpose.

**Shutdown** (Section 65.2) is defined here for purposes including, but not limited to, periodic maintenance, replacement of equipment, or repair. Shutdown does not include the routine rinsing or washing of equipment in batch operation between batches.

**Shutdown** (Section 65.2) means the cessation of operation of a regulated source (for example, chemical manufacturing process unit or a reactor, air oxidation reactor, distillation unit) and equipment required or used to comply with this part, or the emptying and degassing of a storage vessel.

**SI unit** (Section 191.12) means a unit of measure in the International System of Units.

**Side valve engine** (Section 90.3) means an otto-cycle, four stroke engine in which the intake and exhaust valves are located to the side of the cylinder, not within the cylinder head. Such engines are sometimes referred to as "L-head" engines.

**Sievert** (Section 191.12) is the SI unit of effective dose and is equal to 100 rem or one joule per kilogram. The abbreviation is "Sv."

**Signage (including sign posts and supports)** (Section 247.3) is used for identification and directional purposes for public roads and highways, and inside and outside office buildings, museums, parks, and other public places.

**Significant adverse environmental effects** (Section 721.3) means injury to the environment by a chemical substance which reduces or adversely affects the productivity, utility, value, or function of biological, commercial, or agricultural resources, or which may adversely affect a threatened or endangered species. A substance will be considered to have the potential for significant adverse environmental effects if it has one of the following: (1) An acute aquatic EC50 of 1 mg/L or less. (2) An acute aquatic EC50 of 20 mg/L or less where the ratio of aquatic vertebrate 24-hour to 48-hour EC50 is greater than or equal to 2.0. (3) A Maximum Acceptable Toxicant Concentration (MATC) of less than or equal to 100 parts per billion (100 ppb). (4) An acute aquatic EC50 of 20 mg/L or less coupled with either a measured bioconcentration factor (BCF) equal to or greater than 1,000x or in the absence of bioconcentration data a log P value equal to or greater than 4.3.

**Significant adverse reactions** (Section 717.3) are reactions that may indicate a substantial impairment of normal activities, or long-lasting or irreversible damage to health or the environment.

**Significant biological treatment** (Section 133.101) The use of an aerobic or anaerobic biological treatment process in a treatment works to consistently achieve a 30-day average of a least 65 percent removal of BOD5.

**Significant economic loss** (Section 166.3) means that, under the emergency conditions: for a productive activity, the profitability would be substantially below the expected profitability for that activity; or, for other types of activities, where profits cannot be calculated, the value of public or private fixed assets would be substantially below the expected value for those assets. Only losses caused by the emergency conditions, specific to the impacted site, and specific to the geographic area affected by the emergency conditions are included. The contribution of obvious mismanagement to the loss will not be considered in determining loss. In evaluating the significant of an economic loss for productive activities, the Agency will consider whether the expected reduction in profitability exceeds what would be expected as a result of normal fluctuations over a number of years, and whether the loss would affect the long-term financial viability expected from the productive activity. In evaluating the significance of an economic loss for situations other than productive activities, the Agency will consider reasonable measures of expected loss.

**Significant Industrial User** (Section 403.3) (1) Except as provided in paragraph (t)(2) of this section, the term Significant Industrial User means: (i) All industrial users subject to Categorical Pretreatment Standards under 40 CFR 403.6 and 40 CFR chapter I, subchapter N; and (ii) Any other industrial user that: discharges an average of 25,000 gallons per day or more of process wastewater to the POTW (excluding sanitary, noncontact cooling and boiler blowdown wastewater); contributes a process wastestream which makes up 5 percent or more of the average dry weather hydraulic or organic capacity of the POTW treatment plant; or is designated as such by the Control Authority as defined in 40 CFR 403.12(a) on the basis that the industrial user has a reasonable potential for adversely affecting the POTW's operation or for violating any pretreatment standard or requirement (in accordance with 40 CFR 403.8(f)(6)). (2) Upon a finding that an industrial user meeting the criteria in paragraph (t)(1)(ii) of this section has no reasonable potential for adversely affecting the POTW's operation or for violating any pretreatment standard or requirement, the Control Authority (as defined in 40 CFR 403.12(a)) may at any time, on its own initiative or in response to a petition received from an industrial user or POTW, and in accordance with 40 CFR 403.8(f)(6), determine that such industrial user is not a significant industrial user.

**Significant new use notice** (Section 700.43) means any notice submitted to EPA pursuant to section 5(a)(1)(B) of the Act in accordance with part 721 of this chapter.

**Significantly damaged friable miscellaneous ACM** (Section 763.83) means damaged friable miscellaneous ACM where the damage is extensive and severe.

**Significantly damaged friable surfacing ACM** (Section 763.83) means damaged friable surfacing ACM in a functional space where the damage is extensive and severe.

**Significantly more stringent limitation** (Section 133.101) means BOD5 and SS limitations necessary to meet the percent removal requirements of at least 5 mg/l more stringent than the otherwise applicable concentration-based limitations (e.g., less than 25 mg/l in the case of the secondary treatment limits for BOD5 and SS), or the percent removal limitations in Secs. 133.102 and 133.105, if such limits would, by themselves, force significant construction or other significant capital expenditure.

**Silvicultural point source** (Section 122.2) is defined at Sec. 122.27.

**Similar composition** (Section 162.151) refers to a pesticide product which contains only the same active ingredient(s), or combination of active ingredients, and which is in the same category of toxicity, as a federally registered pesticide product.

**Similar emission control systems** (Section 86.1803-01) are engine, fuel metering and emission control system combinations which use the same fuel (e.g., gasoline, diesel, etc.), combustion cycle (e.g., two or four stroke), general type of fuel system (e.g., carburetor or fuel injection), catalyst system (e.g., none, oxidization, three-way plus oxidization, three-way only, etc.), fuel control system (e.g., feedback or non-feedback), secondary air system (e.g., equipped or not equipped) and exhaust gas recirculation (EGR) (e.g., equipped or not equipped).

**Similar product** (Section 162.151) means a pesticide product which, when compared to a federally registered product, has a similar composition and a similar use pattern.

**Similar species** (Section 159.153) means two or more species belonging to the same general taxonomic groups: The general taxonomic groups for purposes of this requirement are: mammals, birds, reptiles, amphibians, fish, aquatic invertebrates, insects, arachnids, aquatic plants (including macrophyte, floating, and submerged plants), and terrestrial (all non-aquatic) plants.

**Similar systems** (Section 86.092-2) are engine, fuel metering and emission control system combinations which use the same fuel (e.g., gasoline, diesel, etc.), combustion cycle (i.e., two or four stroke), general type of fuel system (i.e., carburetor or fuel injection), catalyst system (e.g., none, oxidization, three-way plus oxidization, three-way only, etc.), fuel control system (i.e., feedback or non-feedback), secondary air system (i.e., equipped or not equipped) and EGR (i.e., equipped or not equipped).

**Similar use pattern** (Section 162.151) refers to a use of a pesticide product which, when compared to a federally registered use of a product with a similar composition, does not require a change in precautionary labeling under Sec. 156.10(h) of this chapter, and which is substantially the same as the federally registered use. Registrations involving changed use patterns are not included in this term.

**Simple combustion turbine** (Section 72.2) means a unit that is a rotary engine driven by a gas under pressure that is created by the combustion of any fuel. This term includes combined cycle units without auxiliary firing. This term excludes combined cycle units with auxiliary firing, unless the unit did not use the auxiliary firing from 1985 through 1987 and does not use auxiliary firing at any time after November 15, 1990.

**Simple manufacturing operation** (Section 410.41) shall mean all the following unit processes: Desizing, fiber preparation and dyeing.

**Simple manufacturing operation** (Section 410.61) shall mean the following unit processes: fiber preparation and dyeing with or without carpet backing.

**Simple slaughterhouse** (Section 432.11) shall mean a slaughterhouse which accomplishes very limited by-product processing, if any, usually no more than two of such operations as rendering, paunch and viscera handling, blood processing, hide processing, or hair processing.

**Simultaneous loading** (Section 65.2) means, for a shared control device, loading of regulated materials from more than one transfer arm at the same time so that the beginning and ending times of loading cycles coincide or overlap and there is no interruption in vapor flow to the shared control device.

**Single family structure** (Section 141.2) for the purpose of subpart I of this part only, means a building constructed as a single-family residence that is currently used as either a residence or a place of business.

**Single response** (Section 310.3) means all of the concerted activities conducted in response to a single episode, incident, or threat causing or contributing to a release or threatened release of hazardous substances, or pollutants or contaminants.

**Single stand** (Section 420.101) means those recirculation or direct application cold rolling mills which include only one stand of work rolls.

**Single-seal system** (Section 265.1081) means a floating roof having one continuous seal. This seal may be vapor-mounted, liquid-mounted, or a metallic shoe seal.

**Single-seal system** (Section 65.2) means, for the purposes of subpart C of this part, a floating roof having one continuous seal. This seal may be a vapor-mounted, liquid-mounted, or metallic shoe seal.

**Single-stage topcoat** (Section 59.101) means a topcoat consisting of only one coating.

**Single-unit heavy-duty vehicle** (Section 88.302-93) means a self-propelled motor vehicle with a GVWR greater than 8,500 pounds (3,900 kilograms) built on one chassis which encompasses the engine, passenger compartment, and cargo carrying function, and not coupled to trailered equipment. All buses, whether or not they are articulated, are considered single-unit vehicles.

**Sinking agents** (Section 300.5) means those additives applied to oil discharges to sink floating pollutants below the water surface.

**Site** (Section 122.2) means the land or water area where any "facility or activity" is physically located or conducted, including adjacent land used in connection with the facility or activity.

**Site** (Section 124.2) means the land or water area where any "facility or activity" is physically located or conducted, including adjacent land used in connection with the facility or activity.

**Site** (Section 146.3) means the land or water area where any facility or activity is physically located or conducted, including adjacent land used in connection with the facility or activity.

**Site** (Section 190.02) means the area contained within the boundary of a location under the control of persons possessing or using radioactive material on which is conducted one or more operations covered by this part.

**Site** (Section 191.02) means an area contained within the boundary of a location under the effective control of persons possessing or using spent nuclear fuel or radioactive waste that are involved in any activity, operation, or process covered by this subpart.

**Site** (Section 270.2) means the land or water area where any facility or activity is physically located or conducted, including adjacent land used in connection with the facility or activity.

**Site** (Section 704.3) means a contiguous property unit. Property divided only by a public right-of-way shall be considered one site. There may be more than one plant on a single site. The site for a person who imports a substance is the site of the operating unit within the person's organization which is directly responsible for importing the substance and which controls the import transaction and may in some cases be the organization's headquarters office in the United States.

**Site** (section 710.2) means a contiguous property unit. Property divided only by a public right-of-way shall be considered one site. There may be more than one manufacturing plant on a single site. For the purposes of imported chemical substances, the site shall be the business address of the importer.

**Site** (Section 712.3) means a contiguous property unit. Property divided only by a public right-of-way shall be considered one site. There may be more than one manufacturing plant on a single site.

**Site** (Section 717.3) means a contiguous property unit. Property divided only by a public right-of-way is considered one site. There may be multiple manufacturing, processing, or distribution activities occurring within a single site.

**Site** (Section 721.3) means a contiguous property unit. Property divided only by a public right-of-way is one site. There may be more than one manufacturing plant on a single site.

**Site inspection (SI)** (Section 300.5) means an on-site investigation to determine whether there is a release or potential release and the nature of the associated threats. The purpose is to augment the data collected in the preliminary assessment and to generate, if necessary, sampling and other field data to determine if further action or investigation is appropriate.

**Site lease**, (Section 72.2) as used in part 73, subpart E of this chapter, means a legally-binding agreement signed between a new IPP or a firm associated with a new IPP and a site owner that establishes the terms and conditions under which the new IPP or the firm associated with the new IPP has the binding right to utilize a specific site for the purposes of operating or constructing the new IPP.

**Site** (Section 144.3) means the land or water area where any "facility or activity" is physically located or conducted, including adjacent land used in connection with the facility or activity.

**Site-limited** (section 710.2) means a chemical substance is manufactured and processed only within a site and is not distributed for commercial purposes as a substance or as part of a mixture or article outside the site. Imported substances are never site-limited.

**Site-limited intermediate** (Section 721.3) means an intermediate manufactured, processed, and used only within a site and not distributed in commerce other than as an impurity or for disposal. Imported intermediates cannot be "site-limited."

**Six-minute period** (Section 60.2) means any one of the 10 equal parts of a one-hour period.

**Six-minute period** (Section 63.2) means, with respect to opacity determinations, any one of the 10 equal parts of a 1-hour period.

**Size classes of discharges** (Section 300.5) refers to the following size classes of oil discharges which are provided as guidance to the OSC and serve as the criteria for the actions delineated in subpart D. They are not meant to imply associated degrees of hazard to public health or welfare of the United States, nor are they a measure of environmental injury. Any oil discharge that poses a substantial threat to public health or welfare of the United States or the environment or results in significant public concern shall be classified as a major discharge regardless of the following quantitative measures: (1) Minor discharge means a discharge to the inland waters of less than 1,000 gallons of oil or a discharge to the coastal waters of less than 10,000 gallons of oil.(2) Medium discharge means a discharge of 1,000 to 10,000 gallons of oil to the inland waters or a discharge of 10,000 to 100,000 gallons of oil to the coastal waters. (3) Major discharge means a discharge of more than 10,000 gallons of oil to the inland waters or more than 100,000 gallons of oil to the coastal waters. Size classes of releases refers to the following size classifications which are provided as guidance to the OSC for meeting pollution reporting requirements in subpart B. The final determination of the appropriate classification of a release will be made by the OSC based on consideration of the particular release (e.g., size, location, impact, etc.): (1) Minor release means a release of a quantity of hazardous substance(s), pollutant(s), or contaminant(s) that poses minimal threat to public health or welfare of the United States or the environment. (2) Medium release means a release not meeting the criteria for classification as a minor or major release. (3) Major release means a release of any quantity of hazardous substance(s), pollutant(s), or contaminant(s) that poses a substantial threat to public health or welfare of the United States or the environment or results in significant public concern.

**SLAMS** (Section 58.1) means State or Local Air Monitoring Station(s). The SLAMS make up the ambient air quality monitoring network which is required by Sec. 58.20 to be provided for in the State's implementation plan. This definition places no restrictions on the use of the physical structure or facility housing the SLAMS. Any combination of SLAMS and any other monitors (Special Purpose, NAMS, PSD) may occupy the same facility or structure without affecting the respective definitions of those monitoring station.

**Slaughterhouse** (Section 432.11) shall mean a plant that slaughters animals and has as its main product fresh meat as whole, half or quarter carcasses or smaller meat cuts.

**Slaughterhouse** (Section 432.21) shall mean a plant that slaughters animals and has as its main product fresh meat as whole, half or quarter carcasses or smaller meat cuts.

**Slow Meter Response** (Section 201.1) means that the slow response of the sound level meter shall be used. The slow dynamic response shall comply with the meter dynamic characteristics in paragraph 5.4 of the American National Standard Specification for Sound Level Meters. ANSI S1.4-1971. This publication is available from the American National Standards Institute Inc., 1430 Broadway, New York, New York 10018.

**Slow meter response** (Section 204.2) means the meter ballistics of meter dynamic characteristics as specified by American National Standard S1.4-1971 or subsequent approved revisions.

**Slow sand filtration** (Section 141.2) means a process involving passage of raw water through a bed of sand at low velocity (generally less than 0.4 m/h) resulting in substantial particulate removal by physical and biological mechanisms.

**Sludge** (Section 240.101) means the accumulated semiliquid suspension of settled solids deposited from wastewaters or other fluids in tanks or basins. It does not include solids or dissolved material in domestic sewage or other significant pollutants in water resources, such as silt, dissolved or suspended solids in industrial wastewater effluents, dissolved materials in irrigation return flows or other common water pollutants.

**Sludge** (Section 243.101) means the accumulated semiliquid suspension of settled solids deposited from wastewaters or other fluids in tanks or basins. It does not include solids or dissolved material in domestic sewage or other significant pollutants in water resources, such as silt, dissolved materials in irrigation return flows or other common water pollutants.

**Sludge** (Section 246.101) means the accumulated semiliquid suspension of settled solids deposited from wastewaters or other fluids in tanks or basins. It does not include solid or dissolved material in domestic sewage or other significant pollutants in water resources, such as silt, dissolved material in irrigation return flows or other common water pollutants.

**Sludge** (Section 258.2) means any solid, semi-solid, or liquid waste generated from a municipal, commercial, or industrial wastewater treatment plant, water supply treatment plant, or air pollution control facility exclusive of the treated effluent from a wastewater treatment plant.

**Sludge** (Section 260.10) means any solid, semi-solid, or liquid waste generated from a municipal, commercial, or industrial wastewater treatment plant, water supply treatment plant, or air pollution control facility exclusive of the treated effluent from a wastewater treatment plant.

**Sludge** (Section 272.2) means any solid, semisolid, or liquid waste generated from a municipal, commercial, or industrial wastewater treatment plant, water supply treatment plant, or air pollution control facility or any other such waste having similar characteristics and effect.

**Sludge dryer** (Section 260.10) means any enclosed thermal treatment device that is used to dehydrate sludge and that has a maximum total thermal input, excluding the heating value of the sludge itself, of 2,500 Btu/lb of sludge treated on a wet-weight basis.

**Sludge-only facility** (Section 122.2) means any "treatment works treating domestic sewage" whose methods of sewage sludge use or disposal are subject to regulations promulgated pursuant to section 405(d) of the CWA and is required to obtain a permit under Sec. 122.1(b)(2).

**Sm \3** (Section 464.02) shall mean standard cubic meters.

**Small business** (Section 791.3) refers to a manufacturer or importer whose annual sales, when combined with those of its parent company (if any) are less than $30 million.

**Small business concern** (Section 700.43) means any person whose total annual sales in the person's fiscal year preceding the date of the submission of the applicable section 5 notice, when combined with those of the parent company (if any), are less than $40 million.

**Small diesel refinery** (Section 72.2) means a domestic motor diesel fuel refinery or portion of a refinery that, as an annual average of calendar years 1988 through 1990 and as reported to the Department of Energy on Form 810, had bona fide crude oil throughput less than 18,250,000 barrels per year, and the refinery or portion of a refinery is owned or controlled by a refiner with a total combined bona fide crude oil throughput of less than 50,187,500 barrels per year.

**Small manufacturer or importer** (Section 704.3) means a manufacturer or importer that meets either of the following standards: (1) First standard. A manufacturer or importer of a substance is small if its total annual sales, when combined with those of its parent company (if any), are less than $40 million. However, if the annual production or importation volume of a particular substance at any individual site owned or controlled by the manufacturer or importer is greater than 45,400 kilograms (100,000 pounds), the manufacturer or importer shall not qualify as small for purposes of reporting on the production or importation of that substance at that site, unless the manufacturer or importer qualifies as small under standard (2) of this definition. (2) Second standard. A manufacturer or importer of a substance is small if its total annual sales, when combined with those of its parent company (if any), are less than $4 million, regardless of the quantity of substances produced or imported by that manufacturer or importer. (3) Inflation index. EPA shall make use of the Producer Price Index for Chemicals and Allied Products, as compiled by the U.S. Bureau of Labor Statistics, for purposes of determining the need to adjust the total annual sales values and for determining new sales values when adjustments are made. EPA may adjust the total annual sales values whenever the Agency deems it necessary to do so, provided that the Producer Price Index for Chemicals and Allied Products has changed more than 20 percent since either the most recent previous change in sales values or the date of promulgation of this rule, whichever is later. EPA shall provide Federal Register notification when changing the total annual sales values.

**Small maufacturer or importer** (section 710.2) means a manufacturer or importer whose total annual sales are less than $5,000,000, based upon the manufacturer's or importer's latest complete fiscal year as of January 1, 1978, except that no manufacturer or importer is a "small manufacturer or importer" with respect to any chemical substance which such person manufactured at one site or imported in quantities greater than 100,000 pounds during calendar year 1977. In the case of a company which is owned or controlled by another company, total annual sales shall be based on the total annual sales of the owned or controlled company, the parent company, and all companies owned or controlled by the parent company taken together. Note: The purpose of the exception to the definition is to ensure that manufacturing and importers report production volumes for all chemical substances which they manufactured at one site or imported in quantities equal to or greater than 100,000 pounds during calendar year 1977.

**Small processor** (Section 432.51) shall mean an operation that produces up to 2730 kg (6000 lb) per day of any type or combination of finished products.

**Small quantities** (Section 704.3) solely for research and development (or "small quantities solely for purposes of scientific experimentation or analysis or chemical research on, or analysis of, such substance or another substance, including such research or analysis for the development of a product") means quantities of a chemical substance manufactured, imported, or processed or proposed to be manufactured, imported, or processed solely for research and development that are not greater than reasonably necessary for such purposes.

**Small quantities** (section 710.2) for purposes of scientific experimentation or analysis or chemical research on, or analysis of, such substance or another substance, including any such research or analysis for the development of a product (hereinafter sometimes shortened to small quantities for research and development) means quantities of a chemical substance manufactured, imported, or processed or proposed to be manufactured, imported, or processed that (1) are no greater than reasonably necessary for such purposes and (2) after the publication of the revised inventory, are used by, or directly under the supervision of, a technically qualified individual(s).

**Small quantities** (Section 720.3) solely for research and development (or "small quantities solely for purposes of scientific experimentation or analysis or chemical research on, or analysis of, such substance or another substance, including such research or analysis for the development of a product") means quantities of a chemical substance manufactured, imported, or processed or proposed to be manufactured, imported, or processed solely for research and development that are not greater than reasonably necessary for such purposes.

**Small quantities for research and development** (Section 761.13) means any quantity of PCBs (1) that is originally packaged in one or more hermetically sealed containers of a volume of no more than five (5.0) milliliters, and (2) that is used only for purposes of scientific experimentation or analysis, or chemical research on, or analysis of, PCBs, but not for research or analysis for the development of a PCB product.

**Small quantities solely for research and development** (Section 725.3) (or "small quantities solely for purposes of scientific experimentation or analysis or research on, or analysis of, such substance or another substance, including such research or analysis for development of a product") means quantities of a microorganism manufactured, imported, or processed or proposed to be manufactured, imported, or processed solely for research and development that meet the requirements of Sec. 725.234.

**Small Quantity Generator** (Section 260.10) means a generator who generates less than 1000 kg of hazardous waste in a calendar month.

**Small Quantity Handler of Universal Waste** (Section 273.9) means a universal waste handler (as defined in this section) who does not accumulate 5,000 kilograms or more total of universal waste (batteries, pesticides, thermostats, or lamps, calculated collectively) at any time.

**Small railroad** (Section 92.2) means a railroad that is classified by the Small Business Administration as a small business.

**Small remanufacturer** (Section 92.2) means a remanufacturer that is classified by the Small Business Administration as a small business.

**Small volume engine family** (Section 90.3) means any handheld engine family or any nonhandheld engine family whose eligible production in a given model year are projected at the time of certification to be no more than 5,000 engines.

**Small volume engine manufacturer** (Section 90.3) means, for nonhandheld engines, any engine manufacturer whose total eligible production of nonhandheld engines are projected at the time of certification of a given model year to be no more than 10,000 nonhandheld engines. For handheld engines, the term small volume engine manufacturer means any engine manufacturer whose total eligible production of handheld engines are projected at the time of certification of a given model year to be no more than 25,000 handheld engines.

**Small volume equipment manufacturer** (Section 90.3) means, for nonhandheld equipment, any equipment manufacturer whose production of nonhandheld equipment subject to regulation under this part or powered by engines regulated under this part, does not exceed 5,000 pieces for a given model year or annual production period excluding that equipment intended for introduction into commerce for use in a state that has established its own emission requirements applicable to such equipment or engines in such equipment, pursuant to a waiver granted by EPA under section 209(e) of the Clean Air Act. For handheld equipment, the term small volume equipment manufacturer has the same meaning except that it is limited to 25,000 pieces of handheld equipment rather than 5,000 pieces of nonhandheld equipment.

**Small volume equipment model** (Section 90.3) means, for nonhandheld equipment, any unique model of equipment whose production subject to regulations under this part or powered by engines regulated under this part, does not exceed 500 pieces for a given model year or annual production period excluding that equipment intended for introduction into commerce for use in a state that has established its own emission requirements applicable to such equipment or engines in such equipment, pursuant to a waiver granted by EPA under section 209(e) of the Clean Air Act. For handheld equipment, the term small volume equipment model has the same meaning except that it is limited to 5,000 pieces of handheld equipment, rather than 500 pieces of nonhandheld equipment.

**Small water system** (Section 141.2) for the purpose of subpart I of this part only, means a water system that serves 3,300 persons or fewer.

**Small-sized plants** (Section 428.51) shall mean plants which process less than 3,720 kg/day (8,200 lbs/day) of raw materials.

**Smelter owner and operator** (Section 57.103) means the owner or operator of the smelter, without distinction.

**Smoke** (Section 86.082-2) means the matter in the exhaust emission which obscures the transmission of light.

**Smoke** (Section 87.1) means the matter in exhaust emissions which obscures the transmission of light.

**Smoke** (Section 92.2) means the matter in the engine exhaust which obscures the transmission of light.

**Smoke number (SN)** (Section 87.1) means the dimensionless term quantifying smoke emissions.

**Smooth-one-side (S1S) hardboard** (Section 429.11) means hardboard which is produced by the wet-matting, wet-pressing process.

**Smooth-two-sides (S2S) hardboard** (Section 429.11) means hardboard which is produced by the wet-matting, dry-pressing process.

**Snap beans** (Section 407.71) shall mean the processing of snap beans into the following product styles: Canned and frozen green, Italian, wax, string, bush, and other related varieties, whole, French, fancy, Extra Standard, Standard, and other cuts.

**SO2** (Section 58.1) means sulfur dioxide.

**Soaker hose** (Section 247.3) means a perforated flexible tubing that is used to deliver gentle irrigation to plants;

**Soil** (Section 192.11) means all unconsolidated materials normally found on or near the surface of the earth including, but not limited to, silts, clays, sands, gravel, and small rocks.

**Soil** (Section 268.2) means unconsolidated earth material composing the superficial geologic strata (material overlying bedrock), consisting of clay, silt, sand, or gravel size particles as classified by the U.S. Natural Resources Conservation Service, or a mixture of such materials with liquids, sludges or solids which is inseparable by simple mechanical removal processes and is made up primarily of soil by volume based on visual inspection. Any deliberate mixing of prohibited hazardous waste with soil that changes its treatment classification (i.e., from waste to contaminated soil) is not allowed under the dilution prohibition in Sec. 268.3.

**Soil** (Section 761.123) means all vegetation, soils and other ground media, including but not limited to, sand, grass, gravel, and oyster shells. It does not include concrete and asphalt.

**Soil sample** (Section 745.63) means a sample collected in a representative location using ASTM E1727, "Standard Practice for Field Collection of Soil Samples for Lead Determination by Atomic Spectrometry Techniques," or equivalent method.

**Soil washing** (Section 761.13) means the extraction of PCBs from soil using a solvent, recovering the solvent from the soil, separating the PCBs from the recovered solvent for disposal, and then disposal or reuse of the solvent.

**Sold or distributed** (Section 167.3) means the aggregate amount of a pesticidal product released for shipment by the establishment in which the pesticidal product was produced.

**Sole or principal source aquifer** (Section 146.3) means an aquifer which has been designated by the Administrator pursuant to section 1424 (a) or (e) of the SDWA.

**Sole or Principal Source Aquifer (SSA)** (Section 149.2) means an aquifer which is designated as an SSA under section 1424(e) of the SDWA.

**Solid waste** (Section 243.101) means garbage, refuse, sludges, and other discarded solid materials, including solid waste materials resulting from industrial, commercial, and agricultural operations, and from community activities, but does not include solid or dissolved materials in domestic sewage or other significant pollutants in water resources, such as silt, dissolved or suspended solids in industrial wastewater effluents, dissolved materials in irrigation return flows or other common water pollutants. Unless specifically noted otherwise, the term "solid waste" as used in these guidelines shall not include mining, agricultural, and industrial solid wastes; hazardous wastes; sludges; construction and demolition wastes; and infectious wastes.

**Solid waste** (Section 246.101) means garbage, refuse, sludge, and other discarded solid materials, including solid waste materials resulting from industrial, commercial, and agricultural operations, and from community activities, but does not include solids or dissolved materials in domestic sewage or other significant pollutants in water resources, such as silt, dissolved or suspended solids in industrial wastewater effluents, dissolved materials in irrigation return flows or other common water pollutants. Unless specifically noted otherwise, the term "solid waste" as used in these guidelines shall not include mining, agricultural, and industrial solid wastes; hazardous wastes; sludges; construction and demolition wastes; and infectious wastes.

**Solid waste** (Section 258.2) means any garbage, or refuse, sludge from a wastewater treatment plant, water supply treatment plant, or air pollution control facility and other discarded material, including solid, liquid, semi-solid, or contained gaseous material resulting from industrial, commercial, mining, and agricultural operations, and from community activities, but does not include solid or dissolved materials in domestic sewage, or solid or dissolved materials in irrigation return flows or industrial discharges that are point sources subject to permit under 33 U.S.C. 1342, or source, special nuclear, or by-product material as defined by the Atomic Energy Act of 1954, as amended (68 Stat. 923).

**Solid waste** (Section 260.10) means a solid waste as defined in Sec. 261.2 of this chapter.

**Solid waste** (Section 272.2) means any garbage, refuse, sludge from a waste treatment plant, water supply treatment plant, or air pollution control facility and other discarded material, including solid, liquid, semisolid, or contained gaseous material resulting from industrial, commercial, mining, and agricultural operations, and from community activities, but does not include solid or dissolved materials in domestic sewage, or solid or dissolved material in irrigation return flows or industrial discharges which are point sources subject to permits under section 402 of the Federal Water Pollution Control Act, as amended (86 Stat. 880), or source, special nuclear, or byproduct material as defined by the Atomic Energy Act of 1954, as amended (68 Stat. 923).

**Solid waste incinerator** (Section 72.2) means a source as defined in section 129(g)(1) of the Act.

**Solid waste storage container** (Section 243.101) means a receptacle used for the temporary storage of solid waste while awaiting collection.

**Solid wastes** (Section 240.101) means garbage, refuse, sludges, and other discarded solid materials resulting from industrial and commercial operations and from community activities. It does not include solids or dissolved material in domestic sewage or other significant pollutants in water resources, such as silt, dissolved or suspended solids in industrial wastewater effluents, dissolved materials in irrigation return flows or other common water pollutants.

**Solidification** (Section 437.2) means the addition of sorbents to convert liquid or semi-liquid waste to a solid by means of adsorption, absorption or both. The process is usually accompanied by stabilization.

**Solids control equipment** (Section 435.11) means shale shakers, centrifuges, mud cleaners, and other equipment used to separate drill cuttings and/or stock barite solids from drilling fluid recovered from the wellbore.

**Solution heat treatment** (Section 468.02) shall mean the process introducing a workpiece into a quench bath for the purpose of heat treatment following rolling, drawing or extrusion.

**Solvent extraction operation** (Section 264.1031) means an operation or method of separation in which a solid or solution is contacted with a liquid solvent (the two being mutually insoluble) to preferentially dissolve and transfer one or more components into the solvent.

**Solvent recovery** (Section 437.2) includes fuel blending operations and the recycling of spent solvents through separation of solvent mixtures in distillation columns. Solvent recovery may require an additional, pretreatment step prior to distillation.

**Sorbents (i.e., absorbents and adsorbents)** (Section 247.3) are materials used to retain liquids and gases in a diverse number of environmental, industrial, agricultural, medical, and scientific applications. Absorbents incorporate a substance while adsorbents gather substances on their surfaces.

**Sorbents** (Section 260.10) means a material that is used to soak up free liquids by either adsorption or absorption, or both. Sorbs means to either adsorb or absorb, or both.

**Sorbents** (Section 300.5) means essentially inert and insoluble materials that are used to remove oil and hazardous substances from water through adsorption, in which the oil or hazardous substance is attracted to the sorbent surface and then adheres to it; absorption, in which the oil or hazardous substance penetrates the pores of the sorbent material; or a combination of the two. Sorbents are generally manufactured in particulate form for spreading over an oil slick or as sheets, rolls, pillows, or booms. The sorbent material may consist of, but is not limited to, the following materials: (1) Organic products—(i) Peat moss or straw; (ii) Cellulose fibers or cork; (iii) Corn cobs; (iv) Chicken, duck, or other bird feathers. (2) Mineral compounds—(i) Volcanic ash or perlite; (ii) Vermiculite or zeolite. (3) Synthetic products—(i) Polypropylene; (ii) Polyethylene; (iii) Polyurethane; (iv) Polyester.

**Sound Exposure Level** (Section 201.1) means the level in decibels calculated as ten times the common logarithm of time integral of squared A-weighted sound pressure over a given time period or event divided by the square of the standard reference sound pressure of 20 micropascals and a reference duration of one second.

**Sound Level** (Section 201.1) means the level, in decibels, measured by instrumentation which satisfies the requirements of American National Standard Specification for Sound Level Meters S1.4-1971 Type 1 (or S1A) or Type 2 if adjusted as shown in Table 1. This publication is available from the American National Standards Institute, Inc., 1430 Broadway, New York, New York 10018. For the purpose of these procedures the sound level is to be measured using the A weighting of spectrum and either the FAST or SLOW dynamic averaging characteristics, as designated. It is abbreviated as LA.

**Sound level** (Section 202.10) means the quantity in decibels measured by a sound level meter satisfying the requirements of American National Standards Specification for Sound Level Meters S1.4-1971. This publication is available from the American National Standards Institute, Inc., 1430 Broadway, New York, New York 10018. Sound level is the frequency-weighted sound pressure level obtained with the standardized dynamic characteristic "fast" or "slow" and weighting A, B, or C; unless indicated otherwise, the A-weighting is understood.

**Sound level** (Section 204.2) means the weighted sound pressure level measured by the use of a metering characteristic and weighing A, B, or C as specified in American National Standard Specification for Sound Level Meters S1.4-1971 or subsequent approved revision. The weighting employed must be specified; otherwise A-weighting is understood.

**Sound Level** (Section 205.2) means 20 times the logarithm to base 10 of the ratio of pressure of a sound to the reference pressure. The reference pressure is 20 micropascals (20 micronewtons per square meter). NOTE: Unless otherwise explicitly stated, it is to be understood that the sound pressure is the effective (rms) sound pressure, per American National Standards Institute, Inc., 1430 Broadway, New York, New York 10018.

**Sound Pressure Level (in stated frequency band)** (Section 201.1) means the level, in decibels, calculated as 20 times the common logarithm of the ratio of a sound pressure to the reference sound pressure of 20 micropascals.

**Sound pressure level** (Section 204.2) means, in decibels, 20 times the logarithm to the base ten of the ratio of a sound pressure to the reference sound pressure of 20 micropascals (20 micronewtons per square meter). In the absence of any modifier, the level is understood to be that of a root-mean-square pressure.

**Sound Pressure Level** (Section 205.2) means in decibels, 20 times the logarithm to the base 10 of the ratio of a sound pressure to the reference sound pressure of 20 micropascals (20 micronewtons per square meter). In the absence of any modifier, the level is understood to be that of a root-mean-square pressure. The unit of any sound level is the decibel, having the unit symbol dB.

**Soups** (Section 407.81) shall mean the combination of various fresh and pre-processed meats, fish, dairy products, eggs, flours, starches, vegetables, spices, and other similar raw ingredients into a variety of finished mixes and styles but not including dehydrated soups.

**Source** (Section 129.2) means any building, structure, facility, or installation from which there is or may be the discharge of toxic pollutants designated as such by the Administration under section 307(a)(1) of the Act.

**Source** (Section 174.3) means the donor of the genetic material that encodes a pesticidal substance or leads to the production of a pesticidal substance.

**Source** (Section 66.3) means any source of air pollution subject to applicable legal requirements as defined in paragraph ©.

**Source** (Section 72.2) means any governmental, institutional, commercial, or industrial structure, installation, plant, building, or facility that emits or has the potential to emit any regulated air pollutant under the Act. For purposes of section 502© of the Act, a "source", including a "source" with multiple units, shall be considered a single "facility."

**Source** (Section 96.2) means any governmental, institutional, commercial, or industrial structure, installation, plant, building, or facility that emits or has the potential to emit any regulated air pollutant under the CAA. For purposes of section 502© of the CAA, a "source," including a "source" with multiple units, shall be considered a single "facility."

**Source** (Section 97.2) means any governmental, institutional, commercial, or industrial structure, installation, plant, building, or facility that emits or has the potential to emit any regulated air pollutant under the Clean Air Act. For purposes of section 502© of the Clean Air Act, a "source," including a "source" with multiple units, shall be considered a single "facility."

**Source control action** (Section 300.5) is the construction or installation and start-up of those actions necessary to prevent the continued release of hazardous substances or pollutants or contaminants (primarily from a source on top of or within the ground, or in buildings or other structures) into the environment.

**Source control maintenance measures** (Section 300.5) are those measures intended to maintain the effectiveness of source control actions once such actions are operating and functioning properly, such as the maintenance of landfill caps and leachate collection systems.

**Source separation** (Section 246.101) means the setting aside of recyclable materials at their point of generation by the generator.

**Span** (Section 72.2) means the highest pollutant or diluent concentration or flow rate that a monitor component is required to be capable of measuring under part 75 of this chapter.

**Span gas** (Section 86.082-2) means a gas of known concentration which is used routinely to set the output level of an analyzer.

**Span gas** (Section 86.1803-01) means a gas of known concentration which is used routinely to set the output level of an analyzer.

**Span gas** (Section 86.402-78) means a gas of known concentration which is used routinely to set the output level of any analyzer.

**Span gas** (Section 92.102) means a gas of known concentration which is used routinely to set the output level of an analyzer.

**Special aquatic sites** (Section 230.3) means those sites identified in subpart E. They are geographic areas, large or small, possessing special ecological characteristics of productivity, habitat, wildlife protection, or other important and easily disrupted ecological values. These areas are generally recognized as significantly influencing or positively contributing to the general overall environmental health or vitality of the entire ecosystem of a region. (See Sec. 230.10 (a)(3))

**Special features enabling off-street or off-highway operation and use** (Section 86.1803-01) means a vehicle that has: (1) Four-wheel drive; and (2) At least four of the following characteristics calculated when the automobile is at curb weight, on a level surface, with the front wheels parallel to the vehicle's longitudinal centerline, and the tires inflated to the manufacturer's recommended pressure; approach angle of not less than 28 degrees, breakover angle of not less than 14 degrees, departure angle of not less than 20 degrees, running clearance of not less than 8 inches, and front and rear axle clearances of not less than 7 inches each.

**Special fellow** (Section 46.120) An individual enrolled in an educational program relating to environmental sciences, engineering, professional schools, and allied sciences.

**Special irrigation district** (Section 141.2) means an irrigation district in existence prior to May 18, 1994 that provides primarily agricultural service through a piped water system with only incidental residential or similar use where the system or the residential or similar users of the system comply with the exclusion provisions in section 1401(4)(B)(i)(II) or (III).

**Special irrigation district** (Section 142.2) means an irrigation district in existence prior to May 18, 1994 that provides primarily agricultural service through a piped water system with only incidental residential or similar use where the system or the residential or similar users of the system comply with the exclusion provisions in section 1401(4)(B)(i)(II) or (III).

**Special local need** (Section 162.151) means an existing or imminent pest problem within a State for which the State lead agency, based upon satisfactory supporting information, has determined that an appropriate federally registered pesticide product is not sufficiently available.

**Special Purpose Equipment** (Section 201.1) means maintenance-of-way equipment which may be located on or operated from rail cars including: Ballast cribbing machines, ballast regulators, conditioners and scarifiers, bolt machines, brush cutters, compactors, concrete mixers, cranes and derricks, earth boring machines, electric welding machines, grinders, grouters, pile drivers, rail heaters, rail layers, sandblasters, snow plows, spike drivers, sprayers and other types of such maintenance-of-way equipment.

**Special Purpose Monitor (SPM)** (Section 58.1) is a generic term used for all monitors other than SLAMS, NAMS, PAMS, and PSD monitors included in an agency's monitoring network for monitors used in a special study whose data are officially reported to EPA.

**Special Review** (Section 166.3) refers to any interim administrative review of the risks and benefits of the use of a pesticide conducted pursuant to the provisions of EPA's Rebuttable Presumption Against Registration rules, 40 CFR 162.11(a), or any subsequent version of those rules.

**Special Track Work** (Section 201.1) means track other than normal tie and ballast bolted or welded rail or containing devices such as retarders or switching mechanisms.

**Special wastes** (Section 240.101) means nonhazardous solid wastes requiring handling other than that normally used for municipal solid waste.

**Specialty coatings** (Section 59.101) means adhesion promoters, low-gloss coatings, bright metal trim repair coatings, jambing (cut-in) clearcoats, elastomeric coatings, impact resistant coatings, underbody coatings, uniform finish blenders, and weld-through primers.

**Specialty hot forming operation (or "specialty")** (Section 420.71) applies to all hot forming operations other than "carbon hot forming operations."

**Specialty paper** (Section 763.163) means an asbestos-containing product that is made of paper intended for use as filters for beverages or other fluids or as paper fill for cooling towers. Cooling tower fill consists of asbestos paper that is used as a cooling agent for liquids from industrial processes and air conditioning systems.

**Specialty steel** (Section 420.71) means those steel products containing alloying elements which are added to enhance the properties of the steel product when individual alloying elements (e.g., aluminum, chromium, cobalt, columbium, molybdenum, nickel, titanium, tungsten, vanadium, zirconium) exceed 3% or the total of all alloying elements exceed 5%.

**Species mean acute value (SMAV)** (Section 132.2) is the geometric mean of the results of all acceptable flow-through acute toxicity tests (for which the concentrations of the test material were measured) with the most sensitive tested life stage of the species. For a species for which no such result is available for the most sensitive tested life stage, the SMAV is the geometric mean of the results of all acceptable acute toxicity tests with the most sensitive tested life stage.

**Species mean chronic value (SMCV)** (Section 132.2) is the geometric mean of the results of all acceptable life-cycle and partial life-cycle toxicity tests with the species; for a species of fish for which no such result is available, the SMCV is the geometric mean of all acceptable early life-stage tests.

**Specific chemical identity** (Section 350.1) means the chemical name, Chemical Abstracts Service (CAS) Registry Number, or any other information that reveals the precise chemical designation of the substance. Where the trade name is reported in lieu of the specific chemical identity, the trade name will be treated as the specific chemical identity for purposes of this part.

**Specific emissions** (Section 89.2) means emissions expressed on the basis of observed brake power, using units of g/kW-hr. Observed brake power measurement includes accessories on the engine if these accessories are required for running an emission test (except for the cooling fan). When it is not possible to test the engine in the gross conditions, for example, if the engine and transmission form a single integral unit, the engine may be tested in the net condition. Power corrections from net to gross conditions will be allowed with prior approval of the Administrator.

**Specific emissions** (Section 94.2) means emissions expressed on the basis of observed brake power, using units of g/kW-hr. Observed brake power measurement includes accessories on the engine if these accessories are required for running an emission test (except for the cooling fan). When it is not possible to test the engine in the gross conditions, for example if the engine and transmission form a single integral unit, the engine may be tested in the net condition. Power corrections from net to gross conditions will be allowed with prior approval of the Administrator.

**Specific gravity monitoring device** (Section 65.2) means a unit of equipment used to monitor specific gravity and having a minimum accuracy of [plusmn]0.02 specific gravity units.

**Specific oxygen uptake rate (SOUR)** (Section 503.31) is the mass of oxygen consumed per unit time per unit mass of total solids (dry weight basis) in the sewage sludge.

**Specification** (Section 246.101) means a clear and accurate description of the technical requirements for materials, products or services, identifying the minimum requirements for quality and construction of materials and equipment necessary for an acceptable product. In general, specifications are in the form of written descriptions, drawings, prints, commercial designations, industry standards, and other descriptive references.

**Specification** (Section 247.3) means a description of the technical requirements for a material, product, or service that includes the criteria for determining whether these requirements are met. In general, specifications are in the form of written commercial designations, industry standards, and other descriptive references;

**Specified adjustable range** (Section 92.2) means the range of allowable settings for an adjustable component specified by a certificate of conformity.

**Specified by a certificate of conformity or specified in a certificate of conformity** (Section 92.2) means stated or otherwise specified in a certificate of conformity or an approved application for certification.

**Specified by a certificate of conformity or specified in a certificate of conformity** (Section 94.2) means stated or otherwise specified in a certificate of conformity or an approved application for certification.

**Specified ports and harbors** (Section 300.5) means those ports and harbor areas on inland rivers, and land areas immediately adjacent to those waters, where the USCG acts as predesignated on-scene coordinator. Precise locations are determined by EPA/USCG regional agreements and identified in federal Regional Contingency Plans and Area Contingency Plans.

**Specimen** (Section 792.3) means any material derived from a test system for examination or analysis.

**Specimens** (Section 160.3) means any material derived from a test system for examination or analysis.

**Spectral uncertainty** (Section 211.203) Possible variation in exposure to the noise spectra in the workplace. (To avoid the underprotection that would result from these variations relative to the assumed "Pink Noise" used to determine the NRR, an extra three decibel reduction is included when computing the NRR.)

**Spent acid solution (or spent pickle liquor)** (Section 420.91) means those solutions of steel pickling acids which have been used in the pickling process and are discharged or removed therefrom.

**Spent lubricant** (Section 468.02) shall mean water or an oil-water mixture which is used in forming operations to reduce friction, heat and wear and ultimately discharged.

**Spent nuclear fuel** (Section 191.02) means fuel that has been withdrawn from a nuclear reactor following irradiation, the constituent elements of which have not been separated by reprocessing.

**Spent nuclear fuel** (Section 197.2) means fuel that has been withdrawn from a nuclear reactor following irradiation, the constituent elements of which have not been separated by reprocessing.

**Spill** (Section 761.123) means both intentional and unintentional spills, leaks, and other uncontrolled discharges where the release results in any quantity of PCBs running off or about to run off the external surface of the equipment or other PCB source, as well as the contamination resulting from those releases. This policy applies to spills of 50 ppm or greater PCBs. The concentration of PCBs spilled is determined by the PCB concentration in the material spilled as opposed to the concentration of PCBs in the material onto which the PCBs were spilled. Where a spill of untested mineral oil occurs, the oil is presumed to contain greater than 50 ppm, but less than 500 ppm PCBs and is subject to the relevant requirements of this policy.

**Spill area** (Section 761.123) means the area of soil on which visible traces of the spill can be observed plus a buffer zone of 1 foot beyond the visible traces. Any surface or object (e.g., concrete sidewalk or automobile) within the visible traces area or on which visible traces of the spilled material are observed is included in the spill area. This area represents the minimum area assumed to be contaminated by PCBs in the absence of precleanup sampling data and is thus the minimum area which must be cleaned.

**Spill boundaries** (Section 761.123) means the actual area of contamination as determined by postcleanup verification sampling or by precleanup sampling to determine actual spill boundaries. EPA can require additional cleanup when necessary to decontaminate all areas within the spill boundaries to the levels required in this policy (e.g., additional cleanup will be required if postcleanup sampling indicates that the area decontaminated by the responsible party, such as the spill area as defined in this section, did not encompass the actual boundaries of PCB contamination).

**Spill event** (Section 112.2) means a discharge of oil into or upon the navigable waters of the United States or adjoining shorelines in harmful quantities, as defined at 40 CFR part 110.

**Spill of national significance (SONS)** (Section 300.5) means a spill that due to its severity, size, location, actual or potential impact on the public health and welfare or the environment, or the necessary response effort, is so complex that it requires extraordinary coordination of federal, state, local, and responsible party resources to contain and clean up the discharge.

**Spinach** (Section 407.71) shall mean the processing of spinach and leafy greens into the following product styles: Canned or frozen, whole leaf, chopped, and other related cuts.

**Spitback emissions** (Section 86.096-2) means evaporative emissions resulting from the loss of liquid fuel that is emitted from a vehicle during a fueling operation.

**Spitback emissions** (Section 86.1803-01) means evaporative emissions resulting from the loss of liquid fuel that is emitted from a vehicle during a fueling operation.

**Sponsor** (Section 160.3) means: (1) A person who initiates and supports, by provision of financial or other resources, a study; (2) A person who submits a study to the EPA in support of an application for a research or marketing permit; or (3) A testing facility, if it both initiates and actually conducts the study.

**Sponsor** (Section 790.3) means the person or persons who design, direct and finance the testing of a substance or mixture.

**Sponsor** (Section 792.3) means: (1) A person who initiates and supports, by provision of financial or other resources, a study; (2) A person who submits a study to the EPA in response to a TSCA section 4(a) test rule and/or a person who submits a study under a TSCA section 4 testing consent agreement or a TSCA section 5 rule or order to the extent the agreement, rule or order references this part; or (3) A testing facility, if it both initiates and actually conducts the study.

**Spot allowance** (Section 72.2) means an allowance that may be used for purposes of compliance with a unit's Acid Rain sulfur dioxide emissions limitation requirements beginning in the year in which the allowance is offered for sale.

**Spot auction** (Section 72.2) means an auction of a spot allowance.

**Spot sale** (Section 72.2) means a sale of a spot allowance.

**SPP toxicity** (Section 435.11) as applied to BAT effluent limitations and NSPS for drilling fluids and drill cuttings refers to the bioassay test procedure presented in Appendix 2 of subpart A of this part.

**SPP toxicity** (Section 435.41) as applied to BAT effluent limitations and NSPS for drilling fluids and drill cuttings refers to the bioassay test procedure presented in Appendix 2 of subpart A of this part.

**Spray application** (Section 721.3) means any method of projecting a jet of vapor of finely divided liquid onto a surface to be coated; whether by compressed air, hydraulic pressure, electrostatic forces, or other methods of generating a spray.

**Spray-in-place foam** (Section 247.3) is rigid cellular polyurethane or polyisocyanurate foam produced by catalyzed chemical reactions that hardens at the site of the work. The term includes spray-applied and injected applications;

**Spray-in-place insulation** (Section 247.3) means insulation material that is sprayed onto a surface or into cavities and includes cellulose fiber spray-on as well as plastic rigid foam products;

**sq ft** (Section 401.11) means square foot (feet).

**sq m** (Section 401.11) means square meter(s).

**sq m ["sq ft"]** (Section 413.11) shall mean the area plated expressed in square meters [square feet].

**Squash** (Section 407.71) shall include the processing of pumpkin and squash into canned and frozen styles.

**SS** (Section 133.101) The pollutant parameter total suspended solids.

**Stabilization** (Section 437.2) means a waste process that decreases the mobility of waste constituents by means of a chemical reaction. For the purpose of this rule, chemical precipitation is not a technique for stabilization.

**Stack** (Section 129.2) means any chimney, flue, conduit, or duct arranged to conduct emissions to the ambient air.

**Stack** (Section 51.100) means any point in a source designed to emit solids, liquids, or gases into the air, including a pipe or duct but not including flares.

**Stack** (Section 72.2) means a structure that includes one or more flues and the housing for the flues.

**Stack in existence** (Section 51.100) means that the owner or operator had (1) begun, or caused to begin, a continuous program of physical on-site construction of the stack or (2) entered into binding agreements or contractual obligations, which could not be cancelled or modified without substantial loss to the owner or operator, to undertake a program of construction of the stack to be completed within a reasonable time.

**Stack operating hour** (Section 72.2) means a clock hour during which flue gases flow through a particular stack or duct (either for the entire hour or for part of the hour) while the associated unit(s) are combusting fuel.

**Stack operating hour** (Section 72.2) means any hour (or fraction of an hour) during which flue gases flow through a common stack or bypass stack.

**Stack operating time** (Section 72.2) means the portion of a clock hour during which flue gases flow through a particular stack or duct while the associated unit(s) are combusting fuel. The stack operating time, in hours, is expressed as a decimal fraction, with valid values ranging from 0.00 to 1.00.

**Staging pile** (Section 260.10) means an accumulation of solid, non-flowing remediation waste (as defined in this section) that is not a containment building and that is used only during remedial operations for temporary storage at a facility. Staging piles must be designated by the Director according to the requirements of 40 CFR 264.554.

**Stall barn** (Section 412.11) shall mean specialized facilities wherein producing cows and replacement cows are milked and fed in a fixed location.

**Stand-alone PFPR facility** (Section 455.10) means a PFPR facility where either: No pesticide manufacturing occurs; or where pesticide manufacturing process wastewaters are not commingled with PFPR process wastewaters. Such facilities may formulate, package or repackage or manufacture other non-pesticide chemical products and be considered a "stand-alone" PFPR facility.

**Standard** (Section 171.2) means the measure of knowledge and ability which must be demonstrated as a requirement for certification.

**Standard** (Section 60.2) means a standard of performance proposed or promulgated under this part.

**Standard** (Section 61.02) means a national emission standard including a design, equipment, work practice or operational standard for a hazardous air pollutant proposed or promulgated under this part.

**Standard** (Section 93.101) means a national ambient air quality standard.

**Standard bushel** (Section 406.11) shall mean a bushel of shelled corn weighing 56 pounds.

**Standard bushel** (Section 406.41) shall mean a bushel of wheat weighing 60 pounds.

**Standard conditions** (Section 60.2) means a temperature of 293 K (68F) and a pressure of 101.3 kilopascals (29.92 in Hg).

**Standard conditions** (Section 63.2) means a temperature of 293 K (68 [deg]F) and a pressure of 101.3 kilopascals (29.92 in. Hg).

**Standard conditions** (Section 72.2) means 68 deg.F at 1 atm (29.92 in. of mercury).

**Standard conditions and standard temperature and pressure** (Section 92.102) mean 68 [deg]F (20 [deg]C) and 29.92 in Hg. (101.3 kPa).

**Standard day conditions** (Section 87.1) means standard ambient conditions as described in the United States Standard Atmosphere, 1976, (i.e., Temperature =15 [deg]C, specific humidity =0.00 kg/ H2 O/kg dry air, and pressure =101325 Pa.)

**Standard equipment** (Section 86.082-2) means those features or equipment which are marketed on a vehicle over which the purchaser can exercise no choice.

**Standard equipment** (Section 86.1803-01) means those features or equipment which are marketed on a vehicle over which the purchaser can exercise no choice.

**Standard of performance** (Section 401.11) means any restriction established by the Administrator pursuant to section 306 of the Act on quantities, rates, and concentrations of chemical, physical, biological, and other constituents which are or may be discharged from new sources into navigable waters, the waters of the contiguous zone or the ocean.

**Standard reference material or SRM** (Section 72.2) means a calibration gas mixture issued and certified by NIST as having specific known chemical or physical property values.

**Standard reference material-equivalent compressed gas primary reference material (SRM-equivalent PRM)** (Section 72.2) means those gas mixtures listed in a declaration of equivalence in accordance with section 2.1.2 of the "EPA Traceability Protocol for Assay and Certification of Gaseous Calibration Standards," September 1997, EPA-600/R-97/121.

**Standard sample** (Section 141.2) means the aliquot of finished drinking water that is examined for the presence of coliform bacteria.

**Standard wipe sample** (Section 761.13)means a sample collected for chemical extraction and analysis using the standard wipe test as defined in Sec. 761.123. Except as designated elsewhere in part 761, the minimum surface area to be sampled shall be 100 cm2.

**Standard wipe test** (Section 761.123) means, for spills of high-concentration PCBs on solid surfaces, a cleanup to numerical surface standards and sampling by a standard wipe test to verify that the numerical standards have been met. This definition constitutes the minimum requirements for an appropriate wipe testing protocol. A standard-size template (10 centimeters (cm) x 10 cm) will be used to delineate the area of cleanup; the wiping medium will be a gauze pad or glass wool of known size which has been saturated with hexane. It is important that the wipe be performed very quickly after the hexane is exposed to air. EPA strongly recommends that the gauze (or glass wool) be prepared with hexane in the laboratory and that the wiping medium be stored in sealed glass vials until it is used for the wipe test. Further, EPA requires the collection and testing of field blanks and replicates.

**Standards for sewage sludge use or disposal** (Section 122.2)means the regulations promulgated pursuant to section 405(d) of the CWA which govern minimum requirements for sludge quality, management practices, and monitoring and reporting applicable to sewage sludge or the use or disposal of sewage sludge by any person.

**Standards for sewage sludge use or disposal** (Section 501.2) means the regulations promulgated at 40 CFR part 503 pursuant to section 405(d) of the CWA which govern minimum requirements for sludge quality, management practices, and monitoring and reporting applicable to the generation or treatment of sewage sludge from a treatment works treating domestic sewage or use or disposal of that sewage sludge by any person.

**Starting material** (Section 158.153) means a substance used to synthesize or purify a technical grade of active ingredient (or the practical equivalent of the technical grade ingredient if the technical grade cannot be isolated) by chemical reaction.

**Startup** (Section 264.1031) means the setting in operation of a hazardous waste management unit or control device for any purpose.

**Startup** (Section 60.2) means the setting in operation of an affected facility for any purpose.

**Startup** (Section 61.02) means the setting in operation of a stationary source for any purpose.

**Startup** (Section 63.2) means the setting in operation of an affected source or portion of an affected source for any purpose.

**Startup** (Section 65.2) means the setting into operation of a regulated source (for example, chemical manufacturing process unit or a reactor, air oxidation reactor, distillation unit, a storage vessel after emptying and degassing) and/or equipment required or used to comply with this part. Startup includes initial startup, operation solely for testing equipment, the recharging of equipment in batch operation, and transitional conditions due to changes in product for flexible operation units.

**State** (Section 231.2) means any state agency administering a 404 program which has been approved under section 404(h).

**State** (Section 300.5) means the several states of the United States, the District of

Columbia, the Commonwealth of Puerto Rico, Guam, American Samoa, the U.S. Virgin Islands, the Commonwealth of the Northern Marianas, and any other territory or possession over which the United States has jurisdiction. For purposes of the NCP, the term includes Indian tribes as defined in the NCP except where specifically noted. Section 126 of CERCLA provides that the governing body of an Indian tribe shall be afforded substantially the same treatment as a state with respect to certain provisions of CERCLA. Section 300.515(b) of the NCP describes the requirements pertaining to Indian tribes that wish to be treated as states under CERCLA.

**State** (Section 49.2) means a State, the District of Columbia, the Commonwealth of Puerto Rico, the Virgin Islands, Guam, and American Samoa and includes the Commonwealth of the Northern Mariana Islands.

**State** (Section 55.2) means the State air pollution control agency that would be the permitting authority, a local air pollution permitting agency, or certain Indian tribes which can be the permitting authority for areas within their jurisdiction. State may also be used in the geographic sense to refer to a State, the NOA, or the COA.

**State** (Section 60.2) means all non-Federal authorities, including local agencies, interstate associations, and State-wide programs, that have delegated authority to implement: (1) The provisions of this part; and/or (2) the permit program established under part 70 of this chapter. The term State shall have its conventional meaning where clear from the context.

**State** (Section 61.02) means all non-Federal authorities, including local agencies, interstate associations, and State-wide programs, that have delegated authority to implement: (1) The provisions of this part; and/or (2) The permit program established under part 70 of this chapter. The term State shall have its conventional meaning where clear from the context.

**State** (Section 63.2) means all non-Federal authorities, including local agencies, interstate associations, and State-wide programs, that have delegated authority to implement: (1) The provisions of this part and/or (2) the permit program established under part 70 of this chapter. The term State shall have its conventional meaning where clear from the context.

**State** (Section 65.2) means all non-Federal authorities, including local agencies, interstate associations, and statewide programs, that have delegated authority to implement the provisions of this part; the referencing subparts; and/or the permit program established under part 70 of this chapter. The term State shall have its conventional meaning where clear from the context.

**State** (Section 70.2) means any non-Federal permitting authority, including any local agency, interstate association, or statewide program. The term "State" also includes the District of Columbia, the Commonwealth of Puerto Rico, the Virgin Islands, Guam, American Samoa, and the Commonwealth of the Northern Mariana Islands. Where such meaning is clear from the context, "State" shall have its conventional meaning. For purposes of the acid rain program, the term "State" shall be limited to authorities within the 48 contiguous States and the District of Columbia as provided in section 402(14) of the Act.

**State** (Section 71.2) means any non-Federal permitting authority, including any local agency, interstate association, or statewide program. The term "State" also includes the District of Columbia, the Commonwealth of Puerto Rico, the Virgin Islands, Guam, American Samoa, and the Commonwealth of the Northern Marianas Islands. Where such meaning is clear from the context, "State" shall have its conventional meaning. For purposes of the acid rain program, the term "State" shall be limited to authorities within the 48 contiguous States and the District of Columbia as provided in section 402(14) of the Act.

**State** (Section 96.2) means one of the 48 contiguous States and the District of Columbia specified in Sec. 51.121 of this chapter, or any non-federal authority in or including such States or the District of Columbia (including local agencies, and Statewide agencies) or any eligible Indian tribe in an area of such State or the District of Columbia, that adopts a NOX Budget Trading Program pursuant to Sec. 51.121 of this chapter. To the extent a State incorporates by reference the provisions of this part, the term "State" shall mean the incorporating State. The term "State" shall have its conventional meaning where such meaning is clear from the context.

**State** (Section 97.2) means one of the 48 contiguous States or a portion thereof or the District of Columbia that is specified in Sec. 52.34 of this chapter and in which are located units for which the Administrator makes an effective finding under Sec. 52.34 of this chapter.

**State 404 program or State program** (Section 233.2) means a State program which has been approved by EPA under section 404 of the Act to regulate the discharge of dredged or fill material into certain waters as defined in Sec. 232.2(p).

**State agency** (Section 51.100) means the air pollution control agency primarily responsible for development and implementation of a plan under the Act.

**State Director (Director)** (Section 233.2) means the chief administrative officer of any State or interstate agency operating an approved program, or the delegated representative of the Director. If responsibility is divided among two or more State or interstate agencies, Director means the chief administrative officer of the State or interstate agency authorized to perform the particular procedure or function to which reference is made.

**State Director** (Section 122.2) means the chief administrative officer of any State or interstate agency operating an "approved program," or the delegated representative of the State Director. If responsibility is divided among two or more State or interstate agencies, "State Director" means the chief administrative officer of the State or interstate agency authorized to perform the particular procedure or function to which reference is made.

**State Director** (Section 124.2) means the chief administrative officer of any State, interstate, or Tribal agency operating an approved program, or the delegated representative of the State director. If the responsibility is divided among two or more States, interstate, or Tribal agencies, "State Director" means the chief administrative officer of the State, interstate, or Tribal agency authorized to perform the particular procedure or function to which reference is made.

**State Director** (Section 129.2) means the chief administrative officer of a State or interstate water pollution control agency operating an approved HPDES permit program. In the event responsibility for water pollution control and enforcement is divided among two or more State or interstate agencies, the term State Director means the administrative officer authorized to perform the particular procedure to which reference is made.

**State Director** (Section 133.101) means the chief administrative officer of any State or interstate agency operating an "approved program," or the delegated representative of the State Director.

**State Director** (Section 144.3) means the chief administrative officer of any State, interstate, or Tribal agency operating an "approved program," or the delegated representative of the State director. If the responsibility is divided among two or more States, interstate, or Tribal agencies, "State Director" means the chief administrative officer of the State, interstate, or Tribal agency authorized to perform the particular procedure or function to which reference is made.

**State Director** (Section 146.3) means the chief administrative officer of any State, interstate, or Tribal agency operating an "approved program," or the delegated representative of the State Director. If the responsibility is divided among two or more State, interstate, or Tribal agencies, "State Director" means the chief administrative officer of the State, interstate, or Tribal agency authorized to perform the particular procedure or function to which reference is made.

**State Director** (Section 239.2) means the chief administrative officer of the lead state agency responsible for implementing the state permit program for Subtitle D regulated facilities.

**State Director** (Section 258.2) means the chief administrative officer of the lead state agency responsible for implementing the state permit program for 40 CFR part 257, subpart B and 40 CFR part 258 regulated facilities.

**State Director** (Section 270.2) means the chief administrative officer of any State agency operating an approved program, or the delegated representative of the State Director. If responsibility is divided among two or more State agencies, State Director means the chief administrative officer of the State agency authorized to perform the particular procedure or function to which reference is made.

State means the agency of the State or Tribal government which has jurisdiction over public water systems. During any period when a State or Tribal government does not have primary enforcement responsibility pursuant to section 1413 of the Act, the term "State" means the Regional Administrator, U.S. Environmental Protection Agency.

**State operating permit program** (Section 72.2) means an operating permit program that the Administrator has approved under part 70 of this chapter.

**State or local government official** (Section 1400.2) means—(1) An officer or employee of a State or local government; (2) An officer or employee of an agent or contractor of a State or local government; (3) An individual affiliated with an entity that has been given, by a state or local government, responsibility for preventing, planning for, or responding to accidental releases, such as a member of a Local Emergency Planning Committee (LEPC) or a State EmergencyResponse Commission (SERC), or a paid or volunteer member of a fire or police department; or (4) An officer or employee or an agent or contractor of an entity described in paragraph (e)(3) of this section.

**State or State lead agency** (Section 162.151) as used in this subpart means the State agency designated by the State to be responsible for registering pesticides to meet special local needs under sec. 24© of the Act.

**State primary drinking water regulation** (Section 142.2) means a drinking water regulation of a State which is comparable to a national primary drinking water regulation.

**State Program Director or Director** (Section 501.2) means the chief executive officer of the State sewage sludge management agency.

**State program or permit program** (Section 239.2) means all the authorities, activities, and procedures that comprise the state's system of prior approval and conditions for regulating the location, design, operation, ground-water monitoring, closure, post-closure care, corrective action, and financial assurance of Subtitle D regulated facilities.

**State program revision** (Section 142.2) means a change in an approved State primacy program.

**State regulated waters** (Section 232.2) means those waters of the United States in which the Corps suspends the issuance of section 404 permits upon approval of a State's section 404 permit program by the Administrator under section 404(h). The program cannot be transferred for those waters which are presently used, or are susceptible to use in their natural condition or by reasonable improvement as a means to transport interstate or foreign commerce shoreward to their ordinary high water mark, including all waters which are subject to the ebb and flow of the tide shoreward to the high tide line, including wetlands adjacent thereto. All other waters of the United States in a State with an approved program shall be under jurisdiction of the State program, and shall be identified in the program description as required by part 233.

**State sewage sludge management agency** (Section 501.2) means the agency designated by the Governor as having the lead responsibility for managing or coordinating the approved State program under this part.

**State trading program budget** (Section 96.2) means the total number of NOX tons apportioned to all NOX Budget units in a given State, in accordance with the NOX Budget Trading Program, for use in a given control period.

**State water pollution control agency** (Section 105.2) means the State agency designated by the Governing Authority having responsibility for enforcing State laws relating to the abatement of water pollution.

**State water pollution control agency** (Section 401.11) interstate agency, State, municipality, person, territorial seas, contiguous zone, biological monitoring, schedule of compliance, and industrial user shall be defined in accordance with section 502 of the Act unless the context otherwise requires.

**State/EPA Agreement** (Section 122.2) means an agreement between the Regional Administrator and the State which coordinates EPA and State activities, responsibilities and programs including those under the CWA programs.

**State/EPA agreement** (Section 144.3) means an agreement between the Regional Administrator and the State which coordinates EPA and State activities, responsibilities and programs.

**State/EPA Agreement** (Section 270.2) means an agreement between the Regional Administrator and the State which coordinates EPA and State activities, responsibilities and programs.

**Static loaded radius arc** (Section 86.1803-01) means a portion of a circle whose center is the center of a standard tire-rim combination of an automobile and whose radius is the distance from that center to the level surface on which the automobile is standing, measured with the automobile at curb weight, the wheel parallel to the vehicle's longitudinal centerline, and the tire inflated to the manufacturer's recommended pressure.

**Static sheen test** (Section 435.11) means the standard test procedure that has been developed for this industrial subcategory for the purpose of demonstrating compliance with the requirement of no discharge of free oil. The methodology for performing the static sheen test is presented in Appendix 1 of subpart A of this part.

**Static sheen test** (Section 435.41) means the standard test procedure that has been developed for this industrial subcategory for the purpose of demonstrating compliance with the requirement of no discharge of free oil. The methodology for performing the static sheen test is presented in Appendix 1 of subpart A of this part.

**Station Wagon** (Section 600.022-85) means a passenger automobile with an extended roof line to increase cargo or passenger capacity, cargo compartment open to the passenger compartment, a tailgate, and one or more rear seats readily removed or folded to facilitate cargo carrying.

**Stationary casting** (Section 471.02) is the pouring of molten metal into molds and allowing the metal to cool.

**Stationary compactor** (Section 243.101) means a powered machine which is designed to compact solid waste or recyclable materials, and which remains stationary when in operation.

**Stationary compactor** (Section 246.101) means a powered machine which is designed to compact solid waste or recyclable materials, and which remains stationary when in operation.

**Stationary gas turbine** (Section 72.2) means a turbine that is not self-propelled and that combusts natural gas, other gaseous fuel with a total sulfur content no greater than the total sulfur content of natural gas, or fuel oil in order to heat inlet combustion air and thereby turn a turbine in addition to or instead of producing steam or heating water.

**Stationary source** (Section 1400.2) has the same meaning as provided in 40 CFR part 68 subpart A, Sec. 68.3.

**Stationary source** (Section 60.2) means any building, structure, facility, or installation which emits or may emit any air pollutant.

**Stationary source** (Section 61.02) means any building, structure, facility, or installation which emits or may emit any air pollutant which has been designated as hazardous by the Administrator.

**Stationary source** (Section 63.2) means any building, structure, facility, or installation which emits or may emit any air pollutant.

**Stationary source** (Section 68.3) means any buildings, structures, equipment, installations, or substance emitting stationary activities which belong to the same industrial group, which are located on one or more contiguous properties, which are under the control of the same person (or persons under common control), and from which an accidental release may occur. The term stationary source does not apply to transportation, including storage incident to transportation, of any regulated substance or any other extremely hazardous substance under the provisions of this part. A stationary source includes transportation containers used for storage not incident to transportation and transportation containers connected to equipment at a stationary source for loading or unloading. Transportation includes, but is not limited to, transportation subject to oversight or regulation under 49 CFR parts 192, 193, or 195, or a state natural gas or hazardous liquid program for which the state has in effect a certification to DOT under 49 U.S.C. section 60105. A stationary source does not include naturally occurring hydrocarbon reservoirs. Properties shall not be considered contiguous solely because of a railroad or pipeline right-of-way.

**Stationary source** (Section 70.2) means any building, structure, facility, or installation that emits or may emit any regulated air pollutant or any pollutant listed under section 112(b) of the Act.

**Stationary source** (Section 71.2) means any building, structure, facility, or installation that emits or may emit any regulated air pollutant or any pollutant listed under section 112(b) of the Act.

**Statistical significance** (Section 228.2) shall mean the statistical significance determined by using appropriate standard techniques of multivariate analysis with results interpreted at the 95 percent confidence level and based on data relating species which are present in sufficient numbers at control areas to permit a valid statistical comparison with the areas being tested.

**Statistical Sound Level** (Section 201.1) means the level in decibels that is exceeded in a stated percentage (x) of the duration of the measurement period. It is abbreviated as Lx.

**Steam jet ejector** (Section 65.2) means a steam nozzle that discharges a high-velocity jet across a suction chamber that is connected to the equipment to be evacuated.

**Steam locomotive** (Section 92.2) means a historic locomotive propelled by a steam engine.

**Steam sales agreement** (Section 72.2) is a legally binding agreement between a QF, IPP, new IPP, or firm associated with such facility and an industrial or commercial establishment requiring steam that establishes the terms and conditions under which the facility will supply steam to the establishment.

**Steam stripping operation** (Section 264.1031) means a distillation operation in which vaporization of the volatile constituents of a liquid mixture takes place by the introduction of steam directly into the charge.

**Steel** (Section 464.31) An iron-base alloy containing carbon, manganese, and often other alloying elements. Steel is defined here to include only those iron-carbon alloys containing less than 1.2 percent carbon by weight.

**Steel basis material** (Section 465.02) means cold rolled steel, hot rolled steel, and chrome, nickel and tin coated steel which are processed in coil coating.

**Steep slope** (Section 434.70) means any slope above twenty degrees or such lesser slope as may be defined by the regulatory authority after consideration of soil, climate, and other characteristics of a region or State. This term does not apply to those situations in which an operator is mining on flat or gently rolling terrain, on which an occasional steep slope is encountered and through which the mining operation is to proceed, leaving a plain or predominantly flat area.

**Sterndrive engine** (Section 91.3) means a four stroke marine SI engine that is designed such that the drive unit is external to the hull of the marine vessel, while the engine is internal to the hull of the marine vessel.

**Stipend** (Section 45.115) Supplemental financial assistance, other than tuition and fees, paid directly to the trainee by the recipient organization.

**Stipend** (Section 46.120) Supplemental financial assistance other than tuition, fees, and book allowance, paid directly to the fellow.

**Stock barite** (Section 435.11) means the barite that was used to formulate a drilling fluid.

**Stock barite** (Section 435.41) means the barite that was used to formulate a drilling fluid.

**Stock base fluid** (Section 435.11) means the base fluid that was used to formulate a drilling fluid.

**Stock configuration** (Section 205.165) means that no modifications have been made to the original equipment motorcycle that would affect the noise emissions of the vehicle when measured according to the acceleration test procedure.

**Stock-on-hand** (Section 763.163) means the products which are in the possession, direction, or control of a person and are intended for distribution in commerce. United States has the same meaning as in section 3 of the Act.

**Stoker boiler** (Section 76.2) means a boiler that burns solid fuel in a bed, on a stationary or moving grate, that is located at the bottom of the furnace.

**Storage** (Section 191.02) means retention of spent nuclear fuel or radioactive wastes with the intent and capability to readily retrieve such fuel or waste for subsequent use, processing, or disposal.

**Storage** (Section 197.2) means retention (and any associated activity, operation, or process necessary to carry out successful retention) of radioactive material with the intent or capability to readily access or retrieve such material.

**Storage** (Section 243.101) means the interim containment of solid waste after generation and prior to collection for ultimate recovery or disposal.

**Storage** (Section 246.101) means the interim containment of solid waste after generation and prior to collection for ultimate recovery or disposal.

**Storage** (Section 260.10) means the holding of hazardous waste for a temporary period, at the end of which the hazardous waste is treated, disposed of, or stored elsewhere.

**Storage** (Section 270.2) means the holding of hazardous waste for a temporary period, at the end of which the hazardous waste is treated, disposed, or stored elsewhere.

**Storage** (Section 373.4) means the holding of hazardous substances for a temporary period, at the end of which the hazardous substance is either used, neutralized, disposed of, or stored elsewhere.

**Storage and Retrieval of Aerometric Data (SAROAD) system** (Section 58.1) is a computerized system which stores and reports information relating to ambient air quality. The SAROAD system has been replaced with the AIRS-AQS system; however, the SAROAD data reporting format continues to be used by some States and local air pollution agencies as an interface to AIRS on an interim basis.

**Storage for disposal** (Section 761.13) means temporary storage of PCBs that have been designated for disposal.

**Storm water** (Section 122.2) is defined at Sec. 122.26(b)(13).

**Storm water discharge associated with industrial activity** (Section 122.2) is defined at Sec. 122.26(b)(14).

**Storm-water or wastewater collection system** (Section 280.12) means piping, pumps, conduits, and any other equipment necessary to collect and transport the flow of surface water run-off resulting from precipitation, or domestic, commercial, or industrial wastewater to and from retention areas or any areas where treatment is designated to occur. The collection of storm water and wastewater does not include treatment except where incidental to conveyance.

**Stratum (plural strata)** (Section 144.3) means a single sedimentary bed or layer, regardless of thickness, that consists of generally the same kind of rock material.

**Stratum (plural strata)** (Section 146.3) means a single sedimentary bed or layer, regardless of thickness, that consists of generally the same kind of rock material.

**Strawberries** (Section 407.61) shall mean the processing of strawberries into the following product styles: Canned and frozen, whole, sliced, and pureed.

**Stream design flow** (Section 132.2) is the stream flow that represents critical conditions, upstream from the source, for protection of aquatic life, human health, or wildlife.

**Street motorcycle** (Section 205.151) means: (i) Any motorcycle that: (A) With an 80 kg (176 lb) driver, is capable of achieving a maximum speed of at least 40 km/h (25 mph) over a level paved surface; and (B) Is equipped with features customarily associated with practical street or highway use, such features including but not limited to any of the following: stoplight, horn, rear view mirror, turn signals: or (ii) Any motorcycle that: (A) Has an engine displacement less than 50 cubic centimeters; (B) Produces no more than two brake horse power; (C) With a 80 kg (176 lb) driver, cannot exceed 48 km/h (30 mph) over a level paved surface. (3)Competition motorcycle means any motorcycle designed and marketed solely for use in closed course competition events.

**Street wastes** (Section 243.101) means materials picked up by manual or mechanical sweepings of alleys, streets, and sidewalks; wastes from public waste receptacles; and material removed from catch basins.

**Strip, sheet and plate** (Section 420.91) means those acid pickling operations that pickle strip, sheet or plate products.

**Strip, sheet, and miscellaneous products** (Section 420.121) means steel products other than wire products and fasteners.

**Strong chelating agents** (Section 413.02) is defined as all compounds which, by virtue of their chemical structure and amount present, form soluble metal complexes which are not removed by subsequent metals control techniques such as pH adjustment followed by clarification or filtration.

**Structural fiberboard** (Section 247.3) means a fibrous-felted, homogenous panel made from lignocellulosic fibers (usually wood, cane, or paper) and having a density of less than 31 lbs/ft\3\ but more than 10 lbs/ft\3\. It is characterized by an integral bond which is produced by interfelting of the fibers, but which has not been consolidated under heat or pressure as a separate stage of manufacture;

**Structure** (Section 725.3) means a building or vessel which effectively surrounds and encloses the microorganism and includes features designed to restrict the microorganism from leaving.

**Study** (Section 160.3) means any experiment at one or more test sites, in which a test substance is studied in a test system under laboratory conditions or in the environment to determine or help predict its effects, metabolism, product performance (efficacy studies only as required by 40 CFR 158.640), environmental and chemical fate, persistence and residue, or other characteristics in humans, other living organisms, or media. The term "study" does not include basic exploratory studies carried out to determine whether a test substance or a test method has any potential utility.

**Study** (Section 792.3) means any experiment at one or more test sites, in which a test substance is studied in a test system under laboratory conditions or in the environment to determine or help predict its effects, metabolism, environmental and chemical fate, persistence, or other characteristics in humans, other living organisms, or media. The term "study" does not include basic exploratory studies carried out to determine whether a test substance or a test method has any potential utility.

**Study completion date** (Section 160.3) means the date the final report is signed by the study director.

**Study completion date** (Section 792.3) means the date the final report is signed by the study director.

**Study director** (Section 160.3) means the individual responsible for the overall conduct of a study.

**Study director** (Section 792.3) means the individual responsible for the overall conduct of a study.

**Study initiation date** (Section 160.3) means the date the protocol is signed by the study director.

**Study initiation date** (Section 792.3) means the date the protocol is signed by the study director.

**Stuffing box pressure** (Section 65.2) means the fluid (liquid or gas) pressure inside the casing or housing of a piece of equipment, on the process side of the inboard seal.

**Subconfiguration** (Section 600.022-85) means a unique combination, within a vehicle configuration of equivalent test weight, road-load horsepower, and any other operational characteristics or parameters which the Administrator determines may significantly affect fuel economy within a vehicle configuration.

**Submission** (Section 403.3) means: (1) A request by a POTW for approval of a Pretreatment Program to the EPA or a Director; (2) A request by a POTW to the EPA or a Director for authority to revise the discharge limits in categorical Pretreatment Standards to reflect POTW pollutant removals; or (3) A request to the EPA by an NPDES State for approval of its State pretreatment program.

**Submission** (Section 725.3) means any MCAN or exemption request submitted to EPA under this part.

**Submit or serve** (Section 72.2) means to send or transmit a document, information, or correspondence to the person specified in accordance with the applicable regulation: (1) In person; (2) By United States Postal Service; or (3) By other equivalent means of dispatch, or transmission, and delivery. Compliance with any "submission", "service", or "mailing" deadline shall be determined by the date of dispatch, transmission, or mailing and not the date of receipt.

**Submit or serve** (Section 96.2) means to send or transmit a document, information, or correspondence to the person specified in accordance with the applicable regulation: (1) In person; (2) By United States Postal Service; or (3) By other means of dispatch or transmission and delivery.

**Submit or serve** (Section 97.2) means to send or transmit a document, information, or correspondence to the person specified in accordance with the applicable regulation: (1) In person; (2) By United States Postal Service; or (3) By other means of dispatch or transmission and delivery.

**Submitter** (Section 1601.3) means any person or entity who provides information directly or indirectly to the CSB. The term includes, but is not limited to, corporations, Indian tribal governments, state governments, and foreign governments.

**Submitter** (Section 350.1) means a person filing a required report or making a claim of trade secrecy to EPA under sections 303 (d)(2) and (d)(3), 311, 312, and 313 of Title III.

**Subpart H systems** (Section 141.2) means public water systems using surface water or ground water under the direct influence of surface water as a source that are subject to the requirements of subpart H of this part.

**Subsidence** (Section 146.3) means the lowering of the natural land surface in response to: Earth movements; lowering of fluid pressure; removal of underlying supporting material by mining or solution of solids, either artificially or from natural causes; compaction due to wetting (Hydrocompaction); oxidation of organic matter in soils; or added load on the land surface.

**Subsidence** (Section 147.2902) the lowering of the natural land surface in response to: Earth movements; lowering of fluid pressure; removal of underlying supporting material by mining or solution solids, either artificially or from natural causes; compaction due to wetting (hydrocompaction); oxidation of organic matter in soils; or added load on the land surface.

**Substance** (Section 704.3) means either a chemical substance or mixture unless otherwise indicated.

**Substance** (Section 716.3) means chemical substance as defined at section 3(2)(A) of TSCA, 15 U.S.C. 2602(2)(A).

**Substance** (Section 717.3) means a chemical substance or mixture unless otherwise indicated.

**Substantial business relationship** (Section 280.92) means the extent of a business relationship necessary under applicable state law to make a guarantee contract issued incident to that relationship valid and enforceable. A guarantee contract is issued "incident to that relationship" if it arises from and depends on existing economic transactions between the guarantor and the owner or operator.

**Substantial governmental relationship** (Section 280.92) means the extent of a governmental relationship necessary under applicable state law to make an added guarantee contract issued incident to that relationship valid and enforceable. A guarantee contract is issued "incident to that relationship" if it arises from a clear commonality of interest in the event of an UST release such as coterminous boundaries, overlapping constituencies, common ground-water aquifer, or other relationship other than monetary compensation that provides a motivation for the guarantor to provide a guarantee.

**Substantiation** (Section 350.1) means the written answers submitted to EPA by a submitter to the specific questions set forth in this regulation in support of a claim that chemical identity is a trade secret.

**Substate** (Section 256.06) refers to any public regional, local, county, municipal, or intermunicipal agency, or regional or local public (including interstate) solid or hazardous waste management authority, or other public agency below the State level.

**Substitute data** (Section 72.2) means emissions or volumetric flow data provided to assure 100 percent recording and reporting of emissions when all or part of the continuous emission monitoring system is not functional or is operating outside applicable performance specifications.

**Substitution unit** (Section 72.2) means an affected unit, other than a unit under section 410 of the Act, that is designated as a Phase I unit in a substitution plan under Sec. 72.41.

**Subsurface fluid distribution system** (Section 144.3) means an assemblage of perforated pipes, drain tiles, or other similar mechanisms intended to distribute fluids below the surface of the ground.

**Subsurface fluid distribution system** (Section 146.3) means an assemblage of perforated pipes, drain tiles, or other similar mechanisms intended to distribute fluids below the surface of the ground.

**Subtitle D regulated facilities** (Section 239.2) means all solid waste disposal facilities subject to the revised criteria promulgated by EPA under the authority of RCRA Section 4010©.

**Suitable Substitute Decision** (Section 203.1) means the Administrator's decision whether a product which the Administrator has determined to be a low-noise-emission product is a suitable substitute for a product or products presently being purchased by the Federal Government.

**Sulfide** (Section 425.02) shall mean total sulfide as measured by the potassium ferricyanide titration method described in Appendix A or the modified Monier-Williams method described in Appendix B.

**Sulfite cooking liquor** (Section 430.51) is defined as bisulfite cooking liquor when the pH of the liquor is between 3.0 and 6.0 and as acid sulfite cooking liquor when the pH is less than 3.0.

**Sulfur percentage** (Section 80.2) is the percentage of sulfur in diesel fuel by weight, as determined using the applicable sampling and testing methodologies set forth in Sec. 80.580.

**Sulfur-free generation** (Section 72.2) means the generation of electricity by a process that does not have any emissions of sulfur dioxide, including hydroelectric, nuclear, solar, or wind generation. A "sulfur-free generator" is a generator that is located in one of the 48 contiguous States or the District of Columbia and produces "sulfur-free generation."

**Sulfuric acid pickling** (Section 420.91) means those operations in which steel products are immersed in sulfuric acid solutions to chemically remove oxides and scale, and those rinsing operations associated with such immersions.

**Sulfuric acid plant** (Section 51.100) means any facility producing sulfuric acid by the contact process by burning elemental sulfur, alkylation acid, hydrogen sulfide, or acid sludge, but does not include facilities where conversion to sulfuric acid is utilized primarily as a means of preventing emissions to the atmosphere of sulfur dioxide or other sulfur compounds.

**Sump** (Section 260.10) means any pit or reservoir that meets the definition of tank and those troughs/trenches connected to it that serve to collect hazardous waste for transport to hazardous waste storage, treatment, or disposal facilities; except that as used in the landfill, surface impoundment, and waste pile rules, "sump" means any lined pit or reservoir that serves to collect liquids drained from a leachate collection and removal system or leak detection system for subsequent removal from the system.

**Sunshine Act** (Section 1603.3) means the Government in the Sunshine Act, 5 U.S.C. 552b.

**Superfund Memorandum of Agreement (SMOA)** (Section 300.5) means a nonbinding, written document executed by an EPA Regional Administrator and the head of a state agency that may establish the nature and extent of EPA and state interaction during the removal, pre-remedial, remedial, and/or enforcement response process. The SMOA is not a site-specific document although attachments may address specific sites. The SMOA generally defines the role and responsibilities of both the lead and the support agencies.

**Superfund state contract** (Section 300.5) is a joint, legally binding agreement between EPA and a state to obtain the necessary assurances before a federal-lead remedial action can begin at a site. In the case of a political subdivision-lead remedial response, a three-party Superfund state contract among EPA, the state, and political subdivision thereof, is required before a political subdivision takes the lead for any phase of remedial response to ensure state involvement pursuant to section 121(f)(1) of CERCLA. The Superfund state contract may be amended to provide the state's CERCLA section 104 assurances before a political subdivision can take the lead for remedial action.

**Superphosphoric acid process line** (Section 63.601) means any process line which concentrates wet-process phosphoric acid to 66 percent or greater P<INF>2</INF>O<INF>5</INF> content by weight.

**Supplemental FTP (SFTP)** (Section 86.1803-01) means the additional test procedures designed to measure emissions during aggressive and microtransient driving, as described in Sec. 86.159-00 over the US06 cycle, and also the test procedure designed to measure urban driving emissions while the vehicle's air conditioning system is operating, as described in Sec. 86.160-00 over the SC03 cycle.

**Supplemental FTP, or SFTP** (Section 86.000-2) means the additional test procedures designed to measure emissions during aggressive and microtransient driving, as described in Sec. 86.159-00 over the US06 cycle, and also the test procedure designed to measure urban driving emissions while the vehicle's air conditioning system is operating, as described in Sec. 86.160-00 over the SC03 cycle.

**Supplementary control system (SCS)** (Section 57.103) means any technique for limiting the concentration of a pollutant in the ambient air by varying the emissions of that pollutant according to atmospheric conditions. For the purposes of this part, the term supplementary control system does not include any dispersion technique based solely on the use of a stack the height of which exceeds good engineering practice (as determined under regulations implementing section 123 of the Act).

**Supplier of water** (Section 141.2) means any person who owns or operates a public water system.

**Supplier of water** (Section 142.2) means any person who owns or operates a public water system.

**Supplier of water** (Section 143.2) means any person who owns or operates a public water system.

**Supply-side measure** (Section 72.2) means a measure to improve the efficiency of the generation, transmission, or distribution of electricity, implemented by a utility in connection with its operations or facilities to provide electricity to its customers, and includes the measures set forth in part 73, appendix A, section 2 of this chapter.

**Support** (Section 90.102) means that the operator holds the equipment in position so as to prevent it from falling, slipping or sinking. It is not necessary for the entire weight of the equipment to be borne by the operator.

**Support agency** (Section 300.5) means the agency or agencies that provide the support agency coordinator to furnish necessary data to the lead agency, review response data and documents, and provide other assistance as requested by the OSC or RPM. EPA, the USCG, another federal agency, or a state may be support agencies for a response action if operating pursuant to a contract executed under section 104(d)(1) of CERCLA or designated pursuant to a Superfund Memorandum of Agreement entered into pursuant to subpart F of the NCP or other agreement. The support agency may also concur on decision documents.

**Support agency coordinator (SAC)** (Section 300.5) means the official designated by the support agency, as appropriate, to interact and coordinate with the lead agency in response actions under subpart E of this part.

**Suppressed combustion** (Section 420.41) means those basic oxygen furnace steelmaking wet air cleaning systems which are designed to limit or suppress the combustion of carbon monoxide in furnace gases by restricting the amount of excess air entering the air pollution control system.

**Surface casing** (Section 146.3) means the first string of well casing to be installed in the well.

**Surface coating** (Section 468.02) shall mean the process of coating a copper workpiece as well as the associated surface finishing and flattening.

**Surface collecting agents** (Section 300.5) means those chemical agents that form a surface film to control the layer thickness of oil.

**Surface disposal site** (Section 503.21) is an area of land that contains one or more active sewage sludge units.

**Surface impoundment** (Section 280.12) is a natural topographic depression, man-made excavation, or diked area formed primarily of earthen materials (although it may be lined with man-made materials) that is not an injection well.

**Surface impoundment or impoundment** (Section 260.10) means a facility or part of a facility which is a natural topographic depression, man-made excavation, or diked area formed primarily of earthen materials (although it may be lined with man-made materials), which is designed to hold an accumulation of liquid wastes or wastes containing free liquids, and which is not an injection well. Examples of surface impoundments are holding, storage, settling, and aeration pits, ponds, and lagoons.

**Surface impoundment or impoundment** (Section 272.2) means a facility or part of a facility that is a natural topographic depression, human-made excavation, or diked area formed primarily of earthen materials (although it may be lined with human-made materials), that is designed to hold an accumulation of liquid wastes or wastes containing free liquids and that is not an injection well. Examples of surface impoundments are holding storage, settling, and aeration pits, ponds, and lagoons.

**Surface treatment** (Section 471.02) is a chemical or electrochemical treatment applied to the surface of a metal. Such treatments include pickling, etching, conversion coating, phosphating, and chromating. Surface treatment baths are usually followed by a water rinse. The rinse may consist of single or multiple stage rinsing. For the purposes of this part, a surface treatment operation is defined as a bath followed by a rinse, regardless of the number of stages. Each surface treatment bath, rinse combination is entitled to discharge allowance.

**Surface washing agent** (Section 300.5) is any product that removes oil from solid surfaces, such as beaches and rocks, through a detergency mechanism and does not involve dispersing or solubilizing the oil into the water column.

**Surface water** (Section 141.2) means all water which is open to the atmosphere and subject to surface runoff.

**Surfacing ACM** (Section 763.83) means surfacing material that is ACM.

**Surfacing material** (Section 763.83) means material in a school building that is sprayed-on, troweled-on, or otherwise applied to surfaces, such as acoustical plaster on ceilings and fireproofing materials on structural members, or other materials on surfaces for acoustical, fireproofing, or other purposes.

**Surfactant** (Section 417.91) shall mean those methylene blue active substances amendable to measurement by the method described in "Methods for Chemical Analysis of Water and Wastes," 1971, Environmental Protection Agency, Analytical Quality Control Laboratory, page 131.

**Surge control tank** (Section 264.1031) means a large-sized pipe or storage reservoir sufficient to contain the surging liquid discharge of the process tank to which it is connected.

**Surge control vessel** (Section 65.2) means feed drums, recycle drums, and intermediate vessels. Surge control vessels are used within a process unit (as defined in the specific subpart that references this part) when in-process storage, mixing, or management of flow rates or volumes is needed to assist in production of a product.

**Surrogate pollutant** (Section 439.1) means a regulated parameter that, for the purpose of compliance monitoring, is allowed to serve as a surrogate for a group of specific regulated parameters. Plants would be allowed to monitor for a surrogate pollutant(s), when the other parameters for which it stands are receiving the same degree of treatment as the surrogate pollutant(s) and all of the parameters discharged are in the same treatability class(es) as their respective surrogate pollutant(s). Treatability classes have been identified in Appendix A to this part for both steam stripping and biological treatment technologies, which are the respective technology bases for PSES/PSNS and BAT/NSPS limitations controlling the discharge of regulated organic parameters.

**Susceptibility** (Section 171.2) means the degree to which an organism is affected by a pesticide at a particular level of exposure.

**Suspension** (Section 194.2) means any action taken by the Administrator to withdraw, for a limited period of time, the certification pursuant to section 8(d) (1) of the WIPP LWA.

**SUVA** (Section 141.2) means Specific Ultraviolet Absorption at 254 nanometers (nm), an indicator of the humic content of water. It is a calculated parameter obtained by dividing a sample's ultraviolet absorption at a wavelength of 254 nm (UV 254) (in m =1) by its concentration of dissolved organic carbon (DOC) (in mg/L).

**SW-846** (Section 761.13) means the document having the title "SW-846, Test Methods for Evaluating Solid Waste," which is available from either the National Technical Information Service (NTIS, U.S. Department of Commerce, 5285 Port Royal Rd., Springfield, VA 22161, telephone: (703) 487-4650 or the U.S. Government Printing Office (U.S. GPO, 710 North Capitol St., NW., Washington, DC 20401, telephone: (202) 783-3238.

**Swaging** (Section 471.02) is a process in which a solid point is formed at the end of a tube, rod, or bar by the repeated blows of one or more pairs of opposing dies.

**Sweet water** (Section 417.41) shall mean the solution of 8-10 percent crude glycerine and 90-22 percent water that is a by-product of saponification or fat splitting.

**Switch locomotive** (Section 92.2) means a locomotive designed or used solely for the primary purpose of propelling railroad cars a short distance, and that is powered by an engine with a maximum horsepower rating of 2300 hp or less.

**Switcher Locomotive** (Section 201.1) means any locomotive designated as a switcher by the builder or reported to the ICC as a switcher by the operator-owning-railroad and including, but not limited to, all locomotives of the builder/model designations listed in Appendix A to this subpart.

**Synthetic material** (Section 435.41) as applied to synthetic-based drilling fluid means material produced by the reaction of specific purified chemical feedstock, as opposed to the traditional base fluids such as diesel and mineral oil which are derived from crude oil solely through physical separation processes. Physical separation processes include fractionation and distillation and/or minor chemical reactions such as cracking and hydro processing. Since they are synthesized by the reaction of purified compounds, synthetic materials suitable for use in drilling fluids are typically free of polycyclic aromatic hydrocarbons (PAH's) but are sometimes found to contain levels of PAH up to 0.001 weight percent PAH expressed as phenanthrene. Internal olefins and vegetable esters are two examples of synthetic materials suitable for use by the oil and gas extraction industry in formulating drilling fluids. Internal olefins are synthesized from the isomerization of purified straight-chain (linear) hydrocarbons such as C16-C18 linear alpha olefins. C16-C18 linear alpha olefins are unsaturated hydrocarbons with the carbon to carbon double bond in the terminal position. Internal olefins are typically formed from heating linear alpha olefins with a catalyst. The feed material for synthetic linear alpha olefins is typically purified ethylene. Vegetable esters are synthesized from the acid-catalyzed esterification of vegetable fatty acids with various alcohols. EPA listed these two branches of synthetic fluid base materials to provide examples, and EPA does not mean to exclude other synthetic materials that are either in current use or may be used in the future. A synthetic-based drilling fluid may include a combination of synthetic materials.

**Synthetic material as applied to synthetic-based drilling fluid** (Section 435.11) means material produced by the reaction of specific purified chemical feedstock, as opposed to the traditional base fluids such as diesel and mineral oil which are derived from crude oil solely through physical separation processes. Physical separation processes include fractionation and distillation and/or minor chemical reactions such as cracking and hydro processing. Since they are synthesized by the reaction of purified compounds, synthetic materials suitable for use in drilling fluids are typically free of polycyclic aromatic hydrocarbons (PAH's) but are sometimes found to contain levels of PAH up to 0.001 weight percent PAH expressed as phenanthrene. Internal olefins and vegetable esters are two examples of synthetic materials suitable for use by the oil and gas extraction industry in formulating drilling fluids. Internal olefins are synthesized from the isomerization of purified straight-chain (linear) hydrocarbons such as C16-C18 linear alpha olefins. C16-C18 linear alpha olefins are unsaturated hydrocarbons with the carbon to carbon double bond in the terminal position. Internal olefins are typically formed from heating linear alpha olefins with a catalyst. The feed material for synthetic linear alpha olefins is typically purified ethylene. Vegetable esters are synthesized from the acid-catalyzed esterification of vegetable fatty acids with various alcohols. EPA listed these two branches of synthetic fluid base materials to provide examples, and EPA does not mean to exclude other synthetic materials that are either in current use or may be used in the future. A synthetic-based drilling fluid may include a combination of synthetic materials.

**Synthetic-based** (Section 435.11) means the continuous phase of the drilling fluid is a synthetic material or a combination of synthetic materials.

**Synthetic-based** (Section 435.41) means the continuous phase of the drilling fluid is a synthetic material or a combination of synthetic materials.

**System** (Section 86.082-2) includes any motor vehicle engine modification which controls or causes the reduction of substances emitted from motor vehicles.

**System** (Section 86.402-78) includes any motor vehicle modification which controls or causes the reduction of substances emitted from motor vehicles. Total test distance is defined for each class of motorcycles in Sec. 86.427-78.

**System of records** (Section 1516.2) means a group of any records under the control of the Council from which information is retrieved by the name of the individual or by some identifying number, symbol, or other identifying particular assigned to the individual.

**System with a single service connection** (Section 141.2) means a system which supplies drinking water to consumers via a single service line.

# T

**T, as in "Cyanide, T"** (Section 433.11) shall mean total.

**Tag** (Section 211.203) Stiff paper, metal or other hard material that is tied or otherwise affixed to the packaging of a protector.

**Tailings** (Section 192) means the remaining portion of a metal-bearing ore after some or all of such metal, such as uranium, has been extracted.

**Tailings Closure Plan (Radon)** (Section 192.31) means the Nuclear Regulatory Commission or Agreement State approved plan detailing activities to accomplish timely emplacement of a permanent radon barrier. A tailings closure plan shall include a schedule for key radon closure milestone activities such as wind blown tailings retrieval and placement on the pile, interim stabilization (including dewatering or the removal of freestanding liquids and recontouring), and emplacement of a permanent radon barrier constructed to achieve compliance with the 20 pCi/m2-s flux standard as expeditiously as practicable considering technological feasibility (including factors beyond the control of the licensee).

**Tallow** (Section 432.101) shall mean a product made from beef cattle or sheep fat that has a melting point of 40 [deg]C or greater.

**Tampering** (Section 204.2) means those acts prohibited by section 10(a) (2) of the Act.

**Tampering** (Section 204.51) means those acts prohibited by section 10(a) (2) of the Act. (Secs. 6 and 13, Noise Control Act, Pub. L. 92-574, 86 Stat. 1244; (42 U.S.C. 4912))

**Tampering** (Section 205.151) means the removal or rendering inoperative by any person, other than for purposes of maintenance, repair, or replacement, of any device or element of design incorporated into any product in compliance with regulations under section 6, prior to its sale or delivery to the ultimate purchaser or while it is in use; or the use of a product after such device or element of design has been removed or rendered inoperative by any person.

**Tampering** (Section 205.2) means those acts prohibited by section 10(a)( 2) of the Act.

**Tampering** (Section 205.51) means those acts prohibited by section 10(a) (2) of the Act.

**Tangentially fired boiler** (Section 76.2) means a boiler that has coal and air nozzles mounted in each corner of the furnace where the vertical furnace walls meet. Both pulverized coal and air are directed from the furnace corners along a line tangential to a circle lying in a horizontal plane of the furnace.

**Tangible net worth** (Section 280.92) means the tangible assets that remain after deducting liabilities; such assets do not include intangibles such as goodwill and rights to patents or royalties. For purposes of this definition, "assets" means all existing and all probable future economic benefits obtained or controlled by a particular entity as a result of past transactions.

**Tank** (Section 260.10) means a stationary device, designed to contain an accumulation of hazardous waste which is constructed primarily of non-earthen materials (e.g., wood, concrete, steel, plastic) which provide structural support.

**Tank** (Section 279.1) means any stationary device, designed to contain an accumulation of used oil which is constructed primarily of non-earthen materials, (e.g., wood, concrete, steel, plastic) which provides structural support.

**Tank** (Section 280.12) is a stationary device designed to contain an accumulation of regulated substances and constructed of non-earthen materials (e.g., concrete, steel, plastic) that provide structural support.

**Tank barge** (Section 442.2) means a non-self-propelled vessel constructed or adapted primarily to carry liquid, solid or gaseous commodities or cargos in bulk in cargo spaces (or tanks) through rivers and inland waterways, and may occasionally carry commodities or cargos through oceans and seas when in transit from one inland waterway to another. The commodities or cargos transported are in direct contact with the tank interior. There are no maximum or minimum vessel or tank volumes.

**Tank fuel volume** (Section 86.082-2) means the volume of fuel in the fuel tank(s), which is determined by taking the manufacturer's nominal fuel tank(s) capacity and multiplying by 0.40, the result being rounded using ASTM E 29-67 to the nearest tenth of a U.S. gallon.

**Tank fuel volume** (Section 86.1803-01) means the volume of fuel in the fuel tank(s), which is determined by taking the manufacturer's nominal fuel tank(s) capacity and multiplying by 0.40. The result is rounded to the nearest tenth of a U.S. gallon in accordance with the Rounding-Off Method specified in ASTM E29-93a, Standard Practice for Using Significant Digits in Test Data to Determine Conformance with Specifications (incorporated by reference; see Sec. 86.1)

**Tank system** (Section 260.10) means a hazardous waste storage or treatment tank and its associated ancillary equipment and containment system.

**Tank truck** (Section 442.2) means a motor-driven vehicle with a completely enclosed storage vessel used to transport liquid, solid or gaseous materials over roads and highways. The storage vessel or tank may be detachable, as with tank trailers, or permanently attached. The commodities or cargos transported come in direct contact with the tank interior. A tank truck may have one or more storage compartments. There are no maximum or minimum vessel or tank volumes. Tank trucks are also commonly referred to as cargo tanks or tankers.

**Tank truck** (Section 80.2) means a truck and/or trailer used to transport or cause the transportation of gasoline or diesel fuel, that meets the definition of motor vehicle in section 216(2) of the Act.

**Tank vessel as defined by section 1001 of the OPA** (Section 300.5) means a vessel that is constructed or adapted to carry, or that carries oil or hazardous material in bulk as cargo or cargo residue, and that: (1) is a vessel of the United States; (2) operates on the navigable waters; or (3) transfers oil or hazardous material in a place subject to the jurisdiction of the United States.

**Tankage** (Section 432.101) shall mean dried animal by-product residues used in feedstuffs.

**Target housing** (Section 745.103) means any housing constructed prior to 1978, except housing for the elderly or persons with disabilities (unless any child who is less than 6 years of age resides or is expected to reside in such housing) or any 0-bedroom dwelling.

**Target housing** (Section 745.223) means any housing constructed prior to 1978, except housing for the elderly or persons with disabilities (unless any one or more children age 6 years or under resides or is expected to reside in such housing for the elderly or persons with disabilities) or any 0-bedroom dwelling.

**Taxi/idle (in)** (Section 87.1) means those aircraft operations involving taxi and idle between the time of landing roll-out and final shutdown of all propulsion engines.

**Taxi/idle (out)** (Section 87.1) means those aircraft operations involving taxi and idle between the time of initial starting of the propulsion engine(s) used for the taxi and turn on to duty runway.

**TDS** (Section 401.11) means total dissolved solids.

**Technical grade of active ingredient** (Section 158.153) means a material containing an active ingredient: (1) Which contains no inert ingredient, other than one used for purification of the active ingredient; and (2) Which is produced on a commercial or pilot-plant production scale (whether or not it is ever held for sale).

**Technical Support Document** (Section 66.3) means the Noncompliance Penalties Technical Support Document which accompanies these regulations. The Technical Support Document appears as appendix A to these regulations. All other terms are defined as they are in the Act.

**Technically feasible** (Section 157.21) when applied to child-resistant packaging, means that the technology exists to produce the child-resistant packaging for a particular pesticide.

**Technically qualified individual** (section 710.2) means a person: (1) Who because of his education, training, or experience, or a combination of these factors, is capable of appreciating the health and environmental risks associated with the chemical substance which is used under his supervision, (2) who is responsible for enforcing appropriated methods of conducting scientific experimentation, analysis, or chemical research in order to minimize such risks, and (3) who is responsible for the safety assessments and clearances related to the procurement, storage, use, and disposal of the chemical substance as may be appropriate or required within the scope of conducting the research and development activity. The responsibilities in paragraph (aa)(3) of this section may be delegated to another individual, or other individuals, as long as each meets the criteria in paragraph (aa)(1) of this section.

**Technically qualified individual** (Section 720.3) means a person or persons (1) who, because of education, training, or experience, or a combination of these factors, is capable of understanding the health and environmental risks associated with the chemical substance which is used under his or her supervision, (2) who is responsible for enforcing appropriate methods of conducting scientific experimentation, analysis, or chemical research to minimize such risks, and (3) who is responsible for the safety assessments and clearances related to the procurement, storage, use, and disposal of the chemical substance as may be appropriate or required within the scope of conducting a research and development activity.

**Technically qualified individual** (Section 725.3) means a person or persons: (1) Who, because of education, training, or experience, or a combination of these factors, is capable of understanding the health and environmental risks associated with the microorganism which is used under his or her supervision, (2) Who is responsible for enforcing appropriate methods of conducting scientific experimentation, analysis, or microbiological research to minimize such risks, and (3) Who is responsible for the safety assessments and clearances related to the procurement, storage, use, and disposal of the microorganism as may be appropriate or required within the scope of conducting a research and development activity.

**Temperature monitoring device** (Section 65.2) means a unit of equipment used to monitor temperature and having a minimum accuracy of [plusmn]1 percent of the temperature being monitored expressed in degrees Celsius or [plusmn]1.2 degrees Celsius ( [deg]C), whichever is greater.

**Tempering** (Section 421.61) shall mean the process whereby glass is heated near the melting point and then rapidly cooled to increase its mechanical and thermal endurance.

**Ten (10)-year, 24-hour precipitation event** (Section 440.132) is the maximum 24-hour precipitation event with a probable recurrence interval of once in 10 years as established by the U.S. Department of Commerce, National Oceanic and Atmospheric Administration, National Weather Service, or equivalent regional or rainfall probability information.

**Ten year 24-hour rainfall event** (Section 129.2) means the maximum precipitation event with a probable recurrence interval of once in 10 years as defined by the National Weather Service in Technical Paper No. 40, Rainfall Frequency Atlas of the United States, May 1961, and subsequent amendments or equivalent regional or State rainfall probability information developed therefrom.

**Ten-year 24-hour rainfall event** (Section 418.11) shall mean the maximum 24-hour precipitation event with a probable recurrence interval of once in 10 years as defined by the National Weather Service in technical paper No. 40, "Rainfall Frequency Atlas of the United States", May 1961, and subsequent amendments in effect as of the effective date of this regulation.

**Ten-year 24-hour rainfall event** (Section 422.41) shall mean the maximum precipitation event with a probable recurrence interval of once in 10 years as defined by the National Weather Service in technical paper no. 40, "Rainfall Frequency Atlas of the United States," May 1961, and subsequent amendments or equivalent regional or State rainfall probability information developed therefrom.

**Ten-year 24-hour rainfall event** (Section 422.51) shall mean the maximum precipitation event with a probable recurrence interval of once in 10 years as defined by the National Weather Service in technical paper no. 40, "Rainfall Frequency Atlas of the United States," May 1961, and subsequent amendments or equivalent regional or State rainfall probability information developed therefrom.

**TEQ** (Section 260.10) means toxicity equivalence, the international method of relating the toxicity of various dioxin/furan congeners to the toxicity of 2,3,7,8-tetrachlorodibenzo-p-dioxin.

**TEQ** (Section 62.1201) means toxicity equivalence, the international method of relating the toxicity of various dioxin/furan congeners to the toxicity of 2,3,7,8-tetrachlorodibenzo-p-dioxin.

**Termination under Sec. 280.97(b)(1) and Sec. 280.97(b)(2)** (Section 280.92) means only those changes that could result in a gap in coverage as where the insured has not obtained substitute coverage or has obtained substitute coverage with a different retroactive date than the retroactive date of the original policy.

**Terms and conditions of registration** (Section 154.3) means the terms and conditions governing lawful sale, distribution, and use approved in conjunction with registration, including labeling, use classification, composition, and packaging.

**Terne coating** (Section 420.121) means coating steel products with terne metal by the hot dip process including the immersion of the steel product in a molten bath of lead and tin metals, and the related operations preceding and subsequent to the immersion phase.

**Territorial sea** (Section 230.3) means the belt of the sea measured from the baseline as determined in accordance with the Convention on the Territorial Sea and the Contiguous Zone and extending seaward a distance of three miles.

**Territorial seas** (Section 116.3) means the belt of the seas measured from the line of ordinary low water along that portion of the coast which is in direct contact with the open sea and the line marking the seaward limit of inland waters, and extending seaward a distance of 3 miles.

**Test analyzer** (Section 53.1) means an analyzer subjected to testing as part of a candidate method in accordance with subparts B, C, D, E, or F of this part, as applicable. Test sampler means a PM10 sampler or a PM2.5 sampler subjected to testing as part of a candidate method in accordance with subparts C, D, E, or F of this part.

**Test Compressor** (Section 204.51) means a compressor used to demonstrate compliance with the applicable noise emissions standard.

**Test data** (Section 610.11) means any information which is a quantitative measure of any aspect of the behavior of a retrofit device.

**Test data** (Section 720.3) means data from a formal or informal test or experiment, including information concerning the objectives, experimental methods and materials, protocols, results, data analyses, recorded observations, monitoring data, measurements, and conclusions from a test or experiment.

**Test engine** (Section 86.1002-2001) means an engine in a test sample.

**Test Engine** (Section 86.1002-84) means an engine in a test sample.

**Test engine** (Section 86.1002-97) means an engine in a test sample.

**Test engine** (Section 89.502) means an engine in a test sample.

**Test engine** (Section 90.3) means the engine or group of engines that a manufacturer uses during certification to determine compliance with emission standards.

**Test engine** (Section 91.3) means the engine or group of engines that a manufacturer uses during certification, production line and in-use testing to determine compliance with emission standards.

**Test engine** (Section 94.2) means an engine in a test sample.

**Test exhaust system** (Section 205.165) means an exhaust system in Selective Enforcement Audit test sample.

**Test Facility** (Section 211.203) For this subpart, a laboratory that has been set up and calibrated to conduct ANSI Std S3.19-1974 tests on hearing protective devices. It must meet the applicable requirements of these regulations.

**Test fleet** (Section 89.2) means the engine or group of engines that a manufacturer uses during certification to determine compliance with emission standards.

**Test group** (Section 86.1803-01) means the basic classification unit within a durability group used for the purpose of demonstrating compliance with exhaust emission standards in accordance with Sec. 86.1841-01. The test group is also used as a classification unit for gathering in-use data for the In-Use Verification Program (IUVP) in accordance with Sec. 86.1848-01.

**Test Hearing Protector** (Section 211.203) A hearing protector that has been selected for testing to verify the value to be put on the label, or which has been designated for testing to determine compliance of the protector with the labeled value.

**Test locomotive or locomotive engine** (Section 92.2) means a locomotive or locomotive engine in a test sample.

**Test marketing** (Section 704.3) means the distribution in commerce of no more than a predetermined amount of a chemical substance, mixture, article containing that chemical substance or mixture, or a mixture containing that substance, by a manufacturer or processor, to no more than a defined number of potential customers to explore market capability in a competitive situation during a predetermined testing period prior to the broader distribution of that chemical substance, mixture, or article in commerce.

**Test marketing** (section 710.2) means the distribution in commerce of no more than a predetermined amount of a chemical substance, mixture, or article containing that chemical substance or mixture, by a manufacturer or processor to no more than a defined number of potential customers to explore market capability in a competitive situation during a predetermined testing period prior to the broader distribution of that chemical substance, mixture or article in commerce.

**Test marketing** (Section 712.3) means distributing in commerce a limited amount of a chemical substance or mixture, or article containing such substance or mixture, to a defined number of potential customers, during a predetermined testing period, to explore market capability prior to broader distribution in commerce.

**Test marketing** (Section 720.3) means the distribution in commerce of no more than a predetermined amount of a chemical substance, mixture, or article containing that chemical substance or mixture, by a manufacturer or processor, to no more than a defined number of potential customers to explore market capability in a competitive situation during a predetermined testing period prior to the broader distribution of that chemical substance, mixture, or article in commerce.

**Test method** (Section 63.2) means the validated procedure for sampling, preparing, and analyzing for an air pollutant specified in a relevant standard as the performance test procedure. The test method may include methods described in an appendix of this chapter, test methods incorporated by reference in this part, or methods validated for an application through procedures in Method 301 of appendix A of this part.

**Test product** (Section 204.2) means any product that is required to be tested pursuant to this part.

**Test product** (Section 205.2) means any product that is required to be tested pursuant to this part.

**Test product** (Section 211.102) means any product that must be tested according to regulations published under part 211.

**Test Request** (Section 211.203) A request submitted to the manufacturer by the Administrator that will specify the hearing protector category, and test sample size to be tested according to Sec. 211.212-1, and other information regarding the audit.

**Test rule** (Section 791.3) refers to a regulation ordering the development of data on health or environmental effects or chemical fate for a chemical substance or mixture pursuant to TSCA section 4(a).

**Test sample** (Section 204.51) means the collection of compressors from the same category or configuration which is randomly drawn from the batch sample and which will receive emissions tests.

**Test sample** (Section 205.51) means the collection of vehicles from the same category, configuration or subgroup thereof which is drawn from the batch sample and which will receive noise emissions tests.

**Test sample** (Section 86.1002-2001) means the collection of vehicles or engines of the same configuration which have been drawn from the population of vehicles or engines of that configuration and which will receive emission testing.

**Test Sample** (Section 86.1002-84) means the collection of vehicles or engines of the same configuration which have been drawn from the population of engines or vehicles of that configuration and which will receive exhaust emission testing.

**Test sample** (Section 86.1002-97) means the collection of vehicles or engines of the same configuration which have been drawn from the population of engines or vehicles of that configuration and which will receive exhaust emission testing.

**Test Sample** (Section 86.602-84) means the collection of vehicles of the same configuration which have been drawn from the population of vehicles of that configuration and which will receive exhaust emission testing.

**Test sample** (Section 86.602-98) means the collection of vehicles of the same configuration which have been drawn from the population of vehicles of that configuration and which will receive emission testing.

**Test sample** (Section 89.502) means the collection of engines selected from the population of an engine family for emission testing.

**Test sample** (Section 90.702) means the collection of engines selected from the population of an engine family for emission testing.

**Test sample** (Section 92.2) means the collection of locomotives or locomotive engines selected from the population of an engine family for emission testing or auditing.

**Test sample** (Section 94.2) means the collection of engines or vessels selected from the population of an engine family for emission testing.

**Test sample size** (Section 204.51) means the number of compressors of the same configuration in a test sample.

**Test sample size** (Section 205.51) means the number of vehicles of the same category or configuration in a test sample.

**Test substance** (Section 160.3) means a substance or mixture administered or added to a test system in a study, which substance or mixture: (1) Is the subject of an application for a research or marketing permit supported by the study, or is the contemplated subject of such an application; or (2) Is an ingredient, impurity, degradation product, metabolite, or radioactive isotope of a substance described by paragraph (1) of this definition, or some other substance related to a substance described by that paragraph, which is used in the study to assist in characterizing the toxicity, metabolism, or other characteristics of a substance described by that paragraph.

**Test substance** (Section 790.3) means the form of chemical substance or mixture that is specified for use in testing.

**Test substance** (Section 792.3) means a substance or mixture administered or added to a test system in a study, which substance or mixture is used to develop data to meet the requirements of a TSCA section 4(a) test rule and/or is developed under a TSCA section 4 testing consent agreement or section 5 rule or order to the extent the agreement, rule or order references this part.

**Test system** (Section 160.3) means any animal, plant, microorganism, chemical or physical matrix, including but not limited to soil or water, or subparts thereof, to which the test, control, or reference substance is administered or added for study. "Test system" also includes appropriate groups or components of the system not treated with the test, control, or reference substance.

**Test system** (Section 792.3) means any animal, plant, microorganism, chemical or physical matrix, including but not limited to, soil or water, or components thereof, to which the test, control, or reference substance is administered or added for study. "Test system" also includes appropriate groups or components of the system not treated with the test, control, or reference substance.

**Test vehicle** (Section 205.151) means a vehicle in a Selective Enforcement Audit test sample.

**Test vehicle** (Section 205.51) means a vehicle selected and used to demonstrate compliance with the applicable noise emission standards.

**Test Vehicle** (Section 86.1002-84) means a vehicle in a test sample.

**Test vehicle** (Section 86.1002-97) means a vehicle in a test sample.

**Test Vehicle** (Section 86.602-84) means a vehicle in a test sample.

**Test vehicle** means a vehicle in a test sample.

**Test Weight** (Section 600.022-85) means the weight within an inertia weight class which is used in the dynamometer testing of a vehicle, and which is based on its loaded vehicle weight in accordance with the provisions of part 86.

**Test weight** (Section 86.082-2) means the weight, within an inertia weight class, which is used in the dynamometer testing of a vehicle, and which is based on its loaded vehicle weight in accordance with the provisions of part 86.

**Test Weight** (Section 88.102-94) is defined as the average of the curb weight and the GVWR.

**Test weight basis** (Section 86.094-2) means the basis on which equivalent test weight is determined in accordance with Sec. 86.129-94 of subpart B of this part.

**Test weight basis** (Section 86.1803-01) means the basis on which equivalent test weight is determined in accordance with Sec. 86.129-00 of subpart B of this part.

**Testing agent** (Section 610.11) means any person who develops test data on a retrofit device.

**Testing exemption** (Section 204.2) means an exemption from the prohibitions of section 10(a) (1), (2), (3), and (5) of the Act, which may be granted under section 10(b)(1) of the Act for the purpose of research, investigations, studies, demonstrations, or training, but not including national security where lease or sale of the exempted product is involved.

**Testing exemption** (Section 205.2) means an exemption from the prohibitions of section 10(a) (1), (2), (3), and (5) of the Act, which may be granted under section 10(b)(1) of the Act for the purpose of research, investigations, studies, demonstrations, or training, but not including national security.

**Testing exemption** (Section 211.102) means an exemption from the prohibitions of section 10(a) (1), (2), (3), and (5) of the Act, which may be granted under section 10(b)(1) of the Act for research, investigations, studies, demonstrations, or training, but not for national security.

**Testing exemption** (Section 85.1702) means an exemption which may be granted under section 203(b)(1) for the purpose of research investigations, studies, demonstrations or training, but not including national security.

**Testing exemption** (Section 90.902) means an exemption granted under Sec. 90.1004(b) for the purpose of research, investigations, studies, demonstrations or training, but not including national security.

**Testing exemption** (Section 91.1002) means an exemption which may be granted under Sec. 91.1104 (b) for the purpose of research, investigations, studies, demonstrations or training, but not including national security.

**Testing facility** (Section 160.3) means a person who actually conducts a study, i.e., actually uses the test substance in a test system. "Testing facility" encompasses only those operational units that are being or have been used to conduct studies.

**Testing facility** (Section 792.3) means a person who actually conducts a study, i.e., actually uses the test substance in a test system. "Testing facility" encompasses only those operational units that are being or have been used to conduct studies.

**Tetrachloroethylene (or priority pollutant No. 85)** (Section 420.02) means the value obtained by the standard method Number 610 specified in 44 FR 69464, 69571 (December 3, 1979).

**The BOD5 and SS effluent concentrations consistently achievable through proper operation and maintenance** (Section 133.101) (Sec. 133.101(f)) of the treatment works exceed the minimum level of the effluent quality set forth in Secs. 133.102(a) and 133.102(b), (2) A trickling filter or waste stabilization pond is used as the principal process, and (3) The treatment works provide significant biological treatment of municipal wastewater.

**The term navigable waters of the United States** (Section 112.2) means navigable waters as defined in section 502(7) of the FWPCA, and includes: (1) All navigable waters of the United States, as defined in judicial decisions prior to passage of the 1972 Amendments to the FWPCA (Pub. L. 92-500), and tributaries of such waters; (2) Interstate waters; (3) Intrastate lakes, rivers, and streams which are utilized by interstate travelers for recreational or other purposes; and (4) Intrastate lakes, rivers, and streams from which fish or shellfish are taken and sold in interstate commerce. Navigable waters do not include prior converted cropland. Notwithstanding the determination of an area's status as prior converted cropland by any other federal agency, for the purposes of the Clean Water Act, the final authority regarding Clean Water Act jurisdiction remains with EPA.

**Thermal discharge** (Section 130.2) The discharge of the pollutant heat from a point source that is required to have an NPDES permit.

**Thermal energy** (Section 72.2) means the thermal output produced by a combustion source used directly as part of a manufacturing process but not used to produce electricity.

**Thermal processing** (Section 240.101) means processing of waste material by means of heat.

**Thermal system insulation** (Section 763.83) means material in a school building applied to pipes, fittings, boilers, breeching, tanks, ducts, or other interior structural components to prevent heat loss or gain, or water condensation, or for other purposes.

**Thermal system insulation ACM** (Section 763.83) means thermal system insulation that is ACM.

**Thermal treatment** (Section 260.10) means the treatment of hazardous waste in a device which uses elevated temperatures as the primary means to change the chemical, physical, or biological character or composition of the hazardous waste. Examples of thermal treatment processes are incineration, molten salt, pyrolysis, calcination, wet air oxidation, and microwave discharge. (See also "incinerator" and "open burning".)

**Thermostat** (Section 260.10) means a temperature control device that contains metallic mercury in an ampule attached to a bimetal sensing element, and mercury-containing ampules that have been removed from these temperature control devices in compliance with the requirements of 40 CFR 273.13©(2) or 273.33©(2).

**Thermostat** (Section 273.9) means a temperature control device that contains metallic mercury in an ampule attached to a bimetal sensing element, and mercury-containing ampules that have been removed from these temperature control devices in compliance with the requirements of 40 CFR 273.13© (2) or 273.33© (2).

**Thin-film evaporation operation** (Section 264.1031) means a distillation operation that employs a heating surface consisting of a large diameter tube that may be either straight or tapered, horizontal or vertical. Liquid is spread on the tube wall by a rotating assembly of blades that maintain a close clearance from the wall or actually ride on the film of liquid on the wall.

**Thinner** (Section 59.101) means any solvent used to reduce the viscosity or solids content of a coating.

**Threat of discharge or release** (Section 300.5) see definitions for discharge and release.

**Threat of release** (Section 300.5) see definition for release.

**Three-stage topcoat** (Section 59.101) means a topcoat composed of a pigmented basecoat, a midcoat, and a transparent clearcoat.

**Threshold effect** (Section 132.2) is an effect of a substance for which there is a theoretical or empirically established dose or concentration below which the effect does not occur.

**Threshold planning quantity** (Section 355.20) means, for a substance listed in appendices A and B, the quantity listed in the column "threshold planning quantity" for that substance.

**Threshold quantity** (Section 68.3) means the quantity specified for regulated substances pursuant to section 112®(5) of the Clean Air Act as amended, listed in Sec. 68.130 and determined to be present at a stationary source as specified in Sec. 68.115 of this part.

**Throttle** (Section 86.082-2) means the mechanical linkage which either directly or indirectly controls the fuel flow to the engine.

**Throttle** (Section 86.090-2) means a device used to control an engine's power output by limiting the amount of air entering the combustion chamber.

**Throttle** (Section 86.1803-01) means a device used to control an engine's power output by limiting the amount of air entering the combustion chamber.

**Throttle** (Section 92.2) means the component, or components, which either directly or indirectly controls the fuel flow to the engine.

**Throttle notch** (Section 92.2) means a discrete throttle position for a locomotive with a limited number of throttle positions.

**Throttle notch horsepower** (Section 92.2) means the brake horsepower output of an engine corresponding to each throttle notch position, including dynamic-brake settings.

**Throttle notch speed** (Section 92.2) means the speed of the engine, expressed as the number of revolutions of the crankshaft per unit of time (e.g., rpm), corresponding to each throttle notch position, including dynamic-brake, and hotel power settings.

**Tier 0** (Section 92.2) means relating to emission standards applicable to locomotives originally manufactured before January 1, 2002; or relating to such locomotives.

**Tier 1 engine** (Section 89.2) means an engine subject to the Tier 1 emission standards listed in Sec. 89.112(a).

**Tier 2** (Section 92.2) means relating to emission standards applicable to locomotives originally manufactured on or after January 1, 2005; or relating to such locomotives.

**Tier 2** (Section 94.2) means relating to an engine subject to the Tier 2 emission standards listed in Sec. 94.8.

**Tier 2 engine** (Section 89.2) means an engine subject to the Tier 2 emission standards listed in Sec. 89.112(a).

**Tier 2 HLDT/MDPV** (Section 86.1803-01) means any heavy light-duty truck or medium-duty passenger vehicle, including HEVs and ZEVs, of the 2008 or later model year certified to comply with the Tier 2 FTP exhaust standards contained in Sec. 86.1811-04 including the 0.07 g/mi fleet average NOX standard. The term Tier 2 HLDT/MDPV also includes any heavy light-duty truck or medium-duty passenger vehicle, of any model year, which is certified to Tier 2 FTP exhaust standards for purposes of generating or banking early NOX credits for averaging under Tier 2 requirements, or utilizing alternate phase-in schedules, as allowed in this subpart.

**Tier 2 LDV/LLDT** (Section 86.1803-01) means any light-duty vehicle or light light-duty truck, including HEVs and ZEVs, of the 2004 or later model year certified to comply with the Tier 2 FTP exhaust standards contained in Sec. 86.1811-04 including the 0.07 g/mi fleet average NOX standard. The term Tier 2 LDV/LLDT also includes any light-duty vehicle or light light-duty truck, of any model year, which is certified to Tier 2 FTP exhaust standards for purposes of generating or banking early NOX credits for averaging under Tier 2 requirements, or utilizing alternate phase-in schedules as allowed in this subpart.

**Tier 2 standards** (Section 86.1803-01) means those FTP exhaust emission standards including the 0.07 g/mi full useful life fleet average NOX standard, applicable to new light-duty vehicles and light light-duty trucks that begin a phase-in in the 2004 model year, and those exhaust emission standards including the 0.07 g/mi full useful life fleet average NOX standard, applicable to heavy light-duty trucks and medium-duty passenger vehicles that begin a phase-in in the 2008 model year. These standards are found in Sec. 86.1811-04 of this subpart.

**Tier 2 vehicle** (Section 86.1803-01) means any vehicle certified to comply with the Tier 2 FTP exhaust standards contained in Sec. 86.1811-04 including the 0.07 g/ mi fleet average NOX standard.

**Tier 3 engine** (Section 89.2) means an engine subject to the Tier 3 emission standards listed in Sec. 89.112(a).

**Tier I criteria** (Section 132.2) are numeric values derived by use of the Tier I methodologies in appendixes A, C and D of this part, the methodology in appendix B of this part, and the procedures in appendix F of this part, that either have been adopted as numeric criteria into a water quality standard or are used to implement narrative water quality criteria.

**Tier II values** (Section 132.2) are numeric values derived by use of the Tier II methodologies in appendixes A and C of this part, the methodology in appendix B of this part, and the procedures in appendix F of this part, that are used to implement narrative water quality criteria.

**Tier1** (Section 92.2) means relating to emission standards applicable to locomotives originally manufactured on or after January 1, 2002 and before January 1, 2005; or relating to such locomotives.

**Time period** (Section 51.100) means any period of time designated by hour, month, season, calendar year, averaging time, or other suitable characteristics, for which ambient air quality is estimated.

**Tire** (Section 247.3) means the following types of tires: Passenger car tires, light- and heavy-duty truck tires, high-speed industrial tires, bus tires, and special service tires (including military, agricultural, off-the-road, and slow-speed industrial);

**Title III** (Section 350.1) means Title III of the Superfund Amendments and Reauthorization Act of 1986, also titled the Emergency Planning and Community Right-to-Know Act of 1986.

**Title III** (Section 372.3) means Title III of the Superfund Amendments and Reauthorization Act of 1986, also titled the Emergency Planning and Community Right-To-Know Act of 1986.

**Title V operating permit** (Section 96.2) means a permit issued under title V of the CAA and part 70 or part 71 of this chapter.

**Title V operating permit** (Section 97.2) means a permit issued under title V of the Clean Air Act and part 70 or part 71 of this chapter.

**Title V operating permit regulations** (Section 96.2) means the regulations that the Administrator has approved or issued as meeting the requirements of title V of the CAA and part 70 or 71 of this chapter.

**Title V operating permit regulations** (Section 97.2) means the regulations that the Administrator has approved or issued as meeting the requirements of title V of the Clean Air Act and part 70 or 71 of this chapter.

**Title V permit** (Section 60.2) means any permit issued, renewed, or revised pursuant to Federal or State regulations established to implement title V of the Act (42 U.S.C. 7661). A title V permit issued by a State permitting authority is called a part 70 permit in this part.

**Title V permit** (Section 61.02) means any permit issued, renewed, or revised pursuant to Federal or State regulations established to implement title V of the Act (42 U.S.C. 7661). A title V permit issued by a State permitting authority is called a part 70 permit in this part.

**Title V permit** (Section 63.2) means any permit issued, renewed, or revised pursuant to Federal or State regulations established to implement title V of the Act (42 U.S.C. 7661). A title V permit issued by a State permitting authority is called a part 70 permit in this part.

**Title V permit** (Section 65.2) means any permit issued, renewed, or revised pursuant to Federal or State regulations established under 40 CFR part 70 or 71 to implement title V of the Act (42 U.S.C. 7661).

**Tolerance** (Section 177.3) means: (1) The amount of a pesticide residue that legally may be present in or on a raw agricultural commodity under the terms of a tolerance under FFDCA section 408 or a processed food under the terms of a food additive regulation under FFDCA section 409. Tolerances are usually expressed in terms of parts of the pesticide residue per million parts of the food (ppm), by weight. (2) [Reserved] Subparts B-D [Reserved]

**Tolerance exemption** (Section 176.3) means a formal determination by the Agency pursuant to FFDCA section 408©, 21 U.S.C 346a©, that no tolerance is needed for a given pesticide chemical residue in or on a particular food commodity. For purposes of this part, the term "tolerance" shall include an exemption from the requirement of a tolerance.

**Tolerance** (Section 176.3) means the maximum amount of a pesticide chemical residue that may lawfully be present in or on a raw agricultural commodity, or processed food, or animal feed, expressed as parts per million by weight of the pesticide chemical residue in the food or feed.

**Tolerance with regional registration** (Section 180.1) means any tolerance which is established for pesticide residues resulting from the use of the pesticide pursuant to a regional registration. Such a tolerance is supported by residue data from specific growing regions for a raw agricultural commodity. Individual tolerances with regional registration are designated in separate subsections in 40 CFR 180.101 through 180.999, as appropriate. Additional residue data which are representative of the proposed use area are required to expand the geographical area of usage of a pesticide on a raw agricultural commodity having an established "tolerance with regional registration." Persons seeking geographically broader registration of a crop having a "tolerance with regional registration" should contact the appropriate EPA product manager concerning additional residue data required to expand the use area.

**Tomatoes** (Section 407.61) shall mean the processing of tomatoes into canned, peeled, whole, stewed, and related piece sizes; and processing of tomatoes into the following products and product styles: Canned, peeled and unpeeled paste, concentrate, puree, sauce, juice, catsup and other similar formulated items requiring various other pre-processed food ingredients.

**Tomato-starch-cheese canned specialties** (Section 407.81) shall mean canned specialties resulting from a combination of fresh and pre-processed tomatoes, starches, cheeses, spices, and other flavorings necessary to produce a variety of products similar to but not exclusively raviolis, spaghetti, tamales, and enchiladas.

**Ton or tonnage** (Section 72.2) means any "short ton" (i.e., 2,000 pounds). For the purpose of determining compliance with the Acid Rain emissions limitations and reduction requirements, total tons for a year shall be calculated as the sum of all recorded hourly emissions (or the tonnage equivalent of the recorded hourly emissions rates) in accordance with part 75 of this chapter, with any remaining fraction of a ton equal to or greater than 0.50 ton deemed to equal one ton and any fraction of a ton less than 0.50 ton deemed not to equal any ton.

**Ton or tonnage** (Section 96.2) means any "short ton" (i.e., 2,000 pounds). For the purpose of determining compliance with the NOX Budget emissions limitation, total tons for a control period shall be calculated as the sum of all recorded hourly emissions (or the tonnage equivalent of the recorded hourly emissions rates) in accordance with subpart H of this part, with any remaining fraction of a ton equal to or greater than 0.50 ton deemed to equal one ton and any fraction of a ton less than 0.50 ton deemed to equal zero tons.

**Ton or tonnage** (Section 97.2) means any "short ton" (i.e., 2,000 pounds). For the purpose of determining compliance with the NOX Budget emissions limitation, total tons for a control period shall be calculated as the sum of all recorded hourly emissions (or the tonnage equivalent of the recorded hourly emissions rates) in accordance with subpart H of this part, with any remaining fraction of a ton equal to or greater than 0.50 ton deemed to equal one ton and any fraction of a ton less than 0.50 ton deemed to equal zero tons.

**Too numerous to count** (Section 141.2) means that the total number of bacterial colonies exceeds 200 on a 47-mm diameter membrane filter used for coliform detection.

**Topcoat** (Section 59.101) means any coating or series of coatings applied over a primer or an existing finish for the purpose of protection or beautification.

**Topsoil stockpiling area** (Section 434.80) means the area outside the mined-out area where topsoil is temporarily stored for use in reclamation, including containment berms.

**Total annual sales** (Section 704.3) means the total annual revenue (in dollars) generated by the sale of all products of a company. Total annual sales must include the total annual sales revenue of all sites owned or controlled by that company and the total annual sales revenue of that company's subsidiaries and foreign or domestic parent company, if any.

**Total dissolved solids ("TDS")** (Section 146.3) means the total dissolved (filterable) solids as determined by use of the method specified in 40 CFR part 136.

**Total dissolved solids** (Section 122.2) means the total dissolved (filterable) solids as determined by use of the method specified in 40 CFR part 136.

**Total dissolved solids** (Section 144.3) means the total dissolved (filterable) solids as determined by use of the method specified in 40 CFR part 136.

**Total dissolved solids** (Section 191.22) means the total dissolved (filterable) solids in water as determined by use of the method specified in 40 CFR part 136.

**Total fluorides** (Section 63.601) means elemental fluorine and all fluoride compounds, including the HAP hydrogen fluoride, as measured by reference methods specified in 40 CFR part 60, appendix A, Method 13 A or B, or by equivalent or alternative methods approved by the Administrator pursuant to Sec. 63.7(f).

**Total Hydrocarbon Equivalent** (Section 86.090-2) means the sum of the carbon mass emissions of non-oxygenated hydrocarbons, methanol, formaldehyde or other organic compounds that are separately measured, expressed as gasoline-fueled vehicle hydrocarbons. In the case of exhaust emissions, the hydrogen-to-carbon ratio of the equivalent hydrocarbon is 1.85:1. In the case of diurnal and hot soak emissions, the hydrogen-to-carbon ratios of the equivalent hydrocarbons are 2.33:1 and 2.2:1, respectively.

**Total hydrocarbon equivalent** (Section 86.1803-01) means the sum of the carbon mass emissions of non-oxygenated hydrocarbons, methanol, formaldehyde or other organic compounds that are separately measured, expressed as gasoline-fueled vehicle hydrocarbons. In the case of exhaust emissions, the hydrogen-to-carbon ratio of the equivalent hydrocarbon is 1.85:1. In the case of diurnal and hot soak emissions, the hydrogen-to-carbon ratios of the equivalent hydrocarbons are 2.33:1 and 2.2:1, respectively.

**Total Hydrocarbon Equivalent** (Section 92.2) means the sum of the carbon mass contributions of non-oxygenated hydrocarbons, alcohols and aldehydes, or other organic compounds that are measured separately as contained in a gas sample, expressed as gasoline-fueled vehicle hydrocarbons. The hydrogen-to-carbon ratio of the equivalent hydrocarbon is 1.85:1. Total Hydrocarbon Equivalent is abbreviated THCE.

**Total Hydrocarbon Equivalent** (Section 94.2) means the sum of the carbon mass contributions of non-oxygenated hydrocarbons,alcohols and aldehydes, or other organic compounds that are measured separately as contained in a gas sample, expressed as petroleum-fueled engine hydrocarbons. The hydrogen-to-carbon ratio of the equivalent hydrocarbon is 1.85:1.

**Total maximum daily load (TMDL)** (Section 130.2) A TMDL is a written, quantitative plan and analysis for attaining and maintaining water quality standards in all seasons for a specific waterbody and pollutant. TMDLs may be established on a coordinated basis for a group of waterbodies in a watershed. TMDLs must be established for waterbodies on Part 1 of the list of impaired waterbodies and must include the following eleven elements: (1) The name and geographic location of the impaired waterbody; (2) Identification of the pollutant and the applicable water quality standard; (3) Quantification of the pollutant load that may be present in the waterbody and still ensure attainment and maintenance of water quality standards; (4) Quantification of the amount or degree by which the current pollutant load in the waterbody, including the pollutant load from upstream sources that is being accounted for as background loading, deviates from the pollutant load needed to attain and maintain water quality standards; (5) Identification of source categories, source subcategories or individual sources of the pollutant; (6) Wasteload allocations; (7) Load allocations; (8) A margin of safety; (9) Consideration of seasonal variations; (10) Allowance for reasonably foreseeable increases in pollutant loads including future growth; and (11) An implementation plan. (i) Total Maximum Daily Thermal Load (TMDTL). A TMDTL is a TMDL for impaired waterbodies receiving a thermal discharge.

**Total maximum daily load (TMDL)** (Section 130.2) The sum of the individual WLAs for point sources and LAs for nonpoint sources and natural background. If a receiving water has only one point source discharger, the TMDL is the sum of that point source WLA plus the LAs for any nonpoint sources of pollution and natural background sources, tributaries, or adjacent segments. TMDLs can be expressed in terms of either mass per time, toxicity, or other appropriate measure. If Best Management Practices (BMPs) or other nonpoint source pollution controls make more stringent load allocations practicable, then wasteload allocations can be made less stringent. Thus, the TMDL process provides for nonpoint source control tradeoffs.

**Total maximum daily load (TMDL)** (Section 132.2) is the sum of the individual wasteload allocations for point sources and load allocations for nonpoint sources and natural background, as more fully defined at 40 CFR 130.2(i). A TMDL sets and allocates the maximum amount of a pollutant that may be introduced into a water body and still assure attainment and maintenance of water quality standards.

**Total metal** (Section 413.02) is defined as the sum of the concentration or mass of Copper (Cu), Nickel (Ni), Chromium (Cr) (total) and Zinc (Zn).

**Total of direct and indirect emissions** (Section 93.152) means the sum of direct and indirect emissions increases and decreases caused by the Federal action; i.e., the "net" emissions considering all direct and indirect emissions. The portion of emissions which are exempt or presumed to conform under Sec. 93.153 ©, (d), (e), or (f) are not included in the "total of direct and indirect emissions." The "total of direct and indirect emissions" includes emissions of criteria pollutants and emissions of precursors of criteria pollutants.

**Total Organic Carbon (TOC)** (Section 141.2) means total organic carbon in mg/L measured using heat, oxygen, ultraviolet irradiation, chemical oxidants, or combinations of these oxidants that convert organic carbon to carbon dioxide, rounded to two significant figures.

**Total organic compounds or TOC** (Section 65.2) means those compounds measured according to the procedures specified in Sec. 65.64© and Sec. 65.158(b)(3)(ii)(A), as applicable. Those compounds that the Administrator has determined do not contribute appreciably to the formation of ozone and that are specifically excluded from the definition of volatile organic compound at 40 CFR 51.100(s), as amended, are to be excluded for the purposes of measuring the hourly emission rate as required in Sec. 65.64(f) for process vents subject to subpart III, NNN, or RRR of part 60 of this chapter.

**Total phenols** (Section 464.02) shall mean total phenolic compounds as measured by the procedure listed in 40 CFR part 136 (distillation followed by colorimetric--4AAP).

**Total planned net output capacity** (Section 72.2) means the planned generator output capacity, excluding that portion of the electrical power which is designed to be used at the power production facility, as specified under one or more qualifying power purchase commitments or contemporaneous documents as of November 15, 1990; "Total installed net output capacity" shall be the generator output capacity, excluding that portion of the electrical power actually used at the power production facility, as installed.

**Total residual chlorine (or total residual oxidants for intake water with bromides)** (Section 423.11) means the value obtained using the amperometric method for total residual chlorine described in 40 CFR part 136.

**Total resource effectiveness index value or TRE index value** (Section 65.2) means a calculated value used to determine whether control is required for a process vent. It is based on process vent flow rate, emission rate of regulated material, net heating value, and corrosion properties (halogenated compound content), as quantified by the equations given under Sec. 65.64(h).

**Total solids** (Section 503.31) are the materials in sewage sludge that remain as residue when the sewage sludge is dried at 103 to 105 degrees Celsius.

**Total suspended particulate** (Section 51.100) means particulate matter as measured by the method described in appendix B of part 50 of this chapter.

**Total toxic organics (TTO)** (Section 464.02) shall mean the sum of the mass of each of the toxic organic compounds which are found at a concentration greater than 0.010 mg/l. The specialized definitions for each subpart contain a discrete list of toxic organic compounds comprising TTO for each process segment in which TTO is regulated.

**Total Toxic Organics (TTO)** (Section 464.21) TTO is a regulated parameter under PSES (Sec. 464.25) and PSNS (Sec. 464.26) for the copper subcategory and is comprised of a discrete list of toxic organic pollutants for each process segment where it is regulated, as follows: (1) Casting Quench (Sec. 464.25(a) and Sec. 464.26(a)): 23. chloroform (trichloromethane), 64. pentachlorophenol, 66. bis(2-ethylhexyl)phthalate, 71. dimethyl phthalate, (2) Dust Collection Scrubbers (Sec. 464.25© and 464.26©): 1. acenaphthene, 22. para-chlorometa-cresol,23. chloroform (trichloromethane), 34. 2,4-dimethylphenol, 55. naphthalene, 58. 4-nitrophenol, 64. pentachlorophenol, 65. phenol, 66. bis(2-ethylhexyl)phthalate, 67. butyl benzyl phthalate, 68. di-n-butyl phthalate, 70. diethyl phthalate, 71. dimethyl phthalate, 72. benzo(a)anthracene (1,2-benzanthracene), 74. 3,4-benzofluoranthene, 75. benzo(k) fluoranthene, 76. chrysene, 77. acenaphthylene 78. anthracene, 81. phenanthrene, 84. pyrene, (3) Investment Casting (Sec. 464.25(e) and Sec. 464.26 (e)): 1. acenaphthene, 22. para-chloro meta-cresol, 23. chloroform (trichloromethane), 34. 2,4-dimethylphenol, 55. naphthalene, 58. 4-nitrophenol, 64. pentachlorophenol, 65. phenol, 66. bis (2-ethylhexyl)phthalate, 67. butyl benzyl phthalate, 68. di-n-butyl phthalate, 70. diethyl phthalate, 71. dimethyl phthalate, 72. benzo (a) anthracene (1,2-

benzanthracene), 74. 3,4-benzofluoranthene, 75. benzo(k) fluoranthene, 76. chrysene, 77. acenaphthylene, 78. anthracene, 81. Phenanthrene, 84. pyrene, (4) Melting Furnace Scrubber (Sec. 464.25(f) and Sec. 464.26(f)): 1. acenaphthene, 22. para-chloro meta-cresol, 23. chloroform (trichloromethane), 34. 2,4-dimethylphenol, 55. naphthalene, 58. 4-nitrophenol, 64. pentachlorophenol, 65. phenol, 66. bis (2-ethylhexyl) phthalate, 67. butyl benzyl phthalate, 68. di-n-butyl phthalate, 70. diethyl phthalate, 71. dimethyl phthalate, 72. benzo(a)anthracene (1,2-benzanthracene), 74. 3,4-benzoflouranthene, 75. benzo(k) flouranthene, 76. chrysene, 77. acenaphthylene, 78. anthracene, 81. phenanthrene, 84. pyrene, (5) Mold Cooling (Sec. 464.25(g) and Sec. 464.26(g)): 23. chloroform (trichloromethane), 64. pentachlorophenol, 66. bis (2-ethylhexyl)phthalate, 71. dimethyl phthalate.

**Total Toxic Organics (TTO)** (Section 464.31) TTO is a regulated parameter under PSES (Sec. 464.35) and PSNS (Sec. 464.36) for the ferrous subcategory and is comprised of a discrete list of toxic organic pollutants for each process segment where it is regulated, as follows: (1)Casting Quench (Sec. 464.35(b) and Sec. 464.36(b)): 23. chloroform (trichloromethane), 34. 2,4-dimethylphenol, (2) Dust Collection Scrubber (Sec. 464.35© and Sec. 464.36(b)): 1. acenaphthene, 23. chloroform (trichloromethane), 31. 2,4-dichlorophenol, 34. 2,4-dimethylphenol, 39. fluoranthene, 44. methylene chloride (dichloromethane), 55.naphthalene, 64. pentachlorophenol, 65. phenol, 66.bis (2-ethylhexyl)phthalate, 67. butyl benzyl phthalate, 68. di-n-butyl phthalate, 70. diethyl phthalate, 71.dimethyl phthalate, 72. benzo (a)anthracene (1,2-benzanthracene), 76. chrysene, 77.acenaphthylene, 78. anthracene, 80. fluorine, 81.phenanthrene, 84.pyrene, (3) Investment Casting (Sec. 464.35(e) and Sec. 464.36(e)): 23. chloroform (trichloromethane), 44. methylene chloride (dichloromethane), 66. bis (2-ethylhexyl) phthalate, 77. acenaphthylene, 84. pyrene, (4) Melting Furnace Scrubber (Sec. 464.35(f) and Sec. 464.36(f)): 23. chloroform (trichloromethane), 31. 2,4-dichlorophenol, 34. 2,4-dimethylphenol, 39. fluoranthene, 44. methylene chloride (dichloromethane), 55. naphthalene, 65. phenol, 66. bis (2-ethylhexyl) phthalate, 67.butyl benzyl phthalate, 68.di-n-butyl phthalate, 72. benzo (a)anthracene (1,2-benzanthracene), 76.chrysene, 77.acenaphthylene, 78. anthracene, 80. fluorine, 81.phenanthrene, 84. pyrene, (5) Mold Cooling (Sec. 464.35(g) and Sec. 464.36(g)): 23. chloroform (trichloromethane), 34. 2,4-dimethylphenol, (6) Slag Quench (Sec. 464.35(h) and Sec. 464.36(h)): 34. 2,4-dimethylphenol, 71. dimethyl phthalate, (7) Wet Sand Reclamation (Sec. 464.35(i) and Sec. 464.36(i)): 1. acenaphthene, 34. 2,4-dimethylphenol, 39. fluoranthene, 44. methylene chloride (dichloromethane), 55. naphthalene, 65. phenol, 66. bis (2-ethylhexyl) phthalate, 68. di-n-butyl phthalate, 70.diethyl phthalate, 71.dimethyl phthalate, 72.benzo(a)anthracene (1,2-benzanthracene), 77. acenaphthylene, 84. pyrene, (b) Cast Iron. An iron containing carbon in excess of the solubility in the austentite that exists in the alloy at the eutectic temperature. Cast iron also is defined here to include any iron-carbon alloys containing 1.2 percent or more carbon by weight.

**Total Toxic Organics (TTO)** (Section 464.41) TTO is a regulated parameter under PSES (Sec. 464.45) and PSNS (Sec. 464.46) for the zinc subcategory and is comprised of a discrete list of toxic organic pollutants for each process segment where it is regulated, as follows: (1) Casting Quench (Sec. 464.45(a) and Sec. 464.46(a)): 21.2,4,6-trichlorophenol, 22.para-chloro meta-cresol, 31. 2,4-dichlorophenol, 34. 2,4-dimethylphenol, 39. fluoranthene, 44. methylene chloride (dichloromethane), 65.phenol, 66.bis(2-ethylhexyl) phthalate, 68.di-n-butyl phthalate, 70.diethyl phthalate, 85. tetrachloroethylene, (2) Die Casting (Sec. 465.45(b) and Sec. 464.46(b)): 1. acenaphthene, 21.2,4,6-trichlorophenol, 22.para-chloro meta-cresol, 24.2-chlorophenol, 34. 2,4-dimethylphenol, 44. methylene chloride (dichloromethane), 55. naphthalene, 65.phenol, 66.bis (2-ethylhexyl) phthalate, 68.di-n-butyl phthalate, 70.diethyl phthalate,

85.tetrachloroethylene, 86.toluene, 87.trichloroethylene, (3) Melting Furnace Scrubber (Sec. 464.45© and Sec. 464.46©): 31. 2,4-dichlorophenol, 34. 2,4-dimethylphenol, 39. fluoranthene, 44. methylene chloride (dichloromethane), 55. naphthalene, 65.phenol, 66.bis(2-ethylhexyl) phthalate, 68.di-n-butyl phthalate, 85.tetrachloroethylene, 86.toluene, 87.trichloroethylene, (4) Mold Cooling (Sec. 464.45(d) and Sec. 464.46(d)): 21.2,4,6-trichlorophenol, 22.para-chloro meta-cresol, 31. 2,4-dichlorophenol, 34. 2,4-dimethylphenol, 39. fluoranthene, 44. methylene chloride (dichloromethane), 65.phenol, 66.bis(2-ethylhexyl) phthalate, 68. di-n-butylphthalate, 70.diethyl phthalate, 85. tetrachloroethylene.

**Total Toxic Organics (TTO)** (Section 465.02) shall mean the sum of the mass of each of the following toxic organic compounds which are found at a concentration greater than 0.010 mg/1. 1,1,1-Trichloroethane, 1,1-Dichloroethane, 1,1,2,2-Tetrachloroethane, Bis (2-chloroethyl) ether, Chloroform, 1,1-Dichloroethylene, Methylene chloride (dichloromethane), Pentachlorophenol, Bis (2-ethylhexyl) phthalate, Butyl benzyl-phthalate, Di-N-butyl phthalate, Phenanthrene, Tetrachloroethylene, Toluene.

**Total Toxic Organics (TTO)** (Section 467.02) shall mean the sum of the masses or concentrations of each of the following toxic organic compounds which is found in the discharge at a concentration greater than 0.010 mg/1: p-chloro-m-cresol, 2-chlorophenol, 2,4-dinitrotoluene, 1,2-diphenylhydrazine, ethyblenzene, fluoranthene, isophorone, naphthalene, N-nitro sodi phenyl amine, phenol, benzo(a) pyrene, benzo(ghi)perylene, fluorine, phenanthrene, dibenzo(a,h), anthracene, indeno(1,2,3-c,d)pyrene, pyrene, tetrachloroethylene, toluene, trichloroethylene, endosulfan sulfate, bis(2-ethyl hexyl)phthalate, diethylphthalate, 3,4-benzofluoranthene, benzo(k)fluoranthene, chrysene, acenaphthylene, anthracene, di-n-butyl phthalate, endrin, endrin aldehyde, PCB-1242, 1254, 1221, PCB-1232, 1248, 1260, 1016, acenaphthene, (r) Stationary casting is the pouring of molten aluminum into molds and allowing the metal to air cool.

**Total Toxic Organics (TTO)** (Section 468.02) shall mean the sum of the masses or concentrations of each of the following toxic organic compounds which is found at a concentration greater than 0.010 mg/l. Benzene, 1,1,1-Trichloroethane chloroform, 2,6-Dinitrotoluene ethylbenzene methylene chloride naphthalene, N-nitrosodiphenylamine, anthracene phenanthrene toluene trichloroethylene.

**Total Toxic Organics (TTO)** (Section 469.12) means the sum of the concentrations for each of the following toxic organic compounds which is found in the discharge at a concentration greater than ten (10) micrograms per liter: 1,2,4 Trichlorobenzene chloroform, 1,2 Dichlorobenzene, 1,3, Dichlorobenzene, 1,4, Dichlorobenzene ethylbenzene, 1,1,1 Trichloroethane methylene chloride naphthalene, 2 Nitrophenol phenol bis (2-ethylhexyl) phthalate tetrachloroethylene, toluene trichloroethylene, 2 Chlorophenol, 2,4 Dichlorophenol, 4 Nitrophenol pentachlorophenol di-n-butyl phthalate anthracene, 1,2 Diphenylhydrazine isophorone butyl benzyl phthalate, 1,1 Dichloroethylene, 2,4,6 Trichlorophenol carbon tetrachloride, 1,2 Dichloroethane, 1,1,2 Trichloroethane dichlorobromomethane.

**Total Toxic Organics (TTO)** (Section 469.31) means the sum of the concentrations for each of the following toxic organic compounds which is found in the discharge at a concentration greater than ten (10) micrograms per liter: 1,1,1 Chloroform, Trichloroethane, Methylene chloride, Bis (2-ethylhexyl) phthalate, Toluene, Trichloroethylene.

**Total toxic organics (TTO).** (Section 464.11) TTO is a regulated parameter under PSES (Sec. 464.15) and PSNS (Sec. 464.16) for the aluminum subcategory and is comprised of a discrete list of toxic organic pollutants for each process segment where it is regulated, as follows: (1) Casting Quench (Sec. 464.15(b) and Sec. 464.16(b)): 4. benzene, 21. 2,4,6-trichlorophenol, 22. Para-chloro meta-cresol, 23. chloroform (trichloromethane), 34. 2,4-dimethylphenol,

39. fluoranthene, 44. methylene chloride (dichloromethane), 65. phenol, 66. bis(2-ethylhexyl) phthalate, 67. butyl benzyl phthalate, 84. pyrene, 85. tetrachloroethylene, 87. trichloroethylene, (2) Die Casting (Sec. 464.15© and Sec. 464.16©): 1. acenaphthene, 4. benzene, 7. chlorobenzene, 11. 1,1,1-trichloroethane, 21., 4,6-trichlorophenol, 22. para-chloro meta-cresol, 23. chloroform (trichloromethane), 34. 2,4-dimethylphenol, 39. fluoranthene, 44. methylene chloride (dichloromethane), 55. naphthalene, 65. phenol, 66. bis(2-ethylhexyl) phthalate, 67. butyl benzyl phthalate, 68. di-n-butyl phthalate, 70. diethyl phthalate, 72. benzo (a)anthracene (1,2-enzanthracene), 73. benzo (a)pyrene (3,4-benzopyrene), 76. chrysene, 78. anthracene, 80. fluorine, 81. phenanthrene, 84. pyrene, 85. tetrachloroethylene, 86. toluene, (3) Dust Collection Scrubber (Sec. 464.15(d) and Sec. 464.16(d)): 1.acenaphthene, 21. 2,4,6-trichlorophenol, 23. chloroform (trichloromethane), 34. 2,4-dimethylphenol, 39. fluoranthene, 44. methylene chloride (dichloromethane), 65. phenol, 66. bis (2-ethylhexyl) phthalate, 68. di-n-butyl phthalate, 70. diethyl phthalate, 73. benzo (a)pyrene (3,4-benzopyrene), 84. pyrene, (4) Investment Casting (Sec. 464.15(f) and Sec. 464.16(f)): 11. 1,1,1-trichloroethane, 23. chloroform (trichloromethane), 44. methylene chloride (dichloromethane), 66. bis (2-ethylhexyl) phthalate. 84. pyrene, 85. tetrachloroethylene, 87. trichloroethylene, (5) Melting Furnace Scrubber (Sec. 464.15(g) and Sec. 464.16(g)): 1. acenaphthene, 21. 2,4,6-trichlorophenol, 23. chloroform (trichloromethane), 34. 2,4-dimethylphenol, 39. fluoranthene, 44. methylene chloride (dichloromethane), 65. phenol, 66. bis (2-ethylhexyl) , phthalate, 68. di-n-butyl phthalate, 70. diethyl phthalate, 73. benzo (a)pyrene (3,4-benzopyrene), 84. pyrene, (6) Mold Cooling (Sec. 464.15(h) and Sec. 464.16(h)): 4. benzene, 21. 2,4,6-trichlorophenol, 22. para-chloro meta-cresol, 23. chloroform (trichloromethane), 34. 2,4-dimethylphenol, 39. fluoranthene, 44. methylene chloride, 65. phenol, 66. bis(2-ethylhexyl) phthalate, 67. butyl benzyl phthalate, 84. pyrene, 85. tetrachloroethylene, 87. trichloroethylene.

**Total trihalomethanes (TTHM)** (Section 141.2) means the sum of the concentration in milligrams per liter of the trihalomethane compounds (trichloromethane [chloroform], dibromochloromethane, bromodichloromethane and tribromomethane [bromoform]), rounded to two significant figures.

**Totally chlorine-free (TCF) bleaching** (Section 430.0) Pulp bleaching operations that are performed without the use of chlorine, sodium hypochlorite, calcium hypochlorite, chlorine dioxide, chlorine monoxide, or any other chlorine-containing compound.

**Totally enclosed manner** (Section 761.13) means any manner that will ensure no exposure of human beings or the environment to any concentration of PCBs.

**Totally enclosed treatment facility** (Section 260.10) means a facility for the treatment of hazardous waste which is directly connected to an industrial production process and which is constructed and operated in a manner which prevents the release of any hazardous waste or any constituent thereof into the environment during treatment. An example is a pipe in which waste acid is neutralized.

**Touch-up coating** (Section 59.101) means a coating applied by brush, air-brush, or nonrefillable aerosol can to cover minor surface damage.

**Toxic chemical** (Section 372.3) means a chemical or chemical category listed in Sec. 372.65.

**Toxic pollutant** (Section 122.2) means any pollutant listed as toxic under section 307(a)(1) or, in the case of "sludge use or disposal practices," any pollutant identified in regulations implementing section 405(d) of the CWA.

**Toxic pollutant** (Section 501.2) means any pollutant listed as toxic under section 307(a)(1) or any pollutant identified in regulations implementing section 405(d) of the CWA.

**Toxic pollutants** (Section 131.3) are those pollutants listed by the Administrator under section 307(a) of the Act.

**Toxicity** (Section 171.2) means the property of a pesticide to cause any adverse physiological effects.

**TPQ** (Section 370.2) means the threshold planning quantity for an extremely hazardous substance as defined in 40 CFR part 355.

**Traceable** (Section 50.1) means that a local standard has been compared and certified either directly or via not more than one intermediate standard, to a primary standard such as a National Bureau of Standards Standard Reference Material (NBS SRM), or a USEPA/NBS-approved Certified Reference Material (CRM).

**Traceable** (Section 58.1) means that a local standard has been compared and certified, either directly or via not more than one intermediate standard, to a National Institute of Standards and Technology (NIST)-certified primary standard such as a NIST-Traceable Reference Material (NTRM) or a NIST-certified Gas Manufacturer's Internal Standard (GMIS).

**Tractor** (Section 205.151) means for the purposes of this subpart, any two or three wheeled vehicle used exclusively for agricultural purposes, or for snow plowing, including self-propelled machines used exclusively in growing, harvesting or handling farm produce.

**Trade name product** (Section 372.3) means a chemical or mixture of chemicals that is distributed to other persons and that incorporates a toxic chemical component that is not identified by the applicable chemical name or Chemical Abstracts Service Registry number listed in Sec. 372.65.

**Trade secrecy claim** (Section 350.1) is a submittal under sections 303 (d)(2) or (d)(3), 311, 312 or 313 of Title III in which a chemical identity is claimed as trade secret, and is accompanied by a substantiation in support of the claim of trade secrecy for chemical identity.

**Trade secret** (Section 350.1) means any confidential formula, pattern, process, device, information or compilation of information that is used in a submitter's business, and that gives the submitter an opportunity to obtain an advantage over competitors who do not know or use it. EPA intends to be guided by the Restatement of Torts, Section 757, Comment b.

**Trading** (Section 86.090-2) means the exchange of heavy-duty engine NO<INF>X</INF> or particulate emission credits between manufacturers.

**Trading** (Section 86.1803-01) means the exchange of complete heavy-duty vehicle NOX emission credits between manufacturers.

**Trading** (Section 88.202-94) means the exchange of credits between manufacturers.

**Trading** (Section 91.1302) means the exchange of marine SI engine in-use emission credits between manufacturers and/or brokers.

**Trading** (Section 92.2) means the exchange of locomotive or locomotive engine emission credits between credit holders.

**Trading** (Section 94.2) means the exchange of engine emission credits between credit holders.

**Trainee** (Section 45.115) A student selected by the recipient organization who receives support to meet the objectives in Sec. 45.110.

**Training curriculum** (Section 745.223) means an established set of course topics for instruction in an accredited training program for a particular discipline designed to provide specialized knowledge and skills.

**Training hour** (Section 745.223) means at least 50 minutes of actual learning, including, but not limited to, time devoted to lecture, learning activities, small group activities, demonstrations, evaluations, and/or hands-on experience.

**Training manager** (Section 745.223) means the individual responsible for administering a training program and monitoring the performance of principal instructors and guest instructors.

**Transfer** (Section 92.302) means to convey control of credits generated from an individual locomotive to the purchaser, owner or operator of the locomotive at the time of manufacture or remanufacture; or to convey control of previously generated credits from the purchaser, owner or operator of an individual locomotive to the manufacturer or remanufacturer at the time of manufacture or remanufacture.

**Transfer facility** (Section 260.10) means any transportation related facility including loading docks, parking areas, storage areas and other similar areas where shipments of hazardous waste are held during the normal course of transportation.

**Transfer facility** (Section 270.2) means any transportation-related facility including loading docks, parking areas, storage areas and other similar areas where shipments of hazardous waste are held during the normal course of transportation.

**Transfer facility** (Section 761.13) means any transportation-related facility including loading docks, parking areas, and other similar areas where shipments of PCB waste are held during the normal course of transportation. Transport vehicles are not transfer facilities under this definition, unless they are used for the storage of PCB waste, rather than for actual transport activities. Storage areas for PCB waste at transfer facilities are subject to the storage facility standards of Sec. 761.65, but such storage areas are exempt from the approval requirements of Sec. 761.65(d) and the recordkeeping requirements of Sec. 761.180, unless the same PCB waste is stored there for a period of more than 10 consecutive days between destinations.

**Transfer station** (Section 243.101) means a site at which solid wastes are concentrated for transport to a processing facility or land disposal site. A transfer station may be fixed or mobile.

**Transfer unit** (Section 72.2) means a Phase I unit that transfers all or part of its Phase I emission reduction obligations to a control unit designated pursuant to a Phase I extension plan under Sec. 72.42.

**Transferee** (Section 144.3) means the owner or operator receiving ownership and/or operational control of the well.

**Transferor** (Section 144.3) means the owner or operator transferring ownership and/or operational control of the well.

**Transform** (Section 82.3) means to use and entirely consume (except for trace quantities) a controlled substance in the manufacture of other chemicals for commercial purposes.

**Transformation Credits** (Section 82.3) means those privileges that may be obtained under Sec. 82.9 to produce controlled substances.

**Transfrontier movement** (Section 262.81) means any shipment of wastes destined for recovery operations from an area under the national jurisdiction of one OECD member country to an area under the national jurisdiction of another OECD member country.

**Transhipment** (Section 82.3) means the continuous shipment of a controlled substance, from a foreign state of origin through the United States or its territories, to a second foreign state of final destination, as long as the shipment does not enter into United States jurisdiction. A transhipment, as it moves through the United States or its territories, cannot be re-packaged, sorted or otherwise changed in condition.

**Transient non-community water system or TWS** (Section 141.2) means a non-community water system that does not regularly serve at least 25 of the same persons over six months per year.

**Transit** (Section 93.101) is mass transportation by bus, rail, or other conveyance which provides general or special service to the public on a regular and continuing basis. It does not include school buses or charter or sightseeing services.

**Transit country** (Section 262.51) means any foreign country, other than a receiving country, through which a hazardous waste is transported.

**Transit project** (Section 93.101) is an undertaking to implement or modify a transit facility or transit-related program; purchase transit vehicles or equipment; or provide financial assistance for transit operations. It does not include actions that are solely within the jurisdiction of local transit agencies, such as changes in routes, schedules, or fares. It may consist of several phases. For analytical purposes, it must be defined inclusively enough to: (1) Connect logical termini and be of sufficient length to address environmental matters on a broad scope; (2) Have independent utility or independent significance, i.e., be a reasonable expenditure even if no additional transportation improvements in the area are made; and (3) Not restrict consideration of alternatives for other reasonably foreseeable transportation improvements.

**Transitional low emission vehicle (TLEV)** (Section 86.1702-99) means any vehicle certified to the transitional low emission vehicle standards specified in this subpart.

**Transitional Low-Emission Vehicle** (Section 88.102-94) means any light-duty vehicle or light-duty truck conforming to the applicable Transitional Low-Emission Vehicle standard.

**Transmission Class** (Section 600.022-85) means a group of transmissions having the following common features: Basic transmission type (manual, automatic, or semi-automatic); number of forward gears used in fuel economy testing (e.g., manual four-speed, three-speed automatic, two-speed semi-automatic); drive system (e.g., front wheel drive, rear wheel drive; four wheel drive), type of overdrive, if applicable (e.g., final gear ratio less than 1.00, separate overdrive unit); torque converter type, if applicable (e.g., non-lockup, lockup, variable ratio); and other transmission characteristics that may be determined to be significant by the Administrator.

**Transmission class** (Section 86.082-2) means the basic type of transmission, e.g., manual, automatic, semiautomatic.

**Transmission class** (Section 86.1803-01) means the basic type of transmission, e.g., manual, automatic, semiautomatic.

**Transmission Configuration** (Section 600.022-85) means the Administrator may further subdivide within a transmission class if the Administrator determines that sufficient fuel economy differences exist. Features such as gear ratios, torque converter multiplication ratio, stall speed, shift calibration, or shift speed may be used to further distinguish characteristics within a transmission class.

**Transmission configuration** (Section 86.082-2) means a unique combination, within a transmission class, of the number of the forward gears and, if applicable, overdrive. The Administrator may further subdivide a transmission configuration (based on such criteria as gear ratios, torque convertor multiplication ratio, stall speed and shift calibration, etc.), if he determines that significant fuel economy or exhaust emission differences exist within that transmission configuration.

**Transmission configuration** (Section 86.1803-01) means a unique combination, within a transmission class, of the number of the forward gears and, if applicable, overdrive. The Administrator may further subdivide a transmission configuration (based on such criteria as gear ratios, torque convertor multiplication ratio, stall speed and shift calibration, etc.), if she/he determines that significant fuel economy or exhaust emission differences exist within that transmission configuration.

**Transmissive fault or fracture** (Section 148.2) is a fault or fracture that has sufficient permeability and vertical extent to allow fluids to move between formations.

**Transport vehicle** (Section 260.10) means a motor vehicle or rail car used for the transportation of cargo by any mode. Each cargo-carrying body (trailer, railroad freight car, etc.) is a separate transport vehicle.

**Transport vehicle** (Section 761.13) means a motor vehicle or rail car used for the transportation of cargo by any mode. Each cargo-carrying body (e.g., trailer, railroad freight car) is a separate transport vehicle.

**Transportation** (Section 260.10) means the movement of hazardous waste by air, rail, highway, or water.

**Transportation control measure** (Section 51.100) means any measure that is directed toward reducing emissions of air pollutants from transportation sources. Such measures include, but are not limited to, those listed in section 108(f) of the Clean Air Act.

**Transportation control measure (TCM)** (Section 93.101) is any measure that is specifically identified and committed to in the applicable implementation plan that is either one of the types listed in section 108 of the CAA, or any other measure for the purpose of reducing emissions or concentrations of air pollutants from transportation sources by reducing vehicle use or changing traffic flow or congestion conditions. Notwithstanding the first sentence of this definition, vehicle technology-based, fuel-based, and maintenance-based measures which control the emissions from vehicles under fixed traffic conditions are not TCMs for the purposes of this subpart.

**Transportation equipment cleaning (TEC) process wastewater** (Section 442.2) means all wastewaters associated with cleaning the interiors of tanks including: tank trucks; rail tank cars; intermodal tank containers; tank barges; and ocean/sea tankers used to transport commodities or cargos that come into direct contact with the interior of the tank or container. At those facilities that clean tank interiors, TEC process wastewater also includes wastewater generated from washing vehicle exteriors, equipment and floor washings, TEC-contaminated stormwater, wastewater prerinse cleaning solutions, chemical cleaning solutions, and final rinse solutions. TEC process wastewater is defined to include only wastewater generated from a regulated TEC subcategory. Therefore, TEC process wastewater does not include wastewater generated from cleaning hopper cars, or from food grade facilities discharging to a POTW. Wastewater generated from cleaning tank interiors for purposes of shipping products (i.e., cleaned for purposes other than maintenance and repair) is considered TEC process wastewater. Wastewater generated from cleaning tank interiors for the purposes of maintenance and repair on the tank is not considered TEC process wastewater. Facilities that clean tank interiors solely for the purposes of repair and maintenance are not regulated under this Part.(b) The parameters regulated in this part and listed with approved methods of analysis in Table IB at 40 CFR 136.3, are defined as follows: (1) BOD5 means 5-day biochemical oxygen demand. (2) Cadmium means total cadmium. (3) Chromium means total chromium. (4) Copper means total copper. (5) Lead means total lead. (6) Mercury means total mercury (7) Nickel means total nickel. (8) Oil and Grease (HEM) means oil and grease (Hexane-Extractable Material) measured by Method 1664. (9) Non-polar material (SGT-HEM) means the non-polar fraction of oil and grease (Silica Gel Treated Hexane-Extractable Material) measured by Method 1664. (10) TSS means total suspended solids. (11) Zinc means total zinc. (c) The parameters regulated in this part and listed with approved methods of analysis in Table IC at 40 CFR 136.3, are as follows: (1) Fluoranthene. (2) Phenanthrene.

**Transportation improvement program (TIP)** (Section 93.101) means a staged, multiyear, intermodal program of transportation projects covering a metropolitan planning area which is consistent with the metropolitan transportation plan, and developed pursuant to 23 CFR part 450.

**Transportation plan** (Section 93.101) means the official intermodal metropolitan transportation plan that is developed through the metropolitan planning process for the metropolitan planning area, developed pursuant to 23 CFR part 450.

**Transportation project** (Section 93.101) is a highway project or a transit project.

**Transportation-related and non-transportation-related** (Section 112.2) as applied to an onshore or offshore facility, are defined in the Memorandum of Understanding between the Secretary of Transportation and the Administrator of the Environmental Protection Agency, dated November 24, 1971, 36 FR 24080.

**Transporter** (Section 260.10) means a person engaged in the offsite transportation of hazardous waste by air, rail, highway, or water.

**Transporter** (Section 270.2) means a person engaged in the off-site transportation of hazardous waste by air, rail, highway or water.

**Transporter of PCB waste** (Section 761.13) means, for the purposes of subpart K of this part, any person engaged in the transportation of regulated PCB waste by air, rail, highway, or water for purposes other than consolidation by a generator.

**Transuranic radioactive waste** (Section 191.02) as used in this part, means waste containing more than 100 nanocuries of alpha-emitting transuranic isotopes, with half-lives greater than twenty years, per gram of waste, except for: (1) High-level radioactive wastes; (2) wastes that the Department has determined, with the concurrence of the Administrator, do not need the degree of isolation required by this part; or (3) wastes that the Commission has approved for disposal on a case-by-case basis in accordance with 10 CFR Part 61.

**Trap oxidizer system** (Section 86.1702-99) means an emission control system which consists of a trap to collect particulate matter and a mechanism to oxidize the accumulated particulate.

**TRC (or total residual chlorine)** (Section 420.02) means the value obtained by the iodometric titration with an amperometric endpoint method specified in 40 CFR 136.3.

**Treatability Study** (Section 260.10) means a study in which a hazardous waste is subjected to a treatment process to determine: (1) Whether the waste is amenable to the treatment process, (2) what pretreatment (if any) is required, (3) the optimal process conditions needed to achieve the desired treatment, (4) the efficiency of a treatment process for a specific waste or wastes, or (5) the characteristics and volumes of residuals from a particular treatment process. Also included in this definition for the purpose of the Sec. 261.4 (e) and (f) exemptions are liner compatibility, corrosion, and other material compatibility studies and toxicological and health effects studies. A "treatability study" is not a means to commercially treat or dispose of hazardous waste.

**Treatability Study** (Section 761.13) means a study in which PCB waste is subjected to a treatment process to determine: (1) Whether the waste is amenable to the treatment process; (2) What pretreatment (if any) is required: (3) The optimal process conditions needed to achieve the desired treatment; (4) The efficiency of a treatment process for the specific type of waste (i.e., soil, sludge, liquid, etc.); or, (5) The characteristics and volumes of residuals from a particular treatment process. A "treatability study" is not a mechanism to commercially treat or dispose of PCB waste. Treatment is a form of disposal under this part.

**Treated area** (Section 170.3) means any area to which a pesticide is being directed or has been directed.

**Treatment** (Section 260.10) means any method, technique, or process, including neutralization, designed to change the physical, chemical, or biological character or composition of any hazardous waste so as to neutralize such waste, or so as to recover energy or material resources from the waste, or so as to render such waste non-hazardous, or less hazardous; safer to transport, store, or dispose of; or amenable for recovery, amenable for storage, or reduced in volume.

**Treatment** (Section 270.2) means any method, technique, or process, including neutralization, designed to change the physical, chemical, or biological character or composition of any hazardous waste so as to neutralize such wastes, or so as to recover energy or material resources from the waste, or so as to render such waste non-hazardous, or less hazardous; safer to transport, store, or dispose of; or amenable for recovery, amenable for storage, or reduced in volume.

**Treatment** (Section 437.2) means any method, technique, or process designed to change the physical, chemical or biological character or composition of any metal-bearing, oily, or organic wastes to neutralize such wastes; to render such wastes amenable to discharge; or to recover energy or recover metal, oil, or organic content from the wastes. Treatment does not include (a) the re-use of treated or untreated wastewater in place of potable or pure water in industrial processes such as the use of secondary POTW effluents as non-contact cooling water or storm water in place of process water or (b) the re-use of treated or untreated spent chemicals (such as pickle liquor) as treatment chemicals.

**Treatment facility and treatment system** (Section 434.11) mean all structures which contain, convey, and as necessay, chemically or physically treat coal rine drainage, coal preparation plant process wastewater, or drainage from coal preparation plant associated areas, which remove pollutants regulated by this part from such waters. This includes all pipes, channels, ponds, basins, tanks and all other equipment serving such structures.

**Treatment for destruction** (Section 372.3) means the destruction of a toxic chemical in waste such that the substance is no longer the toxic chemical subject to reporting under EPCRA section 313. Treatment for destruction does not include the destruction of a toxic chemical in waste where the toxic chemical has a heat value greater than 5,000 British thermal units and is combusted in any device that is an industrial furnace or boiler.

**Treatment technique requirement** (Section 142.2) means a requirement of the national primary drinking water regulations which specifies for a contaminant a specific treatment technique(s) known to the Administrator which leads to a reduction in the level of such contaminant sufficient to comply with the requirements of part 141 of this chapter.

**Treatment technology** (Section 300.5) means any unit operation or series of unit operations that alters the composition of a hazardous substance or pollutant or contaminant through chemical, biological, or physical means so as to reduce toxicity, mobility, or volume of the contaminated materials being treated. Treatment technologies are an alternative to land disposal of hazardous wastes without treatment.

**Treatment works treating domestic sewage** (Section 122.2) means a POTW or any other sewage sludge or waste water treatment devices or systems, regardless of ownership (including federal facilities), used in the storage, treatment, recycling, and reclamation of municipal or domestic sewage, including land dedicated for the disposal of sewage sludge. This definition does not include septic tanks or similar devices. For purposes of this definition, "domestic sewage" includes waste and waste water from humans or household operations that are discharged to or otherwise enter a treatment works. In States where there is no approved State sludge management program under section 405(f) of the CWA, the Regional Administrator may designate any person subject to the standards for sewage sludge use and disposal in 40 CFR part 503 as a "treatment works treating domestic sewage," where he or she finds that there is a potential for adverse effects on public health and the environment from poor sludge quality or poor sludge handling, use or disposal practices, or where he or she finds that such designation is necessary to ensure that such person is in compliance with 40 CFR part 503.

**Treatment works treating domestic sewage** (Section 501.2) means a POTW or any other sewage sludge or wastewater treatment devices or systems, regardless of ownership (including Federal facilities), used in the storage, treatment, recycling, and reclamation of municipal or domestic sewage, including land dedicated for the disposal of sewage sludge. This definition does not include septic tanks or similar devices. For purposes of this definition, "domestic sewage" includes waste and waste water from humans or household operations that are discharged to or otherwise enter a treatment works.

**Treatment zone** (Section 260.10) means a soil area of the unsaturated zone of a land treatment unit within which hazardous constituents are degraded, transformed, or immobilized.

**Tribal education agency** (Section 47.105) means a school or community college which is controlled by an Indian tribe, band, or nation, including any Alaska Native village, which is recognized as eligible for special programs and services provided by the United States to Indians because of their status as Indians and which is not administered by the Bureau of Indian Affairs;

**Tributaries of the Great Lakes System** (Section 132.2) means all waters of the Great Lakes System that are not open waters of the Great Lakes, or connecting channels.

**Trihalomethane (THM)** (Section 141.2) means one of the family of organic compounds, named as derivatives of methane, wherein three of the four hydrogen atoms in methane are each substituted by a halogen atom in the molecular structure.

**Trucked batteries** (Section 461.2) shall mean batteries moved into or out of the plant by truck when the truck is actually washed in the plant to remove residues left in the truck from the batteries.

**Trucked batteries** (Section 461.2) shall mean batteries moved into or out of the plant by truck when the truck is actually washed in the plant to remove residues left in the truck from the batteries.

**Trustee** (Section 300.5) means an official of a federal natural resources management agency designated in subpart G of the NCP or a designated state official or Indian tribe or, in the case of discharges covered by the OPA, a foreign government official, who may pursue claims for damages under section 107(f) of CERCLA or section 1006 of the OPA.

**TSCA** (Section 704.3) means the Toxic Substances Control Act, 15 U.S.C. 2601 et seq.

**TSCA** (Section 707.63) means the Toxic Substances Control Act.

**TSCA** (Section 712.3) means the Toxic Substances Control Act, 15 U.S.C. 2601 et seq.

**TSCA** (Section 716.3) means the Toxic Substances Control Act (15 U.S.C. 2601 et seq.).

**TSCA** (Section 745.103) means the Toxic Substances Control Act, 15 U.S.C. 2601.

**TSCA** (Section 761.13) means the Toxic Substances Control Act (15 U.S.C. 2601 et seq.).

**TSCA Experimental Release Application or TERA** (Section 725.3) means an exemption request for a research and development activity, which is not eligible for a full exemption from reporting under Sec. 725.232, 725.234, or 725.238, submitted to EPA in accordance with subpart E of this part.

**TSCA PCB Coordinated Approval** (Section 761.13) means the process used to recognize other Federal or State waste management documents governing the storage, cleanup, treatment, and disposal of PCB wastes. It is the mechanism under TSCA for accomplishing review, coordination, and approval of PCB waste management activities which are conducted outside of the TSCA PCB approval process, but require approval under the TSCA PCB regulations at 40 CFR part 761.

**TSP (total suspended particulates)** (Section 58.1) means particulate matter as measured by the method described in appendix B of part 50 of this chapter, Urban area population means the population defined in the most recent decennial U.S. Census of Population Report.

**TSS (or total suspended solids, or total suspended residue)** (Section 420.02) means the value obtained by the method specified in 40 CFR 136.3.

**TSS** (Section 401.11) means total suspended non-filterable solids.

**TTO** (Section 413.02) shall mean total toxic organics, which is the summation of all quantifiable values greater than 0.01 milligrams per liter for the following toxic organics: Acenaphthene, Acrolein, Acrylonitrile, Benzene, Benzidine, Carbon tetrachloride (tetrachloromethane), Chlorobenzene, 1,2,4-trichlorobenzene, Hexachlorobenzene, 1,2-dichloroethane, 1,1,1-trichloroethane, Hexachloroethane, 1,1-dichloroethane, 1,1,2-trichloroethane, 1,1,2,2-tetrachloroethane, Chloroethane, Bis (2-chloroethyl) ether, 2-chloroethyl vinyl ether (mixed), 2-chloronaphthalene, 2,4,6-trichlorophenol, Parachlorometa cresol, Chloroform (trichloromethane), 2-chlorophenol, 1,2-dichlorobenzene, 1,3-dichlorobenzene, 1,4-dichlorobenzene, 3,3-dichlorobenzidine, 1,1-dichloroethylene, 1,2-trans-dichloroethylene, 2,4-dichlorophenol, 1,2-dichloropropane, 1,3-dichloropropylene (1,3-dichloropropene), 2,4-dimethylphenol, 2,4-dinitrotoluene, 2,6-dinitrotoluene, 1,2-diphenylhydrazine, Ethylbenzene, Fluoranthene, 4-chlorophenyl phenyl ether, 4-bromophenyl phenyl ether, Bis (2-chloroisopropyl) ether, Bis (2-chloroethoxy) methane, Methylene chloride (dichloromethane), Methyl chloride (chloromethane), Methyl bromide (bromomethane), Bromoform (tribromomethane), Dichlorobromomethane, Chlorodibromomethane, Hexachlorobutadiene, Hexachlorocyclopentadiene, Isophorone Naphthalene, Nitrobenzene, 2-nitrophenol, 4-nitrophenol, 2,4-dinitrophenol, 4,6-dinitro-o-cresol, N-nitrosodimethylamine, N-nitrosodiphenylamine, N-nitrosodi-n-propylamine, Pentachlorophenol Phenol, Bis (2-ethylhexyl) phthalate, Butyl benzyl phthalate, Di-n-butyl phthalate, Di-n-octyl phthalate, Diethyl phthalate, Dimethyl phthalate, 1,2-benzanthracene, (benzo(a)anthracene), Benzo(a)pyrene (3,4-benzopyrene), 3,4-Benzofluoranthene (benzo(b)fluoranthene), 11,12-benzofluoranthene (benzo(k)fluoranthene), Chrysene, Acenaphthylene, Anthracene, 1,12-benzoperylene, (benzo(ghi)perylene), Fluorene, Phenanthrene, 1,2,5,6-dibenzanthracene (dibenzo(a,h)anthracene), Indeno (1,2,3-cd) pyrene), (2,3-o-phenylene pyrene), Pyrene Tetrachloroethylene, Toluene, Trichloroethylene, Vinyl chloride (chloroethylene), Aldrin, Dieldrin, Chlordane (technical mixture and metabolites), 4,4-DDT, 4,4-DDE (p,p-DDX), 4,4-DDD (p,p-TDE), Alpha-endosulfan, Beta-endosulfan, Endosulfan sulfate, Endrin, Endrin aldehyde, Heptachlor, Heptachlor epoxide (BHC-hexachlorocyclohexane), Alpha-BHC, Beta-BHC, Gamma-BHC, Delta-BHC, (PCB-polychlorinated biphenyls), PCB-1242 (Arochlor 1242), PCB-1254 (Arochlor 1254), PCB-1221 (Arochlor 1221), PCB-1232 (Arochlor 1232), PCB-1248 (Arochlor 1248), PCB-1260 (Arochlor 1260), PCB-1016 (Arochlor 1016), Toxaphene, 2,3,7,8-tetrachlorodibenzo-p-dioxin (TCDD).

**TTO** (Section 433.11) shall mean total toxic organics, which is the summation of all quantifiable values greater than .01 milligrams per liter for the following toxic organics: Acenaphthene, Acrolein, Acrylonitrile, Benzene, Benzidine, Carbon tetrachloride (tetrachloromethane), Chlorobenzene, 1,2,4-Trichlorobenzene Hexachlorobenzene, 1,2,-Dichloroethane, 1,1,1-Trichloroethane Hexachloroethane, 1,1-Dichloroethane, 1,1,2-Trichloroethane, 1,1,2,2-Tetrachloroethane Chloroethane, Bis (2-chloroethyl) ether, 2-Chloroethyl vinyl ether (mixed), 2-Chloronaphthalene, 2,4,6-Trichlorophenol Parachlorometa cresol Chloroform (trichloromethane), 2-Chlorophenol, 1,2-Dichlorobenzene, 1,3-Dichlorobenzene, 1,4-Dichlorobenzene, 3,3-Dichlorobenzidine, 1,1-Dichloroethylene, 1,2-Trans-dichloroethylene, 2,4-Dichlorophenol, 1,2-Dichloropropane, 1,3-Dichloropropylene (1,3-dichloropropene), 2,4-Dimethylphenol, 2,4-Dinitrotoluene, 2,6-Dinitrotoluene, 1,2-Diphenylhydrazine, Ethylbenzene, Fluoranthene, 4-Chlorophenyl phenyl ether, 4-Bromophenyl phenyl ether, Bis (2-chloroisopropyl) ether, Bis (2-chloroethoxy) methane, Methylene chloride (dichloromethane), Methyl chloride (chloromethane), Methyl bromide (bromomethane), Bromoform (tribromomethane), Dichlorobromomethane, Chlorodibromomethane, Hexachlorobutadiene, Hexachlorocyclopentadiene, Isophorone, Naphthalene, Nitrobenzene, 2-Nitrophenol, 4-Nitrophenol, 2,4-Dinitrophenol, 4,6-Dinitro-o-cresol, N-nitrosodimethylamine, N-nitrosodiphenylamine, N-nitrosodi-n-propylamine, Pentachlorophenol, Phenol, Bis (2-ethylhexyl) phthalate, Butyl benzyl phthalate, Di-n-butyl phthalate, Di-n-octyl phthalate, Diethyl phthalate, Dimethyl phthalate, 1,2-Benzanthracene, (benzo(a)anthracene), Benzo(a)pyrene (3,4-benzopyrene), 3,4-Benzofluoranthene (benzo(b)fluoranthene), 11,12-Benzofluoranthene (benzo(k)fluoranthene), Chrysene, Acenaphthylene, Anthracene, 1,12-Benzoperylene (benzo(ghi)perylene), Fluorene, Phenanthrene, 1,2,5,6-Dibenzanthracene (dibenzo(a,h)anthracene), Indeno(1,2,3-cd) pyrene (2,3-o-phenlene pyrene), Pyrene, Tetrachloroethylene, Toluene, Trichloroethylene, Vinyl chloride (chloroethylene), Aldrin, Dieldrin, Chlordane (technical mixture and metabolites), 4,4-DDT, 4,4-DDE (p,p-DDX), 4,4-DDD (p,p-TDE), Alpha-endosulfan, Beta-endosulfan, Endosulfan sulfate, Endrin, Endrin aldehyde, Heptachlor, Heptachlor epoxide, (BHC-hexachloro-cyclohexane), Alpha-BHC, Beta-BHC, Gamma-BHC, Delta-BHC, (PCB-polychlorinated biphenyls), PCB-1242 (Arochlor 1242), PCB-1254 (Arochlor

1254), PCB-1221 (Arochlor 1221), PCB-1232 (Arochlor 1232), PCB-1248 (Arochlor 1248), PCB-1260 (Arochlor 1260), PCB-1016 (Arochlor 1016), Toxaphene, 2,3,7,8-Tetrachlorodibenzo-p-dioxin (TCDD).

**Tube reducing** (Section 471.02) is an operation which reduces the diameter and wall thickness of tubing with a mandrel and a pair of rolls with tapered grooves.

**Tumbling or barrel finishing** (Section 471.02) is an operation in which castings, forgings, or parts pressed from metal powder are rotated in a barrel with ceramic or metal slugs or abrasives to remove scale, fins, or burrs. It may be done dry or with an aqueous solution.

**Tumbling or burnishing** (Section 468.02) shall mean the process of polishing, deburring, removing sharp corners, and generally smoothing parts for both cosmetic and functional purposes, as well as the process of washing the finished parts and cleaning the abrasion media.

**Turbo-fired boiler** (Section 76.2) means a pulverized coal, wall-fired boiler with burners arranged on walls so that the individual flames extend down toward the furnace bottom and then turn back up through the center of the furnace.

**Two-stage topcoat** (Section 59.101) means a topcoat consisting of a pigmented basecoat and a transparent clearcoat.

**TWTDS** (Section 122.2) means "treatment works treating domestic sewage."

**TWTDS** (Section 501.2) means treatment works treating domestic sewage.

**Type A hybrid electric vehicle** (Section 86.1702-99) means an HEV which achieves a minimum range of 60 miles over the All-Electric Range Test as defined in Sec. 86.1770.

**Type B hybrid electric vehicle** (Section 86.1702-99) means an HEV which achieves a range of 40-59 miles over the All-Electric Range Test as defined in Sec. 86.1770.

**Type C hybrid electric vehicle** (Section 86.1702-99) means an HEV which achieves a range of 0-39 miles over the All-Electric Range test and all other HEVs excluding "Type A" and "Type B" HEVs as defined in Sec. 86.1770.

**Type I Sound Level Meter** (Section 204.2) means a sound level meter which meets the Type I requirements of American National Standard Specification S1.4-1971 for sound level meters. This publication is available from the American National Standards Institute, Inc., 1430 Broadway, New York, New York 10018.

**Type I Sound Level Meter** (Section 205.2) means a sound level meter which meets the Type I requirements of ANSI SI.4-1972 specification for sound level meters. This publication is available from the American National Standards Institute, Inc., 1430 Broadway, New York, New York 10018.

**Type of pesticidal product** (Section 167.3) refers to each individual product as identified by: the product name; EPA Registration Number (or EPA File Symbol, if any, for planned products, or Experimental Permit Number, if the pesticide is produced under an Experimental Use Permit); active ingredients; production type (technical, formulation, repackaging, etc.); and, market for which the product was produced (domestic, foreign, etc.). In cases where a pesticide is not registered, registration is not applied for, or the pesticide is not produced under an Experimental Use Permit, the term shall also include the chemical formulation.

**Typical meteorological conditions** (Section 68.3) means the temperature, wind speed, cloud cover, and atmospheric stability class, prevailing at the site based on data gathered at or near the site or from a local meteorological station.

# U

**U (Uranium)** (Section 440.132) is measured by the procedure discussed in 40 CFR 141.25(b)(2), or an equivalent method.

**U.S. GPO** (Section 761.13) means the U.S. Government Printing Office, 710 North Capitol St., NW., Washington, DC 20401.

**U.S. heavy-duty vehicle sales** (Section 86.1803-01) means sales of heavy-duty vehicles subject to the standards of this subpart, where the sale takes place in any state of the United States except for California (or a state that has adopted California motor vehicle standards for that model year pursuant to section 177 of the Clean Air Act).

**U.S. sales** (Section 86.1803-01) means, unless otherwise specified, sales in any state of the United States except for California or a state that has adopted California motor vehicle standards for that model year pursuant to section 177 of the Clean Air Act. This definition applies only to those regulatory requirements addressing Tier 2 and interim non-Tier 2 vehicles.

**U.S.-directed production** (Section 86.004-2) means the engines and/or vehicles (as applicable) produced by a manufacturer for which the manufacturer has reasonable assurance that sale was or will be made to ultimate purchasers in the United States, excluding engines and/or vehicles that are certified to state emission standards different than the emission standards in this part.

**U.S.-directed production volume** (Section 89.2) means the number of nonroad equipment, vehicle, or marine diesel engine units produced by a manufacturer for which the manufacturer has reasonable assurance that sale was or will be made to ultimate purchasers in the United States.

**U.S.-directed production volume** (Section 94.2) means the number of marine engine units, subject to this part, produced by a manufacturer for which the manufacturer has reasonable assurance that sale was or will be made to ultimate purchasers in the United States.

**UIC** (Section 124.2) means the Underground Injection Control program under Part C of the Safe Drinking Water Act, including an "approved program."

**UIC** (Section 144.3) means the Underground Injection Control program under Part C of the Safe Drinking Water Act, including an "approved State program."

**UIC** (Section 146.3) means the Underground Injection Control program under Part C of the Safe Drinking Water Act, including an "approved program."

**UIC** (Section 270.2) means the Underground Injection Control Program under part C of the Safe Drinking Water Act, including an approved program.

**Ultimate Consumer** (Section 600.022-85) means the first person who purchases an automobile for purposes other than resale or leases an automobile.

**Ultimate purchaser** (Section 205.2) means the first person who in good faith purchases a product for purposes other than resale.

**Ultimate purchaser** (Section 53.1) means the first person or entity who purchases a reference method or an equivalent method for purposes other than resale.

**Ultimate purchaser** (Section 85.1902) shall be given the meaning ascribed to it by section 214 of the Act.

**Ultimate purchaser** (Section 89.2) means, with respect to any new nonroad engine, new nonroad vehicle, or new nonroad equipment, the first person who in good faith purchases such new nonroad engine, nonroad vehicle, or nonroad equipment for purposes other than resale.

**Ultimate purchaser** (Section 90.3) means, with respect to any new nonroad engine or new nonroad vehicle, the first person who in good faith purchases such new nonroad engine or vehicle for purposes other than resale.

**Ultimate purchaser** (Section 91.3) means, with respect to any new marine SI engine the first person who in good faith purchases such new marine SI engine for purposes other than resale.

**Ultimate Purchaser** (Section 94.2) means, with respect to any new engine or vessel, the first person who in good faith purchases such new engine or vessel for purposes other than resale.

**Ultra Low-Emission Vehicle** (Section 88.102-94) means any light-duty vehicle or light-duty truck conforming to the applicable Ultra Low-Emission Vehicle standard, or any heavy-duty vehicle with an engine conforming to the applicable Ultra Low-Emission Vehicle standard.

**Ultra-low emission vehicle (ULEV)** (Section 86.1702-99) means any vehicle certified to the ultra-low emission vehicle standards specified in this subpart.

**Ultrasonic testing** (Section 471.02) is a nondestructive test which applies sound, at a frequency above about 20 HJz, to metal, which has been immersed in liquid (usually water) to locate inhomogeneities or structural discontinuities.

**Unacceptable adverse effect** (Section 231.2) means impact on an aquatic or wetland ecosystem which is likely to result in significant degradation of municipal water supplies (including surface or ground water) or significant loss of or damage to fisheries, shell fishing, or wildlife habitat or recreation areas. In evaluating the unacceptability of such impacts, consideration should be given to the relevant portions of the section 404(b) (1) guidelines (40 CFR part 230).

**Unauthorized dispersion technique** (Section 57.103) refers to any dispersion technique which, under section 123 of the Act and the regulations promulgated pursuant to that section, may not be used to reduce the degree of emission limitation otherwise required in the applicable SIP.

**Uncertainty factor (UF)** (Section 132.2) is one of several numeric factors used in operationally deriving criteria from experimental data to account for the quality or quantity of the available data.

**Uncertified person** (Section 171.2) means any person who is not holding a currently valid certification document indicating that he is certified under section 4 of FIFRA in the category of the restricted use pesticide made available for use. Limited. The following definitions apply only to dealers, dealerships and transactions in States or on Indian Reservations where EPA conducts a Federal Pesticide Applicator Certification Program.

**Uncovered finished water storage facility** (Section 141.2) is a tank, reservoir, or other facility used to store water that will undergo no further treatment except residual disinfection and is open to the atmosphere.

**Under normal circumstances garaged at personal residence** (Section 88.302-94) means a vehicle that, when it is not in use, is normally parked at the personal residence of the individual who usually operates it, rather than at a central refueling, maintenance, and/or business location. Such vehicles are not considered to be capable of being central fueled (as defined in this subpart) and are exempt from the program unless they are, in fact, centrally fueled.

**Under the direct supervision of** (Section 171.2) means the act or process whereby the application of a pesticide is made by a competent person acting under the instructions and control of a certified applicator who is responsible for the actions of that person and who is available if and when needed, even though such certified applicator is not physically present at the time and place the pesticide is applied.

**Underbody coating** (Section 59.101) means a coating designed for protection and sound deadening that is typically applied to the wheel wells and underbody of an automobile.

**Underground area** (Section 280.12) means an underground room, such as a basement, cellar, shaft or vault, providing enough space for physical inspection of the exterior of the tank situated on or above the surface of the floor.

**Underground injection** (Section 144.3) means a "well injection."

**Underground injection** (Section 146.3) means a "well injection."

**Underground injection** (Section 260.10) means the subsurface emplacement of fluids through a bored, drilled or driven well; or through a dug well, where the depth of the dug well is greater than the largest surface dimension. (See also "injection well".)

**Underground injection** (Section 270.2) means a well injection.

**Underground release** (Section 280.12) means any belowground release.

**Underground source of drinking water** (Section 147.2902) an aquifer or its portion: (a) (1) Which supplies any public water system; or (2) Which contains a sufficient quantity of ground water to supply a public water system; and (i) Currently supplies drinking water for human consumption; or (ii) Contains fewer than 10,000 mg/l total dissolved solids; and (b) Which is not an exempted aquifer.

**Underground source of drinking water** (Section 191.22) means an aquifer or its portion which: (1) Supplies any public water system; or (2) Contains a sufficient quantity of ground water to supply a public water system; and (i) Currently supplies drinking water for human consumption; or (ii) Contains fewer than 10,000 milligrams of total dissolved solids per liter.

**Underground source of drinking water** (Section 192) means an aquifer or its portion: (1) (i) Which supplies any public water system as defined in Sec. 141.2 of this chapter; or (ii) Which contains a sufficient quantity of groundwater to supply a public water system; and (A) Currently supplies drinking water for human consumption; or (B) Contains fewer than 10,000 mg/l total dissolved solids; and (2) Which is not an exempted aquifer as defined in Sec. 144.7 of this chapter.
**Administrator's authorized representative** (Section 194.2) means the director in charge of radiation programs at the Agency.

**Underground source of drinking water (USDW)** (Section 144.3) means an aquifer or its portion: (a) (1) Which supplies any public water system; or (2) Which contains a sufficient quantity of ground water to supply a public water system; and (i) Currently supplies drinking water for human consumption; or (ii) Contains fewer than 10,000 mg/l total dissolved solids; and (b) Which is not an exempted aquifer.

**Underground source of drinking water (USDW)** (Section 146.3) means an aquifer or its portion: (1) (i) Which supplies any public water system; or (ii) Which contains a sufficient quantity of ground water to supply a public water system; and (A) Currently supplies drinking water for human consumption; or (B) Contains fewer than 10,000 mg/l total dissolved solids; and (2) Which is not an exempted aquifer.

**Underground source of drinking water (USDW)** (Section 270.2) means an aquifer or its portion: (a) (1) Which supplies any public water system; or (2) Which contains a sufficient quantity of ground water to supply a public water system; and (i) Currently supplies drinking water for human consumption; or (ii) Contains fewer than 10,000 mg/l total dissolved solids; and (b) Which is not an exempted aquifer.

**Underground storage tank or UST** (Section 280.12) does not include any pipes connected to any tank which is described in paragraphs (a) through (i) of this definition.

**Underground storage tank or UST** (Section 280.12) means any one or combination of tanks (including underground pipes connected thereto) that is used to contain an accumulation of regulated substances, and the volume of which (including the volume of underground pipes connected thereto) is 10 percent or more beneath the surface of the ground. This term does not include any: (a) Farm or residential tank of 1,100 gallons or less capacity used for storing motor fuel for noncommercial purposes; (b) Tank used for storing heating oil for consumptive use on the premises where stored; (c) Septic tank; (d) Pipeline facility (including gathering lines) regulated under: (1) The Natural Gas Pipeline Safety Act of 1968 (49 U.S.C. App. 1671, et seq.), or (2) The Hazardous Liquid Pipeline Safety Act of 1979 (49 U.S.C. App. 2001, et seq.), or (3) Which is an intrastate pipeline facility regulated under state laws comparable to the provisions of the law referred to in paragraph (d)(1) or (d)(2) of this definition; (e) Surface impoundment, pit, pond, or lagoon; (f) Storm-water or wastewater collection system; (g) Flow-through process tank; (h) Liquid trap or associated gathering lines directly related to oil or gas production and gathering operations; or (i) Storage tank situated in an underground area (such as a basement, cellar, mineworking, drift, shaft, or tunnel) if the storage tank is situated upon or above the surface of the floor.

**Underground tank** (Section 260.10) means a device meeting the definition of "tank" in Sec. 260.10 whose entire surface area is totally below the surface of and covered by the ground.

**Underlying hazardous constituent** (Section 268.2) means any constituent listed in Sec. 268.48, Table UTS—Universal Treatment Standards, except fluoride, selenium, sulfides, vanadium, and zinc, which can reasonably be expected to be present at the point of generation of the hazardous waste at a concentration above the constituent-specific UTS treatment standards.

**Underutilization** (Section 72.2) means a reduction, during any calendar year in Phase I, of the heat input (expressed in mmBtu for the calendar year) at a Phase I unit below the unit's baseline.

**Undisturbed performance** (Section 191.12) means the predicted behavior of a disposal system, including consideration of the uncertainties in predicted behavior, if the disposal system is not disrupted by human intrusion or the occurrence of unlikely natural events.

**Unexpended Article 5 allowances** (Section 82.3) means Article 5 allowances that have not been used. At any time in any control period a person's unexpended Article 5 allowances are the total of the level of Article 5 allowances the person has authorization under this subpart to hold at that time for that control period, minus the level of controlled substances that the person has produced in that control period until that time.

**Unexpended consumption allowances** (Section 82.3) means consumption allowances that have not been used. At any time in any control period a person's unexpended consumption allowances are the total of the level of consumption allowances the person has authorization under this subpart to hold at that time for that control period, minus the level of controlled substances that the person has produced or imported (not including transhipments and used controlled substances) in that control period until that time.

**Unexpended destruction and transformation credits** (Section 82.3) means destruction and transformation credits that have not been used. At any time in any control period a person's unexpended destruction and transformation credits are the total of the level of destruction and transformation credits the person has authorization under this subpart to hold at that time for that control period, minus the level of controlled substances that the person has produced or imported (not including transhipments and used controlled substances) in that control period until that time.

**Unexpended essential-use allowances** (Section 82.3) means essential-use allowances that have not been used. At any time in any control period a person's unexpended essential-use allowances are the total of the level of essential-use allowances the person has authorization under this subpart to hold at that time for that control period, minus the level of controlled substances that the person has imported or had produced in that control period until that time.

**Unexpended production allowances** (Section 82.3) means production allowances that have not been used. At any time in any control period a person's unexpended production allowances are the total of the level of production allowances he has authorization under this subpart to hold at that time for that control period, minus the level of controlled substances that the person has produced in that control period until that time.

**Unexploded ordnance (UXO)** (Section 266.201) means military munitions that have been primed, fused, armed, or otherwise prepared for action, and have been fired, dropped, launched, projected, or placed in such a manner as to constitute a hazard to operations, installation, personnel, or material and remain unexploded either by malfunction, design, or any other cause.

**Unfit-for use tank system** (Section 260.10) means a tank system that has been determined through an integrity assessment or other inspection to be no longer capable of storing or treating hazardous waste without posing a threat of release of hazardous waste to the environment.

**Uniform finish blender** (Section 59.101) means a coating designed to blend a repaired topcoat into an existing topcoat.

**Unit** (Section 72.2) means a fossil fuel-fired combustion device.

**Unit** (Section 761.13) means a particular building, structure, or cell used to manage PCB waste (including, but not limited to, a building used for PCB waste storage, a landfill, an industrial boiler, or an incinerator).

**Unit** (Section 96.2) means a fossil fuel-fired stationary boiler, combustion turbine, or combined cycle system.

**Unit** (Section 97.2) means a fossil fuel-fired stationary boiler, combustion turbine, or combined cycle system.

**Unit account** (Section 72.2) means an Allowance Tracking System account, established by the Administrator for an affected unit pursuant to Sec. 73.31 (a) or (b) of this chapter.

**Unit load** (Section 72.2) means the total (i.e., gross) output of a unit or source in any calendar year (or other specified time period) produced by combusting a given heat input of fuel, expressed in terms of: (1) The total electrical generation (MWe) for use within the plant and for sale: or (2) In the case of a unit or source that uses part of its heat input for purposes other than electrical generation, the total steam pressure (psia) produced by the unit or source.

**Unit load** (Section 96.2) means the tot l (i.e., gross) output of a unit in any contr l period (or other specified time period) produced by combusting a given heat input of fuel, expressed in terms of: (1) The total electrical generation (MWe) produced by the unit, including generation for use within the plant; or (2) In the case of a unit that uses heat input for purposes other than electrical generation, the total steam pressure (psia) produced by the unit, including steam for use by the unit.

**Unit operating day** (Section 72.2) means a calendar day in which a unit combusts any fuel.

**Unit operating day** (Section 96.2) means a calendar day in which a unit combusts any fuel.

**Unit operating day** (Section 97.2) means a calendar day in which a unit combusts any fuel.

**Unit operating hour** (Section 72.2) means a clock hour during which a unit combusts any fuel, either for part of the hour or for the entire hour.

**Unit operating hour** (Section 72.2) means any hour (or fraction of an hour) during which a unit combusts any fuel.

**Unit operating hour or hour of unit operation** (Section 96.2) means any hour (or fraction of an hour) during which a unit combusts any fuel.

**Unit operating hour or hour of unit operation** (Section 97.2) means any hour (or fraction of an hour) during which a unit combusts any fuel.

**Unit operating quarter** (Section 72.2) means a calendar quarter in which a unit combusts any fuel.

**Unit operating time** (Section 72.2) means the portion of a clock hour during which a unit combusts any fuel. The unit operating time, in hours, is expressed as a decimal fraction, with valid values ranging from 0.00 to 1.00.

**Unit packaging** (Section 157.21) means a package that is labeled with directions to use the entire contents of the package in a single application.

**United States** (Section 302.3) include the several States of the United States, the

District of Columbia, the Commonwealth of Puerto Rico, Guam, American Samoa, the United States Virgin Islands, the Commonwealth of the Northern Marianas, and any other territory or possession over which the United States has jurisdiction.

**United States** (Section 59.101) means the United States of America, including the District of Columbia, Puerto Rico, the Virgin Islands, Guam, American Samoa, and Commonwealth of the Northern Mariana Islands.

**United States** (section 710.2) when used in the geographic sense, means all of the States, territories, and possessions of the United States.

**United States** (Section 720.3) when used in the geographic sense, means all of the States.

**United States** (Section 85.1502) United States includes the Customs territory of the United States as defined in 19 U.S.C. 1202, and the Virgin Islands, Guam, American Samoa and the Commonwealth of the Northern Mariana Islands.

**United States when used in relation to section 311(a)(5) of the CWA** (Section 300.5) means the states, the District of Columbia, the Commonwealth of Puerto Rico, the Northern Mariana Islands, Guam, American Samoa, the United States Virgin Islands, and the Pacific Island Governments. United States, when used in relation to section 101(27) of CERCLA and section 1001(36) of the OPA, includes the several states of the United States, the District of Columbia, the Commonwealth of Puerto Rico, Guam, American Samoa, the United States Virgin Islands, the Commonwealth of the Northern Marianas, and any other territory or possession over which the United States has jurisdiction.

**Universal Waste** (Section 260.10) means any of the following hazardous wastes that are managed under the universal waste requirements of part Sec. 273 of this chapter: (1) Batteries as described in Sec. 273.2 of this chapter; (2) Pesticides as described in Sec. 273.3 of this chapter; (3) Thermostats as described in Sec. 273.4 of this chapter; and (4) Lamps as described in Sec. 273.5 of this chapter.

**Universal Waste** (Section 273.9) means any of the following hazardous waste that are subject to the universal waste requirements of this part 273: (1) Batteries as described in Sec. 273.2, (2) Pesticides as described in Sec. 273.3, (3) Thermostats as described in Sec. 273.4; and (4) Lamps as described in Sec. 273.5.

**Universal Waste Handler** (Section 260.10) (1) Means: (i) A generator (as defined in this section) of universal waste; or (ii) The owner or operator of a facility, including all contiguous property, that receives universal waste from other universal waste handlers, accumulates universal waste, and sends universal waste to another universal waste handler, to a destination facility, or to a foreign destination. (2) Does not mean: (i) A person who treats (except under the provisions of 40 CFR 273.13 (a) or (c), or 273.33 (a) or (c)), disposes of, or recycles universal waste; or (ii) A person engaged in the off-site transportation of universal waste by air, rail, highway, or water, including a universal waste transfer facility.

**Universal Waste Handler** (Section 273.9) Means: (1) A generator (as defined in this section) of universal waste; or (2) The owner or operator of a facility, including all contiguous property, that receives universal waste from other universal waste handlers, accumulates universal waste, and sends universal waste to another universal waste handler, to a destination facility, or to a foreign destination. (b) Does not mean: (1) A person who treats (except under the provisions of 40 CFR 273.13 (a) or (c), or 273.33 (a) or (c)), disposes of, or recycles universal waste; or (2) A person engaged in the off-site transportation of universal waste by air, rail, highway, or water, including a universal waste transfer facility.

**Universal Waste Transfer Facility** (Section 273.9) means any transportation-related facility including loading docks, parking areas, storage areas and other similar areas where shipments of universal waste are held during the normal course of transportation for ten days or less.

**Universal Waste Transporter** (Section 260.10) means a person engaged in the off-site transportation of universal waste by air, rail, highway, or water.

**Universal Waste Transporter** (Section 273.9) means a person engaged in the off-site transportation of universal waste by air, rail, highway, or water.

**Unleaded gasoline** (Section 80.2) means gasoline which is produced without the use of any lead additive and which contains not more than 0.05 gram of lead per gallon and not more than 0.005 gram of phosphorus per gallon.

**Unproven emission control systems** (Section 86.092-2) are emission control components or systems (and fuel metering systems) that do not qualify as proven emission control systems.

**Unproven emission control systems** (Section 86.1803-01) are emission control components or systems (and fuel metering systems) that do not qualify as proven emission control systems.

**Unreasonable adverse effects on the environment** (Section 166.3) means any unreasonable risk to man or the environment, taking into account the economic, social, and environmental costs and benefits of the use of any pesticide.

**Unsanitized** (Section 350.1) means a version of a document from which information claimed as trade secret or confidential has not been withheld or omitted.

**Unsaturated zone or zone of aeration** (Section 260.10) means the zone between the land surface and the water table.

**Unscheduled maintenance** (Section 86.082-2) means any adjustment, repair, removal, disassembly, cleaning, or replacement of vehicle components or systems which is performed to correct a part failure or vehicle (if the engine were installed in a vehicle) malfunction.

**Unscheduled maintenance** (Section 86.1803-01) means any adjustment, repair, removal disassembly, cleaning, or replacement of vehicle components or systems which is performed to correct a part failure or vehicle (if the engine were installed in a vehicle) malfunction which was not anticipated.

**Unstabilized solids** (Section 503.31) are organic materials in sewage sludge that have not been treated in either an aerobic or anaerobic treatment process.

**Unstable area** (Section 503.21) is land subject to natural or human-induced forces that may damage the structural components of an active sewage sludge unit. This includes, but is not limited to, land on which the soils are subject to mass movement.

**Unused military munition is a solid waste when** (Section 266.202) any of the following occurs: (1) The munition is abandoned by being disposed of, burned, detonated (except during intended use as specified in paragraph (a) of this section), incinerated, or treated prior to disposal; or (2) The munition is removed from storage in a military magazine or other storage area for the purpose of being disposed of, burned, or incinerated, or treated prior to disposal, or (3) The munition is deteriorated or damaged (e.g., the integrity of the munition is compromised by cracks, leaks, or other damage) to the point that it cannot be put into serviceable condition, and cannot reasonably be recycled or used for other purposes; or (4) The munition has been declared a solid waste by an authorized military official.

**Upgrade** (Section 280.12) means the addition or retrofit of some systems such as cathodic protection, lining, or spill and overfill controls to improve the ability of an underground storage tank system to prevent the release of product.

**Upgrade** (Section 92.2) means to modify a locomotive or locomotive engine that was originally manufactured prior to January 1, 1973 (or a locomotive or locomotive engine that was originally manufactured on or after January 1, 1973, and that is not subject to the emission standards of this part), such that it is intended to comply with the Tier 0 standards. Upgrading is a type of remanufacturing.

**Uppermost aquifer** (Section 258.2) means the geologic formation nearest the natural ground surface that is an aquifer, as well as, lower aquifers that are hydraulically interconnected with this aquifer within the facility's property boundary.

**Uppermost aquifer** (Section 260.10) means the geologic formation nearest the natural ground surface that is an aquifer, as well as lower aquifers that are hydraulically interconnected with this aquifer within the facility's property boundary.

**Upset** (Section 122.2) is defined at Sec. 122.41(n).

**Uptake** (Section 132.2) is acquisition of a substance from the environment by an organism as a result of any active or passive process.

**Uranium byproduct material** (Section 192.31) means the tailings or wastes produced by the extraction or concentration of uranium from any ore processed primarily for its source material content. Ore bodies depleted by uranium solution extraction operations and which remain underground do not constitute "byproduct material" for the purpose of this subpart.

**Uranium fuel cycle** (Section 190.02) means the operations of milling of uranium ore, chemical conversion of uranium, isotopic enrichment of uranium, fabrication of uranium fuel, generation of electricity by a light-water-cooled nuclear power plant using uranium fuel, and reprocessing of spent uranium fuel, to the extent that these directly support the production of electrical power for public use utilizing nuclear energy, but excludes mining operations, operations at waste disposal sites, transportation of any radioactive material in support of these operations, and the reuse of recovered non-uranium special nuclear and by-product materials from the cycle.

**Urban bus** (Section 85.1402) has the meaning set forth in Sec. 86.091-2 of this chapter.

**Urban bus** (Section 86.091-2) means a heavy heavy-duty diesel-powered passenger-carrying vehicle with a load capacity of fifteen or more passengers and intended primarily for intra-city operation, i.e., within the confines of a city or greater metropolitan area. Urban bus operation is characterized by short rides and frequent stops. To facilitate this type of operation, more than one set of quick-operating entrance and exit doors would normally be installed. Since fares are usually paid in cash or tokens rather than purchased in advance in the form of tickets, urban buses would normally have equipment installed for collection of fares. Urban buses are also typically characterized by the absence of equipment and facilities for long distance travel, e.g., rest rooms, large luggage compartments, and facilities for stowing carry-on luggage. The useful life for urban buses is the same as the useful life for other heavy heavy-duty diesel engines.

**Urban bus** (Section 86.093-2) means a passenger-carrying vehicle powered by a heavy heavy-duty diesel engine, or of a type normally powered by a heavy heavy-duty diesel engine, with a load capacity of fifteen or more passengers and intended primarily for intracity operation, i.e., within the confines of a city or greater metropolitan area. Urban bus operation is characterized by short rides and frequent stops. To facilitate this type of operation, more than one set of quick-operating entrance and exit doors would normally be installed. Since fares are usually paid in cash or tokens, rather than purchased in advance in the form of tickets, urban buses would normally have equipment installed for collection of fares. Urban buses are also typically characterized by the absence of equipment and facilities for long distance travel, e.g., rest rooms, large luggage compartments, and facilities for stowing carry-on luggage. The useful life for urban buses is the same as the useful life for other heavy heavy-duty diesel engines.

**US06** (Section 86.000-2) means the test cycle, described in Sec. 86.159-00 and listed in appendix I, paragraph (g), of this part, which is designed to evaluate emissions during aggressive and microtransient driving.

**US06** (Section 86.1803-01) means the test cycle, described in Sec. 86.159-00 and listed in appendix I, paragraph (g), of this part, which is designed to evaluate emissions during aggressive and microtransient driving.

**USDW** (Section 144.3) means "underground source of drinking water."

**USDW** (Section 146.3) means "underground source of drinking water."

**USDW** (Section 147.2902) underground source of drinking water.

**USDW** (Section 270.2) means underground source of drinking water.

**Use attainability analysis** (Section 131.3) is a structured scientific assessment of the factors affecting the attainment of the use which may include physical, chemical, biological, and economic factors as described in Sec. 131.10(g).

**Use in agricultural or wildlife propagation** (Section 435.51) means that the produced water is of good enough quality to be used for wildlife or livestock watering or other agricultural uses and that the produced water is actually put to such use during periods of discharge.

**Use stream** (Section 721.3) means all reasonably anticipated transfer, flow, or disposal of a chemical substance, regardless of physical state or concentration, through all intended operations of industrial, commercial, or consumer use.

**Used controlled substances** (Section 82.3) means controlled substances that have been recovered from their intended use systems (may include controlled substances that have been, or may be subsequently, recycled or reclaimed).

**Used oil** (Section 260.10) means any oil that has been refined from crude oil, or any synthetic oil, that has been used and as a result of such use in contaminated by physical or chemical impurities.

**Used oil** (Section 279.1) means any oil that has been refined from crude oil, or any synthetic oil, that has been used and as a result of such use is contaminated by physical or chemical impurities.

**Used oil aggregation point** (Section 279.1) means any site or facility that accepts, aggregates, and/or stores used oil collected only from other used oil generation sites owned or operated by the owner or operator of the aggregation point, from which used oil is transported to the aggregation point in shipments of no more than 55 gallons. Used oil aggregation points may also accept used oil from household do-it-yourselfers.

**Used oil burner** (Section 279.1) means a facility where used oil not meeting the specification requirements in Sec. 279.11 is burned for energy recovery in devices identified in Sec. 279.61(a).

**Used oil collection center** (Section 279.1) means any site or facility that is registered/licensed/permitted/recognized by a state/county/municipal government to manage used oil and accepts/aggregates and stores used oil collected from used oil generators regulated under subpart C of this part who bring used oil to the collection center in shipments of no more than 55 gallons under the provisions of Sec. 279.24. Used oil collection centers may also accept used oil from household do-it-yourselfers.

**Used oil filter recycling** (Section 437.2) means crushing and draining of used oil filters of entrained oil and/or shredding and separation of used oil filters.

**Used oil fuel marketer** (Section 279.1) means any person who conducts either of the following activities: (1) Directs a shipment of off-specification used oil from their facility to a used oil burner; or (2) First claims that used oil that is to be burned for energy recovery meets the used oil fuel specifications set forth in Sec. 279.11 of this part.

**Used oil generator** (Section 279.1) means any person, by site, whose act or process produces used oil or whose act first causes used oil to become subject to regulation.

**Used oil processor/re-refiner** (Section 279.1) means a facility that processes used oil.

**Used oil transfer facility** (Section 279.1) means any transportation related facility including loading docks, parking areas, storage areas and other areas where shipments of used oil are held for more than 24 hours and not longer than 35 days during the normal course of transportation or prior to an activity performed pursuant to Sec. 279.20(b)(2). Transfer facilities that store used oil for more than 35 days are subject to regulation under subpart F of this part.

**Used oil transporter** (Section 279.1) means any person who transports used oil, any person who collects used oil from more than one generator and transports the collected oil, and owners and operators of used oil transfer facilities. Used oil transporters may consolidate or aggregate loads of used oil for purposes of transportation but, with the following exception, may not process used oil. Transporters may conduct incidental processing operations that occur in the normal course of used oil transportation (e.g., settling and water separation), but that are not designed to produce (or make more amenable for production of) used oil derived products or used oil fuel.

**Used or fired military munition is a solid waste** (Section 266.202): (1) When transported off range or from the site of use, Where the site of use is not a range, for the purposes of storage, reclamation, treatment, disposal, or treatment prior to disposal; or (2) If recovered, collected, and then disposed of by burial, or landfilling either on or off a range. For purposes of RCRA section 1004(27), a used or fired military munition is a solid waste, and, therefore, is potentially subject to RCRA corrective action authorities under sections 3004(u) and (v), and 3008(h), or imminent and substantial endangerment authorities under section 7003, if the munition lands off-range and is not promptly rendered safe and/or retrieved. Any imminent and substantial threats associated with any remaining material must be addressed. If remedial action is infeasible, the operator of the range must maintain a record of the event for as long as any threat remains. The record must include the type of munition and its location (to the extent the location is known).

**Used solely for competition** (Section 89.2) means exhibiting features that are not easily removed and that would render its use other than in competition unsafe, impractical, or highly unlikely.

**Used solely for competition** (Section 90.3) means exhibiting features that are not easily removed and that would render its use other than in competition unsafe, impractical, or highly unlikely.

**Used solely for competition** (Section 91.3) means exhibiting features that are not easily removed and that would render its use other than in competition unsafe, impractical, or highly unlikely.

**Useful life (Section 85.1502)** A period of time/mileage as specified in part 86 for a nonconforming vehicle which begins at the time of resale (for a motor vehicle or motor vehicle engine owned by the ICI at the time of importation) or release to the owner (for a motor vehicle or motor vehicle engine not owned by the ICI at the time of importation) of the motor vehicle or motor vehicle engine by the ICI after modification and/ or test pursuant to Sec. 85.1505 or Sec. 85.1509.

**Useful life** (Section 85.1902) shall be given the meaning ascribed to it by section 202(d) of the Act and regulations promulgated thereunder.

**Useful Life** (Section 85.2102) means that period established pursuant to section 202(d) of the Act and regulations promulgated thereunder.

**Useful life** (Section 86.001-2) means: (1) For light-duty vehicles, and for light light-duty trucks not subject to the Tier 0 standards of Sec. 86.094-9(a), intermediate useful life and/or full useful life. Intermediate useful life is a period of use of 5 years or 50,000 miles, whichever occurs first. Full useful life is a period of use of 10 years or 100,000 miles, whichever occurs first, except as otherwise noted in Sec. 86.094-9. The useful life of evaporative and/or refueling emission control systems on the portion of these vehicles subject to the evaporative emission test requirements of Sec. 86.130-96, and/or the refueling emission test requirements of Sec. 86.151-2001, is defined as a period of use of 10 years or 100,000 miles, whichever occurs first. For light light-duty trucks subject to the Tier 0 standards of Sec. 86.094-9(a), and for heavy light-duty truck engine families, intermediate and/or full useful life. Intermediate useful life is a period of use of 5 years or 50,000 miles, whichever occurs first. Full useful life is a period of use of 11 years or 120,000 miles, whichever occurs first. The useful life of evaporative emission and/or refueling control systems on the portion of these vehicles subject to the evaporative emission test requirements of Sec. 86.130-96, and/or the refueling emission test requirements of Sec. 86.151-2001, is also defined as a period of 11 years or 120,000 miles, whichever occurs first. (3)

For an Otto-cycle heavy-duty engine family: (i) For hydrocarbon and carbon monoxide standards, a period of use of 8 years or 110,000 miles, whichever first occurs. (ii) For the oxides of nitrogen standard, a period of use of 10 years or 110,000 miles, whichever first occurs. (iii) For the portion of evaporative emission control systems subject to the evaporative emission test requirements of Sec. 86.1230-96, a period of use of 10 years or 110,000 miles, whichever occurs first. (4) For a diesel heavy-duty engine family: (i) For light heavy-duty diesel engines, for hydrocarbon, carbon monoxide, and particulate standards, a period of use of 8 years or 110,000 miles, whichever first occurs. (ii) For light heavy-duty diesel engines, for the oxides of nitrogen standard, a period of use of 10 years or 110,000 miles, whichever first occurs. (iii) For medium heavy-duty diesel engines, for hydrocarbon, carbon monoxide, and particulate standards, a period of use of 8 years or 185,000 miles, whichever first occurs. (iv) For medium heavy-duty diesel engines, for the oxides of nitrogen standard, a period of use of 10 years or 185,000 miles, whichever first occurs. (v) For heavy heavy-duty diesel engines, for hydrocarbon, carbon monoxide, and particulate standards, a period of use of 8 years or 290,000 miles, whichever first occurs, except as provided in paragraph (4)(vii) of this definition. (vi) For heavy heavy-duty diesel engines, for the oxides of nitrogen standard, a period of use of 10 years or 290,000 miles, whichever first occurs. (vii) For heavy heavy-duty diesel engines used in urban buses, for the particulate standard, a period of use of 10 years or 290,000 miles, whichever first occurs.

**Useful life** (Section 86.004-2) means: (1) For light-duty vehicles, and for light light-duty trucks not subject to the Tier 0 standards of Sec. 86.094-9(a), intermediate useful life and/or full useful life. Intermediate useful life is a period of use of 5 years or 50,000 miles, whichever occurs first. Full useful life is a period of use of 10 years or 100,000 miles, whichever occurs first, except as otherwise noted in Sec. 86.094-9. The useful life of evaporative and/or refueling emission control systems on the portion of these vehicles subject to the evaporative emission test requirements of Sec. 86.130-96, and/or the refueling emission test requirements of Sec. 86.151-98, is defined as a period of use of 10 years or 100,000 miles, whichever occurs first. (2) For light light-duty trucks subject to the Tier 0 standards of Sec. 86.094-9(a), and for heavy light-duty truck engine families, intermediate and/or full useful life. Intermediate useful life is a period of use of 5 years or 50,000 miles, whichever occurs first. Full useful life is a period of use of 11 years or 120,000 miles, whichever occurs first. The useful life of evaporative emission and/or refueling control systems on the portion of these vehicles subject to the evaporative emission test requirements of Sec. 86.130-96, and/or the refueling emission test requirements of Sec. 86.151-98, is also defined as a period of 11 years or 120,000 miles, whichever occurs first. (3) For an Otto-cycle HDE family: (i) For hydrocarbon and carbon monoxide standards, a period of use of 10 years or 110,000 miles, whichever first occurs. (ii) For the oxides of nitrogen standard, a period of use of 10 years or 110,000 miles, whichever first occurs. (iii) For the portion of evaporative emission control systems subject to the evaporative emission test requirements of Sec. 86.1230-96, a period of use of 10 years or 110,000 miles, whichever first occurs. (4) For a diesel HDE family: (i) For light heavy-duty diesel engines, for carbon monoxide, particulate, and oxides of nitrogen plus non-methane hydrocarbons emissions standards, a period of use of 10 years or 110,000 miles, whichever first occurs. (ii) For medium heavy-duty diesel engines, for carbon monoxide, particulate, and oxides of nitrogen plus non-methane hydrocarbons emission standards, a period of use of 10 years or 185,000 miles, whichever first occurs. (iii) For heavy heavy-duty diesel engines, for carbon monoxide, particulate, and oxides of nitrogen plus non-methane hydrocarbon emissions standards, a period of use of 10 years or 435,000 miles, or 22,000 hours, whichever first occurs, except as provided in paragraphs (4)(iv) and (4)(v) of this definition. (iv) The useful life limit of 22,000 hours in paragraph (4)(iii) of this definition is effective as a limit to the useful life only when an accurate hours meter is provided by the manufacturer with the engine and only when such hours meter can reasonably be expected to operate properly over the useful life of the engine. (v) For an individual engine, if the useful life hours limit of 22,000 hours is reached before the engine reaches 10 years or 100,000 miles, the useful life shall become 10 years or 100,000 miles, whichever occurs first, as required under Clean Air Act section 202(d). (5) As an option for both light-duty trucks under certain conditions and HDE families, an alternative useful life period may be assigned by the Administrator under the provisions of Sec. 86.094-21(f). Warranty period, for purposes of HDE emissions defect warranty and emissions performance warranty, shall be a period of 5 years/50,000 miles, whichever occurs first, for Otto-cycle HDEs and light heavy-duty diesel engines. For all other heavy-duty diesel engines the aforementioned period shall be 5 years/100,000 miles, whichever occurs first. However, in no case may this period be less than the basic mechanical warranty period that the manufacturer provides (with or without additional charge) to the purchaser of the engine. Extended warranties on select parts do not extend the emissions warranty requirements for the entire engine but only for those parts. In cases where responsibility for an extended warranty is shared between the owner and the manufacturer, the emissions warranty shall also be shared in the same manner as specified in the warranty agreement.

**Useful life** (Section 86.082-2) means: (1) For light-duty vehicles and light-duty trucks a period of use of 5 years or 50,000 miles, whichever first occurs. (2) For gasoline-fueled heavy-duty engines a period of use of 5 years or 50,000 miles of vehicle operation or 1,500 hours of engine operation (or an equivalent period of 1,500 hours of dynamometer operation), whichever first occurs. (3) For diesel heavy-duty engines a period of use of 5 years or 100,000 miles of vehicle operation or 3,000 hours of engine operation (or an equivalent period of 1,000 hours of dynamometer operation), whichever first occurs.

**Useful life** (Section 86.085-2) means: (a) For light-duty vehicles a period of use of 5 years or 50,000 miles, whichever first occurs. (b) For a light-duty truck engine family, a period of use of 11 years or 120,000 miles, whichever occurs first. (c) For a gasoline-fueled heavy-duty engine family (and in the case of evaporative emission regulations, for gasoline-fueled heavy-duty vehicles), a period of use of 8 years or 110,000 miles, whichever first occurs. (d) For a diesel heavy-duty engine family: (1) For light heavy-duty diesel engines, a period of use of 8 years or 110,000 miles, whichever first occurs. (2) For medium heavy-duty diesel engines, a period of use of 8 years or 185,000 miles, whichever first occurs. (3) For heavy heavy-duty diesel engines, a period of use of 8 years or 290,000 miles, whichever first occurs. (e) As an option for both light-duty truck and heavy-duty engine families, an alternative useful life period assigned by the Administrator under the provisions of paragraph (f) of Sec. 86.085-21. (f) The useful-life period for purposes of the emissions defect warranty and emissions performance warranty shall be a period of 5 years/50,000 miles whichever first occurs, for light-duty trucks, gasoline heavy-duty engines, and light heavy-duty diesel engines. For all other heavy-duty diesel engines the aforementioned period is 5 years/100,000 miles, whichever first occurs. However, in no case may this period be less than the manufacturer's basic mechanical warranty period for the engine family.

**Useful life** (Section 86.090-2) means: (a) For light-duty vehicles a period of use of 5 years or 50,000 miles, whichever first occurs. (b) For a light-duty truck engine family, a period of use of 11 years or 120,000 miles, whichever occurs first. (c) For an Otto-cycle heavy-duty engine family, a period of use of 8 years of 110,000 miles, whichever first occurs. (d) For a diesel heavy-duty engine family: (1) For light heavy-duty diesel engines, period of use of 8 years or 110,000 miles, whichever first occurs. (2) For medium heavy-duty diesel engines, a period of use of 8 years or 185,000 miles, whichever first occurs. (3) For heavy heavy-duty diesel engines, a period of use of 8 years or 290,000 miles, whichever first occurs. (e) As an option for both light-duty truck and heavy-duty engine families, an alternative useful life period assigned by the Administrator under the provisions of paragraph (f) of Sec. 86.090-21. (f) The useful-life period for purposes of the emissions defect warranty and emissions performance warranty shall be a period of 5 years/50,000 miles whichever first occurs, for light-duty trucks, Otto cycle heavy-duty engines and light heavy-duty diesel engines. For all other heavy-duty diesel engines the aforementioned period is 5 years/ 100,000 miles, whichever first occurs. However, in no case may this period be less than the manufacturer's basic mechanical warranty period for the engine family.

**Useful life** (Section 86.094-2) means: (a) For light-duty vehicles, and for model year 1994 and later light light-duty trucks not subject to the Tier 0 standards of paragraph (a) of Sec. 86.094-9, intermediate useful life and/or full useful life. Intermediate useful life is a period of use of 5 years or 50,000 miles, whichever occurs first. Full useful life is a period of use of 10 years or 100,000 miles, whichever occurs first, except as otherwise noted in Sec. 86.094-9. (b) For light light-duty trucks subject to the Tier 0 standards of paragraph (a) of Sec. 86.094-9, and for heavy light-duty truck engine families, intermediate and/or full useful life. Intermediate useful life is a period of use of 5 years or 50,000 miles, whichever occurs first. Full useful life is a period of use of 11 years or 120,000 miles, whichever occurs first. (c) For an Otto-cycle heavy-duty engine family, a period of use of 8 years or 110,000 miles, whichever first occurs. (d) For a diesel heavy-duty engine family: (1) For light heavy-duty diesel engines, period of use of 8 years or 110,000 miles, whichever first occurs. (2) For medium heavy-duty diesel engines, a period of use of 8 years or 185,000 miles, whichever first occurs. (3) For heavy-duty diesel engines, a period of use of 8 years or 290,000 miles, whichever first occurs, except as provided in paragraph (d)(4) of this definition. (4) for heavy heavy-duty diesel engines used in urban buses, for the particulate standard, a period of use of 10 years or 290,000 miles, whichever first occurs. (e) As an option for both light-duty trucks under certain conditions and heavy-duty engine families, an alternative useful life period assigned by the Administrator under the provisions of paragraph (f) of Sec. 86.094-21. (f) The useful-life period for purposes of the emissions defect warranty and emissions performance warranty shall be a period of 5 years/50,000 miles, whichever first occurs, for light-duty trucks, Otto-cycle heavy-duty engines and light heavy-duty diesel engines. For all other heavy-duty diesel engines the aforementioned period is 5 years/ 100,000 miles, whichever first occurs. However, in no case may this period be less than the manufacturer's basic mechanical warranty period for the engine family.

**Useful life** (Section 86.096-2) means: (1) For light-duty vehicles, and for light light-duty trucks not subject to the Tier 0 standards of Sec. 86.094-9(a), intermediate useful life and/or full useful life. Intermediate useful life is a period of use of 5 years or 50,000 miles, whichever occurs first. Full useful life is a period of use of 10 years or 100,000 miles, whichever occurs first, except as otherwise noted in Sec. 86.094-9. The useful life of evaporative emission control systems on the portion of these vehicles subject to the evaporative emission test requirements of Sec. 86.130-96 is defined as a period of use of 10 years or 100,000 miles, whichever occurs first. (2) For light light-duty trucks subject to the Tier 0 standards of Sec. 86.094-9(a), and for heavy light-duty truck engine families, intermediate and/or full useful life. Intermediate useful life is a period of use of 5 years or 50,000 miles, whichever occurs first. Full useful life is a period of use of 11 years or 120,000 miles, whichever occurs first. The useful life of evaporative emission control systems on the portion of these vehicles subject to the evaporative emission test requirements of Sec. 86.130-96 is also defined as a period of 11 years or 120,000 miles, whichever occurs first. (3) For an Otto-cycle heavy-duty engine family, a period of use of 8 years or 110,000 miles, whichever occurs first, except for the portion of evaporative emission control systems subject to the evaporative emission test requirements of Sec. 86.1230-96, for which the applicable period of use is 10 years or 110,000 miles, whichever occurs first. (4) For a diesel heavy-duty engine family: (i) For light heavy-duty diesel engines, period of use of 8 years or 110,000 miles, whichever occurs first. (ii) For medium heavy-duty diesel engines, a period of use of 8 years or 185,000 miles, whichever occurs first. (iii) For heavy heavy-duty diesel engines, a period of use of 8 years or 290,000 miles, whichever occurs first, except as provided in paragraph (4)(iv) of this definition. (iv) For heavy heavy-duty diesel engines used in urban buses, for the particulate standard, a period of use of 10 years or 290,000 miles, whichever occurs first. (5) As an option for both light-duty trucks under certain conditions and heavy-duty engine families, an alternative useful life period assigned by the Administrator under the provisions of Sec. 86.094-21(f). (6) The useful-life period for purposes of the emissions defect warranty and emissions performance warranty shall be a period of 5 years/50,000 miles, whichever occurs first, for light-duty trucks, Otto-cycle heavy-duty engines and light heavy-duty diesel engines. For all other heavy-duty diesel engines the aforementioned period is 5 years/ 100,000 miles, whichever occurs first. However, in no case may this period be less than the manufacturer's basic mechanical warranty period for the engine family.

**Useful life** (Section 86.098-2) means: (1) For light-duty vehicles, and for light light-duty trucks not subject to the Tier 0 standards of Sec. 86.094-9(a), intermediate useful life and/or full useful life. Intermediate useful life is a period of use of 5 years or 50,000 miles, whichever occurs first. Full useful life is a period of use of 10 years or 100,000 miles, whichever occurs first, except as otherwise noted in Sec. 86.094-9. The useful life of evaporative and/or refueling emission control systems on the portion of these vehicles subject to the evaporative emission test requirements of Sec. 86.130-96, and/or the refueling emission test requirements of Sec. 86.151-98, is defined as a period of use of 10 years or 100,000 miles, whichever occurs first. (2) For light light-duty trucks subject to the Tier 0 standards of Sec. 86.094-9(a), and for heavy light-duty truck engine families, intermediate and/or full useful life. Intermediate useful life is a period of use of 5 years or 50,000 miles, whichever occurs first. Full useful life is a period of use of 11 years or 120,000 miles, whichever occurs first. The useful life of evaporative emission control systems on the portion of these vehicles subject to the evaporative emission test requirements of Sec. 86.130-96 is also defined as a period of 11 years or 120,000 miles, whichever occurs first. (3) For an Otto-cycle heavy-duty engine family: (i) For hydrocarbon and carbon monoxide standards, a period of use of 8 years or 110,000 miles, whichever first occurs. (ii) For the oxides of nitrogen standard, a period of use of 10 years or 110,000 miles, whichever first occurs. (iii) For the portion of evaporative emission control systems subject to the evaporative emission test requirements of Sec. 86.1230-96, a period of use of 10 years or 110,000 miles, whichever occurs first. (4) For a diesel heavy-duty engine family: (i) For light heavy-duty diesel engines, for hydrocarbon, carbon monoxide, and particulate standards, a period of use of 8 years or 110,000 miles, whichever first occurs. (ii) For light heavy-duty diesel engines, for the oxides of nitrogen standard, a period of use of 10 years or 110,000 miles, whichever first occurs. (iii) For medium heavy-duty diesel engines, for hydrocarbon, carbon monoxide, and particulate standards, a period of use of 8 years or 185,000 miles, whichever first occurs. (iv) For medium heavy-duty diesel engines, for the oxides of nitrogen standard, a period of use of 10 years or 185,000 miles, whichever first occurs. (v) For heavy heavy-duty diesel engines, for hydrocarbon, carbon monoxide, and particulate standards, a period of use of 8 years or 290,000 miles, whichever first occurs, except as provided in paragraph (3)(vii) of this definition. (vi) For heavy heavy-duty diesel engines, for the oxides of nitrogen standard, a period of use of 10 years or 290,000 miles, whichever first occurs. (vii) For heavy heavy-duty diesel engines used in urban buses, for the particulate standard, a period of use of 10 years or 290,000 miles, whichever first occurs.

**Useful life** (Section 86.1803-01) means the period of use or time during which an emission standard applies to light-duty vehicles and light-duty trucks, as described in Sec. 86.1805-01.

**Useful life** (Section 86.402-78) is defined for each class (see Sec. 86.419) of motorcycle: Class I--5.0 years or 12,000 km (7,456 miles), whichever first occurs. Class II--5.0 years or 18,000 km (11,185 miles), whichever first occurs. Class III--5.0 years or 30,000 km (18,641 miles), whichever first occurs. Unscheduled maintenance means any inspection, adjustment, repair, removal, disassembly, cleaning, or replacement of vehicle components or systems which is performed to correct or diagnose a part failure or vehicle malfunction which was not anticipated.

**Useful life** (Section 89.602) A period of time as specified in subpart B of this part which for a nonconforming nonroad engine begins at the time of resale (for a nonroad engine owned by the ICI at the time of importation) or release to the owner (for a nonroad engine not owned by the ICI at the time of importation) of the nonroad engine by the ICI after modification and/or testing pursuant to Sec. 89.605 or Sec. 89.609.

**Useful life** (Section 92.2) means the period during which the locomotive engine is designed to properly function in terms of reliability and fuel consumption, without being remanufactured, specified as work output or miles. It is the period during which a new locomotive or locomotive engine is required to comply with all applicable emission standards.

**Useful life** (Section 94.2) means the period during which an engine is designed to properly function in terms of reliability and fuel consumption, without being remanufactured, specified as hours of operation and years. It is the period during which a new engine is required to comply with all applicable emission standards. (Note: Sec. 94.9(a) specifies minimum requirements for useful life values.)

**USFRS** (Section 266.401) means U.S. Filter Recovery Services, Inc. whose principal place of business for the purposes of these rules is 2430 Rose Place, Roseville, Minnesota.

**USFRS XL Waste** (Section 266.401) means one or more USFRS used water treatment resin canisters and their contents, any associated USFRS pre- or post-resin filters and their containers and their contents from a USFRS XL waste generator located within the State of Minnesota. USFRS XL waste includes the ion exchange resins, the associated pre- and post-resin filters, wastes contained on or within the ion exchange resins and filters and any other wastes contained within the water treatment resin canisters and filter containers. USFRS XL waste also includes spills of XL waste which are handled in accordance with the requirements in this subpart. This definition does not include wastes that were generated prior to the date a generator is added to this USFRS XL Project. USFRS XL waste shall be identified by the waste code XL001.

**USFRS XL Waste Application Form** (Section 266.401) means the form approved by EPA and Minnesota Pollution Control Agency (MPCA) as part of the USFRS XL Waste Project or subsequently modified by USFRS and approved by EPA and MPCA and used for characterization of the chemical constituents of a person's USFRS XL waste. The USFRS XL Waste Application Form shall include all attachments by USFRS or the applicant, including but not limited to, the USFRS Site Engineering Form, Systems Engineering Form and any waste analysis.

**USFRS XL Waste Approved Customer** (Section 266.401) means only those persons located in Minnesota who have properly identified their wastes and processes on the USFRS XL waste application form; have not been excluded by EPA, MPCA or the County Agencies from participation in the USFRS XL waste project; have signed the USFRS XL waste Final Project Agreement (FPA); have certified that they have read and understand the USFRS XL waste training module; and have not generated USFRS XL wastes.

**USFRS XL waste approved transporter** (Section 266.401) means a transporter located within the State of Minnesota who has a satisfactory safety rating from the United States Department of Transportation (USDOT) in the last year; has not been excluded by EPA, MPCA or the County Agencies from participation in the USFRS XL waste project; has signed the USFRS XL waste FPA; and has signed a certification that it has been trained by USFRS on the proper handling of USFRS XL wastes and understands its responsibilities under this subpart.

**USFRS XL Waste Facility or USFRS Facility** (Section 266.401) means the U.S. Filter Recovery Service, Inc. operations located at 2430 Rose Place, Roseville, Minnesota.

**USFRS XL Waste Final Project Agreement (FPA)** (Section 266.401) means the agreement signed by USFRS, EPA, MPCA, the counties of Anoka, Carver, Dakota, Hennepin, Ramsey, Scott and Washington in Minnesota, Pioneer Tank Lines and USFRS XL waste customers, generators and transporters. The FPA may be modified to add or delete participants, subject to the approval of EPA and MPCA. The FPA was signed by EPA, USFRS and MPCA on September 21, 2000.

**USFRS XL Waste Generator** (Section 266.401) means a USFRS XL waste approved customer who generates or generated USFRS XL waste.

**USFRS XL Waste Project, USFRS XL Project or XL Project** (Section 266.401) means the program identified in the Final Project Agreement and this part for the generation, transportation and subsequent treatment, storage and disposal of USFRS XL waste.

**USFRS XL waste training module** (Section 266.401) means the recorded training program approved by EPA and MPCA as part of the USFRS XL Waste Project or subsequently modified by USFRS and approved by EPA and MPCA and developed by USFRS for the purpose of informing USFRS XL waste approved customers, generators and transporters of the special requirements imposed on them by this part and the proper method of handling USFRS XL wastes.

**USFRS XL Waste Transportation Tracking Document** (Section 266.401) means the Transportation Tracking Document developed by USFRS which was approved by EPA and the MPCA as part of the USFRS XL Waste Project or subsequently modified by USFRS and approved by EPA and MPCA; and used when USFRS XL waste is transported off-site from a generator.

**USFRS XL Waste Transporter** (Section 266.401) means USFRS or a USFRS XL waste approved transporter who transports USFRS XL waste.

**UST system or Tank system** (Section 280.12) means an underground storage tank, connected underground piping, underground ancillary equipment, and containment system, if any.

**Utility** (Section 72.2) means any person that sells electricity.

**Utility competitive bid solicitation** (Section 72.2) is a public request from a regulated utility for offers to the utility for meeting future generating needs. A qualifying facility, independent power production facility, or new IPP may be regarded as having been "selected" in such solicitation if the utility has named the facility as a project with which the utility intends to negotiate a power sales agreement.

**Utility regulatory authority** (Section 72.2) means an authority, board, commission, or other entity (limited to the local-, State-, or federal-level, whenever so specified) responsible for overseeing the business operations of utilities located within its jurisdiction, including, but not limited to, utility rates and charges to customers.

**Utility system** (Section 72.2) means all interconnected units and generators operated by the same utility operating company.

**Utility unit** (Section 72.2) means a unit owned or operated by a utility: (1) That serves a generator in any State that produces electricity for sale, or (2) That during 1985, served a generator in any State that produced electricity for sale. (3) Notwithstanding paragraphs (1) and (2) of this definition, a unit that was in operation during 1985, but did not serve a generator that produced electricity for sale during 1985, and did not commence commercial operation on or after November 15, 1990 is not a utility unit for purposes of the Acid Rain Program. (4) Notwithstanding paragraphs (1) and (2) of this definition, a unit that cogenerates steam and electricity is not a utility unit for purposes of the Acid Rain Program, unless the unit is constructed for the purpose of supplying, or commences construction after November 15, 1990 and supplies, more than one-third of its potential electrical output capacity and more than 25 MWe output to any power distribution system for sale.

**Utilization** (Section 72.2) means the heat input (expressed in mmBtu/time) for a unit.

**Utilization** (Section 96.2) means the heat input (expressed in mm Btu/time) for a unit. The unit's total heat input for the control period in each year will be determined in accordance with part 75 of this chapter if the NOX Budget unit was otherwise subject to the requirements of part 75 of this chapter for the year, or will be based on the best available data reported to the Administrator for the unit if the unit was not otherwise subject to the requirements of part 75 of this chapter for the year.

# V

**Valid Emission Performance Warranty Claim** (Section 85.2102) means a claim in which there is no evidence that the vehicle had not been properly maintained and operated in accordance with manufacturer instructions, the vehicle failed to conform to applicable emission standards as measured by an Office Director-approved type of emission warranty test during its useful life and the owner is subject to sanction as a result of the test failure.

**Valid study** (Section 152.83) means a study that has been conducted in accordance with the Good Laboratory Practice standards of 40 CFR part 160 or generally accepted scientific methodology and that EPA has not determined to be invalid.

**Validated test** (Section 154.3) means a test determined by the Agency to have been conducted and evaluated in a manner consistent with accepted scientific procedures.

**Valuable commercial and recreational species** (Section 228.2) shall mean those species for which catch statistics are compiled on a routine basis by the Federal or State agency responsible for compiling such statistics for the general geographical area impacted, or which are under current study by such Federal or State agencies for potential development for commercial or recreational use.

**Value for pesticide purposes** (Section 172.1) means that characteristic of a substance or mixture of substances which produces an efficacious action on a pest.

**Van** (Section 600.022-85) means any light truck having an integral enclosure fully enclosing the driver compartment and load-carrying device, and having no body sections protruding more than 30 inches ahead of the leading edge of the windshield.

**Van** (Section 86.082-2) means a light-duty truck having an integral enclosure, fully enclosing the driver compartment and load carrying device, and having no body sections protruding more than 30 inches ahead of the leading edge of the windshield.

**Van** (Section 86.1803-01) means a light-duty truck or complete heavy-duty vehicle having an integral enclosure, fully enclosing the driver compartment and load carrying device, and having no body sections protruding more than 30 inches ahead of the leading edge of the windshield.

**Vapor balancing system** (Section 65.2) means a piping system that is designed to collect regulated material vapors displaced from tank trucks or railcars during loading and to route the collected regulated material vapors to the storage vessel from which the liquid being loaded originated, or to another storage vessel connected by a common header; or to compress and route to a process or a fuel gas system the collected regulated material vapors.

**Vapor incinerator** (Section 264.1031) means any enclosed combustion device that is used for destroying organic compounds and does not extract energy in the form of steam or process heat.

**Vapor-mounted seal** (Section 265.1081) means a continuous seal that is mounted such that there is a vapor space between the hazardous waste in the unit and the bottom of the seal.

**Vapor-mounted seal** (Section 65.2) means a continuous seal that is mounted so that there is a vapor space between the stored liquid and the bottom of the seal.

**Variance** (Section 122.2) means any mechanism or provision under section 301 or 316 of CWA or under 40 CFR part 125, or in the applicable "effluent limitations guidelines" which allows modification to or waiver of the generally applicable effluent limitation requirements or time deadlines of CWA. This includes provisions which allow the establishment of alternative limitations based on fundamentally different factors or on sections 301©, 301(g), 301(h), 301(i), or 316(a) of CWA.

**Variance** (Section 51.100) means the temporary deferral of a final compliance date for an individual source subject to an approved regulation, or a temporary change to an approved regulation as it applies to an individual source.

**Vector** (Section 240.101) means a carrier, usually an arthropod, that is capable of transmitting a pathogen from one organism to another.

**Vector** (Section 243.101) means a carrier that is capable of transmitting a pathogen from one organism to another.

**Vector attraction** (Section 503.31) is the characteristic of sewage sludge that attracts rodents, flies, mosquitos, or other organisms capable of transporting infectious agents.

**Vegetable oil** (Section 112.2) means a non-petroleum oil or fat of vegetable origin, including but not limited to oils and fats derived from plant seeds, nuts, fruits, and kernels.

**Vegetable tan** (Section 425.02) means the process of converting hides into leather using chemicals either derived from vegetable matter or synthesized to produce effects similar to those chemicals.

**Vegetative reproduction** (Section 174.3) means either: (1) In seed plants, reproduction by apomixis, or (2) In other plants, reproduction by fragmentation, or division of the somatic body.

**Vehicle** (Section 160.3) means any agent which facilitates the mixture, dispersion, or solubilization of a test substance with a carrier.

**Vehicle** (Section 205.151) means any motorcycle regulated pursuant to this subpart.

**Vehicle** (Section 205.51) means any motor vehicle, machine or tractor, which is propelled by mechanical power and capable of transportation of property on a street or highway and which has a gross vehicle weight rating in excess of 10,000 pounds and a partially or fully enclosed operator's compartment.

**Vehicle** (Section 792.3) means any agent which facilitates the mixture, dispersion, or solubilization of a test substance with a carrier.

**Vehicle** (Section 85.2102) means a light duty vehicle or a light duty truck.

**Vehicle** (Section 86.602-84) means any new production light-duty vehicle as defined in subpart A of this part.

**Vehicle Configuration** (Section 600.022-85) means a unique combination of basic engine, engine code, inertia weight class, transmission configuration, and axle ratio within a base level.

**Vehicle configuration** (Section 86.082-2) means a unique combination of basic engine, engine code, inertia weight class, transmission configuration, and axle ratio.

**Vehicle configuration** (Section 86.1803-01) means a unique combination of basic engine, engine code, inertia weight class, transmission configuration, and axle ratio.

**Vehicle curb weight** (Section 86.082-2) means the actual or the manufacturer's estimated weight of the vehicle in operational status with all standard equipment, and weight of fuel at nominal tank capacity, and the weight of optional equipment computed in accordance with Sec. 86.082-24; incomplete light-duty trucks shall have the curb weight specified by the manufacturer.

**Vehicle or Engine Configuration** (Section 85.2113) means the specific subclassification unit of an engine family or certified part application group as determined by engine displacement, fuel system, engine code, transmission and inertia weight class, as applicable.

**Vehicle used for motor vehicle manufacturer product evaluations and tests** (Section 88.302-94) means a vehicle that is owned and operated by a motor vehicle manufacturer (as defined in section 216(1) of the Act), or motor vehicle component manufacturer, or owned or held by a university research department, independent testing laboratory, or other such evaluation facility, solely for the purpose of evaluating the performance of such vehicle for engineering, research and development, or quality control reasons.

**Vented** (Section 264.1031) means discharged through an opening, typically an open-ended pipe or stack, allowing the passage of a stream of liquids, gases, or fumes into the atmosphere. The passage of liquids, gases, or fumes is caused by mechanical means such as compressors or vacuum-producing systems or by process-related means such as evaporation produced by heating and not caused by tank loading and unloading (working losses) or by natural means such as diurnal temperature changes.

**Vertically fired boiler** (Section 76.2) means a dry bottom boiler with circular burners, or coal and air pipes, oriented downward and mounted on waterwalls that are horizontal or at an angle. This definition shall include dry bottom roof-fired boilers and dry bottom top-fired boilers, and shall exclude dry bottom arch-fired boilers and dry bottom turbo-fired boilers.

**Very low sulfur fuel** (Section 72.2) means either: (1) A fuel with a total sulfur content no greater than 0.05 percent sulfur by weight; (2) Natural gas or pipeline natural gas, as defined in this section; or (3) Any gaseous fuel with a total sulfur content no greater than 20 grains of sulfur per 100 standard cubic feet.

**Vessel** (Section 112.2) means every description of watercraft or other artificial contrivance used, or capable of being used as a means of transportation on water, other than a public vessel.

**Vessel** (Section 116.3) means every description of watercraft or other artificial contrivance used, or capable of being used, as a means of transportation on water other than a public vessel;

**Vessel (Section 140.1)** includes every description of watercraft or other artificial contrivance used, or capable of being used, as a means of transportation on waters of the United States;

**Vessel** (Section 1700.3) includes every description of watercraft or other artificial contrivance used, or capable of being used, as a means of transportation on navigable waters of the United States or waters of the contiguous zone, but does not include amphibious vehicles.

**Vessel** (Section 260.10) includes every description of watercraft, used or capable of being used as a means of transportation on the water.

**Vessel** (Section 302.3) means every description of watercraft or other artificial contrivance used, or capable of being used, as a means of transportation on water.

**Vessel** (Section 68.3) means any reactor, tank, drum, barrel, cylinder, vat, kettle, boiler, pipe, hose, or other container.

**Vessel** (Section 89.2) has the meaning given to it in 1 U.S.C. 3.

**Vessel** (Section 94.2) means a marine vessel.

**Vessel as defined by section 101(28) of CERCLA** (Section 300.5) means every description of watercraft or other artificial contrivance used, or capable of being used, as a means of transportation on water; and, as defined by section 311(a)(3) of the CWA, means every description of watercraft or other artificial contrivance used, or capable of being used, as a means of transportation on water other than a public vessel.

**Vibration** (Section 763.83 )means the periodic motion of friable ACBM which may result in the release of asbestos fibers.

**Virgin material** (Section 246.101) means a raw material used in manufacturing that has been mined or harvested and has not as yet become a product.

**Virus** (Section 141.2) means a virus of fecal origin which is infectious to humans by waterborne transmission.

**Visible emission** (Section 63.2) means the observation of an emission of opacity or optical density above the threshold of vision.

**Visible emission** (Section 65.2) means the observation of an emission of opacity or optical density above the threshold of vision.

**Visual inspection for clearance testing** (Section 745.223) means the visual examination of a residential dwelling or a child-occupied facility following an abatement to determine whether or not the abatement has been successfully completed.

**Visual inspection for risk assessment** (Section 745.223) means the visual examination of a residential dwelling or a child-occupied facility to determine the existence of deteriorated lead-based paint or other potential sources of lead-based paint hazards.

**VOC** (Section 58.1) means volatile organic compounds.

**VOC content** (Section 59.101) means the weight of VOC per volume of coating, calculated according to the procedures in Sec. 59.104(a) of this subpart.

**Volatile liquid fuel** (Section 92.2) means any liquid fuel other than diesel or biodiesel.

**Volatile Organic Compound** (Section 60.2) means any organic compound which participates in atmospheric photochemical reactions; or which is measured by a reference method, an equivalent method, an alternative method, or which is determined by procedures specified under any subpart.

**Volatile organic compounds (VOC)** (Section 51.100) means any compound of carbon, excluding carbon monoxide, carbon dioxide, carbonic acid, metallic carbides or carbonates, and ammonium carbonate, which participates in atmospheric photochemical reactions. (1) This includes any such organic compound other than the following, which have been determined to have negligible photochemical reactivity: methane; ethane; methylene chloride (dichloromethane); 1,1,1-trichloroethane (methyl chloroform); 1,1,2-trichloro-1,2,2-trifluoroethane (CFC-113); trichlorofluoromethane (CFC-11); dichlorodifluoromethane (CFC-12); chlorodifluoromethane (HCFC-22); trifluoromethane (HFC-23); 1,2-dichloro 1,1,2,2-tetrafluoroethane (CFC-114); chloropentafluoroethane (CFC-115); 1,1,1-trifluoro 2,2-dichloroethane (HCFC-123); 1,1,1,2-tetrafluoroethane (HFC-134a); 1,1-dichloro 1-fluoroethane (HCFC-141b); 1-chloro 1,1-difluoroethane (HCFC-142b); 2-chloro-1,1,1,2-tetrafluoroethane (HCFC-124); pentafluoroethane (HFC-125); 1,1,2,2-tetrafluoroethane (HFC-134); 1,1,1-trifluoroethane (HFC-143a); 1,1-difluoroethane (HFC-152a); parachlorobenzotrifluoride (PCBTF); cyclic, branched, or linear completely methylated siloxanes; acetone; perchloroethylene (tetrachloroethylene); 3,3-dichloro-1,1,1,2,2-pentafluoropropane (HCFC-225ca); 1,3-dichloro-1,1,2,2,3-pentafluoropropane (HCFC-225cb); 1,1,1,2,3,4,4,5,5,5-decafluoropentane (HFC 43-10mee); difluoromethane (HFC-32); ethylfluoride (HFC-161); 1,1,1,3,3,3-hexafluoropropane (HFC-236fa); 1,1,2,2,3-entafluoropropane (HFC-245ca); 1,1,2,3,3-pentafluoropropane (HFC-245ea); 1,1,1,2,3-pentafluoropropane (HFC-245eb); 1,1,1,3,3-pentafluoropropane (HFC-245fa); 1,1,1,2,3,3-hexafluoropropane (HFC-236ea); 1,1,1,3,3-pentafluorobutane (HFC-365mfc); chlorofluoromethane (HCFC-31); 1 chloro-1-fluoroethane (HCFC-151a); 1,2-dichloro-1,1,2-trifluoroethane (HCFC-123a); 1,1,1,2,2,3,3,4,4-nonafluoro-4-methoxy-butane (C4F9OCH3); 2-(difluoromethoxymethyl)-1,1,1,2,3,3,3-heptafluoropropane ((CF3)2CFCF2OCH3); 1-ethoxy-1,1,2,2,3,3,4,4,4-nonafluorobutane (C4F9OC2H5); 2-(ethoxydifluoromethyl)-1,1,1,2,3,3,3-heptafluoropropane (CF3)2CFCF2OC2H5); methyl acetate and perfluorocarbon compounds which fall into these classes: (i) Cyclic, branched, or linear, completely fluorinated alkanes; (ii) Cyclic, branched, or linear, completely fluorinated ethers with no unsaturations; (iii) Cyclic, branched, or linear, completely fluorinated tertiary amines with no unsaturations; and (iv) Sulfur containing perfluorocarbons with no unsaturations and with sulfur bonds only to carbon and fluorine.(2) For purposes of determining compliance with emissions limits, VOC will be measured by the test methods in the approved State implementation plan (SIP) or 40 CFR part 60, appendix A, as applicable. Where such a method also measures compounds with negligible photochemical reactivity, these negligibility-reactive compounds may be excluded as VOC if the amount of such compounds is accurately quantified, and such exclusion is approved by the enforcement authority. (3) As a precondition to excluding these compounds as VOC or at any time thereafter, the enforcement authority may require an owner or operator to provide monitoring or testing methods and results demonstrating, to the satisfaction of the enforcement authority, the amount of negligibly-reactive compounds in the source's emissions. (4) For purposes of Federal enforcement for a specific source, the EPA shall use the test methods specified in the applicable EPA-approved SIP, in a permit issued pursuant to a program approved or promulgated under title V of the Act, or under 40 CFR part 51, subpart I or appendix S, or under 40 CFR parts 52 or 60. The EPA shall not be bound by any State determination as to appropriate methods for testing or monitoring negligibly-reactive compounds if such determination is not reflected in any of the above provisions. (t)-(w) [Reserved]

**Volatile organic compounds or VOC** (Section 59.101) means any compound of carbon, other than those organic compounds that the Administrator has excluded in 40 CFR part 51, Sec. 51.100 from this definition.

**Volatile organic concentration or VO concentration** (Section 265.1081) means the fraction by weight of the volatile organic compounds contained in a hazardous waste expressed in terms of parts per million (ppmw) as determined by direct measurement or by knowledge of the waste in accordance with the requirements of Sec. 265.1084 of this subpart. For the purpose of determining the VO concentration of a hazardous waste, organic compounds with a Henry's law constant value of at least 0.1 mole-fraction-in-the-gas-phase/mole-fraction-in the liquid-phase (0.1 Y/X) (which can also be expressed as 1.8x10-\6\ atmospheres/gram-mole/m\3\) at 25 degrees Celsius must be included. Appendix VI of this subpart presents a list of compounds known to have a Henry's law constant value less than the cutoff level.

**Volatile solids** (Section 503.31) is the amount of the total solids in sewage sludge lost when the sewage sludge is combusted at 550 degrees Celsius in the presence of excess air.

**Volume of process water used per year** (Section 463.21) is the volume of process water that flows through a cleaning process and comes in contact with the plastic product over a period of one year.

**Volume of process water used per year** (Section 463.31) is the volume of process water that flows through a finishing water process and comes in contact with the plastics product over a period of one year.

**Volumetric flow** (Section 72.2) means the rate of movement of a specified volume of gas past a cross-sectional area (e.g., cubic feet per hour).

**Voluntary Advanced Technology Incentives Program** (Section 430.21) is the program established under Sec. 430.24(b) (for existing direct dischargers) and Sec. 430.25© (for new direct dischargers) whereby participating mills agree to accept enforceable effluent limitations and conditions in their NPDES permits that are more stringent than the "baseline BAT limitations or NSPS" that would otherwise apply, in exchange for regulatory- and enforcement-related rewards and incentives.

**Voluntary agreement (see Sec. 305.25)** (Section 305.3) means a written communication, signed by all the parties or their counsel or representatives, containing an order acceptable to both the Requestor and EPA. A voluntary agreement shall state that, for purposes of this proceeding, EPA consents to the award of a sum certain to the Requestor or such other consideration as the parties deem appropriate. A voluntary agreement is effective without approval of the Presiding Officer and is a final order as defined in this part.

**Voluntary and incentive-based actions, or existing programs, procedures or authorities** (Section 130.2) are acceptable means of demonstrating reasonable assurance if they satisfy the four-part test. Examples of voluntary and incentive-based actions include: State, Territorial, or authorized Tribal programs to audit implementation of agricultural or forestry best management practices; memoranda of understanding between States, Territories, authorized Tribes, and organizations representing categories, subcategories, or individual sources; or State-, Territory-, or authorized Tribe-approved programs for categories, subcategories or individual sources to ensure effectiveness of best management practices. Examples of existing programs, procedures or authorities that may be reliable delivery mechanisms include State, Territorial, and authorized Tribal programs approved by EPA under section 319 of the Clean Water Act; participation in existing United States Department of Agriculture conservation or water quality protection programs; participation in existing programs under the Coastal Zone Act Reauthorization Amendments; regulations; local ordinances; performance bonds; contracts; cost-share agreements; memoranda of understanding; site-specific or watershed-specific voluntary actions; and compliance audits of best management practices.

**Voluntary emission recall** (Section 90.802) means a repair, adjustment, or modification program voluntarily initiated and conducted by a manufacturer to remedy any emission-related defect for which notification of engine owners has been provided.

**Voluntary emission recall** (Section 92.2) means a repair, adjustment, or modification program voluntarily initiated and conducted by a manufacturer or remanufacturer to remedy any emission-related defect for which notification of locomotive or locomotive engine owners has been provided.

**Voluntary emission recall** (Section 94.2) means a repair, adjustment, or modification program voluntarily initiated and conducted by a manufacturer to remedy any emission-related defect for which notification of engine or vessel owners has been provided.

**Voluntary Emissions Recall** (Section 85.1902) shall mean a repair, adjustment, or modification program voluntarily initiated and conducted by a manufacturer to remedy any emission-related defect for which direct notification of vehicle or engine owners has been provided.

**Volunteer** (Section 300.5) means any individual accepted to perform services by the lead agency which has authority to accept volunteer services (examples: See 16 U.S.C. 742f©). A volunteer is subject to the provisions of the authorizing statute and the NCP.

**Vulnerable zone** (Section 1400.2) means the geographical area that could be affected by a worst-case or alternative scenario release from a stationary source, as indicated by the off-site consequence analysis reported by the stationary source in its risk management plan pursuant to the applicable requirements of 40 CFR Part 68. It is defined as a circle, the center of which is the stationary source and the radius of which is the "distance-to-endpoint," or the distance a toxic or flammable cloud, overpressure, or radiant heat would travel after being released and before dissipating to the point that it no longer threatens serious short-term harm to people or the environment.

# W

**Wall-fired boiler** (Section 76.2) means a boiler that has pulverized coal burners arranged on the walls of the furnace. The burners have discrete, individual flames that extend perpendicularly into the furnace area.

**Warning Device** (Section 201.1) means a sound emitting device used to alert and warn people of the presence of railroad equipment.

**Warranty** (Section 204.2) means the warranty required by section 6© (1) of the Act.

**Warranty** (Section 205.151) means the warranty required by section 6(d) (1) of the Act.

**Warranty** (Section 205.2) means the warranty required by section 6©(1) of the Act.

**Warranty Booklet** (Section 85.2102) means a booklet, separate from the owner's manual, containing all warranties provided with the vehicle.

**Warranty period** (Section 90.3) means the period of time the engine or part is covered by the warranty provisions.

**Warranty period** (Section 91.3) means the period of time the engine or part is covered by the warranty provisions.

**Waste** (Section 191.12) as used in this subpart, means any spent nuclear fuel or radioactive waste isolated in a disposal system.

**Waste** (Section 194.2) means the radioactive waste, radioactive material and coincidental material subject to the requirements of part 191 of this chapter.

**Waste** (Section 437.2) includes aqueous, non-aqueous, and solid waste, wastewater, and/or used materials.

**Waste characteristic** (Section 194.2) means a property of the waste that has an impact on the containment of waste in the disposal system.

**Waste component** (Section 194.2) means an ingredient of the total inventory of the waste that influences a waste characteristic.

**Waste determination** (Section 265.1081) means performing all applicable procedures in accordance with the requirements of Sec. 265.1084 of this subpart to determine whether a hazardous waste meets standards specified in this subpart. Examples of a waste determination include performing the procedures in accordance with the requirements of Sec. 265.1084 of this subpart to determine the average VO concentration of a hazardous waste at the point of waste origination; the average VO concentration of a hazardous waste at the point of waste treatment and comparing the results to the exit concentration limit specified for the process used to treat the hazardous waste; the organic reduction efficiency and the organic biodegradation efficiency for a biological process used to treat a hazardous waste and comparing the results to the applicable standards; or the maximum volatile organic vapor pressure for a hazardous waste in a tank and comparing the results to the applicable standards.

**Waste form** (Section 191.12) means the materials comprising the radioactive components of waste and any encapsulating or stabilizing matrix.

**Waste management unit boundary** (Section 258.2) means a vertical surface located at the hydraulically downgradient limit of the unit. This vertical surface extends down into the uppermost aquifer.

**Waste Oil** (Section 761.13) means used products primarily derived from petroleum, which include, but are not limited to, fuel oils, motor oils, gear oils, cutting oils, transmission fluids, hydraulic fluids, and dielectric fluids.

**Waste pile or pile** (Section 272.2) means any noncontainerized accumulation of solid, nonflowing waste that is used for treatment or storage.

**Waste placement** (Section 258.2) in existing units must be consistent with past operating practices or modified practices to ensure good management.

**Waste stabilization** (Section 372.3) means any physical or chemical process used to either reduce the mobility of hazardous constituents in a hazardous waste or eliminate free liquid as determined by a RCRA approved test method for evaluating solid waste as defined in this section. A waste stabilization process includes mixing the hazardous waste with binders or other materials, and curing the resulting hazardous waste and binder mixture. Other synonymous terms used to refer to this process are "stabilization," "waste fixation," or "waste solidification."

**Waste stabilization process** (Section 265.1081) means any physical or chemical process used to either reduce the mobility of hazardous constituents in a hazardous waste or eliminate free liquids as determined by Test Method 9095 (Paint Filter Liquids Test) in "Test Methods for Evaluating Solid Waste, Physical/Chemical Methods," EPA Publication No. SW-846, Third Edition, September 1986, as amended by Update I, November 15, 1992 (incorporated by reference—refer to Sec. 260.11 of this chapter). A waste stabilization process includes mixing the hazardous waste with binders or other materials, and curing the resulting hazardous waste and binder mixture. Other synonymous terms used to refer to this process are "waste fixation" or "waste solidification." This does not include the adding of absorbent materials to the surface of a waste, without mixing, agitation, or subsequent curing, to absorb free liquid.

**Waste treatment systems** (Section 122.2) including treatment ponds or lagoons designed to meet the requirements of CWA (other than cooling ponds as defined in 40 CFR 423.11(m) which also meet the criteria of this definition) are not waters of the United States. This exclusion applies only to manmade bodies of water which neither were originally created in waters of the United States (such as disposal area in wetlands) nor resulted from the impoundment of waters of the United States. [See Note 1 of this section.] Waters of the United States do not include prior converted cropland. Notwithstanding the determination of an area's status as prior converted cropland by any other federal agency, for the purposes of the Clean Water Act, the final authority regarding Clean Water Act jurisdiction remains with EPA.

**Wasteload allocation** (Section 130.2) The portion of a TMDL's pollutant load allocated to a point source of a pollutant for which an NPDES permit is required. For waterbodies impaired by both point and nonpoint sources, wasteload allocations may reflect anticipated or expected reductions of pollutants from other sources if those anticipated or expected reductions are supported by reasonable assurance that they will occur.

**Wasteload allocation (WLA)** (Section 130.2) The portion of a receiving water's loading capacity that is allocated to one of its existing or future point sources of pollution. WLAs constitute a type of water quality-based effluent limitation.

**Wasteload allocation (WLA)** (Section 132.2) is the portion of a receiving water's loading capacity that is allocated to one of its existing or future point sources of pollution, as more fully defined at 40 CFR 130.2(h). In the absence of a TMDL approved by EPA pursuant to 40 CFR 130.7 or an assessment and remediation plan developed and approved in accordance with procedure 3.A of appendix F of this part, a WLA is the allocation for an individual point source, that ensures that the level of water quality to be achieved by the point source is derived from and complies with all applicable water quality standards.

**Wastewater treatment tank** (Section 280.12) means a tank that is designed to receive and treat an influent wastewater through physical, chemical, or biological methods.

**Wastewater treatment unit** (Section 260.10) means a device which: (1) Is part of a wastewater treatment facility that is subject to regulation under either section 402 or 307(b) of the Clean Water Act; and (2) Receives and treats or stores an influent wastewater that is a hazardous waste as defined in Sec. 261.3 of this chapter, or that generates and accumulates a wastewater treatment sludge that is a hazardous waste as defined in Sec. 261.3 of this chapter, or treats or stores a wastewater treatment sludge which is a hazardous waste as defined in Sec. 261.3 of this Chapter; and (3) Meets the definition of tank or tank system in Sec. 260.10 of this chapter.

**Wastewater treatment unit** (Section 270.2) means a device which: (a) Is part of a wastewater treatment facility which is subject to regulation under either section 402 or 307(b) of the Clean Water Act; and (b) Receives and treats or stores an influent wastewater which is a hazardous waste as defined in Sec. 261.3 of this chapter, or generates and accumulates a wastewater treatment sludge which is a hazardous waste as defined in Sec. 261.3 of this chapter, or treats or stores a wastewater treatment sludge which is a hazardous waste as defined in Sec. 261.3 of this chapter; and (c) Meets the definition of tank or tank system in Sec. 260.10 of this chapter.

**Wastewaters** (Section 268.2) are wastes that contain less than 1% by weight total organic carbon (TOC) and less than 1% by weight total suspended solids (TSS).

**Water (bulk shipment)** (Section 260.10) means the bulk transportation of hazardous waste which is loaded or carried on board a vessel without containers or labels.

**Water hold-out coating** (Section 59.101) means a coating applied to the interior cavity areas of doors, quarter panels and rocker panels for the purpose of corrosion resistance to prolonged water exposure.

**Water jet weaving** (Section 410.31) shall mean the internal subdivision of the low water use processing subcategory for facilities primarily engaged in manufacturing woven greige goods through the water jet weaving process.

**Water Management Division Director** (Section 403.3) means one of the Directors of the Water Management Divisions within the Regional offices of the Environmental Protection Agency or this person's delegated representative.

**Water quality limited segment** (Section 130.2) Any segment where it is known that water quality does not meet applicable water quality standards, and/or is not expected to meet applicable water quality standards, even after the application of the technology-based effluent limitations required by sections 301(b) and 306 of the Act.

**Water quality limited segment** (Section 131.3) means any segment where it is known that water quality does not meet applicable water quality standards, and/or is not expected to meet applicable water quality standards, even after the application of the technology-bases effluent limitations required by sections 301(b) and 306 of the Act.

**Water quality management (WQM) plan** (Section 130.2) A State or areawide waste treatment management plan developed and updated in accordance with the provisions of sections 205(j), 208 and 303 of the Act and this regulation.

**Water quality standards** (Section 121.1) means standards established pursuant to section 10© of the Act, and State-adopted water quality standards for navigable waters which are not interstate waters.

**Water quality standards** (Section 131.3) are provisions of State or Federal law which consist of a designated use or uses for the waters of the United States and water quality criteria for such waters based upon such uses. Water quality standards are to protect the public health or welfare, enhance the quality of water and serve the purposes of the Act.

**Water quality standards (WQS)** (Section 130.2) Provisions of State or Federal law which consist of a designated use or uses for the waters of the United States and water quality criteria for such waters based upon such uses. Water quality standards are to protect the public health or welfare, enhance the quality of water and serve the purposes of the Act.

**Water reference level** (Section 159.153) means the level specified in paragraph (1) or (2) of this definition, whichever is lower. (1)

Ten percent of the maximum contaminant level (MCL) established by EPA, or if no MCL has been established by EPA, 10 percent of the most recent draft or final long-term health advisory level (HAL) established by EPA, or if EPA has not published or proposed an MCL or HAL, the lowest detectable amount of the pesticide. (2) The ambient water quality criteria for the protection of aquatic life, established by EPA pursuant to section 304(a) of the Clean Water Act.

**Water-based drilling fluid** (Section 435.11) means the continuous phase and suspending medium for solids is a water-miscible fluid, regardless of the presence of oil.

**Waterbody** (Section 130.2) A geographically defined portion of navigable waters, waters of the contiguous zone, and ocean waters under the jurisdiction of the United States, made up of one or more of the segments of rivers, streams, lakes, wetlands, coastal waters and ocean waters. Identifications of waterbodies should be consistent with the way in which segments are described in State, Territorial, or authorized Tribal water quality standards. (r) List of Impaired Waterbodies or "List." The list of all impaired waterbodies submitted by a State, Territory, or authorized Tribe. This list consists of Parts 1, 2, 3, and 4 described in Sec. 130.27 and the prioritized schedule described in Sec. 130.28. Part 1 of the list consists of the identification of the waterbodies for which TMDLs must be established and a prioritized schedule for establishing TMDLs.

**Waterborne disease outbreak** (Section 141.2) means the significant occurrence of acute infectious illness, epidemiologically associated with the ingestion of water from a public water system which is deficient in treatment, as determined by the appropriate local or State agency.

**Waters of the United States** (Section 230.3) means: (1) All waters which are currently used, or were used in the past, or may be susceptible to use in interstate or foreign commerce, including all waters which are subject to the ebb and flow of the tide; (2) All interstate waters including interstate wetlands; (3) All other waters such as intrastate lakes, rivers, streams (including intermittent streams), mudflats, sandflats, wetlands, sloughs, prairie potholes, wet meadows, playa lakes, or natural ponds, the use, degradation or destruction of which could affect interstate or foreign commerce including any such waters: (i) Which are or could be used by interstate or foreign travelers for recreational or other purposes; or (ii) From which fish or shellfish are or could be taken and sold in interstate or foreign commerce; or (iii) Which are used or could be used for industrial purposes by industries in interstate commerce; (4) All impoundments of waters otherwise defined as waters of the United States under this definition; (5) Tributaries of waters identified in paragraphs (s)(1) through (4) of this section; (6) The territorial sea; (7) Wetlands adjacent to waters (other than waters that are themselves wetlands) identified in paragraphs (s)(1) through (6) of this section; waste treatment systems, including treatment ponds or lagoons designed to meet the requirements of CWA (other than cooling ponds as defined in 40 CFR 423.11(m) which also meet the criteria of this definition) are not waters of the United States. Waters of the United States do not include prior converted cropland. Notwithstanding the determination of an area's status as prior converted cropland by any other federal agency, for the purposes of the Clean Water Act, the final authority regarding Clean Water Act jurisdiction remains with EPA.

**Waters of the United States** (Section 232.2) means: All waters which are currently used, were used in the past, or may be susceptible to us in interstate or foreign commerce, including all waters which are subject to the ebb and flow of the tide. All interstate waters including interstate wetlands. All other waters, such as intrastate lakes, rivers, streams (including intermittent streams), mudflats, sandflats, wetlands, sloughs, prairie potholes, wet meadows, playa lakes, or natural ponds, the use, degradation, or destruction of which would or could affect interstate or foreign commerce including any such waters: Which are or could be used by interstate or foreign travelers for recreational or other purposes; or From which fish or shellfish are or could be taken and sold in interstate or foreign commerce; or Which are used or could be used for industrial purposes by industries in interstate commerce. All impoundments of waters otherwise defined as waters of the United States under this definition; Tributaries of waters identified in paragraphs (g)(1)-(4) of this section; The territorial sea; and Wetlands adjacent to waters (other than waters that are themselves wetlands) identified in paragraphs (q)(1)-(6) of this section. Waste treatment systems, including treatment ponds or lagoons designed to meet the requirements of the Act (other than cooling ponds as defined in 40 CFR 123.11(m) which also meet the criteria of this definition) are not waters of the United States. Waters of the United States do not include prior converted cropland. Notwithstanding the determination of an area's status as prior converted cropland by any other federal agency, for the purposes of the Clean Water Act, the final authority regarding Clean Water Act jurisdiction remains with EPA.

**Waters of the United States** (Section 721.3) has the meaning set forth in 40 CFR 122.2.

**Waters of the United States or waters of the U.S.** (Section 122.2) means: (a) All waters which are currently used, were used in the past, or may be susceptible to use in interstate or foreign commerce, including all waters which are subject to the ebb and flow of the tide; (b) All interstate waters, including interstate "wetlands;" © All other waters such as intrastate lakes, rivers, streams (including intermittent streams), mudflats, sandflats, "wetlands," sloughs, prairie potholes, wet meadows, playa lakes, or natural ponds the use, degradation, or destruction of which would affect or could affect interstate or foreign commerce including any such waters: (1) Which are or could be used by interstate or foreign travelers for recreational or other purposes; (2) From which fish or shellfish are or could be taken and sold in interstate or foreign commerce; or (3) Which are used or could be used for industrial purposes by industries in interstate commerce; (d) All impoundments of waters otherwise defined as waters of the United States under this definition; (e) Tributaries of waters identified in paragraphs (a) through (d) of this definition; (f) The territorial sea; and (g) "Wetlands" adjacent to waters (other than waters that are themselves wetlands) identified in paragraphs (a) through (f) of this definition.

Weather Service in Technical Paper Number 40, "Rainfall Frequency Atlas of the United States", May 1961, and subsequent amendments, or equivalent regional or state rainfall probability information developed therefrom.

weight of more than 6,000 pounds or that has a basic vehicle frontal area in excess of 45 square feet.

**Weighted arithmetic mean** (Section 745.63) means the arithmetic mean of sample results weighted by the number of subsamples in each sample. Its purpose is to give influence to a sample relative to the surface area it represents. A single surface sample is comprised of a single subsample. A composite sample may contain from two to four subsamples of the same area as each other and of each single surface sample in the composite. The weighted arithmetic mean is obtained by summing, for all samples, the product of the sample's result multiplied by the number of subsamples in the sample, and dividing the sum by the total number of subsamples contained in all samples. For example, the weighted arithmetic mean of a single surface sample containing 60 [mu]g/ft2, a composite sample (three subsamples) containing 100 [mu]g/ft2, and a composite sample (4 subsamples) containing 110 [mu]g/ft2 is 100 [mu]g/ ft2. This result is based on the equation [60+(3(Section 745.63)100)+(4(Section 745.63)110)]/(1+3+4).

**Weld-through primer** (Section 59.101) means a primer that is applied to an area before welding is performed, and that provides corrosion resistance to the surface after welding has been performed.

**Well** (Section 144.3) means: A bored, drilled, or driven shaft whose depth is greater than the largest surface dimension; or, a dug hole whose depth is greater than the largest surface dimension; or, an improved sinkhole; or, a subsurface fluid distribution system.

**Well** (Section 146.3) means: A bored, drilled, or driven shaft whose depth is greater than the largest surface dimension; or, a dug hole whose depth is greater than the largest surface dimension; or, an improved sinkhole; or, a subsurface fluid distribution system.

**Well** (Section 147.2902) a bored, drilled, or driven shaft, or a dug hole whose depth is greater than the largest surface dimension.

**Well** (Section 260.10) means any shaft or pit dug or bored into the earth, generally of a cylindrical form, and often walled with bricks or tubing to prevent the earth from caving in.

**Well** (Section 435.61) shall means crude oil producing wells and shall not include gas wells or wells injecting water for disposal or for enchanced recovery of oil or gas.

**Well completion fluids** (Section 435.11) means salt solutions, weighted brines, polymers, and various additives used to prevent damage to the well bore during operations which prepare the drilled well for hydrocarbon production.

**Well completion fluids** (Section 435.41) means salt solutions, weighted brines, polymers, and various additives used to prevent damage to the well bore during operations which prepare the drilled well for hydrocarbon production.

**Well injection** (Section 144.3) means the subsurface emplacement of fluids through a well.

**Well injection** (Section 146.3) means the subsurface emplacement of fluids through a well.

**Well injection** (Section 147.2902) the subsurface emplacement of fluids through a bored, drilled, or driven well; or through a dug well, where the depth of the dug well is greater than the largest surface dimension.

**Well injection** (Section 260.10) (See "underground injection".)

**Well monitoring** (Section 146.3) means the measurement, by on-site instruments or laboratory methods, of the quality of water in a well.

**Well plug** (Section 146.3) means a watertight and gastight seal installed in a borehole or well to prevent movement of fluids.

**Well stimulation** (Section 146.3) means several processes used to clean the well bore, enlarge channels, and increase pore space in the interval to be injected thus making it possible for wastewater to move more readily into the formation, and includes (1) surging, (2) jetting, (3) blasting, (4) acidizing, (5) hydraulic fracturing.

**Well treatment fluids** (Section 435.11) means any fluid used to restore or improve productivity by chemically or physically altering hydrocarbon-bearing strata after a well has been drilled.

**Well treatment fluids** (Section 435.41) means any fluid used to restore or improve productivity by chemically or physically altering hydrocarbon-bearing strata after a well has been drilled.

**Well workover** (Section 147.2902) any reentry of an injection well; including, but not limited to, the pulling of tubular goods, cementing or casing repairs; and excluding any routine maintenance (e.g. re-seating the packer at the same depth, or repairs to surface equipment).

**Well-characterized for introduced genetic material** (Section 725.3) means that the following have been determined: (1) The function of all of the products expressed from the structural gene(s). (2) The function of sequences that participate in the regulation of expression of the structural gene(s). (3) The presence or absence of associated nucleotide sequences and their associated functions, where associated nucleotide sequences are those sequences needed to move genetic material including linkers, homopolymers, adaptors, transposons, insertion sequences, and restriction enzyme sites.

**Western coal mining operation** (Section 434.80) means a surface or underground coal mining operation located in the interior western United States, west of the 100$^{th}$ meridian west longitude, in an arid or semiarid environment with an average annual precipitation of 26.0 inches or less.

**Wet** (Section 420.41) means those steelmaking air cleaning systems that primarily use water for furnace gas cleaning.

**Wet air pollution control scrubbers** (Section 471.02) are air pollution control devices used to remove particulates and fumes from air by entraining the pollutants in a water spray.

**Wet Barking** (Section 430.0) Wet barking operations shall be defined to include hydraulic barking operations and wet drum barking operations which are those drum barking operations that use substantial quantities of water in either water sprays in the barking drums or in a partial submersion of the drums in a "tub" of water.

**Wet bottom** (Section 76.2) means that the ash is removed from the furnace in a molten state. The term "wet bottom boiler" shall include: wet bottom wall-fired boilers, including wet bottom turbo-fired boilers; and wet bottom boilers otherwise meeting the definition of vertically fired boilers, including wet bottom arch-fired boilers, wet bottom roof-fired boilers, and wet bottom top-fired boilers. The term "wet bottom boiler" shall exclude cyclone boilers and tangentially fired boilers.

**Wet desulfurization system** (Section 420.11) means those systems which remove sulfur compounds from coke oven gases and produce a contaminated process wastewater.

**Wet drill cuttings** (Section 435.11) means the unaltered drill cuttings and adhering drilling fluid and formation oil carried out from the wellbore with the drilling fluid.

**Wet drill cuttings** (Section 435.41) means the unaltered drill cuttings and adhering drilling fluid and formation oil carried out from the wellbore with the drilling fluid.

**Wet lot** (Section 412.21) shall mean a confinement facility for raising ducks which is open to the environment with a small portion of shelter area, and with open water runs and swimming areas to which ducks have free access.

**Wet process phosphoric acid process line** (Section 63.601) means any process line manufacturing phosphoric acid by reacting phosphate rock and acid.

**Wet scrubbers** (Section 467.02) are air pollution control devices used to remove particulates and fumes from air by entraining the pollutants in a water spray.

**Wet weather point source** (Section 132.2) means any discernible, confined and discrete conveyance from which pollutants are, or may be, discharged as the result of a wet weather event. Discharges from wet weather point sources shall include only: discharges of storm water from a municipal separate storm sewer as defined at 40 CFR 122.26(b)(8); storm water discharge associated with industrial activity as defined at 40 CFR 122.26(b)(14); discharges of storm water and sanitary wastewaters (domestic, commercial, and industrial) from a combined sewer overflow; or any other stormwater discharge for which a permit is required under section 402(p) of the Clean Water Act. A storm water discharge associated with industrial activity which is mixed with process wastewater shall not be considered a wet weather point source.

**Wet weight** (Section 761.13) means reporting chemical analysis results by including either the weight, or the volume and density, of all liquids. (Sec. 6, Pub. L. 94-469, 90 Stat. 2020 (15 U.S.C. 2605)

**Wetlands** (Section 110.1) means those areas that are inundated or saturated by surface or ground water at a frequency or duration sufficient to support, and that under normal circumstances do support, a prevalence of vegetation typically adapted for life in saturated soil conditions. Wetlands generally include playa lakes, swamps, marshes, bogs and similar areas such as sloughs, prairie potholes, wet meadows, prairie river overflows, mudflats, and natural ponds.

**Wetlands** (Section 122.2) means those areas that are inundated or saturated by surface or groundwater at a frequency and duration sufficient to support, and that under normal circumstances do support, a prevalence of vegetation typically adapted for life in saturated soil conditions. Wetlands generally include swamps, marshes, bogs, and similar areas.

**Wetlands** (Section 230.3) means those areas that are inundated or saturated by surface or ground water at a frequency and duration sufficient to support, and that under normal circumstances do support, a prevalence of vegetation typically adapted for life in saturated soil conditions. Wetlands generally include swamps, marshes, bogs and similar areas.

**Wetlands** (Section 232.2) means those areas that are inundated or saturated by surface or ground water at a frequency and duration sufficient to support, and that under normal circumstances do support, a prevalence of vegetation typically adapted for life in saturated soil conditions. Wetlands generally include swamps, marshes, bogs, and similar areas.

**Wheat** (Section 406.41) shall mean wheat delivered to a plant before processing.

**When it is used to refer to the management of people** (Section 88.302-94) means a person has the authority to direct the activities of another person or employee in a precise situation, such as at the workplace.

**When it is used to refer to the management of vehicles** (Section 88.302-94) means a person has the authority to decide who can operate a particular vehicle, and the purposes for which the vehicle can be operated.

**Whole effluent toxicity** (Section 122.2) means the aggregate toxic effect of an effluent measured directly by a toxicity test. Note: At 45 FR 48620, July 21, 1980, the Environmental Protection Agency suspended until further notice in Sec. 122.2, the last sentence, beginning "This exclusion applies . . ." in the definition of "Waters of the United States." This revision continues that suspension.\1\ \1\ Editorial Note: The words "This revision" refer to the document published at 48 FR 14153, Apr. 1, 1983. (Clean Water Act (33 U.S.C. 1251 et seq.), Safe Drinking Water Act (42 U.S.C. 300f et seq.), Clean Air Act (42 U.S.C. 7401 et seq.), Resource Conservation and Recovery Act (42 U.S.C. 6901 et seq.))

**Whole program** (Section 70.2) means a part 70 permit program, or any combination of partial programs, that meet all the requirements of these regulations and cover all the part 70 sources in the entire State. For the purposes of this definition, the term "State" does not include local permitting authorities, but refers only to the entire State, Commonwealth, or Territory.

**Wholesale purchaser-consumer** (Section 80.2) means any organization that is an ultimate consumer of gasoline, diesel fuel, methanol, natural gas or liquefied petroleum gas and which purchases or obtains gasoline, diesel fuel, natural gas or liquefied petroleum gas from a supplier for use in motor vehicles and, in the case of gasoline, diesel fuel, methanol or liquefied petroleum gas, receives delivery of that product into a storage tank of at least 550-gallon capacity substantially under the control of that organization.

**Wide crosses** (Section 174.3) means to facilitate the formation of viable zygotes through the use of surgical alteration of the plant pistil, bud pollination, mentor pollen, immunosuppressants, in vitro fertilization, pre-pollination and post-pollination hormone treatments, manipulation of chromosome numbers, embryo culture, or ovary and ovule cultures.

**Window trough** (Section 745.63) means, for a typical double-hung window, the portion of the exterior window sill between the interior window sill (or stool) and the frame of the storm window. If there is no storm window, the window trough is the area that receives both the upper and lower window sashes when they are both lowered. The window trough is sometimes referred to as the window "well."

**Wipe sample** (Section 745.63) means a sample collected by wiping a representative surface of known area, as determined by ASTM E1728, "Standard Practice for Field Collection of Settled Dust Samples Using Wipe Sampling Methods for Lead Determination by Atomic Spectrometry Techniques, or equivalent method, with an acceptable wipe material as defined in ASTM E 1792, "Standard Specification for Wipe Sampling Materials for Lead in Surface Dust."

**WIPP** (Section 194.2) means the Waste Isolation Pilot Plant, as authorized pursuant to section 213 of the Department of Energy National Security and Military Applications of Nuclear Energy Authorization Act of 1980 (Pub.L. 96-164; 93 Stat. 1259, 1265).

**WIPP LWA** (Section 194.2) means the Waste Isolation Pilot Plant Land Withdrawal Act of 1992 (Pub.L. 102-579, 106 Stat. 4777).

**Wire products and fasteners** (Section 420.121) means steel wire, products manufactured from steel wire, and steel fasteners manufactured from steel wire or other steel shapes.

**Withdraw specification** (Section 231.2) means to remove from designation any area already specified as a disposal site by the U.S. Army Corps of Engineers or by a state which has assumed the section 404 program, or any portion of such area.

**Wood fiber furnish subdivision mills** (Section 430.111) are those mills where cotton fibers are not used in the production of fine papers.

**Wool** (Section 410.11) shall mean the dry raw wool as it is received by the wool scouring mill.

**Work area** (Section 721.3) means a room or defined space in a workplace where a chemical substance is manufactured, processed, or used and where employees are present.

**Work plan commitments** (Section 35.102) The outputs and outcomes associated with each work plan component, as established in the grant agreement.

**Work plan component** (Section 35.102) A negotiated set or group of work plan commitments established in the grant agreement. A work plan may have one or more work plan components.

**Worker** (Section 170.3) means any person, including a self-employed person, who is employed for any type of compensation and who is performing activities relating to the production of agricultural plants on an agricultural establishment to which subpart B of this part applies. While persons employed by a commercial pesticide handling establishment are performing tasks as crop advisors, they are not workers covered by the requirements of subpart B of this part.

**Working day** (Section 129.2) means the hours during a calendar day in which a facility discharges effluents subject to this part.

**Working day** (Section 1601.3) means a Federal workday that does not include Saturdays, Sundays, or Federal holidays.

**Working day** (Section 350.1) is any day on which Federal government offices are open for normal business. Saturdays, Sundays, and official Federal holidays are not working days; all other days are.

**Working day** (Section 63.2) means any day on which Federal Government offices (or State government offices for a State that has obtained delegation under section 112(l)) are open for normal business. Saturdays, Sundays, and official Federal (or where delegated, State) holidays are not working days.

**Working day (Section 85.1502)** Any day on which Federal government offices are open for normal business. Saturdays, Sundays, and official Federal holidays are not working days.

**Working day** (Section 89.602) Any day on which federal government offices are open for normal business. Saturdays, Sundays, and official federal holidays are not working days.

**Working Level (WL)** (Section 192.11) means any combination of short-lived radon decay products in one liter of air that will result in the ultimate emission of alpha particles with a total energy of 130 billion electron volts.

**Workover fluids** (Section 435.11) means salt solutions, weighted brines, polymers, or other specialty additives used in a producing well to allow for maintenance, repair or abandonment procedures.

**Workover fluids** (Section 435.41) means salt solutions, weighted brines, polymers, or other specialty additives used in a producing well to allow for maintenance, repair or abandonment procedures.

**Workplace** (Section 721.3) means an establishment at one geographic location containing one or more work areas.

**Worst case discharge as defined by section 311(a)(24) of the CWA** (Section 300.5) means, in the case of a vessel, a discharge in adverse weather conditions of its entire cargo, and, in the case of an offshore facility or onshore facility, the largest foreseeable discharge in adverse weather conditions.

**Worst case discharge for an onshore non-transportation-related facility** (Section 112.2) means the largest foreseeable discharge in adverse weather conditions as determined using the worksheets in Appendix D to this part.

**Worst-case release** (Section 68.3) means the release of the largest quantity of a regulated substance from a vessel or process line failure that results in the greatest distance to an endpoint defined in Sec. 68.22(a).

**Written commitment** (Section 93.101) for the purposes of this subpart means a written commitment that includes a description of the action to be taken; a schedule for the completion of the action; a demonstration that funding necessary to implement the action has been authorized by the appropriating or authorizing body; and an acknowledgment that the commitment is an enforceable obligation under the applicable implementation plan.

**Written Instructions for Proper Maintenance and Use** (Section 85.1402) means those maintenance and operation instructions specified in the warranty as being necessary to assure compliance of the retrofit/rebuild equipment with applicable emission standards for the in-use compliance period.

**Written Instructions for Proper Maintenance and Use** (Section 85.2102) means those maintenance and operation instructions specified in the owner's manual as being necessary to assure compliance of a vehicle with applicable emission standards for the useful life of the vehicle that are: (i) In accordance with the instructions specified for performance on the manufacturer's prototype vehicle used in certification (including those specified for vehicles used under special circumstances), and (ii) In compliance with the requirements of 40 CFR 86.094-38 or 86.1808-01 (as appropriate for the applicable model year vehicle/engine classification); and (iii) In compliance with any other regulations promulgated by the Office Director governing maintenance and use instructions.

# XYZ

**X** (Section 409.61) shall mean that fraction of the "net cane" harvested by the advanced harvesting systems.

**Xylenes** (Section 439.1) means a combination of the three isomers: o-xylene, p-xylene, and m-xylene.

**You** (Section 105.2) means the applicant for an award.

**You** (Section 62.1201) means the owner or operator of a hazardous waste combustor.

**Yucca Mountain repository** (Section 197.2) means the excavated portion of the facility constructed underground within the Yucca Mountain site.

**Yucca Mountain site** (Section 197.2) means: (1) The site recommended by the Secretary of DOE to the President under section 112(b)(1)(B) of the Nuclear Waste Policy Act of 1982 (42 U.S.C. 10132(b) (1) (B)) on May 27, 1986; or (2) The area under the control of DOE for the use of Yucca Mountain activities at the time of licensing, if the site designated under the Nuclear Waste Policy Act is amended by Congress prior to the time of licensing.

**Zero (0)** (Section 86.1803-01) miles means that point after initial engine starting (not to exceed 100 miles of vehicle operation, or three hours of engine operation) at which normal assembly line operations and adjustments are completed, and including emission testing, if performed.

**Zero (0) hours** (Section 86.082-2) means that point after normal assembly line operations and adjustments are completed and before ten (10) additional operating hours have been accumulated, including emission testing, if performed.

**Zero (0) miles** (Section 86.082-2) means that point after initial engine starting (not to exceed 100 miles of vehicle operation, or three hours of engine operation) at which normal assembly line operations and adjustments are completed, and including emission testing, if performed.

**Zero air material** (Section 72.2) means either: (1) A calibration gas certified by the gas vendor not to contain concentrations of SO2, NOX, or total hydrocarbons above 0.1 parts per million (ppm), a concentration of CO above 1 ppm, or a concentration of CO2 above 400 ppm; (2) Ambient air conditioned and purified by a CEMS for which the CEMS manufacturer or vendor certifies that the particular CEMS model produces conditioned gas that does not contain concentrations of SO2, NOX, or total hydrocarbons above 0.1 ppm, a concentration of CO above 1 ppm, or a concentration of CO2 above 400 ppm; (3) For dilution-type CEMS, conditioned and purified ambient air provided by a conditioning system concurrently supplying dilution air to the CEMS; or (4) A multicomponent mixture certified by the supplier of the mixture that the concentration of the component being zeroed is less than or equal to the applicable concentration specified in paragraph (1) of this definition, and that the mixture's other components do not interfere with the CEM readings.

**Zero device-miles** (Section 610.11) means the period of time between retrofit installation and the accumulation of 100 miles of automobile operation after installation.

**Zero kilometers** (Section 86.402-78) means that point after normal assembly line operations and adjustments, after normal dealer setup and preride inspection operations have been completed, and before 100 kilometers of vehicle operation of three hours of engine operation have been accumulated, including emission testing if performed.

**Zero-Emission Vehicle** (Section 88.102-94) means any light-duty vehicle or light-dutytruck conforming to the applicable

Zero-Emission Vehicle standard, or any heavy-duty vehicle conforming to the applicable Zero-Emission Vehicle standard.

**Zero-emission vehicle (ZEV)** (Section 86.1702-99) means any vehicle which is certified to produce zero emissions of any criteria pollutants under any and all possible operational modes and conditions. Incorporation of a fuel fired heater shall not preclude a vehicle from being certified as a ZEV provided the fuel fired heater cannot be operated at ambient temperatures above 40 degrees Fahrenheit and the heater is demonstrated to have zero evaporative emissions under any and all possible operational modes and conditions.

**ZEV Mandate** (Section 86.1702-99) means any state regulation or other law that imposes (or purports to impose) obligations on auto manufacturers to produce, deliver for sale, or sell a certain number or percentage of ZEVs.

**Zinc** (Section 415.641) shall mean the total zinc present in the process wastewater stream exiting the wastewater treatment system.

**Zinc** (Section 415.671) shall mean the total zinc present in the process wastewater stream exiting the wastewater treatment system.

**Zinc** (Section 420.02) means total zinc and is determined by the method specified in 40 CFR 136.3.

**Zinc** (Section 437.2) means total zinc.

**Zinc casting** (Section 464.02) The remelting of zinc or zinc alloy to form a cast intermediate or final product by pouring or forcing the molten metal into a mold, except for ingots, pigs, or other cast shapes related to nonferrous (primary) metals manufacturing (40 CFR part 421) and nonferrous metals forming (40 CFR part 471). Processing operations following the cooling of castings not covered under nonferrous metals forming are covered under the electroplating and metal finishing point source categories (40 CFR parts 413 and 433).

**Zone of engineering control** (Section 260.10) means an area under the control of the owner/operator that, upon detection of a hazardous waste release, can be readily cleaned up prior to the release of hazardous waste or hazardous constituents to ground water or surface water.

# APPENDIX: OUTLINE OF 40 CFR

**Section:**

1. Statement of organization and general information

2. Public information

4. Uniform relocation assistance and real property acquisition for Federal and federally assisted programs

5. Nondiscrimination on the basis of sex in education programs or activities receiving Federal financial assistance

6. Procedures for implementing the requirements of the Council on Environmental Quality on the National Environmental Policy Act

7. Nondiscrimination in programs receiving Federal assistance from the Environmental Protection Agency

8. Environmental impact assessment of nongovernmental activities in Antarctica

9. OMB approvals under the Paperwork Reduction Act

10. Administrative claims under Federal Tort Claims Act

11. Security classification regulations pursuant to Executive Order 11652

12. Nondiscrimination on the basis of handicap in programs or activities conducted by the Environmental Protection Agency

13. Claims collection standards

14. Employee personal property claims

15. Reserved

16. Implementation of Privacy Act of 1974

17. Implementation of the Equal Access to Justice Act in EPA administrative proceedings

18. Reserved

19. Adjustment of civil monetary penalties for inflation

20. Certification of facilities

21. Small business

22. Consolidated rules of practice governing the administrative assessment of civil penalties and the revocation/termination or suspension of permits

23. Judicial review under EPA—Administered statutes

24. Rules governing issuance of and administrative hearings on interim status corrective action orders

25. Public participation in programs under the Resource Conservation and Recovery Act, the Safe Drinking Water Act, and the Clean Water Act

26. Protection of human subjects

27. Program fraud civil remedies

28. Reserved

29. Intergovernmental review of Environmental Protection Agency programs and activities

30. Uniform administrative requirements for grants and agreements with institutions of higher education, hospitals, and other non-profit organizations

31. Uniform administrative requirements for grants and cooperative agreements to State and local governments

32. Government wide debarment and suspension (nonprocurement) and government wide requirements for drug-free workplace (grants); Clean Air Act and Clean Water Act ineligibility of facilities in performance of Federal contracts, grants and loans

33. Reserved

34 New restrictions on lobbying

35. State and local assistance

40. Research and demonstration grants

45. Training assistance

46. Fellowships

47. National Environmental Education Act grants

49. Tribal Clean Air Act authority

50. National primary and secondary ambient air quality standards

51. Requirements for preparation, adoption, and submittal of implementation plans

52. Approval and promulgation of implementation plans.

53. Ambient air monitoring reference and equivalent methods

54. Prior notice of citizen suits

55. Outer continental shelf air regulations

56. Regional consistency

57. Primary nonferrous smelter orders

58. Ambient air quality surveillance

59. National volatile organic compound emission standards for consumer and commercial products

60. Standards of performance for new stationary sources.

61. National emission standards for hazardous air pollutants

62. Approval and promulgation of State plans for designated facilities and pollutants

63 National emission standards for hazardous air pollutants for source categories

64. Compliance assurance monitoring

65. Consolidated Federal air rule

66. Assessment and collection of noncompliance penalties by EPA

67. EPA approval of State noncompliance penalty program

68. Chemical accident prevention provisions

69. Special exemptions from requirements of the Clean Air Act

70. State operating permit programs
71. Federal operating permit programs
72. Permits regulation
73. Sulphur dioxide allowance system
74. Sulfur dioxide opt-ins
75. Continuous emission monitoring
76. Acid rain nitrogen oxides emission reduction program
77. Excess emissions
78. Appeal procedures for Acid Rain Program
79. Registration of fuels and fuel additives
80. Regulation of fuels and fuel additives
81. Designation of areas for air quality planning purposes
82. Protection of stratospheric ozone
85. Control of air pollution from mobile sources
87. Control of air pollution from aircraft and aircraft engines
88. Clean-fuel vehicles
89. Control of emissions from new and in-use nonroad compression-ignition engines
90. Control of emissions from nonroad spark-ignition engines
91. Control of emissions from marine spark-ignition engines
92. Control of air pollution from locomotives and locomotive engines
93. Determining conformity of Federal actions to State or Federal implementation plans
94. Control of air pollution from marine compression-ignition engines
95. Mandatory patent licenses
96. NOX Budget Trading Program for State Implementation Plans
97. Federal NOX Budget Trading Program

98-99 [Reserved]

100 [Reserved]

104. Public hearings on effluent standards for toxic pollutants
105. Recognition awards under the Clean Water Act
108. Employee protection hearings
109. Criteria for State, local and regional oil removal contingency plans
110. Discharge of oil
112. Oil pollution prevention
113. Liability limits for small onshore storage facilities
116. Designation of hazardous substances
117. Determination of reportable quantities for hazardous substances

121. State certification of activities requiring a Federal license or permit

122. EPA administered permit programs: The national pollutant discharge elimination system

123. State program requirements

124. Procedures for decisionmaking

125. Criteria and standards for the national pollutant discharge elimination system

129. Toxic pollutant effluent standards

130. Water quality planning and management

131. Water quality standards

132. Water quality guidance for the Great Lakes System

133. Secondary treatment regulation

135. Prior notice of citizen suits

136. Guidelines establishing test procedures for the analysis of pollutants

140. Marine sanitation device standard

141. National primary drinking water regulations

142. National primary drinking water regulations implementation

143. National secondary drinking water regulations

144. Underground injection control program

145. State UIC program requirements

146. Underground injection control program: Criteria and standards

147. State underground injection control programs

148. Hazardous waste injection restrictions

149. Sole source aquifers

150-151 [Reserved]

152. Pesticide registration and classification procedures

153. Registration policies and interpretations

154. Special review procedures

155. Registration standards

156. Labeling requirements for pesticides and devices

157. Packaging requirements for pesticides and devices

158. Data requirements for registration

159. Statements of policies and interpretations

160. Good laboratory practice standards

162. State registration of pesticide products

163. Certification of usefulness of pesticide chemicals

164. Rules of practice governing hearings, under the Federal Insecticide, Fungicide, and Rodenticide Act, arising from refusals to register, cancellations of registrations, changes of

classifications, suspensions of registrations and other hearings called pursuant to section 6 of the Act

166. Exemption of Federal and State agencies for use of pesticides under emergency conditions

167. Registration of pesticide and active ingredient producing establishments, submission of pesticide reports

168. Statements of enforcement policies and interpretations

169. Books and records of pesticide production and distribution

170. Worker protection standard

171. Certification of pesticide applicators

172. Experimental use permits

173. Procedures governing the rescission of State primary enforcement responsibility for pesticide use violations

174. Procedures and requirements for plant-incorporated protectants

176. Time-limited tolerances for emergency exemptions

177. Issuance of food additive regulations

178. Objections and requests for hearings

179. Formal evidentiary public hearing

180. Tolerances and exemptions from tolerances for pesticide chemicals in food 187-189 [Reserved]

190. Environmental radiation protection standards for nuclear power operations

191. Environmental radiation protection standards for management and disposal of spent nuclear fuel, high-level and transuranic radioactive wastes

192. Health and environmental protection standards for uranium and thorium mill tailings

194. Criteria for the certification and re-certification of the Waste Isolation Pilot Plant's compliance with the 40 CFR part 191 disposal regulations

195. Radon proficiency programs

197. Public health and environmental radiation protection standards for Yucca Mountain, Nevada

201. Noise emission standards for transportation equipment; interstate rail carriers

202. Motor carriers engaged in interstate commerce

203. Low-noise-emission products

204. Noise emission standards for construction equipment

205. Transportation equipment noise emission controls

209. Rules of practice governing proceedings under the Noise Control Act of 1972

210. Prior notice of citizen suits

211. Product noise labeling

220. General

221. Applications for ocean dumping permits under section 102 of the Act

222. Action on ocean dumping permit applications under section 102 of the Act

223. Contents of permits; revision, revocation or limitation of ocean dumping permits under section 104(d) of the Act

224. Records and reports required of ocean dumping permittees under section 102 of the Act

225. Corps of Engineers dredged material permits

227. Criteria for the evaluation of permit applications for ocean dumping of materials

228. Criteria for the management of disposal sites for ocean dumping

229. General permits

230. Section 404(b)(1) guidelines for specification of disposal sites for dredged or fill material

231. Section 404© procedures

232. 404 Program definitions; exempt activities not requiring 404 permits

233. 404 State program regulations

238. Degradable plastic ring carriers

239. Requirements for state permit program determination of adequacy

240. Guidelines for the thermal processing of solid wastes

243. Guidelines for the storage and collection of residential, commercial, and institutional solid waste

246. Source separation for materials recovery guidelines

247. Comprehensive procurement guideline for products containing recovered materials

254. Prior notice of citizen suits

255. Identification of regions and agencies for solid waste management

256. Guidelines for development and implementation of State solid waste management plans

257. Criteria for classification of solid waste disposal facilities and practices

258. Criteria for municipal solid waste landfills

259. [Reserved]

260. Hazardous waste management system: General

261. Identification and listing of hazardous waste

262. Standards applicable to generators of hazardous waste

263. Standards applicable to transporters of hazardous waste

264. Standards for owners and operators of hazardous waste treatment, storage, and disposal facilities

265. Interim status standards for owners and operators of hazardous waste treatment, storage, and disposal facilities

266. Standards for the management of specific hazardous wastes and specific types of hazardous waste management facilities

267. [Reserved]

268. Land disposal restrictions

270. EPA administered permit programs: The Hazardous Waste Permit Program

271. Requirements for authorization of State hazardous waste programs

272. Approved State hazardous waste management programs

273. Standards for universal waste management

279. Standards for the management of used oil

280. Technical standards and corrective action requirements for owners and operators of underground storage tanks (UST)

281. Approval of state underground storage tank programs

282. Approved underground storage tank programs

283-299 [Reserved]

300. National Oil and Hazardous Substances Pollution Contingency Plan

302. Designation, reportable quantities, and notification

303. Citizen awards for information on criminal violations under superfund

304. Arbitration procedures for small superfund cost recovery claims

305. Comprehensive Environmental Response, Compensation, and Liability Act (CERCLA) administrative hearing procedures for claims against the superfund

307. Comprehensive Environmental Response, Compensation, and Liability Act (CERCLA) claims procedures

310. Reimbursement to local governments for emergency response to hazardous substance releases

311. Worker protection

350. Trade secrecy claims for emergency planning and community right-to-know information: and trade secret disclosures to health professionals

355. Emergency planning and notification

370. Hazardous chemical reporting: Community right-to-know

372. Toxic chemical release reporting: Community right-to-know

373. Reporting hazardous substance activity when selling or transferring Federal real property

374. Prior notice of citizen suits

375-399 [Reserved]

400. [Reserved]

401. General provisions

402. [Reserved]

403. General pretreatment regulations for existing and new sources of pollution

405. Dairy products processing point source category

406. Grain mills point source category

407. Canned and preserved fruits and vegetables processing point source category

408. Canned and preserved seafood processing point source category

409. Sugar processing point source category

410. Textile mills point source category

411. Cement manufacturing point source category

412. Feedlots point source category

413. Electroplating point source category

414. Organic chemicals, plastics, and synthetic fibers

415. Inorganic chemicals manufacturing point source category

416. [Reserved]

417. Soap and detergent manufacturing point source category

418. Fertilizer manufacturing point source category

419. Petroleum refining point source category

420. Iron and steel manufacturing point source category

421. Nonferrous metals manufacturing point source category

422. Phosphate manufacturing point source category

423. Steam electric power generating point source category

424. Ferroalloy manufacturing point source category

425. Leather tanning and finishing point source category

426. Glass manufacturing point source category

427. Asbestos manufacturing point source category

428. Rubber manufacturing point source category

429. Timber products processing point source category

430. The pulp, paper, and paperboard point source category

431. [Reserved]

432. Meat products point source category

433. Metal finishing point source category

434. Coal mining point source category BPT, BAT, BCT limitations and new source performance standards

435. Oil and gas extraction point source category

436. Mineral mining and processing point source category

437. The centralized waste treatment point source category

439. Pharmaceutical manufacturing point source category

440. Ore mining and dressing point source category

442. Transportation equipment cleaning point source category

443. Effluent limitations guidelines for existing sources and standards of performance and pretreatment standards for new sources for the paving and roofing materials (tars and asphalt) point source category

444. Waste combustors point source category

445. Landfills point source category

446. Paint formulating point source category

447. Ink formulating point source category

454. Gum and wood chemicals manufacturing point source category

455. Pesticide chemicals

457. Explosives manufacturing point source category

458. Carbon black manufacturing point source category

459. Photographic point source category

460. Hospital point source category

461. Battery manufacturing point source category

463. Plastics molding and forming point source category

464. Metal molding and casting point source category

465. Coil coating point source category

466. Porcelain enameling point source category

467. Aluminum forming point source category

468. Copper forming point source category

469. Electrical and electronic components point source category

471. Nonferrous metals forming and metal powders point source category

501. State sludge management program regulations

503. Standards for the use or disposal of sewage sludge

600. Fuel economy of motor vehicles

610. Fuel economy retrofit devices

611-699 [Reserved]

700. General

702. General practices and procedures

704. Reporting and recordkeeping requirements

707. Chemical imports and exports

710. Inventory reporting regulations

712. Chemical information rules

716. Health and safety data reporting

717. Records and reports of allegations that chemical substances cause significant adverse reactions to health or the environment

720. Premanufacture notification

721. Significant new uses of chemical substances

723. Premanufacture notification exemptions

725. Reporting requirements and review processes for microorganisms

745. Lead-based paint poisioning prevention in certain residential structures

747. Metalworking fluids

749. Water treatment chemicals

750. Procedures for rulemaking under section 6 of the Toxic Substances Control Act

761. Polychlorinated biphenyls (PCBs) manufacturing, processing, distribution in commerce, and use prohibitions

763. Asbestos

766. Dibenzo-para-dioxins/dibenzofurans

767-789 [Reserved]

790. Procedures governing testing consent agreements and test rules

791. Data reimbursement

792. Good laboratory practice standards

795. Provisional test guidelines

796. Chemical fate testing guidelines

797. Environmental effects testing guidelines

798. Health effects testing guidelines

799. Identification of specific chemical substance and mixture testing requirements

**CHAPTER IV—ENVIRONMENTAL PROTECTION AGENCY AND DEPARTMENT OF JUSTICE**

1400.Distribution of off-site consequence analysis information

**CHAPTER V—COUNCIL ON ENVIRONMENTAL QUALITY**

1500. Purpose, policy, and mandate

1501. NEPA and agency planning

1502. Environmental impact statement

1503. Commenting

1504. Predecision referrals to the Council of proposed Federal actions determined to be environmentally unsatisfactory

1505. NEPA and agency decisionmaking

1506. Other requirements of NEPA

1507. Agency compliance

1508. Terminology and index

1515. Freedom of Information Act procedures

1516. Privacy Act implementation

1517. Public meeting procedures of the Council on Environmental Quality

Index to parts 1500 through 1508

**CHAPTER VI—CHEMICAL SAFETY AND HAZARD INVESTIGATION BOARD**

1600. [Reserved]

1601. Procedures for disclosure of records under the Freedom of Information Act

1602. Protection of privacy and access to individual records under the Privacy Act of 1974

1603. Rules implementing the Government in the Sunshine Act

1610. Administrative investigations

1611. Testimony by employees in legal proceedings

1612. Production of records in legal proceedings

1613-1699 [Reserved]

**CHAPTER VII—ENVIRONMENTAL PROTECTION AGENCY AND DEPARTMENT OF DEFENSE; UNIFORM NATIONAL DISCHARGE STANDARDS FOR VESSELS OF THE ARMED FORCES**

1700. Uniform National Discharge Standards for vessels of the Armed Forces

# ABOUT THE AUTHOR

Jim has nearly 30 years of experience in quality, environmental, and health and safety management systems consulting.

Jim has refined his management and leadership techniques while being a full time consultant for nearly thirty years. He has assumed the leadership roles in development programs for quality and environmental cost accounting systems, training and auditing both domestically and internationally for a wide variety of industries.

Jim is also a Lead Auditor to ISO 9001:2000, Quality Management System, ISO 14001:1996, Environmental Management System and several Health and Safety management systems.

His experience not only includes corporate management experience, but he has also held leadership positions in startup, and merger operations.

Jim lives in Morgan County, Tennessee and can be reached through his website: **www.JimKing.Biz**.